T0306113

THE ZELDOVICH UNIVERSE:
GENESIS AND GROWTH OF THE COSMIC WEB

IAU SYMPOSIUM 308

COVER ILLUSTRATION:

The skyline of the old town of Tallinn, with its characteristic towers and spires, against the background of a grid rendering of a cosmic density field evolved according to the Zeldovich formalism. Design: Niels Bos & Johan Hidding.

IAU SYMPOSIUM PROCEEDINGS SERIES

Chief Editor
THIERRY MONTMERLE, IAU General Secretary
Institut d'Astrophysique de Paris,
98bis, Bd Arago, 75014 Paris, France
montmerle@iap.fr

Editor
PIERO BENVENUTI, IAU Assistant General Secretary
University of Padua, Dept of Physics and Astronomy,
Vicolo dell'Osservatorio, 3, 35122 Padova, Italy
piero.benvenuti@unipd.it

INTERNATIONAL ASTRONOMICAL UNION

UNION ASTRONOMIQUE INTERNATIONALE

THE ZELDOVICH UNIVERSE: GENESIS AND GROWTH OF THE COSMIC WEB

PROCEEDINGS OF THE 308th SYMPOSIUM OF THE INTERNATIONAL ASTRONOMICAL UNION HELD IN TALLINN, ESTONIA JUNE 23–28, 2014

Edited by

RIEN VAN DE WEYGAERT

Kapteyn Astronomical Institute, University of Groningen, Groningen, the Netherlands

SERGEI SHANDARIN

Department of Physics and Astronomy, Kansas University, Lawrence, Kansas, U.S.A.

ENN SAAR

Tartu Observatory, Toravere, Estonia

and

JAAN EINASTO

Tartu Observatory, Toravere, Estonia

CAMBRIDGE
UNIVERSITY PRESS

University Printing House, Cambridge CB2 8BS, United Kingdom

One Liberty Plaza, 20th Floor, New York, NY 10006, USA

477 Williamstown Road, Port Melbourne, VIC 3207, Australia

314-321, 3rd Floor, Plot 3, Splendor Forum, Jasola District Centre, New Delhi - 110025, India

103 Penang Road, #05-06/07, Visioncrest Commercial, Singapore 238467

Cambridge University Press is part of the University of Cambridge.

It furthers the University's mission by disseminating knowledge in the pursuit of
education, learning and research at the highest international levels of excellence.

www.cambridge.org
Information on this title: www.cambridge.org/9781107078604

© International Astronomical Union 2016

This publication is in copyright. Subject to statutory exception
and to the provisions of relevant collective licensing agreements,
no reproduction of any part may take place without the written
permission of Cambridge University Press.

First published 2016

A catalogue record for this publication is available from the British Library

ISBN 978-1-107-07860-4 Hardback
ISSN 1743-9213

Cambridge University Press has no responsibility for the persistence or
accuracy of URLs for external or third-party internet websites referred to in
this publication, and does not guarantee that any content on such websites is,
or will remain, accurate or appropriate.

IAU Symposium No. 308 and this volume
are dedicated to

YAKOV B. ZELDOVICH (1914-1987)

Table of Contents

CHAPTER 1. Cosmology of Yakov Zeldovich, Historical and Scientific Perspective

CHAPTER 2. Zeldovich legacy: Dynamics & Evolution of the Cosmic Web

CHAPTER 3. Surveys and Observations of the Large Scale Structure of the Universe

CHAPTER 3A. Surveys and Observations: Surveys

CHAPTER 3B. Surveys and Observations: Local Universe

CHAPTER 3C. Surveys and Observations: Filaments

CHAPTER 3D. Surveys and Observations: Groups and Clusters

CHAPTER 4. Cosmic Web Morphology & Identification, Reconstruction & Clustering

CHAPTER 4A. Cosmic Web Morphology & Identification

CHAPTER 4B. Cosmic Web Reconstruction

CHAPTER 4C. Cosmic Web Clustering

CHAPTER 5. Megaparsec Velocity Flows

CHAPTER 6. The Gaseous Cosmic Web

CHAPTER 7. Galaxy Formation & Evolution
in the Cosmic Web

CHAPTER 7A. Galaxy Formation & Evolution

CHAPTER 7B. Galaxy Alignments

CHAPTER 7C. Galaxy Formation and Evolution: Poster papers

CHAPTER 8. Cosmic Voids

CHAPTER 8A. Cosmic Voids:
Structure, Dynamics and Cosmology

CHAPTER 8B. Void Galaxies

CHAPTER 9. Cosmology

CHAPTER 10. Miscellaneous

Preface

On Megaparsec scales, matter and galaxies have aggregated into a complex network of interconnected filaments and walls. This network, which has become known as the *Cosmic Web*, contains structures from a few Megaparsecs up to tens and even hundreds of Megaparsecs of size. It has organized galaxies and mass into a wispy web-like spatial arrangement, marked by highly elongated filaments, flattened wall-like structures and dense compact clusters surrounding large near-empty void regions. Its appearance has been most dramatically illustrated by the maps of the nearby cosmos produced by large galaxy redshift surveys such as the 2dFGRS, the SDSS, and the 2MASS redshift survey, as well as by recently produced maps of the galaxy distribution at larger cosmic depths such as VIPERS.

The Cosmic Web is one of the most striking examples of complex geometric patterns found in nature, and certainly the largest in terms of size. As borne out by a large array of computer simulations of cosmic structure formation, weblike patterns in the overall cosmic matter distribution do represent a universal but possibly transient phase in the gravitationally driven emergence and evolution of cosmic structure. These calculations have shown that weblike patterns defined by prominent anisotropic filamentary and planar features – and with characteristic large underdense void regions – are a natural manifestation of the gravitational cosmic structure formation process. The combination of these theoretical and observational studies have lead to the recognition of the cosmic web as a key aspect of structure in the Universe, marking the transition from the early linear growth of the primordial Gaussian random density fluctuations as it evolves out of the primordial universe towards the emergence of complex patterns, structures and objects.

Instrumental in the development of this view of formation of structure have been the contributions by the Russian physicist and cosmologist Yakov B. Zeldovich (1914-1987). His seminal work paved the way towards a theoretical understanding of the complex weblike patterns observed in our Universe. In the year 2014 it was 100 years ago since his birth. He was born in 1914 in Minsk, Belorussia, at the time in the tsarist Russian Empire, a few months before the outbreak of World War I. In a sense, his life ran parallel to the Soviet Union, of which during the 20th century he became one of its most highly recognized and famous scientists. He played a key role in many areas of physics, with instrumental contributions in - amongst others - chemical physics, shockwave physics, nuclear physics, particle physics, astrophysics and cosmology. Besides his scientific contributions, he left a lasting legacy in terms of an impressive array of students, many of whom have become scientists of great fame.

The centenary year 2014 of Zeldovich has been celebrated by several international physics and astrophysics conferences. In Moscow and Minsk, the Zeldovich 100 conferences addressed the wide range of physical and astrophysical interests that Zeldovich touched upon in his work. In Tallinn, the symposium in honour of Zeldovich that we report in this volume exclusively devoted its attention on the field of which he was one of the founding fathers, the formation of structure in the Universe and the large scale structure of the Universe. The International Astronomical Union recognized that the major developments in the past decades would make a symposium addressing the cosmic web and the formation of structure very timely, and endorsed and sponsored it as IAU Symposium 308. The symposium had the objective of synthesizing the insights from many different observational and theoretical studies relating to the subjects, and to prepare this vibrant field for the host of upcoming surveys and data that will facilitate radically new insights into the cosmic web, the information it entails on pressing cosmological issues and on our understanding of the formation of galaxies.

IAU Symposium 308 pays tribute in more than one way to the historical legacy of the field. While the Zeldovich formalism predicted and preceded the recognition and discovery of the beautiful and complex spatial patterns that large galaxy redshift surveys have uncovered over the past decades, the first observational indications for the reality of a weblike galaxy distribution came along as a result of the collaboration between Zeldovich and the cosmology group at Tartu Observatory lead by Jaan Einasto. In 1977 this culminated in a famous IAU Symposium, No. 79. This symposium was organized in Tallinn, the capital of Estonia, at the time located in the Soviet Union. This conference opened the subject of the large scale structure of the Universe by bringing together cosmologists from the Soviet Union and the Eastbloc with those from the West. The interaction between the different views in East and West of the way structure arose in the Universe started the field as one of the most vibrant and active areas of cosmology. By organizing IAU Symposium 308 in Tallinn, we wish also to pay tribute to the seminal significance of IAU Symposium 79 for the development of the field, and in particular in its instrumental role of exposing western cosmologists to the views of Zeldovich and his students and collaborators.

Following endorsement and sponsoring by IAU Commissions No. 28 (Galaxies) and No. 47 (Cosmology) and Division J (Galaxies and Cosmology), and subsequent approval by the IAU Executive Committe in May 2013, the organization of the symposium was set into motion and the community at large informed. The response was overwhelmingly positive, culminating in a conference that took 6 days, with 22 keynote review papers, 29 invited papers, 41 oral contributions and 62 posters, entertaining 186 participants from 31 countries. Shortly before the symposium, we learnt that one of the organizers, Jaan Einasto, was awarded the Gruber prize in cosmology, along with one of the keynote reviewers, Brent Tully. Given the participation to IAU308 of no less than 7 Gruber cosmology prize winners, on the 5th day a special Gruber prize panel discussion was organized on the subject of the remaining questions in cosmology, actively moderated by Alar Toomre.

Against the beautiful setting of the old and remarkably well preserved medieval Hansa city of Tallinn - a Unesco World Heritage Site - the participants of IAU Symposium 308 gathered from June 23-29, 2014, in the Conference Centre of Tallinn University. Today Tallinn is a vibrant city, priding itself in its modern digital and ICT infrastructure. But, above all, it is the historic heritage that offered IAU308 the serenity and inspiration of past centuries - while wandering through its cobbled streets, along its houses, city walls and towers - to contemplate about the structure of the Universe. IAU308 underlined this with a historic conference banquet in the White Hall of the 16th century House of the Brotherhood of the Black Heads (Schwarzhaupterhaus), where the atmosphere of true medieval brotherhood got framed in a beautiful musical performance by the famed Estonian TV girl's choir, directed by Arne Saluveer.

The symposium was a great success, a midsummer celebration of our fascination with the intricacies of the cosmic web. It provided a wonderful lookback on the giant leaps in understanding and insight that marked the past decades since IAU Symposium 79, and was honoured by the participation of many of whom have been responsible for important contributions. Perhaps even more important were the many new and innovative contributions by young scientists, who traced and defined new unexplored avenues of exploration of the large scale Universe. Just to name a few amongst the many notable contributions and discussions. Advances in observational, computational, as well as analytical work have started to uncover the intimate link between the nature of galaxies and the filamentary or voidlike environments in which they are born. In recent years physical insight into the formation of structure has steeply risen as several studies have opened up 6-dimensional phase space, and representatives of all major contributions along these lines present keynote and invited lectures. The coming surge in available data on the distrition of galaxies and mass has already been marked as a new era, that of big data. Profound new sophisticated computational and statistical formalisms are revolutionizing the way in which we assess such databases and will facilitate a far richer harvest of new insights and accurate measurements of cosmological data. These were presented and discussed at this symposium. There was even much ado about nothing. Amongst the many noteworthy observations was the huge increase in studies on voids. Perhaps in coming years we will conclude IAU308 to have been a watershed in the recognition of the large potential of voids towards answering fundamental cosmological questions.

It is a great pleasure to acknowledge the financial support of our sponsors listed on page xx of these Proceedings, the active support of the members of the LOC in realizing the numerous details always associated with such a symposium. This concerns its chairman Enn Saar, and Jaan Einasto, Elmo Tempel and Antti Tamm. The nice set of conference photographs to be found throughout the book was taken by Antti Tamm (unless stated otherwise). In particular we owe great gratitude to Tiia Lillemaa and Evelyn Silvet for the practical support that made the symposium possible and so enjoyable. Also we wish to thank Niels Bos and Johan Hidding for designing a poster and cover image that honoured both the legacy of Zeldovich' work and the beautiful city of Tallinn. We also wish to thank Lorraine Webb, Elisabeth Woodhouse, and Vince Higgs at CUP for their flexibility, friendliness and patience in the light of the challenges posed by our seemingly never-ending line of requests.

Finally, we wish to state that we are particularly indebted to Tartu Observatory for enabling the success of the symposium and, even more essential, its viability. We therefore wish to thank the director of Tartu Observatory, dr. Anu Reinart, and via her all the members of the institute for enabling the 2nd Tallinn IAU Symposium on Large Scale Structure.

July 2016

Rien van de Weygaert, Sergei Shandarin, Enn Saar and Jaan Einasto

Conference photograph

CONFERENCE PHOTOGRAPH

THE ORGANIZING COMMITTEE

Scientific

J. Bland-Hawthorn (Australia)
J. Richard Bond (Canada)
J. Einasto (co-chair, Estonia)
P. Erdogdu (UK)
C.S. Frenk (UK)
T. Jarrett (South Africa)
B.J.T. Jones (UK, Netherlands)
I. Karachentsev (Russia)

J. Lee (Korea)

A. Nusser (Israel)
N. Padilla (Chile)
V. Sahni (India)
S.F. Shandarin (co-chair, USA)
A. Starobinsky (Russia)
R. Sunyaev (Germany, Russia)
A. Szalay (USA)
R. van de Weygaert (co-chair,
   Netherlands)
J.P. Ying (China)

Local

J. Einasto
A. Tamm

E. Saar (chair)
E. Tempel

Acknowledgements

The symposium is sponsored and supported by the IAU Division J (Galaxies and Cosmology); and by the IAU Commissions No. 28 (Galaxies) and No. 47 (Cosmology).

The Local Organizing Committee operated under the auspices of the
Tartu Observatory, Tõravere, Estonia.

Funding by the
International Astronomical Union,
Tartu Observatory, Tõravere, Estonia
Gambling Tax Council, Estonia,
and
Kapteyn Astronomical Institute, Groningen, the Netherlands
is gratefully acknowledged.

Participants

Miguel **Aragón Calvo**	University of California, Riverside, California	United States
Anna **Aret**	Tartu Observatory, Tõravere	Estonia
Metin **Ata**	Leibniz Institute for Astrophysics, Potsdam	Germany
Lucia **Ayala**	Independent scholar	Spain
Joydeep **Bagchi**	IUCAA, Pune	India
Tobias **Baldauf**	Institute for Advanced Study, Princeton	United States
Eduardo **Battaner**	Fac. Sciences, University of Granada, Granada	Spain
Burcu **Beygu**	Kapteyn Astronomical Institute, Univ. Groningen, Groningen	Netherlands
Davide **Bianchi**	INAF Osservatorio di Brera, Merate	Italy
Monika **Biernacka**	The Jan Kochanowski University, Kielce	Poland
Maciej **Bilicki**	University of Cape Town, Cape Town	South Africa
Joss **Bland-Hawthorn**	Inst. Astronomy, Univ. Sydney, Sydney	Australia
Jonathan **Blazek**	CCAPP, Ohio State University, Columbus, Ohio	United States
Richard **Bond**	CITA/Canadian Institute for Theoretical Astrophysics, Toronto	Canada
Patrick **Bos**	Kapteyn Astronomical Institute, Univ. Groningen, Groningen	Netherlands
Martyn **Bristow**	Liverpool John Moores Univ., Liverpool	United Kingdom
Yan-Chuan **Cai**	ICC, Durham University, Durham	United Kingdom
Russell **Cannon**	AAO/Australian Astronomical Observatory, NSW	Australia
Marius **Cautun**	ICC/Inst. Comput. Cosmology, Durham University, Durham	United Kingdom
Gayoung **Chon**	MPE/Max-Planck-Institut für extraterrestrische Physik, Garching	Germany
Sandrine **Codis**	IAP/Institut d'Astrophysique de Paris, Paris	France
Peter **Coles**	University of Sussex, Brighton	United Kingdom
Christopher **Collins**	Liverpool John Moores Univ., Liverpool	United Kingdom
George James **Conidis**	York University, Toronto	Canada
Martina **Corsi**	Roma TRE University, Rome	Italy
Mousumi **Das**	Indian Institute of Astrophysics, Bangalore	India
Marc **Davis**	UC Berkeley, Berkeley, California	United States
Sylvain **de la Torre**	LAM/Laboratoire d'Astrophysique de Marseille, Marseille	France
Gianfranco **De Zotti**	INAF, Padova and SISSA, Trieste	Italy
Jörg **Dietrich**	Ludwig-Maximilians-Universität, Munich	Germany
Daria **Dobrycheva**	Main Astronomical Observatory NAS Ukraine, Kiev	Ukraine
Rosa **Domínguez-Tenreiro**	Dpt. Theoretical Physics, Universty Autonoma, Madrid	Spain
Le **Duc Thong**	Department of Physics, CNS	Vietnam
Hélène **Dupuy**	IAP/Institut d'Astrophysique, Paris & Institut de Physique Théorique, CEA Saclay	France
Jean-Baptiste **Durrive**	IAS/Institut d'Astrophysique Spatiale, Orsay	France
Jaan **Einasto**	Tartu Observatory, Tõravere	Estonia
Maret **Einasto**	Tartu Observatory, Tõravere	Estonia
Martin **Feix**	Department of Physics, Technion	Israel
Job **Feldbrugge**	Kapteyn Astronomical Institute, Univ. Groningen, Groningen	Netherlands
LongLong **Feng**	Purple Mountain Observatory, Nanjing	China
Piotr **Flin**	Jan Kochanowski University, Kielce	Poland
Estrella **Florido**	University of Granada, Granada	Spain
Gaël **Foëx**	Instituto de Fisica y Astronomia, Valparaiso, Chile	Chile
Jaime **Forero-Romero**	Universidad de los Andes, Bogota	Colombia
Carlos **Frenk**	ICC, Durham University, Durham	United Kingdom
Hirokazu **Fujii**	University of Tokyo, Tokyo	Japan
Bianca **Garilli**	INAF-IASF, Milano	Italy
Carl H **Gibson**	Univ. Calif. San Diego, San Diego	United States
Stefan **Gottloeber**	Leibniz Institute for Astrophysics, Potsdam	Germany
Mirt **Gramann**	Tartu Observatory, Tõravere	Estonia
Bradley **Greig**	Scuola Normale Superiore, Pisa	Italy
Luigi **Guzzo**	Observatory Brera, Merate	Italy
Oliver **Hahn**	ETH, Zürich	Switzerland
Nico **Hamaus**	IAP/Institut d'Astrophysique de Paris, Paris	France
Urmas **Haud**	Tartu Observatory, Tõravere	Estonia
Adam **Hawken**	INAF - Osservatorio di Brera, Merate	Italy
Pekka **Heinämäki**	Tuorla observatory / University of Turku, Piikkiö	Finland
Wojciech **Hellwing**	University of Warsaw (POL), Warsaw & ICC, Univ. Durham, Durham	Poland/United Kingdom
Johan **Hidding**	Kapteyn Astronomical Institute, Univ. Groningen, Groningen	Netherlands
Shaun **Hotchkiss**	University of Sussex, Brighton	United Kingdom
Murat **Hudaverdi**	Yildiz Technical University, Istanbul	Turkey
Mike **Hudson**	University of Waterloo, Waterloo	Canada
Lluís **Hurtado-Gil**	Instituto de Física de Cantabria, Santander	Spain
Gert **Hütsi**	Tartu Observatory, Tõravere	Estonia
Hector **Ibarra-Medel**	INAOE/Instituto Nacional de Astrofisica Optica y Electronica, Tonanzintla	Mexico
Stéphane **Ilic**	IRAP/Institut de Recherche en Astrophysique et Planétologie, Toulouse	France
Ilian **Iliev**	University of Sussex, Brighton	United Kingdom
Tomoaki **Ishiyama**	University of Tsukuba, Tsukuba	Japan
Jaak **Jaaniste**	Estonian University of Life Sciences, Tartu	Estonia
Yipeng **Jing**	Shanghai Jiaotong University, Shanghai	China
Peter **Johansson**	University of Helsinki, Helsinki	Finland
Bernard **Jones**	Kapteyn Astronomical Institute, Univ. Groningen, Groningen	Netherlands
Dennis **Just**	Dept. of Astronomy & Astrophysics, Univ. of Toronto, Toronto	Canada
Nick **Kaiser**	IfA, U. Hawaii, Manoa, Hawaii	United States
Xi **Kang**	Purple Mountain Observatory, Nanjing	China

Kristjan **Kannike**	National Institute of Chemical Physics and Biophysics, Tallinn	Estonia
Igor **Karachentsev**	Special Astrophysical Observatory RAS, Karachai-Cherkessia	Russia
Alexander **Kaurov**	University of Chicago, Chicago	United States
Ryan **Keenan**	ASIAA Academia Sinica Institute of Astronomy and Astrophysics, Taipei	Taiwan
Rain **Kipper**	Tartu Observatory, Tartu University, Tõravere/Tartu	Estonia
Francisco **Kitaura**	Leibniz Institute for Astrophysics, Potsdam	Germany
Katarina **Kovac**	Institute for Astronomy, ETH, Zürich	Switzerland
Kathryn **Kreckel**	MPIA, Heidelberg	Germany
Teet **Kuutma**	Tartu Observatory, Tõravere	Estonia
Florent **Leclercq**	IAP/Institut d'Astrophysique de Paris, Paris	France
Khee-Gan **Lee**	MPIA, Heidelberg	Germany
Jounghun **Lee**	Seoul National University, Seoul	South Korea
Bartosz **Lew**	Nicolaus Copernicus University, Torun	Poland
Guoliang **Li**	Purple Mountain Observatory, Nanhing	China
Noam **Libeskind**	Leibniz Institute for Astrophysics, Potsdam	Germany
Heidi **Lietzen**	IAC, Tenerife	Spain
Lauri Juhan **Liivamägi**	Tartu Observatory, Tõravere	Estonia
Per **Lilje**	University of Oslo, Oslo	Norway
Tiia **Lillemaa**	Tartu Observatory, Tõravere	Estonia
Jon **Loveday**	University of Sussex, Brighton	United Kingdom
Juan **Magaña**	Instituto de Física y Astronomía, Universidad de Valparaíso, Valparaiso	Chile
Christina **Magoulas**	University of Cape Town, Cape Town	South Africa
Dmitry **Makarov**	Special Astrophysical Observatory RAS, Karachai-Cherkessia	Russia
Lidia **Makarova**	Special Astrophysical Observatory RAS, Karachai-Cherkessia	Russia
Henry Joy **McCracken**	IAP/Institut d'Astrophysique de Paris, Paris	France
Stephen **McNeil**	Brigham Young University, Provo, Idaho	United States
Mikhail **Medvedev**	University of Kansas, Lawrence, Kansas	United States
Avery **Meiksin**	Institute for Astronomy, University of Edinburgh, Edinburgh	United Kingdom
Yannick **Mellier**	IAP/Institut d'Astrophysique de Paris, Paris	France
Ziv **Mikulizky**	Technion (Israel Institute of Technology), Haifa	Israel
Alla **Miroshnichenko**	Institute of Radio Astronomy of the NAS of Ukraine, Kharkov	Ukraine
Justyna **Modzelewska**	Copernicus Astronomical Center, Warsaw	Poland
Faizan Gohar **Mohammad**	Osservatorio Astronomico di Brera, Merate	Italy
Joseph **Moody**	Brigham Young University, Provo, Idaho	United States
Volker **Mueller**	Leibniz Institut fuer Astrophysik, Potsdam	Germany
Seshadri **Nadathur**	University of Helsinki, Helsinki	Finland
Masahiro **Nagashima**	Faculty of Education, Bunkyo University, Koshigaya	Japan
Olga **Nasonova**	Special Astrophysical Observatory RAS, Karachai-Cherkessiaa	Russia
Jukka **Nevalainen**	Tartu Observatory, Tõravere	Estonia
Mark **Neyrinck**	Johns Hopkins University, Baltimore	United States
Pasi **Nurmi**	University of Turku/Tuorla observatory, Piikkiö	Finland
Adi **Nusser**	Technion, Haifa	Israel
Nelson **Padilla**	Universidad Catolica de Chile, Santiago	Chile
Isha **Pahwa**	Department of Physics and Astrophysics, University of Delhi, New Delhi	India
Elena **Panko**	V.A. Sukhomlinsky Nikolaev National University, Astronomical Observatory, Nikolaev	Ukraine
Bruce **Partridge**	Haverford College, Haverford, Pennsylvania	United States
John **Peacock**	Royal Observatory, Edinburgh	United Kingdom
Arelie **Penin**	University of KwaZulu-Natal, Durban	South Africa
Andrea **Pezzotta**	Osservatorio di Brera, Merate	Italy
Christophe **Pichon**	IAP/Institut d'strophysique de Paris, Paris	France
Alice **Pisani**	IAP/Institut d'Astrophysique de Paris, Paris	France
Dmitri **Pogosian**	University of Alberta, Edmonton, Alberta	Canada
Laura **Portinari**	Tuorla Observatory, University of Turku, Piikkiö	Finland
Anup **Poudel**	Tuorla Observatory, University of Turku, Piikkiö	Finland
Abhishek **Prakash**	University of Pittsburgh, Pittsburgh, Pennsylvania	United States
Simon **Pustilnik**	Special Astrophysical Observatory RAS, Karachai-Cherkessia	Russia
Antti **Rantala**	University of Helsinki, Helsinki	Finland
John **Regan**	University of Helsinki, Helsinki	Finland
Elena **Ricciardelli**	University of Valencia, Valencia	Spain
Steven **Rieder**	Kapteyn Astronomical Institute, Univ. Groningen, Groningen	Netherlands
Graziano **Rossi**	Sejong University, Seoul	South Korea
Stefano **Rota**	Osservatorio Astronomico di Brera, Merate	Italy
Enn **Saar**	Tartu Observatory, Tõravere	Estonia
Varun **Sahni**	IUCAA, Pune	India
Jorge **Sanchez Almeida**	IAC/Instituto de Astrofisica de Canarias, Tenerife	Spain
Shishir **Sankhyayan**	IISER/Indian Institute of Science Education and Research, Pune, India	India
Prakash **Sarkar**	Tata Institute of Fundamental Research, Mumbai	India
Rudolph **Schild**	CfA/Harvard-Smithsonian Center for Astrophysics, Cambridge, Massachusetts	United States
Carlo **Schimd**	Laboratoire d'Astrophysique de Marseille, Marseille	France
Michael **Schliephake**	SU Stockholms University, Stockholm	Sweden
Marcel **Schmittfull**	UC Berkeley/LBNL, Berkeley, California	United States
Tiit **Sepp**	Tartu University, Tartu	Estonia
Sergei **Shandarin**	University of Kansas, Lawrence, Kansas,	United States
Margarita **Sharina**	Special Astrophysical Observatory RAS, Karachai-Cherkessia	Russia
Cristobal **Sifon**	Leiden Observatory, Univ. Leiden, Leiden	Netherlands
Andrzej **Soltan**	Nicolaus Copernicus Astronomical Center, Warsaw	Poland
Alexei **Starobinsky**	Landau Institute for Theoretical Physics, Moscow	Russia
Ivan **Suhhonenko**	Tartu Observatory, Tõravere	Estonia
Rashid **Sunyaev**	MPA/Max Planck Institut für Astrophysik, Garching	Germany
Paul **Sutter**	IAP/Institut d'Astrophysique de Paris, Paris	France
Erik **Tago**	Tartu Observatory, Tõravere	Estonia
Antti **Tamm**	Tartu Observatory, Tõravere	Estonia
Takayuki **Tatekawa**	University of Fukui, Fukui Prefecture	Japan
Nicolas **Tejos**	UCO/Lick Observatory, Univ. California Santa Cruz, Santa Cruz, California	United States
Elmo **Tempel**	Tartu Observatory, Tõravere	Estonia

Peeter **Tenjes**	Tartu Observatory, Tõravere	Estonia
Alar **Toomre**	MIT, Cambridge, Massachusetts	United States
Jun **Toshikawa**	NAOJ/National Astronomical Observatory of Japan, Mitaka, Tokyo	Japan
Virginia **Trimble**	University of California, Irvine, California & LCOGT	United States
Brent **Tully**	IfA, Univ. Hawaii, Manoa, Hawaii	United States
Cora **Uhlemann**	Arnold Sommerfeld Center for Theoretical Physics / LMU, Munich	Germany
Mathijs **van de Mast**	Kapteyn Astronomical Institute, Univ. Groningen, Groningen	Netherlands
Rien **van de Weygaert**	Kapteyn Astronomical Institute, Univ. Groningen, Groningen	Netherlands
Charlotte **Welker**	IAP, Institut d'Astrophysique de Paris, Paris	France
Jaan **Vennik**	Tartu Observatory, Tõravere	Estonia
David **Wiltshire**	University of Canterbury, Canterbury	New Zealand
Suk-Jin **Yoon**	Yonsei University, Seoul	South Korea
Saleem **Zaroubi**	Kapteyn Astronomical Institute, Univ. of Groningen, Groningen	Netherlands
Paul **Zivick**	The Ohio State University, Columbus, Ohio	United States

Poster of IAU Symposium 308.

Design: Niels Bos & Johan Hidding.

Yakov B. Zeldovich
a collection of historic impressions

Yakov B. Zeldovich
portrait, around 1950.

Yakov B. Zeldovich, Andrei Sacharov and David A. Frank-Kamenetskii,
Sarov, mid 1950s.
Image courtesy: Sacharov-Center.

Yakov B. Zeldovich and David A. Frank-Kamenetskii, 1947.

Zeldovich, after a lecture on cosmology, in 1975.
Image courtesy: A. Starobinsky

Zeldovich, in mid 1970s.

*Yakov B. Zeldovich and astrophysicist Iosif Shklovsky
in 1977.*

*Zeldovich with his wife visiting Estonia, late 1970s.
Image courtesy: Jaan Einasto.*

CHAPTER 1.

Cosmology of Yakov Zeldovich
Historical and Scientific Perspective

Rashid Sunyaev
recalling memories and anecdotes of his supervisor
Yakov Zeldovich

The Zeldovich Universe:
Genesis and Growth of the Cosmic Web
Proceedings IAU Symposium No. 308, 2014
R. van de Weygaert, S. Shandarin, E. Saar & J. Einasto, eds.
© International Astronomical Union 2016
doi:10.1017/S174392131600956X

Zeldovich's legacy in the Discovery and Understanding of the Cosmic Web

Sergei F. Shandarin

Department of Physics and Astronomy, University of Kansas,
1251 Wescoe Hall Drive #1082, Lawrence, KS 66045, USA
email: `sergei@ku.edu`

Abstract. Modeling and understanding the structure of the universe is one the most interesting problems in modern cosmology. It is also an extremely difficult problem. Despite spectacular achievements in observations and computational capabilities at present there is no comprehensive theory of the structure formation. There are successful theoretical models however most of them rely on assumptions and various fits which are not fully physically justified. The Zeldovich Approximation is a rare exception. Suggested more than forty years ago it remains one of the most successful and profound analytic models of structure formation. Some of results stemmed from the Zeldovich Approximation will be briefly reviewed.

Keywords. large-scale structure of universe

1. Introduction

Yakov Borisovich Zeldovich – or Ya.B. as his colleagues and friends called him – celebrated his 70th birthday in 1974. The Academy of Sciences of the USSR made a present to him: it published a selection of his works in two volumes. It was translated into English almost twenty years later (Zeldovich 1993). Ya.B. concluded the second volume by "An Autobiographical Afterword". Among other things Ya.B. included a brief assessment of his contribution to astrophysics and cosmology. Here is his comment on his contribution to the theory of the large-scale structure in the universe.

"This book presents, with commentaries, my papers on astrophysics. It is not reasonable to argue with these commentaries. Today the most important individual work seems to me to be the nonlinear theory of formation of the structure of the Universe or, as it is now called for short the "pancake" theory. The structure of the Universe, its evolution and the properties of the matter which forms the hidden mass, have to this day not been fully established. A major role in this work was played by A. G. Doroshkevich, R. A. Sunyaev, S. F. Shandarin and Ya. E. Einasto. The work continues. However, the "pancake" theory is "beautiful" in and of itself: if the initial assumptions hold, then the theory gives a correct and nontrivial answer. The "pancake" theory is a contribution to synergetics. It was especially pleasant for me to learn that this work in some measure initiated mathematical investigations by V. I. Arnold and others." (Zeldovich 1993, vol. 2, p. 642).

In the last 10 - 15 years of his life Ya.B. had very close relations with V.I. Arnold who described their interactions as follows: "Usually, Yakov Borisovich telephoned me at seven in the morning. 'Doesn't it seem to you ...' he would say, and there would then follow some sort of paradox" (Zeldovich 2004, Part V, p. 195). V.I. Arnold's considered that by formulating the "pancake" theory Ya.B. opened a new topic in mathematics: "Yakov Borisovich's 'pancake theory', in essence, is equivalent to a theory for the simplest so-called Lagrange singularities in symplectic geometry – singularities of projections of

3

Lagrangian manifolds (on which the Poincaré invariant vanishes) from phase space onto a configurational space." (Zeldovich 2004, Part V, p. 197).

Both the great physicist and mathematician expressed very high opinions of the pancake theory. It has been substantiated by numerous studies in the following years. I begin with a brief outline of the mathematical basis of the pancake theory i.e. the Zeldovich approximation (the ZA). Then I will outline some developments that stemmed from the ZA and played important role in understanding of the origin, morphology and evolution of the web. The list of selected topics obviously reflects my personal views and does not pretend to be complete.

2. The Zeldovich approximation in a nutshell

The ZA was formulated for a collisionless fluid which is a good approximation for dark matter (DM) (Zeldovich 1970). It is particularly simple in comoving coordinates $\vec{x} = \vec{r}/a(t)$, where $a(t)$ is the scale factor and $\vec{r}$ are physical coordinates of the fluid particles. It relates the initial Lagrangian coordinates $\vec{q}$ at $t \to 0$ and Eulerian coordinates $\vec{x}$ at time t by an explicit relation

$$\vec{x}(\vec{q}, t) = \vec{q} + D(t)\vec{s}(\vec{q}) \tag{2.1}$$

where $D(t)$ is the linear density growth factor fully specified by the cosmological model and the vector field $\vec{s}(\vec{q})$ is determined by the growing mode of initial density perturbations

$$\nabla \vec{s} = -\frac{1}{D} \left(\frac{\delta \rho}{\bar{\rho}} \right)_{\text{lin,g}}, \tag{2.2}$$

when $\vec{x} \to \vec{q}$. Therefore, the ZA assumes that the initial displacement field $\vec{s}(\vec{q})$ is a potential vector field,

$$\vec{s}_i(\vec{q}) = -\frac{\partial \Psi(\vec{q})}{\partial q_i}, \tag{2.3}$$

where the displacement potential Ψ is directly proportional to the linear perturbation of the gravitational field $\delta\phi_{lin}$ determined by $(\delta\rho/\bar{\rho})_{\text{lin,g}}$

$$\Psi(\vec{q}) = \frac{2}{3Da^2H^2\Omega} \delta\phi_{lin}(\vec{q}). \tag{2.4}$$

In the above equation $H = H(t)$ is the Hubble parameter and Ω is the dimentionless mean total density capable to cluster.

The potential character of the displacement vector is a direct consequence of the potential character of the growing mode of the gravitational instability in an expanding universe. An additional important aspect of the ZA is anisotropic deformation of the mass elements. The deformation is specified by the deformation tensor

$$d_{ij}(\vec{q}) = -\frac{\partial s_i(\vec{q})}{\partial q_j}. \tag{2.5}$$

From equation (2.1) one may easily infer an explicit expression for the density as a function of Lagrangian coordinates and time. If we consider the conservation of mass in differential form (it is worth reminding that $\bar{\rho}$ does not change with time in comoving coordinates)

$$\rho(\vec{x}, t)\mathrm{d}\vec{x} = \bar{\rho}\mathrm{d}\vec{q}, \tag{2.6}$$

the density evolution directly follows from

$$\rho(\vec{x}, t) = \bar{\rho} \left| J \left(\frac{\partial \vec{x}}{\partial \vec{q}} \right) \right|^{-1}, \tag{2.7}$$

where $J(\partial \vec{x}/\partial \vec{q})$ is the Jacobian determinant of the map given by equation 2.1. It is convenient to write equation 2.7 in terms of the eigen values $\lambda_i = (\alpha, \beta, \gamma)$ of the deformation tensor d_{ij}, resulting in an explicit equations for the density as a function of Lagrangian coordinates and time

$$\rho(\vec{q}, t) = \left| \frac{\bar{\rho}}{[1 - D(t)\alpha(\vec{q})][1 - D(t)\beta(\vec{q})][1 - D(t)\gamma(\vec{q})]} \right|, \tag{2.8}$$

where commonly used ordering of the eigen values $\alpha \geqslant \beta$ and $\beta \geqslant \gamma$ is assumed. The eigen values are nonGaussian random fields Doroshkevich (1970).

Expanding the expression for the Lagrangian density (eqn. 2.8) in a Taylor series,

$$\rho(\vec{q}, t) = \bar{\rho} + \bar{\rho}D(t)\left[\alpha(\vec{q}) + \beta(\vec{q}) + \gamma(\vec{q})\right] + \dots, \tag{2.9}$$

clarifies the relation between the known linear expression for the density contrast, δ_{lin} and the eigen values of the deformation tensor

$$\delta_{\text{lin}} = \frac{(\rho_{\text{lin}} - \bar{\rho})}{\bar{\rho}} = D(t)(\alpha + \beta + \gamma). \tag{2.10}$$

Formally speaking, the ZA is valid only in the linear regime when both $|D\alpha| \ll 1$ and $|D\gamma| \ll 1$. However, Zeldovich made a bold prediction: it should be a good qualitative and arguably quantitative approximation up to the beginning of the nonlinear stage, i.e. up to the stage at which $|D\alpha| = 1$ and even a little beyond. In order to compute the density field in Eulerian space one has to apply mapping eqn. 2.1.

One can easily deduce two key features of the density field at the beginning of the nonlinear stage from eqn. 2.8. Firstly the density becomes infinite as soon as $D(t)\alpha = 1$ or $D(t)\beta = 1$ or $D(t)\gamma = 1$. And secondly the collapse of a fluid particle must be anisotropic because the eigenvalues never satisfy the condition $\alpha(\vec{q}) = \beta(\vec{q}) = \gamma(\vec{q})$ for generic $\Psi(\vec{q})$. In addition, the condition

$$D(t)\lambda_i = 1 \tag{2.11}$$

determines caustics i.e. a set of surfaces where density is formally infinite.

It is easy to see that Eulerian linear perturbation theory (ELPT) is a limiting case of the ZA, assuming an additional condition $\vec{x} \approx \vec{q}$. Technically speaking, the ZA is an extrapolation of Lagrangian linear perturbation theory (LLPT) beyond the range of its formal applicability. There are two fundamental differences of LLPT from its Eulerian counterpart ELPT. As may be inferred from equations 2.8 and 2.10, the calculation of the density LLPT uses the full deformation tensor while ELPT relies only on its trace. The second difference is due to the necessity in LLPT of mapping from Lagrangian to Eulerian space in order to evaluate the density field in Eulerian space. Even at small σ_δ, where $\delta = \Delta\rho/\bar{\rho}$, the difference between LLPT and ELPT can be quite noticeable if the scale of the initial i.e. linear velocity field is considerably greater than that of density fluctuations. In this, we assume that σ_δ is evaluated in ELPT. The difference between LLPT and ELPT becomes considerable when both are extrapolated to a larger σ_δ. A particularly obvious problem occurring in ELPT is the emergence of negative densities, i.e. $\rho < 0$, in regions with a large initial density deficit. For example, if ELPT is extrapolated to $\sigma_\delta = 0.5$, the regions with negative densities occupy approximately 2.3% of the volume. This fraction increases to almost 15% at $\sigma_\delta = 1$. Evidently, for a physical model this is an unacceptable circumstance. LLPT is completely free of this problem: at all times it predicts $\rho > 0$, regardless of the magnitude of σ_δ. It is worth mentioning that each of three factors in the denominator of equation 2.8 could be negative. This is a very useful feature of the ZA which is briefly discussed bellow.

3. Further developments
3.1. *N-body simulations and accuracy of the ZA*

Cosmological N-body simulations have played a crucial role in theoretical studies and understanding of the structure in the universe. Despite of well known shortcomings they provide the most realistic data to compare with the observations. The first simulations were carried out in late 1960s. However all N-body simulations conducted in the western countries before 1983 started from Poisson distributions of particles in the simulation box. The only advantage of this initial condition consists in the easiness of its generation. But it has two major shortcomings: the only initial power spectrum of density fluctuations is flat and the smallest scales are in the nonlinear regime from the very beginning.

Doroshkevich, Ryabenkii & Shandarin (1973) were the first who generated initial conditions for a three-dimensional N-body simulation via the ZA. The goal of the study was testing the coherence of the ZA. The simulation involved two groups of particles. One of them moving according the ZA served as a source of the gravitational field. The trajectories of the other much smaller group of particles were integrated in this field. The particles that felt the gravitational field had the siblings in the first group of particles. The pairs of siblings started from identical initial conditions, this diverged with time since the ZA is not an exact solution. The accuracy of the ZA at the nonlinear stage was roughly assessed by comparing the trajectories, velocities and accelerations of each pair of siblings.

The authors also suggested an analytical estimate of the coherence of the ZA by computing the relative difference of two densities. One of which ρ is given by eq. 2.8 and the other $\tilde{\rho}$ is determined by the divergence of the acceleration field obtained from the ZA (eq. 2.1). If the ZA was exact solution then $\rho \equiv \tilde{\rho}$. It was shown that (see also Shandarin & Zeldovich (1989))

$$\frac{\tilde{\rho} - \rho}{\rho} \equiv D^2(-J_2 + 2DJ_3), \qquad (3.1)$$

where $J_2 = \alpha\beta + \alpha\gamma + \beta\gamma$ and $J_3 = \alpha\beta\gamma$. It is interesting that even at the time of the emergence of the pancake at $D = 1/\alpha$ when $\rho \to \infty$ the relative difference between two densities remain finite $(\tilde{\rho} - \rho)/\rho = -\beta/\alpha - \gamma/\alpha + \beta\gamma/\alpha^2$. Equation 3.1 also shows that the ZA is exact in one-dimensional case until shell crossing since $J_2 = J_3 = 0$.

Numerous studies elaborated on the quantitative accuracy and limitations of the approximation, see e.g. Coles, Melott & Shandarin (1993); Melott, Shandarin & Weinberg (1994); Yoshisato, Matsubara & Morikawa (1998); Yoshisato, *et al.* (2006) and references therein.

The first cosmological N-body simulations with Gaussian initial conditions generated via the ZA were carried out by Doroshkevich *et al.* (1980) and (Klypin & Shandarin 1983; Shandarin & Klypin 1984) in two- and three-dimensional cases respectively. This method of the initiation of cosmological simulations has become universal.

3.2. *Relation to Catastrophe Theory*

Ya.B published his paper on pancakes almost ten years prior to first observational evidences for existing highly anisotropic superclusters of galaxies Gregory & Thompson

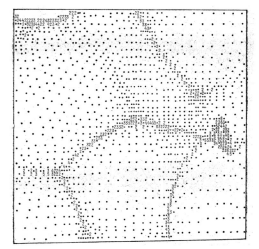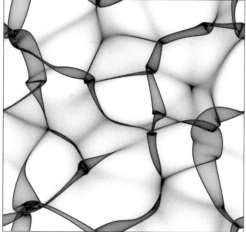

Figure 1. Left Panel: The first theoretical hint of the web structure (Doroshkevich, Zeldovich & Sunyaev 1976). The Russian caption in the original publication says: ' A typical distribution of mass particles obtained by S.F. Shandarin in two-dimensional simulation (see text). Symbols show the number of particles in the mesh cells. Initially particles were placed on a regular grid and slightly perturbed by smooth potential random vector field.' Right Panel: Modern simulation of the structure in 2D using the ZA. (Hidding, Shandarin & van de Weygaert 2014). Gray shades show the density while red and blue lines show the caustics corresponding to two eigen values.

(1978); Chincarini & Rood (1979); for a review see Oort (1983). However most of the cosmologists regarded the prediction of anisotropic structures with conspicuous skepticism for almost twenty years after it was published as described by Shandarin & Sunyaev (2009).

Ya.B. described just one generic type of structure formed by caustics - pancakes which emerged as the first structures at the shell crossing stage. The collapse of a uniform isolated ellipsoid was studied by Lin, Mestel & Shu (1965) who showed that the initially oblate ellipsoid tends toward a disk remaining uniform at all stages. The Zeldovich pancakes formed as a three-stream flow regions bounded by caustic surfaces. The shape of pancakes formed from generic Gaussian initial conditions is very different from ellipses. The first simulation of the ZA in two-dimensional case showed that the structure begins to emerge as a set of isolated pancakes however they weave in essentially single structure very quickly (see left hand side panel Fig. 1).

The explanation of this phenomenon required a dipper understanding of the geometry and topology of the mapping generated by eq. 2.1. Zeldovich turned to Arnold who enthusiastically began working on this problem. Soon the normal form for all generic singularities have been found (Arnold 1982) and the results was applied to cosmology (Arnold, Shandarin & Zeldovich 1982). Arnold, Shandarin & Zeldovich (1982) were able to crudely outline the richness of the web in 2D. A far more detailed study of the two-dimensional web was performed by Hidding, Shandarin & van de Weygaert (2014). Figure 1 (the right hand side panel) illustrates the improvements in both computation and visualization. The visualization of the web in 3D in its full complexity was a serious challenge in 1980s and remain a difficult problem even nowadays. The comprehensive analysis of the ZA in 3D is challenging computationally as well because it requires to study geometry and topology of highly nonGausssian fields which are $\lambda_i(\vec{q})$. Mapping to Eulerian space further complicates the analysis. In order to understand the dark matter

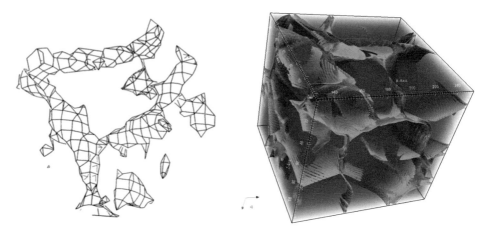

Figure 2. Left Panel: The density contour $\rho = 2.5\bar{\rho}$ obtained in the N-body simulation of the hot dark matter model (Shandarin & Klypin 1984). Right Panel: Prediction of the structure in 3D by the ZA. (Hidding, Shandarin & van de Weygaert 2014). The green surfaces are the α–caustics corresponding to the condition $D(t)\alpha = 1$ and red surfaces are the β-caustics seen only in the openings at the boundaries of the box; γ-caustics are completely obscured by the α- and β-caustics.

structure in the universe we have to understand much simpler structure described by the ZA (see the right hand panel of Fig. 2). It is very likely that the types of singularities in gravitating cold collisionless fluid evolved from potential initial condition will be similar to that in the ZA since the flow in individual streams remains potential as clearly demonstrated by Hahn, Angulo & Abel (2014).

3.3. *Morphology of the web*

Equation (2.1) encapsulates the mapping by the LLPT from Lagrangian space to Eulerian space. Its ramifications can be explored analytically before an overwhelming fraction of mass elements starts to experience shell crossing. The mathematical complications increase rapidly as the number and extent of the multistream flow regions proliferates. In particular interesting are those mass elements which are separated by finite distances in Lagrangian space and end up at the same places in Eulerian space. They belong to different streams and are key manifestations of the dynamically evolving mass distribution and mark the emerging cosmic web. To understand and assess this aspect of the Zeldovich approximation, we require numerical modeling, in particular for the cosmologically relevant situation of random initial conditions. The surfaces separated the regions with different number of streams in Eulerian space are caustics. Their progenitors in Lagrangian space isolate the fluid elements that experienced different number of turns inside out or flip-flops.

Shandarin & Klypin (1984) suggested that the structure emerges in the sequence: pancakes, filaments and finally haloes. The pancakes begin to emerge in the vicinity of $\alpha(\vec{q}) = max$ due to the collapse along the corresponding eigen vector $\vec{n}_\alpha$. Then the following collapse along the eigen vector $\vec{n}_\beta$ results in the emergence of filaments. Finally the collapse along the eigen vector $\vec{n}_\gamma$ leads to the origin of halos. The model obviously represents a substantial simplification of the dynamics in the dark matter component (Hidding, Shandarin & van de Weygaert 2014).

However recently Falck, Neyrinck & Szalay (2012) exploited a very similar idea for developing the ORIGAMI method of identifying structures in cosmological N-body

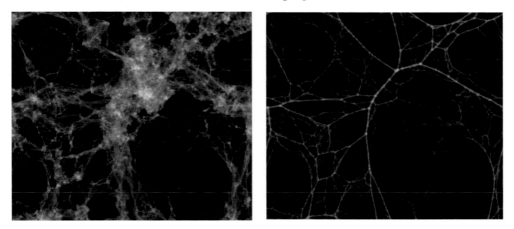

Figure 3. The ZA (left) v.s. Adhesion approximation (right). Courtesy of J. Hidding.

simulations. Their analysis shows a good agreement of the model with cosmological N-body simulations. Another successful numerical algorithm NEXUS for identifying the basic elements of the web (halos, filaments, walls aka pancakes, and voids) that incorporates some essential concepts of the ZA has been recently suggested by Cautun *et al.* (2014).

Klypin and Shandarin (1983, 1984) were puzzled why only the filamentary structure was observed in the simulation of the universe dominated by hot dark matter (see the left hand side panel of Fig. 2) but not Zeldovich's pancakes. They correctly conjectured that the pancakes have too low density contrast to be seen in their simulations. The right hand panel of Fig. 2 demonstrates that in order to see pancakes clearly one has to compute and plot the caustic surfaces or achieve the mass resolution significantly higher than in common cosmological N-body simulations.

3.4. *Topology of the large-scale structure*

The both two and three–dimensional N-body simulations (Doroshkevich *et al.* 1980; Klypin & Shandarin 1983; Shandarin & Klypin 1984) as well as two– and three–dimensional models based on the ZA (Fig. 1 and 2) showed that the filamentary structure spans throughout the entire simulation box. This observation raised the question of the topology of the large-scale structure. Ya.B. initiated the study of the topology of the structure resulted in a series of papers based on percolation theory (Zeldovich 1982; Zeldovich Einasto & Shandarin 1982; Shandarin 1983; Shandarin & Zeldovich 1983, 1984; Klypin 1987; Klypin & Shandarin 1993). Later Gott, Dickinson & Melott (1986) suggested to use the genus statistics for quantifying the topology of the structure. Sahni, Sathyaprakash & Shandarin (1997) compared two techniques and concluded that the percolation method along with ability to distinguish nonGaussian from Gaussian fields also reveals the structure difference in Gaussian fields with different power spectra. The latter property allowed to relate the effects of nature (initial field) and nurture (nonlinear evolution) on the topology of the web (Shandarin, Habib & Heitmann 2010).

At present the term "the large-scale structure" is used less often than "the Cosmic Web" coined by Bond, Kofman & Pogosyan (1996). The difference in the meaning of these two terms is substantial. "The large-scale structure" does not invoke any particular geometry or pattern while "the Cosmic Web" almost inevitably invokes an image of a network made of filaments like a spider-web. Shandarin & Klypin (1984) discussed this issue and and came to the following conclusion: "The regions of high density seem to form

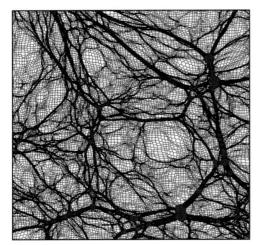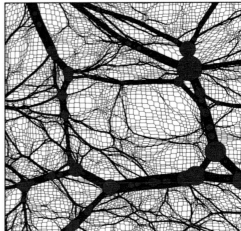

Figure 4. A modern version of the Adhesion approximation. A two-dimensional model shows two stages of the evolution of the web in coming coordinates. One can see the multi scale nature of the web and multiple nature of the halo mergers when small voids are squeezed by expanding large voids. The red circles symbolize the halos which sizes indicate the masses. From Hidding *et al.* (2012).

a single three-dimensional web structure. However, it is not clear from our simulations whether honeycomb structure arises or not." At present the answer is unambiguous: in both the ZA (Fig. 2) and N-body simulations (Ramachandra & Shandarin 2014) the three-stream flow regions form irregular cellular structure made of very thin quasi two-dimensional regions most of which are interconnected in a single percolating structure.

3.5. *The Adhesion model*

Not being accurate the ZA remains qualitatively correct until massive shell crossing happens. The streams continue moving through each other without bounds. This results in unlimited growth of the thickness of the structure as illustrated by the left hand side panel in Fig. 3. The numerical simulations showed that the correctly computed gravitational forces keep the structure very thin. In order to suppress the growth of the thickness of the pancakes Gurbatov, Saichev & Shandarin (1985, 1989) (see also Gurbatov, Saichev & Shandarin (2012) for a recent review) suggested to introduce artificial viscosity into the dynamical equations describing the ZA. Since this modification cannot yield the correct result the form of the viscosity was chosen in the form that results in the Burgers equation well known in theory of turbulence. An important advantage of the Burgers model consists in that it allows an analytic solution. In the limit of the infinitesimal viscosity the structure becomes a two-dimensional surface of a very complicated shape in 3D or a one-dimensional line in 2D as shown in the right hand panel of Fig. 3. The web predicted by the adhesion approximation can be called a skeleton of the web. The accuracy and limitations of the adhesion approximation have been carefully studied (Kofman, Pogosyan & Shandarin 1990; Nusser & Dekel 1990; Kofman *et al.* 1992; Melott, Shandarin & Weinberg 1994). It has also been used for the simulation of deep redshift surveys (Weinberg & Gunn 1990a,b).

A recent very interesting new development of the Adhesion model revealed a deep connection of the relation between Eulerian and Lagrangian space with that between Voronoi and Delaunay tessellations (Hidding *et al.* 2012). It has been found that the walls aka pancakes of the web as edge-like objects in Lagrangian space, whereas filaments have a flattened signature in Lagrangian space. Halos, being the most massive concentrations of mass are therefore most extended progenitors in Lagrangian space. The Adhesion model can be naturally interpreted as a skeleton of the web as illustrated by Fig. 4.

The Adhesion model naturally incorporates the multi scale character of the web, anisotropic accretion of mass on halos from the filaments and to lesser extent from walls, extinguishing of small voids when they pushed to the walls by large voids, simultaneous merger of several halos when a void collapses.

4. Other recent developments

Illustrating the idea of a pancake Ya.B. had a plot (Fig.2 in Zeldovich (1970)) Eulerian coordinate as a function of Lagrangian coordinate at three times: before, after and exactly at the time of birth of a pancake. This is one of the Lagrangian sub-manifolds playing fundamental role in classical mechanics. The distribution of dark matter in $(\vec{q}, \vec{x})$–space, where $\vec{q}$ and $\vec{x}$ are initial and actual comoving coordinates of particles in N-body simulation at time t, is highly degenerate. It occupies a very thin sheet around three-dimensional submanifold due to very low temperature of dark matter.

Recently a very promising technique has emerged based on a triangulation of a three-dimensional submanifold $\vec{x} = \vec{x}(\vec{q}, t)$ in six-dimensional space $(\vec{q}, \vec{x})$ (Shandarin, Habib & Heitmann 2012; Able, Hahn & Kaehler 2012). Based on the triangulation one can compute density, velocity, number of streams, and other fields with considerably higher spatial resolution than other techniques can achieve (Angulo, Hahn & Abel 2013; Hahn, Angulo & Abel 2014; Ramachandra & Shandarin 2014; Shandarin & Medvedev 2014). Therefore this technique allows to visualize and study the web with unprecedented accuracy.

Another extremely interesting example of studies stemmed from the ZA. The ZA was used for computation of the two-point function of the matter and biased tracers (White 2014). The comparison with the N-body simulations and other Lagrangian perturbation theories showed the ZA provides a good fit the N-body results. The ZA was also used to compute the ingredients of the Gaussian streaming model of Reid & White (2011). It was found that this hybrid model, refered to as the Zeldovich streaming model and which involves only simple integrals of the linear theory power spectrum, provides a good match to the N-body measurements down to tens of Mpc.

5. Summary

The ZA continues to inspire cosmologists to create new approaches to the study and understanding of the web: its origin, evolution and morphology. Currently the ZA is 'one of the most successful and insightful analytic models of structure formation' (White 2014). It provides a very sophisticated framework for studies and understanding the complexity of the large-scale structure in the universe. It incoporates a number of necessary concepts and therefore provides language for accounting of the highly nontrivial morphology of the web. It is also a very good approximation for statistical calculations.

References

Abel, T., Hahn, O., & Kaehler R. 2012, *MNRAS*, 427, 61
Angulo, R. E., Hahn, O., & Abel, T. 2013, *MNRAS*, 434, 3337
Arnold, V. I., 1982, *Tr. Semin. imeni I. G. Petrovskogo*, 8, 21

Arnold, V. I., Shandarin, S. F., & Zeldovich, Ya. B., 1982, *Geophys. Astrophys. Fluid Dyn.*, 20, 111

Bond, J. R., Kofman, L., & Pogosyan, D. 1996, *Nature*, 380, 603

Cautun, M., van de Weygaert, R., Jones, B. J. T., & Frenk, C. S. 2014, *MNRAS*, 441, 2923

Chincarini, G. & Rood, H. J. 1979, *ApJ*, 230, 648

Coles, P., Melott, A. L., & Shandarin, S. F., 1993, *MNRAS*, 260, 765

Doroshkevich, A. G. 1970, *Astrophysics*, 6, 320

Doroshkevich, A. G., Ryabenkii, V. S., & Shandarin, S. F., 1973, *Astrophysics*, 9, 144

Doroshkevich, A. G., Zeldovich, Ya. B., & Sunyaev, R. A. 1976, in *The origin and evolution of galaxies and stars*, ed. S.B. Pikelner, Nauka, Moscow, p.77, in Russian

Doroshkevich, A. G., Kotok, E. V., Poliudov, A. N., Shandarin, S. F., Sigov, I. S., & Novikov, I. D., 1980, *MNRAS*, 192, 321

Falck, B. L., Neyrinck, M. C., & Szalay, A. S., 2012 *ApJ.*, 754, 126

Gott, III J. R., Dickinson, M., & Melott, A. L. 1986, *ApJ*, 306, 343

Gregory, S. A. & Thompson, L. A., 1978, *ApJ.*, 222, 784

Gurbatov, S. N., Saichev, A. I., & Shandarin, S. F. 1985, *Sov. Phys. Dokl.*, 20, 921

Gurbatov, S. N., Saichev, A. I., & Shandarin, S. F. 1989, *MNRAS*, 236, 385

Gurbatov, S. N., Saichev, A. I., & Shandarin, S. F. 2012, *Physics Uspekhi*, 55, 223

Hahn, O., Angulo, R., & Abel, T. 2014, *arXiv:1404.2280*

Hidding, J., Shandarin, §. F., & van de Weygaert, R. 2014, *MNRAS*, 437, 3442

Hidding, J., van de Weygaert, R., Vegter, G., & Jones B. J. T. 2012, *arXiv:1211.5385*

Klypin, A. A. 1987, *Sov. Astron.*, 31, 8

Klypin, A. A. & Shandarin, S. F., 1983, *MNRAS*, 204, 891

Klypin, A. A. & Shandarin, S. F., 1993, *ApJ*, 413, 48

Kofman, L., Pogosyan, D., & Shandarin, S. F. 1990, *MNRAS*, 242, 200

Kofman, L., Pogosyan, D., Shandarin, S. F., & Melott, A. L. 1992, *ApJ*, 393, 449

Lin, Mestel & Shu 1965, *ApJ*, 142, L1431

Melott, A. L., Shandarin, S. F., & Weinberg, D., 1994, *ApJ*, 428, 28

Nusser, A., & Dekel, A, 1990*ApJ*, 362, 14

Oort, J. H. 1983, *Ann. Rev., Astron. Astrophys.*, 21, 373

Ramachandra, N. S. & Shandarin, S. F. 2014, *arXiv:1412.7768*

Reid, B. A. & White, M. 2011, *MNRAS* 417, 1913

Sahni, V., Sathyaprakash, B., & Shandarin, S. F., 1997, *ApJ*, 476, L1

Shandarin, S. F., 1983, *Sov. Astron. Lett.*, 9, 104

Shandarin, S. F. & Klypin, A. A., 1984, *Sov. Astron.*, 28, 491

Shandarin, S. F. & Medvedev, M. V. 2014, *arXive:1409.7634*

Shandarin, S. F. & Sunyaev, R. A. 2009, *A & A*, 500, 19

Shandarin, S. F. & Zeldovich, Ya. B., 1983, *Comments on Astrophys.*, 10, 33

Shandarin, S. F. & Zeldovich, Ya. B., 1984, *Phys. Rev. Lett.*, 52, 1418

Shandarin, S. F. & Zeldovich, Ya. B., 1989, *Rev. Mod. Phys.*, 61, 185

Shandarin, S. F., Habib, S., & Heitmann, K. 2010, *Phys. Rev. D*, 81, 103006

Shandarin, S. F., Habib, S., & Heitmann, K. 2012, *Phys. Rev. D*, 85, 083005

Weinberg, D. H., & Gunn, J. E. 2014a, *ApJ*, 352, L25

Weinberg, D. H. & Gunn, J. E. 2014b, *MNRAS*, 247, 260

White, M., 2014, *MNRAS*, 439, 3630

Yosisato, A., Matsubara, T., & Morikawa, M., 1998, *ApJ*, 498, 48

Yosisato, A., Morikawa, M., Gouda, N., & Mouri, H., 2006, *ApJ*, 637, 555

Zeldovich, Ya. B. 1970, *A & A*, 5, 84

Zeldovich, Ya. B. 1982, *Sov. Astron. Lett.*, 8, 102

Zeldovich, Ya. B. 1993, *Selected Works of Yakov Borisovich Zeldovich*, Eds. J.P. Ostriker, G.I. Barenblatt, R.A. Sunyaev, Princeton University Press, Princeton 1993

Zeldovich. Reminiscences, Ed. R.A. Sunyaev, Chapman & Hall/CRC, Boca Raton, London, New York, Washington, D.C. 2004

Zeldovich, Y. B., Einasto, J., & Shandarin, S. F. 1982, *Nature*, 300, 407

The Zeldovich Universe:
Genesis and Growth of the Cosmic Web
Proceedings IAU Symposium No. 308, 2014
R. van de Weygaert, S. Shandarin, E. Saar & J. Einasto, eds.

© International Astronomical Union 2016
doi:10.1017/S1743921316009571

Yakov Zeldovich
and the Cosmic Web Paradigm

Jaan Einasto

Tartu Observatory,
Observatooriumi 1, 61602 Tõravere, Estonia
email: jaan.einasto@to.ee

Abstract. I discuss the formation of the modern cosmological paradigm. In more detail I describe the early study of dark matter and cosmic web and the role of Yakov Zeldovich in the formation of the present concepts on these subjects.

Keywords. cosmology: dark matter, cosmology: large-scale structure of universe, cosmology: theory

1. Formation of the modern cosmological paradigm

The modern classical cosmological paradigm was elaborated step by step during the first part of the 20th Century. It was found that there exist stellar systems outside our Milky Way – external galaxies (Öpik (1922), Hubble (1925)). Next it was found that external galaxies are moving away from us, i.e. the Universe is expanding (Hubble (1929)). On the basis of Einstein relativity theory Friedmann (1922) explained the expansion as a property of the infinite universe. The speed of the expansion can be expressed in terms of the Hubble constant, H_0. Sandage & Tammann (1975) found a value about $H_0 = 50$ km s^{-1} Mpc^{-1}, whereas de Vaucouleurs (1978) and van den Bergh (1972) preferred a value around $H_0 = 100$ km s^{-1} Mpc^{-1}. Due to this uncertainty the Hubble constant is often expressed in dimensionless units h, defined as: $H_0 = 100\ h$ km s^{-1} Mpc^{-1}.

Another basis of the classical cosmological paradigm is the distribution of galaxies and clusters of galaxies. A photographic survey was made using the 48-inch Palomar Schmidt telescope. Abell (1958) used the Palomar survey to compile a catalogue of rich clusters of galaxies for the Northern sky; later the catalogue was continued to the Southern sky (Abell *et al.* (1989)). Zwicky *et al.* (1968) used this survey to compile for the Northern hemisphere a catalogue of galaxies and clusters of galaxies. The galaxy catalogue is complete up to 15.5 photographic magnitude. Both authors noticed that galaxies and clusters of galaxies show a tendency of clustering.

A deeper complete photographic survey of galaxies was made in the Lick Observatory by Shane & Wirtanen(1967). The Lick counts as well as galaxy and cluster catalogues by Zwicky and Abell were analysed by Jim Peebles and collaborators to exclude count limit irregularities (Soneira & Peebles (1978)). These data show the apparent (2-dimensional) distribution of galaxies and clusters on the sky. The basic conclusion from these studies is that galaxies are hierarchically clustered. There exist clusters and superclusters of galaxies, but most galaxies form a more-or-less randomly distributed population of field galaxies.

The mean density due to galaxies was determined using the mean luminosity density, and the mean mass-to-luminosity ratio (M/L) of galaxies. Estimates available in the 1950's indicated a low-density Universe, $\Omega \approx 0.05$.

This complex of data formed the classical cosmological paradigm. However, the theoretical explanation of the data by the Friedman model was mathematical, it did not consider physical processes in the early universe. Thus in the early 1960's in several centres theorists started to think on the physics of the early universe. Most important developments in this direction were made in Princeton by Jim Peebles, and in Moscow by Yakov Zeldovich and their collaborators.

One of the first step in the study of physical processes was the elaboration of the hierarchical clustering scenario of galaxies by Peebles & Yu (1970), Peebles (1971). On the other hand, the Moscow team developed the pancake model of structure formation by Zeldovich (1970), and the theory of the Sunyaev-Zeldovich effect in the Cosmic Microwave Background radiation by Sunyaev & Zeldovich (1969).

To discuss new problems of cosmology and astrophysics Zeldovich organised summer and winter schools. The first of such schools was in the new observatory in Tõravere, 1962; later schools were hold in Caucasus winter resorts. Our Tartu cosmology team was invited to Caucasus winter schools in 1972, 1974 and later. Discussions on winter schools started our collaboration with the Zeldovich team.

Also important observational discoveries were made. Penzias & Wilson (1965) detected the Cosmic Microwave Background radiation. Satellite observatories allowed to detect X-rays from clusters of galaxies, and to find the mass of the hot gas in clusters, as well as the total mass of clusters (Forman *et al.* (1972)).

These observational and theoretical developments were the basis of the formation of the modern cosmological paradigm. In the following I shall discuss in more detail some aspects of the new cosmological paradigm, related the discovery of the cosmic web. Quite unexpectedly we found that the structure of the cosmic web is closely related to another problem – the nature of the dark matter and its role in the formation of the cosmic web.

2. Dark matter

In the middle of 1960's the general attitude of the astronomical community was that the classical cosmological paradigm is in agreement with all observational and theoretical data available. Actually there were some unexplained facts. One of these curious data was the Coma cluster mass paradox. The mass calculated from random motions of galaxies in the cluster was much higher than the expected mass found by adding masses of individual galaxies, as suggested by Zwicky (1933).

Another curious fact was the form of rotation curves of galaxies. As found by Oort (1940), Roberts (1966) and Rubin & Ford (1970), the rotation curves of spiral galaxies are flat on large galactocentric distances. Since the surface brightness of galaxies falls rapidly on the periphery, flat rotation curves mean, that the mass-to-luminosity ratio rapidly increases on large galactocentric distances. Oort (1940) and Roberts (1966) explained this observation with the assumption that on large distances low-mass stars dominate in galaxies. For some unclear reason these observations were ignored by the astronomical community.

Tartu astronomers have studied methods of modelling the structure of galaxies for years. The first dynamical model of the Andromeda galaxy was calculated by Öpik (1922). Kuzmin(1952), Kuzmin (1956) developed more accurate method of modelling galaxies, and applied the method to our Galaxy.

I helped Kuzmin in calculations and was interested to continue the modelling of galaxies, using more observational data on galactic populations. First I studied carefully methods used by previous authors to calculate mass distribution models of galaxies. To my surprise I found that in most models simple conditions of physical reasonability are not satisfied. Most important conditions are: the spatial density must be non-negative and finite, some moments of the density must be finite, in particular moments which define the mass and the effective radius of the model. Thus I found that the only mass distribution profile which satisfies all physical conditions is a generalised exponential model: $\varrho(a) = \varrho_0 \exp\left(-(a/a_c)^{1/N}\right)$, where ϱ_0 is the central density, a is the semi-major axis of the equidensity ellipsoid, a_c is the core radius, and N is the structural parameter, which allows one to vary the shape of the density profile. I used this density profile in my model of the Galaxy (Einasto (1965)), and in models of other galaxies. Presently this profile is known as the "Einasto profile".

The central question in modelling galaxies is the calibration of mass-to-luminosity ratios of populations. This can be done using additional independent data. Most important data are velocity dispersions of open and globular clusters with similar photometric properties, assuming that galactic populations have been formed by dissolution of clusters and star associations. To bring data on populations of different age and composition to a coherent system I developed models of evolution of populations, similar to models by Tinsley (1968). To my surprise I discovered that it is impossible to represent rotation curves of galaxies by the sum of gravitational attraction of known stellar populations. The only way to bring kinematical and photometrical data into agreement was to suppose the presence of a new population — corona — with large radius, mass and M/L ratio.

I calculated models with massive coronas for all major galaxies of the Local Group and the Virgo cluster central galaxy, M87. Results were discussed at the First European Astronomy Meeting in Athens, September 1972 (Einasto (1974)). However, observed rotation curves were not long enough to find the mass and the radius of coronas. Thus I continued to think how to find total masses and radii of coronas. Finally I decided to use companion galaxies as mass tracers of giant galaxies. I collected data needed and found that the mass (and the effective radius) of coronas is about ten times larger than the sum of masses of all known stellar populations. The total cosmological density of matter in galaxies including massive coronas is 0.2 of critical cosmological density (Einasto *et al.* (1974)). A similar density estimate was confirmed by Ostriker *et al.* (1974).

I reported these results in the Arkhõz Winter School in January 1974. My principal conclusion was, that all giant galaxies have massive coronas, and that coronas cannot have stellar nature. Thus the coronal or dark matter is the principal constituent of the universe, and its nature is not clear. After my talk Zeldovich invited me to his room and asked two questions: Can we find data which give some hints to the physical nature of coronas? Can we find observational evidence which can be used to discriminate between various theories of the formation of galaxies?

To discuss the existence and the physical nature of dark matter, we organised in January 1975 a conference in Tallinn, Estonia (Doroshkevich *et al.* (1975)). The rumour on dark matter had spread around the astronomical and physics community and all leading Soviet astronomers and physicists attended. Two basic models were suggested for coronae: faint stars or hot gas. It was found that both models have serious difficulties. Neutrinos were also discussed but excluded since they can form only clumps of rich cluster mass, but coronas of galaxies have thousand time lower masses.

The dark matter problem was discussed also in the Third European Astronomical Meeting in Tbilisi in June 1975. In the dark matter session the principal discussion was between the supporters of the classical paradigm with conventional mass estimates of galaxies, and supportes of the new paradigm with dark matter. The most serious arguments in favour of the classical cosmological paradigm were presented by Materne & Tammann (1976): Big Bang nucleosynthesis suggests a low-density Universe with the density parameter $\Omega \approx 0.05$ (this difficulty was already discussed by Zeldovich in the Tallinn

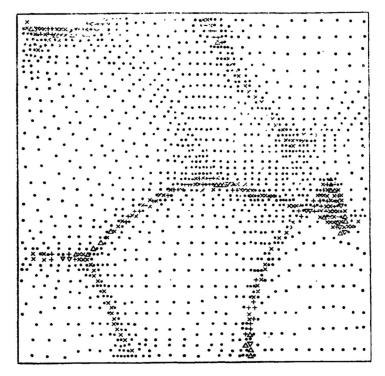

Figure 1. Distribution of particles in simulations according to Zeldovich pancake model, cited by Einasto *et al.*(1980).

conference); the smoothness of the Hubble flow also favours a low-density Universe. It was clear, that the existence of dark matter is in conflict with the classical cosmological paradigm. If it exists, then the density $\Omega \approx 0.2$ must be explained in some other way.

The nature of dark matter and its role in the evolution of the universe was a problem for almost ten years. To solve it data on the distribution of galaxies in space and other new data were needed.

3. Structure of the Universe

When Zeldovich asked the question on the formation of galaxies I had initially no idea how we could find an answer. But soon I remembered our previous experience in the study of galactic populations: their spatial distribution and kinematics evolve slowly. Systems of galaxies are much larger in size, thus their evolution must be even slower. Random velocities of galaxies are of the order of several hundred km/s or less, thus during the whole lifetime of the Universe galaxies have moved from their place of origin only about $1\,h^{-1}$ Mpc. If there exist some regularities in the large-scale distribution of galaxies, these regularities must reflect the conditions in the Universe during the formation of galaxies. Thus we had a leading idea to answer the second Zeldovich question: *We have to study the distribution of galaxies on large scales.*

We started to collect redshift data from all available sources. Since we needed data on large-scale distribution of galaxies, we collected redshifts not only for galaxies, but also for near cluster, both Abell and Zwicky clusters, as well as active galaxies (radio and Markarian galaxies). Our experience showed that clusters and active galaxies are good

Figure 2. Jim Peebles (left), Yakov Zeldovich and Malcolm Longair (right) at the IAU Tallinn Symposium 1977.

tracers of the skeleton of the structure. Redshifts of galaxies and clusters were searched for the whole Northern Hemisphere.

In the middle of 1970's there were two basic structure formation scenarios, the Peebles & Yu (1970) hierarchical clustering scenario and the Zeldovich (1970) pancake scenario. The hierarchical scenario represents well the apparent 2-dimensional distribution of galaxies, seen in Lick maps. Numerical experiments done in the Zeldovich team showed the formation of high-density knots joined by chains of particles to a connected network. Our challenge was to find out whether the real distribution of galaxies shows some similarity with one of the theoretical pictures.

To visualise the three-dimensional distribution we used wedge diagrams at various declination intervals and plots of clusters of galaxies in the Perseus supercluster region at different redshift intervals.

After the Tbilisi Meeting Zeldovich proposed to organise an international symposium devoted solely to cosmology. This suggestion was approved by IAU, and the symposium "Large Scale Structure of the Universe" was hold in Tallinn in September 1977. Two pictures of participants are shown in Figure 2.

The first speaker on the distribution of galaxies was Tully & Fisher (1978), who showed a film of the Local Supercluster. The film showed that the supercluster consists of a number of galaxy chains which branch off from the supercluster's central cluster. No galaxies could be seen in the space between the chains. The presence of voids in the distribution of galaxies was reported also by Tifft & Gregory (1978), and Tarenghi *et al.* (1978) in the Coma and Hercules superclusters, respectively.

In our presentation we showed wedge diagrams and cluster plots obtained in the Perseus supercluster region, see Figure 3. In the Figure left column shows wedge diagrams in three declination zones (Jõeveer & Einasto (1978)). Filled circles are for rich clusters of galaxies, open circles — groups, dots — galaxies, crosses — Markarian galaxies. In right

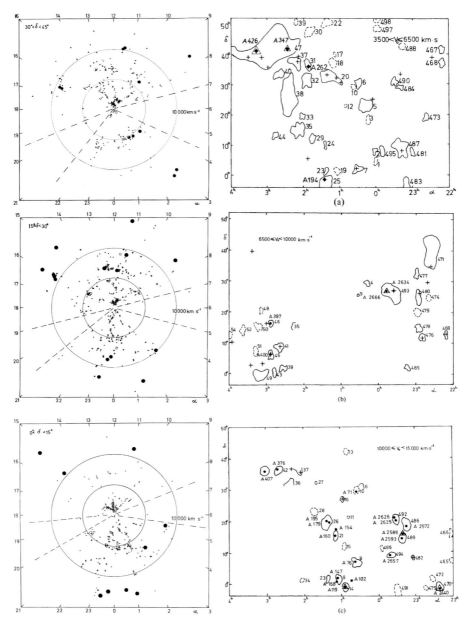

Figure 3. The distribution of galaxies and clusters, see text for explanations.

panels we plot Abell clusters and contours of Zwicky clusters in the Perseus area of sky at three distance intervals (Einasto *et al.* (1980)). When combined these plots show the large-scale three-dimensional distribution of galaxies and systems of galaxies.

These pictures showed a great richness of the distribution of galaxies — instead of random distribution of galaxies and clusters there exists a complicated hierarchical network, which we called "cellular structure". Not only filaments (chains) of galaxies and clusters were seen, but it was clear that galaxy chains form bridges between superclusters, thus there exists an almost continuous network of superclusters and filaments. Some chains are rich and consists of clusters and groups of galaxies, as the main ridge of the Perseus–Pisces supercluster. Filaments of galaxies across large voids are poor and consist only of galaxies and poor Zwicky clusters. In short, the three-dimensional data showed that the structure of the Universe is much richer than believed so far. Presently it is called the cosmic web.

Our picture had some similarity with the simulation made for the Zeldovich pancake scenario. However, the comparison was made only on the basis of visual impression by the comparison of the model and observed distributions of particles/galaxies. Zeldovich (1978) in his talk suggested to compare observations with models using quantitative methods. Thus we started with Zeldovich and his collaborators a search of quantitative methods to investigate properties of the distribution of galaxies.

Our main results were published by Zeldovich *et al.* (1982). Here we used the correlation function, the connectivity of systems of galaxies, the length of the largest system calculated for various linking lengths, and the multiplicity function of systems of galaxies. Comparison was made for a three-dimensional pancake model, hierarchical clustering model, Poisson model, and observations (a volume limited sample of galaxies including the Virgo supercluster). These tests showed that in most tests the pancake model is in good agreement with observations. In contrast, the hierarchical clustering model is in conflict with all tests, see Figure 4.

However, some differences between the pancake model and observations were evident. The most important difference is the lack of systems of intermediate richness in the pancake model, observed in real galaxy samples. As we understood soon, the reason for this disagreement was the assumption that dark matter consists of neutrinos.

4. Astro-particle physics

In early 1980's several important observational and theoretical analyses were made which reinforced the need for a paradigm shift. To understand the nature of dark matter of key value were searches of fluctuations of temperature fluctuations of CMB. From theoretical considerations it was clear that the temperature of CMB cannot be constant, and the expected amplitude of fluctuations was $\delta T/T \approx 10^{-3}$ (assuming the baryonic nature of the hot plasma before recombination). Fluctuations were searched with best radio telescopes available, none was found and the upper limits were much lower than the expected amplitude.

The most important theoretical development was the elaboration of the inflation model of the early universe by Starobinsky (1980), Starobinsky (1982), Guth (1981) and Linde (1982). The inflation model was based on various observational and theoretical considerations. One of the main conclusions of the model is the prediction that the total matter/energy density of the universe must be exactly equal to the critical cosmological density, $\Omega_{tot} = 1$.

The observed density of baryonic matter in the universe is about $\Omega_b = 0.05$, supported by primordial nucleosynthesis considerations (mentioned already by Materne & Tammann (1976) in the dark matter discussion in Tbilisi, 1975). Thus the only way to explain the low level of CMB temperature fluctuations, and conclusions from the inflation theory was to assume, that dark matter is non-baryonic. Non-baryonic matter is very weakly interacting with radiation, and density fluctuations in the non-baryonic matter can start to amplify already during the hot phase of the evolution of the universe.

These problems were discussed in April 1981 in a workshop in Tallinn, where both particle physicists and astronomers attended. A workshop of similar topic was held in

September-October in Vatican. In both workshops one of the main conclusions was the non-baryonic nature of dark matter. These workshops mark the formation of a new area in research – astro-particle physics.

The first natural candidate for the dark matter was neutrino, the only known non-baryonic particle. However, the problems with neutrinos as dark matter candidate were soon realised, as discussed, among others, by Zeldovich *et al.* (1982). Thus astronomers and physicists started to think what would be the alternatives. The main argument against neutrinos was their very high speed, close to the speed of light, which allowed to form only very massive cluster-sized systems. To allow the formation of smaller systems dark matter particles must have higher mass and lover speed. So various hypothetical particles were considered allowing the formation of systems of lower mass. Such particles were commonly called Cold Dark Matter, in contrast to neutrino-based Hot Dark Matter.

In 1983 Adrian Melott has made N-body simulations with density perturbation spectra which corresponded to the hot dark matter as well as to the cold dark matter scenario. He visited Moscow and Tallinn to discuss his results and to compare models with observations. The analysis was made jointly with Moscow and Tartu teams, and was published by Melott *et al.* (1983). Here we applied the same tests as used by Zeldovich *et al.* (1982). Our results showed that the CDM model is in excellent agreement with all quantitative tests. The paper ends with the conclusion, that the formation of the structure starts with the flow of particles to form the filamentary web as in the Zeldovich pancake model, but in the subsequent evolution systems grow as in the hierarchical clustering scenario by Peebles.

The advantages of the CDM model were discussed in detail by Blumenthal *et al.* (1984). Now, finally the presence of dark matter was accepted by leading theorists. A very detailed series of N-body simulations based on CDM and accepting a closed universe with critical density was made by the "Gang of Four" (Efstathiou *et al.* (1985), White *et al.* (1987)).

In 1980's the attention of our cosmology team in Tartu was devoted to quantitative study of the structure of the cosmic web using various tests. In these studies we used initially Melott CDM simulations to compare observation with models of structure formation. But we had the need to have our own simulations to have full control of the model. Enn Saar suggested to develop a model with critical cosmological density with dark matter fraction, $\Omega_{DM} = 0.2$, as we have found from observations (Einasto *et al.* (1974)). The rest of the matter/energy density is in the cosmological Λ term, $\Omega_\Lambda = 1 - \Omega_{DM} = 0.8$. The simulation was made by our gradual student Mirt Gramann.

Our LCDM model was used to investigate various properties of the cosmic web. The first study was devoted to the topology of the cosmic web by Einasto *et al.* (1986). This study shows that the LCDM model fits observational data even better than the standard CDM model with critical density. The need to use the LCDM model was discussed in detail by Efstathiou *et al.* (1990.

Already in our first study of the cosmic web by Jõeveer *et al.* (1977) we had the question: Do galaxies form sheets between filaments as expected in the pancake scenario, or are they formed only in high-density regions as filaments at sheets crossing, and knots at filament crossings. The same question was asked also by Zeldovich *et al.* (1982), and studied on the basis of observational data by Einasto *et al.* (1980). The preliminary answer was – there exists no sheets of galaxies which isolate neighbouring low-density regions between superclusters.

The more detailed study by Einasto *et al.* (1986) showed, that the topology of the web depends on the threshold density level applied to separate low- and high-density regions in simulations. At very low threshold density sheets of particles isolate voids between rich

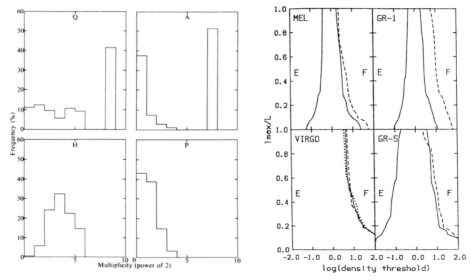

Figure 4. Left: the distribution of galaxies according to the multiplicity of the system. Multiplicity is expressed in powers of 2. Samples are designated: O — observed, A — adiabatic pancake model, H — hierarchical clustering model, P — Poisson model (Zeldovich *et al.* (1982)). Right: the length of the largest system (in units of the box size) versus the density threshold (in units of the mean density of the sample). E is for low density (empty) regions, F high density (filled) regions; MEL — Melott CDM simulation, GR-1 and GR-5 Gramann LCDM simulation at expansion factors 1 and 5.2 (present epoch), and VIRGO the observed sample around the Virgo supercluster. In models solid lines indicate unbiased samples with all test particles included, dashed lines biased samples, where particles in low-density regions have been removed (Einasto *et al.* (1986)).

regions. However, in low-density regions there exists no conditions to form galaxies. The density of the collapsing gas must exceed a threshold about 1.6 of the mean density to have during the Hubble time the possibility to collapse, as shown by Press & Schechter (1974). If we exclude particles from low-density regions (biased galaxy formation) then voids (low-density regions in simulations) form just one large connected region, both in real and model samples — there are no sheets isolating voids, see Figure 4. The length of the largest system depends on the threshold level in all samples. Similar results were obtained by Gott *et al.* (1986).

One aspect of the structure of the cosmic web is its fractal character, as suggested by Mandelbrot (1982), Mandelbrot (1986). During a visit to NORDITA with Enn Saar we investigated the fractal properties of the web in collaboration with Bernard Jones and Vicent Martinez. Ou results showed that both observational and model samples show multi-fractal properties (Jones *et al.* (1988)).

One difficulty with the original pancake scenario by Zeldovich was the shape of objects formed during the collapse. It was assumed that forming systems are flat pancake-like objects, whereas the dominant features of the cosmic web are filaments (Jõeveer & Einasto (1978), Einasto *et al.* (1980)). This discrepancy was explained by Bond *et al.* (1996), who showed that, due to tidal forces, in most cases only essentially one-dimensional structures, i.e. filaments form.

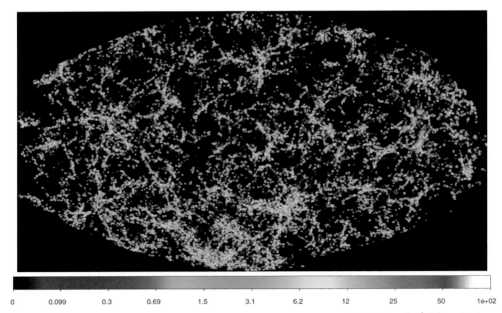

| 0 | 0.099 | 0.3 | 0.69 | 1.5 | 3.1 | 6.2 | 12 | 25 | 50 | 1e+02 |

Figure 5. The luminosity density field of the SDSS in a spherical shell of $10h^{-1}$ Mpc thickness at a distance of $240h^{-1}$ Mpc. The density scale is logarithmic, in units of the mean luminosity density for the whole DR7. The rich complex in the lower area of the picture is part of the Sloan Great Wall; it consists of three very rich superclusters (Suhhonenko *et al.* (2011)).

5. Summary

The contribution of the Zeldovich team to modern cosmology is impressive. Zeldovich was very actively collaborating with other groups, including our Tartu cosmology team. Thank to close collaboration with him and his collaborators we jointly succeeded to get interesting results on the nature of dark matter and the structure of the cosmic web, and the connection between these two phenomena.

What impressed me most in the new cosmological paradigm is its beauty — the Universe is much richer than thought before. The presence of dark matter shows that the Nature of the Universe is richer, in addition to known forms of matter it contains a new population, dark matter, the nature of which is not known even today. The structure of the Universe is also richer, instead of a random background of field galaxies we see now the cosmic web with all its small and large details.

I thank all my collaborators in Tartu and Moscow for very fruitful years of the search of properties of the universe.

References

Abell, G. O. 1958, *ApJS*, 3, 211
Abell, G. O., Corwin, Jr., H. G., & Olowin, R. P. 1989, *ApJS*, 70, 1
Blumenthal, G. R., Faber, S. M., Primack, J. R., & Rees, M. J. 1984, *Nature*, 311, 517
Bond, J. R., Kofman, L., & Pogosyan, D. 1996, *Nature*, 380, 603
de Vaucouleurs, G. 1978, *ApJ*, 224, 710
Doroshkevich, A. G., Joeveer, M., & Einasto, J. 1975, *AZh*, 52, 1113
Efstathiou, G., Davis, M., White, S. D. M., & Frenk, C. S. 1985, *ApJS*, 57, 241

Efstathiou, G., Sutherland, W. J., & Maddox, S. J. 1990, *Nature*, 348, 705

Einasto, J. 1965, *Trudy Astrophys. Inst. Alma-Ata (Tartu Astr. Obs. Teated 17)*, 5, 87

Einasto, J. 1974, in *Stars and the Milky Way System*, ed. L. N. Mavridis, 291

Einasto, J., Gramann, M., Einasto, M., *et al.* 1986, *Tartu Astr. Obs. Preprint*, A-9, 3

Einasto, J., Jõeveer, M., & Saar, E. 1980, *MNRAS*, 193, 353

Einasto, J., Kaasik, A., & Saar, E. 1974, *Nature*, 250, 309

Forman, W., Kellogg, E., Gursky, H., Tananbaum, H., & Giacconi, R. 1972, *ApJ*, 178, 309

Friedmann, A. 1922, *Zeitschrift fur Physik*, 10, 377

Gott, III, J. R., Dickinson, M., & Melott, A. L. 1986, *ApJ*, 306, 341

Guth, A. H. 1981, *Physics Letters D*, 23, 347

Hubble, E. 1929, *Proceedings of the National Academy of Science*, 15, 168

Hubble, E. P. 1925, *ApJ*, 62, 409

Jõeveer, M. & Einasto, J. 1978, in *IAU Symposium*, Vol. 79, *Large Scale Structures in the Universe*, ed. M. S. Longair & J. Einasto, 241–250

Jõeveer, M., Einasto, J., & Tago, E. 1977, *Tartu Astr. Obs. Preprint*, 3

Jones, B. J. T., Martinez, V. J., Saar, E., & Einasto, J. 1988, *ApJL*, 332, L1

Kuzmin, G. 1952, *Tartu Astr. Obs. Publ.*, 32, 211

Kuzmin, G. 1956, *AZh*, 33, 27

Linde, A. D. 1982, *Physics Letters B*, 108, 389

Mandelbrot, B. B. 1982, *The Fractal Geometry of Nature*, ed. Mandelbrot, B. B.

Mandelbrot, B. B. 1986, *Physics Today*, 39, 11

Materne, J. & Tammann, G. A. 1976, in *Stars and Galaxies from Observational Points of View*, ed. E. K. Kharadze, 455–462

Melott, A. L., Einasto, J., Saar, E., *et al.* 1983, *Physical Review Letters*, 51, 935

Oort, J. H. 1940, *ApJ*, 91, 273

Öpik, E. 1922, *ApJ*, 55, 406

Ostriker, J. P., Peebles, P. J. E., & Yahil, A. 1974, *ApJL*, 193, L1

Peebles, P. J. E. 1971, *Physical cosmology* (Princeton Series in Physics, Princeton, N.J.: Princeton University Press, 1971)

Peebles, P. J. E. & Yu, J. T. 1970, *ApJ*, 162, 815

Penzias, A. A. & Wilson, R. W. 1965, *ApJ*, 142, 419

Press, W. H. & Schechter, P. 1974, *ApJ*, 187, 425

Roberts, M. S. 1966, *ApJ*, 144, 639

Rubin, V. C. & Ford, W. K. J. 1970, *ApJ*, 159, 379

Sandage, A. & Tammann, G. A. 1975, *ApJ*, 196, 313

Shane, C. & Wirtanen, C. 1967, *Publ. Lick Obs.*, 22

Soneira, R. M. & Peebles, P. J. E. 1978, *AJ*, 83, 845

Starobinsky, A. A. 1980, *Physics Letters B*, 91, 99

Starobinsky, A. A. 1982, *Physics Letters B*, 117, 175

Suhhonenko, I., Einasto, J., Liivamägi, L., *et al.* 2011, *A&A*, 531, A149

Sunyaev, R. A. & Zeldovich, Y. B. 1969, *Nature*, 223, 721

Tarenghi, M., Tifft, W. G., Chincarini, G., Rood, H. J., & Thompson, L. A. 1978, in *IAU Symposium*, Vol. 79, *Large Scale Structures in the Universe*, ed. M. S. Longair & J. Einasto, 263

Tifft, W. G. & Gregory, S. A. 1978, in *IAU Symposium*, Vol. 79, *Large Scale Structures in the Universe*, ed. M. S. Longair & J. Einasto, 267

Tinsley, B. M. 1968, *ApJ*, 151, 547

Tully, R. B. & Fisher, J. R. 1978, in *IAU Symposium*, Vol. 79, *Large Scale Structures in the Universe*, ed. M. S. Longair & J. Einasto, 214

van den Bergh, S. 1972, *A&A*, 20, 469

White, S. D. M., Frenk, C. S., Davis, M., & Efstathiou, G. 1987, *ApJ*, 313, 505

Zeldovich, Y. B. 1970, *A&A*, 5, 84

Zeldovich, Y. B. 1978, in *IAU Symposium, Vol. 79, Large Scale Structures in the Universe*, ed. M. S. Longair & J. Einasto, 409–420

Zeldovich, Y. B., Einasto, J., & Shandarin, S. F. 1982, *Nature*, 300, 407

Zwicky, F. 1933, *Helvetica Physica Acta*, 6, 110

Zwicky, F., Herzog, E., & Wild, P. 1968, *Catalogue of galaxies and of clusters of galaxies* (Pasadena: California Institute of Technology (CIT), 1961-1968)

The Zeldovich Universe:
Genesis and Growth of the Cosmic Web
Proceedings IAU Symposium No. 308, 2014
R. van de Weygaert, S. Shandarin, E. Saar & J. Einasto, eds.

ⓒ International Astronomical Union 2016
doi:10.1017/S1743921316009583

Ya. B. Zeldovich (1914-1987)
Chemist, Nuclear Physicist, Cosmologist

Varun Sahni

Inter-University Centre for Astronomy and Astrophysics,
Post Bag 4, Ganeshkhind, Pune 411 007, India
email: varun@iucaa.ernet.in

1. Introduction

Yakov Borisovich Zeldovich was remarkably talented.† His active scientific career included major contributions in fields as diverse as chemical physics (adsorption & catalysis), the theory of shock waves, thermal explosions, the theory of flame propogation, the theory of combustion & detonation, nuclear & particle physics, and, during the latter part of his life: gravitation, astrophysics and cosmology (Zeldovich 1992).

Zeldovich made key contributions in all these area's, nurturing a creative and thriving scientific community in the process. His total scientific output exceeds 500 research article and 20 books. Indeed, after meeting him, the famous English physicist Stephen Hawking wrote "Now I know that you are a real person and not a group of scientists like the Bourbaki".

Remarkably Zeldovich never received any formal university education ! He graduated from high school in St. Petersburg at the age of 15 after which he joined the *Institute for Mechanical Processing of Useful Minerals* ('Mekhanabor') to train as a laboratory assistant. The depth of Zeldovich 's questioning and his deep interest in science soon reached senior members of the scientific community and, in 1931, the influential soviet scientist A.F. Ioffe wrote a letter to Mekhanabor requesting that Zeldovich be "released to science". Zeldovich defended his PhD in 1936 and, years later, reminiscenced of the "happy times when permission to defend [a PhD] was granted to people who had no higher education".

In the 1930's, having done pathbreaking work on combustion and detonation (there is a 'Zeldovich number' in combustion theory), Zeldovich moved to nuclear physics writing seminal papers demonstrating the possibility of controlled fission chain reactions among uranium isotopes. This was the time when fascism was on the rise in Germany, and, in an effort for national survival, the Soviet Union was developing its own atomic program of which Zeldovich (then in his mid 20's) quickly became a key member. According to Andrei Sakharov, "from the very beginning of Soviet work on the atomic (and later thermonuclear) problem, Zeldovich was at the very epicenter of events. His role there was completely exceptional" (Sakharov 1988). One might add that Zeldovich 's earlier work on combustion paved the way for creating the internal ballistics of solid-fuel rockets which formed the basis of the Soviet missile program during the 'great patriotic war' and after Ginzburg (1994). After the war Zeldovich went on to do pioneering work in several other aspects of nuclear and particle physics.

† This brief presentation is based on arXiv:1403.1537.

2. Astrophysics

Zeldovich decided to change course midstream and, from about 1964, devoted his phenomenal abilities to problems in astrophysics. Below I (imperfectly) summarize some of Zeldovich's seminal contributions to astrophysics and cosmology.

• In 1962 Zeldovich showed that a black hole could be formed not only during the course of a stellar explosion but by any mechanism which compressed matter to sufficiently high densities. This opened up the possibility of the formation of microscopically small black holes in the early Universe. In 1964 Zeldovich suggested that a black hole may be detected by its influence on the surrounding gas which would accrete onto the hole. Zeldovich also suggested that one could look for a black hole in binary star systems through the holes influence on the motion of its bright stellar companion.

Zeldovich (1971) and his student Starobinsky (1973) showed that under certain conditions a rotating black hole could loose energy via the production of a particle-antiparticle pair. These papers were precursors of later work including Stephen Hawking's famous paper on evaporating black holes published in 1975. Zeldovich and Starobinsky also did a seminal study of particle production in the early universe.

• In 1966 Zeldovich and Gershtein showed that massive neutrino's could play a key cosmological role thus paving the way for the concept of *non-baryonic dark matter*. Zeldovich appreciated the enormous impact that stable topological defects could have within a cosmological setting. He and his colleagues demonstrated that whereas monopoles and domain walls were disastrous for cosmology, cosmic strings might be useful since they would act as 'seeds' onto which matter accreted resulting in the formation of gravitationally bound systems.

• In 1967 Zeldovich applied himself to the issue of the cosmological constant 'Λ'. Originally introduced by Einstein in 1917, the cosmological constant has the unusual property that, unlike other forms of matter, its pressure is *negative* and equal, in absolute terms, to its density ($P = -\rho$). Hence, while the density in normal forms of matter declines in an expanding universe, the density in Λ remains frozen to a constant value $\rho = \Lambda/8\pi G$. After its inception the cosmological constant had fallen into disrepute since it did not seem to be required by observations. Even Einstein distanced himself from it, calling the Λ-term 'my biggest blunder'. Zeldovich radically changed this perspective by persuasively arguing that, within the context of quantum field theory, the prospect of a non-zero value for the Λ-term should be taken extremely seriously. The reason is that the quantum polarization of the vacuum results in a *vacuum energy* which, quite remarkably, has the precise form of a cosmological constant: $\langle T_i^k \rangle = \Lambda \delta_i^k/8\pi G$. By providing a physical basis for the Λ-term, Zeldovich paved the way for future advances including cosmic inflation and dark energy.

• Zeldovich himself felt that his main contribution to cosmology was in the understanding of gravitational instability. In 1972 Zeldovich showed how primordial fluctuations could have a *scale invariant* spectrum. A decade later, the scale invariant spectrum was shown to be a generic prediction of *Inflationary models* of the early Universe.

In 1970 Zeldovich proposed a remarkably simple approximation which could be used to follow a perturbation from its initially linear form into the fully nonlinear regime when it gave rise to the formation of 'pancakes'. Along the way Zeldovich upset a widely prevailing world view according to which the assembly of the first large astrophysical objects in the universe was spherical in nature.

• In 1972, Zeldovich and Sunyaev published a seminal paper in which they showed that photons from the cosmic microwave background would scatter off the hot plasma

Figure 1. Moscow State University. The central building and its wings contained the hostel as well as the department of maths, the main library etc. The right building in the forefront housed the physics department where Zeldovich taught.

trapped in the deep potential wells of clusters. This would alter the brightness of the CMB when viewed in the direction of a cluster.

3. Personal Reminiscences

My first meeting with Zeldovich took place in 1978 when, as a student of physics at Moscow State University (see fig. 2), I was looking for a professor who would guide me for my (pre-MSc) course work. Zeldovich agreed and, to test my skills, gave me a project in general relativity, a subject which I had just started to learn. I soon realized that Zeldovich was an excellent teacher and could explain in very simple language exceedingly complex physical ideas. This was a great boon to students since, despite his extremely busy scientific schedule, Zeldovich always found time to teach courses at Moscow University. So it was that I was initiated into the intricate and beautiful field of cosmology via a twenty two lecture course taught by Zeldovich – each lecture being of roughly 2 hour duration with a ten minute break in between.

Zeldovich continuously modified and expanded his course material taking care to ensure that significantly new developments in the field were covered. (On attending the very same cosmology course 4 years later, I was pleasantly surprised at finding that almost a quarter of its content was new. Zeldovich assiduously incorporated several recent developments including grand unification, topological defects and inflation, into his 'self-revised' syllabus.) Participants at his lectures consisted not only of students, but also senior researchers and professors, many of whom stayed back after class to discuss new science ideas with Zeldovich . Another remarkable quality of Zeldovich was his willingness to learn from others and to acknowledge, often in public, the mistakes which he had made, and what could be learned from them. Thus he admitted in class how he had misunderstood the data regarding the cosmic microwave background in the early 1960's and, to his great regret, had initially advocated the cold big bang model instead of the hot one.

Zeldovich strove to explain complicated ideas simply through numerous entertaining articles and text books (Zeldovich (1992); Zeldovich & Novikov (1971, 1996)) including his excellent monograph 'Higher Mathematics for Beginners' Zeldovich (1970) and Zeldovich (1976) which presented serious mathematical ideas in a form which is accessible to a high school student. He once wrote "the so-called 'strict' proofs and definitions are far more complicated than the intuitive approach to derivatives and integrals. As a result, the mathematical ideas necessary for an understanding of physics reach school-pupils too late. It is like serving the salt and pepper not for lunch, but later – for afternoon tea".

Indeed, the key role played by the science popularisation program for Soviet science cannot be over-emphasised. Talented scientists and mathematicians took great pains in bringing the joy of their disciplines to a much younger audience, including high school students. I believe the great heights reached by the Soviet science school owed much to this popularisation effort, which created an enormous base from which talent could be drawn for the construction of the impressive Soviet science pyramid at the very apex of which belonged Landau, Zeldovich, Kolmogorov and other great thinkers. It might be appropriate to draw here an analogy with the Soviet chess school which also had an enormous grass-root level base, from which rose a formidable range of grandmasters and world champions.

Zeldovich's manner of conducting exams was also quite unusual ! When, towards the end of his cosmology course the time came for exams, I realised that there was no formal time table for a written test, as was usually the case. Instead Zeldovich asked me to meet him outside his office a few days later. When I did, Zeldovich scribbled two problems and asked me to solve them, after which he walked away. Although I had gone through his course material diligently I could not, even after trying hard, figure out how either of these problems could be solved, and so feeling quite dejected I walked back to my hostel a few miles away. This was summer 1979. By winter, with the heavy snows of Moscow having set in, I managed to crack one of the problems and, feeling rather elated, went back to Zeldovich . Zeldovich subjected me to a strenuous viva-voce after which he declared that I had passed with an A and, even more significantly, that my solution to his problem was new and original and could be published as a paper.

I should add that while most people found Zeldovich to be very inspiring, there were some who found his intellectual brilliance rather intimidating. Indeed Zeldovich did not suffer fools lightly and I have seen him demolish in less than a few minutes many a senior scientist propounding a silly idea. On the other hand, Sakharov mentions that Zeldovich's " effect on his pupils was remarkable; he often discovered in them a capacity for scientific creativity which without him would not have been realized or could have been realized only in part and with great difficulty." In this he was following the tradition of the other great Russian *guru*, Landau, who like Zeldovich , fostered and left behind a great scientific legacy in the form of a *school of physics* which was almost unique in scientific method and style.

Around this time I made friends with Lev Kofman. Lev was originally from Tallinn but went to school in Tartu. He had been denied entry into a leading Moscow institution on the pretext that his eyesight was weak. Probably this was not the real reason, but despite this setback Lev was adamant to learn cosmology from the great theorists in Moscow. Lev and I soon became close friends and whenever Lev visited Moscow he stayed with me in the Moscow University hostel. Since Lev did not have a moskovskaya propiska (permission to legally live in Moscow) his stay at the hostel was an act of subterfuge and had to be carefully hidden from the authorities. This meant that he could not enter

Figure 2. With my two gurus Starobinsky (left) and Zeldovich (right) at my PhD defence in Moscow (1985). One of the few times in my life that I wore a suit (which was borrowed).

the university like a normal person but had to climb a four meter high iron fence which surrounded our hostel, a hazardous task especially during winter, when both ground and fence were coated with ice.

Occasionally Lev did get caught. On one such occasion an inspection committee raided my room early one morning and took Lev away to the police precinct. I was very alarmed since I did not want anything nasty to happen to my good friend, so I ran after the police. I rescued Lev by telling the police officer, who happened to be a decent fellow, that we were both students of Academician Zeldovich whose classes would begin that morning, and that it would be a shame if Lev, who had travelled all the way from Estonia for this, should miss them. I guess my persuasive skills were at their best, since the policeman let Lev go, and we immediately ran to the physics building to listen to Zeldovichs lecture on cosmology (without any breakfast). Figure 2 gives an impression of the building complex of Moscow State University.

I might mention here that while Zeldovich was pretty tough when it came to science, on the personal front he was very generous. I experienced Zeldovich empathy on many occasions. Some years after commencing work on my PhD I was shocked to hear that that Zeldovich 's wife had just died of a serious illness. Together with some friends I went to pay my respects and offer condolences. Even before I could utter a word, Zeldovich turned to me and offered his own profuse condolences on the assasination of Indira Gandhi who had lost her life to extremists the very same day that Zeldovich 's wife lost hers to illness. In his eyes I could see how deeply he felt, and I was profoundly moved that he could place the historical anguish of a neighboring nation on the same footing as his own very deep and personal loss.

Although very fond of travel, Zeldovich faced numerous travel restrictions due to his early involvement with the Soviet defence program. In 1982, perhaps because of my political innocence and youthful enthusiasm, I was very keen that Zeldovich visit India, and before embarking home for holidays asked him what I could do to ensure his visit.

V. Sahni

Figure 3. Some members of the Zeldovich school at IUCAA. Alexei Starobinsky and Varun Sahni (top row), Tarun Souradeep (middle), Jatush Sheth, Sanjit Mitra, Amir Hajian and Ujjaini Alam (bottom row).

Looking at me wistfully Zeldovich remarked, "they (the soviet authoroties) will reply to your governments invitation saying Zeldovich is ill and unable to travel, but look at me (Zeldovich flexed his muscles) I am perfectly fit and can travel tomorrow !". Indeed Zeldovich 's visit to India never did materialise and the following year I bore witness to the very bizarre policies of the Soviet government with regard to foreign travel by its eminent scientists. Zeldovich and Starobinsky had both been invited to travel to a famous meeting on general relativity (GR10) scheduled in Padova, Italy in 1983. Indeed, Zeldovich had been invited to organise a special session on the Early Universe while Starobinsky had consented to give a plenary talk on Inflation, which was then a very hot topic. I too was planning to go since my paper (written jointly with Starobinsky and another of his pupils Lev Kofman) had been accepted for presentation. I was very excited by the prospect of travelling to Italy from Moscow (3 days journey by train passing through several countries) and the day before leaving I met Starobinsky who took me to Zeldovich with the remark "Yakov Borisovich here is Varun, the only member of our delegation who is sure of travelling to Italy" ! I was not a little surprised to learn that, with only four days remaining for the meeting, neither Zeldovich nor Starobinsky had as yet received official sanction for their visit. I was broken hearted when, on reaching Padova 4 days later, I learned that neither of my guru's had been allowed to travel, and that the organisers were in a quandary as to how to salvage the early universe session. What was even more disheartening was the fact that while neither Zeldovich nor Starobinsky had been allowed to make the journey to Italy the official Soviet delegation was replete with scientistsc of rather mediocre quality but with deep party connections, who despite their ideological leanings were not ashamed to thoroughly enjoy 'capitalist hospitality'.

Viewed in retrospect, and with the demise of the Soviet Union, it appears quite incomprehensible as to why many of its leading scientists could not travel abroad and enjoy an envigorating science discussion with their contemporaries in the east and west.

4. Back in India

After obtaining my PhD (in 1985 – see fig. 2) I returned to India (via post-docs in the UK and Canada) where I helped in the setting up of a new institution IUCAA. At the time of my return (1991) IUCAA consisted of little more than a tin-roofed shed surrounded by an enormous number of trees, generously populated by birds, snakes and on the ground, even scorpions. We set about building this new institution (under the leadership of Prof. Jayant Narlikar) with great enthusiasm, and in just a few years Sergei Shandarin and I organized the first Zeldovich meeting on large scale structure.

The Zeldovich legacy endures at IUCAA, where Tarun Souradeep (my first PhD student and now an IUCAA faculty) and I have, between us, successfully guided over a dozen cosmology PhDs, several of whom have also been mentored by Alexei Starobinsky, a frequent visitor to India (see fig. 3).

References

Ginsburg, V. L. 1994, Yakov Borissovich Zel'dovich, *Biographical Memoirs of Fellows of the Royal Society* 40, 431-441

Sakharov, A. 1988, A man of universal interests, *Nature*, 331, 671

Zeldovich, Ya. B. 1970, *Higher mathematics for beginners*, Nauka, Moscow

Zeldovich, Ya.B.& Myskis, A. D. 1976 *Elements of Applied Mathematics*, Mir Publishers, Moscow

Zeldovich, Ya.B. 1992, *Selected works of Yakov Borisovich Zeldovich*, Vol. 1 & 2, Princeton University Press

Zeldovich, Ya. B. 1992, *My Universe, selected reviews*, Harwood academic

Zeldovich, Ya.B. & Novikov, I. D. 1971, *Relativistic Astrophysics, Vol. 2: The Structure and Evolution of the Universe*, Chicago, IL: University of Chicago Press

Zeldovich, Ya.B. & Novikov, I. D. 1996, *Relativistic Astrophysics, Vol. 1: Stars and Relativity*, Mineola, NY: Dover Publications

The Zeldovich Universe:
Genesis and Growth of the Cosmic Web
Proceedings IAU Symposium No. 308, 2014
R. van de Weygaert, S. Shandarin, E. Saar & J. Einasto, eds.

© International Astronomical Union 2016
doi:10.1017/S1743921316009595

Why *Planck* (the Satellite) could have been *Zel'dovich*

Bruce Partridge[1]†

[1] Haverford College, 370 Lancaster Ave., Haverford, PA 19041, United States
email: bpartrid@haverford.edu

Abstract. In this brief paper, I cannot provide an overall review of the *Planck* results to date. Instead I will focus on a handful of results from both *Planck* and related cosmic microwave background (CMB) experiments that reflect Ya. B. Zel'dovich's legacy in cosmology. These include the Sunyaev-Zel'dovich effect in clusters of galaxies and in the cosmic web, and a map of the overall distribution of mass in the Universe derived from CMB maps.

Keywords. cosmology, cosmic microwave background, clusters of galaxies

1. Introduction

This brief report is not an attempt to summarize the scientific results to date of the *Planck* mission (see the article by Dick Bond in this volume for further details). Instead, I have tried to select a few topics, several of them astrophysical rather than cosmological, that I believe would have interested Yakov Borisovich Zel'dovich. The selection was not easy, given his wide interests, but in the end I came up with:

(1) *Planck*'s picture of the cosmic web at $z = 1100$,
(2) the Sunyaev-Zel'dovich (SZ) effect in clusters of galaxies,
(3) other loci for the SZ effect,
(4) the cosmic infrared background (CIB; in effect, the light of the cosmic web), and
(5) gravitational lensing maps of the overall mass distribution in cosmic space.

Despite the recent interest in the imprint of early inflation on the cosmic microwave background (CMB) following the BICEP2 results (Ade, Kovacs *et al.* 2014), I skip over that topic since *Planck*'s findings on that subject were not available at the time of the symposium. Instead, I have added two topics that I think might have amused Ya. B.:

(6) using *Planck* to calibrate ground-based radio telescopes and
(7) using *Planck* and related measurements to measure the mass of neutrinos.

Now to explain my title. The cosmic microwave background (CMB) has an almost perfect blackbody spectrum: hence the name *Planck* for the mission. But *Planck* (the satellite) is designed to measure anisotropies in the CMB—the seeds of cosmic structure and a subject of deep interest to Zel'dovich. Given his wide contributions to so much of the science *Planck* was designed to probe, the mission could well have been *Zel'dovich*.

2. *Planck*'s map of the cosmic web at $z \sim 1100$

Images of the CMB like Fig. 1 trace the distribution of matter at the epoch of last scattering at $z \sim 1100$; they are in effect snapshots of the cosmic web at early times. This topic is too rich to treat in detail here (see many other papers in this volume). I will merely say that I regret Zel'dovich did not have the opportunity to see this map, or the equally stunning earlier one made by WMAP.

† In part, on behalf of the Planck Collaboration

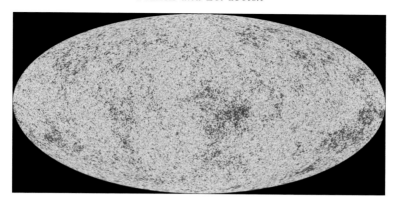

Figure 1. *Planck*'s map (in Galactic co-ordinates); see Planck Collaboration a, 2014. The $\sim 100\ \mu$K temperature variations shown in false color trace faint density perturbations that are the seeds of the cosmic web and all large scale structure.

3. *Planck*'s survey of clusters of galaxies using the Sunyaev-Zel'dovich Effect

Within a few years of the discovery of the CMB, Sunyaev and Zel'dovich (see their 1980 review) had shown that inverse Compton scattering of CMB photons off hot electrons in cosmic plasmas (such as the intergalactic medium in clusters of galaxies) would affect the CMB spectrum. The Sunyaev-Zel'dovich Effect (SZE) is proportional to the pressure of hot electrons, but independent of redshift; hence it has long been recognized as a way to search for clusters at all redshifts. Both *Planck* and ground-based experiments like the South Pole Telescope (SPT; Bleem *et al.* 2014) and the Atacama Cosmology Telescope (ACT; Hasselfield *et al.* 2013) have made surveys of clusters using the SZE, more than doubling the number of known clusters. *Planck*, of course, covers the entire sky, but its broad beams limit sensitivity, so it is best at detecting massive, hot clusters at all redshifts: the 2013 Planck cluster catalog (Planck Collaboration e, 2014) contains all ~ 1200 of the most massive clusters in the Universe. In Fig. 2, I show an example of one of the ground-based surveys, which can reach lower masses than *Planck*.

The number density of clusters as a function of redshift is strongly dependent on cosmological parameters (for instance, the Dark Energy equation of state). A well-defined, mass-limited catalog of clusters thus provides a clear test of such parameters. Fig. 3 shows an example, drawn from work by the SPT Team. The crucial, and still not fully resolved, problem with this method is the scaling relation between the integrated SZE signal (Y) and the total cluster mass (see Fig. 4).

4. Other sites for the SZE

Clusters of galaxies are not alone in producing an SZE signal: any region of hot plasma can do so. That is true, for instance, of the hot gas entrained in the cosmic web, which *Planck* has mapped (Planck Collaboration d, 2014). Another site for the SZE are the gaseous halos of AGN; Gralla *et al.* (2014) have found this signal by stacking ACT maps at the positions of AGN. That work allows an extension of the important scaling relation between cluster mass and the SZE signal to lower masses (Fig. 4, right).

5. The Cosmic Infrared Background (CIB)

Galaxies trace the cosmic web, and the CIB traces galaxies, especially star-forming ones. The CIB, in a sense is the light of the cosmic web. *Planck* has provided detailed

ACT: SZ Selected Galaxy Clusters

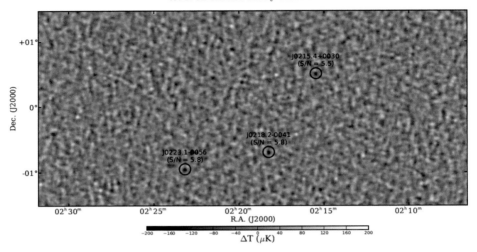

Figure 2. SZE-detected clusters in a small portion of one of the survey fields of ACT. AdvACT, its successor, will detect ∼27,000 clusters in half the sky (all clusters with $M > 2 \times 10^{14}$ Mo).

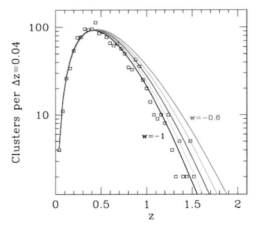

Figure 3. Counts of clusters as a function of redshift, showing the dependence on the Dark Energy equation of state parameter, w (from the SPT Collaboration).

images of the CIB, as well as measurements of its intensity and power spectra (Table 1). Since the CIB spectrum is fairly sharply peaked, observations at different frequencies probe different redshifts, so *Planck* can supply crude tomography of the CIB.

6. Gravitational lensing of the CMB by intervening matter

As CMB photons move through the cosmic web, they are are gravitationally lensed by large scale structure. Measuring this lensing permits a reconstruction of the line-of-sight mass integral across sky (Fig. 5; Planck Collaboration c, 2014). Measuring this lensing also permits an estimate of the lensing power spectrum. I note that the lensing kernel has a broad maximum at $z \sim 2$, just where star formation peaks up.

Band	νI_ν [nW m^{-2} sr^{-1}]	
3000 GHz	13.1	$\pm$ 1.0
857 GHz	7.7	$\pm$ 0.2
545 GHz	2.3	$\pm$ 0.1
353 GHz	0.53	$\pm$ 0.02
217 GHz	0.077	$\pm$ 0.003

Table 1. Intensity of the CIB in the IR (from Planck Collaboration f, 2014).

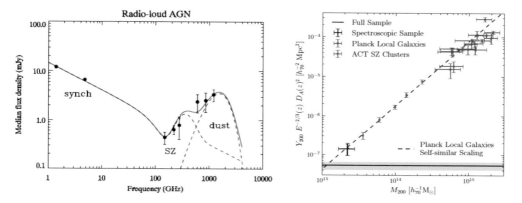

Figure 4. Evidence for the SZE in the gaseous halos of AGN; the result for low mass halos lies along the expected mass-SZE scaling relation for clusters (Gralla, *et al.*, 2014).

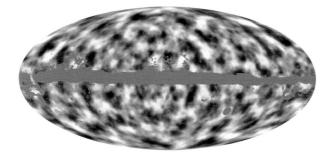

Figure 5. Lensing permits a reconstruction of the line-of-sight integral of all mass in the Universe. That integral is shown here in Galactic coordinates (Planck Collaboration c, 2014).

There is good agreement between the measured lensing power spectrum (Fig. 6) and the value predicted from ΛCDM cosmology (showing that we know how the cosmic web evolves). If anything, the measured lensing is a bit larger than expected.

7. Using *Planck* to calibrate ground-based radio telescopes

The intensity calibration of *Planck* is absolute, pinned to the satellite's motion in the solar system. Furthermore, *Planck* is sensitive enough to observe hundreds of bright radio sources as well as planets. Comparing *Planck* and ground-based measurements allows us to transfer *Planck*'s calibration to the VLA, the Australia Telescope, ALMA, etc. In the case of bright radio sources (and Mars), the variability of the sources can introduce scatter and potentially bias in the *Planck* comparison. We minimize both by making the *Planck* and ground-based observations at roughly the same time, and by including many

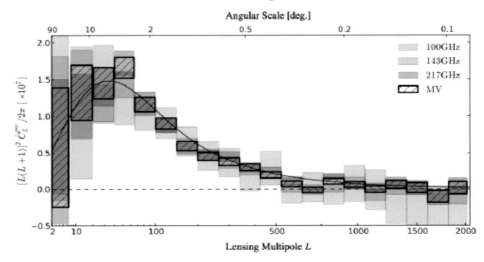

Figure 6. The lensing power spectrum measured by *Planck* at three frequencies (boxes) compared to the expected value derived from the cosmological parameters determined by *Planck*: note the excellent agreement (Planck Collaboration c, 2014).

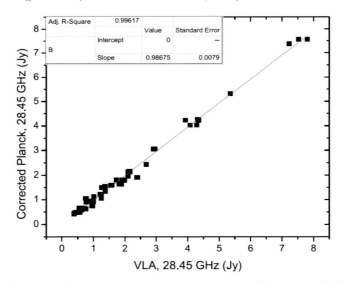

Figure 7. Flux densities of bright radio sources measured by *Planck* and the VLA at 28 GHz. The slope lies within $\sim 1\%$ of unity. The scatter is dominated by variability in these radio sources. An updated version appears in Perley *et al.* (2015).

sources (Perley *et al.*, 2015). An early example is shown in Fig. 7; it demonstrates that the newly adopted microwave flux density scale of Perley and Butler (2013) is consistent with the absolute *Planck* calibration to ~ 1–2% precision at 28 GHz.

8. Using *Planck* to measure the mass of neutrinos

The mass, number and to some extent properties of neutrinos affect various aspects of the power spectrum of anisotropies in the CMB, including the small-angle damping tail (see, for instance, Abazajian t al., 2013). The *Planck* results (Planck Collaboration

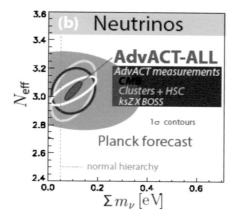

Figure 8. Constraints on neutrino masses expected from the proposed AdvACT experiment.

b, 2014), for instance, are consistent with three (not four) neutrino species, and set a limit of < 0.66 eV on the sum of masses of these three. Adding other astrophysical data allows a lower limit to be set: < 0.23 eV. As an instance of the power of planned CMB experiments to constrain both the number and mass of neutrinos, I show below projections for AdvACTPol.

9. Conclusions

Much of the science I have sketched relies on our increasingly sensitive measurements of anisotropies in the CMB, including *Planck*'s. Even though not all the topics I addressed have a direct link to the cosmic web, I hope I have helped convince you of the value and versatility of current CMB results. I know I do not have to convince you of the range and value of Ya. B. Zel'dovich's contributions to cosmology.

References

Abazajian, K. N, *et al.* 2013, arXiv:1309.5383.

Ade, P., Kovacs, J., *et al.* 2014 *Phys. Rev. Letters* 112, 241101.

Bleem, L. E, *et al.* 2014, arXiv:1409.0850.

Gralla, M., *et al.* 2013, *MNRAS*, submitted; arXiv:1310.8281.

Hasselfield, M. *et al.* 2013, *JCAP* 07, .

Perley, R.A. and Butler, B.J. 2013, *ApJS* 204, 19.

Perley, R. A, Partridge, B., *et al.* 2015 in preparation.

Planck Collaboration a 2014, Planck 2013 results I; arXiv:1303.5062.

Planck Collaboration b 2014, Planck 2013 results XVI; arXiv:1303.5076.

Planck Collaboration c 2014, Planck 2013 results XVII; arXiv:1303.5077.

Planck Collaboration d 2014, Planck 2013 results XXI; arXiv1303.5081.

Planck Collaboration e 2014, Planck 2013 results XXIX; arXiv1303.5089.

Planck Collaboration f 2014, Planck 2013 results XXX; arXiv1309.0382.

Sunyaev, R. A, and Zel'dovich, Ya. B. 1980, *Ann. Rev. Astron. Astrophys.* 18, 537.

The Zeldovich Universe:
Genesis and Growth of the Cosmic Web
Proceedings IAU Symposium No. 308, 2014
R. van de Weygaert, S. Shandarin, E. Saar & J. Einasto, eds.

© International Astronomical Union 2016
doi:10.1017/S1743921316009601

The quest for collapsed/frozen stars in single-line spectroscopic binary systems

Virginia Trimble

University of California Irvine,
Department of Physics and Astronomy
Irvine, California 92697, USA
email: `vtrimble@uci.edu`

Abstract. Black holes are now commonplace, among the stars, in Galactic centers, and perhaps other places. But within living memory, their very existence was doubted by many, and few chose to look for them. Zeldovich and Guseinov were first, followed by Trimble and Thorne, using a method that would have identified HDE 226868 as a plausible candidate, if it had been in the 1968 catalogue of spectroscopic binaries. That it was not arose from an unhappy accident in the observing program of Daniel M. Popper long before the discovery of X-ray binaries and the identification of Cygnus X-1 with that hot, massive star and its collapsed companion.

Keywords. Black holes, collapsed stars, Cygnus X-1, HDE 226868, spectroscopic binaries

---

1. A Contextual Introduction

The context in which this presentation was assembled and written up includes a recent book called *Discovery and Classification in Astronomy: Controversy and Consensus* (Dick 2013) in which the author states at least twice that black holes were not taken seriously by astronomers until the mid 1970's or later. Now the two widely-cited pioneering papers on powering of quasars by accretion on a central supermassive, compact object came from E.E. Salpeter (1964) and Ya. B. Zeldovich & I.D. Novikov (1964b) which had me doubting the claims immediately. This was reinforced by the sudden memory that one of my own papers (Trimble & Thorne 1969) took them seriously rather earlier than the mid-1970's, and that it had its origins in an idea from Zeldovich & Guseinov (1966). There followed a resolve to look back at this topic and offer an oral version of it to the June, 2014 Zeldovich Centenary conference in Moscow. Complexities associated with the June 15th end of the school year at UCI kept me from attending that meeting, and the organizers of IAU Symposium 308 the next week generously made room for "The Quest for Collapsed Stars" just before the conference dinner on the second to last day of the meeting.

2. Cast of Characters

A Newtonian version of the story can be pushed back to publications in the late 18th century, when John Michell and Pierre-Simon de Laplace noted the possibility of astronomical entities with escape velocity equaling or exceeding the speed of light. Relativity enters the picture with a pair of much-cited, but rarely read, papers by Karl Schwarzschild (1916a, b,) that defined what we call the Schwarzschild radius and was already being called the gravitational radius when Oppenheimer and Snyder (1939) wrote "On continued gravitational contraction."

They wrote, and we all now agree, that a sufficiently massive star, having exhausted its thermonuclear sources of energy, must either reduce its mass to of order that of the sun or suffer continuous contraction. Reaching the gravitational radius takes a day or two for an observer riding the star, but forever as seen from infinity, hence the phrase "frozen star" (which came over as "cooled star" in some translations). The last photon that reaches outside observers, however, departs within days, not millennia.

Then there was a war, with very little work on general relativity and its implications being done anyplace for many years. On the far side of that abyss, John Wheeler and a few Princeton students and colleagues began exploring the territory near very massive compact objects, while in the USSR, two critical early papers predicting inescapable collapse were published by Zeldovich & Novikov (1964, 1965). We interrupt this tale to explain that those names are in alphabetical order, in Russian, where the letter that English-speakers transcribe as Z looks sort of like a backwards Greek epsilon and comes early in the alphabet, while N looks like H and comes later.

The oral presentation in Tallinn included a number of images of people, publications, and other pieces of paper not readily reproducible here, and also a Sidney Harris cartoon (caption: It's black, and it looks like a hole. I'd say it's a black hole) still under copyright. Figures 1 and 2 are an inscription from the flyleaf of a 1985 book and a 1986 post-New Years letter, and are included to demonstrate that I have a right to tell Zeldovich stories.

The events referred to near the end of the letter are (1) the GRG meeting in Warsaw in 1963, from which Richard Feynman wrote back to his new wife Gwyneth expressing a considerable distaste for GR and the work being done in its name (reproduced in the volume *Surely You're Joking, Mr. Feynman*), (2) a later GRG meeting in Jena, when Zeldovich and Joe Weber were both elected members of the governing body, but only Joe attended its evening meeting, because at that time Soviet representatives to international organizations could only be those appointed by the government, so Jascha treated Angelica and me to sizable quantities of vodka and heavily whip-creamed desserts, and (3) the General Assembly of the International Astronomical Union in Patras, Greece in 1982, where Zeldovich gave one of the Invited Discourses in a reconditioned ancient Greek theatre. This was his first time in western Europe, marking his achievement of what he called "the second cosmic velocity." The first was enough for a Soviet scientist to travel in Eastern Europe (hence to the Warsaw and Jena meetings), and only the "third cosmic velocity" reached to the United States.

The Patras conference dinner was also the occasion on which Zeldovich expressed to me his puzzlement that Americans, who were allowed to read just about anything, rarely bothered, while Russians would go to a good deal of trouble to read things that were not readily available to them. Indeed we were having a lively discussion about literature (necessarily in English!) when another conference participant came up and asked Yakov Borisovich why he wasn't doing more to help Sakharov. Suddenly his proficiency in the international language of science (broken English, he said) vanished, and he professed not to understand the question.

A later letter expressed unhappiness that we had not been present when he spoke at the US National Academy of Sciences during his first and only visit to the US. I wrote back as soon as possible (letters between the USA and the USSR could take a long time in those days, when they were presumably read by both sides), explaining that merely being on the wrong coast in April would never have kept us away, but that neither Joe nor I was, or would ever be, a member of NAS! I heard from Oktay H. Guseinov once after the advent of email. He died in 2009, having spent the latter years of his career in Azerbaijan and Turkey.

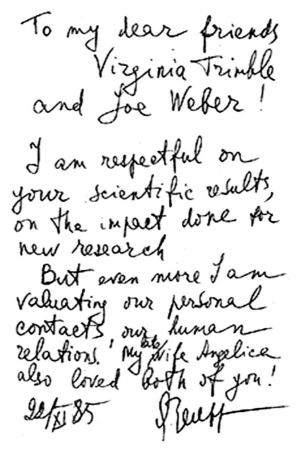

Figure 1. Fly-leaf inscription on a copy of a 1985 volume of reprints of many of Zeldovich's significant papers across many fields, together with a fairly complete list of all his publications up to that time. Nearly all in Russian, of course.

The last preliminary remark was an explanation that John Archibald Wheeler might well have invented the phrase "black hole," but that he could not have done so in 1967 or 1968, as is frequently written, for '"Black Holes" in Space' is the title of a news item by Anne Ewing, published in *Science News Letters* in 1964 (Ewing 1964). The item was a report of the December 1963 meeting of the American Association for the Advancement of Science, held in Cleveland. There is also a rumor around that the phrase was heard at the very first "Texas" symposium in Dallas, earlier in December 1963. Use of white holes to reverse time has never proven practical, so the phrase mush have been coined before those meetings.

3. The Catalogue Searches

Again one could start the story several places, but a good concrete point is the pair of papers (Zeldovich & Novikov 1964, 1965), called "Relativistic Astrophysics I and II" and presaging much of the information later found in the two volumes of their similarly-titled book. The papers (at least the English translations!) talk about neuron stars, continued gravitational collapse, and collapsed or frozen stars (the latter comes out as cooled stars

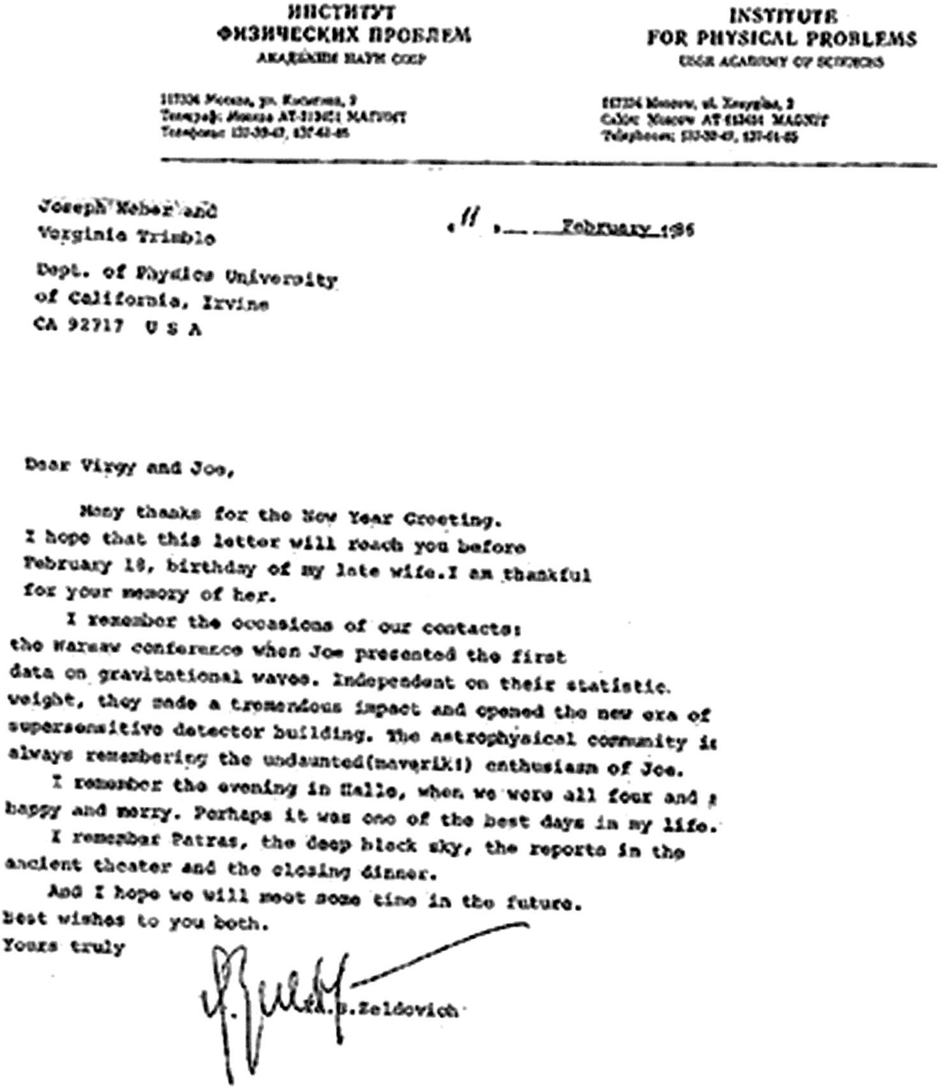

INSTITUTE
FOR PHYSICAL PROBLEMS
USSR ACADEMY OF SCIENCES

Joseph Weber and
Vorginia Trimble

Dept. of Physics University
of California, Irvine
CA 92717 U S A

February 196

Dear Virgy and Joe,

Many thanks for the New Year Greeting.
I hope that this letter will reach you before
February 18, birthday of my late wife.I am thankful
for your memory of her.

I remember the occasions of our contacts:
the Warsaw conference when Joe presented the first
data on gravitational waves. Independent on their statistic.
weight, they made a tremendous impact and opened the new era of
supersensitive detector building. The astrophysical community is
always remembering the undaunted(maverik!) enthusiasm of Joe.

I remember the evening in Halle, when we were all four and :
happy and merry. Perhaps it was one of the best days in my life.

I remember Patras, the deep black sky, the reports in the
ancient theater and the closing dinner.

And I hope we will meet some time in the future.
Best wishes to you both.
Yours truly

Ya.B.Zeldovich

Figure 2. Copy of a February, 1986 letter recalling the three (and only three) occasions on which he and Joe and/or I had been together. A somewhat earlier letter told us of Angelica's death and instructed Joe to take good care of "Vergie." A later one expressed regret that we had not been at his NAS talk, and I have always hoped, though never been quite sure, that our response arrived before he died.

in translation) with some calculations of likely accretion rates and energy release. They cite the Oppenheimer papers, Baade & Zwicky (1934) on neutron star formation as a supernova energy source, and Landau (1938) on whatever you think that paper is about. Zeldovich (1964) which includes the idea of accretion on a neutron star as a source of X-ray is also relevant.

Chapter 13 of "Relativistic Astrophysics II" includes a footnote crediting Guseinov with the idea that detectability of compact stars is increased in binary systems. If you think that I.S. Shklovsky should be part of this story well, so apparently did he. And I refer you to Ginzburg (1990), who outlived the other two and so got the last word, for yet another point of view.

At any rate, armed with the idea that an accreting collapsed star should be an X-ray source but contribute no significant visible light in comparison with a companion of normal luminosity, Guseinov & Zeldovich (1966, submitted 18 October 1965: Zeldovich & Guseinov 1966, submitted 19 November 1965) set out in search of single-line (meaning spectral features from only one star) spectroscopic binaries with invisible companions more massive than the visible stars. They had access only to a 1948 catalogue of SBs and a smattering of later orbits, and found seven candidates. Of the seven, four were very long period systems; none was or is a known X-ray source; and several are probably not even binary systems, but other periodic or quasi-periodic sorts of variables. The earlier paper mentions only collapsed stars as candidates; the latter suggests a neutron star for the seventh system, Alpha Her B.

Five things happened over the next couple of years: (1) the number of known X-ray sources increased, (2) neutron stars as pulsars became part of the inventory of known objects, (3) Batten (1968) compiled a much more extensive and critical catalogue of spectroscopic binaries, (4) Kip Thorne picked up the idea of potential detectability of collapsed stars in SB1s, and (5) Trimble completed her PhD dissertation at Caltech (on motions and structures of filamentary envelope of the Crab Nebula) and had several stray weeks while the thesis was being typed, duplicated, and read by her committee before it could be defended (on 15 April 1968) and she could take off for England as a volunteer postdoc.

KS suggested that VT should go through the Batten catalogue and identify all the SB1 systems with M_2 larger than $1.4M_\odot$ and larger than the mass of the visible star, M_1, as estimated from its spectral type. Barbara A. Zimmerman of the Caltech computing staff provided essential help in writing and executing the program that took the catalogued properties (period and velocity amplitude for the optical star, P and K_1 plus Kepler's third law and turned them into

$$(M_2^3 sin^3 i)/(M_1 + M_2)^2 = const \; X \; PK_1^3 \qquad (3.1)$$

where M_2 is the invisible star; M_1 is the mass you have to guess from its spectral type; and $sin = 1$ yields the lower limit on M_2. It is left as an exercise for the reader to derive this (which applies only to circular orbits, the norm for short-period systems). The presentation erroneously had K_2 in the derivation and expression (3.1) which is customarily called $f(M)$ or the mass function.

We recovered six of the seven Zeldovich & Guseinov (1966) systems and added six more with $M_2 > M_1$ and 38 with $M_2 < M_1$, but greater than $1.4M_\odot$ (the limit for stable white dwarfs). More than 40 years downstream, none of the identified binaries is a known X-ray source (i.e. collapsed star candidate). This was already clear at the time, because the search found about as many systems with $M_2 > M_1$ (etc.) that showed eclipses, meaning that M_2 could not be a compact star.

The paper was written during a two-day stop-over I made in Chicago, where Thorne was then on sabbatical. It is the only paper I ever hand-wrote (typing being so very much easier). And there was a difficulty about who was to be first author. Not I, said I, because it wasn't my idea. Not me, said Thorne, because he hadn't done the necessary bits of arithmetic. Should we, he then asked, call it Zeldovich and Zimmerman? Well perhaps not, and Trimble & Thorne (1969) it then became, and remained sporadically cited for a number of years, because quite a few of the stars turned out to be interesting for other reasons, though some of them weren't even really binaries and were removed from the *Ninth Catalogue* (Batten 2014).

4. What Might Have Been (the first true Z&G system and the last wrong paper)

Three years passed. The Uhuru satellite was launched and, among the strong X-ray sources, Cygnus X-1 proved to be particularly, rapidly, and erratically variable (Schreier *et al.* 1971, Tananbaum *et al.* 1972). Simultaneous flaring in X-ray and radio (Hjellming 1973) permitted optical identification with a previously-known, late O-type supergiant, HDE 226868. Optical astronomers rushed in droves to their telescopes (Bolton 1972, Webster & Murdin 1972; droves were smaller in those days) to check for radial velocity variability. Sure enough, periodic, at $P = 5.6$ days with a velocity half-amplitude of about $75km/sec$, implying a minimum mass of $3.3M_\odot$ for a normal O6 supergiant as the visible star. That limit has crept up over the years (Paczyński 1974 and beyond).

How could the proper response to this tale be anything but wild cheering? Like so. Some time between 1972 and 1999, I was on a conference tour bus seated next to Dan (Daniel Magnes) Popper. Among the anecdotes he related was the fact that he had observed HDE 226868 as part of a radial velocity survey long ago, but had had the bad luck to observe at an integral multiple of the orbit period. In preparing for this talk, I hunted out the relevant paper (Popper 1950). Sure enough, there among 253 O-B6 stars with apparent magnitudes between 8.5 and 11, were color, spectral type, and radial velocity for HD 226868 (between 227704 and 226951). The stellar velocity is tabulated as $-13km/sec$, and of quality A, meaning that two different determinations differed by less than $4km/sec$. The dates of the observing runs appear, and his two spectra were taken 263 days apart, 47 times the orbit period, rather exactly. By ill chance, any two observations could catch an unknown binary at nearly the same phase in its orbit, but I think the constancy to $4km/sec$ out of a full aptitude of $150km/sec$ should happen something less than 10% of the time. Doing a better job of this calculation is also left as an exercise for the reader.

The key point, of course, is that if Popper had prepared a good orbit for the star, it would have been in Batten's catalogue, and the 1972 application of the Zeldovich and Guseinov's (1966) method would have identified it as a black hole candidate.

Meanwhile, as it were, folks interested in the nature of X-ray binaries were very much aware that the mass limit for Cyg X-1 was heavily dependent on that assumed for the visible star on the basis of its spectral type. A very massive star leaving the main sequence is not the only possibility for the combination of high temperature and low atmospheric pressure. Post-asymptotic-giant-branch stars pass through the same conditions (though at much lower luminosity) on their way to becoming white dwarfs. I think I may actually have been the first to think of this possibility for Cygnus X-1, for the fairly obvious reason that I had just calculated models for low mass B stars with low surface gravity (Trimble 1973) meant to describe HZ 22. When I mentioned the idea to a stellar evolution colleague at the University of Maryland, Bill Rose, he said he thought it was a good idea and would like to be an author on the paper. Seeing my fame as author of "Trimble (1973)" fade into the relative obscurity of "Trimble & Rose (1973)" or even "Rose and Trimble (1973)" alphabetically, I co-opted my husband, Joseph Weber, as a third author of Trimble, *et al.* (1973). This is the only paper I have ever had appear in print before the postcard acknowledging receipt of the manuscript arrived.

Once again, optical observers flocked to their telescopes in droves (Margon *et al.* 1973, Bregman *et al.* 1973, droves having become slightly larger in the intervening year) to prove us wrong. This they quickly did, getting distance estimates for the system closer to the $2kpc$ implied by a $20M_\odot$ slightly evolved O supergiant than the $200pc$ required for a post-AGB star. Both distances are far too large for the parallax determinations of the period, and the method used was the determination of the amount of absorption in comparison with that seen for other hot stars in the same direction of the sky.

Later work has, of course, refined knowledge of HDE 226868 = Cygnus X-1, but Trimble *et al.* (1973) was almost certainly the last fundamentally wrong paper on the subject. There is a moral here for theorists: a sufficiently definite prediction of something that can be observed will almost certainly attract the attention of colleagues wanting to prove you wrong. Unfortunately, most of the topics discussed at IAUS 308 are too complex for such definitive predictions to be made at present.

Acknowledgments

I am grateful to Alan Batten and Roger Griffin for e-discussions of the status of some of the Z&G systems, to the late Dan Popper for confiding his unhappy tale long ago, to the organizers of IAUS 308 for finding time on the program for what I had to say, and to Alison Lara for her usual expert keyboarding of my usual untidy typescript.

References

Baade, W. & Zwicky, F. 1934, *Proc. USNAS*, 20, 254
Batten, A. H. 1968, *Publ. Dom. Ap. Obs.* 8, 119
Batten, A. H. 2014, *personal e-mail communication*
Bolton, C. T. 1972, *Nature* 235, 271; *Nature Phys. Sci.* 240, 124
Bregman, J. *et al.* 1973, *ApJL* 186, L117
Dick, S. J. 2013, *Discovery and Classification in Astronomy: Controversy and Consensus*, Cambridge Univ. Press
Ewing, A. 1964, *Science News Letters* 8, 39 (January 18)
Ginzburg, V. L. 1990, *ARA&A* 20, 1
Guseinov, O. H. & Zeldóvich, Ya. B. 1966, *A. Zh* 43; 303 Sov. Astr. AJ 10, 251
Hjellming, R. M. 1973, *ApJL* 182, L29
Landau, L. 1932, *Phys. Z. Sowjetunion* 1, 285
Landau, L. 1938, *Nature* 141, 333
Margon, B. 1973, *ApJL* 185, L143
Oppenheimer, J. R. & Snyder, H. 1939, *PR* 56 455
Oppenheimer, J. R. & Volkoff, G. 1939, *PR* 55, 374
Paczyński, B. 1974, *A&A* 46, 513
Popper, C. M. 1950, *ApJ* 111, 495
Salpeter, E. E. *et al.* 1971, *ApJ* 140, 796
Schreier, E. *et al.* 1971, *ApJ* 170, 121
Schwarzschild, K. 1916, *Sitzber. Preus. Deut. Akad. Wiss. Berlin, KL Math.Phys.* 189-196 & 424-434
Tananbaum, H. *et al.* 1972, *ApJL* 177, L5
Trimble, V. 1973, *A&A* 23, 81
Trimble, V., Rose, W. K., & Weber, J. 1973, *MNRAS* 162, 1p
Trimble, V. L. & Thorne, K. S. 1969, *ApJ* 156, 1013
Webster, V. L. & Murdin, P. 1972, *Nature* 235, 37
Zeldóvich, Ya. B. 1964, *Sov. Phys. Dokl.* 9, 165
Zeldóvich, Ya. B. & Guseinov, O. H. 1966, *ApJ* 144, 840; Sov. Phys. Dokl. 162, 791
Zeldóvich, Ya. B. & Novikov, I. D. 1964b, *Sov. Phys. Dokl.* 9, 834

Zeldovich, Ya. B. & Novikov, I. D. 1964, *Uspekhi Phys. Nauk.* 84, 377 (= 7,763 in translation)
Zeldovich, Ya. B. & Novikov, I. D. 1965, *Uspekhi Phys. Nauk.* 86, 447 (= 8,522, in translation)
Zeldovich, Ya. B. & Novikov, I. D. 1966, *Nuovo Cimento.* I 4, 840

CHAPTER 2.

Zeldovich legacy:

Dynamics & Evolution of the Cosmic Web

An organizer's discussion:
Jaan Einasto, Sergei Shandarin and Rien van de Weygaert.

A selection of participants at IAU308.

The Zeldovich Universe:
Genesis and Growth of the Cosmic Web
Proceedings IAU Symposium No. 308, 2014
R. van de Weygaert, S. Shandarin, E. Saar & J. Einasto, eds.

© International Astronomical Union 2016
doi:10.1017/S1743921316009613

Understanding the cosmic web

Marius Cautun[1,2], Rien van de Weygaert[2], Bernard J. T. Jones[2] and Carlos S. Frenk[1]

[1]Department of Physics, Institute for Computational Cosmology, University of Durham,
South Road, Durham DH1 3LE, UK, email: m.c.cautun@durham.ac.uk
[2]Kapteyn Instituut, Rijksuniversiteit Groningen, P.O. Box 800, 9700 AV Groningen,
The Netherlands

Abstract. We investigate the characteristics and the time evolution of the cosmic web from redshift, $z = 2$, to present time, within the framework of the NEXUS+ algorithm. This necessitates the introduction of new analysis tools optimally suited to describe the very intricate and hierarchical pattern that is the cosmic web. In particular, we characterising filaments (walls) in terms of their linear (surface) mass density, which is very good in capturing the evolution of these structures. At early times the cosmos is dominated by tenuous filaments and sheets, which, during subsequent evolution, merge together, such that the present day web is dominated by fewer, but much more massive, structures. We show also that voids are more naturally described in terms of their boundaries and not their centres. We illustrate this for void density profiles, which, when expressed as a function of the distance from void boundary, show a universal profile in good qualitative agreement with the theoretical shell-crossing framework of expanding underdense regions.

Keywords. large-scale structure of universe, dark matter

1. Introduction

On megaparsec scales the matter distribution of the Universe is not uniform, but it forms an intricate pattern which is known as the *Cosmic Web* (Bond, Kofman & Pogosyan 1996). This is the most salient feature of the anisotropic gravitational collapse of matter, the motor behind the formation of structure in the Universe. Identifying and characterising the cosmic web network, in both numerical simulations and observations, is very challenging due to the overwhelming complexity of the individual structures, their connectivity and their intrinsic multiscale nature. It is even more difficult to follow the time evolution of the cosmic web, since the dominant components and scales change rapidly with redshift. This necessitates the use of scale- and user-free methods that naturally adapt to the complex geometry of the web and that extract the maximum information available among the components of this network.

2. Simulation and methods

We follows the evolution of the cosmic web using the high resolution Millennium and Millennium-II dark matter simulations (Springel *et al.* 2005; Boylan-Kolchin *et al.* 2009), which describe the formation of structure in a periodic box of length 500 and 100 h^{-1}Mpc, respectively. We use the Delaunay Tessellation Field Estimator (Schaap & van de Weygaert 2000; van de Weygaert & Schaap 2009; Cautun & van de Weygaert 2011) to obtain continuous density and velocity fields. These are used as input for the NEXUS+ algorithm (Cautun, van de Weygaert & Jones 2013), which identifies the cosmic web

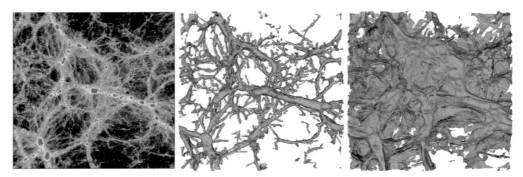

Figure 1. The complexity and multiscale character of the cosmic web as identified by NEXUS+. It shows the density field (left), the filaments (centre) and the walls (right) in a $100 \times 100 \times 10 \ (h^{-1}\mathrm{Mpc})^3$ slice trough the Millennium-II Simulation.

components. The outcome of applying NEXUS+ to the density field is illustrated in Fig. 1. Based on the Multiscale Morphology Filter (MMF) method of Aragón-Calvo *et al.* (2007b), we developed NEXUS+, and its sister method, NEXUS, to use a multitude of tracer fields for classifying the cosmic web environments, from density, to tidal, velocity shear and velocity divergence fields. In addition, NEXUS and NEXUS+ have the unique feature of employing natural and self-consistent criteria for identifying the cosmic web, in contrast to the less clear percolation threshold criteria used in MMF. These approaches are optimally suited for identifying the morphological environments since these methods are multiscale, parameter-free and designed to fully account for the anisotropic nature of gravitational collapse.

In particular, NEXUS+ uses the mass distribution, as traced by the density field, to identify clusters, filaments, walls and voids. The dominant morphological signal is extracted from a 4-dimensional scale-space representation of the matter distribution. The fourth dimension is constructed by filtering the logarithm of the density with a Gaussian kernel, for a range of smoothing scales. The environments are classified on the basis of the dominant local morphological signature of the filtered density field.

For this study, NEXUS+ has two main advantages. Firstly, it determines in a self-consistent way all the different morphological components. And, secondly, the multiscale character of the method make it ideal not only for identifying both prominent and tenuous environments, but also for studying the time evolution of the cosmic web without having to choose a user-defined scale.

3. The evolution of the cosmic web

Fig. 2 shows the evolution of filamentary environments starting with a redshift of $z = 1.9$ down to the present time. At early times, the filaments form a complex network that pervades most of the cosmic volume, with the exception of the most underdense regions. While the network has a few thick structures, it is dominated by small scale filaments. These thin filaments seem to be packed much more tightly close to prominent structures, suggesting that overdense regions have a higher richness of filaments. By $z = 1$, we find that most of the tenuous structures have disappeared and that we can more easily see the pronounced filaments. Going forward in time, to $z = 0.5$ and 0, we find that the evolution of the cosmic web significantly slows down, with only minor changes after $z = 0.5$. Though not shown, a similar evolution can be seen for the wall network too. For a more in depth analysis see Cautun *et al.* (2014).

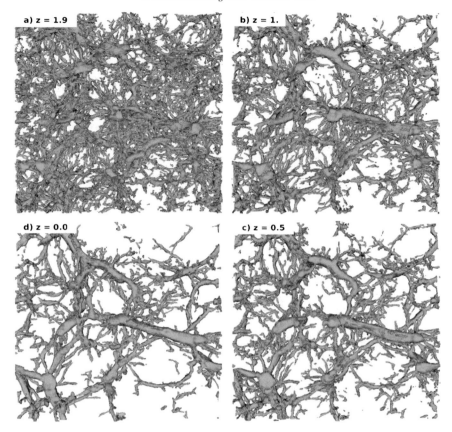

Figure 2. The evolution of the filamentary network as identified by NEXUS+ in a $100 \times 100 \times 10 \ (h^{-1}\,\mathrm{Mpc})^3$ slice trough the Millennium-II Simulation.

3.1. *The mass distribution along filaments and walls*

One of the most widely employed methods to study the cosmic web properties involve the use of global properties, like mass and volume filling fraction of each component (e.g. Aragón-Calvo *et al.* 2007b; Hahn *et al.* 2007; Forero-Romero *et al.* 2009). When applied to our study, such an approach leads to the results presented in top row of Fig. 3. For example, it shows that the volume occupied by filaments decreases since high redshift, probably due to the merging of the thin and tenuous filaments with the more prominent structures. But, more importantly, such a simple analysis cannot characterise the complex evolution seen in Fig. 2. For example, it cannot tell which components, tenuous or prominent, contained the most mass and how this mass distribution evolves in time. To do so, one needs a more complex analysis framework that takes into account the geometry of the various cosmic components.

Such a framework has been introduced in Cautun *et al.* (2014, see also Aragón-Calvo, van de Weygaert & Jones 2010) and takes advantage that, to a first approximation, filaments and walls can be seen as 1-dimensional lines and 2-dimensional surfaces, respectively. These represent the spine of filaments and the central plane of sheets; and can be computed using the techniques described in Aragón-Calvo, van de Weygaert & Jones (2010) and Cautun *et al.* (2014). This reduces the complex filamentary network to a simpler distribution of interconnected curves, with all the mass and halo distribution

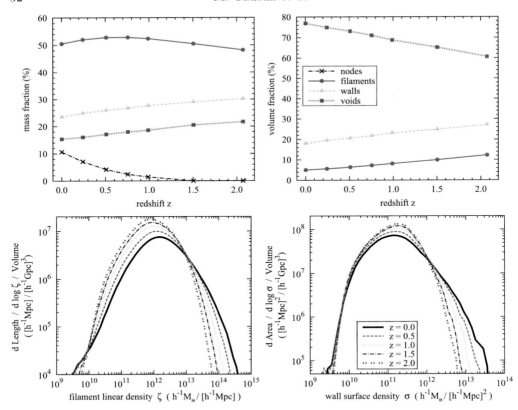

Figure 3. *Top row:* the time evolution of the mass and volume fraction in each cosmic web environment. *Bottom row:* the distribution of linear mass density of filaments (left) and the surface mass density of walls (right) at different redshifts.

of the filament compressed to its spine. Following this, one can move along the resulting curves and compute local quantities, like the mean linear density and the mean diameter of filaments (for details see Cautun *et al.* 2014).

The bottom-left panel of Fig. 3 show the results of such an analysis. It gives the length of filaments, per unit volume, that have a given linear mass density, when measured within a window of $2\ h^{-1}\mathrm{Mpc}$. The figure shows that there are few very low or very high mass filaments and that most of the length of the filamentary network is given by segments with linear densities of $\sim 10^{12}\,\mathrm{M_\odot}/\mathrm{Mpc}$. This distribution evolves in time, to show that segments with high mass become more common at late times, while at the same time there are fewer low mass segments. Even more telling, is the shift to the right in the peak of the distribution, showing that time evolution leads to an increase in the mass of the typical filament segment.

In contrast to filaments, the typical sheet regions become less massive at present time, as shown in bottom-right panel of Fig. 3. The decrease in wall surface density is seen as the shift in the peak of the distribution towards lower σ_{wall} values at later times. It shows that the decrease in the mass fraction of walls seen in the top-left panel takes place via two processes. First, as we just argued, typical sheet stretches become less massive. And secondly, the extent of the wall network reduces at later times, as seen in the decreasing peak values of the σ_{wall} distribution.

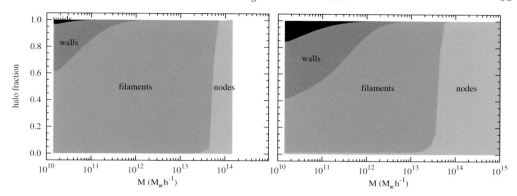

Figure 4. The fraction of haloes in each cosmic web environment as a function of halo mass. We shows results for redshift, $z = 2$ (left), and for the present time (right).

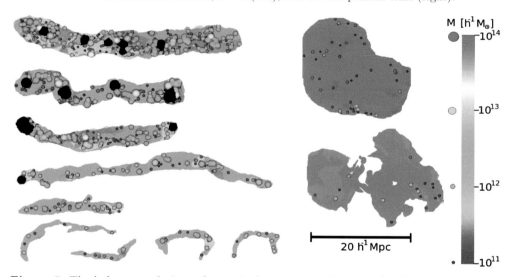

Figure 5. The haloes populating a few typical cosmic web filaments (left) and walls (right). The black points show haloes found in node environments.

3.2. *The halo distribution among the web environments*

The distribution of haloes across web environments plays a key role. First, the morphology of the cosmic web determines the preferential directions of accretion and thus influences the shape and angular momentum of haloes, inducing large scale alignments (e.g. Aragón-Calvo *et al.* 2007a; Hahn *et al.* 2007). And secondly, to identify and study the cosmic web in observations one makes use of galaxies, which are hosted in haloes. So knowing how haloes populate the morphological components is crucial to understand the web environments seen in observations. To this end, we show in Fig. 4 the distribution of haloes in the cosmic web at both $z = 2$ and 0. The figure shows a clear segregation of haloes across environments, with the most massive ones living in nodes and prominent filaments. Walls typically host $\sim 10^{12}~h^{-1}\mathrm{M}_\odot$ and lower mass objects, while void regions are populated with even lower mass haloes.

More interestingly, the halo population varies not only with environment, but also between structures with similar morphological features. This is shown in the left panel of Fig. 5, where we illustrate the haloes found in several filamentary branches. From top

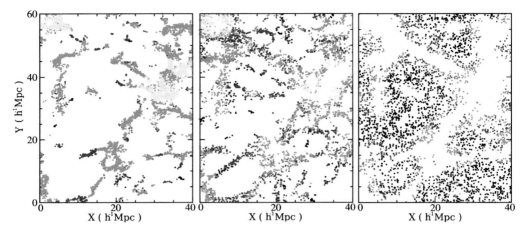

Figure 6. The mass transport across the web components. It shows dark matter particles identified at $z = 2$ as part of filaments (left), walls (centre) and voids (right). These particles are coloured according to their web classification at $z = 0$: node (light-grey), filament (grey), wall (dark-grey) and void (black).

to bottom, we show progressively less prominent filamentary branches, with the bottom most ones corresponding to filaments found in voids. The panel shows a clear trend of the halo distribution with filament properties. Thicker filaments, which are typically outstretched between cluster pairs, are populated with more massive haloes which are also more tightly packed together. In contrast, haloes in tenuous filaments are typically low mass, similar to the ones in walls, and are widely spaced apart.

Fig. 5 illustrates that while prominent filaments are easy to find observationally, since they host many bright galaxies, detecting more tenuous structures presents many observational challenges, given that these systems are mostly inhabited by low luminosity galaxies far apart. Thus, such tenuous objects are not conspicuous features in the spatial distribution of galaxies. The configuration of three aligned galaxies inside a void found by Beygu *et al.* (2013) is probably an example of such a thin filament (Rieder *et al.* 2013).

3.3. *The mass transport across the cosmic web*

By comparing the cosmic web identified at different redshifts, we can trace the path taken by the anisotropic collapse of matter and study the transport of mass among the different web environments. According to the gravitational instability theory, the matter distribution follows a well defined path, with mass flowing from voids into sheets, from sheets into filaments and only in the last step into the cosmic nodes. We illustrate this with the help of Fig. 6 that shows the dark matter particles in a thin $2\ h^{-1}\mathrm{Mpc}$ slice at redshift $z = 2$. Each panel gives the particles identified as part of filaments, walls and voids at $z = 2$, with the particles coloured according to the environments they are found in at the present time.

Fig. 6 shows some of the most important characteristics of the mass transport across the cosmic web, with these conclusions supported by a more in-depth and quantitative analysis of (Cautun *et al.* 2014). Among others, it shows that nodes form at the intersections of the filamentary network and that most of the mass in them has been accreted from regions that correspond to $z = 2$ filaments. For filaments, most of the mass found in these objects at $z = 2$ is also found in filaments at present time. In contrast, walls loose more than half of their $z = 2$ mass to mostly filaments, with the rest remaining in walls. And finally, voids also loose around half of their $z = 2$ mass, which flows into walls

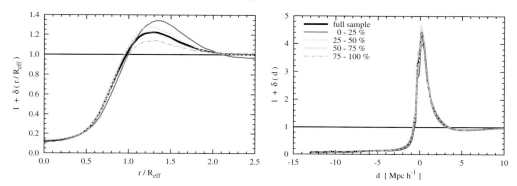

Figure 7. The density profile, $1 + \delta$, of voids for the full population and for subsamples split according to the volume of the void, from small to large. The left panel show the spherical profile of voids. The right panel shows the void profile as a function of the distance from the void boundary. We use the convention that negative d values correspond to the inside of voids, while positive ones correspond to the profile outside the void.

and filaments. In addition, Fig. 6 illustrates some of the limitations of the analysis. A small fraction of $z = 2$ filaments and walls are identified as present day walls and voids, respectively. This is restricted to minor filaments and walls and it is due to the difficulty of identifying such tenuous structures, especially across multiple time steps.

4. A natural void profile

The filaments and walls are not the only ones that have complex shapes and morphologies. Voids have it too. The simple picture of an expanding underdensity in a uniform background suggests that voids become more spherical as they evolve (Icke 1984; Sheth & van de Weygaert 2004). But in reality, voids are not isolated and this simple picture does not hold. There are two major factors that affect the evolution of voids. Firstly, since voids are nearly empty the force field that dictates their growth is dominated by external forces. This external tidal field determines the anisotropic expansion of voids and manifest itself as large scale correlations, over distances larger than $30\ h^{-1}\mathrm{Mpc}$, between the shapes of neighbouring voids (Platen, van de Weygaert & Jones 2008). Secondly, as voids expand, they encounter neighbouring voids resulting in a packing problem. These effects lead to voids that have highly aspherical and complex shapes (Platen, van de Weygaert, & Jones 2007; Platen, van de Weygaert & Jones 2008).

Up to now, voids were characterised in terms of spherical profiles with respect to the void's center, typically the barycentre. Such a methodology results in the void profiles shown in the left panel of Fig. 7. These profiles have been rescaled following the prescription of Hamaus, Sutter & Wandelt (2014), where R_{eff} is the effective radius of the void corresponding to a spherical void with the same volume as the real object. This results in void profiles that are not fully universal and where one needs to follow a complex procedure to rescale voids of different sizes (for details see Hamaus, Sutter & Wandelt 2014; Nadathur *et al.* 2014). More worryingly, the resulting profile does not show the large density caustic at their boundary resulting from the shell-crossing of expanding underdense shells (Sheth & van de Weygaert 2004). So what is the reason for the mismatch?

The discrepancy between spherical void profiles and the shell-crossing predictions of Sheth & van de Weygaert (2004) arise from the fact that voids are very far from having a spherical shape. Thus, using spherically averaged profiles does not lead to an accurate description of void structure. In fact, it is easier to determine the void boundary, where most of the mass and galaxies reside, than the void center, which is devoid of tracers. In fact, this very fact is the cornerstone of Watershed-based void finders (Platen, van de Weygaert, & Jones 2007; Neyrinck 2008). Thus, it is more natural to describe voids with respect to their boundary than with respect to their center.

It suggests that void profiles, and in general void properties, should also be computed with respect to the void boundary. For this, we define the boundary distance field $d_i(\mathbf{x})$ as the minimum distance between point $\mathbf{x}$ and the boundary of void i. In addition, to distinguish between points inside and outside the void, we take the convention that $d_i(\mathbf{x})$ is negative if $\mathbf{x}$ is inside void i and positive otherwise. The resulting density profiles, as a function of the void boundary distance, are shown in the right panel of Fig. 7. First, we find the sharp increase in the density profile at $d{\sim}0$, corresponding to the caustic resulting from shell-crossing. And secondly, voids of different size show a much more similar and universal profile. This method gives a better and more natural description of not only density profiles, but also of void velocity profiles (Cautun *et al.* in prep.).

Acknowledgements

MC and CSF acknowledge the support of the ERC Advanced Investigator grant COS-MIWAY [grant number GA 267291]. RvdW acknowledges support by the John Templeton Foundation, grant [#FP05136-O].

References

Aragón-Calvo, M. A., Jones, B. J. T., van de Weygaert, R., & van der Hulst, J. M., 2007a, *ApJ*, 655, L5
Aragón-Calvo, M. A., Jones, B. J. T., van de Weygaert, R., & van der Hulst, J. M., 2007b, *A&A*, 474, 315
Aragón-Calvo, M. A., van de Weygaert, R., & Jones, B. J. T., 2010, *MNRAS*, 408, 2163
Beygu, B., Kreckel, K., van de Weygaert, R., van der Hulst, J. M., & van Gorkom, J. H., 2013, *AJ*, 145, 120
Bond, J. R., Kofman, L., & Pogosyan, D., 1996, *Nature*, 380, 603
Boylan-Kolchin, M., Springel, V., White, S. D. M., Jenkins, A., & Lemson G., 2009, *MNRAS*, 398, 1150
Cautun, M., van de Weygaert, R., & Jones, B. J. T., 2013, *MNRAS*, 429, 1286
Cautun, M., van de Weygaert, R., Jones, B. J. T., & Frenk, C. S., 2014, *MNRAS*, 441, 2923
Cautun, M. C. & van de Weygaert, R., 2011, preprint arXiv:1105.0370
Forero-Romero, J. E., Hoffman, Y., Gottlöber, S., Klypin, A., & Yepes G., 2009, *MNRAS*, 396, 1815
Hahn, O., Carollo, C. M., Porciani, C., & Dekel, A., 2007, *MNRAS*, 375, 489
Hamaus, N., Sutter, P. M., & Wandelt, B. D., 2014, *Physical Review Letters*, 112, 251302
Icke, V., 1984, *MNRAS*, 206, 1P
Nadathur, S., Hotchkiss, S., Diego, J. M., Iliev, I. T., Gottlöber S., Watson, W. A., & Yepes, G., 2014, preprints arXiv:1407.1295
Neyrinck, M. C., 2008, *MNRAS*, 386, 2101
Platen, E., van de Weygaert, R., & Jones, B. J. T., 2007, *MNRAS*, 380, 551
Platen, E., van de Weygaert, R., & Jones, B. J. T., 2008, *MNRAS*, 387, 128
Rieder, S., van de Weygaert, R., Cautun, M., Beygu, B., & Portegies Zwart S., 2013, *MNRAS*
Schaap, W. E. & van de Weygaert, R., 2000, *A&A*, 363, L29
Sheth, R. K. & van de Weygaert, R., 2004, *MNRAS*, 350, 517
Springel, V. *et al.*, 2005, *Nature*, 435, 629
van de Weygaert, R. & Schaap, W., 2009, in Lecture Notes in Physics, Berlin Springer Verlag, Vol. 665, Data Analysis in Cosmology, Martínez V. J., Saar E., Martínez-González E., Pons-Bordería M.-J., eds., pp. 291–413

The Zeldovich Universe:
Genesis and Growth of the Cosmic Web
Proceedings IAU Symposium No. 308, 2014
R. van de Weygaert, S. Shandarin, E. Saar & J. Einasto, eds.

© International Astronomical Union 2016
doi:10.1017/S1743921316009625

The Peak/Dip Picture of the Cosmic Web

Graziano Rossi

Department of Astronomy and Space Science, Sejong University, Seoul, 143-747, Korea
email: graziano@sejong.ac.kr

Abstract. The initial shear field plays a central role in the formation of large-scale structures, and in shaping the geometry, morphology, and topology of the cosmic web. We discuss a recent theoretical framework for the shear tensor, termed the 'peak/dip picture', which accounts for the fact that halos/voids may form from local extrema of the density field – rather than from random spatial positions; the standard Doroshkevich's formalism is generalized, to include correlations between the density Hessian and shear field at special points in space around which halos/voids may form. We then present the 'peak/dip excursion-set-based' algorithm, along with its most recent applications – merging peaks theory with the standard excursion set approach.

Keywords. Methods: analytical, statistical, numerical; cosmology: theory, large-scale structure of universe.

1. The Cosmic Web: Geometry, Morphology, Topology

The 'cosmic web', a complex large-scale spatial organization of matter, is the result of the anisotropic nature of gravitational collapse (Zel'Dovich 1970; Peebles 1980; Bardeen *et al.* 1986; Bond *et al.* 1991). Figure 1 exemplifies this intricate pattern of filaments, halos, voids and sheets, in its gaseous component at $z = 3$ (top left panel) – when structures are still forming – and at the present epoch ($z = 0$; top right panel), when the geometry, morphology, and topology of the web are well-delineated; these snapshots are extracted from a $25h^{-1}$Mpc box-size low-resolution hydrodynamical simulation, assuming a Planck (2013) reference cosmology (Borde *et al.* 2014; Rossi *et al.* 2014). In the same figure, the bottom panels show the corresponding distribution of the internal energy at the two different redshifts considered. The cosmic web has been confirmed by several observations, in particular using data from the 2dF Galaxy Redshift Survey (Colless *et al.* 2003) and the Sloan Digital Sky Survey (York *et al.* 2000), and most recently by a spectacular three-dimensional detection of a cosmic web filament in Lyman-α emission at $z \simeq 2.3$, discovered during a search for cosmic gas fluorescently illuminated by background bright quasars (Cantalupo *et al.* 2014). Along with observational efforts, numerical studies are essential and unavoidable in order to understand all the complicated physical phenomena involved in the nonlinear formation of structures, requiring larger simulated volumes and yet accuracy in resolution – e.g., see Cautun *et al.* (2014) and Hidding *et al.* (2014) for recent developments. Some analytic work is also helpful in guiding and interpreting results from numerical studies. To this end, we discuss here a theoretical framework which patches together several ingredients: the fact that dark matter halos and voids are clearly not spherical in shape, but at the very least triaxial; the observation, supported by simulation results, that halos and voids tend to form in or around local maxima/minima of the density field – rather than at random spatial locations; the indication that an 'embryonic cosmic web' is already present in the primordial density field (Bond *et al.* 1991); the evidence that there is a good correspondence between peaks/dips in the initial conditions and halos/voids at later times; the crucial role of the tidal field in shaping the cosmic web, and the effects of its correlation with the density Hessian; and the fact that

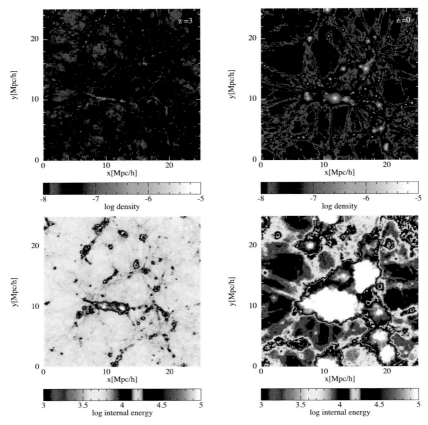

Figure 1. Snapshots of the gaseous component of the cosmic web, from a cosmological hy-drodynamical simulation with $25h^{-1}$Mpc box-size and resolution $N_{\rm p} = 192^3$ particles/type, assuming a reference Planck (2013) cosmology. Top panels are full projections of the density field in the x and y directions, at redshifts $z = 3$ (left) and $z = 0$ (right). Bottom panels show the corresponding internal energy distributions, reflecting the temperature of the gas.

the standard excursion set theory, which determines the initial conditions and is based on the statistics of Gaussian fields, only considers random spatial positions. The aim is to provide a more realistic theoretical language for describing the morphology of the cosmic web, in closer support of numerical studies. We briefly elaborate on these points in the next sections, while details can be found in Rossi et al. (2011) and in Rossi (2012, 2013).

2. Initial Shear Field and Dynamical Evolution

The initial shear field – rather than the density Hessian – is a major player in shaping the geometrical structure of the cosmic web, and originates the characteristic observed pattern of filaments, halos, voids and sheets. The morphology of halos and voids departs from sphericity, as the statistics of Gaussian fields imply that spherically symmetric initial configurations should be a set of measure zero (Doroshkevich 1970); the observed triaxial-ity has several important implications in determining the assembly histories, kinematics, clustering and fundamental structural properties of halos and voids. Rossi et al. (2011) describe a simple dynamical model able to link the final shapes of virialized halos to their initial shapes, by combining the physics of the ellipsoidal collapse with the excursion set

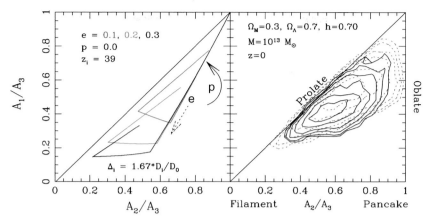

Figure 2. A simple model for describing the shapes of dark matter halos (Rossi *et al.* 2011), based on the combination of excursion set theory and ellipsoidal collapse. The left panel shows the evolution of halos in the 'axis ratio plane', as a function of their ellipticity e and prolateness p, when $p = 0$ and $e = 0.1, 0.2, 0.3$, respectively. The right panel compares the final axial ratio distributions predicted from the model (black solid lines) with corresponding results from numerical simulations (blue dashed lines), for 10,000 halos of mass $M = 10^{13} M_\odot$.

theory. An illustration is provided in Figure 2: in a useful planar representation of halo axial ratios, the left panel shows the effect of the initial values of ellipticity and prolateness (e, p) – which parametrize the surrounding shear field – in determining the future evolution of an object of given mass, for three representative values of e when $p = 0$; the right panel compares the distribution of final axial ratios in the model (solid black lines) with results from numerical simulations (dashed blue lines), when 10,000 halos of mass $M = 10^{13} M_\odot$ are considered – in a flat cosmology with $\Omega_M = 0.3$ and $h = 0.7$.

3. The Peak/Dip Formalism: Theory and Applications

Although idealized, the previous model provides some useful insights into the nonlinear collapse of structures, and in particular it highlights the fact that a collapsing patch will eventually become a filament, pancake, or halo depending on its initial shape and overdensity. Hence, the eigenvalues of the initial shear field are a key ingredient in determining the final destiny of an object, and more generally in shaping the morphology of the cosmic web. In 1970, Doroshkevich derived the joint 'unconditional' probability distribution of an ordered set of tidal field eigenvalues corresponding to a Gaussian potential. Recently, akin in philosophy to that of van de Weygaert & Bertschinger (1996), Rossi (2012) provided a set of analytic expressions which extend the work of Doroshkevich (1970) and Bardeen *et al.* (1986), to incorporate the density peak/dip constraint into the statistical description of the initial shear field. In this generalized formalism, termed the 'peak/dip picture of the cosmic web', the probability of observing a tidal field $\mathbf{T}$ for the gravitational potential, given a curvature $\mathbf{H}$ for the density field and a correlation strength γ is given by (Rossi 2012):

$$p(\mathbf{T}|\mathbf{H}, \gamma) = \frac{15^3}{16\sqrt{5}\pi^3} \frac{1}{\sigma_T^6(1-\gamma^2)^3} \exp\left[-\frac{3}{2\sigma_T^2(1-\gamma^2)}(2K_1^2 - 5K_2)\right]. \qquad (3.1)$$

The previous equation has lead to the 'peak/dip excursion-set-based' algorithm, able to sample the constrained eigenvalues of the initial shear field associated with Gaussian statistics at positions which correspond to peaks or dips of the correlated density field

　　　　　　　　　　　　　　G. Rossi

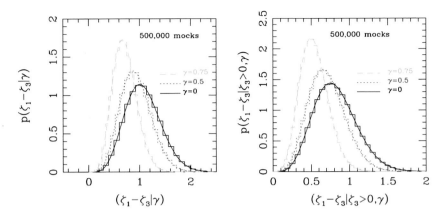

Figure 3. Conditional distributions of eigenvalues in the 'peak/dip picture', relevant for the conditional shape distributions (Rossi 2012, 2013), for different values of the correlation parameter γ – as specified in the panels. Theoretical predictions are confronted with results from 500,000 mock simulations (histograms), obtained with the 'peak-dip excursion-set-based algorithm'.

(Rossi 2013); the algorithm can be readily inserted into the excursion set framework (Peacock & Heavens 1990; Bond *et al.* 1991; Lacey & Cole 1993) to account for a subset of spatial points from where halos/voids may form – hence merging the peaks theory description with the excursion set approach. Along these lines, it is possible to extend the standard distributions of shape parameters (i.e. ellipticity and prolateness) in the presence of the density peak/dip constraint (Rossi 2013, and Figure 3), which generalize some previous literature work and combine the formalism of Bardeen *et al.* (1986) – based on the density field – with that of Bond & Myers (1996) – based on the shear field.

Acknowledgments

This work and the participation to the IAU Symposium 308 'The Zeldovich Universe: Genesis and Growth of the Cosmic Web' (June 2014) in Tallinn, Estonia, were supported by the faculty research fund of Sejong University in 2014. G.R. is also supported by the National Research Foundation of Korea (SGER Grant 2014055950). It is a pleasure to thank Sergei Shandarin and Rien van de Weygaert for the superb organization, along with the scientific and local organizing committees and the secretariat.

References

Bardeen, J. M., Bond, J. R., Kaiser, N., & Szalay, A. S. 1986, *ApJ*, 304, 15
Bond, J. R. & Myers, S. T. 1996, *ApJS*, 103, 1
Bond, J. R., Cole, S., Efstathiou, G., & Kaiser, N. 1991, *ApJ*, 379, 440
Cautun, M., van de Weygaert, R., Jones, B. J. T., & Frenk, C. S. 2014, *MNRAS*, 441, 2923
Doroshkevich, A. G. 1970, *Astrofizika*, 6, 581
Hidding, J., Shandarin, §. F., & van de Weygaert, R. 2014, *MNRAS*, 437, 3442
Peebles, P. J. E. 1980, Princeton University Press, 1980. 435 p.
Rossi, G., Palanque-Delabrouille, N., Borde, A., et al. 2014, *A&A*, 567, A79
Rossi, G. 2013, *MNRAS*, 430,1486
Rossi, G. 2012, *MNRAS*, 421, 296
Rossi, G., Sheth, R. K., & Tormen, G. 2011, *MNRAS*, 416, 248
van de Weygaert, R. & Bertschinger, E. 1996, *MNRAS*, 281, 84
Zel'Dovich, Y. B. 1970, *A&A*, 5, 84

The Zeldovich Universe:
Genesis and Growth of the Cosmic Web
Proceedings IAU Symposium No. 308, 2014
R. van de Weygaert, S. Shandarin, E. Saar & J. Einasto, eds.

© International Astronomical Union 2016
doi:10.1017/S1743921316009637

Non Gaussian Minkowski functionals and extrema counts for CMB maps

Dmitri Pogosyan[1,2,3], Sandrine Codis[2] and Christophe Pichon[2]

[1]Department of Physics, University of Alberta, 11322-89 Avenue, Edmonton, Alberta, T6G 2G7, Canada
[2]CNRS, UPMC, Institut d'astrophysique de Paris, 98 bis boulevard Arago, 75014, Paris, France
[3]CNRS, Institut Lagrange de Paris, 98 bis boulevard Arago, 75014, Paris, France

Abstract. In the conference presentation we have reviewed the theory of non-Gaussian geometrical measures for 3D Cosmic Web of the matter distribution in the Universe and 2D sky data, such as Cosmic Microwave Background (CMB) maps that was developed in a series of our papers. The theory leverages symmetry of isotropic statistics such as Minkowski functionals and extrema counts to develop post Gaussian expansion of the statistics in orthogonal polynomials of invariant descriptors of the field, its first and second derivatives. The application of the approach to 2D fields defined on a spherical sky was suggested, but never rigorously developed. In this paper we present such development treating the effects of the curvature and finiteness of the spherical space S_2 exactly, without relying on flat-sky approximation. We present Minkowski functionals, including Euler characteristic and extrema counts to the first non-Gaussian correction, suitable for weakly non-Gaussian fields on a sphere, of which CMB is the prime example.

Random fields are ubiquitous phenomena in physics appearing in areas ranging from turbulence to the landscape of string theories. In cosmology, the sky-maps of the polarized Cosmic Microwave Background (CMB) radiation – a focal topic of current research – is a prime example of such 2D random fields, specified on S_2 spherical space. Modern view of the cosmos, developed primarily through statistical analysis of these fields, points to a Universe that is statistically homogeneous and isotropic with a hierarchy of structures arising from small Gaussian fluctuations of quantum origin. While the Gaussian limit provides the fundamental starting point in the study of random fields, non-Gaussian features of the CMB fields are of great interest. Indeed, CMB inherits a high level of Gaussianity from initial fluctuations, but small non-Gaussian deviations may provide a unique window into the details of processes in the early Universe. The search for the best methods to analyze non-Gaussian random fields is ongoing.

In the paper Pogosyan, Gay & Pichon (2009) the general invariant based formalism for computing topological and geometrical characteristics of non Gaussian fields was presented. The general formulae for the Euler characteristic to all orders has been derived, which encompasses the well known first correction Matsubara (2003) and which was later confirmed to the next order by Matsubara (2010). This work was followed by the detailed exposition of the theory in 2D and 3D flat (Cartesian) space in Pogosyan, Pichon & Gay (2011) and Gay, Pichon & Pogosyan (2012), and generalized to the 3D redshift space where isotropy is broken in Codis *et al.* (2013).

The goal of this paper is to extend these results to the fields defined on a finite curved spherical space S_2 without reliance on the flat field (small angle) approximation. While these proceedings were being prepared, similar work has been done for statistics of peaks in the Gaussian case within Marcos-Caballero *et al.* (2016). Here our focus is on non-Gaussian corrections. We discuss how to compute exact Minkowski functionals for the excursion sets of a scalar field on a S_2 sphere to all orders in non-Gaussian expansion and provide an explicit expression for the Euler characteristic to the first order. Expressions for the total extrema counts to the first non-Gaussian order are also given, while analytical formulas for differential extrema counts to the same order will be published elsewhere due to their length. These results have a direct relevance to CMB data analysis.

1. Joint distribution function of the field and its derivatives on S_2 sphere.

The statistics of Minkowski functionals, including the Euler number, as well as extrema counts requires the knowledge of the one-point joint probability distribution function (JPDF) $P(x, x_i, x_{ij})$ of the field x, its first, x_i, and second, x_{ij}, derivatives of the field x. Let us consider random field x defined on a 2D sphere S_2 of radius R represented as the expansion in spherical harmonics

$$x(\theta, \phi) = \sum_{l=0}^{\infty} \sum_{m=-l}^{l} a_{lm} Y_{lm}(\theta, \phi) \tag{1.1}$$

where for the Gaussian statistically homogeneous and isotropic field random coefficients a_{lm} are uncorrelated with m-independent variances C_l of each harmonic

$$\langle a_{lm} a_{l'm'}^* \rangle = C_l \delta_{ll'} \delta_{mm'} \tag{1.2}$$

The variance of the field is then given by

$$\sigma^2 \equiv \langle x^2 \rangle = \frac{1}{4\pi} \sum_l C_l (2l+1) \tag{1.3}$$

When considering derivatives in the curved space, we use covariant derivatives $x_{;\theta}$, $x_{;\phi}$, $x^{;\theta}_{;\theta}$, $x^{;\phi}_{;\phi}$, $x^{;\theta}_{;\phi}$ where it will be seen immediately that mixed version for the second derivatives is the most appropriate choice. The 2D rotation-invariant combinations of derivatives are

$$q^2 = x_{;\phi} x^{;\phi} + x_{;\theta} x^{;\theta} , \quad J_1 = \left(x^{;\theta}_{;\theta} + x^{;\phi}_{;\phi} \right)^2 , \quad J_2 = \left(x^{;\theta}_{;\theta} - x^{;\phi}_{;\phi} \right)^2 + 4 x^{;\theta}_{;\phi} x^{;\phi}_{;\theta} \tag{1.4}$$

where J_1 is linear in the field and q^2 and J_2 are quadratic, always positive, quantities. The derivatives are also random Gaussian variables, which variances are easily computed

$$\sigma_1^2 \equiv \langle q^2 \rangle = \frac{1}{4\pi R^2} \sum_l C_l l (l+1)(2l+1) \tag{1.5}$$

$$\sigma_2^2 \equiv \langle J_1^2 \rangle = \frac{1}{4\pi R^4} \sum_l C_l l^2 (l+1)^2 (2l+1) \tag{1.6}$$

$$\sigma_2'^2 \equiv \langle J_2 \rangle = \frac{1}{4\pi R^4} \sum_l C_l (l-1) l (l+1)(l+2)(2l+1) \tag{1.7}$$

where the fundamental difference between a sphere and the 2D Cartesian space is in the fact that $\sigma_2' \neq \sigma_2$. Among the cross-correlations the only non-zero one is between the the field and its Laplacian $\left\langle x \left(x^{;\theta}_{;\theta} + x^{;\phi}_{;\phi} \right) \right\rangle = -\sigma_1^2$.

From now on we rescale all random quantities by their variances, so that rescaled variables have $\langle x^2 \rangle = \langle J_1^2 \rangle = \langle q^2 \rangle = \langle J_2 \rangle = 1$. Introducing $\zeta = (x + \gamma J_1)/\sqrt{1 - \gamma^2}$ (where the spectral parameter $\gamma = -\langle x J_1 \rangle = \sigma_1^2/(\sigma \sigma_2)$) leads to the following JPDF for the Gaussian 2D field

$$G_{2D} = \frac{1}{2\pi} \exp\left[-\frac{1}{2}\zeta^2 - q^2 - \frac{1}{2}J_1^2 - J_2\right]. \tag{1.8}$$

In Pogosyan, Gay & Pichon (2009) we have observed that for non-Gaussian JPDF the invariant approach immediately suggests a Gram-Charlier expansion in terms of the orthogonal polynomials defined by the kernel G_{2D}. Since ζ, q^2, J_1 and J_2 are uncorrelated variables in the Gaussian limit, the resulting expansion is

$$P_{2D}(\zeta, q^2, J_1, J_2) = G_{2D}\left[1+\right.$$

$$\sum_{n=3}^{\infty} \sum_{i,j,k,l=0}^{i+2j+k+2l=n} \frac{(-1)^{j+l}}{i!\,j!\,k!\,l!} \left\langle \zeta^i q^{2j} J_1^{\,k} J_2^{\,l} \right\rangle_{GC} H_i\left(\zeta\right) L_j\left(q^2\right) H_k\left(J_1\right) L_l\left(J_2\right)\right], \tag{1.9}$$

where terms are sorted in the order of the field power n and $\sum_{i,j,k,l=0}^{i+2j+k+2l=n}$ stands for summation over all combinations of non-negative i, j, k, l such that $i + 2j + k + 2l$ adds to the order of the expansion term n. H_i are (*probabilists'*) Hermite and L_j are Laguerre polynomials. The coefficients of expansion

$$\left\langle \zeta^i q^{2j} J_1^k J_2^l \right\rangle_{GC} = \frac{j!\,l!}{(-1)^{j+l}} \left\langle H_i\left(\zeta\right) L_j\left(q^2\right) H_k\left(J_1\right) L_l\left(J_2\right)\right\rangle. \tag{1.10}$$

are related (and for the first non-Gaussian order n=3 are equal) to the moments of the field and its derivatives (see Gay, Pichon & Pogosyan (2012) for details).

Up to now our considerations are practically identical to the theory in the Cartesian space, which facilitates using many of the Cartesian calculations. We stress again the only, but important difference being $\sigma_2' \neq \sigma_2$. We shall see in the next sections how this difference plays out. Here we introduce the spectral parameter β that describes this difference

$$\beta \equiv 1 - \frac{\sigma_2'^2}{\sigma_2^2} = 2\frac{\sum_l C_l l(l+1)(2l+1)}{\sum_l C_l l^2(l+1)^2(2l+1)} \tag{1.11}$$

Let us review the scales and parameters that the theory has. As in the flat space, we have two scales $R_0 = \sigma/\sigma_1$ and $R_* = \sigma_1/\sigma_2$ and the spectral parameter $\gamma = R_*/R_0$ (which also describes correlation between the field and its second derivatives). On a sphere we have a third scale, the curvature radius R. The meaning of β becomes clear if we notice that $\sigma_2^2 - \sigma_2'^2 = 2\sigma_1^2/R^2$, thus $\beta = 2R_*^2/R^2$, i.e describes the ratio of the correlation scale R_* to the curvature of the sphere. As with the γ, β varies from 0 to 1, with $\beta = 0$ corresponding to the flat space limit. From Eq. (1.11) we find that $\beta = 1$ is achieved when the field has only the monopole and the dipole in its spectral decomposition.

2. Minkowski functionals on S_2 beyond the Gaussian limit

There are three Minkowski functionals that are defined for the excursion set above threshold ν of a 2D field, namely the filling factor, $f_V(\nu)$, i.e the volume fraction occupied by the region above the threshold ν, the length (per unit volume) of isofield contours, $\mathcal{L}(\nu)$ and Euler characteristic $\chi(\nu)$. Statistics of the first two do not depend on second derivatives of the field, and thus are identical on S_2 and the 2D Cartesian space. Here, for completeness, we reproduce the non-Gaussian expansions for these quantities from

Gay, Pichon & Pogosyan (2012)

$$f(\nu) = \frac{1}{2}\mathrm{Erfc}\left(\frac{\nu}{\sqrt{2}}\right) + \frac{1}{\sqrt{2\pi}}e^{-\frac{\nu^2}{2}}\sum_{n=3}^{\infty}\frac{\langle x^n\rangle_{\mathrm{GC}}}{n!}H_{n-1}(\nu). \tag{2.1}$$

$$\mathcal{L}(\nu) = \frac{1}{2\sqrt{2}R_0}e^{-\frac{\nu^2}{2}}\left(1 + \frac{1}{2\sqrt{\pi}}\sum_{n=3}^{\infty}\sum_{i,j}^{i+2j=n}\frac{(-1)^{j+1}}{i!j!}\frac{\Gamma(j-\frac{1}{2})}{\Gamma(j+1)}\langle x^i q^{2j}\rangle_{\mathrm{GC}}H_i(\nu)\right). \tag{2.2}$$

Euler characteristic density of the region above a threshold $x = \nu$ is a more interesting case. It is given by the average of the determinant of the Hessian matrix of the second derivatives of the field at the points where the first derivatives vanish Adler (1981), Longuet-Higgins (1957)

$$\chi(\nu) = \int_\nu^\infty \mathrm{d}x \int \mathrm{d}^3 x_{ij}\, P(x, x_i = 0, x_{ij})\det(x_{ij}). \tag{2.3}$$

It has been argued in Pogosyan, Gay & Pichon (2009) that on S_2 the determinant should be that of the Hessian of the mixed covariant derivatives $\det(x^{\cdot i}_{\cdot j})$. It is this choice that provides the density relative to the invariant volume element $R^2 \sin^2\theta \mathrm{d}\theta \mathrm{d}\phi$ and has a scalar trace equal to the Laplacian of the field. Using scaled invariant variables

$$\det(x^{\cdot i}_{\cdot j}) = \frac{\sigma_2^2}{4}\left(J_1^2 - (1-\beta)J_2\right) \equiv \sigma_2^2 I_2 , \tag{2.4}$$

where we have introduced another scaled quadratic invariant I_2. In terms of the eigenvalues of the Hessian, $\sigma_2^2 I_2 = \lambda_1 \lambda_2$, while $\sigma_2^2 J_1 = \lambda_1 + \lambda_2$ and $\sigma_2^2(1-\beta)J_2 = (\lambda_1 - \lambda_2)^2$.

In the Gaussian limit the Euler characteristic density becomes

$$\chi(\nu) = \frac{\sigma_2^2}{8\pi^2\sigma_1^2}\int_{-\infty}^{\infty}\mathrm{d}J_1\int_0^\infty \mathrm{d}J_2 \int_{\frac{\nu+\gamma J_1}{\sqrt{1-\gamma^2}}}^{\infty}\mathrm{d}\zeta\,\exp\left[-\frac{1}{2}\zeta^2 - \frac{1}{2}J_1^2 - J_2\right]\left(J_1^2 - (1-\beta)J_2\right) \tag{2.5}$$

It evaluates to

$$\chi(\nu) = \frac{\gamma^2}{4\pi\sqrt{2\pi}R_*^2}\nu e^{-\frac{\nu^2}{2}} + \frac{\beta}{8\pi R_*^2}\mathrm{erfc}\left(\frac{\nu}{\sqrt{2}}\right) \tag{2.6}$$

which differ from the well known Cartesian result by the $\beta \neq 0$ term. On a sphere which has a finite volume $4\pi R^2$ it is appropriate to quote the total Euler characteristic in the whole volume, which, recalling the relation between γ, β, R and R_* becomes

$$4\pi R^2\chi(\nu) = \frac{R^2}{\sqrt{2\pi}R_0^2}\nu e^{-\frac{\nu^2}{2}} + \mathrm{Erfc}\left(\frac{\nu}{\sqrt{2}}\right) \tag{2.7}$$

which explicitly demonstrates that if $\nu = -\infty$. i.e the whole space is included in the excursion set, the total Euler characteristic is equal to that of a sphere, $4\pi R^2\chi(-\infty) = 2$, as expected.

Evaluation of the non-Gaussian expansion for $\chi(\nu)$ entails integration Eq. (2.3) with distribution function P_{2D} given by the Eq. (1.9). The procedure is similar to that in Cartesian space as elaborated in detail in Gay, Pichon & Pogosyan (2012) and which led to the complete expression for the Euler characteristic to all orders first reported in Pogosyan, Gay & Pichon (2009). Indeed, the quantity I_2 that we average over all the range of J_2 can be rewritten as $H_2(J_1) + \beta + (1-\beta)L_1(J_2)$. Thus only $l = 0, 1$ terms of the expansion, i.e containing $L_0(J_2)$ or $L_1(J_2)$, do not vanish after integration. Here we should limit ourselves with presenting only the result of the most practical use - up to the first, cubic in the field, non-Gaussian correction

$$\chi(\nu) = \frac{\beta}{8\pi R_*^2} \text{Erfc}\left(\frac{\nu}{\sqrt{2}}\right) + \frac{1}{4\pi\sqrt{2\pi}R_*^2}\exp\left(-\frac{\nu^2}{2}\right)$$

$$\times \left[\gamma^2 H_1(\nu) + 2\gamma\langle q^2 J_1\rangle + 4\langle xI_2\rangle - (\gamma^2\langle xq^2\rangle + \gamma\langle x^2 J_1\rangle) H_2(\nu) + \frac{\gamma^2}{6}\langle x^3\rangle H_4(\nu)\right.$$

$$\left. + \beta\left(-\langle xq^2\rangle H_0(\nu) + \frac{1}{6}\langle x^3\rangle H_2(\nu)\right)\right], \tag{2.8}$$

where the Gram-Charlier moments of the non-primary variables x and I_2 are understood as correspondent combinations of Gram-Charlier moments of the expansion variables ζ, J_1 and J_2. The first term of Eq. (2.8) is the Gaussian result on the sphere that is responsible for the total Euler number of the excursion set to be that of the total sphere when $\nu = -\infty$. The last $\propto \beta$ terms is a correction to non-Gaussian result due to the curvature of the sphere. In conclusion we as well write explicitly the result for the total Euler number above threshold ν

$$4\pi R^2\chi(\nu) = \text{Erfc}\left(\frac{\nu}{\sqrt{2}}\right) + \frac{2}{\sqrt{2\pi}}\exp\left(-\frac{\nu^2}{2}\right)\left(\frac{1}{6}\langle x^3\rangle H_2(\nu) - \langle xq^2\rangle H_0(\nu)\right)$$

$$+ \frac{R^2}{\sqrt{2\pi}R_0^2}\exp\left(-\frac{\nu^2}{2}\right) \tag{2.9}$$

$$\times \left[H_1(\nu) + \frac{2}{\gamma}\langle q^2 J_1\rangle + \frac{4}{\gamma^2}\langle xI_2\rangle - \left(\langle xq^2\rangle + \frac{1}{\gamma}\langle x^2 J_1\rangle\right) H_2(\nu) + \frac{1}{6}\langle x^3\rangle H_4(\nu)\right].$$

3. Extrema counts on S_2 beyond the Gaussian limit

The number density of extrema above a threshold ν is given by an integral very similar to the Euler characteristic Adler (1981), Longuet-Higgins (1957)

$$n_{\text{ext}}(\nu) = \int_\nu^\infty dx \int d^3x_{ij} P(x, x_i = 0, x_{ij})|x_{ij}|\Theta_{ext}(\lambda_m). \tag{3.1}$$

where the theta function $\Theta_{ext}(\lambda_m)$ chooses the regions of integration in the space of second derivatives with appropriate to the particular extremum type signs of the Hessian eigenvalues. In 2D, assuming $\lambda_1 \geqslant \lambda_2$, $\Theta_{ext}(\lambda_m) = \Theta(-\lambda_1)$ for maxima, $= \theta(\lambda_2)$ for minima, and $= \theta(\lambda_1)\theta(-\lambda_2)$ for saddle points. Indeed, we have a well-known topological relation

$$\chi(\nu) = n_{\max}(\nu) - n_{\text{sad}}(\nu) + n_{\min}(\nu) \tag{3.2}$$

The integral Eq. (3.1) has a very transparent form when the Hessian is described in invariant variables. It is equivalent to Eq. (2.5), except that the limits of integration over J_1 are partitioned into the regions of the fixed sign of the determinant I_2. Namely, maxima correspond to the range $J_1 \in (-\infty, -\sqrt{(1-\beta)J_2})$, minima to $J_1 \in (\sqrt{(1-\beta)J_2}, \infty)$ and saddle points to $J_1 \in (-\sqrt{(1-\beta)J_2}, \sqrt{(1-\beta)J_2})$.

Calculations for differential density of extremal points, $\partial n_{\text{ext}}/\partial(\nu)$ can be carried out analytically even for the general expression Eq, (1.9) (see discussion in Gay, Pichon & Pogosyan (2012) for the flat case). The resulting expressions are cumbersome, and here we limit ourselves to presenting results for the total density of extrema in the first non-Gaussian order only. The total number density of maxima is given by

$$n_{\max} = \frac{\sigma_2^2}{4\sigma_1^2}\int_0^\infty dJ_2\int_{-\infty}^{-\sqrt{(1-\beta)J_2}} dJ_1 P_{2D}(q^2 = 0, J_1, J_2)\left|J_1^2 - (1-\beta)J_2\right| \tag{3.3}$$

and, similarly, for the minima and the saddle points. The result is

$$n_{\max/\min} = \frac{(1-\beta)^{3/2} + \beta\sqrt{3-\beta}}{8\pi\sqrt{3-\beta}R_*^{\,2}} \pm \frac{6(3-\beta)\left\langle q^2 J_1\right\rangle - (5-3\beta)\left\langle J_1^3\right\rangle + 6(1-\beta)\left\langle J_1 J_2\right\rangle}{6\pi\sqrt{2\pi}R_*^{\,2}(3-\beta)^2},$$

$$n_{\text{sad}} = \frac{(1-\beta)^{3/2}}{4\pi\sqrt{3-\beta}R_*^{\,2}}, \tag{3.4}$$

where we immediately see that $n_{\max} + n_{\min} - n_{\text{sad}} = \beta/(4\pi R_*^2) = 2/(4\pi R^2)$ as expected. The total number of saddles, as well as of all the extremal points, $n_{\max} + n_{\min} + n_{\text{sad}}$, are preserved in the first order (the latter following from the former), but the symmetry between the minima and the maxima is broken.

It is instructive to look how the Gaussian extrema counts are modified by the properties of spherical space when the curvature radius is large relative to the typical extrema separation scale R_*, i.e when β is small. Up to the first order in β

$$n_{\max/\min} \sim \frac{1}{8\sqrt{3}\pi R_*^2} + \frac{9-4\sqrt{3}}{36\pi R^2} \approx \frac{1}{8\sqrt{3}\pi R_*^2}\left(1 + 0.8 R_*^2/R^2\right) \tag{3.5}$$

$$n_{\text{sad}} \sim \frac{1}{4\sqrt{3}\pi R_*^2} - \frac{2}{3\sqrt{3}\pi R^2} \approx \frac{1}{4\sqrt{3}\pi R_*^2}\left(1 - 2.7 R_*^2/R^2\right) \tag{3.6}$$

This shows that being on a sphere, increases the number density of maxima and minima, but decrease (and in more significant way) the number of saddles. Incidently, assuming large-angle CMB power spectrum, truncated at $l = 30$ gives $\beta \approx 1/170$, i.e 1% correction to the count of extrema relative to the flat-sky approximation.

References

Adler, R. J. *The Geometry of Random Fields*. The Geometry of Random Fields, Chichester: Wiley, 1981.

Bardeen, J. M., Bond, J. R., Kaiser, N., & Szalay, A. S. The statistics of peaks of Gaussian random fields. *ApJ*, 304:15–61, May 1986.

Codis, S., Pichon, C., Pogosyan, D., Bernardeau, F., & Matsubara, T. Non-Gaussian Minkowski functionals and extrema counts in redshift space. *MNRAS*, 435:531–564, October 2013.

Doroshkevich, A. G. The space structure of perturbations and the origin of rotation of galaxies in the theory of fluctuation. *Astrofizika*, 6:581–600, 1970.

Gay, C., Pichon, C., & Pogosyan, D. Non-Gaussian statistics of critical sets in 2D and 3D: Peaks, voids, saddles, genus, and skeleton. *Phys. Rev. D*, 85(2):023011, January 2012.

Longuet-Higgins, M. S. The statistical analysis of a random, moving surface. *Royal Society of London Philosophical Transactions Series A*, 249:321–387, February 1957.

Marcos-Caballero, A., Fernández-Cobos, R., Martínez-González, E., & Vielva, P. The shape of CMB temperature and polarization peaks on the sphere. *JCAP*, 4:058, April 2016.

Matsubara, T. Statistics of Smoothed Cosmic Fields in Perturbation Theory. I. Formulation and Useful Formulae in Second-Order Perturbation Theory. *ApJ*, 584:1–33, February 2003.

Matsubara, T. Analytic Minkowski functionals of the cosmic microwave background: Second-order non-Gaussianity with bispectrum and trispectrum. *Phys. Rev. D*, 81(8):083505, April 2010.

Pogosyan, D., Gay, C., & Pichon, C. Invariant joint distribution of a stationary random field and its derivatives: Euler characteristic and critical point counts in 2 and 3D. *Phys. Rev. D*, 80(8):081301, October 2009.

Pogosyan, D., Gay, C., & Pichon, C. Erratum: Invariant joint distribution of a stationary random field and its derivatives: Euler characteristic and critical point counts in 2 and 3D [Phys. Rev. D 80, 081301 (2009)]. *Phys. Rev. D*, 81(12):129901, June 2010.

Pogosyan, D., Pichon, C., & Gay, C. Non-Gaussian extrema counts for CMB maps. *Phys. Rev. D*, 84(8):083510, October 2011.

The Zeldovich Universe:
Genesis and Growth of the Cosmic Web
Proceedings IAU Symposium No. 308, 2014
R. van de Weygaert, S. Shandarin, E. Saar & J. Einasto, eds.

© International Astronomical Union 2016
doi:10.1017/S1743921316009649

Large-scale structure non-Gaussianities with modal methods

Marcel Schmittfull

Berkeley Center for Cosmological Physics,
Department of Physics and Lawrence Berkeley National Laboratory,
University of California, Berkeley, CA 94720, USA
email: `mschmittfull@lbl.gov`

Abstract. Relying on a separable modal expansion of the bispectrum, the implementation of a fast estimator for the full bispectrum of a 3d particle distribution is presented. The computational cost of accurate bispectrum estimation is negligible relative to simulation evolution, so the bispectrum can be used as a standard diagnostic whenever the power spectrum is evaluated. As an application, the time evolution of gravitational and primordial dark matter bispectra was measured in a large suite of N-body simulations. The bispectrum shape changes characteristically when the cosmic web becomes dominated by filaments and halos, therefore providing a quantitative probe of 3d structure formation. Our measured bispectra are determined by ~ 50 coefficients, which can be used as fitting formulae in the nonlinear regime and for non-Gaussian initial conditions. We also compare the measured bispectra with predictions from the Effective Field Theory of Large Scale Structures (EFTofLSS).

Keywords. cosmology: large-scale structure of universe, methods: n-body simulations, methods: data analysis, methods: statistical

---

Due to non-linear gravitational collapse, the probability distribution function (pdf) of large-scale structure (LSS) dark matter (DM) density fluctuations deviates from a Gaussian pdf even if the initial conditions are Gaussian. This non-Gaussianity is characterized by non-vanishing higher-order n-point functions, e.g. the bispectrum, which is the Fourier transform of the $3-$point function and corresponds to the probability of finding a third overdensity mode given two other overdensity modes. An additional source of non-Gaussianity is primordial non-Gaussianity of the initial conditions that can be generated by certain inflation models. Measuring these non-Gaussianities in observations can help to break degeneracies present at the level of 2-point statistics, e.g. between linear bias b_1 and the normalization of fluctuations σ_8, or to constrain inflation models.

While it is numerically straightforward to sample initial conditions from a Gaussian pdf and to estimate the power spectrum of a given density, sampling from a non-Gaussian pdf and estimating the bispectrum of a given density perturbation are numerically rather challenging tasks. We address these issues by expanding the bispectrum in separable basis functions which are particularly suited for efficient numerical evaluation.

1. Non-Gaussian initial conditions for N-body simulations

To draw a realization from a pdf with power P_Φ and bispectrum $f_{\mathrm{NL}} B_\Phi$, we add $\Phi^B = \int_{\mathbf{k}'} W(k, k', |\mathbf{k} - \mathbf{k}'|) \Phi^G_{\mathbf{k}'} \Phi^G_{\mathbf{k}-\mathbf{k}'}$ to a Gaussian field $\Phi^G_{\mathbf{k}}$, where (Wagner *et al.* (2012))

$$W(k, k', |\mathbf{k} - \mathbf{k}'|) \equiv \frac{f_{\mathrm{NL}}}{2} \frac{B_\Phi(k, k', |\mathbf{k} - \mathbf{k}'|)}{P_\Phi(k) P_\Phi(k') + 2\,\mathrm{perms}}. \qquad (1.1)$$

In practice, Φ^B can only be computed efficiently if W is product-separable, i.e. consisting of terms of the form $f_1(k)f_2(k')f_3(k'')$, because then Φ^B reduces to a convolution of filtered Gaussian fields. Unfortunately, the symmetrized denominator in (1.1) often destroys separability. As a general solution to this problem, the kernel W can be expanded in product-separable basis functions following Fergusson *et al.* (2012), which we show in Regan *et al.* (2012) to perform well in simulations.

2. Separable bispectrum estimation

Estimating the bispectrum of a density perturbation is computationally expensive due to the large number of possible triangle configurations and integrals similar to the expression for Φ^B above. Rather than estimating the bispectrum individually for every triangle, Fergusson *et al.* (2012) propose to estimate the amplitude of many independent separable basis bispectrum templates and reconstruct the full bispectrum by summing up the contributions. We restrict ourselves to a finite separable monomial basis that covers all theoretically motivated bispectra and implicitly uses all triangles. This reduces the computational cost of bispectrum estimation dramatically, e.g. the full bispectrum of a 1024^3 grid is estimated in only one hour on six cores.

We implemented this method and tested it on a large suite of N-body simulations with Gaussian and different types of non-Gaussian initial conditions in Schmittfull *et al.* (2013). All bispectrum measurements agree with perturbation theory on large scales, validating our framework and implementation. In the non-linear regime, we find that the bispectrum shape characterizes the 3d dark matter structures, e.g. pancake-like structures correspond to flattened bispectra, while filaments and clusters enhance equilateral contributions to the bispectrum. The bispectrum characterizes these structures in a quantitative way which can both be modeled and extracted from simulations.

In our approach, the full bispectrum information is compressed to $n_{\max} = \mathcal{O}(50)$ amplitudes of (orthonormalized) basis bispectrum shapes. The first ten modes already contain 99% of the information, which we exploit to construct fitting formulae for the dark matter bispectrum with ten parameters at every redshift and $k_{\max}$. The effect of primordial non-Gaussianity can be modeled by a time shift compared to gravitational evolution in a universe with Gaussian initial conditions, which leads to simple fitting formulae for the excess bispectrum due to primordial non-Gaussianity of various types (see Schmittfull *et al.* (2013) for details).

In another application of the separable bispectrum estimation method, we compare the measured bispectra against predictions of Effective Field Theory of Large-Scale Structure (EFTofLSS) in Angulo *et al.* (2014) (see also Baldauf *et al.* (2014)), finding that EFTofLSS significantly improves upon standard perturbation theory.

Acknowledgements I thank my collaborators for the work summarized here, R. Angulo, J. Fergusson, S. Foreman, D. Regan, L. Senatore and P. Shellard.

References

Angulo, R., Foreman, S., Schmittfull, M., & Senatore, L. 2014, *arXiv* 1406.4143
Baldauf, T., Mercolli, L., Mirbabayi, M., & Pajer, E. 2014, *arXiv* 1406.4135
Fergusson, J., Regan, D., & Shellard, E. P. S.. 2012, *Phys. Rev. D* 86, 063511, 1008.1730
Regan, D., Schmittfull, M., Shellard, E. P. S., & Fergusson, J. 2012, *Phys. Rev. D* 86, 123524, 1108.3813
Schmittfull, M., Regan, D., & Shellard, E. P. S.. 2013, *Phys. Rev. D* 88, 063512, 1207.5678
Wagner, C. & Verde, L. 2012, *JCAP* 3, 2, 1102.3229

The Zeldovich Universe:
Genesis and Growth of the Cosmic Web
Proceedings IAU Symposium No. 308, 2014
R. van de Weygaert, S. Shandarin, E. Saar & J. Einasto, eds.

© International Astronomical Union 2016
doi:10.1017/S1743921316009650

The Zeldovich & Adhesion approximations and applications to the local universe

Johan Hidding[1], Rien van de Weygaert[1] and Sergei Shandarin[2]

[1]Kapteyn Astronomical Institute, University of Groningen,
Postbus 800, NL-9700AV, Groningen, the Netherlands
email: `johannes.hidding@gmail.com`

[2]Department of Physics and Astronomy, University of Kansas,
1082 Malott,1251 Wescoe Hall Dr., Lawrence, KS 66045-7582

Abstract. The Zeldovich approximation (ZA) predicts the formation of a web of singularities. While these singularities may only exist in the most formal interpretation of the ZA, they provide a powerful tool for the analysis of initial conditions. We present a novel method to find the skeleton of the resulting cosmic web based on singularities in the primordial deformation tensor and its higher order derivatives. We show that the A_3-lines predict the formation of filaments in a two-dimensional model. We continue with applications of the adhesion model to visualise structures in the local ($z < 0.03$) universe.

---

1. The Zeldovich approximation

The Zeldovich Approximation (ZA) (Zeldovich 1970, Shandarin & Zeldovich 1989) describes structure formation in the form of a deceptively simple equation

$$\boldsymbol{x}(\boldsymbol{q}, t) = \boldsymbol{q} - D_+(t) \boldsymbol{\nabla} \Phi_0(\boldsymbol{q}). \tag{1.1}$$

Rather than describing just ballistic motion, this equation hides a formalism of Lagrangian collision-free fluid mechanics. In the context of emerging interest in phase-space folding descriptions of structure formation (Shandarin *et al.* 2012, Falck *et al.* 2012, Abel *et al.* 2012), it becomes all the more relevant to understand this expression at a much deeper level. We can see why there is more to the ZA than inertial motion, if we compute particle densities from the above expression. Density increases or decreases locally as a fluid element contracts or expands. Taking a fluid element from Lagrangian location $\boldsymbol{q}$, we can quantify its deformation in terms of the deformation tensor $d_{ij} = -\partial_i \partial_j \Phi_0$. This tensor is best studied locally in the eigenvector frame of reference $\{\boldsymbol{e}_\lambda\}$, where d_{ij} becomes diagonal. The density is then

$$\delta(\boldsymbol{x}) + 1 = \sum_{q \in \{q^\star\}} \left| \det \frac{\partial x_i}{\partial q_j} \right|_q^{-1} = \sum_{q \in \{q^\star\}} \left| \prod_i (1 - D_+ \lambda_i) \right|_q^{-1}, \tag{1.2}$$

where $\{\boldsymbol{q}^\star\}$ is the set of Lagrangian locations solving for $\boldsymbol{x}$ in equation 1.1, and λ_i are the eigenvalues of the deformation tensor d_{ij}. Note that this expression for the density has singularities whenever for one of the eigenvalues we have

$$\lambda_i = 1/D_+(t).$$

This notion gives us the traditional interpretation of ZA, namely one where gravitational collapse occurs in three possible stages. First objects form along the major axis of collapse, making pancakes; then along the second eigenvector filaments form; and finally if and when all three eigenvalues have passed a singularity, a cluster forms. This

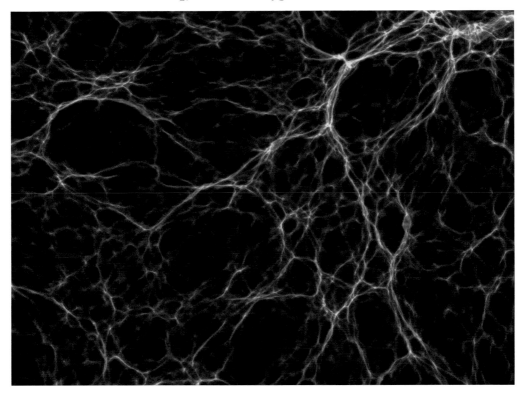

Figure 1. The Zeldovich approximation. Density map of the cosmic structure following the evolution according to the Zeldovich formalism. The cosmic web is sharply rendered, with most of the structures residing just before or around shell crossing. From Hidding 2016.

corresponds to the *Morse theory* view of nodes and saddles exploited in many structure finders. We show that, even in the narrow confines of the ZA, this interpretation is not complete; even that it is wrong on the account of the formation of the first filaments (Hidding *et al.* 2014). Taking the mathematics of Morse theory a step further, we arrive at Lagrangian *singularity theory* (Arnold *et al.* 1982, Arnold 1986). This theory shows how to predict the evolution of *folds, cusps, swallow-tails, butterflies* and *umbilics* directly from the initial velocity potential Φ_0. Due to the relative complexity of this method we are forced to restrict our further discussion to the two-dimensional case.

Formation of pancakes. A *fold* is the simplest kind of singularity we have. It is the caustic that separates single-stream from multi-stream regions and is also found under the cryptic name A_2 †. At any moment in time, pancakes can be identified as the locations of A_2 folds. At a fold the phase-space sheet (see Fig. 2) is tangent to line of projection. In the case of ZA, this happens when $\lambda = 1/D_+$, identifying the level-sets of λ as the Lagrangian progenitors of folds. Two folds may connect at a *cusp*. A fold being a line of tangency on the phase-space sheet, there exists points where the fold line itself is tangent to the projection, these points are the cusps (see Fig. 2). In the tensor field d_{ij}, a cusp is found where a level-set of λ is tangent to the corresponding eigenvector $\boldsymbol{e}_\lambda$, or

$$\boldsymbol{\nabla}\lambda \cdot \boldsymbol{e}_\lambda = 0. \qquad (1.3)$$

† A_2 refers to the ADE classification of singularities introduced by Arnold. In this paper we also deal with A_3 for cusp, A_4 for swallow-tail, and D_4 for umbilic singularities.

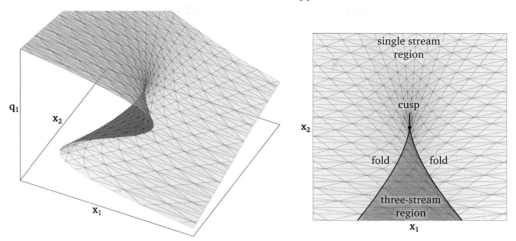

Figure 2. *The cusp singularity.* Singularities arise if we project a smooth manifold down one or more directions. In this case we show one Lagrangian direction as the z-axis and the two remaining axes in Eulerian space. On the right is the projected view, with the fold and cusp locations marked. Between the folds we find a three-stream region.

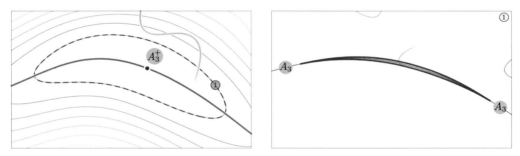

Figure 3. *Pancake genesis.* On the left we see in Lagrangian space the contours of the first eigenvalue around an A_3^+-point. The Eulerian counterpart of the dashed contour is shown on the right. It shows the pancake at its prime, highly elongated with a cusp on each end.

Finding all A_3-points for each level-set of λ traces an A_3-line. The set of A_3-lines trace the entire network of filaments formed in the ZA, throughout time. It can be shown that all maxima and saddle points of the function $\lambda(q)$ also lie on an A_3-line. Lowering a level-plane down on the function $\lambda(q)$, we can see that at the maxima of λ (A_3^+-points) two cusps are created, while at the saddle points (A_3^--points) they annihilate, merging two pancakes. A_3-lines terminate only in D_4 *umbilic* points, but we choose to also truncate them where $\lambda = 0$, beyond which they loose their physical significance.

Splitting of pancakes. A pancake may branch by creating two new cusps at a fold. Typically one of these cusps remains within the confines of the present pancake and the other dashes out to create a subsidiary pancake, often to merge later with another pancake at an A_3^--point, creating a three-legged structure. The point at which a pancake branches is called a swallow-tail, denoted A_4. An A_4 singularity is found in Lagrangian space at points where an A_3-line is tangent to the corresponding eigenvector, or equivalently at local maxima of λ limited to the A_3-line. Important to note here, is that we don't need to involve the second eigenvalue to create a node in the network of caustics. Moving this discussion to the three-dimensional case, we don't strictly need to collapse along the

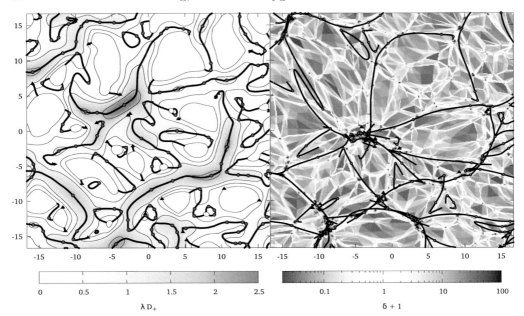

Figure 4. *Comparison with N-body.* On the left: ZA on filtered initial conditions; contours show the first eigenvalue, A_3-lines in red, and triangles showing D_4 points. On the right: the density contrast resulting from a 2D PM code, with the Eulerian displaced A_3-lines over-plotted.

second eigenvector to create a filament-like structure. This also suggests the existence of different possible late-time morphologies for filaments.

Scaling and comparison with N-body. We computed the A_3-lines for a set of initial conditions and compared the result with those of a 2D N-body code. A non-linear time evolution may be approximated by truncating power of the initial conditions at scales smaller than the scale of non-linearity. It is well known that observable filaments have a density contrast around unity, so this method should give realistic results. We find good agreement for all relevant $P(k) \propto k^n$ power-spectra, in an eye-ball comparison of filaments predicted by ZA with density fields from a 2D PM code. An example is given in Fig. 4.

2. The Adhesion approximation

We showed how the emergence of caustics in the ZA allows us to trace the formation of cosmic structures in a formal, yet physically meaningful way. However, the ZA suffers from a major flaw in that it doesn't allow for gravitational interaction, and therefore hierarchical structure formation. This is because the ZA is solely based on local analysis of the velocity potential and its chain of derivatives. The adhesion model moves beyond local considerations, which makes it computationally more intensive than the ZA (though still much faster than N-body). Still, results are computed from initial conditions directly and with complete accountability. The adhesion model is arrived at by taking the source-free (hence collision-free) Euler equation, and adding an artificial viscosity term to *emulate* the effects of gravity (Gurbatov & Saichev 1984, Shandarin & Zeldovich 1989). The resulting equation

$$\partial_t \boldsymbol{u} + (\boldsymbol{u} \cdot \boldsymbol{\nabla})\boldsymbol{u} = \nu\nabla^2\boldsymbol{u}, \qquad (2.1)$$

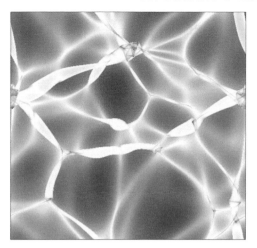
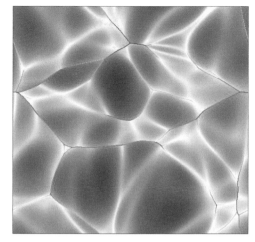

Figure 5. *ZA and adhesion compared.* Outside multi-stream regions the results from the ZA and adhesion are identical. Adhesion contains an artificial viscosity term that only 'activates' when streams cross. Multi-stream regions are thus collapsed to infinitesimally thin structures.

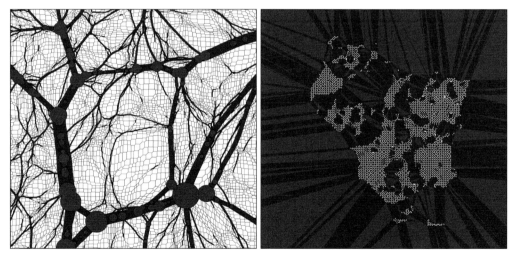

Figure 6. *Dual structures in adhesion.*

is known as Burgers' equation; in the limit where $\nu \to 0$, it has the exact solution

$$\Phi(\boldsymbol{x}, t) = \max_{q} \left(\Phi_0(\boldsymbol{q}) - \frac{(\boldsymbol{x} - \boldsymbol{q})^2}{2D_+(t)} \right). \tag{2.2}$$

The *global maximum* in this solution can be computed efficiently using either a Legendre transform, convex hull (Vergassola *et al.* 1994) or a weighted Voronoi diagram (Hidding *et al.* 2012, Hidding *et al.* 2016a (2016)). One condition for reaching an extremum is that the first derivative of the maximised quantity should vanish. Performing this test reduces above equation to the ZA as presented in equation 1.1. The global maximisation guarantees that the resulting map from Lagrangian to Eulerian space stays *monotonic* always. Where and whenever shell-crossing occurs in the ZA, adhesion creates a solid structure. Matter inside these structures is confined to stay inside, which is the reason people may refer to the adhesion model as having "sticky particles". Outside collapsed

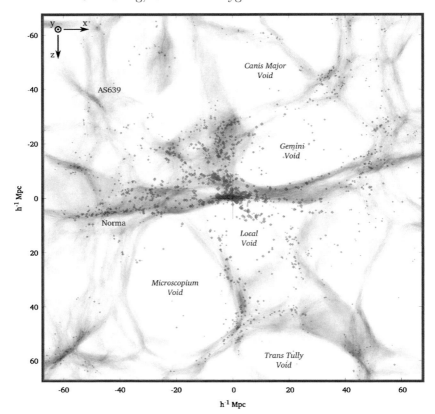

Figure 7. *Local Universe adhesion reconstruction.* Structure of the lcoal Universe. The reconstruction of the weblike structure in the local Universe, sampled by the 2MRS survey, has been obtained on the basis of the adhesion formalism applied to a set of 25 constrained Bayesian KIGEN realizations of the primordial density and velocity field in the Local Universe. The image shows the density field in a 10 Mpc thick slice perpendicular to the plane of the Local Supercluster. Note that the density field concerns the dark matter distribution. The red dots are the 2MRS galaxies in the same volume. From Hidding 2016, Hidding *et al.* 2016b.

structures the results from the ZA and adhesion are identical; caustics from ZA are compressed to infinitesimally thin structures (see Fig. 5). This unifies Zeldovich' idea of collapsed structures in terms of shell crossing with a hierarchical formation model.

Dual geometry. The solution to Burgers' equation given in expression 2.2 is identical to the definition of the *weighted Voronoi tessellation*, weighted by the potential. The Voronoi cell of a Lagrangian point $q \in \mathcal{L}$ occupies an area in Eulerian space $\mathcal{E}$ given by

$$V_q = \left\{ \boldsymbol{x} \in \mathcal{E} \;\middle|\; (\boldsymbol{x} - \boldsymbol{q})^2 + w_q \leqslant (\boldsymbol{x} - \boldsymbol{p})^2 + w_p, \; \forall \, \boldsymbol{p} \in \mathcal{L} \right\}. \qquad (2.3)$$

Taking $w(\boldsymbol{q}) = 2D_+ \Phi_0(\boldsymbol{q})$ as the weights in this expression reduces it to the given solution in equation 2.2. Where there is a Voronoi tessellation, there is its dual: the *Delaunay triangulation*. It is the latter that gives us the origin and mass of matter residing in the nodes, edges and faces of the Voronoi tessellation (see Fig. 6).

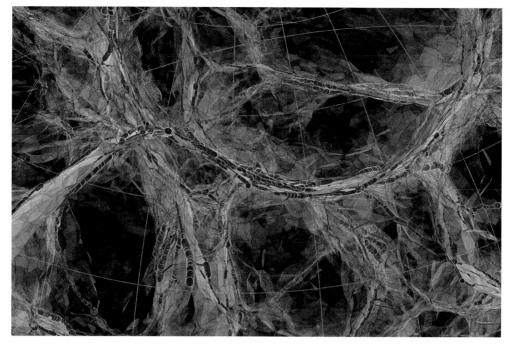

Figure 8. *The Pisces-Perseus Supercluster.* A 3-D isodensity surface rendering of the intricate filamentary structure around the Pisces-Perseus supercluster. It is based on the adhesion reconstruction of the local Cosmic Web, based on constrained realizations of the local primordial density and velocity field implied by the 2MRS galaxy redshift survey. From Hidding 2016, Hidding *et al.* 2016b.

3. The Local Universe

One application of the adhesion model is the detection of walls, filaments and nodes in cases where some form of an initial potential is available. We ran our adhesion code on a set of 25 constrained initial conditions *reconstructed* (Kitaura 2013, Heß *et al.* 2013) to produce structures in our local universe ($z < 0.03$) (Hidding 2016, Hidding *et al.* 2016b (2016)). This reconstruction is based on the 2MASS redshift catalog (Huchra *et al.* 2012), which covers the full sky except for galactic lattitudes $|b| < 5°$.

Figure 7 provides a remarkably detailed reconstruction of the cosmic web in the 2MRS volume. It shows the (surface) density of the weblike structures in the Local Universe. These are the result of adhesion simulations by Hidding 2016 and Hidding *et al.* 2016b (2016), based on the the constrained Bayesian KIGEN reconstruction by Kitaura 2013 of the initial conditions in the local volume traced by the 2MRS redshift survey. For a given Gaussian primordial field, the adhesion formalism allows the accurate reconstruction of the rich pattern of weblike features that emerge in the same region as a result of gravitational evolution. The adhesion formalism was applied to 25 constrained realizations of the 2MRS based primordial density field (Hidding *et al.* 2012, Hidding 2016). The mean of these realizations gives a reasonably accurate representation of the significant filamentary and wall-like features in the Local Universe. Most outstanding is the clear outline of the void population in the local Universe. The reconstruction also includes the velocity flow in the same cosmic region. It reveals the prominent nature of the outflow from the underdense voids, clearly forming a key aspect of the dynamics of the Megaparsec scale universe.

The Local Universe structure in figure 7 presents a telling image of a void dominated large scale Universe. Many of the voids in the adhesion reconstruction can be identified with the void nomenclature proposed by Fairall (Fairall (1998)), who mainly identified these voids by eye from the 6dFGRS survey. It is interesting to see that the socalled Tully void appears to be a richly structured underdense region, containing at least the Microscopium Void, the Local Void and the "Trans Tully Void".

In the same reconstruction, we are studying the intricate filamentary network in and around the Pisces-Perseus supercluster. The image in figure 8 provides a nice impression of the complex 3-dimensional structure and connectivity along the main ridge of the Pisces-Perseus supercluster. It also shows how the main ridge connects to several neighbouring filaments, connecting near massive clusters along the ridge, and how these filaments surround a lower density planar structure. Interesting is to note the clustering and alignment of the small (whitish) filamentary tendrils in and around the main arteries of the Cosmic Web. Analysis of this weblike structures region is under progress and will be first reported in Hidding 2016.

References

Abel, T., Hahn, O., & Kaehler, R. 2012, *MNRAS*, 427, 61-76
Arnold, V. I. 1986, *Journal of Soviet Mathematics*, 32:03, 229-258
Arnold, V. I., Shandarin, S. F., & Zeldovich, Ya. B. 1982, *Geophysical and Astrophysical Fluid Dynamics*, 20, 111-130
Fairall A. P. 1998, *Large-scale structures in the Universe* (Wiley)
Falck, B. L., Neyrinck, M. C., & Szalay, A. S. 2012, *ApJ*, 754, 126
Gurbatov, S. N. & Saichev, A. I. 1984, *Radiophysics and Quantum Electronics*, 27:4, 303-313
Heß, S., Kitaura, F.-S., & Gottlöber, S. 2013, *MNRAS* 435, 2065-2076
Hidding, J., Van de Weygaert, R., Vegter, G., Jones, B. J. T., & Teillaud, M. 2012, *Proc. of the 28th SoCG*, doi:10.1145/2261250.2261316
Hidding, J., Shandarin, S. F., & Van de Weygaert, R. 2014, *MNRAS*, 437, 3442-3472
Hidding, J., 2016, *in preparation* PhD thesis, Univ. Groningen
Hidding, J., Van de Weygaert, R., & Vegter, G. 2016, *in preparation*.
Hidding, J., Van de Weygaert, R., Kitaura F.-S. & Hess S. 2016, *in preparation*.
Huchra, J. P., Macri, L. M., Masters, K. L. *et al.* 2012, *ApJS*, 199, 26
Kitaura, F.-S. 2013, *MNRAS*, 429, L84-L88
Shandarin, S. F. & Zeldovich, Ya. B. 1989, *Rev. Mod. Phys.*, 61, 185-220
Shandarin, S. F., Habib, S., & Heitmann, K. 2012, *Phys. Rev. D*, 85, 083005
Vergassola, M., Dubrulle, B., Frisch, U., & Noullez, A. 1994, *A&A*, 289, 325-356
Zeldovich, Ya. B. 1970, *A&A*, 5, 84-89

The Zeldovich Universe:
Genesis and Growth of the Cosmic Web
Proceedings IAU Symposium No. 308, 2014
R. van de Weygaert, S. Shandarin, E. Saar & J. Einasto, eds.
© International Astronomical Union 2016
doi:10.1017/S1743921316009662

Dynamics of The Tranquil Cosmic Web

Adi Nusser

Technion- Israel Institute of Technology
32000 Haifa, Israel
email: adi@physics.technion.ac.il

Abstract. The phase space distribution of matter out to ~ 100Mpc is probed by two types of observational data: galaxy redshift surveys and peculiar motions of galaxies. Important information on the process of structure formation and deviations from standard gravity have been extracted from the accumulating data. The remarkably simple Zel'dovich approximation is the basis for much of our insight into the dynamics of structure formation and the development of data analyses methods. Progress in the methodology and some recent results is reviewed.

Keywords. cosmology: large-scale structure of universe

1. Introduction

Merging and star formation activities in the galaxy population have calmed gown by the current epoch ($z = 0$). This lead to the establishment of a) a tight relation between the distributions of galaxies and the underlying mass of the dark matter, and b) relations between galaxy intrinsic properties, allowing for measurements of distance. Therefore, the $z \sim 0$ Large Scale Structure is an excellent laboratory for probing cosmological models. Two complementary observational sets are our main window to the phase space distribution of matter. The first, surveys of galaxy redshifts, cz, and apparent magnitudes, m. The second, distance measurements d_e, and hence peculiar motions v_p, of galaxies obtained via intrinsic relations such as the Tully-Fisher (TF). Distance measurements are more difficult to obtain than just cz and m and hence the number of galaxies with observed peculiar motions is significantly smaller than in redshift surveys. An example of the first set is Two Micron All Sky Redshift Survey (2MRS) (Huchra *et al.* 2012), of about 45000 galaxies with a mean redshift of ~ 8000kms^{-1} and the deeper SDSS containing about half a million galaxies but with partial sky coverage. The second type of data include the SFI++ catalog (Masters *et al.* 2006) of TF measurements of ~ 4000 galaxies, and the Cosmic Flows 2 (CF2)(Tully *et al.* 2013) catalog of ~ 8000. Fig. 1 is a visual representation of the data in the Super-galactic (SG) plane. Note the patchiness and sparseness of galaxies in the CF2 catalogue (left) compared to the 2MRS (right).

While the galaxy distribution is a biased tracer of the underlying mass density field of the dominant dark matter, the equivalence principle implies that galaxies are comoving with the dark matter on large scales away from non-gravitational forces. But the peculiar velocity field (as a function of the measured distance, d_e) derived from the noisy data suffers from inhomogeneous Malmquist biases (Lynden-Bell *et al.* 1988), resulting from the systematic difference between d_e and the mean of true distances of galaxies with the same d_e. It is very difficult to correct for this bias because of its dependence on the unknown distribution of galaxies in true distance space. In contrast, galaxy biasing is likely to be well approximated by a simple linear relation $\delta_{gal} = b\delta_{dm}$ between the galaxy and dark matter density fluctuations, as seen in Fig. 2.

In the standard paradigm, the observed structure has grown by Gravitational Instability (GI) from tiny initial fluctuations. Neglecting gas related effects, the equations of

A. Nusser

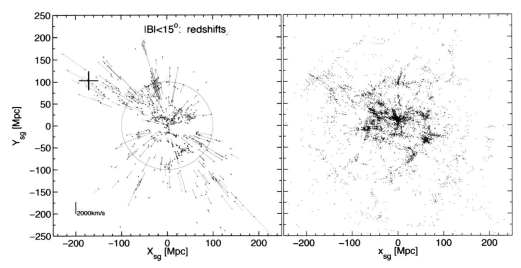

Figure 1. *Left:* The measured v_p (in the CMB frame) of galaxies within $15°$ of the SG plane in CF2. Black dots indicate the observed redshifts in the CMB frame. Red and blue arrows correspond to peculiar motions pointing away and towards the observer, respectively. The large cross plus sign indicates the location of the Shapley supercluster. *Right:* The distribution of galaxies within 20Mpc of the SG plane.

motion (EoM) of the perturbations are the usual Euler, Poisson and continuity equations in an expanding background. Supplemented with initial conditions appropriate for cosmological perturbations, the full solution to the EoM is possible only via numerical simulations which have achieved an impressive dynamical range from small galaxies to a significant fraction of the Hubble volume. Nonetheless, approximate solutions have been and will remain the basis for observational analyses methods and a physical understanding the numerical results. The simplest approximate solution is provided by linear theory which yields $\delta(\mathbf{x}, t) = \delta_0(\mathbf{x})D^+(t) + \delta_-(\mathbf{x})D^-(t)$ where D^- describes a decaying mode and the growing mode obeys $\ddot{D} + 2H\dot{D} - 3\Omega H^2 D/2 = 0$. Linear theory also yields

$$\delta = -\frac{1}{Hf}\mathbf{\nabla}\cdot\mathbf{v},\qquad(1.1)$$

where $f = d\ln D/d\ln a \approx \Omega^\gamma$ is the growth rate and $\mathbf{v}(\mathbf{x})$ is the 3D peculiar velocity field. The index γ depends on the cosmology (e.g. through the dark energy model) and the underlying theory for gravity. Accurate determination of γ is one of the goals of future large surveys of galaxies. This $\delta - \mathbf{v}$ has been used extensively for the prediction of velocity fields associated with the distribution of galaxies in a given redshift survey. It is the basis for modeling redshift distortions of correlation functions from redshift surveys on large scales.

Within GI the two independent data sets can be analyzed in several ways: *a*) Correlation functions (and power spectra) have been estimated from the distribution of galaxies in redshift surveys. These correlations can be compared with predictions of cosmological models. Further, $cz = Hd + v_p$ implies that correlations in redshift space indirectly probe v_p through the fingers of god effect on small scales and the enhancement of clustering on large scales (Davis & Peebles 1983; Kaiser 1988). *b*) Correlation analysis of the observed peculiar velocity field have also been done. However, this analysis is intrinsically plagued with inhomogeneous Malmquist biases. Quantitative conclusions from this type of analysis should always be examined critically. *c*) Comparison of low order moments of

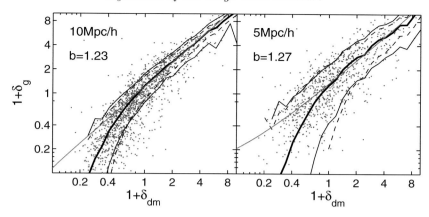

Figure 2. A scatter plot (logarithmic scale) of the galaxy versus the dark matter over-densities in 2MRS mock galaxy catalogs De Lucia & Blaizot (2007). The left and right panels correspond to densities in cubic cells of $10\,h^{-1}\mathrm{Mpc}$ and $5\,h^{-1}\mathrm{Mpc}$ on the side, respectively. The thick solid curve in each panel is the mean of $1 + \delta_{\mathrm{g}}$ at a given $1 + \delta_{\mathrm{dm}}$. The two thin solid curves are $\pm 1\sigma$ scatter computed from points above and below the mean. Dashed curves are the expected $\pm 1\sigma$ Poisson (shot-noise) scatter. The nearly straight red lines show $\delta_{\mathrm{g}} = b\delta_{\mathrm{dm}} + const$, where b (indicated in the figure) are determined using linear regression from points in the range $-0.5 < \delta_{\mathrm{dm}} < 4$.

the peculiar velocity field, e.g. the bulk flow, with predictions of cosmological models. *d*) Testing GI by assessing the alignment of the gravitational force field (or the peculiar velocity) derived from redshift survey with the observed v_p of galaxies in the peculiar velocity catalogs. This comparison is particularly important since it minimizes cosmic variance in the estimation of the cosmological parameters.

2. The Zel'dovich approximation

Full analytic solutions to the EoM are available only for initial conditions with a high degree of symmetry, e.g. self-similar collapse/expansion with planar, cylindrical or spherical symmetry. Zel'dovich (1970) proposed a remarkably simple approximation for the evolution of generic cosmological perturbations in the quasi-linear regime (laminar flow). The Zel'dovich approximation (ZA) states that the current position $\mathbf{x}(t)$ and the initial Lagrangian coordinate, $\mathbf{q}$, of a particle are related by

$$\mathbf{x} = \mathbf{q} + D(t)\boldsymbol{\psi}(\mathbf{q}) \,. \tag{2.1}$$

In the paper, Zel'dovich considered only baryons and argued that the natural perturbation scale is the Silk damping mass scale, $M_{\mathrm{S}} \approx 10^{12} M_{\odot}$. Further, the probability distribution of the eigenvalues of $\partial_i \psi_j$, revealed a preference for planar-like perturbations. Hence, ZA was the basis for the top-down pancake paradigm for structure formation. The approximation (2.1) is an exact solution to the full EoM for planar perturbations up to the onset of shell crossing (in collision-less fluids). The proof appears in Zeldovich & Novikov (1983), but not in the 1970 paper. Although not highly accurate the ZA has given us fantastic physical insight into the working of nonlinear dynamics, e.g. the growth of angular momentum of galaxies (Doroshkevich 1970; White 1984), nonlinear density power spectrum with and without redshift distortions (Schneider & Bartelmann 1995; Fisher & Nusser 1996; Taylor & Hamilton 1996; White 2014) and the probability distribution of the density field Kofman *et al.* (1994).

2.1. *Extension: Lagrangian Perturbation Theory*

Here, the displacement is expanded in a Taylor series in an appropriate parameter (Buchert & Ehlers 1993) which can be taken as the linear growth factor, D. Therefore, $\mathbf{x} = \mathbf{q} + \sum_s D^s \mathbf{\Psi}^{(s)}(\mathbf{q})$, where the ZA term ($s = 1$) is entirely fixed by the initial conditions, while the EoM dictate all $s > 1$ terms via a recurrence relation involving lower order terms only (Zheligovsky & Frisch 2014). Unfortunately these recurrence relations become messy for $s > 2$. One of the reasons for that is the emergence of non-vanishing Lagrangian vorticity $\mathbf{\nabla}_q \times \mathbf{\Psi}^{(s)} \neq 0$ for $s > 2$ †. Second order Lagrangian perturbation (2LPT) ($s = 2$), gives the density as

$$\frac{1}{\rho_{2\mathrm{LPT}}} = 1 + D\mathbf{\nabla} \cdot \mathbf{\Psi}^{(1)} + \frac{4}{7}D^2(\mu_1\mu_2 + \mu_1\mu_3 + \mu_2\mu_3) . \tag{2.2}$$

where μ_i are the eigenvalues of $\partial_i \psi_j^{(1)}$. The continuity equation yields the widely known expression for the density in the ZA,

$$\frac{1}{\rho_{\mathrm{zel}}} = 1 + D\mathbf{\nabla} \cdot \mathbf{\Psi}^{(1)} + D^2(\mu_1\mu_2 + \mu_1\mu_3 + \mu_2\mu_3) + D^3\mu_1\mu_2\mu_3 . \tag{2.3}$$

The ZA and 2LPT share the algebraic form of the second order term (D^2), but with different coefficients. Thus, ZA is inaccurate even to second order. However, simulations show that The ZA provides a better match to the density derived from the velocity in (Gramann 1993) in high density regions, although 2LPT is better for low densities.

One of the most important common application of these approximation is the generation of particle displacements and velocities to be used as initial conditions for N-body simulations. The ZA and 2LPT are traditionally used but the latter yields is more accurate for this purpose.

3. Peeble's action method

We are given the positions of mass tracers (galaxies) today. What are the tracers' paths from the nearly homogeneous early Universe to the current configuration? This is *a boundary value problem* where a solution to the equations of motion is sought for *boundary conditions* (BC) at two different times‡. Its solution allows a reconstruction of the velocity field associate with the observed distribution of tracers and also the initial density field which lead to this distribution through gravitational interactions.

ZA and 2LPT can be employed to derive approximate solutions to the orbits. However, the most general (and elegant) method to do that has been designed by Peebles (1989) based on the least action principle. Orbits, $x(t)$, obeying the EoM also render the action stationary with respect to variations, $\delta x(t)$, satisfying the BC $p\delta x = 0$ at the limiting times t_1 and $t_2 > t_1$, where p is the momentum. In the cosmological problem, $\delta x(t_2) = 0$ is naturally imposed. The initial positions are unknown, but Peebles noted that for the growing mode of cosmological perturbations the momentum vanishes as $t_1 \to 0$. Hence, the solutions to the cosmological boundary value problem can be obtained by minimizing the action with respect to trial orbits constructed to satisfy $a)$ known current positions and $b)$ $p \to 0$ near the Big Bang. For sufficiently general trial functions the orbits should

---

† A vanishing Eulerian vorticity as a function of time (for an initial irrotational flow) is protected by Kelvin's circulation theorem until the onset of orbit-mixing. In other words, Eulerian vorticity remains zero for any order in perturbation theory.

‡ It is closely tied to transport problems where displacements from a clumpy into a uniform distribution are sought by minimizing a cost function (Frisch *et al.* 2002). One still needs a dynamical prescription, e.g. ZA, to get the orbits from total displacements.

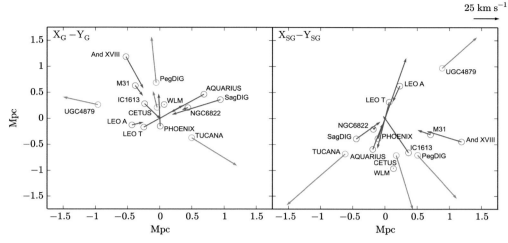

Figure 3. Galaxies in the LG in the Galactic (left) and SG (right) planes the arrow represent observed radial velocities transformed to the frame of reference comoving with the LG. Curtsey of Ziv Mikulsy.

be a solution to the full equations satisfying the BC, should such a solutions exist. In general, boundary value problems allow for multiple solutions or no solutions at all. Consider for example the linear oscillator $\ddot{x} + x = 0$ subject to the BC, $x(t_1 = 0) = 0$ and $x(t_2 = 2\pi) = 0$. There is an infinite number of solutions: $x(t) = A\sin(t)$ for any A. There are also BC where the action has no extremum orbit. An example is $x(0) = 0$ and $x(2\pi) = 1$. There is no physical orbit which satisfies these BC. In this case, it is easy to see that action can acquire infinitely large (positive and negative) values for certain choices of the orbits. For mixed BC where one of the conditions is $p = 0$, the situation can even be more intriguing. The oscillator equation of motion constrained to $\dot{x}(0) = 0$ and $x(2\pi) = 1$, is solved by $x(t) = \cos(t)$. Let us compute the action $S = \int_0^{2\pi} dt(\dot{x}^2 - x^2)/2$ for the following choice for the perturbed orbits: $x_p = \cos(t) + A\cos(wt)$. These orbits satisfy the BC for $w = (2n+1)/4$, but not the equation of motion. It is easy to see that $S = A^2\pi(w^2 - 1)/2$, i.e. the extremum point is a maximum for $w < 1$ and a minimum for $w > 1$.

3.1. *Application to the Local Group (LG) of galaxies: masses of MW & M31*

The LG contains about a dozen known galaxies within a distance ~ 1.5 Mpc. Although not a virialized object (see below), it is gravitationally bound and detached from the expansion. Galaxy members (excluding satellites) of the LG are shown in Fig. 3. The MW and M31 are by far the most luminous (M31 is 100 more luminous than NGC6822-the third most luminous galaxy shown in the figure). The radial velocities in the LG frame are represented by the arrow in the figure. The (radial) velocity dispersion is $\sim 60\,\mathrm{kms}^{-1}$ and the flow pattern clearly reveals a non-virialized system that is most likely is on a first infall. Thus, the virial theorem will over-estimate the total mass of the LG. To constraint the masses of MW and M31, we apply the action principle to the nearby galaxies. The application is basically a generalization of the timing argue (TA) of Woltjer and Kahn who considered only the MW and M31. By treating the two galaxies as point particles with zero angular momentum (relative to the center of mass), TA constrains the total mass $M_{MW} + M_{31}$ by demanding that the two galaxies originated from zero separation a Hubble time ago, reaching their current separation and relative velocity today. The action method breaks the degeneracy between the masses by including kinematical observations of the smaller members of the LG. Although dynamically unimportant, the observed distances and velocities of these galaxies will allow us to resolve the individual masses M_{MW} and M_{31}.

Galaxy orbits which render the action stationary are found iteratively using standard techniques. They are verified as solutions to the EoM in a leapfrog approximation. Since the solutions are non-unique, different choices of initial trial orbits will generally give different solutions for the galaxy paths. We define a χ^2 measure of fit for all relevant observables, from which a best-fit solution can be selected. Maps of χ^2 in four different scenarios, all with $H_0 = 67$ and $\Omega_0 = 0.27$, are shown in Fig. 4. At upper left are the contours in χ^2 generated from a simplified catalog consisting of only MW and M31, to check the consistency of the action method with the Timing Argument. As expected, we find a well-defined constraint on the sum $M_{MW} + M_{M31}$. The upper right panel, shows the results from a reduced version of our catalog which includes the LG actors but excludes external galaxies. The additional dynamical actors has broken the degeneracy in the TA, giving independent masses of $2.5 \pm 1.5 \times 10^{12} M_\odot$ for the MW and $3.5 \pm 1.0 \times 10^{12} M_\odot$ for M31. With the addition of the external galaxies (Fig. 4, lower left), the best mass for the MW increases to $3.5 \pm 1.0 \times 10^{12} M_\odot$. This is consistent at the lower end with previous TA measurements of the total LG mass and the individual MW mass. When the transverse velocity constraints on M31, LMC, M33, IC10, and LeoI are added (lower right), the confidence intervals are broadened and the best-fit mass for MW decreases slightly, to $3.0 \pm 1.5 \times 10^{12} M_\odot$, reflecting the fact that lower masses for MW are correlated to lower transverse velocities for M31 and other nearby galaxies. These values are to be understood as the masses contained within roughly half the separation between the two galaxies. For the MW, the value is more than twice what stellar motions yield for its virial mass. This could pose a challenge to the standard model since such an increase of the mass is not seen in N-body simulations.

4. Cosmological constraints on 20-100 Mpc scales

Restricting the analysis to these scales, away from non-linearities and hydrodynamical effects, greatly simplifies observational analyses of observations. Linear theory dynamical relations are, by and large, adequate on these scales but with the inclusion of scatter as a result from the presence of small scale modes. Further, large scale galaxy biasing is well described by a linear relation, as seen in Fig. 2.

4.1. Velocity-velocity comparison

Davis *et al.* (2011) have shown that GI passes a very important test: an excellent agreement between the velocity field predicted from the distribution of galaxies via linear theory and the observed motions of galaxies obtained from the TF measurements of spiral galaxies. The beauty of this test is that the effects of cosmic variance are minimized (in principles two good measurements are enough). The contribution of Marc Davis to these proceedings offers more details.

4.2. The recovery of the CMB dipole, i.e. the motion of the LG with respect to the CMB

The CMB temperature dipole together with astronomical estimation of the LG motion relative to the Sun, provide $V_{lg} = 627 \pm 22 \mathrm{km\,s^{-1}}$ toward $(l, b) = (276° \pm 3°, 30° \pm 3°)$ for the LG motion relative to the CMB frame. The LG is accelerated by the cumulative gravitational pull of the surrounding large scale structure. Therefore, an important probe of the GI paradigm is whether the observed large scale structure, as traced by the galaxy

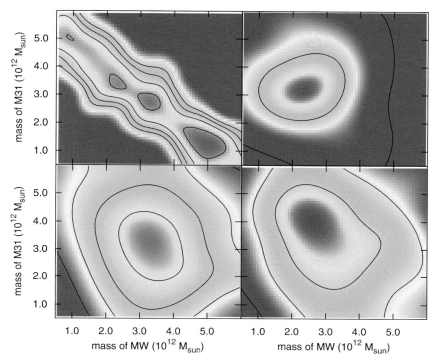

Figure 4. Contours in χ^2 for different values of MW and M31 masses. Upper left: results from the two-body problem of MW + M31. Upper right: LG actors only. Lower left: LG actors + four external groups. Lower right: same as lower left, with transverse velocity constraints included for five nearby galaxies. The first contour level (solid black line) marks the region of 95% confidence. From Phelps *et al.* (2013).

distribution, could indeed account for the LG motion. To do that, we need to compute the gravitational force on the LG from an all sky survey of galaxies. According to linear theory, this force should be proportional to the LG motion. We use the 2MRS which is the deepest nearly-all sky survey of angular positions and spectroscopic galaxy redshifts limited to $K_s = 11.75$ and arguably the best sample of objects to estimate the LG motion.

A source of uncertainty is the error in the linear theory dynamical reconstruction of the velocity from a given density field (equation 1.1). To assess this error we have estimated the LG motion from the full dark matter out to the largest possible outer radius in the simulation, i.e. $R_{out} = 250$ h^{-1}Mpc. The corresponding 1σ error, ~ 90km s^{-1}, is substantially smaller that the typical error in linear reconstruction of the peculiar velocity of a generic observer in the Universe. The reason for this is the strict criteria we have applied in selecting the "LG observer" in the mock catalogs, aimed at matching the quietness and moderate density environment of the observed LG. Removing these selection criteria boosts the error to $\gtrsim 300$km s^{-1}, consistent with previous studies (Nusser & Branchini 2000). Nonlinear dynamical reconstruction methods can potentially reduce the dynamical error. However, because the particular environment of the LG, errors due to linear reconstruction are subdominant compared to the total error budget.

For a more realistic assessment of the recovery of the LG motion in real data we resort to mock catalogs designed to match the 2MRS with $K_s < 11.75$ (as in Fig. 3). The results for various cases are shown in Fig. 5, as described in the caption. There is a dramatic decrease in the scatter when matter within $R_{out} = 250$ h^{-1}Mpc is included. Still the

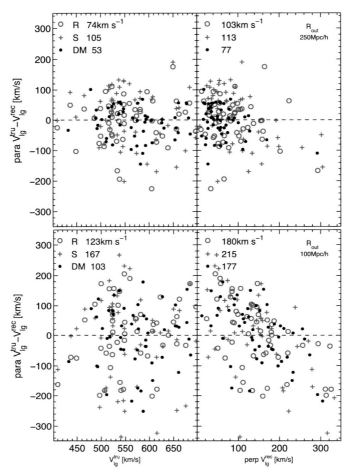

Figure 5. A scatter plot showing the velocity residual in the parallel and perpendicular directions from mock catalogs. The blue dots are in real space and the red crosses are in redshift space, while black dots show recovery from the full dark matter density field in real space. The *rms* values of the parallel and perpendicular residuals are listed in the left and right panels, respectively. Top and bottom panels correspond to velocity reconstruction with only data within $R_{\mathrm{out}} = 250 \ \mathrm{h}^{-1}\mathrm{Mpc}$ and $R_{\mathrm{out}} = 100 \ \mathrm{h}^{-1}\mathrm{Mpc}$, respectively. The *rms* of the parallel and perpendicular residuals are indicated, respectively, in the left and right panels.

residuals are at the level of $70 - 100 \mathrm{km \, s}^{-1}$ depending on the case considered. This is consistent with Bilicki *et al.* (2011) who derived a similar result analytically. Current all sky data do not allow a reliable assessment of the contribution of fluctuations beyond $100 \ \mathrm{h}^{-1}\mathrm{Mpc}$. Much of that is because of the Kaiser rocket effect (Kaiser 1987) (see Nusser *et al.* (2014) for further details).

4.3. *The bulk flow*

The bulk flow, $\mathbf{B}(r)$, is the mean motion of a sphere of a given radius, r. Computing $\mathbf{B}$ from the sparse and noisy peculiar velocity catalogs is very challenging. Improper handling of the data may easily lead to artificially large flow. Nusser & Davis (2011) have estimated $\mathbf{B}(r)$ from the SFI++ survey. They have discarded *a*) the fainter galaxies which do not obey the TF relation and *b*) galaxies beyond $100 \ \mathrm{h}^{-1}\mathrm{Mpc}$ which are likely to suffer from systematics related to the measurements of TF parameters. They found $\mathbf{B}(r)$ to be consistent with the standard ΛCDM model and derived the constraint: $\sigma_8(\Omega_m/0.266)^0.28 = 0.86 \pm 0.11$, which leads to a σ_8 higher (although consistent with) than the WMAP value. However, taking $\Omega_m = 0.317$ from the recent Planck results, yields a best fit value $\sigma_8 = 0.819$, very close to the result reported by the Planck collaboration. Despite earlier claims of anomalous bulk flows, the emergent consensus is that bulk flow measurements from numerous data (intrinsic relations, SN and kSZ), is consistent with the standard ΛCDM model (Nusser & Davis 2011; Dai *et al.* 2011; Colin *et al.* 2011; Hong *et al.* 2014; Feix *et al.* 2014; Watkins & Feldman 2014; Ma & Pan 2014; Planck Collaboration 2014).

5. Summary

The gravitational instability model for structure formation with its current ΛCDM incarnation described the nearby large scale structure very well. Whatever corrections for this model should be small as far as large scales are concerned. Probing deviations from this model on large scales maybe possible only with next generation redshift surveys. Modifications on smaller scales ($\sim$ a few Mpcs) are a different matter (e.g. Peebles & Nusser 2010) but are not the subject of this contribution.

Upcoming redshift surveys will contain a large number of galaxies to allow constraints on the cosmological velocity field independently of the classical redshift distortion the correlation functions. The idea is that galaxy redshifts cz depend on v_p. Hence, using cz instead of true distances, d, in order to estimate galaxy luminosities (from the observed apparent magnitudes) will introduce spatial variations of the luminosity function. Assuming negligible environmental dependence in the luminosity function, these variations can put constraints on the velocity field. (e.g. Yahil *et al.* 1980; Nusser *et al.* 2012a; Feix *et al.* 2014). Basically this method assumes that galaxy luminosity is a standard candle where the very large distance error is beaten by the large number of galaxies. The method can be applied to surveys with photometric redshifts (Bilicki *et al.* 2014) and is not restricted to spectroscopic surveys. The method can constrain galaxy biasing and the the linear growth rate is a velocity model based on the actual galaxy distribution is used to model the luminosity variations. The contribution of Martin Feix addresses potential caveats and presents an application to the SDSS.

Another potential probe is offered by Gaia. There could be a large number of galaxies detected as point sources by Gaia (Nusser *et al.* 2012b). For example, the nuclei of M87 and N5121 (both at d=17.8Mpc) should be detected with an end of mission accuracy of $600 \, \mathrm{km \, s^{-1}}$ in the transverse motion. The surface brightness profiles of the Carnegie-Irvine Galaxy Survery show that 70% of galaxies in this survey could be detected by Gaia. The majority of these nearby galaxies will be detected if placed at $\gtrsim 500$Mpc (early type) and $\gtrsim 250$Mpc (late type). Of course, parallax distance errors increase quadratically with distance. Therefore, the error on Gaia's distances for extragalactic objects will be huge and cannot be used to get the transverse velocities from the measured proper motions. But, at such distances, the redshifts can be used as proxies for the true distance without introducing a significant error in the transverse velocities.

References

Bilicki, M., Chodorowski, M., Jarrett, T., & Mamon, G. A. 2011, *ApJ*, 741, 31
Bilicki, M., Jarrett, T. H., Peacock, J. A., Cluver, M. E., & Steward, L. 2014, *ApJ. S*, 210, 9
Buchert, T. & Ehlers, J. 1993, *MNRAS*, 264, 375

Colin, J., Mohayaee, R., Sarkar, S., & Shafieloo, A. 2011, *MNRAS*, 414, 264

Dai, D., Kinney, W. H., & Stojkovic, D. 2011, ArXiv:1102.0800

Davis, M., Nusser, A., Masters, K. L., Springob, C., Huchra, J. P., & Lemson, G. 2011, *MNRAS*, 413, 2906

Davis, M. & Peebles, P. J. E. 1983, *ApJ*, 267, 465

De Lucia, G. & Blaizot, J. 2007, *MNRAS*, 375, 2

Doroshkevich, A. G. 1970, *Astrophysics*, 6, 320

Feix, M., Nusser, A., & Branchini, E. 2014, *JCAP*, 9, 19

Fisher, K. B. & Nusser, A. 1996, *MNRAS*, 279, L1

Frisch, U., Matarrese, S., Mohayaee, R., & Sobolevski, A. 2002, *Nature*, 417, 260

Gramann, M. 1993, *ApJ*, 405, 449

Hong, T., Springob, C. M., Staveley-Smith, L. *et al.* 2014, *MNRAS*, 445, 402

Huchra, J. P., Macri, L. M., Masters, K. L. *et al.* 2012, *ApJ. S*, 199, 26

Kaiser, N. 1987, *MNRAS*, 227, 1

—. 1988, *MNRAS*, 231, 149

Kofman, L., Bertschinger, E., Gelb, J. M., Nusser, A., & Dekel, A. 1994, *ApJ*, 420, 44

Lynden-Bell, D., Faber, S. M., Burstein, D. *et al.* 1988, *ApJ*, 326, 19

Ma, Y.-Z. & Pan, J. 2014, *MNRAS*, 437, 1996

Masters, K. L., Springob, C. M., Haynes, M. P., & Giovanelli, R. 2006, *ApJ*, 653, 861

Nusser, A. & Branchini, E. 2000, *MNRAS*, 313, 587

Nusser, A., Branchini, E., & Davis, M. 2012a, *ApJ*, 744, 193

—. 2012b, *ApJ*, 755, 58

Nusser, A. & Davis, M. 2011, *ApJ*, 736, 93

Nusser, A., Davis, M., & Branchini, E. 2014, *ApJ*, 788, 157

Peebles, P. J. E. 1980, The large-scale structure of the universe (Princeton University Press)

—. 1989, *ApJL*, 344, L53

Peebles, P. J. E. & Nusser, A. 2010, *Nature*, 465, 565

Phelps, S., Nusser, A., & Desjacques, V. 2013, *ApJ*, 775, 102

Planck Collaboration. 2014, *A&A*, 561, A97

Schneider, P. & Bartelmann, M. 1995, *MNRAS*, 273, 475

Taylor, A. N. & Hamilton, A. J. S. 1996, *MNRAS*, 282, 767

Tully, R. B. *et al.* 2013, *Astronomical. J*, 146, 86

Watkins, R. & Feldman, H. A. 2014, ArXiv e-prints

White, M. 2014, *MNRAS*, 439, 3630

White, S. D. M. 1984, *ApJ*, 286, 38

Yahil, A., Sandage, A., & Tammann, G. A. 1980, *ApJ*, 242, 448

Zeldovich, I. B. & Novikov, I. D. 1983, Relativistic astrophysics. Volume 2 - The structure and evolution of the universe (Chicago, IL, University of Chicago Press, 1983)

Zel'dovich, Y. B. 1970, *A&A*, 5, 84

Zheligovsky, V. & Frisch, U. 2014, *Journal of Fluid Mechanics*, 749, 404

The Zeldovich Universe:
Genesis and Growth of the Cosmic Web
Proceedings IAU Symposium No. 308, 2014
R. van de Weygaert, S. Shandarin, E. Saar & J. Einasto, eds.

© International Astronomical Union 2016
doi:10.1017/S1743921316009674

Collisionless Dynamics and the Cosmic Web

Oliver Hahn

Department of Physics, ETH Zurich,
CH-8093 Zürich, Switzerland
email: `hahn@phys.ethz.ch`

Abstract. I review the nature of three-dimensional collapse in the Zeldovich approximation, how it relates to the underlying nature of the three-dimensional Lagrangian manifold and naturally gives rise to a hierarchical structure formation scenario that progresses through collapse from voids to pancakes, filaments and then halos. I then discuss how variations of the Zeldovich approximation (based on the gravitational or the velocity potential) have been used to define classifications of the cosmic large-scale structure into dynamically distinct parts. Finally, I turn to recent efforts to devise new approaches relying on tessellations of the Lagrangian manifold to follow the fine-grained dynamics of the dark matter fluid into the highly non-linear regime and both extract the maximum amount of information from existing simulations as well as devise new simulation techniques for cold collisionless dynamics.

Keywords. cosmology: theory, dark matter, large-scale structure of universe, methods: numerical

1. Introduction

Zeldovich's legendary formula (Zeldovich 1970) describes cosmological structure formation and in particular the emergence of singularities under gravitational collapse that emerge from critical curves in the velocity perturbation field. It arises rather simply from the Lagrangian motion of pressure-free fluid parcels under self gravity. The relation between Lagrangian and Eulerian space can at all times be described by the position $\mathbf{x}$ and velocity $\mathbf{v}$ of a fluid parcel labeled by its three-dimensional Lagrangian coordinate $\mathbf{q}$ as formally given by

$$\mathbf{x_q}(t) = \mathbf{q} + \mathbf{L}(\mathbf{q}, t), \quad \text{and} \quad \mathbf{v_q}(t) = \dot{\mathbf{x}}_\mathbf{q} = \dot{\mathbf{L}}(\mathbf{q}, t). \quad (1.1)$$

The displacement field $\mathbf{L}$ at linear perturbative order is simply proportional to the growing mode of the initial velocity perturbation field. It evolves with the growth factor of linear density perturbations

$$\mathbf{L}(\mathbf{q}, t) \propto D_+ \mathbf{v_q}(0), \quad \text{and} \quad \dot{\mathbf{L}}(\mathbf{q}, t) \propto \dot{D}_+ \mathbf{v_q}(0), \quad (1.2)$$

which *is* the Zeldovich approximation, implying that the fluid elements move on straight lines as determined by their initial velocity vectors. The density then becomes obviously (simply through the Jacobian of the transformation from $\mathbf{q}$ to $\mathbf{x}$)

$$1 + \delta = \prod_{i=0}^{3} (1 + D_+ \lambda_i)^{-1}, \quad \text{where} \quad \lambda_i = \mathrm{eig}\left\{ T_{ij} \equiv \partial_{ij}\phi \right\}, \quad (1.3)$$

and ϕ is the velocity potential (i.e. $\mathbf{v_q}(t) \propto \dot{D}_+ \boldsymbol{\nabla}\phi$), which is, at first order, proportional to the gravitational potential. Since T_{ij} is a symmetric and real tensor, its eigenvalues are real numbers and thus can be arranged $\lambda_1 \leqslant \lambda_2 \leqslant \lambda_3$. This however implies that δ can undergo a maximum of three singularities over time, depending on the number of negative

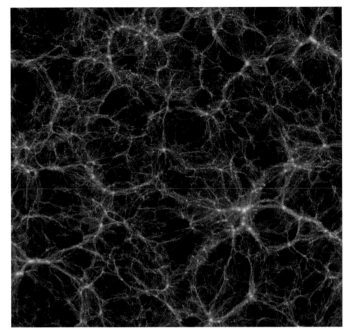

Figure 1. The cosmic web as formed in an N-body simulation of the ΛCDM cosmology at $z = 0$. The width of the image is $250\,h^{-1}\mathrm{Mpc}$, shown is the logarithmic overdensity $\log(1 + \delta)$. Zel'dovich's first order perturbation theory predicts the formation of pancakes, filaments, halos and voids in the cosmic matter distribution. Since in CDM perturbation exist down to very small scales, filaments are of course made up of haloes, and pancakes of filaments.

eigenvalues λ_i. If $\lambda_1 < 0$, then at some finite time t_1, when $D_+(t_1)\lambda_1 = -1$, the fluid element will undergo a singularity along the eigenvector associated with λ_1, followed by the other axes assuming they correspond to negative eigenvalues. Thus, the first objects that form are two-dimensional structures (the infamous Zeldovich pancakes). Then the collapse along a second axis happens and a one-dimensional structure emerges (a filament), before also the third axis collapses and finally a (roughly) spherical structure forms. Of course this is a somewhat simplistic picture and in general the singularities/catastrophes that can arise have a multitude of structures (see Hidding *et al.* 2014 for an extensive discussion of catastrophes in the Zel'dovich approximation). Since density perturbations exist on various scales, this pancake collapse scenario occurs simultaneously on different scales leading to the multi-scale nature of the cosmic web. On the smallest scales sit the halos, embedded in larger perturbations for which the final direction has not collapsed yet, corresponding to filaments. In turn, the filaments are embedded in even larger scale perturbations for which two axes have not collapsed yet, corresponding to the pancakes. Naturally, if any eigenvalue is smaller than zero, the corresponding axis will never collapse in this approximation. In Figure 1, we show the cosmic web as it emerges in a density map of an N-body simulation of a ΛCDM cosmology.

As is well known, the approximation of eq. (1.2) breaks down in two cases: (1) after shell-crossing in one-dimension, when the gravitational force acting on a fluid parcel reverses its direction, and (2) in multiple dimensions if higher derivatives of the gravitational perturbation field (e.g. $\partial_{ijk}\phi$) become non-negligible compared to the tidal (i.e. $\partial_{ij}\phi$) term (e.g. Crocce *et al.* 2006). We will give point (1) more attention next.

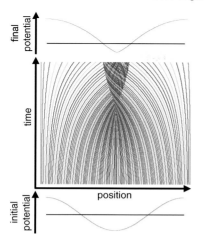

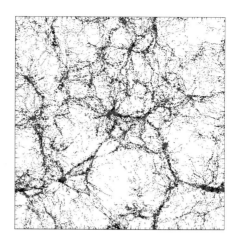

Figure 2. Left: Collapse of a plane wave under self gravity. Shell-crossing leads to a gravita-tionally bound structure forming in the centre. Regions of tidal expansion in the final potential correspond to regions outside the bound structure and regions of tidal compression to the inside. **Right:** Haloes from an N-body ΛCDM simulation classified by the signature of the tidal ten-sor eigenvalues smoothed on $2\,h^{-1}$Mpc splitting them into clusters (red: -,-,-), filaments (blue: +,-,-), walls (green: +,+,-) and voids (yellow: +,+,+) according to the method of Hahn *et al.* (2007b). The image is $180\,h^{-1}$Mpc wide.

2. The Zeldovich approximation, non-linear structure formation and identification of dynamically distinct structures of the cosmic web

In Figure 2, we show the collapse of a collisionless self-gravitating plane wave, i.e. the collapse of a sinusoidal perturbation $\mathbf{L}(\mathbf{q},0) \propto \sin(\mathbf{k}\cdot\mathbf{q})$, or equivalently $\phi \propto -\cos(\mathbf{k}\cdot\mathbf{q})$, where $\mathbf{k}$ is the wave vector of the perturbation. We plot the fully non-linear solution in a slightly different way than is done usually. The bottom panel shows the initial gravitational potential, the middle panel shows the trajectories of particles over time (a given curve corresponds to a fixed $\mathbf{q}$), and the top panel shows the gravitational potential at the final time. Time is plotted logarithmically in the middle panel to highlight better the dynamics around the time of first shell-crossing. Several important observations can be made from this simple diagram, which is perfectly reproduced by the Zeldovich approximation for all points outside of the outer caustics of the multi-stream region. In the fully non-linear setting, repeated shell-crossing happens in the center due to self-gravity. We observe that (1) the initial tidal field is given by $\lambda = \partial_{ij}\phi \propto \cos(\mathbf{k}\cdot\mathbf{q})$ implying that trajectories in the central region are convergent ($\lambda < 0$), and trajectories outside are divergent ($\lambda > 0$). Of course, the outer trajectories still collapse onto the central structure, but the Eulerian volume which had $\lambda > 0$ initially remains divergent and becomes underdense. (2) the final potential has not changed qualitatively from the initial. Obviously, it still has a convergent ($\lambda < 0$) central part and divergent ($\lambda > 0$) outer parts (in both cases separated by the horizontal black lines which are drawn to intersect at the inflection points of ϕ). Furthermore, the convergent region agrees well with the shell-crossed region. This latter observation provides a strong motivation to use the final tidal field to classify the cosmic web into dynamically distinct regions as suggested by Hahn *et al.* (2007b).

Naturally, in CDM, almost all matter is in dark matter halos so that a simple classi-fication by the final potential would identify halos as dips in the potential rather than describe the large-scale structure. For this reason, Hahn *et al.* (2007b) smoothed the potential on about twice the virial scale of M_* halos of $\sim 2\,h^{-1}\mathrm{Mpc}$ at $z=0$ and on correspondingly smaller scales at high redshift (Hahn *et al.* 2007a). The resulting smoothed gravitational potential ψ can then be classified into distinct regions according to the signature of the eigenvalues of $\partial_{ij}\psi$, just as in eq. (1.3). Regions of three-dimensional compression $(-,-,-)$ correspond then to tidal compression on the smoothed scale and thus can be identified with clusters (or nodes of the web), and corresondingly $(+,-,-)$ with filaments, $(+,+,-)$ with walls and $(+,+,+)$ with voids. The performance of such a classification can be seen in the right panel of Figure 2. Very clearly, the method is able to identify the nodes of the cosmic web and the larges filaments very well and assigns the finer filaments to walls. However, One can immediately see from this classification that the volume assigned to voids is very small, which subsequently has led Forero-Romero *et al.* (2009) to argue that the signature should not be evaluated w.r.t. to zero as this describes only the asymptotic behavior, i.e. whether $1+D_+\lambda$ vanishes within a finite time in eq. (1.3). They have thus introduced an additional (free) parameter λ_{th} with respect to which the eigenvalues are compared, i.e. the signature of $(\lambda_i + \lambda_{th})$ is determined instead. Another (arguable) shortcoming is the fixed smoothing scale, which e.g. identifies all nodes of the cosmic only on one scale. The combination of multi-scale filtering with an evaluation of the tidal eigenvalues has been discussed by Cautun *et al.* (2013).

During the linear stages of structure formation – as expressed by the Zeldovich approximation – the velocity field and the gradient of the gravitational potential are proportional to each other (cf. eq. 1.2). This implies that at early times a classification by the eigenvalues of the tidal tensor eig $\partial_{ij}\psi$ is equivalent to a classification by the velocity deformation tensor eig $\partial_i v_j$. At late times, this equality is however broken by non-linear terms so that Hoffman *et al.* (2012) have proposed to use the latter when classifying structures of the cosmic web. In particular, while primordial vorticity is usually neglected since it is a decaying perturbation, at late time vorticity arises in the non-linear cosmic velocity field (see the discussion in the next section). In addition to these approaches inspired by the evolution of large-scale structure in the Zeldovich approximation, also density fields and logarithmic density fields have been used (most notably Aragón-Calvo *et al.* 2007a, 2010; Sousbie 2011; Cautun *et al.* 2013). With such a multitude of classifiers available, the question what needs to be classified arises, since all definitions are to a large degree arbitrary. What all methods have achieved is to provide a clear prediction for the alignment of halo spins and shapes with the cosmic web (e.g. Aragón-Calvo *et al.* 2007b; Hahn *et al.* 2007a; Codis *et al.* 2012, among others). It still remains to be seen whether there are other observable properties of galaxies that depend on the cosmic web environment rather than just the halo mass and the central vs. satellite distinction.

3. The Zeldovich approximation, the Lagrangian manifold and Lagrangian tesselations

Another interesting property of the mapping between Lagrangian and Eulerian space is that it has a manifold structure and directly describes the distribution function of a perfectly cold fluid (Arnold *et al.* 1982; Shandarin and Zeldovich 1989). This means that the mapping between Lagrangian space and Eulerian phase space varies smoothly between points close in Lagrangian space, i.e. the mapping is differentiable,

$$\mathbf{q} \mapsto (\mathbf{x_q}(t), \mathbf{v_q}(t)), \quad \Rightarrow \quad T_{\mathbf{q}} = (\boldsymbol{\nabla_q}\mathbf{x}, \boldsymbol{\nabla_q}\mathbf{v}), \tag{3.1}$$

where the latter defines the three-dimensional space tangent to the cold distribution function in six-dimensional phase space. If the cold distribution function is initially suf-

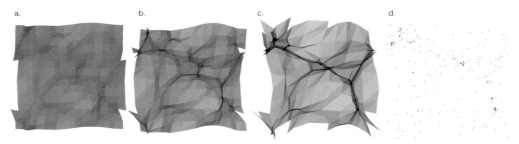

Figure 3. Foldings of the triangulated phase space sheet in 2+2 dimensional phase space over time. Panels (a) to (c) show the time evolution of a two-dimensional N-body simulation of only 16^2 particles where the density field is given by triangles connecting particles neighbouring in Lagrangian space. The resulting density estimate is defined everywhere in space. Panel (d) shows the positions of the N-body particles at the same time as panel (c).

ficiently smooth, it will remain so by virtue of the collisionless Vlasov-Poisson system of equations. The density of dark matter in the vicinity of **q** is given by the determinant of the spatial part, i.e.

$$\rho = m_p \left| \det \frac{\partial x_i}{\partial q_j} \right|^{-1}, \qquad (3.2)$$

where m_p is the particle mass. Notably, as we had already seen in eq. (1.3), the density becomes singular in caustics, where this determinant vanishes, i.e. where the distribution function has a tangent perpendicular to configuration space. The evolution of the tangent space can be described in terms of a geodesic deviation equation around point **q** (Vogelsberger *et al.* 2008) and allows one to study in simulations e.g. how often a particle goes through a caustic (White and Vogelsberger 2009).

More recently however, Abel *et al.* (2012) and Shandarin *et al.* (2012) noted that the tangent space can be easily approximated in simulations by combining information from neighboring points in Lagrangian space. This can be achieved by determining the neighbors of point **q** by a Delaunay triangulation of the regular lattice that the N-body particles form in Lagrangian space. Thus, the discrete points $\mathbf{q}_i$ (i.e. the N-body particles) can be associated with the vertices of a tetrahedral mesh that covers all of Lagrangian space, and thus also all of Eulerian configuration space. Each tetrahedron uniquely defines an approximation to $T_{\mathbf{q}}$, i.e. one can calculate $\partial x_i/\partial q_j$ and $\partial v_i/\partial q_j$ uniquely for each tetrahedron. This in turn enables one to calculate the density of a tetrahedron according to eq. (3.2). This is the single-stream density. Since after shell crossing several tetrahedra may overlap a given point **x** in configuration space, the densities of all tetrahedra overlapping that point need to be added up to obtain the configuration space density. We illustrate the procedure in two dimensions in Figure 3. The left-most panel shows the initial triangular mesh. Before shell-crossing occurs, the triangles do not overlap. Density perturbations come from particles that are displaced between Lagrangian and Eulerian space leading to density perturbations between the triangles consistent with Zeldovich's formula. Over time, the perturbations grow, shell-crossing occurs and several triangles can overlap the same point in configuration space. Clearly the filamentary cosmic web emerges already from this simple demonstration. For comparison, the right panel shows the locations of the N-body particles where the filamentary structure is by far not as clearly visible as in the triangulated density estimate that tracks explicitly the anisotropic local deformation tensors.

The approximation of the phase space sheet by finite volume tetrahedral elements provides access to a wide range of properties of dark matter flows that could not be measured before. Most notably, the tetrahedral elements allow to measure density fields with significantly reduced noise without the need for any smoothing. This allows for rather breath-taking possibilities for visualization (Abel *et al.* 2012; Kaehler *et al.* 2012) or gravitational lensing simulations with suppressed noise (Angulo *et al.* 2014) and allows to, e.g., explicitly count the number of streams overlaying points in configuration space (Shandarin *et al.* 2012; Abel *et al.* 2012). They also provide a new way to determine whether structures are collapsed (and thus folded) along three, two, one or no directions which allows one to more directly identify halos, filaments, walls and voids (Neyrinck and Shandarin 2012; Falck *et al.* 2012).

Furthermore, this Lagrangian tessellation approach provides excellent estimates of cosmic mean velocity fields and their derivatives (Hahn *et al.* 2014). Since the single stream density is known, a very accurate estimate of the bulk velocity field can be obtained. It is defined as the density weighted mean velocity

$$\langle \mathbf{v} \rangle = \frac{\sum_{s \in S} \mathbf{v}_s(\mathbf{x}) \, \rho_s(\mathbf{x})}{\sum_{s \in S} \rho_s(\mathbf{x})}, \tag{3.3}$$

where S stands for all streams that contain a point $\mathbf{x}$ in configuration space. This allows one to define the *exact* differentials of the collisionless mean multi-stream velocity field as

$$\boldsymbol{\nabla} \cdot \langle \mathbf{v} \rangle = \langle (\boldsymbol{\nabla} \log \rho) \cdot (\mathbf{v} - \langle \mathbf{v} \rangle) \rangle + \langle \boldsymbol{\nabla} \cdot \mathbf{v} \rangle \tag{3.4}$$

and

$$\boldsymbol{\nabla} \times \langle \mathbf{v} \rangle = \langle (\boldsymbol{\nabla} \log \rho) \times (\mathbf{v} - \langle \mathbf{v} \rangle) \rangle, \tag{3.5}$$

which hold if the derivative is not taken across caustics. Most notably, the curl of the mean field vanishes in single stream regions since gravity cannot generate vortical modes, but it can arise in multi-stream regions as a 'collective' phenomenon (see also Pichon and Bernardeau 1999). We furthermore see that the velocity divergence also has a 'collective' term, which depends on alignment of local flows with the density gradient, in addition to the single stream divergence.

In Figure 4, we show a slice through the density field (left panel) and the velocity divergence field (right panel) calculated from an N-body simulation of halo in a warm dark matter cosmology. We see that underdense regions have positive velocity divergence before shell-crossing, while overdense regions have negative divergence. After shell-crossing, overdense regions also show a positive divergence. The correlation before shell crossing follows directly from a series expansion of eq. (1.3) which yields $1 + \delta \simeq 1 - A \sum \text{eig} \, \partial v_i / \partial q_j$ (where A is a positive constant, see also Kitaura *et al.* 2012). The reversal after shell crossing must thus come from the 'collective' term. Care has to be taken when comparing these results to velocity fields which are sampled on a mesh and then differentiated on that mesh, since in that case the discontinuous velocity jumps across caustics completely dominate the differential fields so that one essentially only measures the motion of caustics. The exact derivatives that can be computed using the triangulated sheet do not suffer from these inherent problems.

Another exciting possibility is to attempt to evolve the N-body system self-consistently using the tetrahedral approximation to the distribution function (Hahn *et al.* 2013). In such an approach, the vertices become tracers of the flow and the mass is, unlike in the N-body approach, not centered at the vertices, but is contained in the tetrahedral volume element. Hahn *et al.* (2013) have shown that this "TetPM" approach significantly reduces discreteness effects known for N-body simulations (e.g. Wang and White 2007).

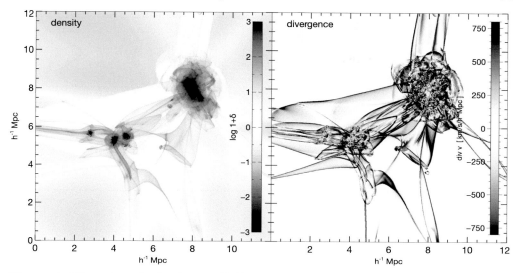

Figure 4. Slices through the density (left) and bulk velocity divergence field (right panel) from a warm dark matter simulation. Before shell-crossing underdense regions are divergent, and overdense regions are convergent. After shell-crossing, overdense regions are predominantly divergent as well which reflects that the volume of the dark matter sheet is growing over time. [Adapted from Hahn *et al.* (2014)].

We show the result of such a simulation, where the mass distribution due to the tetrahedra is approximated by sampling the tetrahedra with mass-tracing pseudo-particles, in the top panels of Figure 5. The top left panel shows the well known artificial fragmentation of filaments in warm dark matter simulations if the N-body method is used at a high force to low mass resolution ratio. The top right panel shows that the tetrahedral approach spreads the mass smoothly between the flow tracers leading to sharp well-defined filaments that show no sign of fragmentation.

These fragments have made it inherently difficult if not impossible to measure the mass function of halos in warm dark matter (WDM) cosmologies robustly (e.g. Schneider *et al.* 2013) and their very presence might well invalidate many other predictions from such simulations. Using the TetPM method Angulo *et al.* (2013) were able to measure the WDM halo mass function in the absence of such fragments and found a distinctly different picture of large-scale structure and halo formation than in CDM: smooth pancakes permeated by filaments with very dense cores connecting the nodes of the cosmic web. They found that halos do not form below the truncation scale of the power spectrum, but come only into existence once the last axis collapses, quite in accord with the prediction of Zeldovich's equation (1.3).

The rapid (phase and possibly chaotic) mixing of dark matter as it orbits in the potential wells of dark matter halos leads to a growth of the dark matter sheet that cannot be tracked by the Lagrangian motion of the tracer particles alone. The approximation that the tetrahedra describe the neighbourhood of a Lagrangian point $\mathbf{q}$ well breaks down as a consequence. The practical effect of this is that mass gets preferentially assigned to the center of halos making them more and more dense (Hahn *et al.* 2013). This limits the applicability of a tessellation approach based on a fixed number of Lagrangian tracers in the inner parts of dark matter halos where strong mixing occurs. This problem can however be resolved by inserting new vertices in the tessellation, i.e. by splitting the tetrahedra adaptively, in order to keep following the evolution of the dark matter sheet

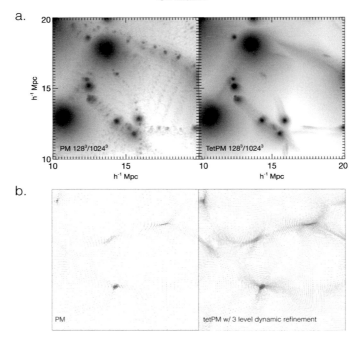

Figure 5. Top: Magnitude of the gravitational force in a simulation of structure formation in a warm dark matter cosmology. The N-body method (top left) shows artificial fragmentation of the filaments into clumps due to inherent discreteness effects. Spreading the mass between particles using tetrahedral elements strongly suppresses such discreteness effects. **Bottom:** Adaptive refinement of the phase space sheet (right) will allow to track its rapid super-Lagrangian growth in the inner of halos. [Top panel adapted from Hahn *et al.* (2013)]

through all its complicated foldings at late times. The effect of such a refinement can be seen in the bottom panels of Figure 5. In the left panel a standard N-body particle distribution is shown, while the right panel shows the distribution of flow tracers after tetrahedra were allowed to split into smaller units. The newly inserted vertices clearly trace out the regions of strongest deformation of the sheet. An article discussing the feasibility of such a adaptive dynamic refinement technique of the tessellation is currently in preparation (Hahn and Angulo 2015). Refinement will allow to track the full evolution of the fine-grained distribution function in the deeply non-linear regime.

4. Summary

In this article, we have attempted to present an overview of the connection between Zeldovich's groundbreaking work from 1970 describing the evolution of Lagrangian fluid elements in an expanding universe (Zeldovich 1970), predicting a hierarchy of structures of pancakes, filaments and clusters, and modern work to dissect the cosmic web into distinct components. We have described how this work has inspired methods to classify the large-scale structure into dynamically distinct structure through the signature of eigenvalues of the tidal tensor or the velocity deformation tensor. Arguably the most interesting result of these attempts has been the discovery of (mass-dependent) alignments of the spin and shape of dark matter halos with the surrounding cosmic web.

In the second part, we described how Zeldovich's description of the Lagrangian motion of fluid elements and their tidal deformation is equivalent to recent attempts to decompose the dark matter sheet (or Lagrangian manifold) into finite volume elements with the help of tessellations. The resulting tetrahedra, spanned by four tracer particles (which can be the N-body particles of a cosmological simulation) are finite difference approximations to Zeldovich's famous formula and describe the same dynamical behaviour allowing for shell-crossing along three distinct directions. We gave a brief overview of the published applications of these tessellation approaches to a finite volume Zeldovich approximation in terms of using them to provide new insights into the phase space structure of N-body simulations. Finally we discussed how they can be used to create alternative simulation methods for self-gravitating dark matter in the cold limit that evolve the tessellated phase space sheet self-consistently in phase-space. The rapid wrapping of the phase space sheet inside of halos where mixing occurs necessitates the adaptive refinement of the tessellation. Adaptively refined tesselations are a method which is currently in development and will provide new exciting insights into structure formation since the adaptive refinement allows to follow the full evolution of the fine-grained distribution function of dark matter as it is warped through gravitational collapse from pancakes to filaments and finally folded up in a halo. Information which is unaccessible for N-body methods and can only approximately be reconstructed without dynamic refinement.

Acknowledgments

OH acknowledges support from the Swiss National Science Foundation through the Ambizione fellowship. The author thanks Tom Abel, Raul Angulo, Cristiano Porciani and Ralf Kaehler for many discussions surrounding the topics of this article, as well as Sergei Shandarin, Rien van de Weygaert, Enn Saar and Jaan Einasto for the invitation to speak at such a wonderful conference on the occasion of Zeldovich's 100th birthday.

References

Abel, T., Hahn, O., & Kaehler, R. Tracing the dark matter sheet in phase space. MNRAS, 427: 61–76, November 2012.

Angulo, R. E., Hahn, O., & Abel, T. The warm dark matter halo mass function below the cut-off scale. MNRAS, 434:3337–3347, October 2013.

Angulo, R. E., Chen, R., Hilbert, S., & Abel, T. Towards noiseless gravitational lensing simulations. MNRAS, 444:2925–2937, November 2014.

Aragón-Calvo, M. A., Jones, B. J. T., van de Weygaert, R., & van der Hulst, J. M.. The multiscale morphology filter: identifying and extracting spatial patterns in the galaxy distribution. A&A, 474:315–338, October 2007a.

Aragón-Calvo, M. A., van de Weygaert, R., B. J. T. Jones, & van der Hulst, J. M.. Spin Alignment of Dark Matter Halos in Filaments and Walls. ApJ, 655:L5–L8, January 2007b.

Aragón-Calvo, M. A., van de Weygaert, R., & B. J. T. Jones. Multiscale phenomenology of the cosmic web. MNRAS, 408:2163–2187, November 2010.

Arnold, V. I., Shandarin, S. F., & Zeldovich, I. B. The large scale structure of the universe. I - General properties One- and two-dimensional models. *Geophysical and Astrophysical Fluid Dynamics*, 20:111–130, 1982.

Cautun, M., van de Weygaert, R., & Jones, B. J. T. NEXUS: tracing the cosmic web connection. MNRAS, 429:1286–1308, February 2013.

Codis, S., Pichon, C., Devriendt, J., Slyz, A., Pogosyan, D., Dubois, Y., & Sousbie, T. Connecting the cosmic web to the spin of dark haloes: implications for galaxy formation. MNRAS, 427: 3320–3336, December 2012.

Crocce, M., Pueblas, S., & Scoccimarro, R. Transients from initial conditions in cosmological simulations. MNRAS, 373:369–381, November 2006.

Falck, B. L., Neyrinck, M. C., & Szalay, A. S. ORIGAMI: Delineating Halos Using Phase-space Folds. ApJ, 754:126, August 2012.

Forero-Romero, J. E., Hoffman, Y., Gottlöber, S., Klypin, A., & Yepes, G. A dynamical classi-
    fication of the cosmic web. MNRAS, 396:1815–1824, July 2009.

Hahn, O., Carollo, C. M., Porciani, C., & Dekel, A. The evolution of dark matter halo properties
    in clusters, filaments, sheets and voids. MNRAS, 381:41–51, October 2007a.

Hahn, O., Porciani, C., Carollo, C. M., & Dekel, A. Properties of dark matter haloes in clusters,
    filaments, sheets and voids. MNRAS, 375:489–499, February 2007b.

Hahn, O., Abel, T., & Kaehler, R. A new approach to simulating collisionless dark matter fluids.
    MNRAS, 434:1171–1191, September 2013.

Hahn, O., Angulo, R. E., & Abel, T. The Properties of Cosmic Velocity Fields. *ArXiv e-prints*,
    April 2014.

Hidding, J., Shandarin, S. F., & van de Weygaert, R.. The Zel'dovich approximation: key to
    understanding cosmic web complexity. MNRAS, 437:3442–3472, February 2014.

Hoffman, Y., Metuki, O., Yepes, G., S. Gottlöber, J. E. Forero-Romero, Libeskind, N. I., &
    Knebe, A. A kinematic classification of the cosmic web. MNRAS, 425:2049–2057, September
    2012.

Kaehler, R., Hahn, O., & Abel, T. A Novel Approach to Visualizing Dark Matter Simulations.
    ArXiv e-prints, August 2012.

Kitaura, F.-S., Angulo, R. E., Hoffman, Y., & S. Gottlöber. Estimating cosmic velocity fields
    from density fields and tidal tensors. MNRAS, 425:2422–2435, October 2012.

Neyrinck, M. C. and Shandarin, S. F. Tessellating the cosmological dark-matter sheet: origami
    creases in the universe and ways to find them. *ArXiv e-prints*, July 2012.

Pichon, C. and Bernardeau, F. Vorticity generation in large-scale structure caustics. A&A, 343:
    663–681, March 1999.

Schneider, A., Smith, R. E., & Reed, D. Halo mass function and the free streaming scale.
    MNRAS, 433:1573–1587, August 2013.

Shandarin, S., Habib, S., & Heitmann, K. Cosmic web, multistream flows, and tessellations.
    Phys. Rev. D, 85(8):083005, April 2012.

Shandarin, S. F. and Zeldovich, Y. B. The large-scale structure of the universe: Turbulence,
    intermittency, structures in a self-gravitating medium. *Reviews of Modern Physics*, 61:
    185–220, April 1989.

Sousbie, T. The persistent cosmic web and its filamentary structure - I. Theory and implemen-
    tation. MNRAS, 414:350–383, June 2011.

Vogelsberger, M., White, S. D. M., Helmi, A., & Springel, V. The fine-grained phase-space
    structure of cold dark matter haloes. MNRAS, 385:236–254, March 2008.

Wang, J. and White, S. D. M. Discreteness effects in simulations of hot/warm dark matter.
    MNRAS, 380:93–103, September 2007.

White, S. D. M. and Vogelsberger, M. Dark matter caustics. MNRAS, 392:281–286, January
    2009.

Zeldovich, Y. B. Gravitational instability: An approximate theory for large density perturba-
    tions. A&A, 5:84–89, March 1970.

The Zeldovich Universe:
Genesis and Growth of the Cosmic Web
Proceedings IAU Symposium No. 308, 2014
R. van de Weygaert, S. Shandarin, E. Saar & J. Einasto, eds.

ⓒ International Astronomical Union 2016
doi:10.1017/S1743921316009686

An Origami Approximation to the Cosmic Web

Mark C. Neyrinck

Department of Physics and Astronomy, The Johns Hopkins University, Baltimore, MD 21211
email: neyrinck@pha.jhu.edu

Abstract. The powerful Lagrangian view of structure formation was essentially introduced to cosmology by Zel'dovich. In the current cosmological paradigm, a dark-matter-sheet 3D manifold, inhabiting 6D position-velocity phase space, was flat (with vanishing velocity) at the big bang. Afterward, gravity stretched and bunched the sheet together in different places, forming a cosmic web when projected to the position coordinates.

Here, I explain some properties of an origami approximation, in which the sheet does not stretch or contract (an assumption that is false in general), but is allowed to fold. Even without stretching, the sheet can form an idealized cosmic web, with convex polyhedral voids separated by straight walls and filaments, joined by convex polyhedral nodes. The nodes form in 'polygonal' or 'polyhedral' collapse, somewhat like spherical/ellipsoidal collapse, except incorporating simultaneous filament and wall formation. The origami approximation allows phase-space geometries of nodes, filaments, and walls to be more easily understood, and may aid in understanding spin correlations between nearby galaxies. This contribution explores kinematic origami-approximation models giving velocity fields for the first time.

In the current structure-formation paradigm, walls, filaments, and cosmic-web nodes in the Universe form somewhat like the origami-folding of a sheet. In paper origami, a 2D non-stretchy sheet is folded in three dimensions. In cosmological structure-formation, the sheet is a stretchable 3D manifold, which folds up in symplectic 6D position-velocity phase space. As in paper origami, the sheet is continuous and cannot tear, and also cannot pass through itself in 6D. This concept is essentially a Lagrangian fluid-dynamics framework (following mass elements, not using a fixed spatial coordinate system). If the matter were not collisionless (dark), Lagrangian patches could not pass through each other in position space to form folded structures.

Fig. 1 shows an example cosmic web folded from a collisionless dark-matter sheet, whose vertices have been displaced according to the Zel'dovich approximation. This is a projection of a 2D dark-matter sheet, residing in 4D phase space, down to position space.

In catastrophe theory, singularities (caustics, or folds) that can occur in this sheet can be formally classified (Arnold *et al.* 1982; Hidding *et al.* 2014). However, these descriptions apply only to the local vicinity of caustics. Here, we aim for a description of collections of caustics, that occur at 'nodes' that approximate the outer caustic geometry of haloes.

The origami approximation imposes the following assumptions on the displacement field $\mathbf{\Psi}(\mathbf{q})$, giving the comoving displacement of a mass element from the initial to final conditions: $\mathbf{\Psi}(\mathbf{q}) \equiv \mathbf{x}(\mathbf{q}) - \mathbf{q}$, where $\mathbf{q}$ is a Lagrangian position coordinate, and $\mathbf{x}(\mathbf{q})$ is the position at some later time. The first two assumptions are thought to hold rather well in reality, but the third is unrealistically strong in general. See Neyrinck (2014) for a complementary discussion of this model, aimed at origami researchers.

(*a*) After applying a sufficiently large smoothing filter to the displacement field, all of space is single-stream (*void*; Falck *et al.* 2012; Falck and Neyrinck 2014; Shandarin and Medvedev 2014). Also, at any resolution being considered, void regions exist.

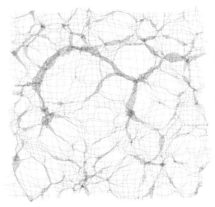

Figure 1. A dark-matter sheet in a $200\,h^{-1}$ Mpc 2D patch, distorted and folded according to the Zel'dovich approximation. The darkness of the color at each position gives the number of streams there. Initially, all vertices were nearly on a regular lattice. Since then, gravity has distorted the mesh, causing regions with a bit more matter than average to accumulate more matter around them. Nodes correspond to galaxies or clusters of galaxies.

(*b*) In void regions, $\boldsymbol{\Psi}(\mathbf{q})$ is irrotational. This is reasonable, since any initial vorticity decays with the expansion of the Universe, and since gravity is a potential force. However, in multi-stream regions (where $\mathbf{x}(\mathbf{q})$ is many-to-one), the flow, averaged among streams, often carries vorticity (e.g. Pichon and Bernardeau 1999).

(*c*) The sheet does not inhomogeneously stretch, i.e. $\mathbf{x}(\mathbf{q})$ is piecewise-isometric, or $\nabla_{\text{Lagrangian}} \cdot \boldsymbol{\Psi} = 0$, except at caustics, where it is undefined. This is the only assumption that is manifestly broken in reality; however, as Hahn *et al.* (2014) have found in warm-dark-matter simulations, the velocity divergence is perhaps surprisingly uniform. It remains largely positive even in filaments, except where it is undefined at caustics. Despite its inaccuracy, we expect that the piecewise-isometric assumption still allows some correct conclusions to be drawn.

2. Collapsed structures in 1D and 2D

In 1D, collapsed structures are delineated by their outer caustics; for example, the canonical phase-space spiral in 2D phase space. From here on, we focus on the outer caustics of structures, by looking at 'simple' collapsed regions, i.e. without inner phase-space windings.

In 2D, the simplest collapsed structure is a *filament*, an extrusion of a 2D node. With the piecewise-isometry assumption, a filament must consist of straight caustics (Demaine and O'Rourke 2008). Furthermore, the caustics must be parallel, since two reflections produced by non-parallel caustics would cause neighboring voids to be rotated with respect to each other. Here I focus on the outer caustics, and ignore substructure within collapsed regions, but a filament in 2D can form with multiple parallel caustics too; for example, it can form an intricate phase-space spiral in cross-section.

'Polygonal collapse' occurs at a polygonal *node* (intersection of filaments). At a vertex of the polygon, i.e. intersection of caustics, Kawasaki's theorem (Kawasaki 1989) (both alternating sums of vertex angles add to 180°) implies that all angles between caustics are acute. Thus, nodes are convex. Each vertex also joins an even number $\geqslant 4$ of caustics. Thus, nodes cannot form in isolation; they must form in conjunction with e.g. filaments. Kawasaki's theorem, with some simple geometry, also implies that angles at which filaments come off of the node's edges must equal each other (e.g. Kawasaki (1997)). From Lagrangian to Eulerian space, polygonal collapse is a rotation of a convex polygon, producing filaments simultaneously.

Note that circular (spherical) collapse is impossible without stretching the dark-matter sheet, because caustics must be straight lines. Circular (spherical) collapse can be seen as

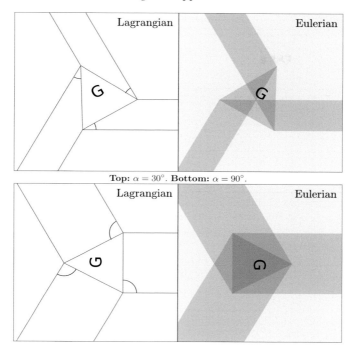

Figure 2. Triangular-collapse models, with different rotation angles α. The bottom panels show irrotational collapse, the closest analog to spherical collapse if stretching of the dark-matter sheet is allowed.

one special case of the collapse of a region, with isotropy around the node, and in which the sheet stretches substantially. Polygonal (polyhedral) collapse can be seen as another special case, with anisotropy, but no stretching.

Fig. 2 shows two examples of polygonal collapse. At left are the various regions and their caustics in Lagrangian space; at right they appear in Eulerian space. In both, the collapsing polygon is an equilateral triangle, but they differ in the angle α (indicated by arcs). From Lagrangian to Eulerian space, the triangle rotates by 2α. The bottom panels show irrotational triangular collapse, which can be considered a parity inversion of all elements, since a rotation by $180°$ is a 2D reflection. Irrotational ($\alpha = 90°$) collapse is the closest analog to spherical collapse in the origami approximation.

For a triangular node forming along with filaments, regions can exist with 1 (void), 3 (filament), 5, and 7 streams. In Lagrangian space, the node could be defined simply as the central polygon, but in Eulerian space, it is ambiguous how to delineate the node. In 2D, the ORIGAMI algorithm (Falck *et al.* 2012) (not to be confused with the present origami approximation) defines voids, filaments and nodes to have undergone collapse along 0, 1, and 2 orthogonal axes. For these particular models, ORIGAMI classifies 1, 3, and ($\geqslant 5$)-stream models as voids, filaments, and nodes, respectively. According to this definition, the central triangle is not classified as a node automatically, unless it rotates by $> 90°$ (with $\alpha > 45°$). 3-stream regions of the triangle exist in Eulerian space, classified as filament instead of node. At bottom, the entire hexagon is classified as a node. The 'outer caustic' of the node is propeller-shaped at top, and hexagonal at bottom.

The origami approximation also imposes restrictions on the topology of cosmic-web components. Since filaments must consist of straight, parallel lines, they are either infinite, or terminate at nodes. Thus, all voids are either infinite, or are convex polygons,

Lagrangian v Eul momentum, $\Sigma_{\text{streams}} v_i$ Eul vel, $(\Sigma_{\text{streams}} v_i)/N_{\text{streams}}$

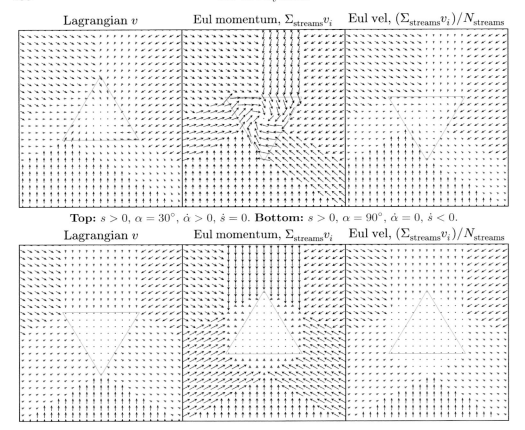

Top: $s > 0$, $\alpha = 30°$, $\dot{\alpha} > 0$, $\dot{s} = 0$. **Bottom:** $s > 0$, $\alpha = 90°$, $\dot{\alpha} = 0$, $\dot{s} < 0$.

Lagrangian v Eul momentum, $\Sigma_{\text{streams}} v_i$ Eul vel, $(\Sigma_{\text{streams}} v_i)/N_{\text{streams}}$

Figure 3. Kinematic triangular-collapse models, with different values of (the angle) α and (the scale parameter) s, and their time-derivatives. **Left:** Lagrangian space. **Middle, Right:** Eulerian space, showing both momentum and mean velocity. **Top:** purely rotational collapse, in which only s increases with time. **Bottom:** irrotational collapse, where only s increases with time. At the current snapshots, these correspond (if tilted by 90°) to the models in Fig. 2.

forming a tessellation (e.g. Gjerde 2008). This idealized property does not seem to hold in reality; we (Falck and Neyrinck 2014) found that voids in an N-body simulation generally percolate through space, not forming such convex structures. These properties generalize to 3D, as well.

3. Kinematic models

Velocity fields can be defined in a kinematic model, with time-varying model parameters. A polygonal-collapse model with fixed shape has two parameters: s, the scale of the model (e.g. the side length of the triangle); and α, the angle. Velocities may be defined from $s(t)$ and $\alpha(t)$. Thus far, we have not discussed the requirement that the dark matter sheet cannot cross itself in phase space; this may be tested by checking that at each position, velocities on all streams differ.

Fig. 3 shows two polygonal-collapse models with different sets of parameters: purely rotational, and purely irrotational models. As expected, there is obvious vorticity in the rotational model in the central node, but not elsewhere. Curiously, the mean velocities (right panels) in multistream regions are lower than in the voids. In the filaments, this

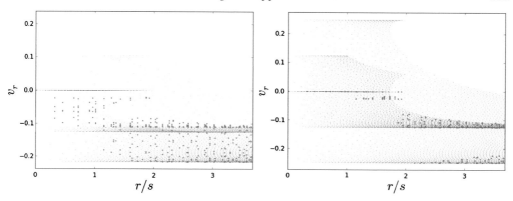

Figure 4. Radial velocities v_r in triangular-collapse models, as a function of radius r/s, where s is the incircle radius of triangle (the distance from its center to the nearest edge). Small blue dots show v_r of individual particles, often on different streams, on a fine Lagrangian mesh. Larger, red dots show v_r at the Eulerian positions of arrows in Fig. 3, averaged over streams. **Left**: rotational (corresponding to Fig. 3, top); **Right**: irrotational (corresponding to Fig. 3, bottom).

is because the velocity is the average over three streams, two of which have rather small components pointing toward the node. The Eulerian momentum, on the other hand, is large, as expected, in filaments.

One use of such velocity fields is in practically relating the positions of actual outer caustics to the velocity field around haloes. Diemer and Kravtsov (2014) have recently proposed a minimum in $v_r(r)$ to delineate outer caustics. Fig. 4 shows v_r explicitly, for the models in Fig. 3. In these plots, blue dots show v_r as a function of Eulerian separation from the node center r for particles on a regular Lagrangian lattice. Larger, red dots show stream-averaged velocities at Eulerian positions.

For $r \lesssim 3.5s$ (where s is the incircle radius of the triangular node), in both the rotational and irrotational cases, there are particles moving both inward (with negative v_r) and outward. In the rotational case, all Eulerian positions show net inward movement, whereas in the irrotational case, the interior of the node has exactly zero velocity for $r < s$. So, rather surprisingly, a rotational model displays more consistent inward movement than the irrotational model. In this idealized case of a single node, there is no transition to cosmic expansion, which is what would form a minimum in $v_r(r)$, as Diemer and Kravtsov (2014) find. This behavior might occur if a mean expansion of the sheet is included in a kinematic model. This expansion would be incorporated in a model including several neighboring nodes that remain in place in comoving space, but which grow in mass.

4. Polyhedral collapse in 3D

Finally, in Fig. 5 we make a preliminary foray into 3D, looking at the truly 3D collapse of a tetrahedral node; see Neyrinck (2014) for further details of the model. In the origami approximation, although voids and walls are irrotational, filaments can (and in general do) rotate about their axes, correlating the spins of nodes linked by filaments together. The spins of filaments intersecting at nodes are related to each other; in Fig. 5, for instance, the top filament rotates counter-clockwise by $60°$, while the smaller, bottom filaments rotate clockwise by $120°$. This can be conceptualized as a set of rods joined by gears in the central node; some of them turn one direction, and others turn oppositely.

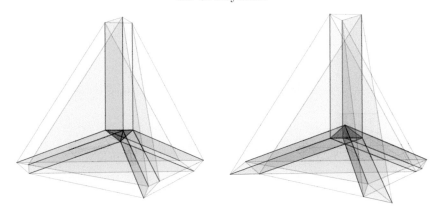

Figure 5. A rotational tetrahedral-collapse model. Filament caustics (green) are triangular tubes, intersecting at the central node. Wall caustics (blue) extend from filament edges through the thin lines drawn between filaments. Node caustics are in red. **Left:** Pre-collapse (Lagrangian). **Right:** Post-collapse (Eulerian). Walls invert along their central planes; the filaments rotate; and the central node both rotates and inverts. The top filament rotates counter-clockwise by 60°, while the smaller, bottom filaments rotate clockwise by 120°. See `http://skysrv.pha.jhu.edu/~neyrinck/TetCollapse` for an interactive model.

We will explore the quantitative restrictions this puts on neighboring cosmic-web nodes, comparing to simulations, in future work.

References

Arnold, V. I., Shandarin, S. F., & Zeldovich, I. B.: 1982, *Geophysical and Astrophysical Fluid Dynamics* **20**, 111

Demaine, E. & O'Rourke, J.: 2008, *Geometric Folding Algorithms: Linkages, Origami, Polyhedra*, Cambridge University Press

Diemer, B. & Kravtsov, A. V.: 2014, *ApJ* **789**, 1

Falck, B. & Neyrinck, M. C.: 2014, *MNRAS,* submitted, arXiv:1410.4751

Falck, B. L., Neyrinck, M. C., & Szalay, A. S.: 2012, *ApJ* **754**, 126

Gjerde, E.: 2008, *Origami tessellations: awe-inspiring geometric designs*, A K Peters

Hahn, O., Angulo, R. E., & Abel, T.: 2014, *MNRAS,* submitted, arXiv:1404.2280

Hidding, J., Shandarin, S. F., & van de Weygaert, R.: 2014, *MNRAS* **437**, 3442

Kawasaki, T.: 1989, in *Proceedings of the 1st International Meeting of Origami Science and Technology*, pp 229–237

Kawasaki, T.: 1997, in K. Miura (ed.), *Origami Science and Art: Proceedings of the Second International Meeting of Origami Science and Scientific Origami*, pp 31–40

Neyrinck, M. C.: 2014, *submitted for refereeing to the Proceedings of the 6th International Meeting on Origami in Science, Mathematics, and Education*

Pichon, C. and Bernardeau, F.: 1999, *A&A* **343**, 663

Shandarin, S. F. & Medvedev, M. V.: 2014, arXiv:1409.7634

The Zeldovich Universe:
Genesis and Growth of the Cosmic Web
Proceedings IAU Symposium No. 308, 2014
R. van de Weygaert, S. Shandarin, E. Saar & J. Einasto, eds.

© International Astronomical Union 2016
doi:10.1017/S1743921316009698

Disentangling the Cosmic Web
with Lagrangian Submanifold

Sergei F. Shandarin and Mikhail V. Medvedev

Department of Physics and Astronomy, University of Kansas, Lawrence, KS 66045

Abstract. The Cosmic Web is a complicated highly-entangled geometrical object. Remarkably it has formed from practically Gaussian initial conditions, which may be regarded as the simplest departure from exactly uniform universe in purely deterministic mapping. The full complexity of the web is revealed neither in configuration no velocity spaces considered separately. It can be fully appreciated only in six-dimensional (6D) phase space. However, studies of the phase space is complicated by the fact that every projection of it on a three-dimensional (3D) space is multivalued and contained caustics. In addition phase space is not a metric space that complicates studies of geometry. We suggest to use Lagrangian submanifold i.e., $\mathbf{x} = \mathbf{x}(\mathbf{q})$, where both $\mathbf{x}$ and $\mathbf{q}$ are 3D vectors instead of the phase space for studies the complexity of cosmic web in cosmological N-body dark matter simulations. Being fully equivalent in dynamical sense to the phase space it has an advantage of being a single valued and also metric space.

Keywords. voids, dark matter, N-body simulations, excursion set, large-scale structure

1. Introduction

Modern redshift surveys reveal an intricate 3D structure in the spatial distribution of galaxies, often called a Cosmic Web. Generic building blocks of the large-scale structure (LSS) are: halos, filaments, walls, and voids. Historically, halos have attracted the most of attention in theoretical studies of the LSS formation. From the observational point of view, halos are most closely related to galaxies, galaxy groups and clusters of galaxies, which provide most of information about our universe. However, direct modeling of galaxy formation based on fundamental laws of physics is precluded by enormous complexity of the physical processes involved, such as the highly nonlinear gravitational evolution of collisionless dark matter (DM) together with the hydrodynamical and thermal processes in baryons including star formation and the stellar wind feedback, shocks and supernovae explosions, gas accretion onto black holes in active galactic nuclei and the feedback via relativistic jets, and others. Hence various semi-empirical models of galaxy formation have been suggested, e.g., that galaxies are formed in the host DM halos of corresponding masses. The DM halos themselves are formed in a chain of mergers of smaller DM halos which may start from tiny halos of a planet mass (Diemand, *et al.* 2005). When two or more halos merge their remnants may survive for a long time as subhalos and/or streams within the resultant halo. Therefore, DM halos are likely to have a hierarchical structure resembling a Russian doll or "Matryoshka", where each subhalo includes a number of even smaller subhalos down to the smallest halos allowed by the initial power spectrum (Diemand, *et al.* 2005; Ghigna, *et al.* 1998).

In cosmological N-body simulations, DM halos were loosely defined as compact concentrations of the simulation particles in configuration space. Numerous sophisticated methods to identify halos, subhalos and other elements of the LSS have been developed over decades (see e.g. Shandarin (1983); Davis, *et al.* (1985); Vogelsberger & White (2011); Knebe, *et al.* (2013); Hoffmann, *et al.* (2014) and references therein). Most of

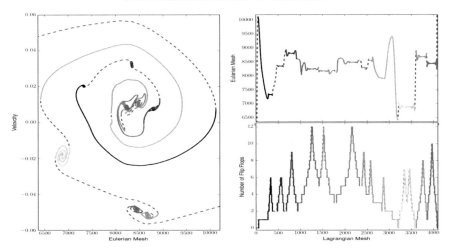

Figure 1. (*left panel*) The phase space of a one-dimensional halo simulated from random but smooth initial condition. The individual subhalos are shown by different colors. (*right panel*) Fields $x(q)$ and $n_{\mathrm{ff}}(q)$ are plotted in the top and bottom panels respectively.

the methods use only the configuration space information. However, they suffer from the projection effect that causes dynamically distinct structures in phase space to overlap in configuration space (for illustration, see the left panel of Fig. 1). Unfortunately, using all dynamical information provided by phase space is complicated by the fact that it is not a metric space (Ascasibar & Binney 2005). Here we propose an novel technique which does not suffer from all the above problems.

2. Lagrange Submanifold

We propose to identify the elements of the LSS by analyzing the mapping — the epimorphism — namely, $\mathbf{x} = \mathbf{x}(\mathbf{q}, t)$, where $\mathbf{x}$ and $\mathbf{q}$ are the coordinates of the particles in Eulerian and Lagrangian spaces respectively, (Shandarin & Medvedev 2012). Topologically, this mapping, referred to as the Lagrangian submanifold, is a 3D sheet in the the the six-dimensional $(\mathbf{q}, \mathbf{x})$ space. Our method is based on a concept of a DM sheet $\mathbf{v} = \mathbf{v}(\mathbf{x}, t)$, which major difference from the conventional interpretation of N-body simulations is in a different role of simulation particles. In contrast to the common interpretation of particles as carriers of mass, the new approach treats them as massless markers of the vertices in a tessellation of the 3D DM sheet in 6D phase space. The particles' mass is uniformly distributed inside each tetrahedra of the tessellation (Shandarin, *et al.* 2012; Abel, *et al.* 2012). Once the tessellation is built in the initial state of the simulation, it must remain intact through the whole evolution because of the Liouville's theorem, as long as the thermal velocities of the DM particles are vanishing. This requirement differs markedly from the Delaunay tessellation approach proposed by Schaap & van de Weygaert (2000) for estimating the density from particle distributions.

We stress that whereas both $(\mathbf{x}, \mathbf{v})$ and $(\mathbf{q}, \mathbf{x})$ spaces contain all the information about a dynamical system, the latter is a *metric* space and hence superior to the non-metric phase space. Moreover, the Lagrangian submanifold mapping, $\mathbf{x} = \mathbf{x}(\mathbf{q})$, is a single-valued function, unlike the phase-space mappings $\mathbf{v} = \mathbf{v}(\mathbf{x})$ or $\mathbf{x} = \mathbf{x}(\mathbf{v})$ which are multivalued.

To illustrate the above statements, Fig. 1 shows the phase space of a halo evolved in a one-dimensional (1D) universe from some smooth random initial condition. A halo can be naturally defined as the region in Eulerian space where the number of streams is greater

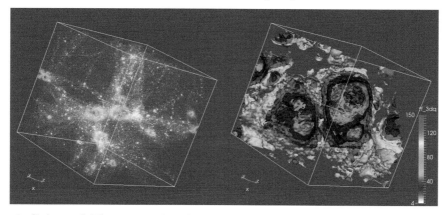

Figure 2. (*left panel*) The scatter plot of the structure in Euler space, $\mathbf{x}$, in the simulation of 1 Mpc/h box in the ΛCDM cosmology at $z = 0$. The colors from white to red correspond to the range $n_{\mathrm{ff}} \geqslant 4$. (*right panel*) The contour plot of $n_{\mathrm{ff}}(\mathbf{q})$ field in the Lagrange space, $\mathbf{q}$, cut with a 2D cross-section plane through the center of the simulation cube. The same colors are is used.

than one. The number of stream changes by two at caustics where the tangent to the phase space curve becomes vertical and the density in the corresponding stream becomes formally infinite. Subhalos and streams are shown by different colors. It is obvious from the figure that identifying individual subhalos in the configuration (Eulerian) space is difficult even in a simple 1D model due to projection effects and the presence of tidal streams. The situation with 3D simulations is even more complex.

In our approach, the initial (Lagrangian) coordinates q_i of the particles, which are in essence their IDs (which are known *exactly*, by the way), increase monotonically. However, their final (Eulerian) coordinates x_i are not monotonic, see top-right panel of Fig. 1, i.e., there are DM fluid elements with $x_{i+1} < x_i$ while $q_{i+1} > q_i$. We call a swap of the Eulerian coordinates of the two neighboring Lagrange particles as a flip-flop. In 3D, this corresponds to a formal change of the sign of the Jacobian $J(\mathbf{q}, t) = |\partial x_i / \partial q_j|$. The total number of flip-flops at a given time is shown in the bottom panel of Fig. 1. Colors show individual peaks of the flip-flop field in Lagrangian coordinates. The correspondence of the flip-flop peaks in Lagrangian space to the individual subhalos in the phase space is striking. The tidal streams and their progenitor halos are also easily, unambiguously and robustly identified via the flip-flop field, see the bottom panel of Fig. 1.

3. Implications

Now, we demonstrate our technique in real DM-only N-body simulations using an appropriately modified GADGET code (Springel 2005). The initial conditions were generated with N-GenIC code with the standard ΛCDM cosmology, $\Omega_m = 0.3, \Omega_\Lambda = 0.7, \Omega_b = 0, \sigma_8 = 0.9, h = 0.7$ and the initial redshift $z = 50$. For illustration purposes, we show here a small zoomed-in simulation with 256^3 DM particles in a $1h^{-1}$ Mpc (comoving) box. The the main purpose of this example is to demonstrate that the flip-flop field of halos in a highly nonlinear dynamic state, which still retains rich information about the substructure in haloes.

Fig. 2 (left panel) shows the map of the excursion set $n_{\mathrm{ff}}(\mathbf{q}, z = 0) \geqslant 4$ to Eulerian space and the flip-flop field in Lagrangian space. The colors — white-blue-green-red — of the particles represent the number of flip-flops. Fig. 2 (right panel) shows the corresponding flip-flop field in Lagrangian space with the same color coding. Obviously,

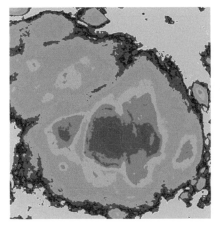

Figure 3. The contour plot of $n_\mathrm{ff}(\mathbf{q})$ field, i.e., the number of flip-flops in the Lagrange space, for the same simulation shown in Fig. 2. Right panel show the entire simulation cube and the left panel show one of six faces of the cube. Much finer contour levels demonstrate the complex "Matrioshka"-type structure of halos/subhalos and of the entire Cosmic Web.

this flip-flop field in Lagrangian space traces the distribution of matter in the universe extremely accurately. One can clearly see that halos (and the LSS, in general) form a large number of distinct flip-flop peaks in Lagrangian space. In order to reveal the much greater richness and complexity of the structure of LSS in the flip-flop field, we also plot a full 3D cube and its one face in the Lagrangian space in Fig. 3. This figure shows a complex hierarchy of peaks in much greater detail. We stress that the regions in Lagrange space with zero flip-flops are, by definition, voids. Thus, the flip-flop field $n_\mathrm{ff}(\mathbf{q}, z)$, along with the number of streams field $n_\mathrm{str}(\mathbf{x}, z)$ (Shandarin, *et al.* 2012) are the superior void detectors in N-body simulations, as they do not suffer from the poor density contrast or other issues. Note, it differs from another useful "sOrigami" method (Neyrinck 2012), as our method does not contain any free parameters. Our method is also "Diophantine" as it deals with integer numbers of flip-flops, so it does not suffer from numerical accuracy errors. Therefore, we can call our universal 'flip-flop' approach to cosmological LSS formation the "Diophantine cosmology".

References

T. Abel, O. Hahn, & R. Kaehler, *MNRAS* 427, 61 (2012).

Y. Ascasibar & J. Binney, *MNRAS* 356, 872 (2005).

M. Davis, G. Efstathiou, C. S. Frenk, & S. D. M. White, *ApJ* 292, 371 (1985).

J. Diemand, B. Moore, & J. Stadel, *Nature* 433, 389 (2005).

S. Ghigna *et al.*, *MNRAS* 300, 146 (1998).

K. Hoffmann *et al.*, *MNRAS* 442, 1197 (2014).

A. Knebe *et al.*, *MNRAS* **435**, 1618 (2013).

M. C. Neyrinck, *MNRAS* **427**, 494 (2012).

W. E. Schaap & R. van de Weygaert, *A&A* 363, L29 (2000).

S. F. Shandarin, *Soviet Astronomy Letters* 9, 104 (1983).

S. F. Shandarin, S. Habib, & K. Heitmann, *Phys. Rev. D* 385, 083005 (2012).

S. F. Shandarin & M. V. Medvedev, *submitted*; ArXiv:1409.7634 (2014).

V. Springel, *MNRAS* **364**, 1105 (2005).

M. Vogelsberger & S. D. M. White, *MNRAS* 413, 1419 (2011).

The Zeldovich Universe:
Genesis and Growth of the Cosmic Web
Proceedings IAU Symposium No. 308, 2014
R. van de Weygaert, S. Shandarin, E. Saar & J. Einasto, eds.

© International Astronomical Union 2016
doi:10.1017/S1743921316009704

Statistics of Caustics
in Large-Scale Structure Formation

Job L. Feldbrugge, Johan Hidding and Rien van de Weygaert

Kapteyn Instituut, University of Groningen,
Postbus 800, NL-9700AD, Groningen, the Netherlands
email: feldbrug@astro.rug.nl

Abstract. The cosmic web is a complex spatial pattern of walls, filaments, cluster nodes and underdense void regions. It emerged through gravitational amplification from the Gaussian primordial density field. Here we infer analytical expressions for the spatial statistics of caustics in the evolving large-scale mass distribution. In our analysis, following the quasi-linear Zel'dovich formalism and confined to the 1D and 2D situation, we compute number density and correlation properties of caustics in cosmic density fields that evolve from Gaussian primordial conditions. The analysis can be straightforwardly extended to the 3D situation. We moreover, are currently extending the approach to the non-linear regime of structure formation by including higher order Lagrangian approximations and Lagrangian effective field theory.

Keywords. Cosmology, large-scale structure, Zel'dovich approximation, catastrophe theory, caustics

1. Introduction

The large-scale structure of the universe contains clusters, filaments and walls. This cosmic web owes its intricate structure to the density fluctuations of the very early universe, as observed in the cosmic microwave background radiation field. For discussions on the observational and theoretical aspects of the cosmic web we refer to Shandarin & Zel'dovich (1989), Bond, Kofman & Pogosyan (1996), van de Weygaert & Bond (2008), Aragón-Calvo *et al.* (2010a), and Cautun *et al.* (2014).

In this paper we present a framework to analytically quantify the cosmic web in terms of the statistics of these initial density fluctuations. In a Lagrangian description of gravitational structure growth, we see that caustics form where shell-crossing is occurring. We link the caustic features of the large-scale structure to singular points and curves in the initial conditions. We subsequently study the statistics of these singularities. We will concentrate on the linear Lagrangian approximation, known as the Zel'dovich approximation (see Zel'dovich (1970)). We restrict our description to one and two spatial dimensions. It is straightforward to generalize the approach to spaces with higher dimensions. Moreover, we may forward it to the nonlinear regime by means of higher order corrections computed in Lagrangian effective field theory.

2. The Zel'dovich approximation

The gravitational collapse of small density fluctuations in an expanding universe can be modeled in multiple ways. In the Eulerian approach, we analyze the evolution of the (smoothed) density and velocity field. The resulting set of equations of motion is relatively concise, and leads to a reasonably accurate description of the mean flow field in a large fraction of space. However, due to its limitation to the mean velocity, the Eulerian formalism fails to describe the multistream structure of the matter distribution.

In the Lagrangian approach, we assume that every point in space consists of a mass element. The evolution of the mass elements is described by means of a displacement field. The Zel'dovich approximation (Zel'dovich (1970)) is the first order approximation of the displacement field of mass elements in the density field. It entails a product of a (universal) temporal factor and a spatial field of perturbations. The linear density growth factor $D_+(t)$ encapsulates all global cosmological information. The spatial component is represented by the gradient of the linearized velocity potential $\Psi(\mathbf{q})$, and expresses the effect of the initial density fluctuation field. The Zel'dovich approximation thus translates into a ballistic motion, expressed in terms of the displacement $\mathbf{s}(\mathbf{q},t)$ of a mass element with Lagrangian coordinate $\mathbf{q}$, at time t,

$$\mathbf{s}(\mathbf{q},t) = \mathbf{x}(\mathbf{q},t) - \mathbf{q} = -D_+(t)\nabla_{\mathbf{q}}\Psi(\mathbf{q}), \tag{2.1}$$

with $\mathbf{x}(\mathbf{q},t)$ the Eulerian position of the mass element. Up to linear order the linearized velocity potential $\Psi(\mathbf{q})$ is proportional to the linearly extrapolated gravitational potential at the current epoch $\phi_0(\mathbf{q})$,

$$\Psi(\mathbf{q}) = \frac{2}{3\Omega_0 H_0^2}\phi_0(\mathbf{q}), \tag{2.2}$$

with Hubble constant H_0 and density parameter Ω_0.

The Lagrangian approach is aimed at the displacement of mass elements. We can express the density in terms of the displacement fields, i.e.

$$\rho(\mathbf{x}',t) = \sum_{\mathbf{q}\in A(\mathbf{x}',t)} [\rho_u + \delta\rho_i(\mathbf{q})] \left\| \frac{\partial\mathbf{x}(\mathbf{q},t)}{\partial\mathbf{q}} \right\|^{-1}, \tag{2.3}$$

with $A(\mathbf{x}',t) = \{\mathbf{q} \mid \mathbf{x}(\mathbf{q},t) = \mathbf{x}'\}$ the pre-image of the function $\mathbf{x}(\mathbf{q},t)$, the average initial density ρ_u, and the initial density fluctuation $\delta\rho_i(\mathbf{q})$. The latter is formally equal to zero at the initial time $t = 0$, because at the recombination epoch the average density ρ_u dominates the fluctuations $\delta\rho_i(\mathbf{q})$. For this reason the density evolution according to the Zel'dovich approximation is given as

$$\rho(\mathbf{x}',t) = \sum_{\mathbf{q}\in A(\mathbf{x}',t)} \frac{\rho_u}{(1 - D_+(t)\lambda_1(\mathbf{q}))(1 - D_+(t)\lambda_2(\mathbf{q}))\dots(1 - D_+(t)\lambda_d(\mathbf{q}))}, \tag{2.4}$$

with λ_i the ordered eigenvalue fields, $\lambda_1(\mathbf{q}) \geqslant \lambda_2(\mathbf{q}) \geqslant \dots \geqslant \lambda_d(\mathbf{q})$ and corresponding eigenvector fields $\{\mathbf{n}^{\lambda_i}(\mathbf{q})\}$, of the deformation tensor

$$\psi_{ij}(\mathbf{q}) = \frac{\partial^2\Psi(\mathbf{q})}{\partial q_i \partial q_j}, \qquad i,j = 1,2,\dots,d, \tag{2.5}$$

or equivalently the tidal tensor (the Hessian of the gravitational potential ϕ_0). From this, we see that the evolution of the density field is completely determined by the eigenvalue fields of the deformation tensor of the linear velocity potential and, through eqn. 2.2, by the tidal field.

3. Caustics and catastrophe theory

The most prominent features of equation 2.4 are its poles. When at a given Lagrangian location $\mathbf{q}$ at least one of the eigenvalue fields is positive, the density ρ of that mass element can attains an infinite value at some time t. This circumstance occurs when mass elements temporarily accumulate in an infinitesimal volume. This defines a caustic. Physically we recognize these as the sites where matter streams start to cross (shell

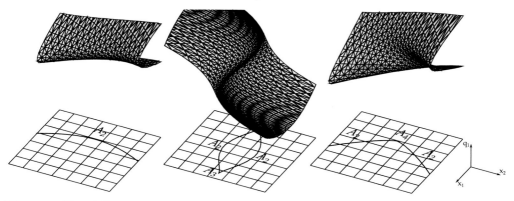

Figure 1. Visual illustrations of catastrophe classes A_2, A_3 and A_4 (Hidding *et al.* (2014)). The surfaces represent a Lagrangian manifold in the phase space $(\mathbf{q}, \mathbf{x})$ defined by the Lagrangian and Eulerian coordinates of mass elements, with on the horizontal axis the (final) Eulerian positions and on the vertical axis the (initial) Lagrangian positions. Note that we show only one of the two Lagrangian coordinates. The manifold is projected onto the Eulerian space. This results in several catastrophes. Left: an A_2 line catastrophe. Center: a Zel'dovich pancake singularity, consisting of two A_2 fold arcs and two A_3 cusp catastrophes. Right: a A_4 singularity.

crossing). In mathematics, caustics have been extensively studied and are also known as catastrophes. We can identify them with the locations in configuration space (i.e, the space of final positions of the mass elements), where the projection of the Lagrangian phase-space † submanifold $(\mathbf{q}, \mathbf{x})$ onto the configuration space leads to an infinite density by the accumulation of mass elements, i.e. where

$$\left\| \frac{\partial \mathbf{x}}{\partial \mathbf{q}} \right\| = 0 \tag{3.1}$$

For a visual appreciation of this, we refer to the illustration in figure 1.

Arnol'd (1972) developed a classification of these (Lagrangian) catastrophes up to local coordinate transformations. Here we shortly summarize the main features of this classification. In the one-dimensional Zel'dovich approximation, poles can only (stably‡) occur in two manifestations, the so-called *fold* and *cusp* catastrophes. In the classification scheme of Arnol'd, these are denotes by A_2 and A_3. In two-dimensional space, there are two additional classes of (stable) catastrophes, to a total of four catastrophe classes. These are the *swallowtail* and *umbilic* catastrophes, denoted by A_4 and D_4. Finally, in three-dimensional space, we have a total of seven catastrophe classes. These include the additional A_5, D_5 and E_5 catastrophes. To obtain a visual impression of these catastrophes, figure 1 contains illustrations of the classes A_2, A_3 and A_4. The surfaces represent a Lagrangian manifold in the phase space $(\mathbf{q}, \mathbf{x})$, defined by the Lagrangian and Eulerian coordinates of mass elements. The projection of these surfaces on to configuration space $\mathbf{x}$ reveals itself in the spatial identity of the corresponding singularities.

---

† We consider a phase space $(\mathbf{q}, \mathbf{x})$ consisting of the (initial) Lagrangian $(\mathbf{q})$ and (final) Eulerian positions $(\mathbf{x})$ of the mass elements.

‡ Stable is understood in the sense that a small fluctuation of the Lagrangian manifold does not remove the caustic, but only shifts in in space and time. Note that the singularities of highest dimension only exist at a point in time. The A_2 and A_3 singularities in figure 1 move in time while the A_4 singularity occurs in a transition of the singularity.

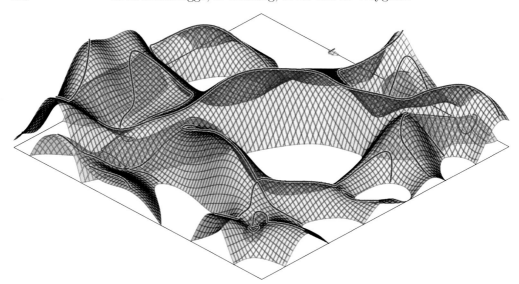

Figure 2. Iso-eigenvalue surfaces of an initial fluctuation field (in Lagrangian space). The yellow surface corresponds to the first eigenvalue λ_1 while the blue surface corresponds to the second eigenvalue field λ_2. Superimposed are a set of curves. The red and blue lines represent the A_3-lines corresponding to the first and second eigenvalue fields respectively. The eigenvalue fields of the (initial) Lagrangian density fluctuations completely determine the evolution in the Zel'dovich approximation. The caustic lines and points characterize the generation of singularities.

4. Zel'dovich and the cosmic skeleton

Arnol'd, Shandarin and Zel'dovich (1982) linked the classification of catastrophes to the geometry of the deformation tensor eigenvalue fields in the one- and two-dimensional Zel'dovich approximation. In one dimension, the fold catastrophes correspond to level crossings of the eigenvalue field λ, whereas the cusp catastrophes correspond to maxima and minima of this eigenvalue field.

4.1. Catastrophes in the 2D Zel'dovich formalism

On the basis of catastrophe theory, we may classify the caustics that arise in a cosmic density field that evolves according to the Zel'dovich approximation. The classification is based on the spatial characteristics of the field of the first and second eigenvalues λ_1 and λ_2 of the deformation tensor ψ_{ij} (eqn 2.5). Following Arnol'd *et al.* (1982), we may then identify the corresponding set of A_3-lines, and A_3, A_4 and D_4 points (see figure 2, from Hidding *et al.* (2014), for an illustration). These caustic points and curves describe the regions where infinite densities arise in the Zel'dovich approximation.

In two dimensions, the fold catastrophes correspond to isocontours of the first or second eigenvalue fields λ_1 or λ_2, known as A_2-lines defined as $A_2^i(\lambda) = \{\mathbf{q} \mid \lambda_i(\mathbf{q}) = \lambda\}$ (see figure 2). The cusp catastrophes are located on the A_2-lines at which the eigenvector $\mathbf{n}^{\lambda_i}$ is orthogonal to the gradient of the corresponding eigenvalue field,

$$\mathbf{n}^{\lambda_i}(\mathbf{q}) \cdot \nabla \lambda_i(\mathbf{q}) = 0, \qquad i = 1, 2. \tag{4.1}$$

Generically, these points form a piecewise smooth curve, known as the A_3-line defined as $A_3^i = \{\mathbf{q} \mid \mathbf{n}^{\lambda_i}(\mathbf{q}) \cdot \nabla \lambda_i(\mathbf{q}) = 0\}$ (see figure 2).

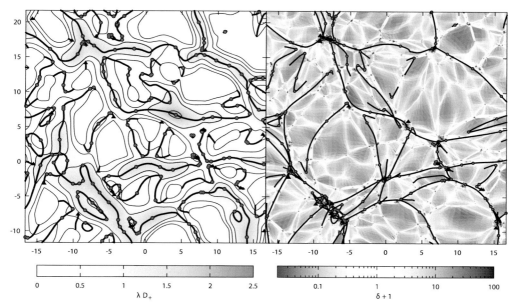

Figure 3. Spatial distribution of singularities in Lagrangian and Eulerian Cosmic Web. The figure compares the spine of the cosmic web with the mass distribution in a 2-D N-body simulation. Left panel: initial field of density fluctuations and the skeleton of identified singularities/catastrophes. Right panel: density field of an evolved 2D cosmological N-body simulation, in which the Lagrangian skeleton of singularities is mapped by means of the Zel'dovich approximation. From Feldbrugge *et al.* 2014.

Note that the caustic points on the A_3- line assume their A_3 singularity state at different times: starting at a maximum on the A_3 line, the location of the cusps moves along the line towards lower values of the eigenvalue λ_i. Hence, we observe that the maxima and saddle points of the eigenvalue field are special cusp catastrophes. In the Zel'dovich approximation, the maxima mark the points at which the first infinite densities emerge. Subsequently, as time proceeds, the caustics at the maxima become Zel'dovich pancakes consisting of two A_2 fold arcs (see figure 1). The cusps at the tips of the pancake are defined by the corresponding points on the A_3-line. Within this context, we observe the merging of two pancakes at the saddle points in the eigenvalue field.

The A_4 and D_4 catastrophes occur at the singularities of the A_3-line. The swallowtail catastrophe occurs when the tangent of the A_3-line becomes parallel to the isocontour of the corresponding eigenvalue field. The umbilic catastrophes occur in points where the two eigenvalue fields coincide. We refer to Hidding *et al.* (2014) for a more detailed study of the geometric and dynamic nature of the catastrophes in the two-dimensional Zel'dovich approximation.

4.2. *Catastrophes and the cosmic skeleton*

Although the Zel'dovich approximation is only accurate in the linear and quasi-nonlinear regime, the emerging caustics turn out to be manifest themselves in the present-day nonlinearly evolved large-scale structure. They are a proxy for the cosmic skeleton of the evolving weblike mass distribution. This can be directly appreciated from figure 3. It compares a two-dimensional dark matter N-body simulation with the distribution of caustic curves and points inferred from the Zel'dovich approximation.

The initial Gaussian random density field has a power-law power spectrum. The left panel of figure 3 contains an isocontour map of the first eigenvalue field λ_1. Superimposed on this map we find the corresponding A_3-lines and A_3 and D_4 points. The depicted caustics are the ones obtained from the initial density field filtered with a Gaussian kernel, such that the fluctuation amplitude in the linearly extrapolated density slightly exceeds unity. In this manner, the extracted skeleton corresponds to structures that have just entered the collapse phase.

The mass distribution in the right panel is the result of the gravitational growth of the initial Gaussian random field. To follow this, we evolved the initial density and velocity field by a two-dimensional dark matter N-body simulation. The resulting density field is marked by a weblike pattern of clusters, filaments and voids. Superimposed on the density map are the A_3-lines, A_3, A_4 and D_4 points, mapped from their Lagrangian towards their Eulerian location by means of the Zel'dovich approximation (eqn. 2.1).

Comparison between both panels reveals a close match between the weblike structure in the density field and the spatial distribution of the catastrophe points and lines. The point catastrophes are mostly located in the dense cluster nodes, while the A_3-lines are found to closely trace the filaments. It demonstrates the fact the spine of the cosmic web is already outlined by the catastrophes in the Gaussian initial conditions in the Lagrangian volume. In this process, the main structure remains largely intact. It forms a justification for describing the cosmic web in terms of the defining caustic singularities, as may be inferred from detailed studies of the evolving mass and halo distribution in and around the cosmic web (see e.g. Aragón-Calvo et al. (2010a), Cautun et al. (2014) and Robles (2014)).

It is also good to emphasize the dynamical nature of this definition of the spine of the cosmic web. In this, it differs from the skeleton of the cosmic web identified on the basis of structural and topological aspects of the density field (e.g. Novikov et al. (2006), Aragón-Calvo et al. (2007), Sousbie (2008), Aragón-Calvo et al. (2010b), Sousbie (2011) and Cautun et al. (2013)). On the other hand, it establishes a close link to the reecent studies by Shandarin et al. (2012), Abel et al. (2012) and Neyrinck (2012), who assessed the phase-space structure of the cosmic mass distribution to identify the various morphological elements of the cosmic web.

5. Statistics of caustics

As we argued in the previous sections, the catastrophes of the one- and two-dimensional Zel'dovich approximation can be linked to local properties of the deformation tensor eigenvalue fields in the (initial) Lagrangian density field. We assume that this density field closely resembles a Gaussian random field. So far, no counter-evidence for this (i.e., non-Gaussianities) has been found in observations of the cosmic microwave background radiation field. Moreover, the assumption follows naturally from both inflation theory and the central limit theorem.

Gaussian random fields have been extensively studied. Doroshkevich (1970) was among the first to study the eigenvalue field of Gaussian random fields. Bardeen et al. (1986) and Adler (1981) studied the statistics of Gaussian random fields near critical points (also see van de Weygaert & Bertschinger (1996), Catelan & Porciani (2001), Desjacques & Smith (2008), Rossi (2012)).

A Gaussian random field is completely characterized by the second order moment of the fluctuation field, i.e., by its autocorrelation function or, equivalently, its power spectrum. For any finite number of points $\mathbf{q}_1, \mathbf{q}_2, \ldots, \mathbf{q}_n \in \mathbb{R}^d$, the probability distribution that the density field $f : \mathbb{R}^d \to \mathbb{R}$ assumes the values $f_i \in \mathbb{R}$ in $f(\mathbf{q}_i)$ for $i = 1, 2, \dots, n$ is given by

$$p(f(t_1), \dots, f(t_n)) = \frac{\exp\left[-\frac{1}{2}\sum_{i,j} f(t_i)(M^{-1})_{ij} f(t_j)\right]}{[(2\pi)^n \det M]^{1/2}}, \tag{5.1}$$

in which the matrix elements M_{ij} express the spatial 2pt correlation of the field $f(\mathbf{q})$,

$$M_{ij} = \langle f(\mathbf{q}_i) f(\mathbf{q}_j)\rangle. \tag{5.2}$$

Using equation 5.2, we can now calculate statistical properties of the caustics in the one- and two-dimensional Zel'dovich approximation.

Here we evaluate the number density of critical points - maxima, minima and saddle points - and catastrophe points, as well as the average line length of catastrophe lines in (initial) Lagrangian space. It is straightforward to extend this description to the curves and points in Eulerian space.

In order to calculate the density of points in random fields we use Rice's formula (Rice (1944), Rice (1945)). The number density of points $\mathbf{q}$ for which a function $f = (f_1, f_2)$: $\mathbb{R}^2 \to \mathbb{R}^2$ takes a value $\mathbf{y} = (y_1, y_2) \in \mathbb{R}^2$ is given by

$$\mathcal{N} = \left\langle \delta^{(2)}(f - \mathbf{y}) \det(f_{i,j})\right\rangle = \int p(f_1 = y_1, f_2 = y_2, f_{i,j}) \det(f_{i,j}) \mathrm{d}f_{1,1} \dots \mathrm{d}f_{2,2}, \tag{5.3}$$

with $f_{i,j} = \frac{\partial f_i}{\partial q_j}$. By using the properties of the eigenvalue fields at the caustics, the density of the A_2 fold and A_3 cusp catastrophes in the one-dimensional case can be expressed as

$$\mathcal{N}_{A_2}(\alpha) = \int |\lambda_{1,1}| p(\lambda_1 = \alpha, \lambda_{1,1}) \mathrm{d}\lambda_{1,1}, \tag{5.4}$$

$$\mathcal{N}_{A_3}(\alpha) = \int |\lambda_{1,11}| p(\lambda_1 = \alpha, \lambda_{1,1} = 0, \lambda_{1,11}) \mathrm{d}\lambda_{1,11}. \tag{5.5}$$

Note that in equation 5.4 we integrate over $\mathbb{R}$ whereas in equation 5.5 we integrate over $(-\infty, 0)$ for the maxima and $(0, \infty)$ for the minima of λ_1. In the two-dimensional Zel'dovich approximation, the density of D_4 points can be computed by evaluating the integral

$$\mathcal{N}_{D_4}(\alpha) = \int |\lambda_{1,1}\lambda_{2,2} - \lambda_{1,2}\lambda_{2,1}| p(\lambda_1 = \alpha, \lambda_2 = \alpha, \lambda_{1,1}, \lambda_{1,2}, \lambda_{2,1}, \lambda_{2,2}) \mathrm{d}\lambda_{1,1} \mathrm{d}\lambda_{1,2} \mathrm{d}\lambda_{2,1} \mathrm{d}\lambda_{2,2}. \tag{5.6}$$

The density of A_3 and A_4 points in the two-dimensional case are obtained in an analogous fashion.

For the A_3-lines in the two-dimensional case, we study the curve length density, by adapting the statistical analysis of Longuet-Higgins (1957). The average length of the iso-contour of a function $f : \mathbb{R}^2 \to \mathbb{R}$ at level y is given by

$$\mathcal{L}(y) = \left\langle \delta^{(1)}(f - y)\sqrt{f_{,1}^2 + f_{,2}^2}\right\rangle = \int p(f = y, f_{,1}, f_{,2})\sqrt{f_{,1}^2 + f_{,2}^2}\ \mathrm{d}f_{,1}\mathrm{d}f_{,2}. \tag{5.7}$$

By using applying the eigenvalue conditions of the A_3-lines, we obtain the differential A_3-line length with respect to the first eigenvalue field λ_1,

$$\mathcal{L}_{A_3}(\lambda_1) = \pi \int \sqrt{\lambda_{1,11}^2 + \lambda_{1,12}^2}\ p(\lambda_1, \lambda_2, \lambda_{1,1} = 0, \lambda_{1,11}, \lambda_{1,12})(\lambda_1 - \lambda_2)\mathrm{d}\lambda_{1,11}\mathrm{d}\lambda_{1,12}\mathrm{d}\lambda_2. \tag{5.8}$$

Other local properties, such as the curvature or the correlation function between caustics, can be determined analogously.

6. Conclusion

Here we have presented a formalism to describe the spatial statistics of caustics in the one- and two-dimensional Zel'dovich approximation, for a given power spectrum of the initial random Gaussian density field. The visual comparison of the spatial distribution of these caustics with the pattern of the mass distribution in N-body simulations demonstrates that the caustics define the spine of the cosmic web. It reflects the strong correspondence between catastrophe lines and points and the emerging weblike structures in the cosmic mass distribution. In other words, the skeleton of the cosmic web appears to be defined by the spatial properties of the tidal force and deformation field in the initial Gaussian mass distribution.

It is straightforward to extend the one- and two-dimensional formalism presented here to three dimensions. Moreover, currently we are looking into how to extend the formalism to more advanced stages of dynamical evolution, using Lagrangian effective field theory.

References

Abel, T. , Hahn, O., & Kaehler, R., 2012, *MNRAS*, 427, 61

&Adler, R. J., 1981, *The Geometry of Random Fields*, Wiley

Adler, R. J., & Taylor, J. E., 2007, *Random Fields and Geometry*, Springer

Aragón-Calvo, M. A., Jones, B. J. T., van de Weygaert,R., & van der Hulst, J. M., 2007, *Astron. Astrophys*, 474, 315

Aragón-Calvo, M. A., van de Weygaert, R. , & Jones, B. J. T.., 2010, *MNRAS*, 408, 2163

Aragón-Calvo, M. A., Platen, E., van de Weygaert,R., & Szalay, A. S., 2010, *Astrophys. J.*, 723, 364

Arnol'd, V. I., 1972, *Funct. Anal. and its Appl.*, 6, 254

Arnol'd, V. I., Shandarin, S. F., & Zeldovich, Ia.B., 1982, *Geophys. and Astrophys. Fluid Dynamics*, 20, no. 1-2, 111

Bardeen, J. M., Bond, J. R., Kaiser, N., & Szalay, A. S., 1986, *Astrophys. J.*, 304, 15

Bond, J. R., Kofman, L., & Pogosyan, D., 1996, *Nature*, 380, 603

Catelan, P. & Porciani, C., 2001, *MNRAS*, 323, 713

Cautun, M., van de Weygaert, R. , & Jones, 2013, *MNRAS*, 429, 1286

Cautun, M., van de Weygaert, R. , Jones, B. J. T., & Frenk, C. S., 2014, *MNRAS*, 441, 2923

Desjacques, V. & Smith, R. E., 2008, *Phys. Rev. D*, 78, 023527

&Doroshkevich, A. G., 1970, *Astrophysics*, 6, Issue 4, 320

Hidding, J., Shandarin, S. F., & van de Weygaert, R., 2014, *MNRAS*, Vol. 437, , 3442

&Longuet-Higgins, M. S., 1957, *Philosophical Trans. of the Royal Society of London* Vol. 249, 321

&Neyrinck, M., 2012, *MNRAS*, 427, 494

Novikov, D., Colombi, S., & Doré, O., 2006, *MNRAS*, 366, 1201

&Rice, S. O., 1944, *Bell Systems Tech. J.*, 23, 282

&Rice, S. O., 1945, *Bell Systems Tech. J.*, 24, 46

Robles, S., Domínguez-Tenreiro R., Oñorbe J., & Martínez-Serrano F., 2014, *NMRAS*, subm.

Rossi, G., 2012, *MNRAS*, 421, 296

Shandarin, S. F., Habib, S., & Heitmann, K., 2012, *Phys. Rev. D*, 85, 083005

Sousbie, T., Pichon, C., Colombi, S., Novikov, D., & Pogosyan, D., 2008, *MNRAS*, 383, 1655

Sousbie, T., 2011, *MNRAS*, 414, 350

Shandarin, S. F. & Zel'dovich, Ia.B., 1989, *Rev. Mod. Phys.*, 61, 185

van de Weygaert, R. & Bertschinger, E., 1996, *MNRAS*, 281, 84

van de Weygaert, R. & Bond, J. R., 2008, *Lecture Notes in Physics*, 740, 335

Zel'dovich Ia.B., 1970, *Astron. Astrophys.*, 5, 84

The Zeldovich Universe:
Genesis and Growth of the Cosmic Web
Proceedings IAU Symposium No. 308, 2014
R. van de Weygaert, S. Shandarin, E. Saar & J. Einasto, eds.

© International Astronomical Union 2016
doi:10.1017/S1743921316009716

Beyond single-stream
with the Schrödinger method

Cora Uhlemann and Michael Kopp

Excellence Cluster Universe, Boltzmannstr. 2, D-85748 Garching
Arnold Sommerfeld Center for Theoretical Physics, LMU, Theresienstr. 37, D-80333 Munich

Abstract. We investigate large scale structure formation of collisionless dark matter in the phase space description based on the Vlasov-Poisson equation. We present the Schrödinger method, originally proposed by Widrow and Kaiser, 1993 as numerical technique based on the Schrödinger Poisson equation, as an analytical tool which is superior to the common standard pressureless fluid model. Whereas the dust model fails and develops singularities at shell crossing the Schrödinger method encompasses multi-streaming and even virialization.

Keywords. large scale structure, cold dark matter, cosmic web, halo formation

---

Introduction. The standard model of large-scale structure and halo formation is based on collisionless cold dark matter (CDM), a particle species that for this purpose can be assumed to interact only gravitationally and to be cold or initially single-streaming. We are therefore interested in the dynamics of a large collection of identical point particles that via gravitational instability evolve from initially small density perturbations into eventually bound structures, like halos that are distributed along the cosmic web.

1. Phase space description of cold dark matter

The dynamics of CDM with mass m is described by the one-particle phase space density $f(\boldsymbol{x}, \boldsymbol{p}, t)$ which fulfills the Vlasov-Poisson equation

$$\partial_t f = -\frac{\boldsymbol{p}}{a^2 m} \cdot \boldsymbol{\nabla}_x f + m \boldsymbol{\nabla}_x V \cdot \boldsymbol{\nabla}_p f \quad , \quad \Delta V = \frac{4\pi G \rho_0}{a} \left(\int d^3 p \, f - 1 \right). \tag{1.1}$$

This description is valid in the absence of irreducible two-body correlations, which is the case for a smooth matter distribution. The cosmology dependence is encoded in the scale factor a, today's background matter density ρ_0 and the initial conditions f_{ini}.

Cumulants. The cumulants $C^{(n)}$ of the phase space distribution are in practice the physical quantities of interest – observationally accessible via redshift space distortions and peculiar velocities or numerically determinable from N-body simulations. They encode the number density $n(\boldsymbol{x}) = \exp C^{(0)}$, the velocity $u_i(\boldsymbol{x}) = C_i^{(1)}$ and the velocity dispersion $\sigma_{ij} = C_{ij}^{(2)}$. They can be calculated from f as

$$G[\boldsymbol{J}] = \int d^3 p \, \exp\left[i \boldsymbol{p} \cdot \boldsymbol{J}\right] f \quad , \quad C_{i_1 \cdots i_n}^{(n)} := (-i)^n \left. \frac{\partial^n \ln G[\boldsymbol{J}]}{\partial J_{i_1} \ldots \partial J_{i_n}} \right|_{\boldsymbol{J}=0}. \tag{1.2}$$

For a general f it is impossible to perform the integration over momentum analytically. Therefore we would like to resort to an ansatz with a specific $\boldsymbol{p}$-dependence.

Vlasov hierarchy. The Vlasov hierarchy is constituted by the evolution equations for the cumulants $C^{(n)}$ determined from the Vlasov equation (1.1)

$$\partial_t C^{(n)} = -\frac{1}{a^2 m} \left\{ \nabla \cdot C^{(n+1)} + \sum_{|S|=0}^{n} C^{(n+1-|S|)} \cdot \nabla C^{(|S|)} \right\} - \delta_{n1} m \nabla V. \tag{1.3}$$

	Schrödinger method $f_{\mathrm{H}}(\boldsymbol{x},\boldsymbol{p},t)$	dust model $f_{\mathrm{d}}(\boldsymbol{x},\boldsymbol{p},t)$
degrees of freedom	$1\times\mathbb{C}\colon \psi=\sqrt{n}\exp[i\phi/\hbar]$	$2\times\mathbb{R}\colon n_{\mathrm{d}},\phi_{\mathrm{d}}$
equations of motion	Schrödinger-Poisson equation (2.1)	fluid equations (1.5)
Vlasov equation (1.1) solved	approximately $(\hbar,\sigma_x)$	exactly
shell-crossing	well-behaved	singularities
multi-streaming, virialization	✓, ✓	✗, ✗
closed-form cumulants	✓, $C^{(n\geqslant2)}\neq0$	(✓), $C_{\mathrm{d}}^{(n\geqslant2)}\equiv0$

Table 1. Comparison between the Schrödinger method and the dust model

It is an infinite coupled hierarchy: in order to determine the time-evolution of the n-th cumulant, the $(n+1)$-th is required. The dust model with $C^{(n\geqslant2)}=0$ is the only consistent truncation but fails after shell-crossing where all cumulants become equally important as demonstrated by [Pueblas and Scoccimarro, 2009].

Dust model. The dust model is an ansatz for the phase-space distribution function with trivial $\boldsymbol{p}$-dependence

$$f_{\mathrm{d}}(\boldsymbol{x},\boldsymbol{p},t)=n_{\mathrm{d}}(\boldsymbol{x},t)\delta\Big(\boldsymbol{p}-\boldsymbol{\nabla}\phi_{\mathrm{d}}(\boldsymbol{x},t)\Big). \tag{1.4}$$

The cumulants are given by the density $C_{\mathrm{d}}^{(0)}=\ln n_{\mathrm{d}}(\boldsymbol{x},t)$ and a curl-free velocity $\boldsymbol{C}_{\mathrm{d}}^{(1)}=\boldsymbol{\nabla}\phi_{\mathrm{d}}(\boldsymbol{x},t)/m$ since all higher cumulants vanish identically $C_{\mathrm{d}}^{(n\geqslant2)}=0$. Therefore solving the Vlasov equation for f_{d} is equivalent to solving the coupled fluid system consisting of continuity and Bernoulli equation for n_{d} and ϕ_{d}.

$$\partial_t n_{\mathrm{d}}=-\frac{\boldsymbol{\nabla}\cdot(n_{\mathrm{d}}\boldsymbol{\nabla}\phi_{\mathrm{d}})}{a^2 m}\ ,\ \ \partial_t\phi_{\mathrm{d}}=-\frac{1}{2}\frac{(\boldsymbol{\nabla}\phi_{\mathrm{d}})^2}{a^2 m}-mV_{\mathrm{d}}\ ,\ \ \Delta V_{\mathrm{d}}=\frac{4\pi G\,\rho_0}{a}\Big(n_{\mathrm{d}}-1\Big). \tag{1.5}$$

2. The Schrödinger method

The Schrödinger method (ScM), originally proposed by [Widrow and Kaiser, 1993] as numerical technique to study CDM dynamics, is a special ansatz for the distribution function that is based on the Schrödinger Poisson equation

$$i\hbar\partial_t\psi=-\frac{\hbar^2}{2a^2 m}\Delta\psi+mV(\boldsymbol{x})\psi\ \ ,\ \ \ \Delta V=\frac{4\pi G\,\rho_0}{a}\left(|\psi|^2-1\right). \tag{2.1}$$

If a wavefunction ψ fulfils (2.1) then the Husimi distribution function f_{H}

$$f_{\mathrm{H}}(\boldsymbol{x},\boldsymbol{p},t)=N\left\{\int d^3y\ \exp\left[-\frac{(\boldsymbol{x}-\boldsymbol{y})^2}{4\sigma_x{}^2}-\frac{i}{\hbar}\boldsymbol{p}\cdot\left(\boldsymbol{y}-\frac{1}{2}\boldsymbol{x}\right)\right]\psi(\boldsymbol{y},t)\right\}^2, \tag{2.2}$$

where $\hbar$ and σ_x are free parameters and $N(\hbar,\sigma_x)$ is a normalization constant, approximately fulfils the coarse grained Vlasov equation [Takahashi, 1989] obtained by a Gaussian smoothing of f over σ_x and $\sigma_p=\hbar/(2\sigma_x)$. Therefore, physical processes taking place at scales larger than σ_x and σ_p can be modeled with arbitrary precision.

Key features. The advantages of the Schrödinger method compared to the standard dust model are summarized in Tab. 1. The special $\boldsymbol{p}$-dependence of (2.2) allows to compute cumulants analytically. All cumulants are nonzero and can be expressed as Gaussian smoothed functions of n and $\boldsymbol{\nabla}\phi$ and their derivatives which allows for closing the Vlasov hierarchy, see [Uhlemann, Kopp and Haugg, 2014].

Numerical example: Pancake collapse. In Fig. 1 we show the standard toy example of plane parallel (or pancake) collapse, whose exact solution in the case of dust is given by

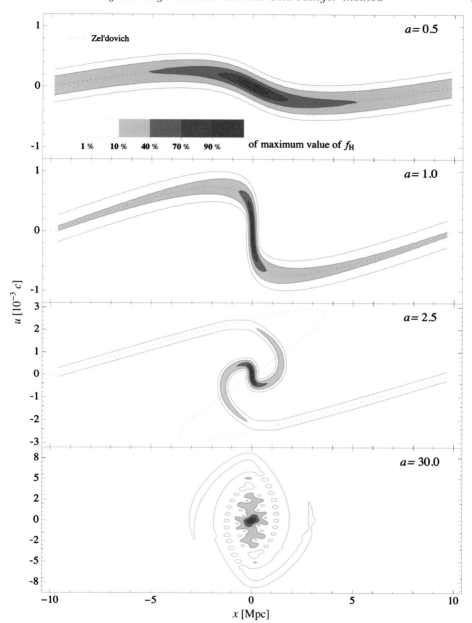

Figure 1. *shaded* Schrödinger method phase space density $f_{\rm H}$, *dotted* exact dust solution.

the Zel'dovich approximation [Zel'dovich, 1970]. We therefore have analytic expressions for $n_{\rm d}$ and $\phi_{\rm d}$. Nearly cold initial conditions can be implemented by choosing the initial wave function at some early time where shell crossings have not yet occurred as

$$\psi_{\rm ini}(x) = \sqrt{n_{\rm d}(a_{\rm ini}, x)} \exp\left[i\phi_{\rm d}(a_{\rm ini}, x)/\hbar\right] . \tag{2.3}$$

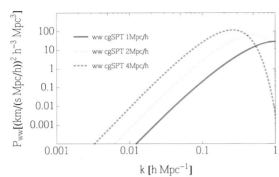

The coarse-graining naturally leads to a mass-weighted velocity thereby generating large-scale vorticity which is also observed in N-body measurements [Hahn, Angulo and Abel, 2014] optimizing the agreement for a smoothing scale of $\sigma_x = 1\,\mathrm{Mpc}$.

Figure 2. Power spectrum of vorticity $\boldsymbol{w} = \boldsymbol{\nabla} \times \boldsymbol{v}$ in 1-loop Eulerian perturbation theory for coarse-grained dust (cgSPT) and three different smoothing scales.

3. Coarse-grained dust model

The coarse-grained dust model studied in [Uhlemann and Kopp, 2014] is limiting case of the Schrödinger method when $\hbar \to 0$ given by

$$\bar{f}_{\mathrm{d}}(\boldsymbol{x}, \boldsymbol{p}) = \int \frac{d^3 x' d^3 p'}{(2\pi\sigma_x\sigma_p)^3} \exp\left[-\frac{(\boldsymbol{x} - \boldsymbol{x}')^2}{2\sigma_x{}^2} - \frac{(\boldsymbol{p} - \boldsymbol{p}')^2}{2\sigma_p{}^2} \right] f_{\mathrm{d}}(\tilde{\boldsymbol{x}}, \tilde{\boldsymbol{p}}). \qquad (3.1)$$

It is much closer to the distribution extracted from N-body simulations, which necessarily involves averaging over phase space cells of width σ_x and σ_p. Indeed, implementing the coarse-graining in this way results in a resummation in the large scale parameter of the macroscopic model suggested by [Dominguez, 2000] when the corresponding fluid-type equations are expressed in terms of coarse grained quantities.

4. Prospects

Correlation functions of the phase space density are necessary for analyzing observations of large scale structure. Of particular interest is the 2-point correlation function in redshift space $1 + \xi(\boldsymbol{s}) = \langle (1 + \delta(\boldsymbol{s}_1))(1 + \delta(\boldsymbol{s}_2)) \rangle$ for biased tracers, like halos or galaxies, relevant to observations made in galaxy surveys. This is investigated for the coarse-grained dust model in [Kopp and Uhlemann et al., 2014].

The universality of halo density profiles may be understood by determining stationary complex solutions of the Schrödinger-Poisson equation. Since the Schrödinger method allows for virialization, it could prove useful in further analytical understanding of violent relaxation [Lynden-Bell, 1967] that leads to universal density profiles [Navarro, Frenk and White, 1997]. These properties might be derived from an entropy principle for collisionless self-gravitating systems as described in [He, 2012].

References

Dominguez, 2000, *Phys.Rev.*, D62:103501.
Hahn, Angulo & Abel, 2014, arXiv:1404.2280.
He, 2012, *MNRAS*, 419:1667–1681, 1103.5730.
Kopp & Uhlemann *et al.*, 2014, in preparation.
Lynden-Bell, 1967, *MNRAS*, 136:101.
Melott, Pellman & Shandarin, 1994, *MNRAS*, 269:626, arXiv:astro-ph/9312044.
Navarro, Frenk & White, *ApJ*, 490:493, arXiv:astro-ph/9611107.
Pueblas & Scoccimarro, 2009, *Phys.Rev.*, D80:043504, arXiv:0809.4606.
Takahashi, 1989, *Progress of Theoretical Physics Supplement*, 98:109–156.
Uhlemann & Kopp, 2014, arXiv:1407.4810.
Uhlemann, Kopp & Haugg, 2014, *Phys.Rev.*, D90:023517, arXiv:1403.5567.
Widrow & Kaiser, 1993, *Ap. Lett.*, 416:L71.
Zel'dovich, 1970, *A&A*, 5:84–89.

The Zeldovich Universe:
Genesis and Growth of the Cosmic Web
Proceedings IAU Symposium No. 308, 2014
R. van de Weygaert, S. Shandarin, E. Saar & J. Einasto, eds.

© International Astronomical Union 2016
doi:10.1017/S1743921316009728

Higher-order Lagrangian perturbative theory for the Cosmic Web

Takayuki Tatekawa[1,2] and Shuntaro Mizuno[3]

[1] Center for Infromation Initiative, University of Fukui,
3-9-1 Bunkyo, Fukui, Fukui, 910-8507, Japan
email: `tatekawa@u-fukui.ac.jp`

[2] Research Institute for Science and Engineering, Waseda University,
3-4-1 Okubo, Shinjuku, Tokyo, 169-8555, Japan

[3] Waseda Institute for Advanced Study, Waseda University,
1-6-1 Nishi-Waseda, Shinjuku, Tokyo, 169-8050, Japan

Abstract. Zel'dovich proposed Lagrangian perturbation theory (LPT) for structure formation in the Universe. After this, higher-order perturbative equations have been derived. Recently fourth-order LPT (4LPT) have been derived by two group. We have shown fifth-order LPT (5LPT) In this conference, we notice fourth- and more higher-order perturbative equations. In fourth-order perturbation, because of the difference in handling of spatial derivative, there are two groups of equations. Then we consider the initial conditions for cosmological N-body simulations. Crocce, Pueblas, and Scoccimarro (2007) noticed that second-order perturbation theory (2LPT) is required for accuracy of several percents. We verify the effect of 3LPT initial condition for the simulations. Finally we discuss the way of further improving approach and future applications of LPTs.

Keywords. large-scale structure of universe, methods: analytical

1. Derivation of 5LPT equations and their fitting formulae

The formation of the cosmic web is one of most important problems in modern cosmology. It is considered that the primordial density fluctuation grows by its-self gravitational instability. As one of theoretical approach for the evolution, Lagrangian perturbation has been considered for long time. At first, Zel'dovich (1970) proposed Lagrangian perturbation theory (LPT). Although the perturbation remains linear stage, quasi nonlinear evolution of the density fluctuation is described well. In 1990s, second- and third-order LPTs had been derived (Bouchet *et al.* (1992), Buchert (1992), Buchert & Ehlers (1993), Buchert (1994), Bouchet *et al.* (1995), Sasaki & Kasai (1998)). Recently, fourth-order LPT (4LPT) have been derived (Rampf & Buchert (2012), Tatekawa (2013)). Furthermore, the recursive formula for the derivation of higher-order LPT have been proposed (Rampf (2012)).

In 4LPT, because of difference of treatment for spatial derivative, there are two groups of perturbative equations. In this conference, we have derived perturbative equations for 5LPT based on the procedure by Rampf & Buchert (2012). 5LPT equations consist of 15 longitudinal modes and 11 transverse modes.

In Einstein-de Sitter Universe model, we can derive analytical form for temporal parts up to fifth-order LPT. However in LCDM model, we can derive analytical form for temporal parts only in 1LPT. Therefore we consider fitting formula for temporal parts. Peebles (1984) and Bouchet *et al.* (1995) derived fitting formula up to third-order LPT. We have tried to improve past formulae and derive new formula up to 5LPT. First, we define logarithmic derivative of the temporal parts:

$$f_n \equiv \frac{a}{g_n} \frac{\mathrm{d}g_n}{\mathrm{d}a} \,, \tag{1.1}$$

where a means the scale factor. g_n means the temporal parts in n-th order LPT. Following past formulae, we assume the formula in nLPT as

$$f_n \simeq n\Omega_M^\alpha \,, \tag{1.2}$$

where Ω_M means the density parameter. For derivation, we apply least-squares method for derivation of new formula in $0.1 \leqslant \Omega_M \leqslant 1$. In preliminary results, the formulas are seldom improved. We should consider other approach for the derivation of the fitting formulae.

Before considering application of 5LPT, we should verify two groups of 4LPT equations (Rampf & Buchert (2012), Tatekawa (2013)). After verification, we will consider 3-loop correction for the power spectrum using 5LPT solutions.

2. Initial condition problem for cosmological N-body simulations

For analyses in strongly nonlinear stage, cosmological N-body simulations have been carried out. As initial conditions for the simulations, 1LPT has been applied for long time (for example, Ma & Bertschinger (1995)). Recently, several groups point out that the higher-order LPT requires for the initial condition. Crocce, Pueblas, & Scoccimarro (2006) pointed out if 2LPT is ignored in the initial conditions, the statistical quantities such as non-Gaussianity in the density field at low-z region deviate with several percent.

A new question arises here. Even if the effect of 2LPT seems important for late time, we cannot decide whether 2LPT initial condition is enough or not. If the effect of 3LPT initial condition is negligible, 2LPT initial condition is enough for cosmological N-body simulations. We have analyzed the effect of 3LPT initial conditions (Tatekawa & Mizuno(2007), Tatekawa (2014)).

By the analysis of the evolution for the power spectrum and non-Gaussianity in the density field, we conclude when we require sub percent accuracy, we should consider 3LPT initial conditions. Then the effect of the transverse mode in 3LPT seems negligible.

References

Buchert, T. 1992, *Mon. Not. R. Astron. Soc.*, 254, 729
Barrow, J. D. & Saich, P. 1993, *Class. Quantum Grav.*, 10, 79
Bouchet, F. R., Juszkiewicz, R., Colombi, S., & Pellat, R. 1992, *Astrophys. J.*, 394, L5
Buchert, T. & Ehlers, J. 1993, *Mon. Not. R. Astron. Soc.*, 264, 375
Buchert, T. 1994, *Mon. Not. R. Astron. Soc.*, 267, 811
Bouchet, F. R., Colombi, S., Hivon, E., & Juszkiewicz, R. 1995 *Astron. Astrophys.*, 296, 575
Catelan, P. 1995, *Mon. Not. R. Astron. Soc.*, 276, 115
Crocce, M., Pueblas, S., & Scoccimarro, R. 2006 *Mon. Not. R. Astron. Soc.*, 373, 369
Ma, C.-P. & Bertschinger, E. 1995 *Astrophys. J.*, 455, 7
Peebles, P. J. E. 1984, *Astrophys. J.*, 284, 439
Rampf, C. & Buchert, T. 2012, *J. Cosmol. Astropart. Phys.*, 06, 021
Rampf, C. 2012, *J. Cosmol. Astropart. Phys.*, 12, 004
Sasaki, M. & Kasai, M. 1998 *Prog. Theor. Phys.*, 99, 585
Tatekawa, T. & Mizuno, S. 2007, *J. Cosmol. Astropart. Phys.* **12** 014
Tatekawa, T. 2013, *Prog. Theor. Exp. Phys.*, 013E03
Tatekawa, T. 2014, *J. Cosmol. Astropart. Phys.*, 04, 025
Zel'dovich, Ya. B. 1970, *Astron. Astrophys.*, 5, 84

The Zeldovich Universe:
Genesis and Growth of the Cosmic Web
Proceedings IAU Symposium No. 308, 2014
R. van de Weygaert, S. Shandarin, E. Saar & J. Einasto, eds.

© International Astronomical Union 2016
doi:10.1017/S174392131600973X

Non-linear description of massive neutrinos in the framework of large-scale structure formation

Hélène Dupuy[1,2]

[1]Institut de Physique Théorique, CEA, IPhT, URA 2306 du CNRS, F-91191 Gif-sur-Yvette, France
[2]Institut d'Astrophysique de Paris, UMR 7095 du CNRS, Université Pierre et Marie Curie, 98 bis bd Arago, 75014 Paris, France
email: `helene.dupuy@cea.fr` ; `dupuy@iap.fr`

Abstract. There is now no doubt that neutrinos are massive particles fully involved in the non-linear growth of the large-scale structure of the universe. A problem is that they are particularly difficult to include in cosmological models because the equations describing their behavior in the non-linear regime are cumbersome and difficult to handle. In this manuscript I present a new method allowing to deal with massive neutrinos in a very simple way, based on basic conservation laws. This method is still valid in the non-linear regime. The key idea is to describe neutrinos as a collection of single-flow fluids instead of seeing them as a single hot multi-flow fluid. In this framework, the time evolution of neutrinos is encoded in fluid equations describing macroscopic fields, just as what is done for cold dark matter. Although valid up to shell-crossing only, this approach is a further step towards a fully non-linear treatment of the dynamical evolution of neutrinos in the framework of large-scale structure growth.

Keywords. cosmology: theory and large-scale structure of universe, neutrinos

1. Introduction

Observational cosmology has been particularly fruitful recently, as evidenced by the great success of the Planck mission (Planck Collaboration 2013). From the theoretical point of view, such an achievement naturally opens perspectives. Any component of the universe involved in large-scale structure formation has to be examined minutely. In particular, the discovery of the massiveness of neutrinos has triggered a considerable effort in theoretical and numerical cosmology to infer the impact of this mass on the evolution of cosmological perturbations.

The first study in which massive neutrinos are properly treated in the linear theory of gravitational perturbations dates back from 1994 (Ma and Bertschinger 1994). Then the connection between neutrino masses and cosmology has been investigated in full detail (Lesgourgues and Pastor 2006). To a large extent, current and future cosmology projects aim at exploiting this connection to put constraints on neutrino masses. Such surveys are sensitive to the non–linear growth of structure, whence the importance of studying the non-linear regime in cosmological perturbation theory. In this manuscript, I summarize the results of a study in which the authors design analytical tools to explore the impact of massive neutrinos on large-scale structure growth within the non-linear regime (Dupuy and Bernardeau 2014).

2. The standard linear description of neutrinos

The strategy usually adopted to describe neutrinos, massive or not, is calqued on that used to describe the radiation fluid: neutrinos are considered as a single hot multi-stream fluid whose evolution is dictated by the behavior of its distribution function in phase-space f. The key equation is the Boltzmann equation. For neutrinos, contrary to radiation, it is taken in the collisionless limit since neutrinos do not interact with ordinary matter (neither at the time of recombination nor after). It leads to the Vlasov equation, $\frac{df}{d\eta} = 0$, where η is a time coordinate.

The standard linear description of neutrinos is based i) on a decomposition of the phase-space distribution function into a homogeneous part and a first-order inhomogeneous contribution in the Vlasov equation and ii) on a decomposition of the latter into harmonic functions. It leads to the standard Boltzmann hierarchy. Finally, relevant physical quantities can be built out of the coefficients of the harmonic decomposition (see e.g. Ma and Bertschinger 1994). This approach is powerful and largely used in the litterature. But, unfortunately, it is valid in the linear regime only.

To get a non-linear description, which is essential to do precision cosmology, one can e.g compute the fully non-linear moments of the Vlasov equation (Van de Rijt 2012). Although interesting because fully non-linear, the resulting hierarchy is unfortunately very difficult to handle. As a matter of fact, the non-linear equations one can find in the litterature to describe neutrinos are generally not exploitable because of their complexity. An analytical description of neutrinos in the non-linear regime is therefore still missing.

3. The non-linear description of cold dark matter, a model to follow

Cold dark matter is described as a collection of identical point particles, non-relativistic and sensitive to gravitational interaction only (for notations and more details, see Bernardeau 2013). Using the Newtonian approximation, its time evolution is encoded in the Vlasov-Poisson system, which leads to the continuity and Euler equations

$$\frac{\partial \delta(\mathbf{x},t)}{\partial t} + \frac{1}{a}[(1 + \delta(\mathbf{x},t))u_i(\mathbf{x},t)]_{,i} = 0, \tag{3.1}$$

$$\frac{\partial u_i(\mathbf{x},t)}{\partial t} + \frac{\dot{a}}{a}u_i(\mathbf{x},t) + \frac{1}{a}u_j(\mathbf{x},t)u_i(\mathbf{x},t)_{,j} = -\frac{1}{a}\Phi(\mathbf{x},t)_{,i} - \frac{(\rho(\mathbf{x},t)\sigma_{ij}(\mathbf{x},t))_{,j}}{a\rho(\mathbf{x},t)}. \tag{3.2}$$

A major simplification comes from *the single-flow approximation*, which consists in neglecting the velocity dispersion term σ_{ij} because, in a cold fluid, it is extremely small compared to the velocity gradients induced by density fluctuations. This assumption breaks down as soon as shell-crossing starts. Very little analytical results are known in presence of shell-crossing so description is then purely numerical.

Besides, the Euler equation shows that, in the single-flow approximation, the velocity field is a gradient. Equations can thus be much simplified when introducing the velocity divergence θ. Those two specificities of cold dark matter allow to write the motion equations on a compact form (see Bernardeau 2013 for notations),

$$\frac{\partial \Psi_a(\mathbf{k},\eta)}{\partial \eta} + \Omega_a{}^b(\eta)\Psi_b(\mathbf{k},\eta) = \gamma_a{}^{bc}(\mathbf{k}_1,\mathbf{k}_2)\Psi_b(\mathbf{k}_1,\eta)\Psi_c(\mathbf{k}_2,\eta), \tag{3.3}$$

where $\Psi_a(\mathbf{k},\eta) \equiv (\delta(\mathbf{k},\eta), -\theta(\mathbf{k},\eta))$. It makes the writing of a formal solution possible

$$\Psi_a(\mathbf{k},\eta) = g_a{}^b(\eta)\Psi_b(\mathbf{k},\eta_0) + \int_{\eta_0}^{\eta} d\eta' g_a{}^b(\eta,\eta')\gamma_b{}^{cd}(\mathbf{k}_1,\mathbf{k}_2)\Psi_c(\mathbf{k}_1,\eta')\Psi_d(\mathbf{k}_2,\eta'). \tag{3.4}$$

This is a very good starting point for the development of theories allowing to establish a connection with some interesting observables. The formalism that has to be developed to do this is explained in detail in Bernardeau 2013. It is now well-established and it can be considered as a model to follow when it comes to study the impact of neutrinos on large-scale structure formation in the non-linear regime.

4. A non-linear alternative to the standard description of neutrinos

The leading idea of Dupuy and Bernardeau 2014 is to describe neutrinos as a collection of single-flow fluids instead of considering them as a single multi-flow fluid in order to take advantage of the single-flow approximation. This approximation can not be applied directly to the overall neutrino fluid because, unlike cold dark matter, there is no reason why velocity dispersion should be small for neutrinos. To circumvent the problem, the overall neutrino fluid is splitted into several flows in this study. Each flow, or each neutrino fluid, is defined as the collection of all the particles that have a given initial velocity. Such fluids are thus actually single-flow fluids and remain so until shell-crossing starts. In practice, it starts when neutrino velocities become low enough for gravity to make neutrinos turn around. So, at that time, neutrinos are not relativistic anymore and behave as cold dark matter. Describing this phenomenon is beyond the scope of this study. So far, it is also beyond the scope of the study of cold dark matter.

The first equation of motion is simply the conservation of the number of particles in each neutrino fluid, which gives in a perturbed Friedmann-Lemaître metric (written in the conformal Newtonian gauge)

$$\partial_\eta n - (1 + 2\phi + 2\psi)\partial_i \left(\frac{P_i}{P_0} n \right) = 3n(\partial_\eta \phi - \mathcal{H}) + n(2\partial_i \psi - \partial_i \phi)\frac{P_i}{P_0}, \qquad (4.1)$$

where n is the proper number density and P_i is the comoving momentum field. To obtain this, only linear terms in the metric perturbations ψ and ϕ have been taken into account. Indeed, observations demonstrate that what matters in the non-linear regime is not metric-metric coupling but the non-linear growth of the fields. Moreover, for a single-flow fluid, the conservation of the energy-momentum tensor combined with the conservation of the number of particles gives an equation for P_i only,

$$\partial_\eta P_i - (1 + 2\phi + 2\psi)\frac{P_j}{P_0}\partial_j P_i = P_0\partial_i \psi + \frac{P_j P_j}{P_0}\partial_i \phi. \qquad (4.2)$$

This equation shows in particular that the linear part of P_i is a gradient. This property is interesting since the fact that the velocity field is a gradient is one of the key ingredients of the cold dark matter description. The set of equations (4.1)-(4.2) is a relativistic generalization of the motion equations of cold dark matter. It is fully equivalent to the non-linear moments of the Boltzmann equation (Van de Rijt 2012). The advantage is that it is much more manageable in the non-linear regime than an infinite non-linear hierarchy. Some numerical tests have been realized in order to compare this approach to the standard one. The goal was to compute the overall (i.e. integrated over all the neutrino fluids) multipole energy distribution and to check that it gives the same values as integration of the linear Boltzmann hierarchy. Calculations are thus perfomed in the linear regime only. As illustrated on Fig.1, the agreement between both approaches is extremely good. By its specificity, the multi-fluid approach allows also to show the convergence of the number density contrast and of the velocity divergence of each flow to the ones of the cold dark matter component. As an illustration of this phenomenon, the time evolution of the velocity divergences of several neutrino fluids is presented on Fig. 2.

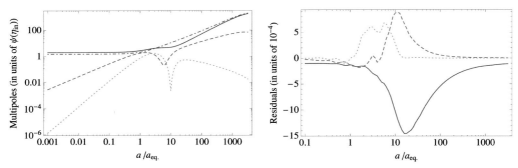

Figure 1. Time evolution of the energy density contrast (solid line), velocity divergence (dashed line) and shear stress (dotted line) of the neutrinos. The dot-dashed line is presented for comparison and corresponds to the density contrast of the dark matter component. Left panel: the quantities are computed with the multi–fluid approach. Right panel: residuals (defined as the relative differences) when the two methods are compared. Numerical integration has been done with 40 neutrino fluids, k is set to $k_{\rm eq} \approx 0.01h/{\rm Mpc}$ and the neutrino mass m is set to 0.3 eV.

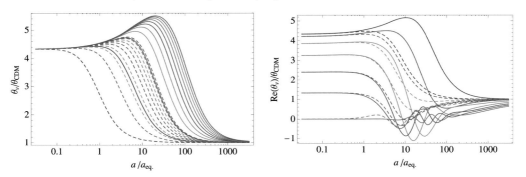

Figure 2. Time evolution of the velocity divergence. Left panel: values of the initial velocity moduli, denoted τ, range from $0.45\,k_B T_0$ (bottom lines) to $9\,k_B T_0$ (top lines) with $\mu = 0$ (μ being the cosine of the angle between the initial velocity vector and the wave vector). Right panel is for $\tau = 3.6\,k_B T_0$ and μ ranging from $\mu = 0$ (top lines) to $\mu = 1$ (bottom lines). The time evolution of the velocity divergence of each flow is plotted in units of the dark matter velocity divergence. The wave number is set to $k_{\rm eq}$, the solid lines correspond to a 0.05 eV neutrino mass and the dashed lines to a 0.3 eV neutrino mass.

References

Bernardeau, F. 2013, *ArXiv e-prints*, 1311.2724

Dupuy, H. & Bernardeau, F. 2014, *JCAP*, 1:30

Lesgourgues, J. & Pastor, S. 2006, *Phys. Rept.*, 429:307–379

Ma, C.-P. & Bertschinger, E. 1994, *ApJ*, 429:22–28

Planck Collaboration 2013, *ArXiv e-prints*, 1303.5062

Planck Collaboration 2013, *ArXiv e-prints*, 1303.5076

Van de Rijt, N. 2012, *PhD thesis "Signatures of the primordial universe in large-scale structure surveys"*, Ecole Polytechnique & Institut de Physique Théorique, CEA Saclay

CHAPTER 3.

Surveys and Observations
of the Large Scale Structure of the Universe

Jaan Einasto during his lecture.

Participants following one of the presentations.
In the center, Varun Sahni and Bernard Jones.

CHAPTER 3A.

Surveys and Observations:
Surveys

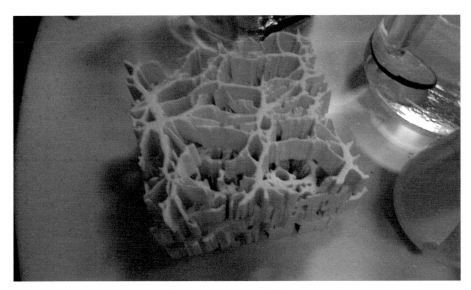

Zeldovich pancakes in a state-of-the-art presentation:
3-D print of the Spine of the Cosmic Web by Miguel Aragon-Calvo.
Photo courtesy: Peter Coles

The Zeldovich Universe:
Genesis and Growth of the Cosmic Web
Proceedings IAU Symposium No. 308, 2014
R. van de Weygaert, S. Shandarin, E. Saar & J. Einasto, eds.

© International Astronomical Union 2016
doi:10.1017/S1743921316009741

The cosmic web:
a selective history and outlook

John A. Peacock

Institute for Astronomy, University of Edinburgh
Royal Observatory, Edinburgh EH9 3HJ, UK
email: jap@roe.ac.uk

Abstract. In the Century since Slipher's first observations, roughly three million galaxy redshifts have been measured. The resulting maps of large-scale structure have taught us much of central importance in cosmology, ranging from the matter content of the universe to the study of the primordial density fluctuations. This talk aims to review some of the key observational and theoretical milestones on this journey, and to speculate about what the future may bring.

1. Introduction: the pre-history of redshift surveys

The large-scale structure seen in the galaxy distribution is one of the most important probes of fundamental cosmology, undoubtedly representing a relic of the earliest times and highest-energy physics. It's sobering to think that all of this was undreamed-of only within living memory, with V.M. Slipher publishing the first galaxy radial velocity in 1913 – the same year in which my Father was born. It's remarkable how little fuss was made of the recent Centenary of Slipher's revolutionary work, although at least a celebratory conference was held at the Lowell Observatory, to mark Slipher's achievements there. For a decade, he had the field of galaxy spectroscopy to himself, and by 1917 had already gathered sufficient data to prove that galaxies had to be redshifted on average (see e.g. Peacock 2013 for details of these achievements).

Beyond about 1923, the frontier of redshift measurement moved to the larger telescopes at Mt Wilson, but things still advanced slowly because of the limitations of photographic plates for recording spectroscopic information. More than three decades of effort yielded the 620 redshifts listed by Humason, Mayall & Sandage (1956), and this logarithmic pace of progress continued even into the mid-1970s, at which point the total number of known redshifts was not greatly above 1000.

Remarkably, a good deal had nevertheless been learned about galaxy clustering by this stage, thanks to careful analysis of the angular clustering seen in projected galaxy catalogues – which were painfully assembled by visual inspection of wide-area photographic survey plates (Shane & Wirtanen 1954). Much of the wisdom of this heroic era was gathered in the classic textbook by Peebles (1980). But by this stage, the field had already embarked on a series of technological revolutions, which have collectively brought us to the present happy glut of data.

2. Selected observational revolutions

2.1. *The CfA survey*

The first big step on the road to modern studies of the cosmic web was taken when the CfA survey applied the tools of the electronic revolution: image-intensifier tubes as detectors instead of photographic plates, and digital computers to undertake semi-automated analysis of the data. As a result, Huchra *et al.* (1983) were able to assemble a coherent sample of 2401 redshifts, suitable for the first true analyses of the 3D galaxy distribution. The CfA efforts continued, leading to the iconic slice picture of the cosmic web (de Lapparent, Geller & Huchra 1986), which was the first clear demonstration of the void-filament network with which we are now so familiar (although the void phenomenon was already known: Gregory & Thompson 1978). By the time they terminated, in 1995, the CfA surveys had accumulated over 20,000 redshifts.

2.2. *The LCRS*

The CfA efforts were impressive, but the rate of progress was limited by the need to observe galaxies one at a time. Even a larger telescope and more sensitive instrumentation only help to a degree, since there are unavoidable overheads associated with changing from one target to the next. What was required was a means of observing many galaxies in parallel. A number of approaches were tried, starting with objective prisms applied to early Schmidt photographic plates, but such an approach is unsatisfactory both through object overlap and non-rejection of the sky background. The latter problem can be beaten by applying a slit mask, but spectral overlap remains; in addition, it is hard to cover a very large area in this way. Thus the real revolution in redshift surveys came with the application of fibre optics. A number of groups developed this technique, particularly at the Steward Observatory and the Anglo-Australian Observatory (see e.g. Hill 1988 and Gray 1988 for reviews of the early work). The largest amount of early astronomical results from fibre instrumentation certainly came from the AAO, especially when the initial approach of plugging fibres into a pre-drilled plate (FOCAP) gave way to the more flexible approach where fibres could be placed by a robot at arbitrary positions on the focal plane (AUTOFIB).

But in terms of cosmological information, it is important for redshift surveys to cover a large area of sky, and the AAO's initial efforts were limited by the restricted Cassegrain field of view. Thus, the first true wide-area redshift survey to use fibre multiplexing was the Las Campanas Redshift Survey. Shectman *et al.* (1996) give details of this effort, which measured 26418 redshifts over the period 1988–1994. The LCRS fibre system used 112 fibres over a field 1.5 degrees square, employing pre-drilled plug plates, to cover an eventual area totalling 700 deg^2. The main achievement of this survey was, in their own words, "the end of greatness". Previous redshift surveys had seen features that were comparable to the size of the survey (the 'great wall' being the imaginative name given to one such structure), and it was fair to question whether these studies as yet constituted a fair sample of the universe. But the LCRS observed equal fields in the Northern and Southern hemispheres, and the statistical character of these two sub-surveys were closely comparable. This was a major achievement, setting the scene for the work that followed.

2.3. *The 2dFGRS*

The success of the AAO fibre experiments showed that the power of the technique on the Anglo-Australian 4m could be increased by moving to a wider field. This required the construction of a new corrector to allow the fibres to work at the prime focus. The initial planning started around 1988, with formal approval given in 1990 for the project to construct the 'two-degree field' system: a 2-degree diameter corrected field with 400 fibres placed by a robot. 1994 saw the formation of UK-Australian consortia to carry out major galaxy (and quasar) redshift surveys with the new facility, for which observing eventually started in 1996. The galaxy survey (2dFGRS) was based on an input catalogue derived from the UK Schmidt photographic plates scanned at Cambridge with the APM

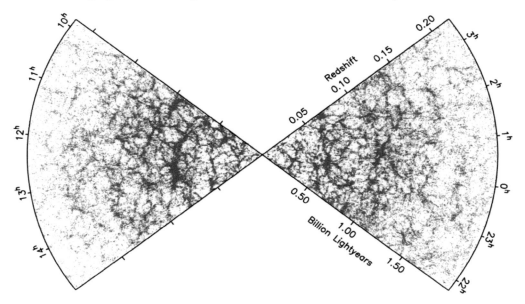

Figure 1. The 2dFGRS view of the local universe, based on 221,414 redshifts. This is the highest-fidelity picture we have of a large expanse of the local cosmic web.

machine (blue 'J' plates) and at Edinburgh with SuperCOSMOS (red 'F' plates). By 2003, redshifts had been measured for 221,414 galaxies over approximately 2000 deg^2, to an extinction-corrected depth of $B_J < 19.45$ (Colless *et al.* 2003). The resulting image of the local cosmic web (Figure 1) remains the highest fidelity picture we have (the deepest truly wide-area fully-sampled survey).

This advance in size of nearly an order of magnitude with respect to the LCRS, with corresponding improvement in volume spanned, is what allowed 2dFGRS to pass the threshold in precision needed to make qualitatively new discoveries. These include accurate measurements of the power spectrum, yielding a precise matter density and the detection of Baryon Acoustic Oscillations (Percival *et al.* 2001; Cole *et al.* 2005), and the measurement of Redshift-Space Distortions (Peacock *et al.* 2001). The latter was originally seen as a further means of measuring Ω_m; but as explained below, RSD has now emerged as a leading means of testing theories of gravity.

It should be pointed out that the impact of a given project depends on timing, and the context of developments elsewhere. 2dFGRS made especially strong advances because of parallel developments in CMB anisotropies; both classes of study complemented each other, tying down degrees of freedom that would otherwise have been degenerate (e.g. the shape of the power spectrum largely measures $\Omega_m h$, not Ω_m, and it has a further degeneracy with the slope of the primordial spectrum, n_s). As a result, LSS information from 2dFGRS played a major part in, for example, the cosmological interpretation of the initial results from WMAP (Spergel *et al.* 2003).

2.4. *SDSS*

All the surveys listed so far suffered from one common disadvantage: they were only part-time tenants of their telescopes. So time allocation committees had to perform the delicate juggling act of giving sufficient time to allow the surveys to make worthwhile progress, while not stifling the community wanting to use the telescope for diverse smaller projects. For example, the 2dFGRS used 272 AAT nights over a 5-year period – less than

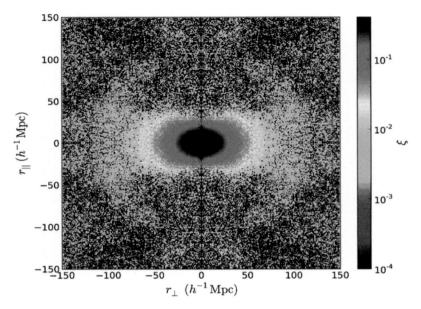

Figure 2. The redshift-space correlation function of BOSS LRGs (Samushia *et al.* 2014). The apparent radial coordinate is affected by peculiar velocities, yielding the marked apparent deviations from circular symmetry around $20\,h^{-1}$ Mpc. This figure also nicely shows the ring around $100\,h^{-1}$ Mpc corresponding to the Baryon Acoustic Oscillation feature.

1/3 of the available dark/grey time. In contrast, the SDSS was able to operate with a dedicated telescope, albeit one split between imaging and spectroscopy. The number of fibres went up slightly: 600 at the time of the 1998 commissioning, rising to 1000 in SDSS-III, and the field size was 3 degrees. But with a much smaller 2.5m mirror, SDSS would have been a less powerful facility than 2dF without the ability to make use of all observing time. Indeed, the main galaxy redshift survey, limited at $r = 17.77$, is somewhat less deep than 2dFGRS as a result.

But the most powerful step taken by SDSS was to exploit its own multicolour imaging, which was used to generate the input catalogue. This permitted the selection of a probe uniquely suited to accurate statistical measurement of LSS properties: the Luminous Red Galaxies. Given the limited light grasp of the SDSS mirror, efficient surveying to large redshifts requires the pre-selection of unusually luminous galaxies – and these tend to be the ellipticals, which can be picked out by their lack of short-wavelength emission. Moreover, such objects have strong spectral features, allowing successful determination of redshifts even when the spectra are of low S/N. For a final bonus, such objects are strongly biased, so the clustering signal is enhanced. As a result, large volumes could be surveyed in a dilute way while yielding accurate measurements of clustering. The initial LRG selection worked to $z < 0.5$, and the first LRG results were already sufficiently powerful to detect the BAO signal with only 46,748 redshifts (Eisenstein *et al.* 2005). Subsequent work has refined the selection method, pushing to $z < 0.7$ in the 'BOSS' sub-project of SDSS-III, yielding sample sizes of 690,826 LRGs in the 'CMASS' sample at $z > 0.43$, supplemented by 313,780 at lower redshifts (Anderson *et al.* 2014; Samushia *et al.* 2014).

As a result of this large survey volume, BOSS currently sets the standard for the most precise measurements of the key LSS statistics, including the redshift-space correlation

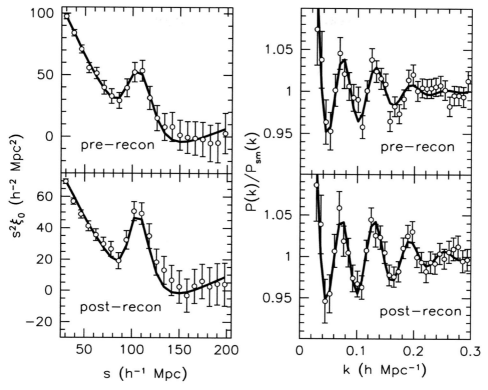

Figure 3. The BAO feature as seen in configuration space and in Fourier space by the BOSS LRG survey (Anderson *et al.* 2014). The precision of the measurement is improved by 'reconstruction', in which an approximate correction is made for weakly nonlinear displacements at the BAO scale. It is striking to compare the quality of this measurement with the first bare detections of only nine years previously.

function (Figure 2), and the location of the BAO scale, with the latter measurement probing the empirical $D(z)$ relation and hence parameters relevant to the cosmic expansion history (Figures 3 & 4).

3. Revolutions in theory

Observational advances have tended to proceed in step with important refinements of theory and/or statistical treatment of data. It is natural that better data should stimulate improved methods, but all the major surveys of recent years have been directed at theoretical targets, even if in some cases the target emerged after the survey had commenced.

In so many ways, studies of large-scale structure rest on the titanic foundations of the book by Peebles (1980). Almost all the statistical concepts and tools used today are set out there, at least in embryo. All that was really missing was a specific theoretical model on which the general methods could be focused – and the CDM model was not long in being developed (Peebles 1982). Next to Peebles, probably the majority of the most important ideas that are in use today derive from three big ideas developed in classic 1980s papers by Kaiser:

(1) **Bias**. It was known empirically that different galaxies displayed different amplitudes of correlations, but this remained a puzzle until Kaiser (1984) showed that

regions of density high enough to collapse by the present will inevitable show enhanced correlations. Although this breakthrough is commonly cited as the invention of biased galaxy formation, this is incorrect. Initially, it was thought that this phenomenon could arise via a threshold in density below which star formation in galaxies would be suppressed (Bardeen *et al.* 1986), which turned out to be a blind alley. Eventually, it was realised that the correct way to view the Kaiser result was in terms of a bias that depended on the mass of dark-matter haloes (e.g. Mo & White 1996), which then only needs a mass-dependent halo occupation number for a full understanding of the clustering of different classes of galaxy (Seljak 2000; Peacock & Smith 2000).

(2) **Sparse sampling**. Many redshift surveys, from the early CfA surveys to the 2dFGRS, used a simple 'shoot everything that moves' strategy: apply a magnitude limit and obtain redshifts for all galaxies brighter than this over as large an area as possible. But Kaiser (1986) pointed out that this is wasteful. The precision with which cosmological statistics such as the power spectrum can be measured is limited by a mixture of cosmic variance and shot noise, and for the best results with fixed telescope time one should optimise the balance between these. Kaiser found that a fully-sampled survey generally overkills shot-noise suppression: if galaxies all lived in clumps of size N, then most of the clumps have been found once a fraction $\sim 1/N$ of the redshifts have been taken. Beyond this, discreteness noise declines slowly, and one is better covering more area at the same sampling. This strategy was used effectively in the IRAS-selected QDOT survey (Lawrence *et al.* 1999), and is implicitly what has permitted the outstanding success of the SDSS LRG strategy.

(3) **Redshift-space distortions**. The final 1980s revolution concerned the issue that the radial coordinate deduced from redshifts was imperfect through the modification of the observed redshift by peculiar velocities. Although this had long been clear in a general sense, Kaiser (1987) supplied the quantitative analysis of the large-scale linear anisotropic clustering that resulted: $P_s(k, \mu) = P_r(k)(b + f_g\mu^2)^2$, where the s subscript denotes redshift space; μ is the cosine of the angle between the wavevector and the line of sight; b is a bias parameter, and f_g is the logarithmic growth rate of density fluctuations with respect to scale factor, $f_g = d\ln\delta/d\ln a$. This equation has been hugely influential in provoking observational attempts to measure the growth rate, as discussed further below.

3.1. *BAO and dark energy*

Probably the main aspect of current LSS studies that was not put in place during the 1980s is the Baryon Acoustic Oscillations – although these were well understood in the context of the CMB during this time, and it was known that the effects on the matter power spectrum would be small if dark matter dominated. One strong stimulus to thinking seriously about BAO effects was the impressive analytical work on the form of the power spectrum by Eisenstein & Hu (1998); it also became clear around the same time that the small baryonic features would survive nonlinear evolution (Meiksin, White & Peacock 1999), so there was a strong theoretically motivated signal to search for.

It should therefore not have come as much of a surprise when evidence for BAO features in the galaxy power spectrum started to become available. The first claim of a detection was made in the initial 2dFGRS power spectrum measurements by Percival *et al.* (2001), with a greater significance in the final data (Cole *et al.* 2005), at the same time as the the initial SDSS LRG results of Eisenstein *et al.* (2005). It should be noted that all these pieces of evidence were model-dependent: the ΛCDM theory made a well-specified prediction of BAO features, and this model was statistically preferred to a featureless power spectrum – even though an empirical featureless spectrum was not formally inconsistent with the data. This is exactly the position we occupied with the CMB prior to about 2000: the data at that time constrained very accurately the height and location of the main acoustic peak at $\ell \simeq 220$, even though a spectrum with no peak at all was consistent with the measurements. This represents the additional information injected by having a strong theoretical prior.

In fact, what might be considered surprising about the BAO features is that they are so small. Right from the first calculations of the evolution of cosmological perturbations, it was clear that the resulting transfer function should contain strong oscillatory features as a result of sound waves in the primordial matter+radiation fluid (Peebles & Yu 1970; Sunyaev & Zeldovich 1970). The amplitude of these order unity fluctuations becomes damped only in the presence of collisionless dark matter, which cannot support sound waves (e.g. Bond & Szalay 1983). At the time of decoupling, the matter-radiation fluid has order unity BAO, whereas the CDM has none at all; subsequently, with a much reduced sound speed, the two components fall towards each other. Both have the same phase of Fourier perturbations, but with different transfer functions with respect to some initial fluctuation:

$$\delta_{\text{tot}} = \delta_i \left[f_1 T_1(k) + f_2 T_2(k) \right],$$

where f_1 and f_2 are the fractions in the two components. Since about 20% of the matter is baryonic, the overall amplitude of the BAO features in the total power spectrum are correspondingly reduced. Thus, from the moment that data revealed a relatively smooth large-scale galaxy power spectrum (e.g. Peacock & Dodds 1994), we were in possession of rather general evidence for the existence of collisionless dark matter.

In any case, it was quickly realised that the important application of the BAO signal was not so much in constraining cosmological parameters from the shape of the power spectrum; rather, it lay in using the acoustic scale to provide a standard ruler suitable for geometrical cosmology (Blake & Glazebrook 2003; Seo & Eisenstein 2003). One can treat the BAO scale as something completely empirical, observed at different redshifts to yield an angle subtended or the corresponding radial increment of redshift – respectively probing the comoving angular-diameter distance $D_C(z)$ or the epoch-dependent Hubble parameter $H(z)$. In practice, it is normal to combine these into a single effective distance measure derived from averaging over radial shells:

$$D_V(z) \equiv \left(D_C^2(z) \, cz/H(z) \right)^{1/3}.$$

One can then measure $D_V(z)$ subject to an overall normalization, which already places important limits on $H(z)$ and thus the equation of state of dark energy via the Friedmann relation $H^2(z) \propto \Omega_m a^{-3} + (1 - \Omega_m)a^{-3(1+w)}$ for a flat universe. But more precise limits can be set if we are prepared to adopt the ΛCDM prediction of the acoustic scale. This is the origin of the impressive plot shown in Figure 4 (Anderson *et al.* 2014), where the BAO data are seen to be consistent with a prediction involving no free parameters (or, rather, the parameters of the simplest flat ΛCDM model can all be determined from the CMB). The precision of this agreement is what permits the current very tight limits on the evolution of dark energy: $w = -1.006 \pm 0.045$ (Planck Consortium 2015).

This has been a great success story for LSS, and there seems no reason why it should not continue. The physics that determines the BAO scale in CMB and LSS is often described as 'simple' – perhaps a little unfair when one considers the decades of detailed work needed to generate a precise theory for the effect – but certainly there is a good case that the signal is robust and independent of the details of galaxy formation. Nonlinear

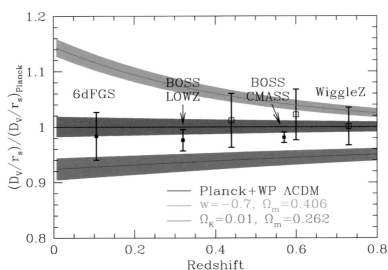

Figure 4. The BAO length scale is calculable given the cosmological parameters of the ΛCDM model. On the assumption of flatness, these can be determined very accurately from the CMB alone, and the BOSS BAO measurements give a completely independent test of the validity of this prediction, which is passed extremely well (figure from Anderson *et al.* 2014). If the model is then generalized to open other degrees of freedom (curvature, or dark energy equation of state $w \neq -1$), these are heavily constrained.

evolution of structure does potentially alter the power spectrum on BAO scales, but to an extent that can be modelled so that systematics in the BAO scale are $\sim 0.1\%$ (Eisenstein *et al.* 2007). Since the sensitivity of distance to w is $d\ln D/dw \lesssim 0.2$, this implies that future experiments should be able to attain 1%–level precision on w, and seeking to measure whether dark energy evolves at this level is a realistic goal for the next decade.

3.2. *RSD and modified gravity*

Kaiser's work on redshift-space distortions revealed a quadrupole anisotropy that could be used as a diagnostic of the growth-rate of cosmological inhomogeneities. From an early stage, it was clear that this linear-theory expression required supplementing with allowance for small-scale virialized motions, which can be treated by a radial convolution. In Fourier space, this amounts to multiplying by a function of the radial component of **k** (Peacock & Dodds 1994), which gave rise to the widely-used 'dispersion model':

$$P_s(k, \mu) = P_m(k) \frac{(b + f_g \mu^2)^2}{1 + k^2 \mu^2 \sigma_p^2/2} .$$

There is a choice of whether the matter power spectrum, P_m, should be taken from linear theory; in practice, a better match to data arises if it is the full nonlinear real-space spectrum.

From the shape of this equation, the distortion parameter $\beta \equiv f_g/b$ can be extracted. This is complicated by the unknown bias, but this is constrained by the amplitude of the observed large-scale galaxy clustering, $b^2 P_m$ (this real-space quantity can be measured by projecting the 2D redshift-space correlation function along the radial direction). Thus e.g. $b^2 \sigma_8^2$ is observable, so the bias-independent quantity $f_g \sigma_8$ can be constructed. If this is measured as a function of redshift, it is an informative combination of the differential and integrated growth of fluctuations.

Initially, the RSD phenomenon was seen as an independent way to weigh the universe via the approximate relation $f_g \simeq \Omega_m(z)^{0.6}$, and the first detailed measurement of the effect by the 2dFGRS was focused on measuring Ω_m (Peacock *et al.* 2001). But in more recent years, it has been appreciated that the deeper significance of f_g is that it gives a way no probe modified theories of gravity. This change of perspective was promoted by Guzzo *et al.* (2008), who were also the first to measure RSD at high redshifts ($z \simeq 0.8$). At high redshifts, the Kaiser flattening becomes partially degenerate with the 'Alcock-Paczynski' geometrical flattening which can arise through incorrect choice of the cosmological geometry when calculating the 2D correlations. But with sufficiently good data, this degeneracy can be broken – and the geometrical effect then becomes another useful constraint on the cosmological model (Ballinger *et al.* 1998).

Once we decide to move beyond Einstein's original relativistic theory of gravity, a range of possibilities becomes available (e.g. Clifton *et al.* 2012). From the point of view of LSS, it is simplest to condense this variety into two parameters that express deviations from Einstein gravity on linear scales. We start by writing the metric in terms of two scalar potentials:

$$ds^2 = a^2(t)\left[(1+2\Psi)d\eta^2 - (1-2\Phi)(dx^2 + dy^2 + dz^2)\right] ;$$

in Einstein gravity, $\Phi = \Psi$ is the Newtonian potential, which obeys Poisson's equation. Note that, in moving away from this model, we are not really abandoning general relativity, despite the common claim that this approach tests GR: there is still a metric, and we seek a description of gravity in terms of relativistic invariants. All that changes is that we abandon the simplest gravitational Lagrangian $\mathcal{L} \propto R + 2\Lambda$ in favour of a more complicated invariant based on the metric (such as the common $f(R)$ models). In the linear limit of small deviations from Einstein, we can parameterise the slip (relation of Ψ and Φ) and the effective strength of gravity in Poisson's equation:

$$\Psi = (1+\alpha)\Phi$$
$$\nabla^2\Phi = 4\pi(1+\beta)Ga^2\bar{\rho}\delta.$$

There is no standard notation for these parameters, as discussed by e.g. Daniel *et al.* (2010).

Rather than working with slip and effective G, a more directly physical parameterisation is to describe the change in the acceleration felt by non-relativistic particles (dictated by Ψ) and relativistic particles (dictated by $\Psi + \Phi$):

$$\Psi = (1+\mu)\Psi_E$$
$$\Psi + \Phi = (1+\Sigma)(\Psi+\Phi)_E = 2(1+\Sigma)\Phi_E,$$

where the E subscript (often written 'GR', which as discussed earlier is inappropriate) stands for Einstein gravity, in which $\nabla^2\Phi_E = 4\pi Ga^2\bar{\rho}\delta$. These parameters are related to first order to the slip and G parameters by $\mu = \alpha + \beta$; $\Sigma = \alpha/2 + \beta$. In this parameterisation, μ is probed by redshift-space distortions, while Σ is probed by gravitational lensing. But even so, this decomposition leaves open the extent to which the gravitational modification depend on scale and on epoch. The scale dependence is normally ignored, so that we are implicitly determining the properties of gravity on the $\sim$ 100-Mpc scales of LSS. As for epoch dependence, modifications of gravity gain a strong motivation from the acceleration of the universe, with the conjecture that changing gravity may produce this acceleration without the need for dark energy as a physical substance. Acceleration

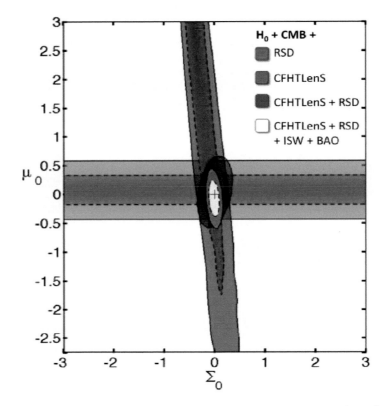

Figure 5. Limits on the two principal linear modified-gravity parameters, from Simpson *et al.* (2013). Both non-relativistic accelerations (from RSD) and relativistic accelerations (from gravitational lensing) seem to be within ~ 20% of their standard values.

is a late-time phenomenon, so it is reasonable to assume that modifications of gravity are negligible at early times (apart from anything else, we would prefer not to spoil the good agreement between the standard model and the CMB or primordial nucleosynthesis). Thus it is usual to scale $(\mu, \Sigma) \propto a^3$ or $\propto \Omega_v(a)$.

As we have seen, the μ parameter influences the growth rate of density fluctuations: this is often parameterised as $f_g = \Omega_m(a)^\gamma$, where the standard value of $\gamma \simeq 0.55$ changes when gravity is modified. According to Linder & Cahn (2007), this index is well approximated by

$$\gamma = \frac{3}{5 - 6w}\left(1 - w - \frac{\mu(a)}{\Omega_v(a)}\right),$$

where $\mu(a)/\Omega_v(a)$ will generally be assumed to be nearly constant in order that the modifications vanish at high z, so it can be taken to be μ_0/Ω_v. The sensitivity to μ_0 is therefore

$$\frac{d\ln f_g}{d\mu_0} = \frac{-3\ln\Omega_m(a)}{(5 - 6w)\Omega_v}.$$

For data at $z = 1$, $\Omega_m(a) \simeq 0.77$, so $d\ln f_g/d\mu_0 \simeq 0.10$ (as compared to 0.47 at $z = 0$), so high-redshift measurements probe μ_0 less effectively (although this comes entirely from the model assumptions, and it is still of interest to look at the high-redshift f_g in order to be sensitive to a broader range of possibilities). The highest redshift probed to date is

$\bar{z} = 0.8$ via the VIPERS project (de la Torre *et al.* 2013), which measured $f_g \sigma_8$ to 17% precision. Combining lower-redshift BAO data (especially from BOSS), together with measurements of gravitational lensing, we can set limits to the parameters governing modified gravity, specified by the current values μ_0 and Σ_0. Current constraints of this sort are shown in Figure 5, where we see that standard gravity in both its relativistic and non-relativistic respects is consistent with current data at about the 20% level.

Future experiments will change this sensitivity to the sub-% level; do we have any reason to expect that a deviation from Einstein will be seen? Certainly there is no compelling model that leads us to expect a signal at this level, so a null result is probably the favoured prior expectation. Nevertheless, carrying out such tests has been a hugely positive development for cosmology. In previous generations, the correctness of Einstein gravity had to be assumed in order to reach cosmological conclusions; but now we can validate this fundamental assumption, rather than having to take it on trust.

4. Summary and conclusions

In this brief overview, we have taken stock of the growth in understanding of the large-scale structure of the universe from its inception almost exactly a century ago, to the beautiful precision of present-day measurements. Over 2 million galaxy redshifts are known, a total that would probably have been unimaginable to the early pioneers. As summarised above, this advance has been possible through a series of technological revolutions, which have allowed the stockpile of redshifts to advance exponentially with time over more than three decades. As illustrated in Figure 6, the doubling time of this advance has been a mere 3.5 years; at this rate, less than 50 years remains before spectroscopic information for the visible universe would be complete.

Naive exponential extrapolation is guaranteed to generate comical errors in due course. As a child, I owned an early-1960s book that looked ahead correctly to human landings on the Moon in 1969 – but the same book also informed the reader that landings on Mars would follow in 1984. Similarly, one can list several reasons why the spectroscopic exploration of the universe may not continue to grow at its recent astonishing rate. The fundamental problem is that there is no new technology: efficiency of digital detectors has saturated; future degrees of spectroscopic multiplex will be only moderately larger; primary mirrors for spectroscopy will become slightly larger. This does not add up to a revolution. It is somewhat disappointing to think that DESI, as the leading spectroscopic survey project around 2020 (Levi *et al.* 2013), will have the same mirror size and 'only' about 10 times the number of fibres of 2dF – an instrument 25 years older. Moreover, the cost of such an instrument can tend to increase linearly with the number of fibres, as this increases the number of spectrographs to be replicated. DESI will also increase its power over 2dF by becoming a dedicated instrument that uses all the time on its telescope; this factor, together with the increase in numbers of fibres and a lowering of S/N target will amount to two orders of magnitude in terms of the numbers of redshifts gathered with respect to 2dFGRS. But the cost has arisen substantially, also. 2dF cost about \$10M for the hardware, which dominates the cost of 2dFGRS; DESI's budget is in excess of \$100M, of which about 1/3 comes from supporting the sheer number of nights required. Even though the cost per redshift has fallen if we assign the entire cost of 2dF to 2dFGRS, this approach is becoming expensive. Add to this the fact that increasing depth of imaging surveys really requires a larger mirror, and we can see that limits are being reached. A spectroscopic survey 'dream machine' with 10m mirror, 10,000 fibres over a 4-5 degree field, would clearly be in the same billion-dollar league as e.g. LSST,

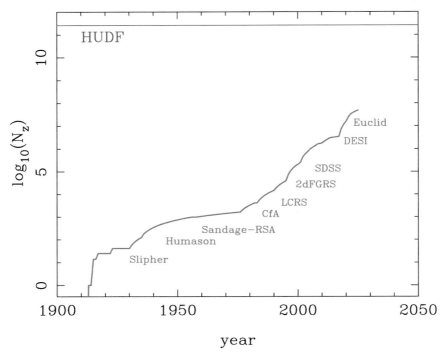

Figure 6. The approximate cumulative measurement history of galaxy spectroscopic redshifts, with selected major landmark studies indicated. The upper line marked 'HUDF' corresponds to the Hubble Ultra Deep Field, extrapolated to all sky (i.e. every galaxy in the visible universe). Over several decades, we have sustained a 'MooreZ Law', in which the stock of redshifts doubles every 3.5 years. On this basis, spectroscopic exploration of the universe of galaxies may be expected to cease in about 2060.

and thus would probably take the same 20+ years to come to fruition. This cannot be the way ahead.

The nearest thing to a new technology on offer is the approach taken by Euclid, measuring redshifts via slitless spectroscopy. This is of course also a very old approach, given a new lease of life because the sky background is very low in space. But this too has its limitations: object overlap is a major headache unless the spectral resolution is made unacceptably low. Also, working in space inevitably escalates the cost, pricing Euclid's $\sim 25M$ expected redshifts at several times \$100M. Thus, both on ground and in space, it is hard to see how to take current approaches to the next level of $\sim 10^9$ redshifts. Two possible ways ahead aim to achieve the equivalent science goals by sacrificing some element of detailed spectroscopy on individual objects. The first approach is extreme photometric redshifts, using narrow-band filters to estimate redshifts with a precision of around 0.5% in z. Two competing Spanish-led projects are pursuing this approach (J-PAS and PAU) (Benitez *et al.* 1014; Martí *et al.* 2014). This approach should work well for BAO studies, although the redshift precision is equivalent to a velocity of worse than $1000\,\mathrm{km\,s^{-1}}$, so RSD studies will be challenging. Also, the restriction to optical bands means that the accessible redshift range will substantially overlap the volume already studied by BOSS, and subject to the same cosmic-variance effects.

Alternatively, we can ask why we bother to measure redshift from individual objects, when we only care about how the emission is distributed on 100-Mpc scales. The philosophy of intensity-mapping experiments is to measure this directly by sacrificing spatial resolution in exchange for sky coverage. In order to avoid being degraded by sky background emission, this approach is most effective in the radio. The first experiment likely to demonstrate this approach in practice is CHIME – the Canadian HI Mapping Experiment (Bandura *et al.* 2014). This will map a hemisphere with about 0.4° resolution (tens of Mpc), with a frequency resolution of 0.1% over $z = 0.8 - 2.5$. The lack of angular resolution will make removal of foregrounds more challenging, but this could be a revolutionary approach. The frequency precision is good enough that the results should be useable for RSD as well as BAO.

But how far the field will be able to make use of expanded datasets depends on theoretical and modelling developments. We have seen that it is already hard to extract the growth rate correctly at the 1–2% level from existing data, and improving this precision will be a challenge. The BAO systematics are simpler, so probably there is no reason that this method cannot be pushed robustly well below the current 1% level. There is good reason to be hopeful in both areas, since future datasets will yield large samples of very different galaxies, so there will be a strong motivation to continue theoretical efforts until the correspondingly improved deductions about dark energy and gravity are available. It's hard to know whether we should really expect to find a deviation from the ΛCDM model in this way; but it is profoundly important that these fundamentals of the model can be tested rather than assumed. To have achieved so much in this direction in only a century of effort should be a source of great pride to this community.

References

Anderson, L. *et al.* 2014, *MNRAS*, 439, 83 [arXiv: 1312.4877]
Ballinger, W. E., Peacock, J. A. & Heavens, A. F. 1996, *MNRAS*, 282, 877
Bandura, K. *et al.* 2014, arXiv:1406.2288
Benitez, N. *et al.* 2014, arXiv:1403.5237
Blake, C. & Glazebrook, K., 2003, *ApJ*, 594, 665
Bond, J. R. & Szalay, A. S., 1983, *ApJ*, 274, 443
Clifton, T., Ferreira, P. G., Padilla, A., & Skordis C., 2012, *Physics Reports*, 513, 1
Colless, M. *et al.* 2003, arXiv:astro-ph/0306581
Cole, S. *et al.* 2005, *MNRAS*, 362, 505
Daniel, S. F. *et al.*, 2010, Phys. Rev. D, 81, 123508
de Lapparent, V., Geller, M. J., & Huchra, J. P. 1986, *ApJ*, 302, L1
de la Torre, S. *et al.*, 2013, *A&A*, 557, A54
Eisenstein, D. J. & Hu, W. 1998, *ApJ*, 496, 605
Eisenstein, D. J. *et al.*, 2005, *ApJ*, 633, 560
Eisenstein, D. J., Seo, H.-J., & White, M., 2007, *ApJ*, 664, 660
Hill, J. M. 1988, *ASP Conference 'Fiber Optics in Astronomy'*, 3, 285
Gregory, S. A. & Thompson, L. A. 1978, *ApJ*, 222, 784
Guzzo, L., Pierleoni, M., Meneux, B., *et al.* 2008, *Nature*, 451, 541
Hill, J. M. 1988, *ASP Conference 'Fiber Optics in Astronomy'*, 3, 77
Huchra, J., Davis, M., Latham, D., & Tonry, J. 1983, *ApJS*, 52, 89
Humason, M. L., Mayall, N. U., & Sandage, A. R. 1956, *AJ*, 61, 97
Kaiser, N. 1984, *ApJ*, 284, L9
Kaiser, N. 1986, *MNRAS*, 219, 785
Kaiser, N. 1987, *MNRAS*, 227, 1
Lawrence, A. *et al.* 1999, *MNRAS*, 308, 897
Levi, M. *et al.*, 2013, arXiv:1308.0847
Linder, E. & Cahn, R., 2007, *Astroparticle Physics*, 28, 481
Martí, P. *et al.* 2014, arXiv:1402.3220
Meiksin, A., White, M., & Peacock, J. A. 1999, *MNRAS*, 304, 851

Mo, H. J. & White, S. D. M. 1996, *MNRAS*, 282, 347

Peacock, J. A. & Dodds, S. J. 1994, *MNRAS*, 267, 1020

Peacock, J. A. & Smith, R. E. 2000, *MNRAS*, 318, 1144

Peacock, J. A. *et al.* 2001, *Nature*, 410, 169

Peacock, J. A. 2013, arXiv:1301.7286

Peebles, P. J. E. & Yu, J. T., 1970, *ApJ*, 162, 815

Peebles, P. J. E., 1980, *The Large-Scale Structure of the Universe*, Princeton

Peebles, P. J. E., 1982, *ApJ*, 263, L1

Percival, W. J. *et al.* 2001, *MNRAS*, 327, 1297

Planck Consortium (Ade *et al.*), 2015, arXiv:1502.01589

Reid, B. A. *et al.* 2014, arXiv:1404.3742

Samushia, L. *et al.* 2014, *MNRAS*, 439, 3504 [arXiv: 1312.4899]

Seljak, U. 2000, *MNRAS*, 318, 203

Seo, H.-J. & Eisenstein, D. J., 2003, *ApJ*, 598, 720

Shane, C. D. & Wirtanen, C. A. 1954, *AJ*, 59, 285

Shectman, S. A., Landy, S. D., Oemler, A., *et al.* 1996, *ApJ*, 470, 172

Simpson, F., Heymans, C., Parkinson, D., *et al.* 2013, *MNRAS*, 429, 2249

Spergel, D. N. *et al.* 2003, *ApJ suppl*, 148, 175

Sunyaev, R. A. & Zeldovich, Y. B., 1970, *Astrophysics & Space Science*, 7, 3

The Zeldovich Universe:
Genesis and Growth of the Cosmic Web
Proceedings IAU Symposium No. 308, 2014
R. van de Weygaert, S. Shandarin, E. Saar & J. Einasto, eds.

© International Astronomical Union 2016
doi:10.1017/S1743921316009753

Mapping the Cosmic Web
with the largest all-sky surveys

Maciej Bilicki[1,2], John A. Peacock[3], Thomas H. Jarrett[1],
Michelle E. Cluver[1] and Louise Steward[1]

[1]Department of Astronomy, University of Cape Town, South Africa
email: `maciek(at)ast.uct.ac.za`

[2]Kepler Institute of Astronomy, University of Zielona Góra, Poland

[3]Institute for Astronomy, University of Edinburgh, United Kingdom

Abstract. Our view of the low-redshift Cosmic Web has been revolutionized by galaxy redshift surveys such as 6dFGS, SDSS and 2MRS. However, the trade-off between depth and angular coverage limits a systematic three-dimensional account of the entire sky beyond the Local Volume ($z < 0.05$). In order to reliably map the Universe to cosmologically significant depths over the full celestial sphere, one must draw on multiwavelength datasets and state-of-the-art photometric redshift techniques. We have undertaken a dedicated program of cross-matching the largest photometric all-sky surveys – 2MASS, WISE and SuperCOSMOS – to obtain accurate redshift estimates of millions of galaxies. The first outcome of these efforts – the 2MASS Photometric Redshift catalog (2MPZ, Bilicki *et al.* 2014a) – has been publicly released and includes almost 1 million galaxies with a mean redshift of $z = 0.08$. Here we summarize how this catalog was constructed and how using the WISE mid-infrared sample together with SuperCOSMOS optical data allows us to push to redshift shells of $z \sim 0.2$–0.3 on unprecedented angular scales. Our catalogs, with ~ 20 million sources in total, provide access to cosmological volumes crucial for studies of local galaxy flows (clustering dipole, bulk flow) and cross-correlations with the cosmic microwave background such as the integrated Sachs-Wolfe effect or lensing studies.

Keywords. catalogs, surveys, galaxies: distances and redshifts, techniques: photometric, methods: data analysis, (cosmology:) large-scale structure of universe, cosmology: observations

1. The need for all-sky galaxy surveys in three dimensions

A complete picture of the Cosmic Web we live in would only be known if we could observe the entire extragalactic sky (4π sr) up to the surface of last scattering and with full redshift information. This is of course observationally unachievable, in part due to our Galaxy creating the so-called Zone of Avoidance (ZoA) and the instrumental limitations which force us to choose between wide angular coverage and the depth of a redshift survey. There is, however, a reason and need to observe as much of the sky as possible in three dimensions, as only this enables the framework for comprehensively testing cosmological models. For instance, cosmic microwave background (CMB) observations provide a consistent picture of an early Universe which was very homogeneous and isotropic; such an assumption is also applied to the late-time Universe when modeling it with a simple Friedman-Lemaître metric. However, validity of this Copernican Principle (CP) needs to be verified observationally, and this is only possible if we observe the deep cosmos in all possible directions, and at various redshifts: cosmological tests of the CP require access to the entire celestial sphere in 3D. For instance, analyses of the CMB data have brought to light several 'anomalies' such as the quadrupole-octopole alignment or low observed variance of the CMB signal (Planck collaboration XXIII 2013). A fundamental question is whether these anomalies are confirmed as today's anisotropy and/or inhomogeneity in the galaxy distribution; some studies indicate this is not the case (Hirata 2009; Pullen & Hirata 2010). The issue, however, requires more scrutiny using more comprehensive low-redshift catalogs. One possible manifestation of late-time violation of the CP would be high-amplitude large-scale flows of galaxies and indeed claims of their existence have been made (Kashlinsky *et al.* 2008; Watkins *et al.* 2009), although several independent analyses show otherwise (e.g. Nusser & Davis 2011; Turnbull *et al.* 2012). Also the issue of which structures, and on what scales, contribute to the pull on the Local Group of galaxies has still not been settled; despite some claims of the convergence of the 'clustering dipole' at distances ~ 100 Mpc (Rowan-Robinson *et al.* 2000; Erdogdu *et al.* 2006), other analyses suggest influence on our motion from much larger depths (Kocevski & Ebeling 2006; Bilicki *et al.* 2011; see also Nusser *et al.* 2014). Finally, several other important cosmological signals – such as the integrated Sachs-Wolfe effect (ISW), CMB lensing on the large-scale structure (LSS) or baryon acoustic oscillations (BAO) – do not require full-sky coverage, but are only detectable with sufficiently wide-angle galaxy surveys probing large volumes.

State of the art in all-sky surveys of galaxies. Presently, the largest all-sky catalog of extended sources is the Two Micron All Sky Survey Extended Source Catalog (2MASS XSC, Jarrett *et al.* 2000). It contains approx. 1.5 million galaxies, of which 1 million are within the completeness limit of the survey, $K_s < 13.9$ mag. This catalog is however only photometric, and the largest all-sky 3D spectroscopic dataset is its subsample, the 2MASS Redshift Survey of 44,000 galaxies (2MRS, Huchra *et al.* 2012), with median redshift of only $\langle z \rangle = 0.03$. There have been attempts to expand 2MRS by adding redshifts of 2MASS galaxies available from other surveys such as 6dFGS or SDSS, which resulted in the the the 2M++ compilation by Lavaux & Hudson (2011) of 70,000 galaxies; the depth of such a sample is however non-uniform and limited to what is available from 2MRS on the parts of the sky not covered by other redshift surveys.

In addition to 2MASS, there exist two other major photometric all-sky datasets, namely the Wide-field Infrared Survey Explorer (WISE, Wright *et al.* 2010) and optical SuperCOSMOS (SCOS for short, Hambly *et al.* 2001). These samples include hundreds of millions sources, mostly galaxies at high latitudes. WISE mid-infrared data were gathered by an orbiting telescope, which makes them free of such issues as seeing or atmospheric glow. WISE currently provides only a point-source catalog with no 'XSC' release; however, its limited angular resolution will not allow to resolve sources beyond the depths already probed by 2MASS (Cluver *et al.* 2014). SCOS on the other hand was produced from digitized and calibrated photographic plate data, which are of limited accuracy compared to modern CCD-based surveys. Despite such shortcomings, these datasets will remain the largest all-sky astronomical catalogs at least until Gaia, which will however be dominated by stars. We have thus decided to take advantage of the information contained in 2MASS, WISE and SCOS, focusing on their usefulness for extragalactic science. Specifically, we have undertaken a dedicated program of combining them into multi-wavelength galaxy datasets and adding the third dimension through the methodology of photometric redshifts.

2. 2MASS Photometric Redshift catalog (2MPZ)

At the moment, about 30% of 2MASS galaxies have spectroscopic redshifts from other surveys, such as 2MRS, 2dFGRS, 6dFGS or SDSS. This percentage is likely to improve in the coming years, notably in the southern hemisphere with the planned TAIPAN survey, we cannot, however, hope for *complete* spectroscopic coverage of all the 2MASS galaxies in

Figure 1. All-sky Aitoff projection of 1 million galaxies in the 2MASS Photometric Redshift catalog (2MPZ) in Galactic coordinates, color-coded by photometric redshifts.

the coming years. Presently, the only way to obtain 3D information for this sample – even if of limited accuracy – is to estimate redshifts based on other information, such as fluxes,. This is the widely used technique of *photometric redshifts* (photo-z's), which so far has not been very popular for low-z samples because of limited availability of multiwavelength data. The situation has though changed mostly due to WISE, which together with SCOS allowed us to add the third dimension to the 2MASS XSC sample. For that purpose we used 2MASS J, H and K_s near-IR bands, together with mid-IR WISE ($W1$, 3.4 μm and $W2$, 4.6 μm) and optical B, R, I from SCOS. This gave us 8-band coverage for almost 95% of 2MASS galaxies (with incompleteness mostly at low Galactic latitudes), and the availability of comprehensive spectroscopic subsamples allowed to apply empirical (machine-learning) methodology to calculate photo-z's. We employed an artificial neural network algorithm (ANNz, Collister & Lahav 2004), trained on a representative spec-z subsample of 350,000 galaxies. As a result, we have produced the 2MASS Photometric Redshift catalog (2MPZ, Bilicki *et al.* 2014a), containing 940,000 galaxies with $\langle z \rangle = 0.07$, covering most of the sky. This dataset was publicly released in December 2013 and is available for download from `http://surveys.roe.ac.uk/ssa/TWOMPZ`. The photometric redshifts are well constrained: 1σ scatter in δz is $\sigma_{\delta z} = 0.015$, typical photo-z error is 13% and there are only 3% outliers beyond $3\sigma_{\delta z}$. Fig. 1 presents the all-sky distribution of 2MPZ sources, color-coded by photometric redshifts. Despite the tendency of the photo-z's to dilute radial information, the 3D cosmic web is evident, with major structures such as the Shapley Superconcentration prominent in the maps. Note that by using infrared-selected galaxies we are able to probe deep into the ZoA, although photo-z's at very low latitudes are of limited accuracy due to uncertainties in extinction and large star density compromising the photometric multiwavelength data.

First cosmological results from 2MPZ. The 2MPZ catalog has been already applied to several cosmological tests, of which some have been published, and others are in preparation. Appleby & Shafieloo (2014) used it to test the degree of isotropy in the Local Universe through the K_s-band luminosity function (LF), while Xu *et al.* (2014) showed that the sample is appropriate for galaxy cluster identification. In a forthcoming paper (Steward *et al.* 2014, in prep.) we will present the constraints on the ISW obtained by cross-correlating 2MPZ with Planck CMB data. Working in redshift shells (cf. Francis & Peacock 2010, FP10), we have found only mild preference for the ISW over no signal: less than 2σ (odds 3.5:1). This is an improvement over FP10 (who had odds 1.5:1), possible thanks to our much more accurate photo-z's, the lack of significant detection is however hardly surprising. 2MASS is too shallow to probe the full ISW signal, and the constraints would not be much more significant even if we had full spectroscopic coverage.

Other applications in progress of the 2MPZ data include a detailed study of the near-IR LF at $z \sim 0$ (cf. Branchini *et al.* 2012) from the largest sample to date (Feix *et al.*, in prep.), as well as constraining the bulk flow in a sphere of ~ 300 Mpc/h from LF variations (cf. Nusser *et al.* 2011; Feix *et al.* 2014 and Feix *et al.* in this volume). The catalog is now also analyzed in terms of its applicability to various clustering-related analyses: looking for the transition to homogeneity through the angular correlation function (cf. Alonso *et al.* 2014) or more general attempts to recover the 3D power spectrum and correlation function from photometric redshifts, setting the ground for upcoming and future surveys such as DES or Euclid. Many more applications of the catalog are of course expected.

3. Beyond 2MASS: 20 million galaxies from WISE×SuperCOSMOS

The depth of 2MPZ is limited by the shallowest of the three cross-matched samples, i.e. 2MASS. Both SCOS and WISE probe the Cosmic Web to much larger distances: the former is about 3 times deeper than 2MASS, while the latter reaches even farther. In addition, both contain a significant number of quasars, some at high redshifts (especially WISE). By pairing up WISE with SCOS we were thus able to access sources and redshift shells not available with 2MPZ. Here the sample is limited by the depth and accuracy of the optical data: we adopted $B < 21$ and $R < 19.5$ (AB) as the reliability limits, requiring also source detections in both these bands, while WISE was preselected to have measurements in $W1$ and $W2$. Cross-matching the two samples at $|b_{\mathrm{Gal}}| > 10°$ gives 170 million sources, however such a catalog is dominated by stars at the bright end and at low latitudes, so further cuts were needed to clean up the sample. Using Galaxy And Mass Assembly (GAMA, Driver *et al.* 2009) and SDSS DR10 (Ahn *et al.* 2014) spectroscopic data, we defined simple color cuts to purify the sample of stars and high-z quasars, which together with SCOS morphological information gave us about 20 million galaxies over $> 3\pi$ sr. The quasars present in our sample, which were treated as contamination here, will be a subject of a future study; we estimate that there might be even 2 million of them in the WISE×SCOS cross-match. A forthcoming paper (Krakowski *et al.* 2014) will present a more sophisticated method of source classification in our catalog, namely with the use of Support Vector Machines (e.g. Solarz *et al.* 2012; Małek *et al.* 2013) trained on SDSS DR10 spectroscopic data.

The next step was to calculate photometric redshifts for the galaxies in the sample. We used five photometric bands for that purpose: optical B, R and infrared $W1, W2, W3$. As in 2MPZ, we took the empirical approach (ANNz, Collister & Lahav 2004) but here the training sets were the most recent GAMA and SDSS DR10 data. The former of the two spectroscopic catalogs is crucial here, as it is complete in 3 equatorial fields to $r < 19.8$, which makes it deeper than our sample, hence representative for photo-z calibration. The resulting photometric redshifts have a median $\langle z \rangle \sim 0.2$, the redshift distribution is however broad and the sample probes the LSS reliably up to $z \sim 0.35$, with some information even at $z \gtrsim 0.4$. Normalized 1σ scatter of photo-z errors is $\sigma_{\delta z} = 0.03$, median error is 12% and there are 3% outliers over $3\sigma_{\delta z}$. With the present catalog this is the best performance attainable, as there is no other all-sky data of sufficient depth that could bring additional bands to our full sample. Further details regarding the

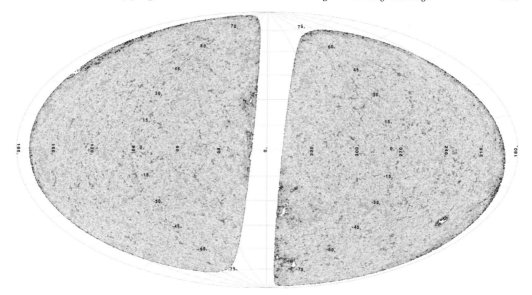

Figure 2. The Cosmic Web 2.5 Gyr ago: the large-scale structure of the Universe at $z = 0.2$. The plot shows 1.7 million galaxies from the WISE×SCOS photometric redshift catalog in a shell of $0.19 < z_{\rm phot} < 0.21$, in Aitoff projection in Supergalactic coordinates. The gaps are due to our Galaxy and the LMC obscuring the view.

WISE×SCOS photometric redshift catalog construction will be provided in a forthcoming paper (Bilicki *et al.* 2014b, in prep.). Fig. 2 shows an example of what the WISE×SCOS photo-z sample gives access to: the Cosmic Web as it was 2.5 Gyr ago. Pictured here are 1.7 million galaxies in a redshift shell of $0.19 < z_{\rm phot} < 0.21$; note that there are more sources in this narrow shell than in the entire 2MASS galaxy catalog.

Possible applications of the WISE×SuperCOSMOS photo-z catalog. The availability of the deep, almost all-sky WISE×SCOS data opens new possibilities of LSS analysis, as this is the first 3D galaxy catalog which probes the redshifts of $z = 0.2 \sim 0.3$ on 75% of the sky. In particular, it will be suitable for similar studies as done with 2MPZ, but in more distant redshift shells and in much larger volumes. Most generally, we expect it to allow for testing the Copernican Principle of isotropy and homogeneity of the Universe up to $z \sim 0.4$. It should help constrain the contributions to the bulk flows and pull on the Local Group from scales of > 1 Gpc, if there are any. It is also very promising in terms of various correlation analyses: it should provide a higher S/N of ISW than possible with 2MPZ; it can also be cross-correlated with the CMB lensing signal to constrain e.g. non-Gaussianity (Giannantonio & Percival 2014). Some other possible applications include those planned with such major photometric redshift surveys as DES, for instance measuring angular BAOs in redshifts shells to constrain expansion history (see e.g. Blake & Bridle 2005 for an early discussion of such applications). This is certainly not an exhaustive list; as we are planning to publicly release the final photometric redshift catalog, the community will be able to apply many other ideas to these data.

...and beyond: quasars and deep WISE data. Except for the WISE×SCOS $z \sim 0.2$ galaxy catalog described here, we are also planning to compile a quasar photo-z sample based on the same data. Our present estimates show that we should be able to identify about 2 million quasars on $> 3\pi$ sr (Krakowski *et al.* 2014, in prep.); it remains to verify how accurate quasar photo-z's will be possible with the limited photometric

information we have. In addition, as far as WISE itself is concerned, its all-sky 5σ limit is at ~ 17 mag (Vega) in the $W1$ band, which makes it useful to probe additional redshift layers beyond those available through the cross-match with SCOS. WISE itself is also a great AGN/quasar repository as already shown by several authors. All these properties give great promise for future cosmological studies that we are envisaging to undertake.

Acknowledgments. We thank the Wide Field Astronomy Unit at the IfA, Edinburgh for archiving the 2MPZ catalog, which can be accessed at `http://surveys.roe.ac.uk/ssa/TWOMPZ`. We made use of data products from the Two Micron All Sky Survey, Wide-field Infrared Survey Explorer, SuperCOSMOS Science Archive, Sloan Digital Sky Survey and Galaxy And Mass Assembly catalogs. Special thanks to Mark Taylor for his wonderful TOPCAT software, `http://www.starlink.ac.uk/topcat/`. The financial assistance of the South African National Research Foundation (NRF) towards this research is hereby acknowledged. MB was partially supported by the Polish National Science Center under contract #UMO-2012/07/D/ST9/02785.

References

Ahn, C. P., Alexandroff, R., Allende Prieto, C., Anders, F., *et al.* 2014, *ApJS*, 211, 2
Alonso, D., Bueno Belloso, A., Sánchez, F. J., García-Bellido, J., *et al.* 2014, *MNRAS*, 440, 1
Appleby, S. & Shafieloo, A. 2014, *arXiv:1405.4595*
Bilicki, M., Chodorowski, M., Jarrett, T., & Mamon, G. A. 2011, *ApJ*, 741, 1
Bilicki, M., Jarrett, T. H., Peacock, J. A., Cluver, M. E., & Steward, L. 2014, *ApJS*, 210, 9
Bilicki, M., Peacock, J. A., Jarrett, T. H., Maddox, N., Steward, L., & Cluver, M. E. 2014, in prep.
Blake, C. & Bridle, S. 2005, *MNRAS*, 363, 4
Branchini, E., Davis, M., & Nusser, A. 2012, *MNRAS*, 424, 1
Cluver, M. E., Jarrett, T. H., Hopkins, A. M., Driver, S. P., Liske, J., *et al.* 2014, *ApJ*, 782, 2
Collister, A. A. & Lahav, O. 2004, *PASP*, 116, 818
Driver, S. P., Norberg, P., Baldry, I. K., Bamford, S. P., Hopkins, A. M., *et al.* 2009, *A&G*, 50, 5
Erdogdu, P., Huchra, J. P., Lahav, O., Colless, M., Cutri, R. M., *et al.* 2006, *MNRAS*, 368, 4
Feix, M., Nusser, A., & Branchini, E. 2014, *arXiv:1405.6710*
Francis, C. L. & Peacock, J. A. 2010, *MNRAS*, 406, 1
Giannantonio, T. & Percival, W. J. 2014, *MNRAS*, 441, 1
Hambly, N. C., MacGillivray, H. T., Read, M. A., Tritton, S. B., *et al.* 2001, *MNRAS*, 326, 4
Hirata, C. M. 2009, *JCAP*, 09, 011
Huchra, J. P., Macri, L. M., Masters, K. L., Jarrett, T. H., Berlind, P., *et al.* 2012, *ApJS*, 199, 2
Jarrett, T. H., Chester, T., Cutri, R., Schneider, S., Skrutskie, M., & Huchra, J. P. 2000, *AJ*, 119, 5
Kashlinsky, A., Atrio-Barandela, F., Kocevski, D., & Ebeling, H. 2008, *ApJ*, 686, 2
Kocevski, D. D. & Ebeling, H. 2006, *ApJ*, 645, 2
Krakowski, T., *et al.* 2014, in prep.
Lavaux, G. & Hudson, M. J. 2011, *MNRAS*, 416, 4
Małek, K., Solarz, A., Pollo, A., Fritz, A., Garilli, B., Scodeggio, M., *et al.* 2013, *A&A*, 557, A16
Nusser, A., Branchini, E., & Davis, M. 2011, *ApJ*, 735, 2
Nusser, A. & Davis, M. 2011, *ApJ*, 736, 2
Nusser, A., Davis, M., & Branchini, E. 2014, *ApJ*, 788, 2
Planck Collaboration: Ade, P. A. R., Aghanim, N., *et al.* 2013, *arXiv:1303.5083*
Pullen, A. R. & Hirata, C. M. 2010, *JCAP*, 05, 027
Rowan-Robinson, M., Sharpe, J., Oliver, S. J., Keeble, O., *et al.* 2000, *MNRAS*, 314, 2
Solarz, A., Pollo, A., Takeuchi, T. T., Pępiak, A., Matsuhara, H., *et al.* 2012, *A&A*, 541, A50
Steward, L., Bilicki, M., Peacock, J. A., Jarrett, T. H., & Cluver, M. E. 2014, in prep.
Turnbull, S. J., Hudson, M. J., Feldman, H. A., Hicken, M., *et al.* 2012, *MNRAS*, 420, 1
Watkins, R., Feldman, H. A., & Hudson, M. J. 2009, *MNRAS*, 392, 2
Wright, E. L., Eisenhardt, P. R. M., Mainzer, A. K., Ressler, M. E., *et al.* 2010, *AJ*, 140, 6
Xu, W. W., Wen, Z. L., & Han, J. L. 2014, *arXiv:1406.0943*

The Zeldovich Universe:
Genesis and Growth of the Cosmic Web
Proceedings IAU Symposium No. 308, 2014
R. van de Weygaert, S. Shandarin, E. Saar & J. Einasto, eds.

© International Astronomical Union 2016
doi:10.1017/S1743921316009765

Measuring Large-Scale Structure at $z \sim 1$ with the VIPERS galaxy survey

Luigi Guzzo†

INAF - Osservatorio Astronomico di Brera, 20122 Milano, Italy
email: luigi.guzzo@brera.inaf.it

Abstract. The VIMOS Public Extragalactic Redshift Survey (VIPERS) is the largest redshift survey ever conducted with the ESO telescopes. It has used the Very Large Telescope to collect nearly $100,000$ redshifts from the general galaxy population at $0.5 < z < 1.2$. With a combination of volume and high sampling density that is unique for these redshifts, it allows statistical measurements of galaxy clustering and related cosmological quantities to be obtained on an equal footing with classic results from local redshift surveys. At the same time, the simple magnitude-limited selection and the wealth of ancillary photometric data provide a general view of the galaxy population, its physical properties and the relation of the latter to large-scale structure. This paper presents an overview of the galaxy clustering results obtained so far, together with their cosmological implications. Most of these are based on the $\sim 55,000$ galaxies forming the first public data release (PDR-1). As of January 2015, observations and data reduction are complete and the final data set of more than $90,000$ redshifts is being validated and made ready for the final investigations.

Keywords. catalogs, surveys, galaxies, distances and redshifts, techniques: spectroscopic, methods: data analysis, (cosmology:) large scale structure of universe, cosmology: observations

1. Motivation, design and status of VIPERS

Statistical measurements of large-scale structure in the galaxy distribution represent one of the pillars upon which the current "standard" model of cosmology rests. The power spectrum of this distribution contains a wealth of information on the history of cosmological fluctuations, which in principle allows precise constraints on cosmological parameters like the mean density of matter Ω_m and the baryon fraction. Additionally, galaxy clustering has emerged more recently as a powerful means to understand the origin of the apparent acceleration of cosmic expansion, discovered at the end of the last century (Riess *et al.* 1998, Perlmutter *et al.* 1999). The tiny "baryonic wiggles" in the shape of the power spectrum are the mark of a specific spatial scale, the remnant of the sound horizon size at the baryon drag epoch. One of the major discoveries of the past decade has been that there are enough baryons in the cosmic mixture to influence the dominant dark-matter fluctuations (Cole *et al.* 2005; Eisenstein *et al.* 2005) and leave a visible signature in the galaxy distribution. Such Baryonic Acoustic Oscillations (BAO) represent a formidable standard ruler to measure the expansion history of the Universe

† *On behalf of the VIPERS team* (http://vipers.inaf.it): **TO BE UPDATED**, U. Abbas, C. Adami, S. Arnouts, J. Bel, M. Bolzonella, D. Bottini, E. Branchini, A. Burden, A. Cappi, J. Coupon, O. Cucciati, I. Davidzon, S. de la Torre, G. De Lucia, C. Di Porto, P. Franzetti, A. Fritz, M. Fumana, B. Garilli, B. R. Granett, L. Guennou, A. Hawken, O. Ilbert, A. Iovino, J. Krywult, V. Le Brun, O. Le Fèvre, D. Maccagni, K. Malek, A. Marchetti, C. Marinoni, F. Marulli, H. J. McCracken, Y. Mellier, L. Moscardini, R. C. Nichol, L. Paioro, J. A. Peacock, W. J. Percival, S. Phleps, M. Polletta, A. Pollo, S. Rota, H. Schlagenhaufer, M. Scodeggio, A. Solarz, L. A. M. Tasca, R. Tojeiro, D. Vergani, M. Wolk, G. Zamorani **and** A. Zanichelli

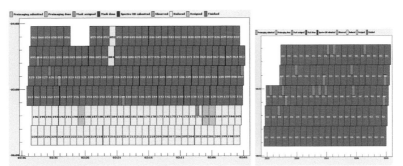

Figure 1. Pictorial view of the mosaic of 288 VIMOS pointings that cover the VIPERS survey area, extracted from the survey monitoring web page. At the date of writing this paper (January 2015) all fields have been observed. Colours here indicate where redshifts are already in the database (red) or data have been reduced and need to be quality checked before final acceptance (yellow and orange).

$H(z)$, similar to what has been originally done using supernovae (see e.g. Anderson *et al.* 2014). Additionally, the apparent anisotropy of clustering induced by the contribution of peculiar velocities to the measured redshifts, what we call Redshift Space Distortions (RSD, Kaiser 1987), provides us with a precious complementary tool which probes the growth rate of structure. This key information can break the degeneracy in explaining the observed expansion history, between the presence of a cosmological constant in Einstein's equations or a rather more radical revision of these equations, which we call "modified gravity". While RSD are a well-known phenomenon since long, their potential in the context of understanding the origin of cosmic acceleration has been recognized only in recent times (Guzzo *et al.* 2008).

Translating galaxy clustering observations into precise and accurate cosmological measurements, however, requires careful modelling of the effects of non-linear evolution, galaxy bias (i.e. how galaxies trace mass) and redshift-space distortions themselves. Galaxy surveys collecting broad samples of the galaxy population with a simple, well-defined selection function and extensive photometric information are crucial to ease this. In general, they allow us to play with different classes of tracers of the underlying matter density field (e.g. red and blue galaxies, groups, clusters), which are differently affected by the above effects. This has been the case of the two largest surveys of the local Universe, i.e. the Sloan Digital Sky Survey (SDSS, York *et al.* 2000) and the 2dF Galaxy Redshift Survey (2dFGRS, Colless *et al.* 2001).

The VIMOS Public Extragalactic Redshift Survey (VIPERS) was conceived to extend this concept, with comparable statistical accuracy, to the $z \sim 1$ Universe, pursuing the following main goals:

• To precisely and accurately measure galaxy clustering up to scales $\sim 100\,h^{-1}$ Mpc at $z \sim 1$ and obtain cosmological constraints from the power spectrum and correlation function, at an epoch when the Universe was about half its current age.

• To measure the growth of structure through RSD out to $z \sim 1$, possibly using different populations of tracers, as allowed by the broad selection function.

• To characterize the density field at such redshifts, tracing the evolution and non-linearity of galaxy bias and identifying non-linear structures as groups and clusters.

• To precisely measure statistical properties of the galaxy population (luminosity, colour, stellar mass) and their relationship with large-scale structure.

To achieve this, VIPERS exploits the unique capabilities of the VIMOS multi-object spectrograph at the ESO VLT. This is used to measure redshifts for $\sim 10^5$ galaxies over

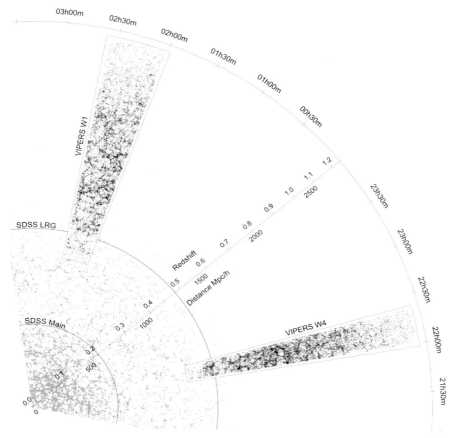

Figure 2. The large-scale distribution of galaxies within the two VIPERS survey volumes, integrated over the declination direction (1.84 and 1.54 degrees thick, respectively for W1 and W4). The new data are matched to the SDSS main and LRG samples at lower redshift (for which a 4-degree-thick slice is shown). Note how VIPERS pushes into a new epoch, with detailed sampling on all scales, similarly to SDSS Main below $z \sim 0.2$. Conversely, sparse samples like the SDSS LRG are excellent statistical probes of the largest scales, but (by design) fail to register the details of the underlying nonlinear structure.

~ 24 square degrees and located at $0.5 < z < 1.2$. Galaxy targets are selected from the W1 and W4 fields of the CanadaFranceHawaii Telescope Legacy Survey Wide catalogue (CFHTLS–Wide), which provides high-quality photometry in five bands $(ugriz)$.† The survey area is tiled with a mosaic of 288 VIMOS pointings, producing the geometry shown schematically in Fig. 1. The target sample is magnitude-limited to $i_{AB} = 22.5$, and further constrained to $z > 0.5$ through a thoroughly-tested $ugri$ colour pre-selection (Guzzo *et al.* 2014). Discarding the $z < 0.5$ population allows us to nearly double the sampling density within the high-redshift volume of interest, reaching an average sampling $> 40\%$. At the same time, the area and depth correspond to a volume of 5×10^7 h^{-3} Mpc^3, comparable to that of the 2dFGRS at $z \sim 0$ (Colless *et al.* 2001). Such a combination of sampling and volume is unique among redshift surveys at $z > 0.5$. VIPERS has used 372 hours of multi-object spectroscopy, plus 68.5 hours of pre-imaging, corresponding to an effective total investment of about 55 nights of VLT time. These and more details on the survey

† http://terapix.iap.fr/rubrique.php?id_rubrique=252

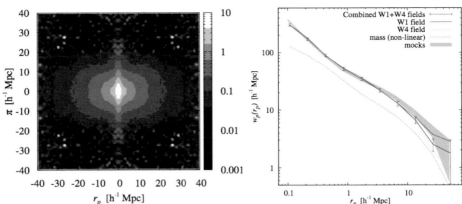

Figure 3. Redshift-space two-point correlation function over the full $0.5 < z < 1.0$ range, from the VIPERS PDR-1 catalogue. Left: $\xi(r_p, \pi)$, showing the well-defined signature of linear redshift distortions, i.e. the oval shape of the contours (de la Torre *et al.*, 2013). Right: the projected correlation function $w_p(r_p)$, obtained by integrating $\xi(r_p, \pi)$ along the π direction, for W1 and W4 fields separately and for the total sample. These are compared to the best-fitting Cold Dark Matter model for the mass (dotted line, prediction using the HALOFIT code, Smith *et al.* 2003). The shaded area corresponds to the $1-\sigma$ error corridor, computed from the scatter in the measurements of a large set of mock surveys, custom built for VIPERS (see de la Torre *et al.* 2013).

construction and the properties of the sample can be found in Guzzo *et al.* (2014) and Garilli *et al.* (2014).

2. The cosmic web at $0.5 < z < 1.2$

The simplest, yet striking result of a redshift survey like VIPERS can be appreciated in the the maps of the galaxy distribution shown in Fig. 2. The comparison of the two VIPERS wedges to the data of the local Universe evidences the remarkable combination of volume and dynamical range (in terms of scales sampled), which is a unique achievement of VIPERS at these redshifts. The two slices are about 2 degrees thick and display with great detail the cellular structure with which we have become familiar from local surveys and numerical simulations, but at an epoch when the Universe was about half its current age and for which no such detailed and extended map existed so far.

3. Redshift-space clustering and the growth of structure

One of the original goals of VIPERS is the measurement of the amplitude and anisotropy of the redshift-space two-point correlation function at redshift approaching unity. $\xi(r_p, \pi)$ and the corresponding projected function $w_p(r_p)$ from the first data release (PDR-1) catalogue are shown in the two panels of Fig. 3 (de la Torre *et al.* 2013). As we discuss in detail in that paper, crucial for these measurements is an accurate knowledge of several ancillary pieces of information from the survey, such as the photometric and spectro-scopic angular selection masks, the target sampling rate and the spectroscopic success rate. These allow us to assign a weight to every observed galaxy in the survey to correct for the overall incompleteness introduced by the different observational limitations. When the goal is to push systematic effects to percent values, this is not a trivial operation and very subtle effects need to be accounted for (see e.g. Pezzotta *et al.*, these proceedings).

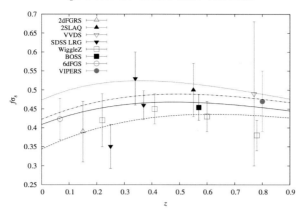

Figure 4. The growth rate of structure at $z \sim 0.8$ estimated from the VIPERS PDR-1 data (de la Torre *et al.* 2013). This is expressed as the product $f\sigma_8$ and compared to a compilation of recent measurements, respectively from 2dFGRS (Hawkins *et al.* 2003), 2SLAQ (Ross *et al.* 2007), VVDS (Guzzo *et al.* 200)8, SDSS LRG (Cabre & Gaztanaga 2009), WiggleZ (Blake *et al.* 2011), BOSS (Reid *et al.* 2012) and 6dFGS, (Beutler *et al.* 2012). Model predictions correspond to GR in a ΛCDM model with WMAP9 parameters (solid), while the dashed, dotted, and dot-dashed curves are respectively Dvali-Gabadaze-Porrati (dashed, Dvali *et al.* 2000), $f(R)$ (dotted), and coupled dark energy (dot-dashed) model expectations. For these latter two models, the analytical growth rate predictions from Di Porto *et al.* (2012) have been used.

The fingerprint of RSD is evident in the flattening of $\xi(r_p, \pi)$ along the line-of-sight direction (left panel of Fig. 3). The right panel shows the projected correlation function $w_p(r_p)$, obtained by projecting $\xi(r_p, \pi)$ along the line-of-sight direction, for the full PDR-1 data and for W1 and W4 separately. The agreement in shape and amplitude of these two independent sets is remarkable, although for $r > 10$ h^{-1} Mpc the measurement from W4 drops more rapidly.

Modelling $\xi(r_p, \pi)$, we have obtained a first estimate of the mean growth rate of structure at an effective redshift $z = 0.8$, $f\sigma_8 = 0.47 \pm 0.08$ (de la Torre *et al.* 2013). Our measurement, conventionally expressed as the product of the growth rate f with the *rms* amplitude of matter fluctuations, σ_8, is compared to a selection of literature results and models in Fig. 4, where the models have been self-consistently normalised to the Planck estimate of σ_8. The VIPERS value is in agreement with the predictions of general relativity within the current error bars. We expect the final VIPERS catalogue to push the error bars down to $\sim 10\%$, or to allow splitting the measurement into two bins with error comparable with the current one. We are also exploring the combined use of different tracers of RSD (e.g. blue and LRG galaxies, or groups – see Mohammad *et al.*, these proceedings), as a way to reduce statistical and systematic errors, a specific advantage allowed by the high sampling and broad selection function of VIPERS.

4. Constraints on cosmological parameters at $z \sim 1$

4.1. *Counts-in-cells two-point statistics: the Clustering Ratio*

Complementarily to computing directly the likelihood of the observed correlation function against a Λ-CDM model, in Bel *et al.* (2014), we applied a novel statistics based on the combination of two-point quantities, the *clustering ratio* $\eta_{g,R}$. This statistics is defined by smoothing the density field in spheres of a given radius R and then computing the ratio of the correlation function of the smoothed field on a multiple of this scale, $\xi(nR)$, and its variance $\sigma^2(R)$ (Bel & Marinoni 2014) . This is an indirect probe of the power spectrum shape, which in the linear regime is by definition insensitive to linear bias and linear redshift-space distortions of the specific galaxy tracers. In Bel *et al.* (2014), $\eta_{g,R}$ has been applied to the PDR-1 data, pushing it to mildly non-linear scales, verifying the impact of non-linearities and of the survey mask through our set of mock samples. The results show the robustness of this statistics, thanks to its definition as a ratio of quantities that are similarly affected by non-linearities, redshift distortions and bias. At an effective redshift $z = 0.93$, we use $R = 5 \, h^{-1}$ Mpc and $n = 3$ to estimate $\eta_{g,R}(15) = 0.141 \pm 0.013$. Assuming the best available priors on H_0, n_s and baryon fraction in a flat ΛCDM scenario, this yields a value $\Omega_{m,0} = 0.270^{+0.029}_{-0.025}$ for the current mean density parameter. This agrees very well with an estimate at $z \simeq 0.3$ performed with the same technique on the SDSS-LRG sample (Bel & Marinoni 2014). The two measurements together produce a combined estimate $\Omega_{m,0} = 0.274 \pm 0.017$. The clustering ratio is now being tested against its ability to detect the specific spectral feature of massive neutrinos, with promising results (Bel *et al.* 2015, in preparation).

4.2. *The VIPERS three-dimensional power spectrum*

The power spectrum in Fourier-space provides a complementary view to the correlation function of the spatial distribution of galaxies. Ideally, in Fourier space one could estimate the power spectrum in bins corresponding to independent wave modes k. In practice, the survey geometry and mask, especially in cases like VIPERS when the size in one direction is significantly smaller than the other two, give rise to a window function in Fourier-space which damps the power and couples modes. The specific window function must be accurately accounted for to compare models with the observations.

We present the VIPERS monopole power spectrum measurement in Rota et al (2015; see also these proceedings). To account for the variation in galaxy bias over the extended redshift range of VIPERS, we split the sample into two redshift bins $0.6 \leqslant z \leqslant 0.9$ and $0.9 \leqslant z \leqslant 1.1$. The power spectra are shown in the left panel of Fig. 5. To properly compare the measurements to theory, we must model the window functions separately for the two fields and the two redshift bins. We carry out a joint likelihood analysis of the four measurements containing the galaxy bias in each sample, the matter density and the baryon fraction. We compare the constraints found from the full sample $0.6 \leqslant z \leqslant 1.1$ with those derived from the joint analysis of the two redshift sub samples, as shown in the right panel of Fig. 5. We find that the degeneracy in the $\Omega_m - f_B$ plane is considerably lessened by splitting the sample into two redshift ranges. In Rota et al (2015) we explicitly show that the systematic dependence of the likelihood on the choice of the maximum wave number k_{max} to be included, is weaker for the high-redshift bin, where clustering is more linear. This bin is also providing a much tighter statistical constraint on the two considered quantities, driving the global likelihood contours. The right panel of Fig. 5 can be directly compared to Fig. 24 in Cole *et al.* (2005), to notice how the constraining power of VIPERS, under the same assumptions, is similar to that of the 2dFGRS at low redshift. When, in analogy to the clustering ratio measurement, we include the best available priors on H_0, n_s (from Planck) and $\Omega_b h^2$ (from BBN), we obtain a value $\Omega_{m,0} = 0.272^{+0.027}_{-0.030}$. This agrees within 1% with the result obtained on the same PDR-1 data with the clustering ratio (based on counts-in-cells). Such an agreement between two estimates in Fourier and configuration space is not at all obvious, given the way systematic effects can enter these computations.

4.3. *The combined CFHTLS-VIPERS power spectrum*

The VIPERS targeting algorithm may be extended over the full ~ 130 square degree area of the CFHTLS photometric sample. Thus, by combining the three-dimensional

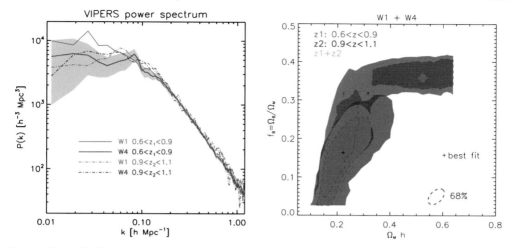

Figure 5. Left: Estimates of the three-dimensional power spectrum of the galaxy distribution from the VIPERS data. The four curves correspond to two redshift bins for the two separated fields W1 and W4, which have slightly different window functions. Right: corresponding constraints on the mean matter density and baryonic fraction. This plot can be directly compared to that at $z \sim 0.1$ from the 2df galaxy survey by Cole *et al.* (2005). See Rota *et al.* (2015) for details.

clustering from VIPERS with a projected clustering measurements from CFHTLS, we may improve the constraining power of the two individual surveys.

In this spirit, in a work that pre-dated the 2013 results from the PDR-1, we measured the projected power spectrum of the CFHTLS using the VIPERS target selection at $0.5 < z < 1.2$ (Granett *et al.* 2012). The redshift distribution from an early VIPERS catalogue was then used to deproject the angular distribution and directly estimate the three-dimensional power spectrum. Using a maximum likelihood estimator, we obtained an estimate of the galaxy linear bias for a VIPERS-like sample, on scales $k < 0.2 \, h^{-1}$ Mpc averaged over $0.5 < z < 1.2$, $b_g = 1.38 \pm 0.05$ (assuming $\sigma_8 = 0.8$). This agrees well with the more recent direct estimate from the PDR-1 sample, shown in Fig. 6 for three different redshift slices (Di Porto *et al.* 2014). Considering three photometric redshift slices, and marginalising over the bias factors while keeping other Λ-CDM parameters fixed, we find a value $\Omega_m = 0.30 \pm 0.06$ for the matter density parameter. These measurements were then used to constrain separately the total neutrino mass $\sum m_\nu$ and the effective number of neutrino species N_{eff} (Xia *et al.* 2012). The combination of our CFHTLS-VIPERS power spectrum with WMAP7 and a prior on the Hubble constant provided an upper limit of $\sum m_\nu < 0.29 \, \text{eV}$ and $N_{\text{eff}} = 4.17^{+1.62}_{-1.26}$ ($2 \, \sigma$ confidence levels). Combining with other large scale structure probes further improves these constraints, showing the complementarity of higher-redshift samples such as CFHTLS to other cosmological probes of the neutrino mass (see Granett *et al.* 2012 and Xia et al. 2012 for details).

5. Density field reconstruction, biasing and higher-order correlations

5.1. *Evolution and non-linearity of galaxy bias*

We already mentioned the recent paper based on the PDR-1 data set (Di Porto *et al.* 2014), in which we have studied the biasing relation and its nonlinearity over the redshift range covered by VIPERS. This analysis is based on an improved version of the method

L. Guzzo

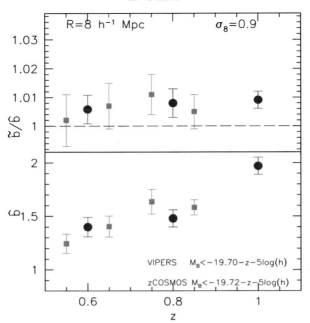

Figure 6. Evolution of the linear bias parameter (bottom) and of the bias non-linearity (top), for luminous galaxies in the VIPERS survey (circles), compared to previous measurements from the zCOSMOS survey (squares, Kovač et al. 2011). See Di Porto et al. (2014) for details.

originally proposed and developed by Dekel & Lahav (1999) and Sigad et al. (2000). It entails estimating the 1-point PDF of galaxies from counts in cells of a given size and, assuming a model for the mass PDF, inferring the full bias relation. We detect small ($\lesssim 3$ %), but significant deviations from linear bias, with the mean biasing function being close to linear in regions above the mean density. The linear bias parameter increases both with luminosity and with redshift, with a strong bias evolution only for $z > 0.9$, in agreement with some, but not all, previous studies. Fig. 6 shows the trend with redshift of the linear bias $\hat{b}$ and the bias non-linearity $\tilde{b}/\hat{b}$, compared to the previous result from zCOSMOS Kovač et al. (2011). For the first time out to $z = 1$, a significant increase of the bias with scale, from 4 to 8 h^{-1} Mpc, is also seen. The amount of nonlinearity in the bias depends on redshift, luminosity and scale, but no monotonic trend is detected. The VIPERS results demonstrate that the observed tension between the zCOSMOS and VVDS-Deep bias estimates at $z \sim 1$ is fully accounted for by cosmic variance.

5.2. Higher-order clustering

Starting from the measured PDF obtained from counts-in-cells, in a very recent paper (Cappi et al., 2015), we have measured the volume-averaged two-, three- and four-point correlation functions and the normalised skewness S_{3g} and kurtosis S_{4g}. This work is based on volume-limited subsamples of the PDR-1 catalogue and represents the first measurement of high–order correlation functions at $z \sim 1$ obtained from a spectroscopic survey. We find the expected hierarchical scaling between correlation functions of different order, over the whole range of scales and redshifts explored. There is no significant dependence of S_{3g} on luminosity. S_{3g} and S_{4g} do not depend on scale, except beyond $z \sim 0.9$, where they have higher values for $R \geqslant 10$ h^{-1} Mpc. We show this to be arising from only one of the two VIPERS fields. Once this is accounted for, neither

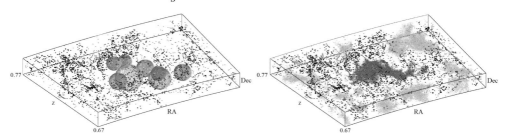

Figure 7. Identifying cosmic voids in the VIPERS PDR-1 W1 volume (see Micheletti *et al.* 2014 for details). This region includes the largest maximal sphere in the catalogue, which has a radius of 31 h^{-1} Mpc. The left-hand panel shows this sphere and the other six maximal spheres detected in this void region. The right-hand panel shows, in red, the centres of the overlapping significant spheres that make up the void, other void regions within this area of the survey are shown in orange. The grey and black points in both panels are the isolated and unisolated galaxies, respectively.

significant evolution with redshift of these quantities, nor of the variance is detected. The corresponding values and non-linearity of the bias function agree with the independent analysis of the same data set by Di Porto *et al.* (2014).

6. Cosmic voids

VIPERS probes the galaxy density field to $z \sim 1$ with unprecedented fidelity, allowing us to experiment with additional tests of cosmological and galaxy formation models. Cosmic voids are one such promising probe, still being scrutinised as to evaluate their real potential beyond possible systematics. Void statistics and profiles have gained popularity in recent times, thanks to the increased volumes sampled by redshift surveys (e.g. Sutter *et al.* 2012). The specific sampling of a redshift survey, i.e. its "resolution" in defining cosmic structures into the non-linear regime, however, is another important ingredient in order to properly define void catalogues. VIPERS, despite its nearly 2D "slice" geometry is one such survey, where the sampling density is nearly two orders of magnitude higher than surveys of much larger volume reaching similar redshifts (as e.g. the WiggleZ survey, Drinkwater *et al.* 2010).

Voids have been identified in VIPERS over the range $0.55 < z < 0.9$, using search method based upon the identification of empty spheres that fit between galaxies, as shown in Fig. 7. This allows us to characterise cosmic voids despite the presence of complex survey boundaries and internal gaps. The void size distribution and the void-galaxy correlation function are found to agree well with the equivalent functions in Λ-CDM mock catalogues. Details can be found in Micheletti *et al.* (2014). The anisotropy of the void-galaxy cross correlation function indicates that galaxies are outflowing from voids. Modelling this can be used to put loose constraints on the growth rate of structure (see Hawken *et al.*, this proceedings).

7. The interplay of large-scale structure and galaxy properties

7.1. *Dependence of clustering on galaxy luminosity and stellar mass*

The dependence of galaxy clustering on the luminosity and stellar mass of the galaxy tracers is a well-established effect at low redshifts (e.g. Norberg *et al.* (2001). At $z \sim 1$ results have been so far less conclusive, although both DEEP-2 (Coil *et al.* 2006) and VVDS-Deep (Pollo *et al.* 2006) suggest a steepening of the two-point correlation function,

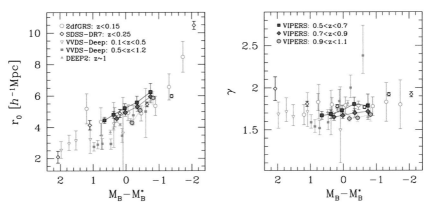

Figure 8. The dependence of clustering on galaxy luminosity (B-band absolute magnitude) as seen in VIPERS, compared to previous measurements at low and high redshift. The best-fit values of the correlation length r_0 and slope γ, for a power-law fit to the two-point correlation function, are plotted for three redshift slices of the PDR-1 catalogue.

when compared to local samples. This analysis has been repeated on the PDR-1 catalogue by Marulli *et al.* (2013) and the resulting dependence on luminosity is shown in Fig. 8. In the figure, the amplitude and slope derived from the projected function $w_p(r_p)$ are plotted and confronted with previous results. While the trend of r_0 seems to be consistent among surveys, both at low and high redshift, the strong steepening evidenced by VVDS-Deep at $z \sim 1$ (large value of γ, right panel), is not confirmed by the VIPERS data (green circles). In Marulli *et al.* (2013) we see similar trends when samples are selected based on ranges of stellar mass. In the same paper we provide a thorough discussion of the difficulties and biases that are intrinsic in studies of the clustering dependence on stellar mass, when using data that are flux-limited in origin.

7.2. *Reconstruction of the galaxy density field*

Properly accounting for the correlations between intrinsic galaxy properties and their spatial distribution can lead to improved estimates of the underlying matter clustering statistics and tighter constraints on cosmological parameters. To carry out a joint analysis, considering full covariances, we have recently turned to Bayesian methods (Granett *et al.* 2015). We have modelled the VIPERS galaxy counts as a function of luminosity and colour with a multivariate Gaussian distribution parameterised by the matter power spectrum, matter density, galaxy number density and galaxy bias. The high-dimensional parameter space is explored with Monte Carlo Gibbs sampling using the Wiener filter to reconstruct the density field. The Markov chain provides simultaneous reconstructions of the redshift-space density field and power spectrum as well as the galaxy biasing and luminosity functions. A slice through the galaxy distribution is shown in Fig. 9 alongside a Wiener density field reconstruction. The results of the Bayesian analysis, fully accounting for the covariances between galaxy properties and environment, give constraints on redshift-space clustering and galaxy intrinsic properties consistent with the individual VIPERS analyses, building a coherent picture of the cosmic density field (Granett et al. 2015).

8. Conclusions

VIPERS is the first redshift survey reaching beyond $z = 1$ while providing simultaneously volume and sampling that are comparable to the best surveys of the local Universe.

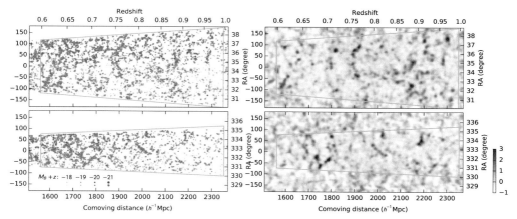

Figure 9. VIPERS cone diagrams for the fields W1 (top) and W4 (bottom). The left panels show the redshift-space positions of observed galaxies. The marker colour indicates the blue or red colour class and the marker size scales with B-band luminosity. The depth of the slice is $10\,h^{-1}$ Mpc. The orange line traces the field boundaries cut in the redshift direction at $0.6 < z < 1.0$. At right we show a slice of the density field taken from one step in the Markov chain. It represents the anisotropic Wiener reconstruction from the weighted combination of galaxy tracers. The field is filled with a constrained Gaussian realisation. The field has been smoothed with a Gaussian kernel with full-width-half-max $10\,h^{-1}$ Mpc. The colour scale gives the over-density value.

We are currently (January 2015) in the phase of validating the last $\sim 35,000$ redshifts collected during the just concluded, final observing season. These will bring the final redshift sample close to 90,000 redshifts, out of $\sim 95,000$ targeted galaxies (plus 2500 contaminating stars). This completed sample will be used to produce definitive results on several of the large-scale structure investigations described here. In addition, it will enable specific new analyses, which have been waiting for the largest possible survey volume. These include, e.g., the construction of a group/cluster catalogue (Iovino *et al.*, in preparation) or the combination of weak lensing and RSD.

The parallel, very important results obtained with the VIPERS PDR-1 data in the field of galaxy evolution have not been discussed in this review, which focuses on large-scale structure. Among these, the precise measurement of the stellar mass function evolution over the $0.5 < z < 1.2$ range needs to be mentioned. This study benefits of the large volume sampled by VIPERS, to provide an unprecedented sampling of the massive end at $z \sim 1$ (Davidzon *et al.* 2013). Also, a thorough study of the evolution of the Colour-Magnitude diagram and the galaxy luminosity function over the same range has been instead presented in Fritz *et al.* (2012).

Finally, VIPERS has distinguished itself for a very early public release of more than half of the survey data (the PDR-1 catalogue), in October 2013, i.e. only six months after submission of the first series of scientific papers. A similar policy will be followed for the complete survey data, for which we foresee a public distribution in 2016 (please monitor the survey web site, `http://vipers.inaf.it`, for related announcements). As a consequence, we expect (and encourage), a large number of further and important analyses, both on large-scale structure and galaxy evolution, to be produced by the scientific community at large.

References

Anderson, L., *et al.* 2014, *MNRAS*, 441, 24

Bel, J., VIPERS Team 2014, *A&A*, 563, A37

Bel, J. & Marinoni, C. 2014, *A&A*, 563, A36

Beutler, F., *et al.* 2012, *MNRAS*, 423, 3430

Blake, C., *et al.* 2011, *MNRAS*, 415, 2876

Cabrè, A. & Gaztanñaga, E. 2009, *MNRAS*, 393, 1183

Cappi, A., VIPERS Team 2015, *A&A*, submitted

Coil, A. L., *et al.* 2006, *ApJ*, 644, 671

Cole, S., *et al.* 2005, *MNRAS*, 362, 505

Colless, M., *et al.* 2001, *MNRAS*, 328, 1039

Davidzon, I., VIPERS Team 2013, *A&A*, 558, A23

de la Torre, S., Guzzo, L., Peacock, J. A., *et al.* 2013, *A&A*, 557, A54

Di Porto, C., Amendola, L., & Branchini, E. 2012, *MNRAS*, 419, 985

Di Porto, C., VIPERS Team 2014, *A&A*, submitted (arXiv:1406.6692)

Drinkwater, M. J., *et al.* 2010, *MNRAS*, 401, 1429

Dvali, G., Gabadadze, G., & Porrati, M. 2000, *Phys. Lett. B*, 485, 208

Eisenstein, D. J., *et al.* 2005, *ApJ*, 633, 560

Fritz, A., VIPERS Team 2014, 563, A92

Garilli, B., VIPERS Team 2014, 562, A23

Guzzo, L., Pierleoni, M., Meneux, B., *et al.* 2008, *Nature*, 451, 541

Guzzo, L., VIPERS Team 2014, *A&A*, 566, A108

Granett, B. R., VIPERS Team 2012, *MNRAS*, 421, 251

Granett, B. R., VIPERS Team 2015, *A&A*, submitted

Hawkins, E., *et al.* 2003, *MNRAS*, 346, 78

Kaiser, N. 1987, *MNRAS*, 227, 1

Kovač, K., *et al.* 2011, *ApJ*, 731, 102

Marulli, F., VIPERS Team 2013, *A&A*, 557, A17

Micheletti, D., VIPERS Team 2014, *A&A*, 570, A106

Norberg, P., *et al.* 2001, *MNRAS*, 328, 64

Perlmutter, S., *et al.* 1999, *ApJ*, 517, 565

Pollo, A., VIPERS Team 2006, *A&A*, 451, 409

Reid, B. A., *et al.* 2012, *MNRAS*, 426, 2719

Riess, A. G., *et al.* 1998, *AJ*, 116, 1009

Ross, N. P., *et al.* 2007, *MNRAS*, 381, 573

Rota, S., VIPERS Team 2015, *A&A*, submitted

Samushia, L., *et al.* 2013, *MNRAS*, 29, 1514

Smith, R. E., *et al.* 2003, *MNRAS*, 341, 1311

Sutter, P. M., *et al.* 2012, *ApJ*, 761, 44

York, D. G., *et al.* 2000, *AJ*, 120, 1579

Xia, J. Q., *et al.* 2012, *JCAP*, 6, 10

The Zeldovich Universe:
Genesis and Growth of the Cosmic Web
Proceedings IAU Symposium No. 308, 2014
R. van de Weygaert, S. Shandarin, E. Saar & J. Einasto, eds.
© International Astronomical Union 2016
doi:10.1017/S1743921316009777

Tracing high redshift cosmic web with quasar systems

Maret Einasto

Tartu Observatory,
Observatooriumi 1, 61602 Tõravere, Estonia
email: `maret.einasto@to.ee`

Abstract. We study the cosmic web at redshifts $1.0 \leqslant z \leqslant 1.8$ using quasar systems based on quasar data from the SDSS DR7 QSO catalogue. Quasar systems were determined with a friend-of-friend (FoF) algorithm at a series of linking lengths. At the linking lengths $l \leqslant 30\,h^{-1}$ Mpc the diameters of quasar systems are smaller than the diameters of random systems, and are comparable to the sizes of galaxy superclusters in the local Universe. The mean space density of quasar systems is close to the mean space density of local rich superclusters. At larger linking lengths the diameters of quasar systems are comparable with the sizes of supercluster complexes in our cosmic neighbourhood. The richest quasar systems have diameters exceeding $500\,h^{-1}$ Mpc. Very rich systems can be found also in random distribution but the percolating system which penetrate the whole sample volume appears in quasar sample at smaller linking length than in random samples showing that the large-scale distribution of quasar systems differs from random distribution. Quasar system catalogues at our web pages (`http://www.aai.ee/ maret/QSOsystems.html`) serve as a database to search for superclusters of galaxies and to trace the cosmic web at high redshifts.

Keywords. Cosmology: large-scale structure of the Universe; quasars: general

1. Introduction

According to the contemporary cosmological paradigm the cosmic web formed and evolved from tiny density perturbations in the very early Universe by hierarchical growth driven by gravity (van de Weygaert & Schaap (2009) and references therein). To understand how the cosmic web formed and evolved we need to describe and quantify it at low and high redshifts. Large galaxy redshift surveys like SDSS enable us to describe the cosmic web in our neighbourhood in detail. One source of information about the cosmic structures at high redshifts is the distribution of quasars — energetic nuclei of massive galaxies. Already decades ago several studies described large systems in quasar distribution (Webster(1982), Clowes & Campusano (1991), Clowes *et al.* (2012), Clowes *et al.* (2013)) which are known as Large Quasar Groups (LQGs). LQGs may trace distant galaxy superclusters (Komberg *et al.* 1996). The large-scale distribution of quasar systems gives us information about the cosmic web at high redshifts which are not yet covered by large and wide galaxy surveys.

The aim of our study is to study the high redshift cosmic web using data about quasar systems. We find quasar systems and analyse their properties and large-scale distribution at redshifts $1.0 \leqslant z \leqslant 1.8$ using quasar data from Schneider *et al.* (2010) catalogue of quasars, based on the Sloan Digital Sky Survey Data Release 7.

We select from this catalogue a subsample of quasars in the redshift interval $1.0 \leqslant z \leqslant 1.8$, and apply i-magnitude limit $i = 19.1$. In order to reduce the edge effects of our analysis, we limit the data in the area of SDSS sky coordinate limits $-55 \leqslant \lambda \leqslant 55$ degrees and $-33 \leqslant \eta \leqslant 35$ degrees. Our final sample contains data of 22381 quasars.

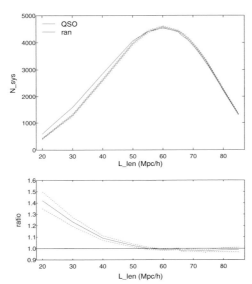

Figure 1. The number of quasar and random systems (upper panel) and the ratio of the numbers of quasar and random systems (lower panel) vs. the linking length. Red solid line denote quasar systems, and blue lines denote random systems, black line shows ratio 1.

The mean space density of quasars is very low, approximately $1.1 \cdot 10^{-6} (\mathrm{h}^{-1} \mathrm{Mpc})^{-3}$, therefore it is important to understand whether their distribution differs from random distribution. To compare quasar and random distributions we generated random samples with the same number of points, and sky coordinate and redshift limits as quasar samples.

We assume the standard cosmological parameters: the Hubble parameter $H_0 = 100\ h$ km s^{-1} Mpc^{-1}, the matter density $\Omega_{\mathrm{m}} = 0.27$, and the dark energy density $\Omega_\Lambda = 0.73$.

2. Results

We determined quasar and random systems with the friend-of-friend (FoF) algorithm at a series of linking lengths and present catalogues of quasar systems. FoF method collects objects into systems if they have at least one common neighbour closer than a linking length. At each linking length we found the number of systems in quasar and random samples with at least two members, calculated multiplicity functions of systems, and analysed the richness and size of systems. For details we refer to Einasto *et al.* (2014).

Up to the linking lengths approximately 50 h^{-1} Mpc the number of quasar systems is larger than the number of systems in random catalogues (Fig. 1), at larger linking lengths the number of systems becomes similar to that in random catalogue. The number of systems in both quasar and random catalogues reaches maximum at 60 h^{-1} Mpc. At higher values of the linking lengths systems begin to join into larger systems and the number of systems decreases. Multiplicity functions in Fig. 2 show that at the linking length $l = 85\ h^{-1}$ Mpc about half of quasars join the richest quasar system — a percolation occurs. In random catalogues the richest system is much smaller than the richest quasar system. Therefore FoF analysis shows that the distribution of quasars and the properties of quasar systems differ from random at small and large linking lengths.

In Fig. 3 we compare the distribution of diameters (maximum distance between quasar pairs in a system) of quasar and random systems at the linking lengths 30 and 70 h^{-1} Mpc. At the linking length 30 h^{-1} Mpc in the whole diameter interval the number of quasar

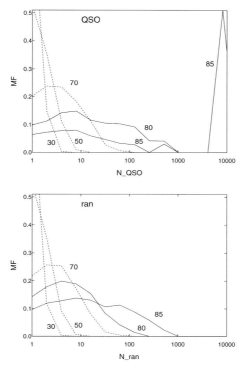

Figure 2. Multiplicity functions MF (the fraction of systems of different richness) of quasar (upper panel) and random (lower panel) data at the linking lengths 30, 50, 70, 80, and 85 h^{-1} Mpc.

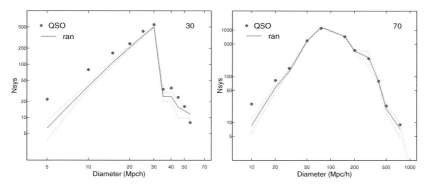

Figure 3. Number of quasar (red dots) and random (blue lines) systems of different diameter for linking lengths 30 and 70 h^{-1} Mpc (upper and lower panel, correspondingly).

systems with a given diameter is higher than that of random systems, the difference is statistically highly significant. At larger linking lengths ($l \geqslant 40\ h^{-1}$ Mpc; we show this for 70 h^{-1} Mpc) the number of quasar systems with diameters up to 20 h^{-1} Mpc is always larger than the number of random systems at these diameters. From diameters $\approx 30\ h^{-1}$ Mpc the number of systems of different diameter in quasar and random catalogues becomes similar. Among both quasar and random systems there are several very large systems with diameters larger than 500 h^{-1} Mpc.

In Fig. 4 we show the median, and minimum and maximum values of quasar system diameters vs. their richness at the linking lengths 50 and 70 h^{-1} Mpc. At the linking

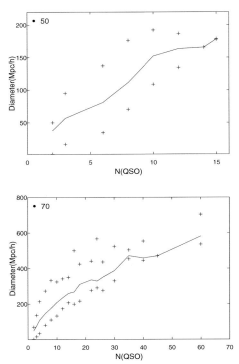

Figure 4. System richness N_{QSO} vs. their diameter D_{max} for quasars for linking lengths 50 and 70 h^{-1} Mpc. Lines show median values of diameters, crosses denote the smallest and the largest diameters.

length 50 h^{-1} Mpc the sizes of the richest quasar systems, $\approx 200\ h^{-1}$ Mpc, are comparable to the sizes of the richest superclusters in the local Universe (Einasto $et\ al.$ 1994). The mean space density of quasar systems of order of $10^{-7}(h^{-1}\mathrm{Mpc})^{-3}$, this is close to the mean space density of local rich superclusters (Einasto $et\ al.$ 1997).

The sizes of the largest quasar systems at $l = 70\ h^{-1}$ Mpc, $500 - 700\ h^{-1}$ Mpc, are comparable with the sizes of supercluster complexes in the local Universe (Einasto $et\ al.$ (2011c), Liivamägi $et\ al.$ (2012)). At this linking length we obtain systems of the same size also from the random catalogues.

We show in Fig. 5 the distribution of quasars in systems of various richness at linking length 70 h^{-1} Mpc in cartesian coordinates x, y, and z (see Einasto $et\ al.$ (2014)):

$$x = -d\sin\lambda,$$
$$y = d\cos\lambda\cos\eta, \qquad\qquad (2.1)$$
$$z = d\cos\lambda\sin\eta,$$

where d is the comoving distance, and λ and η are the SDSS survey coordinates. We plot in the figure also quasars from the richest systems at $l = 20\ h^{-1}$ Mpc (quasar triplets).

Visual inspection of Fig. 5 shows that very rich quasar systems form a certain pattern. In some areas of the figure there are underdense regions between rich quasar systems with diameters of about 400 h^{-1} Mpc (e.q. in the upper panel between $-1000 < x < 1000\ h^{-1}$ Mpc). The size of underdense regions in this figure is much larger than the sizes of typical large voids in the local Universe (see Einasto $et\ al.$ (2011a)) but is close to the sizes of the largest voids covered by SDSS survey (Einasto $et\ al.$ (2011b), Park

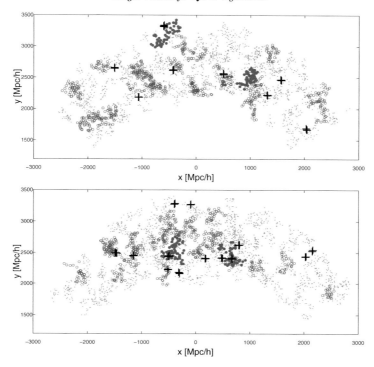

Figure 5. Distribution of QSO systems at the linking length 70 h^{-1} Mpc in x and y coordinates in two slices by z coordinate (upper panel: $z \leqslant 0\ h^{-1}$ Mpc, lower panel: $z > 0\ h^{-1}$ Mpc). Grey dots denote quasars in systems with $10 \leqslant N_{QSO} \leqslant 24$, blue circles denote quasars in systems with $25 \leqslant N_{QSO} \leqslant 49$, and red filled circles denote quasars in systems with $N_{QSO} \geqslant 50$. Black crosses denote quasar triplets at a linking length 20 h^{-1} Mpc.

et al.(2012)). Very rich systems were found also from random catalogues but the percolation analysis shows that the large-scale distribution of quasar systems differs from random distribution. We shall analyse the large scale distribution of quasar systems in detail in another study.

The richest system at the linking length $l = 70\ h^{-1}$ Mpc at $x \approx 1000\ h^{-1}$ Mpc and $y \approx 2500\ h^{-1}$ Mpc is the Huge-LQG described in Clowes *et al.* (2013). The presence of very rich systems as supercluster complexes is an essential property of the cosmic web, and do not violate homogeneity of the universe at very large scales, as claimed by Clowes *et al.* (2013).

3. Summary

We determined quasar systems at a series of linking lengths, and found that at small linking lengths their diameters and space density are similar to those of rich galaxy superclusters in the local Universe. At the linking lengths $l \geqslant 50\ h^{-1}$ Mpc the diameters of the richest quasar systems are comparable with the sizes of supercluster complexes in our cosmic neighbourhood, exceeding 500 h^{-1} Mpc. Systems of similar richness were determined also in random catalogues but the large-scale distribution of quasar systems differs from random distribution. We may conclude that quasar systems as markers of galaxy superclusters and supercluster complexes give us a snapshot of the high-redshift cosmic web. Quasar system catalogues serve as a database to search for high-redshift superclusters of galaxies and to trace the cosmic web at high redshifts.

I thank my coauthors Erik Tago, Heidi Lietzen, Changbom Park, Pekka Heinämäki, Enn Saar, Hyunmi Song, Lauri Juhan Liivamägi Jaan Einasto for enjoyable and fruitful collaboration.

The present study was supported by ETAG project IUT26-2, and by the European Structural Funds grant for the Centre of Excellence "Dark Matter in (Astro)particle Physics and Cosmology" TK120.

References

Clowes, R. G. & Campusano, L. E. 1991, *MNRAS*, 249, 218

Clowes, R. G., Campusano, L. E., Graham, M. J., & Söchting, I. K. 2012, *MNRAS*, 419, 556

Clowes, R. G., Harris, K. A., Raghunathan, S., *et al.* 2013, *MNRAS*, 429, 2910

Einasto, M., Einasto, J., Tago, E., Dalton, G. B., & Andernach, H. 1994, *MNRAS*, 269, 301

Einasto, M., Tago, E., Jaaniste, J., Einasto, J., & Andernach, H. 1997, *A&AS*, 123, 119

Einasto, M., Liivamägi, L. J., Tempel, E., *et al.* 2011c, *ApJ*, 736, 51

Einasto, M., Liivamägi, L. J., Tago, E., *et al.* 2011b, *A&A*, 532, A5

Einasto, J., Suhhonenko, I., Hütsi, G., *et al.* 2011a, *A&A*, 534, A128

Einasto, M. and Tago, E. and Lietzen, H. *et al.* 2014, *A&A*, 568, A46

Komberg, B. V., Kravtsov, A. V., & Lukash, V. N. 1996, *MNRAS*, 282, 713

Liivamägi, L. J., Tempel, E., & Saar, E. 2012, *A&A*, 539, A80

Park, C., Choi, Y.-Y., Kim, J., *et al.* 2012, *ApJL*, 759, L7

Schneider, D. P., Richards, G. T., Hall, P. B., *et al.* 2010, *AJ*, 139, 2360

van de Weygaert, R. & Schaap, W. 2009, in Lecture Notes in Physics, Berlin Springer Verlag, Vol. 665, Data Analysis in Cosmology, ed. V. J. Martínez, E. Saar, E. Martínez-González, & M.-J. Pons-Bordería, 291–413

Webster, A. 1982, *MNRAS*, 199, 683

The Zeldovich Universe:
Genesis and Growth of the Cosmic Web
Proceedings IAU Symposium No. 308, 2014
R. van de Weygaert, S. Shandarin, E. Saar & J. Einasto, eds.

© International Astronomical Union 2016
doi:10.1017/S1743921316009789

Improved correction of VIPERS angular selection effects in clustering measurements

A. Pezzotta[1,2], B.R. Granett[2], J. Bel[2], L. Guzzo[1,2], S. de la Torre[3] and VIPERS team[*]

[1]Università di Milano Bicocca, Milano, Italy
[2]INAF - Osservatorio Astronomico di Brera, Merate(LC), Italy
[3]Aix, Marseille Université, CNRS, LAM, Marseille, France
email: pezzotta@brera.inaf.it
[*]see http://vipers.inaf.it for the full member list

Abstract. Clustering estimates in galaxy redshift surveys need to account and correct for the way targets are selected from the general population, as to avoid biasing the measured values of cosmological parameters. The VIMOS Public Extragalactic Redshift Survey (VIPERS) is no exception to this, involving slit collisions and masking effects. Pushed by the increasing precision of the measurements, e.g. of the growth rate f, we have been re-assessing these effects in detail. We present here an improved correction for the two-point correlation function, capable to recover the amplitude of the monopole of the two-point correlation function $\xi(r)$above 1 h^{-1} Mpc to better than 2 %.

Keywords. Large-scale structure, clustering, cosmological parameters

---

VIPERS has used the VIsible Multi-Object Spectrograph (VIMOS) at the ESO VLT to collect $\sim 90,000$ spectra for galaxies in the redshift range $0.5 < z < 1.2$. VIMOS spectra are collected within four distinct quadrants on the sky, corresponding to four metal masks on which slits are carved. With the VIPERS selection function and a single-pass strategy (see Guzzo *et al.* 2014), on average about 40% of the galaxies in the full parent sample get targeted by a slit. This fraction depends on the specific pointing/quadrant, defining a Target Sampling Rate TSR$_i$(Q)). Simulations show that this introduces a constant 10% systematic damping in the amplitude of $\xi(r)$ on scales above 1 h^{-1} Mpc. This is essentially due to low-density regions being more represented than high-density ones in the overall pair counting. A second, interconnected effect happens at small angular separation due to "proximity bias", i.e. to the fact that two galaxies closer than a typical slit dimension cannot be both observed spectroscopically in a single visit. This effect, if not corrected, introduces a 50% systematic damping on scales below 1 h^{-1} Mpc.

To correct such effects, de la Torre *et al.*(2013) introduced appropriate weights when computing two-point correlations. Large-scale TSR variations were accounted for assigning to each galaxy a weight $w^Q = 1/\text{TSR(Q)}$. Then, each galaxy pair was then up-weighted using an angular completeness function (Fig. 1), defined as the ratio between the angular correlation functions of the observed parent samples. With such correction, the original amplitude of the (projected) correlation function above 1 h^{-1} Mpc could be recovered, for galaxies in the distant redshift bin of VIPERS ($z > 0.75$), but a residual lack of amplitude $\sim -4\%$ remained at smaller redshifts. We present here an improvement over this early scheme.

To better account for the non-homogeneous sampling introduced by the target selection, we first introduce a galaxy up-weighting based on a more local value of the TSR. Rather than using a value averaged over a single quadrant, we define it over a smaller aperture of given shape and dimension around each galaxy. This is similar to what Pollo

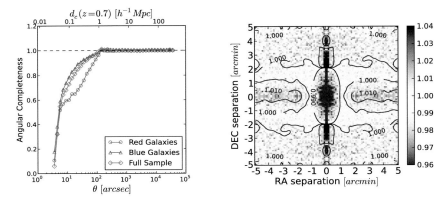

Figure 1. 1d (left) and 2d (right) angular completeness functions computed using VIPERS mock samples. It is remarkable how the spectrum masking effect is able to reduce the completeness below a typical scale of about 100 arcsec.

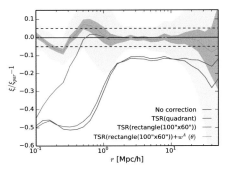

Figure 2. Fractional error on measurements of the real-space monopole $\xi(r)$. Shaded regions stand for the dispersion on the mean value (26 mocks) and on the single measurement for our final correction. w^A marks the small-scale correction using the completeness function.

et al. (2005) developed for the VVDS-Deep survey. We test the impact of two different patterns, i.e. circular and rectangular. The latter shape is actually important as it mimics the "shadow" of galaxy spectra, with its anisotropy between RA and DEC. The dimension of these regions are calibrated using the already cited angular completeness function (1d and 2d respectively, Fig. 1). We couple this "local TSR" correction to the usual smaller-scale one based on the angular correlation functions. The results (Fig. 2) are excellent on a large set of statistical tools (projected correlation function, multipoles, power spectrum) and also for different colour classes and redshift bins. Specifically, we recover the clustering amplitude above 1 h^{-1} Mpc to better than $\sim$2% using our improved TSR correction and to better than 5% below this scale, through the angular correlation function ratio.

References

Guzzo, L., the VIPERS team, 2014, *A& A*, 566, A108

de la Torre, S., the VIPERS team, 2013, *A& A*, 557, A54

Pollo, A., *et al*, 2005, *A& A*, 439, 887-900

The Zeldovich Universe:
Genesis and Growth of the Cosmic Web
Proceedings IAU Symposium No. 308, 2014
R. van de Weygaert, S. Shandarin, E. Saar & J. Einasto, eds.

© International Astronomical Union 2016
doi:10.1017/S1743921316009790

Measuring the VIPERS galaxy power spectrum at $z \sim 1$

Stefano Rota[1], Julien Bel[1], Ben Granett[1], Luigi Guzzo[1] and VIPERS Team

[1]INAF - Osservatorio astronomico di Brera,
Via Bianchi 46, I-23807, Merate (LC), Italy
email: stefano.rota@brera.inaf.it

Abstract. The VIMOS Public Extragalactic Redshift Survey [VIPERS, Guzzo *et al.* (2014)] is using the VIMOS spectrograph at the ESO VLT to measure redshifts for $\sim 100,000$ galaxies with $I_{AB} < 22.5$ and $0.5 < z < 1.2$, over an area of 24 deg^2 (split over the W1 and W4 fields of CFHTLS). VIPERS currently provides, at such redshifts, the best compromise between volume, number of galaxies and dense spatial sampling. We present here the first estimate of the power spectrum of the galaxy distribution, P(k), at redshifts $z \sim 0.75$ and $z \sim 1$, obtained from the $\sim 55,000$ redshifts of the PDR-1 data release. We discuss first constraints on cosmological quantities, as the matter density and the baryonic fraction, obtained for the first time at an epoch when the Universe was about half its current age.

1. Overview

In the estimation of the galaxy power spectrum, the survey shape and volume play a dominant role. The observed power P_{obs} is, in fact, a convolution of the true one, P_t, with a window function (WF), W, that depends on the geometry of the survey:

$$P_{obs}(\mathbf{k}) = \int \frac{d_3 k}{(2\pi)^3} P_t(\mathbf{k}') |W(\mathbf{k} - \mathbf{k}')|^2 \qquad (1.1)$$

The narrower is the dimension of a survey along a given direction, the broader is the WF in Fourier space, leading to a strong mixing of modes through equation 1.1. The effect of the WF on the recovered power is a strong suppression on small-k modes (large scales). These effects are particularly severe in the case of VIPERS, which has a relatively small angular aperture along the declination, even more so at the current stage, with the PDR-1 public release covering $\sim 65\%$ of the final sample. To understand and overcome the impact of the WF, we have used a suite of 26 mock samples accurately reproducing the VIPERS selection function and footprint, built from the MultiDark simulation [Prada *et al.* (2012)] and described in de la Torre *et al.* (2013). A series of accurate tests have been performed, including the VIPERS sampling strategy and other subtleties (Rota *et al.*, 2015, in preparation). The results of this work convincingly demonstrate that we are able to correctly model the effects of sampling and survey geometry, such that unbiased cosmological information can be extracted from the measured P(k). Briefly, this is obtained by matching the model power spectrum to the observed P_{obs}, after the former has been brought to redshift space and then convolved with the WF (see Rota *et al.* 2015 for details). Particular care has been taken in including the effect of redshift-space distortions (RSD). This has been done *before* performing the convolution with the WF, an approach usually avoided due to the time-consuming nature of the required 3D convolution.

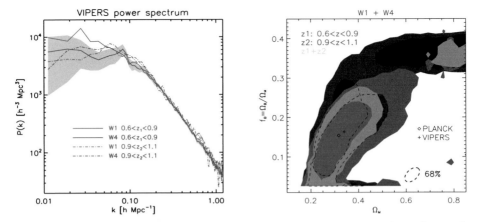

Figure 1. *Left:* measured power spectrum in the two redshift bins, [0.6-0.9] and [0.9-1.1] from the two independent W1 and W4 fields of VIPERS. *Right:* chi-square contours on the $f_B - \Omega_M$ plane for the two redshift bins (low redshift red contours, high redshift blue contours). The green region corresponds to a joint likelihood between z_1 and z_2.

To test the robustness of our approach, the measured power spectrum from each of the 26 VIPERS mock catalogues is fitted with the model, using the standard chi-square technique including an estimate of the covariance matrix (obtained from a larger suite of 200 mock surveys, built using the Pinocchio code [Monaco *et al.* (2002)], and then re-normalized to match the variance of the "primary" mocks). The averaged redshift-space power spectrum from the 26 mocks is measured separately for the W1 and W4 areas, each split in turn into two redshift bins, $0.6 < z_1 < 0.9$ and $0.9 < z_2 < 1.1$. We vary the matter density parameter and the bias factor (assumed to be linear and scale independent), and fix all the other parameters to the known MultiDark values. In all cases we recover the true value Ω_M of the MultiDark simulation within statistical errors.

2. Results

We split the VIPERS data as done with the mocks, building four independent samples. These four data sets are characterised by two slightly different bias values and growth factors (at different redshifts) and two slightly different window functions (W1 and W4). We can handle these differences at best by treating each of them separately. As shown in the left panel of Figure 1, at $k > 0.1\,h^{-1}$ Mpc, where the effect of the WF is negligible, the combined effect of bias/growth is visible, with a slightly larger amplitude for the two higher-z samples (higher bias due to larger mean luminosity). When accounting for this, on large scales ($k < 0.1\,h^{-1}$ Mpc) the four power spectra are all compatible to each other within statistical errors, showing that small differences between the four WFs are smaller than statistical fluctuations.

Finally, in the right panel we show the results of a first model fit to the VIPERS measured power spectra in the range $0.01 < k < 0.3\,[h\,\mathrm{Mpc}^{-1}]$. In analogy with the measurement at $z \sim 0$ from the 2dFGRS [Percival *et al.* (2001), Cole *et al.* (2005)], we impose flat priors on the baryonic fraction, $f_B = \Omega_B/\Omega_M$, the matter density, Ω_M and the bias, while fix the other cosmological parameters to Planck values [Planck Collaboration *et al.* (2014)]. In order to estimate the marginalised (over the bias) posterior likelihood distribution, we run a MCMC on the combined W1 and W4 measurements (accounting for the different WF), but treating separately the two redshift bins, given the different bias parameters (blue and red contours). The green contours show the joint likelihood between z_1 and z_2. **The best-fit values correspond to** $f_B = 0.16^{+0.08}_{-0.12}$ **and** $\Omega_M = 0.34^{+0.11}_{-0.14}$. These values agree well with the Planck ones, **although they are not completely independent**.

References

Guzzo, L., Scodeggio, M., Garilli, B., *et al.* 2014, A& A, 566, AA108

de la Torre, S., Guzzo, L., Peacock, J. A., *et al.* 2013, A& A, 557, A54

Prada, F., Klypin, A. A., Cuesta, A. J., Betancort-Rijo, J. E., & Primack, J. 2012, MNRAS, 423, 3018

Monaco, P., Theuns, T., & Taffoni, G. 2002, MNRAS, 331, 587

Percival, W. J., Baugh, C. M., Bland-Hawthorn, J., *et al.* 2001, MNRAS, 327, 1297

Cole, S., Percival, W. J., Peacock, J. A., *et al.* 2005, MNRAS, 362, 505

Planck Collaboration, Ade, P. A. R., Aghanim, N., *et al.* 2014, A& A, 571, AA16

CHAPTER 3B.

Surveys and Observations: Local Universe

IAU308 symposium gathers for the official opening reception.
Photo courtesy: Peter Coles

Caustics in the Local Universe:
Johan Hidding & Job Feldbrugge in front of an Estonian farmhouse
Photo courtesy: Steven Rieder

The Zeldovich Universe:
Genesis and Growth of the Cosmic Web
Proceedings IAU Symposium No. 308, 2014
R. van de Weygaert, S. Shandarin, E. Saar & J. Einasto, eds.
© International Astronomical Union 2016
doi:10.1017/S1743921316009807

Dynamics of galaxy structures in the Local Volume

I. D. Karachentsev

Special Astrophysical Observatory,
Russian Academy of Sciences N.Arkhyz, KChR, 369167, Russia
email: ikar@sao.ru

Abstract. I consider a sample of 'Updated Nearby Galaxy Catalog' that contains eight hundred objects within 11 Mpc. Environment of each galaxy is characterized by a tidal index Θ_1 depending on separation and mass of the galaxy Main Disturber (=MD). The UNGC galaxies with a common MD are ascribed to its 'suite' and ranked according to their Θ_1. Fifteen the most populated suites contain more than half of the UNGC sample. The fraction of MDs among the brightest galaxies is almost 100% and drops to 50% at $M_B = -18$ mag. The observational properties of galaxies accumulated in UNGC are used to derive orbital masses of giant galaxies via motions of their satellites. The average orbital-to-stellar mass ratio for them is $M_{orb}/M_* \simeq 30$, corresponding to the mean local density of matter $\Omega_m \simeq 0.09$, i.e 1/3 of the global cosmic one. The dark-to-stellar mass ratio for the Milky Way and M31 is typical for other neighboring giant galaxies.

Keywords. cosmology: observations - dark matter - galaxies: groups: general

1. Introduction and the data sample

As observational data show, the bulk of galaxies inhabit the groups, like our Local Group. Due to their cosmic abundance, the galaxy groups make a main contribution to the average density of matter in the universe. Most of the groups are concentrated in the filaments and sheets, forming a large-scale 'cosmic web' (Zeldovich 1970, Shandarin *et al.* 2004, Einasto *et al.* 2011). Until recently, the scarcity of data on the distances of even the closest galaxies was the major obstacle in the development of observational cosmology in the Local universe.

The recently published Updated Nearby Galaxy Catalog (= UNGC, Karachentsev *et al.*, 2013) contains a summary of data on radial velocities, distances and other observable parameters of about 800 galaxies located within a 11 Mpc radius around us. More than 300 galaxies of this sample have accurate distance measurements with a $(5 - 10)\%$ accuracy obtained by the Tip of the Red Giant Branch from observations with the Hubble Space Telescope. Due to the proximity of the UNGC- objects, the kinematic data density in the catalog proves to be one order higher than in the sample of Sloan Digital Sky Survey (Abazajian *et al.* 2009). This circumstance, and the presence of individual distance measurements in many UNGC galaxies allows us to investigate the structure of nearby groups and their vicinities with unprecedented detail.

2. Neighboring giants and their suites

Possessing the data on the distances and luminosities of eight hundred galaxies of the Local Volume, Karachentsev *et al.* (2013) have determined for each galaxy its tidal index

$$\Theta_1 = \max[\log(M_n^*/D_n^3)] + C, \ \ n = 1, 2, ...N,$$

175

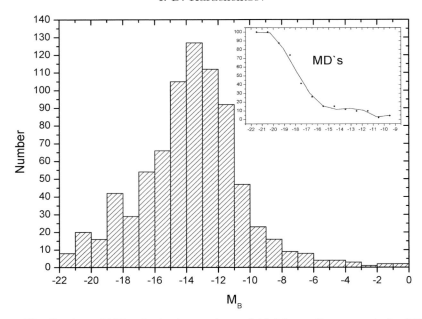

Figure 1. Distribution of 795 galaxies in a sphere of 11 Mpc radius around the Milky Way on absolute B-magnitudes, corrected for the internal and external extinction. The inset shows what percentage of these galaxies in each bin act as the Main Disturber.

where M^* is the stellar mass of the neighboring galaxy, and D_n is its spatial separation from the considered galaxy. The stellar mass of the galaxy was assumed to be equal to its K-band luminosity at $M/L_K = 1 \times M/L$. Ranking the surrounding galaxies by the magnitude of their tidal force, $F_n \sim M^*/D^3$, allowed to find the most influential neighbor, called the Main Disturber (= MD). Here the ratio of the total mass of the galaxy to its stellar mass was considered to be constant regardless of the luminosity and morphology of galaxies. The constant $C = -10.96$ was chosen so that the galaxy with $\Theta_1 = 0$ was located at the zero velocity sphere relative to its MD. In other words, the galaxy with $\Theta_1 \geqslant 0$ was regarded as gravitationally related to its MD as their crossing time was shorter than the age of the universe.

The galaxies which have a common MD can be combined in a certain association, or a MD suite. At that, an aggregate of suite members with positive values is quite consistent with the notion of a physically bound group of galaxies. The most massive MDs possess the most populous suites. The total number of companions around 15 most massive galaxies makes up about a half of the total population of the Local Volume.

Fig. 1 represents the distribution of galaxies of the Local Volume by the absolute B-band magnitude. The inset picture shows what fraction of the MDs as function of the absolute magnitude. The relative number of MDs among the brightest galaxies is close to 100%. The fraction of MDs decreases towards the low-luminosity galaxies, dropping below 50% at $M_B - 18$ mag.

The distribution of 351 companions by the radial velocity difference and projection separation relative to their main galaxies is presented in three panels of Fig.2. The upper panel of the figure shows the $\{\Delta_V, R_p\}$ diagram for 31 companions of the Milky Way = MW (squares) and 39 members of the M 31 = Andromeda suite (diamonds). The companions of massive galaxies with the positive tidal index, considered to be physical, are represented by closed symbols, while the members of the suites with $\Theta_1 = (0, -0.5)$

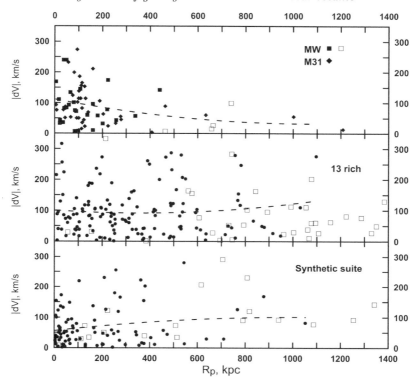

Figure 2. Line-of-sight velocity of the suite members relative to the main galaxy as a function of their projected linear separation.

are shown by the open symbols. The extension of the companion sample by the objects with slightly negative values of was done not to miss some possible physical members of the group, in which the distances are as yet measured with low accuracy. The objects in this boundary category may appear to be both the real companions of main galaxies or belong to the population of general field.

The middle panel of Fig.2 shows the $\{\Delta V, R_p\}$ distribution for 174 members of rich suites around 13 other massive nearby galaxies. Prospective physical companions (N = 142) are also marked here by solid symbols. In addition to 15 rich suites, the Local Volume comprises a lot of small suites, where the radial velocities are measured in one or several presumed companions. We have combined these small suites in a composite suite. The $\{\Delta V, R_p\}$ diagram for 107 companions uniting small suites is represented on the lower panel of Fig. 2. The dashed lines in all the three panels show quadratic regressions of the velocity difference on the projection separation of companions.

3. Orbital and stellar masses

If the group is dominated by a massive galaxy, surrounded by a set of test particles with random orientation of their orbits, one can use the mass estimate (Karachentsev,2005):

$$M_{orb} = (16/\pi) \times G^{-1} \times \langle \Delta V_{12}^2 \times R_{p12} \rangle,$$

where G is the gravitational constant, ΔV_{12} and R_{p12} are the velocity difference and the projected separation of companions relative to the main galaxy. The prevailing orbit

Table 1. Basic properties of the nearby suites.

MD	N	N_v	$\langle R_p \rangle$ kpc	$\langle \Delta V \vert \rangle$ km s^{-1}	M_{MD}^* $dex(10)M_\odot$	$\langle M_{orb} \pm \rangle$ $dex(12)M_\odot$
MW	38	27	121	90	3.5	1.44±0.46
M31	42	39	198	93	5.4	1.76±0.33
M81	53	26	219	116	8.5	4.89±1.41
N5128	37	15	343	110	8.1	6.71±2.09
N4594	32	6	577	153	20.0	28.47±17.80
N3368	31	20	408	150	6.8	17.00±4.30
N4258	31	11	316	96	8.7	3.16±1.01
N4736	31	14	515	50	4.1	2.67±0.90
N5236	28	10	294	57	7.2	1.06±0.28
N253	25	7	500	51	11.0	1.51±0.59
N3115	12	6	215	82	8.9	3.43±2.00
M101	11	6	167	76	7.1	1.47±0.67
IC342	10	8	321	66	4.0	1.810±.82
N3627	8	7	254	69	10.2	1.45±0.39
N6946	8	6	163	60	5.8	0.66±0.34

The columns contain: (1) name of the suite/group by its main galaxy, (2) the total number of satellites, (3) the number of physical companions with measured radial velocities, (4) the average projection separation of the companions from the MD, (5) the mean absolute value of the radial velocity difference of the companions relative to the MD, (6) the main galaxy stellar mass, (7) the value of orbital mass of the group.

eccentricity is assumed to be $1/\sqrt{2}$. The basic characteristics of the considered suites are presented in Table 1.

4. The Milky Way and Andromeda suites as compared with others.

Modeling the structure and kinematics of galaxy groups within the ΛCDM paradigm, many authors (Libeskind *et al.* 2010, Knebe *et al.* 2011) choose the Local Group to make a comparison with the observational data. Previously Karachentsev *et al.* (2014) noted that judging on some morphological features the groups of galaxies around the MW and M31 are not quite typical.

Six histograms of Fig.3 represent the distributions of 15 most populated suites in the Local Volume based on their global parameters. The groups of galaxies around the MW and Andromeda are marked with 'M' and 'A', respectively. The histogram data show that based on their stellar masses, and orbital masses, both MW and M 31 do not get in the top ten most massive galaxies of the Local Volume, whereas based on the orbital-to-stellar mass ratio ~ 30 both groups are not significantly different from the rest.

5. Masses derived from Hubble flows around the nearby groups.

A high density of observational data on the radial velocities and distances of galaxies in the Local Volume gives an opportunity to determine the masses of nearby groups not only by the virial motions, but also by perturbations of the Hubble flow around them. This idea was proposed by Lynden-Bell (1981), and is based on the measurement of the radius of the zero velocity sphere, R_0, which separates a group from the surrounding volume that expands. In the standard cosmological model with the parameters $H_0 = 73$ km s^{-1}Mpc^{-1} and $\Omega_m = 0.24$ the total mass of a spherical overdensity is expressed as

$$M_T/M_\odot = 2.12 \times 10^{12} \times (R_0/\text{Mpc})^3.$$

An important circumstance here is that the estimate of the total mass of a group corresponds to the scale of R_0, which is ~ 3.7 times larger than its virial radius. The analysis of observational data on radial velocities and separations of galaxies in the

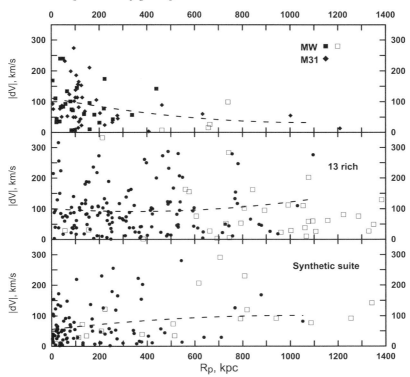

Figure 3. The distributions of the 15 richest nearby suites according to: a) the mean projected separation of physical companions, b) radial velocity dispersion, c) logarithm of stellar mass of the main galaxy, d) logarithm of the orbital mass, e) the mean orbital-to-stellar mass ratio, f) the mean crossing time for the components in units of the global cosmic time T_0. The Milky Way suite and the Andromeda (M31) suite are depicted by 'M' and 'A', respectively.

Table 2. Total masses of nearby groups via internal and external motions.

Group	$\log(M_{orb})$	R_0,(Mpc)	$\log(M_{tot})$	$\log(M_T/M_{orb})$
MW+M31	12.52 ± 0.08	0.98 ± 0.03	12.30 ± 0.05	-0.22 ± 0.09
IC342	12.26 ± 0.21	0.90 ± 0.10	12.19 ± 0.14	-0.07 ± 0.25
M81	12.69 ± 0.13	1.05 ± 0.07	12.39 ± 0.09	-0.30 ± 0.16
N5128+N5236	12.89 ± 0.14	1.26 ± 0.15	12.63 ± 0.15	-0.23 ± 0.21
N253	12.18 ± 0.18	0.70 ± 0.10	11.86 ± 0.18	-0.32 ± 0.25
N4736	12.43 ± 0.15	1.04 ± 0.20	12.38 ± 0.24	-0.05 ± 0.28

vicinity of the Local Group and other nearby groups was done by different authors. A summary for six groups is presented in Table 2. The references to derived masses are given in Karachentsev & Kudrya (2014).

In general, the estimates of mass by two independent methods agree with each other quite well. However, a moderate systematic difference of mass estimates in favour of the orbital masses is noteworthy. For six groups the mean log-difference amounts to -0.20 ± 0.05. This paradoxical result lying in the fact that the estimates of the total mass of the groups on the scale of $R_0 \sim 3.7R_v$ are lower than the orbital mass estimates on the scale of the virial radius R_v can have a simple interpretation. Chernin *et al.* (2013) noted that the estimate of the total mass of a group includes two components: $M_T = M_M + M_{DE}$, where M_M is the mass of dark and baryonic matter, and $M_{DE} =$

$(8\pi/3) \times \rho_{DE} \times R^3$ is the mass, negative in magnitude, determined by the dark energy with the density of ρ_{DE}. On the scale of R_v the contribution of this component in the group mass is small, not exceeding 1%. But in the sphere of R_0 radius, the role of this 'mass defect' becomes significant, reaching about 40%. A correction to the total mass by a factor of 1.4 can almost completely eliminate the observed discrepancy between the group mass estimates at different scales.

In turn, such an agreement of mass estimates by the internal and external motions after the correction for the dark energy component can be interpreted as another empirical evidence for the existence of the dark energy itself appearing in the dynamics of nearby groups.

6. Concluding remarks.

The high-density data on the distances and radial velocities of ~ 800 most nearby galaxies from the UNGC catalog provides an unique opportunity to investigate the distribution of light and dark matter in the Local Volume in outstanding detail.

For the mass of dark halo around the MW and around M 31, we have obtained the values of (1.35 ± 0.47) and (1.76 ± 0.33) in the units of $10^{12} M_\odot$, respectively. Within the Local Volume, there are 15 rich groups containing more than 50% of the total Local Volume population. The typical ratio of orbital-to-stellar mass for them is ~ 30. This quantity is confirmed by independent estimates of total masses derived from Hubble flows around the groups. The ratio of $M_{orb}/M^* \simeq 30$ corresponds to the mean local density of matter $\Omega_m \simeq 0.09$, which is only $1/3$ of the global cosmic density of matter.

Acknowledgements

My co-laborators in this work are Yuri Kudrya, Dmitry Makarov and Elena Kaisina. This work was supported by the grant of Russian Foundation for Basic Research 13–02–90407 Ukr-f-a and the grant of the Ukraine F53.2/15.

References

Abazajian, K. N., Adelman-McCarthy, J. K. & Agüeros, M. A. *et al.* 2009, *ApJS* 182, 543

Chernin, A. D., Bisnovatyi-Kogan, G. S., Teerikorpi, P., Valtonen, M. J, Byrd, G. G. & Merafina, M. 2013, *A&A* 553, 101

Einasto, J., Hütsi, G., Saar, E., Suhhonenko, I., Liivamgi, L. J, Einasto, M., Müller, V., Starobinsky, A. A., Tago, E. & Tempel, E. 2011, *A&A* 531A, 75

Karachentsev, I. D. & Kudrya, Y.N. 2014, *AJ* 148, 50

Karachentsev, I. D., Kaisina, E. I. & Makarov, D.I. 2014, *AJ* 147, 13

Karachentsev, I. D., Makarov, D. & Kaisina, E. 2013, *AJ* 145, 101 (=UNGC)

Karachentsev I. D. 2005, *AJ* 129, 178

Knebe, A., Libeskind, N. I., Doumler, T., Yepes, G., Gottlöber, S. & Hoffman, Y. 2011, *MNRAS* 417L, 56

Libeskind N. I., Yepes G., Knebe A., Gottlöber, S., Hoffman, Y. & Knollmann, S.R. 2010, *MNRAS* 401, 1889

Lynden-Bell, D. 1981, *Observatory* 101, 111

Shandarin, S. F., Sheth, J. V. & Sahni V. 2004, *MNRAS* 353, 162

Zeldovich, Ya. B., 1970, *A&A* 5, 84

The Zeldovich Universe:
Genesis and Growth of the Cosmic Web
Proceedings IAU Symposium No. 308, 2014   © International Astronomical Union 2016
R. van de Weygaert, S. Shandarin, E. Saar & J. Einasto, eds.   doi:10.1017/S1743921316009819

The place of the Local Group in the cosmic web

Jaime E. Forero-Romero[1] and Roberto González[2,3]

[1]Departamento de Física, Universidad de los Andes,
Cra. 1 No. 18A-10, Edificio Ip
Bogotá, Colombia
email: `je.forero@uniandes.edu.co`

[2]Instituto de Astrofísica, Pontificia Universidad Católica de Chile
Av. Vicuña Mackenna 4860
Santiago, Chile

[3]Centro de Astro-Ingeniería, Pontificia Universidad Católica de Chile
Av. Vicuña Mackenna 4860
Santiago, Chile
email: `regonzar@astro.puc.cl`

Abstract. We use the Bolshoi Simulation to find the most probable location of the Local Group (LG) in the cosmic web. Our LG simulacra are pairs of halos with isolation and kinematic properties consistent with observations. The cosmic web is defined using a tidal tensor approach. We find that the LG's preferred location is regions with a dark matter overdensity close to the cosmic average. This makes filaments and sheets the preferred environment. We also find a strong alignment between the LG and the cosmic web. The orbital angular momentum is preferentially perpendicular to the smallest tidal eigenvector, while the vector connecting the two halos is strongly aligned along the the smallest tidal eigenvector and perpendicular to the largest tidal eigenvector; the pair lies and moves along filaments and sheets. We do not find any evidence for an alignment between the spin of each halo in the pair and the cosmic web.

Keywords. cosmology: large-scale structure of universe; cosmology:dark matter; cosmology: simulations; Galaxy: formation

---

1. Introduction

The kinematic configuration of the Local Group (LG) is not common in the cosmological context provided by the Λ Cold Dark Matter (CDM) model.

The LG is dominated by the presence of two spiral galaxies, the Milky Way (MW) and M31. It is relatively isolated of other massive structures; the next most luminous galaxy is 10 times less massive than M31, with several dwarf galaxies around a sphere of $\sim$ 3Mpc and the closest massive galaxy cluster, the Virgo Cluster, is 16.5 Mpc away. The velocity vector of M31 has a low tangential component and is consistent with a head-on collision toward the MW, and the velocity dispersion of nearby galaxies up to $\sim$ 8 Mpc is relatively low.

These features make LG analogues uncommon in numerical simulations. Only about of 2% of MW-sized halos reside in a pair similar to the MW-M31 in a similar environment. Additionally, the strong alignments of dwarf satellites of the MW and M31 in the form of polar and planar structures calls asks for a detailed explanation of the large scale structure environment that can breed such halos (González & Padilla(2010), González et al. (2014), González et al. (2013), Forero-Romero et al. (2011), Forero-Romero et al. (2013)).

181

Here we present results of the study of the large scale environment of LG analogues in the context of ΛCDM. We use a cosmological N-body simulation (Bolshoi) to infer the most probable place of the LG in the cosmic web and its alignments. A detailed description of these results can found in Forero-Romero & González (2014).

2. Finding the cosmic web in numerical simulations

We use a web finding algorithm based on the tidal tensor computed from the gravitational potential field computed over a grid. We define the tensor as:

$$T_{ij} = \frac{\partial^2 \phi}{\partial r_i \partial r_j},\qquad(2.1)$$

where the index $i = 1, 2, 3$ refers to the three spatial directions in euclidean space and ϕ is a normalized gravitational potential that satisfies the following Poisson equation $\nabla^2 \phi = \delta$, where δ is the dark matter overdensity.

The algorithm finds the eigenvalues of this tensor, $\lambda_1 > \lambda_2 > \lambda_3$, and use them to classify each cell in the grid as a peak, filament, sheet or void if three, two, one or none of the eigenvectors is larger than a given threshold λ_{th}. Each eigenvalue has associated to it an eigenvector (e_1, e_2, e_3) which are the natural basis to define local directions in the web. Details describing the algorithm can be found in Forero-Romero *et al.* (2009). The dark matter density is interpolated over a grid with cells of size $0.97\ h^{-1}$Mpc and smoothed with a Gaussian filter with the same spatial scale. We use a threshold of $\lambda_{th} = 0.25$ to define the different cosmic web environments.

We use the Bolshoi simulation with a ΛCDM cosmology described by the parameters $\Omega_{\mathrm{m}} = 1 - \Omega_{\Lambda} = 0.27$, $H_0 = 70\,\mathrm{km/s/Mpc}$, $\sigma_8 = 0.82$, $n_s = 0.95$ (Klypin *et al.* (2011)). The simulation followed the evolution of dark matter in a $250h^{-1}$Mpc box with spatial resolution of $\approx 1h^{-1}$ kpc and mass resolution of $m_{\mathrm{p}} = 1.35 \times 10^8$ M$_\odot$. Halos are identified with the Bound Density Maxima (BDM) algorithm (Klypin & Holtzman(1997)). The data for the halos and the cosmic web are available through a public database located at http://www.cosmosim.org/. A detailed description of the database structure was presented by Riebe *et al.* (2013).

3. Local Groups in cosmological simulations

We construct a sample of MW-M31 pairs at $z \sim 0$ by using multiple snapshots from the simulation asking for consistency with the following criteria:

• Relative distance. The distance between the center of mass of each halo in the pair cannot be larger than 1.3 Mpc.

• Individual halo mass. Each halo has a mass in the mass range $5 \times 10^{11} < M_{200c} < 5 \times 10^{13}M_\odot$.

• Isolation. No neighboring halos more massive than either pair member can be found within 5Mpc.

• Isolation from Virgo-like halos. No dark matter halos with mass $M_{200c} > 1.5 \times 10^{14}M_\odot$ within 12Mpc.

With these selection criteria we select close to 6×10^3 pairs to build a General Sample (GS). From the GS we select two sub-samples according to the tolerance in kinematic constraints. These sub-samples are named 2σ and 3σ, they correspond to a tolerance of two and three times the observational errors in the radial velocity, tangential velocity and separation. The number of pairs in each sample is 46 and 120, respectively.

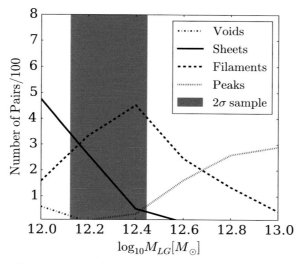

Figure 1. Number of pairs in a preferred environment as a function of the LG total mass. The lines show the results for the General Sample, the shaded region shows the mass interval spanned by the 2σ sample. The preferred environments are filaments and sheets.

4. The place of the Local Group in the Cosmic Web

We find that the pairs in the general sample are located across different environments with a strong dependence on the total pair mass. Figure 1 summarizes this correlation between environment and mass. Each line represents the distribution of pairs in the four different environments. High mass pairs are mostly located in peaks and filaments while less massive ones in voids and sheets. The shaded regions represent the 68% confidence intervals of the mass distributions in the 2σ sample. By and large the LGs in the 2σ and 3σ samples are located in filaments and sheets. In both samples, $\sim 50\%$ of the pairs can be found in filaments while $\sim 40\%$ are in sheets.

We can also characterize the preferred place of the pair samples in terms of the web overdensity. Figure 2 shows the values of the overdensity as a function of the total pair mass. The GS and its uncertainties are represented by the shaded region. The symbols represent the results for the 2σ and 3σ samples. We see that the range of values for the restricted samples samples are completely expected from the mass constraint alone. Higher mass pairs are located in high density regions. The 2σ and 3σ samples have narrower mass range and are located within a narrower range of overdensities $0.0 < \delta < 2.0$ peaking at $\delta \sim 1$.

5. Alignments with the cosmic web

There is wide evidence showing that DM halo formation properties only depend on environment through the local DM density. Whether they are located in sheet or a filament is irrelevant as long the local density is the same.

However there is long history of alignment measurement (shape, spin, peculiar velocities) of individual halos with the cosmic web. In a recent paper Forero-Romero *et al.* (2014) presented a study using the same simulation and cosmic web definition we use here. They also presented a comprehensive review of all the previous results from simulations that also inspected the alignment of halo shape and spin. We refer the reader to the paper for a complete list of references.

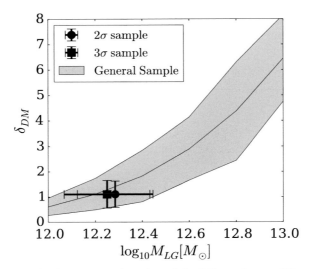

Figure 2. Dark matter overdensity as a function of the LG total mass. The shaded region show the results for the General Sample. There is a strong correlation where more massive pairs sit in denser regions. The symbols represent the 2σ and 3σ samples. The preferred overdensity is in the interval $0 < \delta < 2$.

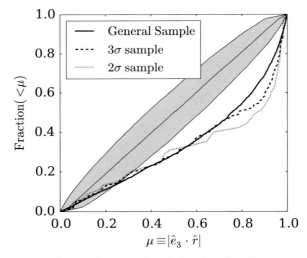

Figure 3. Cumulative distribution for $\mu = |\hat{e}_3 \cdot \hat{r}|$ showing the alignment of the vector joining the two halos in the LG with the smallest tidal eigenvector. The shaded regions represents the expected result from a distribution of vectors randomly placed in space and the corresponding 5% and 75% percentiles for distributions with the same number of points as the 2σ sample.

The main results from the study in Forero-Romero *et al.* (2014) is that the halo shape presents the strongest alignment signal. In this case the DM halo major axis lies along the smallest eigenvector e_3, regardless of the web environment. This alignments is stronger for higher halo masses. Concerning spin alignment the simulations show a weak anti-alignment with respect to e_3 for halo masses larger that $10^{12} M_\odot$, and no alignment signal for masses below that threshold. The peculiar velocities show a strong alignment signal along e_3 for all masses.

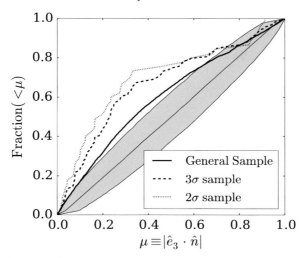

Figure 4. Same as Figure 3 for the vector $\hat{n}$ that indicates the direction of the orbital angular momentum of the pair.

In our case we test for the alignment of the vector connecting the two halos $(\hat{r})$, the orbital angular momentum of the pair $(\hat{n})$ and the spin of each halo. We quantify the alignment using the absolute value of the cosine of the angle between two vectors $\mu = |\hat{e}_3 \cdot \hat{n}|$ or $\mu = |\hat{e}_3 \cdot \hat{r}|$.

The results for the first two alignments are summarized in Figures 3 and 4. We find that the vector $\hat{r}$ is strongly aligned with $\hat{e}_3$, along filaments and sheets. This trend is already present in the GS and gets stronger for the 2σ and 3σ samples. Concerning the vector $\hat{n}$ we have a strong anti-correlation with $\hat{e}_3$, again this tendency is stronger for the 2σ and 3σ samples.

6. Conclusions

Here, we have presented results on the expected place of the Local Group in the cosmic web. Our results are based on cosmological N-body simulations and the tidal web method to define the cosmic web. We constructed different Local Groups samples from dark matter halo pairs that fulfill observational kinematic constraints.

We found a tight correlation of the LG pairs' total mass with the local overdensity. For the LG pairs closer to the observational constraints their total mass is in the range $1 \times 10^{12} M_\odot < M_{LG} < 4 \times 10^{12} M_\odot$ preferred overdensity value is constrained to be in the range $0 < \delta < 2$. This restricts the preferred environment to be filaments and sheets.

We also found strong alignments of the pairs with the cosmic web. The strongest alignment is present for the vector joining the two LG halos. This vector is aligned with the lowest eigenvector and anti-aligned with the highest eigenvector. This trend is already present in wide sample of pairs and becomes stronger as the kinematic constraints are closer to their observed values.

These results raise the need to use observations to constraint the alignments of LG pairs with their cosmic web environment. There are many algorithms available to reconstruct the DM distribution from large galaxy surveys to tackle this task. This would allow a direct quantification of how common are the LG alignments we have found, providing a potential new test for ΛCDM.

J.E.F-R Acknowledges the IAU and the Local Organizing Committee for providing the financial support to attend this meeting.

References

Forero-Romero, J. E., & González, R. E., *ApJ* accepted, ArXiv:1408.3166

Forero-Romero, J. E., Contreras, S., & Padilla, N. 2014, MNRAS, 443, 1090

Forero-Romero, J. E., Hoffman, Y., Bustamante, S., Gottlöber, S., & Yepes, G. 2013, ApJL, 767, L5

Forero-Romero, J. E., Hoffman, Y., Yepes, G., Gottlöber, S., Piontek, R., Klypin, A., & Steinmetz, M. 2011, MNRAS, 417, 1434

Forero-Romero, J. E., Hoffman, Y., Gottlöber, S., Klypin, A., & Yepes, G. 2009, MNRAS, 396, 1815

González, R. E., Kravtsov, A. V., & Gnedin, N. Y. 2013, ApJ, 770, 96

—, 2014, ApJ, 793, 91

González, R. E. & Padilla, N. D. 2010, MNRAS, 407, 1449

Hoffman Y., Metuki O., Yepes G., Gottlöber S., Forero-Romero J. E., Libeskind N. I., Knebe A., 2012, MNRAS, 425, 2049

Klypin, A. A., Trujillo-Gomez, S., & Primack, J. 2011, ApJ, 740, 102

Klypin, A. & Holtzman, J. 1997, ArXiv:9712217

Riebe, K., Partl, A. M., Enke, H., Forero-Romero, J., Gottlöber, S., Klypin, A., Lemson, G., Prada, F., Primack, J. R., Steinmetz, M., & Turchaninov, V. 2013, *Astronomische Nachrichten*, 334, 691

CHAPTER 3C.

Surveys and Observations:
Filaments

line-up, from left to right:
Jaan Einasto, Alar Toomre, Tiia Lillemaa, Rien van de Weygaert and
dr. Anu Reinart, director of Tartu Observatory

line-up at the Gruber Prize panel discussion on future challenges in cosmology.
from left to right: Gert Hütsi, Marc Davis, Dick Bond, Noam Libeskind,
Jaan Einasto, Brent Tully, Rashid Sunyaev and Alar Toomre (moderator)

The Zeldovich Universe:
Genesis and Growth of the Cosmic Web
Proceedings IAU Symposium No. 308, 2014
R. van de Weygaert, S. Shandarin, E. Saar & J. Einasto, eds.

© International Astronomical Union 2016
doi:10.1017/S1743921316009820

Structure and kinematics
of the Bootes filament

O. Nasonova[1], I. Karachentsev[1] and V. Karachentseva[2]

[1]Special Astrophysical Observatory of RAS, Nizhnij Arkhyz, Russia
[2]Main Astronomical Observatory of NASU, Kyiv, Ukraine

Abstract. Bootes filament of galaxies is a dispersed chain of groups residing on sky between the Local Void and the Virgo cluster. We consider a sample of 361 galaxies inside the sky area of RA $= 13^h.0...18^h.5$ and Dec $= -5°... + 10°$ with radial velocities $V_{LG} < 2000$ km/s to clarify its structure and kinematics. In this region, 161 galaxies have individual distance estimates. We use these data to draw the Hubble relation for galaxy groups, pairs as well as the field galaxies, and to examine the galaxy distribution on peculiar velocities. Our analysis exposes the known Virgo-centric infall at RA $< 14^h$ and some signs of outflow from the Local Void at RA $> 17^h$. According to the galaxy grouping criterion, this complex contains the members of 13 groups, 11 pairs and 140 field galaxies. The most prominent group is dominated by NGC 5846. The Bootes filament contains the total stellar mass of $2.7 \times 10^{12} M_\odot$ and the total virial mass of $9.07 \times 10^{13} M_\odot$, having the average density of dark matter to be $\Omega_m = 0.09$, i.e. a factor three lower than the global cosmic value.

Keywords. galaxies: distances and redshifts, cosmology: large-scale structure of universe

1. Introduction

Oservational data on distances and velocities of galaxies in the Local Supercluster have been enriched significantly by recent optical and HI surveys, allowing to obtain the nearby field of peculiar motions and hence to study the local distribution of dark matter. Previously, we considered motions of galaxies in some sky areas neighbouring the Virgo cluster as the Local Supercluster centre. In the Virgo Southern Extension filament (Karachentsev & Nasonova 2013) and the Ursa Majoris cloud (Karachentsev *et al.* 2013) we derived a low mean matter density: $\Omega_m = 0.11$ and 0.08, respectively. But in the Coma I region we suggested the existence of a dark attractor with the total mass of $\sim 2 \times 10^{14} M_\odot$ (Karachentsev *et al.* 2011).

2. Observational data

The initial sample of galaxies was selected from the Lyon Extragalactic Database = LEDA (http://leda.univ-lyon1.fr) and limited by radial velocities $V_{LG} \leqslant 2000$ km/s and equatorial coordinates RA $= 13^h.0 ... 18^h.0$ and Dec $= -5° ... + 10°$. The major part of the considered strip is covered by SDSS survey (Abazajian *et al.* 2009) as well as HIPASS (Zwaan *et al.* 2003) and ALFALFA (Haynes *et al.* 2011) HI surveys. It allowed us to perform an independent morphological classification of galaxies and to determine distances for many galaxies from the Tully & Fisher (1977) relation between luminosity of a galaxy and its HI line width W_{50}. The resulting list of 361 galaxies includes 161 galaxies with individual distance estimates.

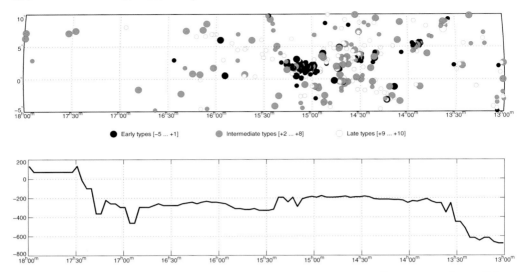

Figure 1. Upper panel: distribution of galaxies in the Bootes strip. Bottom panel: a poliline represents the running median of peculiar velocity with a window of $0^h.5$.

3. Structure and sub-structures

The Bootes filament of galaxies is a scattered chain of galaxy groups residing between the Local Void (RA = $19^h.0$, Dec = $+3°$) and the Virgo cluster (RA = $12^h.5$, Dec = $+12°$). The kinematics of this structure should be influenced both by galaxies infalling towards the Virgo cluster as well as more eastern galaxies moving away from the expanding Local Void (Nasonova & Karachentsev 2011).

The upper panel of Fig. 1 presents the distribution of 361 galaxies in the Bootes strip. The galaxies of different morphological types are shown by circles of different density. According to the galaxy grouping criterion (Makarov & Karachentsev 2011), the Bootes strip contains 13 groups, 11 pairs and 140 field galaxies. The most notable feature in the strip is the compact group around NGC 5846 (just right from the strip center), which numbers 74 members with measured radial velocities. The bottom panel of Fig. 1 shows the running median of the peculiar velocity, $V_{pec} = V_{LG} - 72 D_{Mpc}$, with a window of $0^h.5$ along RA. The most common value is $V_{pec} = -250$ km/s, which remains almost flat from 14^h to 17^h. Galaxies in the Virgo infall zone (RA < 14^h) demonstrate clearly a droop of the V_{pec} median. To the contrary, in the vicinity of the Local Void the median rises which is quite expectable since galaxies move away from the void centre.

4. Local density of matter

The virial (projected) mass distribution of galaxy groups (squares) and pairs (triangles) in the considered strip versus their total stellar mass is depicted in Fig. 2. There is a positive correlation between virial and stellar masses, well-known from other data. While masses are small, the significant vertical scatter is caused mainly by projection factors.

According to Jones et al.(2006), the mean density of stellar matter in the universe is $4.6 \times 10^8 M_\odot/\mathrm{Mpc}^3$. The global matter density $\Omega_m = 0.28$ is equivalent to dark-to-stellar matter ratio of $M_{DM}/M_* = 97$. This value is shown as a diagonal in Fig. 2. All the groups and all the pairs except one in the Bootes strip are situated under this line. The Bootes filament contains the total stellar mass of $2.7 \times 10^{12} M_\odot$ and the total virial mass of $9.07 \times 10^{13} M_\odot$, having the average ratio $\sum M_p / \sum M_* = 33$ or the local density of

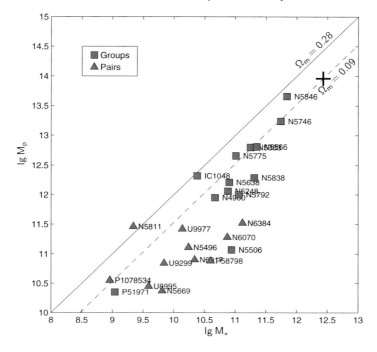

Figure 2. The relation between projected (virial) mass and total stellar mass.

dark matter $\Omega_m = 0.09$. This is a factor three lower than the global cosmic value. The sum of virial masses-to-sum of stellar masses ratio for all the groups and pairs is plotted in Fig. 2 as a cross. The dashed line drawn through the cross indicates the mean mass density Ω_m (Bootes) $\simeq 0.09$. Considering field galaxies should only reduce slightly this proportion, as they contribute evidently both to numerator and denominator of the ratio $\sum M_p / \sum M_*$. Thus, the observational data on galaxy motions in the Bootes strip give us an argument that this filamentary structure does not contain a large amount of dark matter. This statement is based on internal (virial) motions of galaxies, but supposing that 3–5 times larger mass is hidden in the Bootes filament between the groups, then the velocity dispersion for centres of groups and pairs should be considerably larger than what is observed.

This work is supported by the Russian Science Foundation (project No. 14-12-00965), Russian Foundation for Basic Research (grant No. 13-02-90407) and the State Fund for Fundamental Researches of Ukraine (grant No. F53.2/15). Olga Nasonova thanks the non-profit Dmitry Zimins Dynasty Foundation for the financial support. This research has made use of NASA/IPAC Extragalactic Database (http://ned.ipac.caltech.edu), HyperLeda database (http://leda.univ-lyon1.fr) and SDSS archive (http://www.sdss.org).

References

Abazajian, K. N., Adelman-McCarthy, J. K., Agueros, M. A., *et al.* 2009, *ApJS*, 182, 54
Haynes, M. P., Giovanelli, R., Martin, A. M., *et al.* 2011, *AJ*, 142, 170
Jones, D. H., Peterson, B. A., Colless, M., & Saunders, W. 2006, *MNRAS*, 369, 25
Karachentsev, I. D., Nasonova, O. G., & Courtois, H. M. 2011, *ApJ*, 743, 123
Karachentsev, I. D. & Nasonova, O. G. 2013, *MNRAS*, 429, 2677

Karachentsev, I. D., Nasonova, O. G., & Courtois, H. M. 2013, *MNRAS*, 429, 2264

Karachentsev, I. D., Tully, R. B., Shaya, E. J., *et al.* 2014, *ApJ*, 782, 4

Makarov, D. I. & Karachentsev, I. D. 2011, *MNRAS*, 412, 2498

Nasonova, O. G. & Karachentsev, I. D. 2011, *Astrophysics*, 54, 1

Tully R. B. & Fisher R. J. 1977, *A&A*, 54, 661

Zwaan, M. A., Staveley-Smith, L., Koribalski, B. S., *et al.* 2003, *AJ*, 125, 2842

The Zeldovich Universe:
Genesis and Growth of the Cosmic Web
Proceedings IAU Symposium No. 308, 2014
R. van de Weygaert, S. Shandarin, E. Saar & J. Einasto, eds.

ⓒ International Astronomical Union 2016
doi:10.1017/S1743921316009832

The Dark Matter filament between Abell 222/223

Jörg P. Dietrich[1], Norbert Werner[2], Douglas Clowe[3], Alexis Finoguenov[4], Tom Kitching[5], Lance Miller[6] and Aurora Simionescu[2]

[1]Universitäts-Sternwarte München, Ludwig-Maximilians-Universität München, Scheinerstr. 1, 81679 München, Germany
email: dietrich@usm.lmu.de

[2]Kavli Institute for Particle Astrophysics and Cosmology, Stanford University, 382 Via Pueblo Mall, Stanford, CA 94305-4060, USA

[3]Dept. of Physics & Astronomy, Ohio University, Clippinger Lab 251B, Athens, OH 45701, USA

[4]Max-Planck-Institut fur extraterrestrische Physik, Giessenbachstraße, 85748 Garching b. München, Germany

[5]Institute for Astronomy, The University of Edinburgh, Royal Observatory, Blackford Hill, Edinburgh EH9 3HJ, U.K.

[6]Department of Physics, University of Oxford, The Denys Wilkinson Building, Keble Road, Oxford OX1 3RH, U.K.

Abstract. Weak lensing detections and measurements of filaments have been elusive for a long time. The reason is that the low density contrast of filaments generally pushes the weak lensing signal to unobservably low scales. To nevertheless map the dark matter in filaments exquisite data and unusual systems are necessary. SuprimeCam observations of the supercluster system Abell 222/223 provided the required combination of excellent seeing images and a fortuitous alignment of the filament with the line-of-sight. This boosted the lensing signal to a detectable level and led to the first weak lensing mass measurement of a large-scale structure filament. The filament connecting Abell 222 and Abell 223 is now the only one traced by the galaxy distribution, dark matter, and X-ray emission from the hottest phase of the warm-hot intergalactic medium. The combination of these data allows us to put the first constraints on the hot gas fraction in filaments.

Keywords. gravitational lensing: weak, galaxies: clusters: individual (Abell 222, Abell 223), large-scale structure of universe

It is a firm prediction of the concordance Cold Dark Matter (CDM) cosmological model that galaxy clusters live at the intersection of large-scale structure filaments (Bond *et al.* 1996). The thread-like structure of this "cosmic web" has been traced by galaxy redshift surveys for decades (e.g. Joeveer *et al.* 1978; Geller & Huchra 1989). More recently the Warm-Hot Intergalactic Medium (WHIM) residing in low redshift filaments has been observed in emission (Werner *et al.* 2008) and absorption (Buote *et al.* 2009; Fang *et al.* 2010). However, a reliable direct detection of the underlying Dark Matter skeleton, which should contain more than half of all matter (Aragón-Calvo *et al.* 2010), remained elusive for much longer, as earlier candidates for such detections (Kaiser *et al.* 1998; Gray *et al.* 2002; Dietrich *et al.* 2005) were either falsified (Gavazzi *et al.* 2004; Heymans *et al.* 2008) or suffered from low signal-to-noise ratios (Kaiser *et al.* 1998; Dietrich *et al.* 2005) and unphysical misalignements of dark and luminous matter (Gray *et al.* 2002; Dietrich *et al.* 2005).

Abell 222 and Abell 223, the latter a double galaxy cluster in itself, form a supercluster system of three galaxy clusters at a redshift of $z \sim 0.21$ (Dietrich *et al.* 2002), separated on the sky by $\sim 14'$. Gravitational lensing distorts the images of faint background galaxies as their light passes massive foreground structures. The foreground mass and its distribution can be deduced from measuring the shear field imprinted on the shapes of the background galaxies. The mass reconstruction in Figure 1 shows a mass bridge connecting A 222 and the southern component of A 223 (A 223-S) at the 4.1σ significance level. This mass reconstruction does not assume any model or physical prior on the mass distribution.

To show that the mass bridge extending between A 222 and A 223 is not caused by the overlap of the cluster halos but in fact due to additional mass, we also fit parametric models to the three clusters plus a filament component. The clusters were modelled as elliptical Navarro-Frenk-White (NFW) profiles (Navarro *et al.* 1997) with a fixed mass-concentration relation (Dolag *et al.* 2004). We used a simple model for the filament, with a flat ridge line connecting the clusters, exponential cut-offs at the filament end points in the clusters, and a King profile (King 1966) describing the radial density distribution, as suggested by previous studies (Colberg *et al.* 2005; Mead *et al.* 2010). We showed in the original publication of this work (Dietrich *et al.* 2012) that the exact ellipticity has little impact on the significance of the filament.

The best fit parameters of this model were determined with a Monte-Carlo Markov Chain (MCMC) and are shown in Fig. 2. The likelihood-ratio test prefers models with a filament component with 96.0% confidence over a fit with three NFW halos only. A small degeneracy exists in the model between the strength of the filament and the virial radii of A 222 and A 223-S. The fitting procedure tries to keep the total amount of mass in the supercluster system constant at the level indicated by the observed reduced shear. Thus, it is not necessarily the case that sample points with a positive filament contribution indeed have more mass in the filament area than a 3 clusters only model has. The reason is that the additional filament mass might be compensated for with lower cluster masses. We find that the integrated surface mass density along the filament ridge line exceeds that of the clusters only model in 98.5% of all sample points. This indicates that the data strongly prefers models with additional mass between A 222 and A 223-S and that this preference is stronger than the confidence level derived from the likelihood-ratio test. The difference is probably due to the oversimplified model, which is not a good representation of the true filament shape.

The virial masses inferred from the MCMC are lower than those reported earlier for this system (Dietrich *et al.* 2005), which were obtained from fitting a circular two-component NFW model to A 222 and A 223. Compared to this approach, our more complex model removes mass from the individual supercluster constituents and redistributes it to the filament component. Reproducing the two-component fit with free concentration parameters, which was used in the previous study, we find $M_{200}(\text{A } 222) = (2.7^{+0.8}_{-0.7}) \times 10^{14} \, M_{\odot}$, which is in good agreement, and $M_{200}(\text{A } 223) = (3.4^{+1.3}_{-1.0}) \times 10^{14} \, M_{\odot}$, which overlaps the 1σ error bars of the earlier study. Here and in the following, all error bars are single standard deviations.

The detection of a filament with a dimensionless surface mass density of $\kappa \sim 0.03$ is unexpected. Simulations generally predict the surface mass density of filaments to be much lower (Dietrich *et al.* 2005) and not to be detectable individually (Mead *et al.* 2010). These predictions, however, are based on the assumption that the longer axis of the filament is aligned with the plane of the sky and that we look through the filament along its minor axis. If the filament were inclined with respect to the line-of-sight and we were to look almost along its major axis, the projected mass could reach the observed

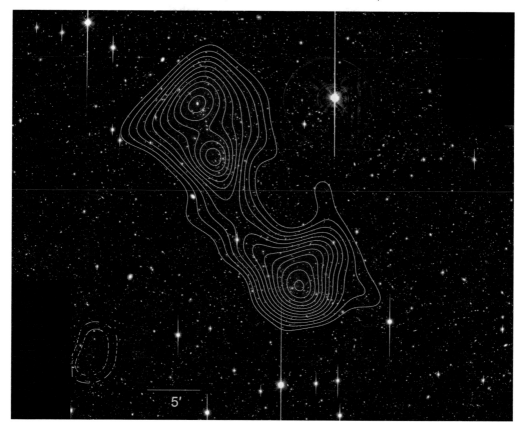

Figure 1. Mass reconstruction of A 222/223. The background image is a three colour-composite SuprimeCam image based on observations with the 8 m Subaru telescope during the nights of Oct. 15 (A 222) and 20 (A 223), 2001 in V-, R_c- and i'-bands. We obtained the data from the SMOKA science archive (http://smoka.nao.ac.jp/). The FWHM of the stellar point-spread function varies between $0''.57$ and $0''.70$ in our final co-added images. Overlayed are the reconstructed surface mass density (blue) above $\kappa = 0.0077$, corresponding to $\Sigma = 2.36\,M_\odot\,\mathrm{Mpc}^{-2}$, and significance contours above the mean of the field edge, rising in steps of 0.5σ and starting from 2.5σ. Dashed contours mark underdense regions at the same significance levels. Supplementary Figure 1 shows the corresponding B-mode map. The reconstruction is based on 40,341 galaxies whose colours are not consistent with early type galaxies at the cluster redshift. The shear field was smoothed with a $2'$ Gaussian. The significance was assessed from the variance of 800 mass maps created from catalogues with randomised background galaxy orientation. We measured the shapes of these galaxies primarily in the R_c-band, supplementing the galaxy shape catalogue with measurements from the other two bands for galaxies for which no shapes could be measured in the R_c-band, to estimate the gravitational shear (Miller *et al.* 2007; Kitching *et al.* 2008). A 222 is detected at $\sim 8.0\sigma$ in the south, A 223 is the double-peaked structure in the north seen at $\sim 7\sigma$.

level. A timing argument (Kahn & Woltjer 1959; Sandage 1986) can be made to show that the latter scenario is more plausible in the A 222/3 system. In this argument we treat A 223 as a single cluster and neglect the filament component, such that we have to deal only with two bodies, A 222 and A 223. The redshifts of A 222 and A 223 differ by $\Delta z = 0.005$, corresponding to a line-of-sight separation of 18 Mpc if the redshift difference is entirely due to Hubble flow. Let us assume for a moment that the difference is caused only by peculiar velocities. Then at $z = \infty$, the clusters were at the same location in the

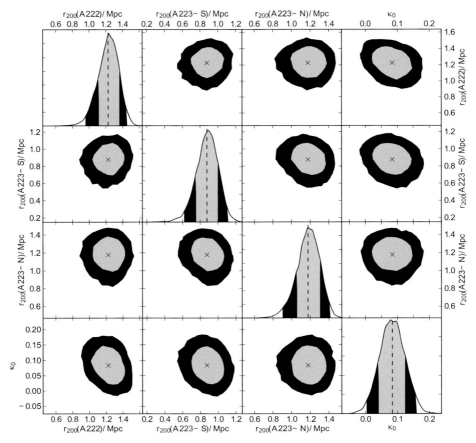

Figure 2. Posterior probability distributions for cluster virial radii and filament strength. Shown are the 68% and 95% confidence intervals on the cluster virial radii $r_{200}(\cdot)$ and the filament strength κ_0. The confidence intervals are derived from 30,000 MCMC sample points. The filament model is described by $\kappa(\theta, r) = \kappa_0 \left\{ 1 + \exp\left[(|\theta| - \theta_1)/\sigma\right] + (r/r_{\rm c})^2 \right\}^{-1}$, where the coordinate θ runs along the filament ridge line and r is orthogonal to it. This model predicts the surface mass density at discrete grid points from which we computed our observable, the reduced shear, via a convolution in Fourier space. The data cannot constrain the steepness of the exponential cut-off at the filament endpoints σ and the radial core scale $r_{\rm c}$. These were fixed at their approximate best-fit values of $\sigma = 0.45\,\mathrm{Mpc}$ and $r_{\rm c} = 0.54\,\mathrm{Mpc}$. The data also cannot constrain the cluster ellipticity and orientation. These were held fixed at the values measured from the isodensity contours of early-type galaxies (Dietrich *et al.* 2002). The ratios of minor/major axes and the position angles of the ellipses are $(0.63, 0.69, 0.70)$ and $(65°, 34°, 3°)$ for A 222, A 223-S, and A 223-N, respectively. We further explore the impact of cluster ellipticity on the filament detection in the supplementary information.

Hubble flow. We let them move away from each other with some velocity and inclination angle with respect to the line-of-sight and later turn around and approach each other. The parameter space of total system mass and inclination angle that reproduces the observed configuration at $z = 0.21$ is completely degenerate. Nevertheless, in order to explain the observed configuration purely with peculiar velocity, this model requires a minimum mass of $(2.61 \pm 0.05) \times 10^{15}\,M_\odot$ with an inclination angle of 46 degrees, where the error on the mass is caused solely by the uncertainty of the Hubble constant. Since this is more than 10 standard deviations above our mass estimate for the sum of both

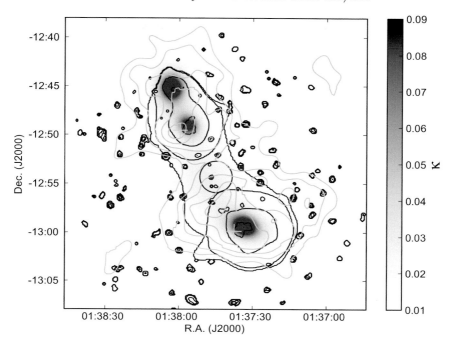

Figure 3. Surface mass density of the best fit parametric model. The surface mass density distribution of the best parameters in Fig. 2 was smoothed with a 2′ Gaussian to have the same physical resolution as the mass reconstruction in Fig. 1. The yellow crosses mark the end points of the filament model. These were determined from the visual impression of the filament axis in Fig. 1. The MCMC is not able to constrain their location. In the model, the filament ridge line is not aligned with the axis connecting the centers of A 222 and A 223-S. This is a fairly common occurrence (∼ 9%) for straight filament but may also indicate some curvature, which occurs in ∼ 53% of all intercluster filaments (Colberg *et al.* 2005) and is not included in our simple model. Overlayed are X-ray contours from XMM-Newton observations (Werner *et al.* 2008) (red) and significance contours of the colour-selected early-type galaxy density (Dietrich *et al.* 2005) (beige), showing the alignment of all three filament constituents. The black circle marks the region inside which the gas mass and the filament mass were estimated.

clusters, we infer that at least part of the observed redshift difference is due to Hubble flow, and that we are looking along the filament's major axis.

The combination of our weak-lensing detection with the observed X-ray emission of 0.91 ± 0.25 keV WHIM plasma (Werner *et al.* 2008) lets us constrain the hot gas fraction in the filament. Assuming that the distribution of the hot plasma is uniform and adopting a metallicity of $Z = 0.2$ Solar, the mass of the X-ray emitting gas inside a cylindrical region with radius 330 kpc centred on (01:37:45.00, 12:54:19.6) with a length along our line-of-sight of $l = 18$ Mpc, as suggested by our timing argument, is $M_{\mathrm{gas}} = 5.8 \times 10^{12} \, M_{\odot}$. The assumption of uniform density is certainly a strong simplification. Because the X-ray emissivity depends on the average of the squared gas density, a non-uniform density distribution can lead to strong changes in the X-ray luminosity. Thus, if the filament consists of denser clumps embedded into lower density gas (as has been observed in the outskirts of the Perseus Cluster (Simionescu *et al.* 2011)), or even if there is a smooth non-negligible density gradient within the region used for spectral extraction, then our best fit mean density will be overestimated. The quoted gas mass should therefore be considered as an upper limit, and the true mass may be lower by up to a factor of 2–3.

We estimated the total mass of the filament from the reconstructed surface mass-density map and the model fits within the same region where we measured the gas mass. In the reconstructed κ-map, the mass inside the extraction circle is $M_{\mathrm{fil}} = (6.5 \pm 0.1) \times 10^{13}\,M_{\odot}$, where the error is small due to the highly correlated noise of the smoothed shear field inside the extraction aperture. For the parametric model fit, the inferred mass is higher but consistent within one standard deviation, $M_{\mathrm{fil}} = (9.8 \pm 4.4) \times 10^{13}\,M_{\odot}$. The corresponding upper limits on the hot gas fractions vary between $f_{\mathrm{X}} = 0.06 - 0.09$, a value that is lower than the gas fraction in galaxy clusters (Allen *et al.* 2008). This is consistent with the expectation that a significant fraction of the WHIM in filaments is too cold to emit X-rays detectable by XMM-Newton (Davé *et al.* 2001).

The results reported in this proceeding were first published in Dietrich *et al.* (2012).

References

Allen, S. W., Rapetti, D. A., Schmidt, R. W., *et al.* 2008, *Mon. Not. R. Astron. Soc.*, 383, 879

Aragón-Calvo, M. A., van de Weygaert, R., & Jones, B. J. T. 2010, *Mon. Not. R. Astron. Soc.*, 408, 2163

Bond, J. R., Kofman, L., & Pogosyan, D. 1996, *Nature*, 380, 603

Buote, D. A., Zappacosta, L., Fang, T., *et al.* 2009, *Astrophys. J.*, 695, 1351

Colberg, J. M., Krughoff, K. S., & Connolly, A. J. 2005, *Mon. Not. R. Astron. Soc.*, 359, 272

Davé, R., Cen, R., Ostriker, J. P., *et al.* 2001, *Astrophys. J.*, 552, 473

Dietrich, J. P., Clowe, D. I., & Soucail, G. 2002, *Astron. Astrophys.*, 394, 395

Dietrich, J. P., Schneider, P., Clowe, D., Romano-Díaz, E., & Kerp, J. 2005, *Astron. Astrophys.*, 440, 453

Dietrich, J. P., Werner, N., Clowe, D., *et al.* 2012, *Nature*, 487, 202

Dolag, K., Bartelmann, M., Perrotta, F., *et al.* 2004, *Astron. Astrophys.*, 416, 853

Fang, T., Buote, D. A., Humphrey, P. J., *et al.* 2010, *Astrophys. J.*, 714, 1715

Gavazzi, R., Mellier, Y., Fort, B., Cuillandre, J.-C., & Dantel-Fort, M. 2004, *Astron. Astrophys.*, 422, 407

Geller, M. J. & Huchra, J. P. 1989, *Science*, 246, 897

Gray, M. E., Taylor, A. N., Meisenheimer, K., *et al.* 2002, *Astrophys. J.*, 568, 141

Heymans, C., Gray, M. E., Peng, C. Y., *et al.* 2008, *Mon. Not. R. Astron. Soc.*, 385, 1431

Ilbert, O., Arnouts, S., McCracken, H. J., *et al.* 2006, *Astron. Astrophys.*, 457, 841

Joeveer, M., Einasto, J., & Tago, E. 1978, *Mon. Not. R. Astron. Soc.*, 185, 357

Kahn, F. D. & Woltjer, L. 1959, *Astrophys. J.*, 130, 705

Kaiser, N., Wilson, G., Luppino, G., *et al.* 1998, astro-ph/9809268

King, I. R. 1966, *Astron. J.*, 71, 64

Kitching, T. D., Miller, L., Heymans, C. E., van Waerbeke, L., & Heavens, A. F. 2008, *Mon. Not. R. Astron. Soc.*, 390, 149

Mead, J. M. G., King, L. J., & McCarthy, I. G. 2010, *Mon. Not. R. Astron. Soc.*, 401, 2257

Miller, L., Kitching, T. D., Heymans, C., Heavens, A. F., & van Waerbeke, L. 2007, *Mon. Not. R. Astron. Soc.*, 382, 315

Navarro, J. F., Frenk, C. S., & White, S. D. M. 1997, *Astrophys. J.*, 490, 493

Sandage, A. 1986, *Astrophys. J.*, 307, 1

Simionescu, A., Allen, S. W., Mantz, A., *et al.* 2011, *Science*, 331, 1576

Werner, N., Finoguenov, A., Kaastra, J. S., *et al.* 2008, *Astron. Astrophys.*, 482, L29

CHAPTER 3D.

Surveys and Observations:
Groups & Clusters

The Zeldovich Universe:
Genesis and Growth of the Cosmic Web
Proceedings IAU Symposium No. 308, 2014
R. van de Weygaert, S. Shandarin, E. Saar & J. Einasto, eds.

© International Astronomical Union 2016
doi:10.1017/S1743921316009844

Characterising large-scale structure with the REFLEX II cluster survey

Gayoung Chon

Max-Planck-Institut für extraterrestrische Physik
Garching 85748, Germany
email: gchon@mpe.mpg.de

Abstract. We study the large-scale structure with superclusters from the REFLEX X-ray cluster survey together with cosmological N-body simulations. It is important to construct superclusters with criteria such that they are homogeneous in their properties. We lay out our theoretical concept considering future evolution of superclusters in their definition, and show that the X-ray luminosity and halo mass functions of clusters in superclusters are found to be top-heavy, different from those of clusters in the field. We also show a promising aspect of using superclusters to study the local cluster bias and mass scaling relation with simulations.

Keywords. cosmology:large-scale structure of universe, X-rays: galaxies: clusters, methods: n-body simulations

1. Introduction

Superclusters are the largest overdense structures found in the galaxy and galaxy cluster surveys, defined in general as groups of two or more galaxy clusters above a certain density enhancement (Bahcall (1988)). Unlike clusters they have not reached a quasi-equilibrium configuration, hence the definition of superclusters must be made clear such that their properties can be studied quantitatively. One solution to this problem is to include the future evolution of superclusters into the definition of the object, selecting only those structures that will collapse in the future. This allows us to obtain a more homogeneous class of superclusters. We have been exploring this approach observationally using an appropriate selection criterion to construct an X-ray supercluster catalogue (Chon *et al.* (2013), Chon *et al.* (2014)). It is based on the REFLEX X-ray cluster survey, which provides the largest and homogeneous X-ray cluster sample to date in the southern sky (Böhringer *et al.* (2013), Chon & Böhringer (2012)). Since the REFLEX survey has a well-defined selection function, we can apply equivalent criteria to dark matter halos in cosmological N-body simulations to construct superclusters, which allows us to study superclusters more quantitatively.

2. Defining superclusters

To identify a region that will collapse in the future, we approximate the overdense regions by homogeneous density spheres, which has been successfully used in the literature for many applications. We can then model the evolution of the overdense region with respect to the expansion of the background cosmology with reference to Birkhoff's theorem. This allows us to describe the evolution of both the overdense and background regions by the respective values of the local and global Hubble constant, H, matter density, Ω_m, and Ω_Λ corresponding to a cosmological constant. We evolve both regions from a starting redshift of 500, which results in an accuracy well below 1% in the final calculation. We solve the Friedmann equations for a spherical collapse model iteratively where the local matter density is enhanced at the starting redshift in the overdense region. We then obtain an criterion for R, which is the ratio between the minimally required local overdensity to the mean density of the universe today. This density ratio, for example, is 7.858 for the flat ΛCDM cosmology with $\Omega_m=0.3$ and $h=0.7$. Since we use a friends-of-friends (fof) algorithm to build superclusters from clusters, the linking length must reflect this required density ratio. Given that the linking length is inversely proportional to a third power of the local density of clusters, we find that the overdensity parameter, f, used in the fof algorithm has to be about 25 for a cluster bias of 2-3. In Chon $et\ al.$ (2013) we adopted a slightly more generous value of ten for the nominal catalogue together with that built with $f=50$ for comparison so as to collect a slightly larger regions than just the core of the collapsing superclusters. We recover a number of known superclusters including Shapley, Hydra-Centaurus, and Aquaris B as well as a number of new X-ray superclusters. Based on our physically driven choice of the overdensity parameter, one of the consequences is that, for example, the Shapley supercluster is fragmented into three smaller mass concentrations. A further study of the radial profile of Shapley mass concentration implies that in fact approximately the central 11 h^{-1} Mpc of Shapley is under-going a collapse currently, while regions outside 13 h^{-1} Mpc will not collapse in the future despite the fact that the outskirts of Shapley are rich with clusters (Chon $et\ al.$, submitted).

3. Superclusters as dark matter tracers

The fact that the REFLEX II supercluster sample has been constructed by means of a statistically well-defined sample of closely mass-selected clusters motivates us to search for a more precise physical characterisation of the superclusters in simulations. We achieve this by applying criteria equivalent to those used in our X-ray selection, and regard halos as clusters by imposing a mass limit to halos in the Millennium simulation (Springel (2005)). Our nominal lower mass limit is $10^{14}h^{-1}M_\odot$ corresponding to a typical mass limit of a cluster survey, and we construct superclusters with the same overdensity parameter that was used for the REFLEX II superclusters.

One of the interesting aspects in the study of superclusters is how superclusters trace the underlying dark matter. Since we have almost no direct access to determined their masses, an indirect mass estimate would still be helpful. Since we have mass estimates of the member clusters through mass-observable scaling relations, the total supercluster mass can be estimated if we can calibrate the cluster mass fraction in superclusters. Similarly we would be able to determine the dark matter overdensity traced by a supercluster if the mass fraction relation or the overdensity bias could be calibrated. Hence we consider two quantities, the mass fraction represented by the total clusters mass in a supercluster, and the bias of clusters in superclusters. A better knowledge of the former would allow us to make a prediction of the underlying total mass of a superclusters through a mass scaling relation like that of clusters.

The left panel of Fig. 1 shows the total cluster mass as a function of the total mass of superclusters with the best fit to a power-law shown in a solid line, which is enclosed by two dashed lines representing its two sigma scatter. The total mass of a supercluster is defined to be the sum of all halos in the volume with a correction factor that takes the particles which do not contribute to the halo mass, M_{200}, assuming that they are distributed throughout the volume in an unbiased way. The fact that the total cluster mass is closely correlated with the total halo mass of a supercluster gives a first encouraging

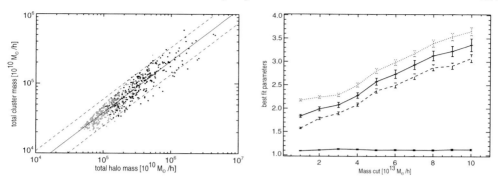

Figure 1. (Left) Total mass of a supercluster probed by the total mass of member clusters. The halos of the lower halo mass limit of $10^{14} h^{-1} M_\odot$ are considered here. (Right) Fitted slope and amplitude for the mass bias as a function of the mass limit in the cluster catalogue. In comparison we show the fitted amplitude of the pair (dotted) and richer (dashed) superclusters separately where the slope is fixed to that for the entire sample.

evidence that there is a potential to use the cluster observables to trace the dark matter distribution in the regions of superclusters.

The power spectrum of clusters of galaxies measures the density fluctuation amplitude of the distribution of clusters as a function of a scale, where the amplitude ratio of this power spectrum in comparison with the power spectrum of the dark matter is interpreted as a bias that clusters have. Analogous to the power spectrum of clusters, we take the number overdensity of clusters in superclusters as a measure of bias against the dark matter overdensity, for which we take again the halo mass overdensity as a tracer. This approach makes use of the observable, the number overdensity of clusters, so it can be calibrated against a quantity from simulations, the dark matter mass overdensity. In this case we consider a continuous range of lower halo mass limits from $10^{13} h^{-1} M_\odot$ to $10^{14} h^{-1} M_\odot$ to form ten cluster catalogues from the simulation and construct ten supercluster samples. We fit a power-law to the cluster number density as a function of the mass density of halos in superclusters. The best fit parameters are shown in the right panel of Fig. 1 where dotted and dashed lines represent those superclusters with two member clusters and richer superclusters. The uncertainties are calculated by one thousand bootstrappings of the sample. The fitted slope of this relation turns out to be nearly constant, ~ 1.1, for all mass ranges of consideration, hence we interpret the fitted amplitude as a bias. We see that the bias increases with an increasing lower mass limit in a cluster catalogue, which is also expected. We note that for the $10^{14} h^{-1} M_\odot$ mass limit, the local bias is 3.36, and for the $10^{13} h^{-1} M_\odot$ it decreases to 1.83. This result is in line with the biases obtained from the power spectrum analysis, the former being 2.1 and the latter 3.3 (Balaguera-Antolinez *et al.* (2011)). The fair agreement between the cluster biases calculated locally and globally is encouraging because the cluster number overdensity clearly extends into non-linear regime whereas the calculated bias of the power spectrum is mainly based on linear theory. This motivates the next section where we will quantitatively test how much this non-linear environment differs from the field.

4. Supercluster environment

Cluster mass functions can be predicted on the basis of cosmic structure formation models such as the hierarchical clustering model for CDM universe (Press & Schechter (1974)). Due to the close relation of X-ray luminosity and gravitational mass of clusters

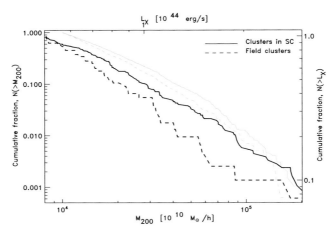

Figure 2. Measured cumulative X-ray luminosity function from REFLEX shown in thick black lines in comparison to the cumulative mass function of the superclusters in simulations in grey lines. In both cases clusters in superclusters are denoted by a solid line, while field clusters by dashed lines.

the mass function is reflected by the X-ray luminosity function (XLF) of clusters. Thus we use the observed XLF as a substitute for the mass function. The clusters in superclusters are used separately to form their own XLF in comparison to those in the field, and to better illustrate the difference we consider a normalised cumulative luminosity function as shown in Fig. 2. The black lines represent the volume-limited sample of the REFLEX II clusters.

We see a clear effect that we have a more top-heavy luminosity function for the clusters in superclusters compared to those in the field. Since the cumulative distribution is unbinned, we quantify the difference with a Kolmogorov-Smirnov (KS) test yielding the probability, P, that both luminosity distributions come from the same parent distribution with a value of 0.03. With the Millennium simulation, we form the same cumulative distribution directly with masses shown in grey lines in Fig. 2. In this case the probability from the KS test is practically zero. Hence both in our REFLEX data and simulation that there are over-abundance of more luminous clusters in the superclusters than in the field.

References

Bahcall, N. 1988, *ARAA*, 26, 631B
Balaguera-Antolinez, A. 2011, *MNRAS*, 413, 386B
Böhringer, H., Chon, G., Collins, C., Guzzo, L., Nowak, N., & Bobrovsky, S. 2013, *A&A* 555, 30
Böhringer, H., Chon, G., & Collins, C. A. (2014) 2014, *A&A* in press
Böhringer, H., Chon, G., Bristow, M., & Collins, C. A. 2014, *A&A* submitted
Chon, G. & Böhringer, H. 2013, *A&A*, 538, 35
Chon, G., Böhringer, H., & Nowak, N. 2013, *MNRAS*, 429, 3272
Chon, G., Böhringer, H., Collins, C., & Krause, M. 2014, *A&A*, 567, 144
Chon, G., Böhringer, H., & Zaroubi, S. 2014, *A&A* submitted
Press, W. H. & Schechter, P. 1974, *ApJ*, 187, 425
Springel, V. 2005, *MNRAS*, 364, 1105

The Zeldovich Universe:
Genesis and Growth of the Cosmic Web
Proceedings IAU Symposium No. 308, 2014
R. van de Weygaert, S. Shandarin, E. Saar & J. Einasto, eds.
© International Astronomical Union 2016
doi:10.1017/S1743921316009856

Galaxy Group Properties
in Observations and Simulations

Pasi Nurmi

University of Turku,
Department of Physics and Astronomy, Tuorla Observatory, Finland
email: pasnurmi@utu.fi

Abstract. In this project, we compare different properties of galaxy groups in cosmological N-body simulations and SDSS galaxy group catalogs. In the first part of the project (Nurmi et al. 2013) we compared the basic properties of the groups like the luminosity functions, group richness and velocity dispersion distributions and studied how good is the agreement between the mock group catalog and the SDSS group catalog. Here we continue the earlier study and use updated galaxy group catalog (SDSS DR10) and new simulation data (Guo *et al.* 2013). We reanalyse earlier group properties and include new properties in the analysis like group environment, star formation rates and group masses. Our analysis show that there are clear differences between the simulated and observed properties of galaxy groups, especially for small groups with a few members. Also, the high luminosities are clearly overestimated in the simulations compared with the SDSS group data.

Keywords. methods: numerical – methods: statistical – galaxies: clusters: general – cosmology: miscellaneous – large-scale structure of Universe

1. Introduction

Many numerical and analytical studies of the cluster scale dark matter halos agree well with observed cluster abundances. However, group environment influences the galaxy formation and galaxies have different properties in different large-scale environments. In our earlier study (Nurmi *et al.* 2013) we compared mock galaxy group catalogs against the galaxy groups obtained from the observations (SDSS, DR7). All group property distributions had similar shapes and amplitudes for richer groups, but for smaller groups and galaxy pairs there were clear differences. This indicates that we don't fully understand the galaxy formation in group environment. We continue the previous study and use updated simulation data and observational group catalogue.

2. Data and analysis

Our comparison data consists of two samples: galaxy groups from observations (SDSS, DR10) and galaxy groups from the simulations. The simulated galaxies are taken from the Millennium simulation database and the simulation is based on the updated version of the Millennium simulation (Springel *et al.* 2005). The simulation follows the evolution of 2150^3 particles from the redshift $z = 127$ in a box of $500\ h^{-1}$Mpc on a side and the cosmological parameters of the simulation agree with WMAP7 cosmology: $\Omega_{\mathrm{m}} = \Omega_{\mathrm{dm}} + \Omega_{\mathrm{b}} = 0.27$, $\Omega_{\mathrm{b}} = 0.045$, $h = 0.70$, $\Omega_{\lambda} = 0.73$, $n = 0.96$, and $\sigma_8 = 0.81$. The simulated galaxy evolution is based on merger trees and galaxy properties are obtained by using semi-analytical galaxy formation models, where the star formation and its regulation by feedback processes are parameterized in terms of analytical physical models (Guo *et al.* 2013). From their catalogue, we use the r-band magnitudes that include the dust

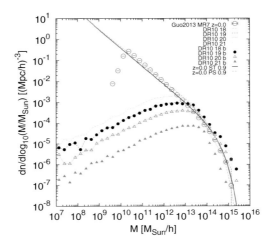

Figure 1. The mass functions of groups in different volume limited samples. Two different mass estimations of groups are shown. Guo2013 refers to mass function calculated from the virial masses of the DM halos in simulations. Also analytical mass functions by Sheth & Tormen (1999) and Press & Schechter (1974) are given.

extinction. The observed galaxies and galaxy groups come from the SDSS DR10 (Tempel *et al.* 2014).

3. Results

As a first test, we calculated the luminosity functions of all galaxies in the simulations and observations. The agreement is good except for large luminosities that are notably overestimated in the simulations. The agreement has been improved from one simulation to another, but clear differences are still seen for galaxies with $M < -22.5$. The Fig. 1 shows the mass functions of the galaxy groups obtained from the observations. These are shown together with the virial mass function of DM halos obtained from the Quo *et al.* (2013) simulation. Although the luminosity functions do not agree, the mass functions are in better agreement for cluster size objects. Group masses start to deviate from the halo mass function due to the incompleteness of the observed galaxies. This occurs for groups with $M_{tot} < 10^{13} h^{-1} M_{Sun}$ that may be the border beyond which all the groups are not virialized and the assumptions fail. We also studied the richness distribution of groups in DR10 and mock data. For galaxy pairs and small groups with < 10 members the abundances agree well, but in the simulation there are less rich groups (n of galaxies > 10). This can be due to the problems in the grouping-algorithm or then the galaxies in the SAM groups are distributed in the different way.

References

Guo, Q., White, S., Angulo, R. E., *et al.* 2013, *MNRAS*, 428, 1351
Nurmi, P., Heinämäki, P., Sepp, T., *et al.* 2013, *MNRAS*, 439, 380
Press, W. H. & Schechter, P. 1974, *ApJ*, 187
Sheth, R. & Tormen, G. 1999, *MNRAS*, 308
Springel, V., White, S., Jenkins, A., *et al.* 2005, *Nature*, 435, 629
Tempel, E., Tamm, A., Gramann, M., *et al.* 2014, *A&A*, 566, A1T

The Zeldovich Universe:
Genesis and Growth of the Cosmic Web
Proceedings IAU Symposium No. 308, 2014
R. van de Weygaert, S. Shandarin, E. Saar & J. Einasto, eds.

© International Astronomical Union 2016
doi:10.1017/S1743921316009868

3D structure of nearby groups of galaxies

L. Makarova[1], D. Makarov[1], A. Klypin[2] and S. Gottlöber[3]

[1]Special Astrophysical Observatory, Nizhniy Arkhyz, Karachai-Cherkessia 369167, Russia
email: `lidia@sao.ru`

[2]Astronomy Department, New Mexico State University, MSC 4500, P.O.Box 30001, Las Cruces, NM, 880003-8001, USA

[3]Leibniz-Institut für Astrophysik (AIP), An der Sternwarte 16, D-14482 Potsdam, Germany

Abstract. Using high accuracy distance estimates, we study the three-dimensional distribution of galaxies in five galaxy groups at a distance less than 5 Mpc from the Milky Way. Due to proximity of these groups our sample of galaxies is nearly complete down to extremely small dwarf galaxies with absolute magnitudes $M_B = -12$. We find that the average number-density profile of the groups shows a steep power-law decline $dn/dV \sim R^{-3}$ at distances R=(100–500) kpc consistent with predictions of the standard cosmological model. We also find that there is no indication of a truncation or a cutoff in the density at the expected virial radius: the density profile extends at least to 1.5 Mpc. Vast majority of galaxies within 1.5 Mpc radius around group centres are gas-rich star-forming galaxies. Early-type galaxies are found only in the central ~ 300 kpc region. Lack of dwarf spheroidal and dwarf elliptical galaxies in the field and in the outskirts of large groups is a clear indication that these galaxies experienced morphological transformation when they came close to the central region of forming galaxy group.

Keywords. (cosmology:) distance scale, (cosmology:) dark matter, galaxies: distances and redshifts, galaxies: dwarf

---

We use the Updated Nearby Galaxy Catalog (Karachentsev, Makarov, Kaisina, 2013, AJ, 145, 101), selecting seven nearest well-known groups of galaxies. Each of the group contains one main central galaxy and a significant family of satellite galaxies. We surveyed the groups around the Milky Way galaxy, M31 (Andromeda), M81, Centaurus A, M83, Canes Venatici I cloud (M94), and Sculptor (NGC253).

The luminosity function (LF) (the left panel of the Fig. 1) is constructed using adaptive binning with 10 galaxies per bin. The combined luminosity function does not show any indication of cut off on the bright end of the distribution. A simple power law $n \sim L^\alpha$ fits it well with a slope $\alpha = -1.24 \pm 0.02$. The LF of all galaxies within 7 Mpc is fitted by a Schechter function with a slope $\alpha = -1.26 \pm 0.04$. The faint end of the LF for combined group is in good accordance with the whole 7 Mpc sample.

The observed density profile (the right panel of the Fig. 1) does not show any statistically significant drop or any other feature around the virial radius. The virial radius for these type of galaxies is expected to be in the range of $\sim 200 - 400$ kpc. The density profiles smoothly declines from large (~ 1000) overdensities in the central ~ 100 kpc region and gradually approaches the background at ~ 1 Mpc. The dashed curve shows the Navarro-Frenk-White profile with parameters compatible with LCDM predictions: virial mass $2.5 \times 10^{12} M_\odot$, concentration C = 10. The virial radius for this halo is shown in the plot. The average density of the Universe is added to the Navarro-Frenk-White profile.

The distribution of galaxies in the considered groups by morphological types is demonstrated in the Fig. 2. Note, that dSphs present about 40% of the whole sample, while outside the virialized zone they consist only 20% of galaxies. It illustrates the fact that

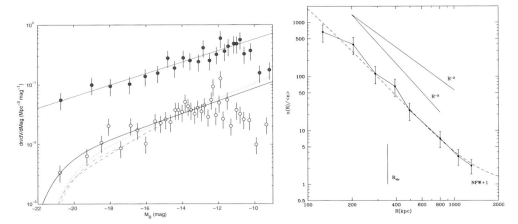

Figure 1. Left panel: the LF of the considered groups is shown with filled circles. Open symbols display the LF of all Local Volume galaxies within 7 Mpc from UNGC. For comparison we plot the global LFs from SDSS (Blanton *et al.* 2005, ApJ, 631, 208) as long dashed line and from 2dFGRS (Norberg *et al.* 2002, *MNRAS*, 336, 907) as dashed-dotted line. The dotted line illustrates it using the Schechter approximation of nearby galaxies with fixed cut-off $M_B^* = 20.5$, which roughly corresponds to the global value. **Right panel:** the full curve with error bars shows the average number-density of galaxies in four isolated centrally dominated groups: Cen A, M 83, M 81, and NGC 253. The number-density was normalized to the average number-density of galaxies brighter than $M_B = 12$ inside 10 Mpc region. The errors are poissonian errors.

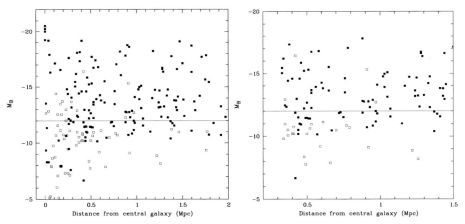

Figure 2. Left panel: the distribution of the absolute magnitudes of galaxies, corrected for galactic and internal absorption as a function of the distance of the galaxy from the central object in the group. Open squares represent giant and dwarf ellipticals and dwarf spheroidal galaxies, whereas filled squares are giant spirals, S0, irregular and dwarf irregular galaxies. **Right panel:** the same galaxies, but located in the non-virialized zone of each of the group, that is, $0.3 < r \leqslant 1.5$ Mpc. The resulting subsample contains 50% of the original one.

most of dSphs located inside $0.3 - 0.5$ Mpc from the central galaxies, while dIrs are distributed more widely.

Acknowledgements. We acknowledge the support from RFBR grant 13-02-00780 and Research Program OFN–17 of the Division of Physics, Russian Academy of Sciences, and the grant of Russian Scientific Foundation 14–12–00965. Our collaboration has been supported by the German Science Foundation (DFG).

The Zeldovich Universe:
Genesis and Growth of the Cosmic Web
Proceedings IAU Symposium No. 308, 2014
R. van de Weygaert, S. Shandarin, E. Saar & J. Einasto, eds.

© International Astronomical Union 2016
doi:10.1017/S174392131600987X

Structure of
the Canes Venatici I cloud of galaxies

Dmitry I. Makarov, Lidia N. Makarova and Roman I. Uklein

Special Astrophysical Observatory, Russian Academy of Sciences, Nizhnii Arkhyz, 369167
Russia
email: *dim@sao.ru, lidia@sao.ru, uklein@sao.ru*

Abstract. We study the spatial distribution of the sparse cloud of galaxies in the Canes Venatici constellation. We determined distances of 30 galaxies using the tip of the red giant branch (TRGB) method. This homogeneous sample allows us to distinguish the zone of chaotic motions around the center of the system. A group of galaxies around M94 is characterized by the mass-luminosity ratio of $M/L_B = 159$ $(M/L)_\odot$. It is significantly higher than the typical ratio $M/L_B \sim 30$ $(M/L)_\odot$ for the nearby groups of galaxies. The CVn I cloud of galaxies contains 4–5 times less luminous matter compared with the well-known nearby groups, such as the Local Group, M 81 and Centaurus A. The central galaxy M 94 is at least 1 mag fainter than any other central galaxy of these groups. However, the concentration of galaxies in the Canes Venatici may have a comparable total mass.

Keywords. galaxies: distances and redshifts

Prominent fuzzy concentration of galaxies in the Canes Venatici constellation with line-of-sight velocity around $V_{\rm LG} = 300$ km s^{-1} forms so called the Canes Venatici I cloud (CVn I) of galaxies (see left panel of Fig. 1). This feature has been noted by Karachentsev (1966) and de Vaucouleurs (1975). This cloud clearly differs from the other nearby galaxy groups, such as the Local Group. CVn I shows absence of clearly prominent gravitational center and looks diffuse. This complex is mostly populated by dwarf galaxies of late morphological types.

We redefined the distances for 30 galaxies of CVn I using the deep images from the Hubble Space Telescope archive with the WFPC2 and ACS cameras. We carried out a high-precision stellar photometry of the resolved stars in these galaxies using the HST-phot (Dolphin 2000) and DOLPHOT (Dolphin 2002) software. Resulting color-magnitude diagrams allowed us to determine photometric distances by TRGB method using an advanced technique (Makarov *et al.* 2006) and modern calibrations (Rizzi *et al.* 2007). Comparison of our distance estimates with measurements by other authors (Karachentsev *et al.* 2002, Karachentsev *et al.* 2003, Karachentsev *et al.* 2006, Tully *et al.* 2006, Dalcanton *et al.* 2009) shows a very good agreement with a generally better accuracy.

M 94 is a giant spiral galaxy located within the CVn I cloud and could be claimed as its gravitating center. The luminosity function of RGB stars in the M 94 galaxy appears to be much more complex than that for the normal dwarf galaxies. It has a long, extended plateau near the cut-off. This is probably related with the complex history of star formation and metal enrichment in giant spiral M 94. We estimated the distance modulus of 28.14 ± 0.08 mag, which corresponds to the linear distance of $D = 4.25 \pm 0.15$ Mpc.

The Hubble diagram of nearby galaxies in the Canes Venatici is shown in right panel of Fig. 1. The new high precision of distance determinations allow us to identify an area of chaotic motions around the center of the system. The group of galaxies around M 94 is characterized by a median velocity of 287 km s^{-1}, median distance of 4.28 Mpc,

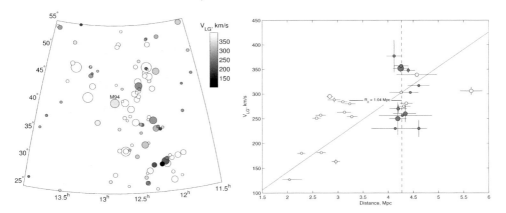

Figure 1. Left panel represents the distribution on the sky of galaxies in the Canes Venatici constellation. The distance–velocity diagram is shown on the right panel. The solid line corresponds to linear Hubble law with $H_0 = 71$ km/(s×Mpc). The galaxies within 1 Mpc from the center of the system are marked in dark gray.

and the total luminosity of $L_B = 1.61 \times 10^{10}\ L_\odot$. The virial mass amounts to $M_{\mathrm{vir}} = 1.93 \times 10^{12}\ M_\odot$, which corresponds to $M/L_B = 120\ (M/L)_\odot$. The projection estimation gives $M_{\mathrm{proj}} = 2.56\times10^{12}\ M_\odot$ and $M/L_B = 159\ (M/L)_\odot$. If we assume that specific 'wave' on Hubble diagram in front of CVn I is result of gravitational attraction to the M 94 group of galaxies, we can estimate the radius of the zero velocity sphere of 1.04 ± 0.15 Mpc. It corresponds to the mass $M_{\mathrm{R_0}} = 2.38 \times 10^{12}\ M_\odot$.

Our mass-luminosity estimate, $M/L_B = 120$–$160\ (M/L)_\odot$ for the CVn I cloud greatly exceeds the typical ratio ~ 30 for the nearby groups of galaxies, such as the Local Group $M/L_B = 15$–$20\ (M/L)_\odot$ and M 81 group $M/L_B = 19$–$32\ (M/L)_\odot$. Note that compared with the well-known nearby groups, such as the Local Group, M 81, and Centaurus A, the CVn I cloud contains 4–5 times less luminous matter, and M 94 is at least 1 mag fainter than any other central galaxy of these groups. However, the concentration of galaxies in the Canes Venatici may have a comparable total mass.

Acknowledgements

The authors thank prof. I. D. Karachentsev for constructive discussions. We are thankful for the support of the Russian Science Foundation grant 14–12–00965. We acknowledge the usage of the HyperLeda database (`http://leda.univ-lyon1.fr`).

References

J. J. Dalcanton, B. F. Williams, A. C. Seth, *et al.* 2009, *ApJS*, 183, 67
G. de Vaucouleurs, *Nearby Groups of Galaxies* (University of Chicago Press, Chicago, 1975), p. 557
A. E. Dolphin 2000, *PASP*, 112, 1383
A. E. Dolphin 2002, *MNRAS*, 332, 91
I. D. Karachentsev 1966, *Astrophysics*, 2, 39
I. D. Karachentsev, M. E. Sharina, D. I. Makarov, *et al.* 2002, *A&A*, 389, 812
I. D. Karachentsev, M. E. Sharina, A. E. Dolphin, *et al.* 2003, *A&A*, 398, 467
I. D. Karachentsev, A. Dolphin, R. B. Tully, *et al.* 2006, *AJ*, 131, 1361
D. Makarov, L. Makarova, L. Rizzi, *et al.* 2006, *AJ*, 132, 2729
L. Rizzi, R. B. Tully, D. Makarov, *et al.* 2007, *ApJ*, 661, 815
R. B. Tully, L. Rizzi, A. E. Dolphin, *et al.* 2006, *AJ*, 132, 729

The Zeldovich Universe:
Genesis and Growth of the Cosmic Web
Proceedings IAU Symposium No. 308, 2014
R. van de Weygaert, S. Shandarin, E. Saar & J. Einasto, eds.
© International Astronomical Union 2016
doi:10.1017/S1743921316009881

Studying structure formation and evolution with strong-lensing galaxy groups

Gaël Foëx[1], Veronica Motta[1], Marceau Limousin[2], Tomas Verdugo[3] and Fabio Gastaldello[4]

[1]Instituto de Física y Astronomía, Universidad de Valparaíso,
Avda. Gran Bretaña 1111, Valparaíso, Chile
email: gael.foex@uv.cl

[2]Aix Marseille Université, CNRS, Laboratoire d'Astrophysique de Marseille UMR 7326,
13388, Marseille, France

[3]Centro de Investigaciones de Astronomía,
AP 264, Mérida 5101-A, Venezuela

[4]INAF - IASF Milano,
via E. Bassini 15, I-20133 Milano, Italy

Abstract. We present the analysis of a sample of strong-lensing galaxy group candidates. Our main findings are: confirmation of group-scale systems, complex light distributions, presence of large-scale structures in their surroundings, and evidence of a strong-lensing bias in the mass-concentration relation. We also report the detection of the first 'Bullet group'.

Keywords. Dark matter, gravitational lensing, galaxy groups.

1. The SARCS sample

The Strong Lensing Legacy Survey (SL2S, Cabanac *et al.* 2007) is a semi-automated search of strong-lensing systems on the full Canada-France-Hawaii Telescope Legacy Survey (CHFTLS). Using the ARCFINDER algorithm, More *et al.* (2012) compiled the SL2S-ARCS sample (SARCS) made of group- and cluster-scale lens candidates. Basically, ARCFINDER searches for elongated and contiguous features of pixels above a given intensity threshold, and tags the most promising features as arc candidates according to their width, length, area, and curvature. Roughly 1000 candidates/deg^2 were found, which were then inspected visually to reduce the sample to 413 systems ($\sim$ 2.75 candidates/deg^2). The most promising candidates with an arc radius $R_A \gtrsim 2$" were kept, leading to a total of 127 objects (More at al., 2012). The sample spans a redshift range $z \in [0.2\text{-}1.2]$ and peaks at z $\sim$ 0.5. The SARCS distribution of image separation is located between the galaxy-scale SLACS sample and the massive cluster MACS sample, thus corresponding mostly to groups and poor clusters of galaxies.

2. Weak-lensing and optical analyses

For each SARCS candidate, we the fitted the measured shear profile by the singular isothermal sphere (SIS) mass model to estimate the velocity dispersion σ_v. We also constructed luminosity maps using the bright galaxies populating the red sequence. We combined these two analyses to build a sample of 80 most secure lens candidates, characterized by a positive weak-lensing detection ($\sigma_v - \sigma > 0$) and a clear light over-density associated to the strong-lensing feature (Foëx *et al.* 2013). With this reduced sample, we investigated the optical scaling relations of strong-lensing galaxy groups. Despite a large scatter (up to 35%), we found correlations between the SIS σ_v and the optical richness and luminosities. We combined the SARCS sample with a sample of massive galaxy clusters (Foëx *et al.*, 2012) to derive scaling laws consistent with the expectations of the hierarchical model of structure formation and evolution.

The morphological study of the luminosity maps revealed that a significant fraction of groups present a complex light distribution: $\sim 42\%$ with highly-elongated luminosity contours, $\sim 16\%$ with a multimodal structure (Foëx *et al.*, 2013). These results suggest that galaxy groups are dynamically-young objects, a picture consistent with a temporary stage towards the formation of more massive clusters. We also inspected the groups' luminosity map at larger scales. We found 10 systems with crowded environments made of several light over-densities not randomly distributed, suggesting the presence of large-scale filamentary structures (Foëx *et al.*, in prep.).

In a second paper (Foëx *et al.*, 2014), we performed a stacked weak-lensing analysis to constrain the $c(M)$ mass-concentration relation of strong lenses. We found an average concentration $c_{200} = 8.6 \pm 1.8$ for an average $M_{200} = (0.73 \pm 0.1) \times 10^{14}\,\mathrm{M}_\odot$, a concentration in disagreement at the 3σ level with the predictions from numerical simulations (Duffy *et al.*, 2008). We combined our composite strong lenses with massive strong-lensing galaxy clusters to derive the $c(M)$ over nearly two decades in mass. We found a relation much steeper than expected, resulting from projections effects of highly-elongated haloes with a major axis close to the line of sight.

3. The 'Bullet Group'

A deeper investigation of the group SL2S J08544-0121 revealed a separation of 124 ± 20 kpc between the X-ray emission peak and the mass centers of this bi-modal system (Gastaldello *et al.* 2014). Such a separation between the collisional gas and the collisionless galaxies and dark matter is characteristic of merging systems in the plan of the sky, as the so-called 'Bullet Cluster'. The estimated mass of the system is $M_{200} = (2.4 \pm 0.6) \times 10^{14}\,\mathrm{M}_\odot$ from a M-T scaling relation, and $M_{200} = (2.2 \pm 0.5) \times 10^{14}\,\mathrm{M}_\odot$ from the weak-lensing analysis, which makes it the lowest mass bullet-like object found to date. We used this 'Bullet Group' to derive an upper limit on the dark matter self-interaction cross-section of $10\,\mathrm{cm}^2\mathrm{g}^{-1}$. We showed in a parallel study based on numerical simulations (Fernandez-Trincado *et al.* 2014) that bullet groups are more numerous than massive bullet clusters. Therefore, with this first detection of a low-mass bullet-like system, we prove the possibility of using galaxy groups to perform a statistical study of the dark matter cross-section.

G.F. acknowledges funds from FONDECYT grant #3120160. V.M. acknowledges funds from FONDECYT grant #1120741. GF, VM, ML acknowledge funds from ECOS-CONICYT C12U02.

References

Cabanac, R., *et al.* 2007, *A&A*, 461, 813
Duffy, A. R., *et al.* 2008, *MNRAS*, 390, 64
Gastaldello, F., *et al.* 2014, *MNRAS*, 442, 76
Fernandez-Trincado, J. G., *et al.* 2014, *ApJ*, 787, 34
Foëx, G., *et al.* 2012, *A&A*, 546, 106
Foëx, G., *et al.* 2013, *A&A*, 559, 105
Foëx, G., *et al.* 2014, *ArXiv*, 1409.5905
More, A., *et al.* 2012, *ApJ*, 749, 38

The Zeldovich Universe:
Genesis and Growth of the Cosmic Web
Proceedings IAU Symposium No. 308, 2014
R. van de Weygaert, S. Shandarin, E. Saar & J. Einasto, eds.

© International Astronomical Union 2016
doi:10.1017/S1743921316009893

The Adopted Morphological Types of 247 Rich PF Galaxy Clusters

Elena Panko[1], Katarzyna Bajan[2], Piotr Flin[3] and Alla Gotsulyak[4]

[1]Kalinenkov Astronomical Observatory, Nikolaev National University, Nikolaev, Ukraine
email: panko.elena@gmail.com

[2]Institute of Physics, Pedagogical University, Cracow, Poland,
[3]Institute of Physics, Jan Kochanowski University, Kielce, Poland
[4]Astronomical Department, Odessa National University, Odessa, Ukraine

Abstract. Morphological types were determined for 247 rich galaxy clusters from the PF Catalogue of Galaxy Clusters and Groups. The adopted types are based on classical morphological schemes and consider concentration to the cluster center, the signs of preferential direction or plane in the cluster, and the positions of the brightest galaxies. It is shown that both concentration and preferential plane are significant and independent morphological criteria.

Keywords. Galaxies: clusters: morphological types.

1. Introduction

The classification of galaxy clusters at optical wavelengths is carried out using several different parameters: cluster richness (number of galaxies within a specific limiting magnitude), degree of central concentration, the presence of bright galaxies in the center of the cluster, etc. The prevalent Bautz-Morgan (BM) (Bautz & Morgan, 1970) and Rood-Sastry (Rood & Sastry, 1971) classification schemes are in agreement and complement each other. López-Cruz *et al.* (1997) introduced the definition of a cD cluster, the complement of which is called a non-cD cluster.

2. Observational Data

A Catalogue of Galaxy Clusters and Groups (Panko & Flin, 2006, hereafter PF) was constructed from the Münster Red Sky Survey Galaxy Catalogue (Ungrue, Seitter & Duerbeck, 2003, hereafter MRSS) mainly for statistical analysis of properties for large-scale structures. Unfortunately, so far we have only been able to study the cluster parameters from the morphology for 1056 PF clusters that are coincident with those in the ACO catalogue (Abell, Corvin & Olovin, 1989). Similarly, only 247 PF clusters with richness $N \geqslant 100$ have assumed BM morphological types according to the ACO catalogue. Those morphological types permit us to find alignments of the brightest galaxy relative to the parent clusters for BM type I (Panko, Juszczyk & Flin, 2009). The Binggeli effect (Binggeli, 1982) is strongest for BM type I clusters, as well presents for BM III clusters (Flin *et al.*, 2011). Moreover, Godłowski *et al.* (2010), using data for 97 PF galaxy clusters, found a weak dependence of galaxy velocity dispersion with BM type for the parent cluster. Other morphology schemes for PF clusters were not used. Presently we adopted the prevalent morphological systems for the MRSS observational data and determined our morphological types for PF galaxy clusters using a $2D$ distribution of galaxies in rectangular coordinates relative to the cluster center for each.

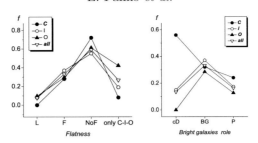

Figure 1. Variation of frequencies for flatness signs (left) and BCM role in C-I-O and all clusters.

3. The adopted morphological scheme

For input data we established adopted morphological types based on 3 parameters: concentration, signs of flatness, and bright cluster members (BCM) positions. The adopted types correspond to concentration (C - compact, I - intermediate, and O - open), flatness (L - line, F - flat, and no symbol if no indication of flatness is present), and the role of bright galaxies (cD or BG if the BCM role is significant). Other peculiarities are noted as P. The details of the approach are described and justified elsewhere Panko (2013). The designations can be combined, for example CFcD or ILP.

For 247 rich PF clusters with BM types from the ACO comparison, we determined the adopted morphological types and analyzed the frequencies of each. The sign of the flatness type is independent of concentration class: those for L and F types are similar in C-I-O groups, as shown in Fig. 1, left panel. In contrast, the role of BCMs is strongly connected with cluster concentration: the number of cD clusters is greatest in C-type (Fig. 1, right panel). Note, CcD type corresponds to BM I type. For L and F clusters we found a correlation between position angle for the major axes of the best-fit ellipse and the direction of the preferred plane.

4. Conclusions

From $2D$ maps of 247 rich PF galaxy clusters we determined their adopted morphological types. It is shown that concentration and flatness are independent morphological criteria. The direction of the major axis of the best-fitting ellipse for a cluster (calculated in the PF catalogue) is close to the direction determined by the L or F regions; the difference between the two directions increases for O-type galaxy clusters.

References

Abell, G. O., Corwin, H. G., & Olowin, R. P. 1989, *ApJS*, 70, 1
Bautz, P. & Morgan, W. W. 1970, *ApJ*, 162, L149
Binggeli, B. 1970, *A& A*, 162, L149
Flin, P., Biernacka, M., Godlowski, W., *et al.* 2011, *Balt. Astr.*, 20, 251
Godłowski, W., Piwowarska, P., Panko, E., *et al.* 2011, *ApJ*, 723, 985
López-Cruz, O., Yee, H. K. C., Brown, J. P., *et al.* 1997, *ApJ*, 475, L97
Panko, E. 2013, *Odessa Astr. Publ.*, 26, 90
Panko, E. & Flin, P. 2006, *Jornal of Astronomical Data*, 12, 1
Panko, E., Juszczyk, T., & Flin, P. 2009, *AJ*, 138, 1709
Rood, H. J. & Sastry, G. N. 1971, *PASP*, 83, 313
Ungrue, R., Seitter, W. C., & Duerbeck, H. W. 2003, *Jornal of Astronomical Data*, 9, 1

The Zeldovich Universe:
Genesis and Growth of the Cosmic Web
Proceedings IAU Symposium No. 308, 2014
R. van de Weygaert, S. Shandarin, E. Saar & J. Einasto, eds.

© International Astronomical Union 2016
doi:10.1017/S174392131600990X

Galaxy and Mass Assembly (GAMA): Selection of the Most Massive Clusters.

Héctor J. Ibarra-Medel[1,2], Maritza Lara-López[3], Omar López-Cruz[1] and the GAMA Team.

[1] National Institute of Astrophysics, Optics and Electronic,
Luis Enrique Erro No 1, Tonantzintla, Puebla, México
email: `omarlx@inaoep.mx`

[2] email: `ibarram@inaoep.mx`

[3] Institute of Astronomy, National Autonomous University of México,
Box 70-264, México City, México
email: `maritza@astro.unam.mx`

Abstract. We have developed a galaxy cluster finding technique based on the Delaunay Tessellation Field Estimator (DTFE) combined with caustic analysis. Our method allows us to recover clusters of galaxies within the mass range of 10^{12} to 10^{16} $\mathcal{M}_\odot$. We have found a total of 113 galaxy clusters in the Galaxy and Mass Assembly survey (GAMA). In the corresponding mass range, the density of clusters found in this work is comparable to the density traced by clusters selected by the thermal Sunyaev Zel'dovich Effect; however, we are able to cover a wider mass range. We present the analysis of the two-point correlation function for our cluster sample.

Keywords. methods: data analysis, surveys, galaxies: clusters: general, cosmology: miscellaneous, cosmology: observations

1. Introduction

The GAMA survey (http://www.gama-survey.org/) is a multi-wavelength spectroscopic survey that covers ~ 360 deg^2, which includes $\sim 400,000$ galaxy redshifts down to a magnitude limit of $r_{AB} = 19.8$ (Driver *et al.* 2011). We chose three stripes within GAMA that cover ~ 144 deg^2 with $\sim 110,000$ galaxy spectra. These three equatorial sky stripes are centred at 9h, 12h and 14.5h (Driver *et al.* 2011).

We have implemented a new cluster finding technique to find overdensities and estimate cluster masses, simultaneously. We find number galaxy overdensities by using an adaptive method based on the Delaunay Tessellation Field Estimator (DTFE, Schaap & van de Weygaert 2000, Platen 2009), mass estimation is done using caustic analysis (e.g., Serra *et al.* 2011, Alpaslan *et al.* 2012). We use this method to detect clusters of galaxies within the mass range of 10^{12} to 10^{16} $\mathcal{M}_\odot$, up to $z = 0.3$.

2. Overview and Results

We have found 113 cluster within GAMA. For this sample we have estimated positions, cluster redshifts, velocity dispersions, cluster sizes, and cluster integrated luminosity. Our algorithm has been tested using the GAMA mock catalogs (Robotham *et al.* 2011). The calculation the cluster luminosities have been generated by using the individual cluster-galaxy luminosity functions (LF) corrected for completeness. We have evaluated the cluster selection function by the application of a simple halo occupation distribution (HOD) model. We want to stress that the density of clusters found by mass selection methods (e.g., the Atacama Cosmology Telescope (ACT), Menanteau *et al.* 2013) is

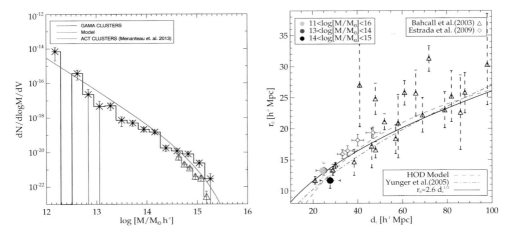

Figure 1. The left panel present the comoving density traced by the cluster found in this study (indicated by filled stars), we have compare our results those of Menanteau *et al.* (2013) generated from a sample of massive clusters selected by the Sunyaev-Zel'dovich effect (open triangles), the continuous dashed line is a simple halo occupation model (HOD). In the right panel we present a compilation of previous results on the characteristic scale for the two-point cluster correlation as a function of cluster separation, the filled big dots represents the results for our sample. We find agreement with previous studies and models

comparable to one found in this work; however, we have covered a larger mass range by more than three orders of magnitude. In addition, we have generated the two-point correlation for clusters of galaxies for our sample. We find broad agreement previous observations and predictions (Estrada *et al.* 2009). We have generated the mass-to-light ratio (M/L) for the clusters and BCGs in our sample, we find that a single power law $\mathcal{L} \propto \mathcal{M}^{\eta}$ can describe . We found $\eta = 0.6 - 1$ for clusters and $\eta_{BGC} = 0.1 - 0.4$ for BCGs. These relations agree with the results of Lin *et al.* (2004) and Lin & Mohr (2004).

The sample found in this study can be used for further studies in galaxy evolution and its relation with environment. We have shown that optical surveys such as GAMA can be used to select cluster by mass. A sample of cluster selected by our method can be used to traced baryon acoustic oscillations using in a survey in which galaxies are selected in the same fashion as GAMA but covering a larger volume.

References

Alpaslan, M., Robotham, A. S. G., Driver, S., *et al.* 2012, *MNRAS*, 3012
Bahcall, N. A., Dong, F., Hao, L., *et al.* 2003, *ApJ*, 599, 814
Driver, S. P., Hill, D. T., *et al.* 2011, *MNRAS*, 413, 971
Estrada, J., Sefusatti, E., & Frieman, J. A. 2009, *ApJ*, 692, 265
Robotham, A. S. G., Norberg, P., Driver, S. P., *et al.* 2011, *MNRAS*, 416, 2640
Menanteau, F, *et al.* 2013, *ApJ*, 765, 67
Platen, E. 2009, Ph.D. Thesis
Schaap, W. E. & van de Weygaert, R. 2000, *A& A*, 363, L29
Serra, A. L., Diaferio, A., Murante, G., & Borgani, S. 2011, *MNRAS*, 412, 800
Lin, Y.-T., Mohr, J. J., & Stanford, S. A. 2004, *ApJ*, 610, 745
Lin, Y.-T. & Mohr, J. J. 2004, *ApJ*, 617, 879
Younger, J. D., Bahcall, N. A., & Bode, P. 2005, *ApJ*, 622, 1

The Zeldovich Universe:
Genesis and Growth of the Cosmic Web
Proceedings IAU Symposium No. 308, 2014
R. van de Weygaert, S. Shandarin, E. Saar & J. Einasto, eds.

© International Astronomical Union 2016
doi:10.1017/S1743921316009911

The Evolving Shape of Galaxy Clusters

Dennis W. Just[1], H. K. C. Yee[1], Adam Muzzin[2], Gillian Wilson[3], David G. Gilbank[4] and Michael Gladders[5]

[1] Department of Astronomy and Astrophysics, University of Toronto,
50 St. George St., Toronto, ON, M5S 3H4, Canada
email: dwjust@gmail.com

[2] Leiden Observatory, Leiden University,
P.O. Box 9513, 2300 RA Leiden, The Netherlands
email: muzzin@strw.leidenuniv.nl

[3] Department of Physics and Astronomy, University of California, Riverside,
Pierce Hall, Riverside, CA 92521, USA
email: gillian.wilson@ucr.edu

[4] South African Astronomical Observatory,
PO Box 9, 7935, South Africa
email: gilbank@saao.ac.za

[5] The Department of Astronomy and Astrophysics, and the Kavli Institute for Cosmological
Physics, The University of Chicago
5640 South Ellis Avenue, Chicago, IL 60637, USA
email: gladders@oddjob.uchicago.edu

Abstract. We present the first measurement of the evolution of the apparent projected shape of galaxy clusters from $0.2 \lesssim z \lesssim 2$. We measure the ellipticities (ϵ_{cl}) of homogeneously selected galaxy clusters over this wide redshift range. We confirm the predictions of N-body simulations that clusters are more elongated at higher redshift, finding the mean projected ellipticity changes linearly from 0.36 ± 0.01 to 0.25 ± 0.01 over that range. The fraction of relaxed clusters (defined as having $\epsilon_{cl} < 0.2$) is $9^{+5}_{-3}\%$ at $z \sim 1.8$, steadily increasing to $42^{+7}_{-6}\%$ by $z \sim 0.3$. Because more spherical clusters have a higher degree of virialization, our result shows significant evolution in the degree of cluster virialization over cosmic time.

Keywords. galaxies: clusters: general

1. Introduction

Galaxy clusters are the largest virialized structures in the Universe, and play important roles in our understanding of both cosmology and galaxy evolution. The aspherical shapes of clusters and groups has long been recognized tending to be preferentially prolate with cluster ellipticities (ϵ_{cl}) of ~ 0.2–0.5. In a hierarchical structure formation scenario, clusters are predicted to exhibit more elongated shapes when infall and merging rates are higher, as they are at earlier times in the Universe. With galaxy infall preferentially occurring along a few filaments that feed the cluster, the distribution of galaxies takes an elongated shape. As the Universe becomes more rarefied, clusters find themselves cut off from significant inflow and have the opportunity to gravitationally relax to a more spherical shape with a higher degree of virialization (Tsai & Buote 1996).

In this paper, we present the first measurement that tracks the mean cluster ellipticity from $0.2 \lesssim z \lesssim 2$ using homogeneously selected galaxy clusters. The low redshift clusters come from the Red-Sequence Cluster Survey-2 (RCS-2; Gilbank *et al.* 2011), with higher redshift clusters from the Spitzer Adaptation of the RCS (SpARCS; Wilson *et al.* 2009),

D. W. Just *et al.*

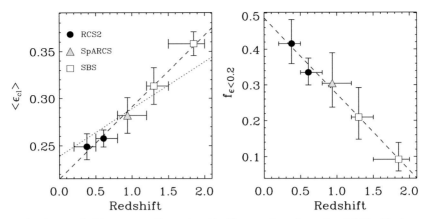

Figure 1. (*Left panel*) The mean ellipticity ($\langle \epsilon_{cl} \rangle$) as a function of redshift. Horizontal error bars show the bin sizes. The dotted line uses the slope from Hopkins *et al.* (2005), with the intercept fitted to the data: $\langle \epsilon_{cl} \rangle = 0.240 + 0.05z$. The dashed line is the best-fit based on the data. (*Right panel*) The fraction of clusters with $\epsilon < 0.2$ as a function of redshift.

including clusters identified with the stellar bump sequence method (SBS; Muzzin *et al.* 2009).

2. Results

We determine projected cluster shapes by assigning a probability that a given galaxy belongs to the cluster red sequence based on its color (Gladders *et al.* 2000, Muzzin *et al.* 2013). We construct probability maps by smoothing the spatial distribution of galaxies with a 0.5 Mpc gaussian kernel and measure ϵ_{cl} from the second moment of the probability contours containing 60% of the peak probability in the map.

In Fig. 1 we present $\langle \epsilon_{cl} \rangle$ and the fraction of "relaxed" clusters (defined by $\epsilon_{cl} < 0.2$; $f_{\epsilon_{cl} < 0.2}$) as a function of redshift. The former quantity tracks how the shape of the clusters changes over redshift, while the latter quantifies the mix of "relaxed" and more elongated clusters over cosmic time. We compare to the N-body simulation of Hopkins *et al.* (2005).

Both are strong functions of redshift, with the best linear fit to both trends having slopes > 5 standard deviations from zero. The rate of change in $\langle \epsilon_{cl} \rangle$ is steeper than predicted, although the difference in measuring cluster shape between observations and simulations can account for this. Our result shows the evolving degree of cluster virialization over ≈ 7.5 Gyr of cosmic time and agreement with the ΛCDM cosmology.

References

Gilbank, D. G., Gladders, M. D., Yee, H. K. C., & Hsieh, B. C. 2011, *AJ*, 141, 94

Gladders, M. D. & Yee, H. K. C. 2000, *AJ*, 120, 2148

Hopkins, P. F., Bahcall, N. A., & Bode, P. 2005, *ApJ*, 618, 1

Muzzin, A., Wilson, G., Lacy, M., Yee, H. K. C., & Stanford, S. A. 2008, *ApJ*, 686, 966

Muzzin, A., Wilson, G., Demarco, R., *et al.* 2013, *ApJ*, 767, 39

Tsai, J. C. & Buote, D. A. 1996, *MNRAS*, 282, 77

Wilson, G., Muzzin, A., Yee, H. K. C., *et al.* 2009 *ApJ*, 698, 1943

CHAPTER 4.

Cosmic Web
Morphology & Identification
Reconstruction & Clustering

View over the skyline of the medieval old town of Tallinn,
UNESCO world heritage site.
Photo: Rien van de Weijgaert.

Part of the medieval walls around the old town of Tallinn,
the stretch of walkway between the Nun and Sauna Tower.
Photo: Rien van de Weijgaert.

CHAPTER 4A.

Cosmic Web
Morphology & Identification

View towards St. Olav Church, old town Tallinn.
The church was the tallest building in Europe between 1549 and 1625.
Photo: Rien van de Weijgaert.

One of the towers of the medieval defensive wall of Tallinn.
Photo: Rien van de Weijgaert.

The Zeldovich Universe:
Genesis and Growth of the Cosmic Web
Proceedings IAU Symposium No. 308, 2014
R. van de Weygaert, S. Shandarin, E. Saar & J. Einasto, eds.

© International Astronomical Union 2016
doi:10.1017/S1743921316009923

The structural elements of the cosmic web

Bernard J.T. Jones[1] and Rien van de Weygaert[2]†

[1]Kapteyn Astronomical Institute, University of Groningen, The Netherlands
email: jones@astro.rug.nl

[2] Kapteyn Astronomical Institute, University of Groningen, The Netherlands
email: weygaert@astro.rug.nl

Abstract. In 1970 Zel'dovich published a far-reaching paper presenting a simple equation describing the nonlinear growth of primordial density inhomogeneities. The equation was remarkably successful in explaining the large scale structure in the Universe that we observe: a Universe in which the structure appears to be delineated by filaments and clusters of galaxies surrounding huge void regions. In order to concretise this impression it is necessary to define these structural elements through formal techniques with which we can compare the Zel'dovich model and N-body simulations with the observational data.

We present an overview of recent efforts to identify voids, filaments and clusters in both the observed galaxy distribution and in numerical simulations of structure formation. We focus, in particular, on methods that involve no fine-tuning of parameters and that handle scale dependence automatically. It is important that these techniques should result in finding structures that relate directly to the dynamical mechanism of structure formation.

1. A Short Historical Introduction

In the early 1960's there was an ongoing debate between the proponents of two different cosmological models. There was also a debate as to whether or not galaxy clusters were stable agglomerations of galaxies. The first of these was to be resolved in 1965 with the discovery of the cosmic background radiation (CBR) by Penzias & Wilson (1965) and its interpretation by Dicke *et al.* (1965) in terms of the relict radiation from a hot big bang origin of the Universe. The second debate would take several decades to resolve in terms of the existence of non-luminous "dark matter" in the Universe. Nonetheless, progress on describing the origin of cosmic structure started very soon after the CBR discovery with seminal papers by Peebles (1965), Sachs & Wolfe (1967), Silk (1967), Zel'dovich (1970), Sunyaev& Zeldovich (1970), Peebles & Yu (1970), among others.

1.1. *The Zel'dovich Legacy*

Zel'dovich (1970) presented a simple model describing the non-linear evolution of structure in a co-expanding frame of reference described by a coordinate system $\{\mathbf{x}\} : \mathbf{x} = \mathbf{r}/a(t)$. $a(t)$ is the usual cosmic expansion factor and $\mathbf{r}$ is the position of the particle in the expanding space. The Zel'dovich *ansatz* is that a particle starting from initial position $\mathbf{x}(t_0) = \mathbf{q}$ would move ballistically according to

$$\mathbf{x} = \mathbf{q} - D(t)\mathbf{v}(\mathbf{q}), \quad \text{with} \quad \mathbf{v} = \frac{\partial \Phi}{\partial q_j} \tag{1.1}$$

for some scalar velocity potential $\Phi(\mathbf{q})$. Here, $\mathbf{v}(\mathbf{q})$ is a pseudo-velocity vector with which the particle is projected from the point $\mathbf{q}$ and $D(t)$ is the so-called *linear growth factor*

---

† This article reflects some of the work of the **Groningen Cosmic Web Group**, which has at various times included thesis work and subsequent papers written with Miguel Aragon-Calvo, Marius Cautun and Erwin Platen.

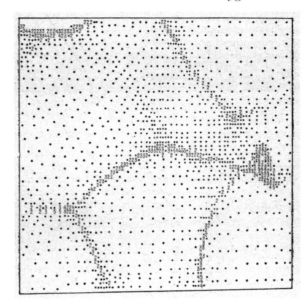

Figure 1. First 2-D simulation of the formation of structure using the Zel'dovich approximation (Doroshkevich & Shandarin 1978). This was done in 1976 at a time when filaments and voids had not been recognized as key elements in a web-like structure. The paper introduced use of the Hessian in defining local structural morphology.

that determines the time dependence of the particle motion. $D(t)$ mimics the effects of gravity in such a way that the growth of small amplitude density fluctuations accords with linear perturbation theory (Peebles 1965) . It is the separation of the displacement into a temporal component, $D(t)$, and a spatial component $\mathbf{v}(\mathbf{q})$, depending only on the initial particle positions that underlies the elegance and power of this model.

A small element of volume $d^3\mathbf{q}$ in the initial configuration is carried by equation 1.1 into a a small element $d^3\mathbf{x}$ by the flow. If there is no flux of material into or out of these elemental volumes, the ratio of these two volumes gives the evolution of the density. Likewise, the relative shear and rotation of the volumes describes the local kinematics. This is expressed in terms of the *Zel'dovich deformation tensor* $\mathbf{Z} = \{Z_{ij}\}$ which has components

$$Z_{ij} = \frac{\partial x_i}{\partial q_j} \tag{1.2}$$

The density evolution in the physical space is

$$\frac{\rho(\mathbf{x}, t)}{\rho_0(t)} = \det \mathbf{Z} = \left| \frac{\partial x_i}{\partial q_j} \right| = \frac{1}{(1 - D(t)\gamma_1)(1 - D(t)\gamma_2)(1 - D(t)\gamma_3)} \tag{1.3}$$

where the γ_i are the eigenvalues of $\mathbf{Z}$, ordered so that $\gamma_1 \geqslant \gamma_2 \geqslant \gamma_3$.

A key point about this approximation is that the terms in the denominator all have the same time dependence: $D(t)$. So the first term to become singular is the one involving the largest eigenvalue, γ_1. Hence the first structure to form is was a flattened structure that became known as a *pancake*. This would be followed by the formation of a linear structure (a *filament*) when the term involving γ_2 became singular and finally a point-like structure (a *cluster*). The pancakes would surround regions referred to as *voids*. The first numerical simulation of this process (Fig. 1) was due to (Doroshkevich & Shandarin 1978). This

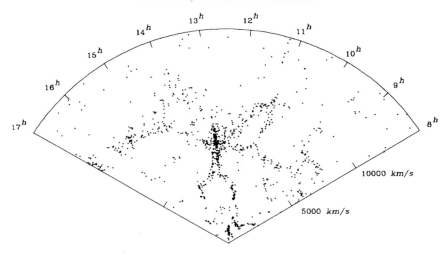

Figure 2. The CfA redshift slice de Lapparent *et al.* (1986). Although voids had earlier been recognized in surveys centered on rich galaxy clusters, this study demonstrated the ubiquity of voids. Co-author John Huchra is famously quoted as saying *"If we are right, these bubbles fill the universe just like suds filling the kitchen sink"* (Sullivan 1986).

was to be followed by more than a decade of increasingly sophisticated simulations of the Zel'dovich process.

This was all done within the framework of a cosmological model dominated by hot dark matter in which the large scales would be the first to collapse. Smaller scale structure would have to form via a fragmentation process taking place in the baryonic component. The astrophysical consequences of equation 1.3 were discussed at length by Doroshkevich & Shandarin (1974) and in terms of the gravitational and thermodynamic instability of the pancake structures by Jones *et al.* (1981).

1.2. *Evidence for a cosmic web*

The first evidence for such large scale structures came from the seminal paper of de Lapparent *et al.* (1986) describing the CfA Redshift Slice (widely known as the *deLapparent Slice*, see Figure 2). Previous to that others had identified voids by doing redshift surveys in the direction of rich galaxy clusters: in particular we think of the study of the Boötes void and of the Perseus-Pisces chain. Subsequent ever larger redshift surveys served to enhance the first impression given by the de Lapparent slice. In particular the *2dF Redshift Survey* (Colless *et al.* 2001) not only showed the ubiquity of this structure (see Figure 3), but also set the pattern in terms of the sample definition and analysis procedures for future redshift surveys.

1.3. *The cosmic web in numerical simulations*

One of the questions that arose in the late 1980's, with the popularization of the notion that the dark matter in the Universe might be cold (Frenk *et al.* 1985), was whether such visually dominant structures could in fact arise in Cold Dark Matter (CDM) models where the structure was dominated by the very smallest scales. Davis *et al.* (1985) addressed this issue, at the same time introducing into simulations the important notion that the particles in the simulations that were to be identified with galaxies were only located where there were significant density peaks on galaxy scales (Kaiser 1984; Bardeen *et al.* 1986). This was referred to as *biasing*: it was a simple mechanism for identifying

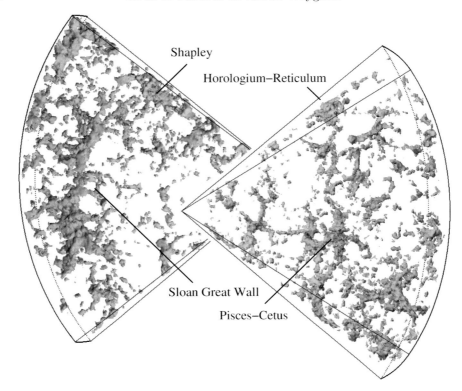

Figure 3. The distribution of galaxies in the 2dF Redshift Survey (2dFGRS) (Colless *et al.* 2001). The survey had redshifts for some 250,000 galaxies obtained with a 400-fibre multi-object spectrograph covering a 2 □° field on the Anglo Australian Telescope. The data was quickly made public, thereby enhancing the science value of the project.

This smooth iso-density version of the survey map was produced from the original point distribution using the Delaunay Triangulation Field Estimator ('DTFE') of van de Weygaert& Schaap (2007). DTFE provides an efficient mass-preserving mechanism for interpolating the field density at any point of the sample space. Some of the most famous very large scale features are identified on the map.

where the luminous matter might be found and for highlighting the structures we see. With this mechanism, in those early models, the web-like structure was visible but not particularly striking.

Larger simulations with improved models for identifying the sites of luminous galaxies did, however, bring the structure to light (Springel *et al.* 2005, The *Millenium Simulation*). Figure 4 shows the structure in the galaxies that have been located in that simulation using semi-analytic models ('SAM') for galaxy evolution to identify haloes that host luminous galaxies.

2. Identification of Cosmic Structures

The human brain is good at picking out structures from complex scenes, so we have little or no trouble in appreciating the presence of and identifying structural entities from pictures of galaxy distributions. However, we guess, but do not know *a priori*, that these structures arose as a consequence of the kind of dynamics suggested by the Zel'dovich model. The issue is to relate what we perceive to dynamical models.

Figure 4. The cosmic web as seen through filamentary distribution of galaxies in the Virgo consortium's 10^{10}-particle *Millenium Simulation* (Springel *et al.* 2005). The subsequent decade has seen even larger simulations, some dealing with enhanced gas dynamic flow modeling (see the presentation of Pichon in these proceedings).

To do that effectively we must have tools that will systematically and automatically isolate the structures that we perceive on the basis of point patterns taken either from galaxy redshift catalogs or numerical simulations. With that we can then go further and discuss galaxy evolution in different environments. We can also design observational projects to test our surmises about galaxy evolution in those various environments.

2.1. *Structure finding scenarios*

How you find structure in a distribution of particles depends on the information available in the source data. There are the following possibilities to consider:

6-dimensional

phase space in which both velocities and positions are specified for every particle.

3-dimensional

positions of all the particles

2+1 dimensions

2D positions on the sky + radial velocity

All of these three views are directly available from cosmological N-body models, whereas what we observe is only the "2+1" view.

It is not possible to review or even to list all published approaches to cosmological image segmentation in this short article. However, many of these techniques are covered by other contributors to this volume.

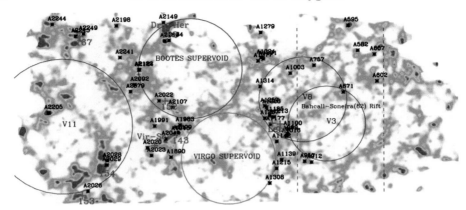

Figure 5. Super-voids found in the DR7 release of the Sloan Digital Sky Survey (SDSS) using the Watershed Void Finder of Platen *et al.* (2007, 2008). The super-void locations are depicted by a circle having a radius equal to the effective void radius. The boundaries of the super-voids are in places well defined by aggregates (super-clusters) of rich galaxy clusters which are marked on the figure by black crosses.

The 'Virgo super-void' is so-called because it lies in the direction of and far behind the Virgo cluster at a distance of $\sim 140h^{-1}$Mpc. This is Void 7 in Table 1 of Tully (1986), the other V-numbers are from that table. These super-voids voids range in diameter assigned by the WVF method from $\sim 82h^{-1}$Mpc for V3 to $\sim 148h^{-1}$Mpc for V11.

Sizes are typically $\sim 100h^{-1}$Mpc, which tempts us to identify them with the BAO scales.

2.2. *General Comments: Voids, Walls, Filaments and clusters*

The data we are presented with is almost always a set of points in one of the data-spaces described above. The points are merely a statistical sample of the underlying structure that we wish to discover and analyze. Most methods worked on a smoothed, grid-based, version of the given point set. The idea of smoothing is that it somehow compensates or corrects for the statistical nature of the data we are presented with. Smoothing of this underlying field was generally done with isotropic smoothing kernels. This was fine for finding clusters of points, but vitiated against finding thin structures such as walls and filaments. Nor was it very successful at finding voids, which, by definition, were regions where there were very few, if any, data points (see, for example, Colberg *et al.* (2008)).

The alternative to smoothing is re-sampling the given data onto a grid. One of the central tools to do this was developed by van de Weygaert& Schaap (2007): the Delaunay Triangulation Field Estimator ("DTFE") or its dual, the Voronoi Tesseleation Field Estimator ("VTFE"). The DTFE mechanism provides a continuous field on a grid that is free of structural bias and closely reflects the original point data. Any noise present in the original data is still present in the gridded data, and so there is no cost resulting from assumptions about either the noise or the underlying data.

3. Void Finders

The best way of finding voids appears to be the Watershed based methods, as first introduced by Platen *et al.* (2007, 2008, The Watershed Void Finder, WVF) , and Neyrinck (2008, ZOBOV).†

---

† For an excellent and recent review of watershed and other void finding methods see Way *et al.* (2014). See also the Void Finder comparison test of Colberg *et al.* (2008).

The watershed method is parameter-free and has been widely used as the basis for identifying the other cosmic structures: walls, filaments and clusters. Since the voids are thought to be space-filling, their intersections can be identified as walls, while the intersections of the walls are the filaments, and the meeting points of the filaments are the clusters: in fact, just as envisaged in the original pancake theory (Doroshkevich & Shandarin 1974).

An example of this is the *Spineweb* method of Aragón-Calvo *et al.* (2010). However, one of the practical disadvantages of this approach is that filaments necessarily connect clusters, whereas in the real world we see filaments that emanate from clusters to terminate inside a void. We also appear to see filaments within the walls (Rieder *et al.* 2013).

3.1. *WVF: Watershed Void Finder*

The Watershed Void Finder (WVF) of Platen *et al.* (2007, 2008) has a data preprocessing stage which involves re-sampling data onto a grid using the DTFE, and then smoothing the sampled data on different scales R_f ranging from $1h^{-1}$ Mpc to $10h^{-1}$. Importantly, the data is then median filtered several times to remove small scale fluctuations that may be due to remaining smaller scale structure than the filter R_f. This yields a clear basin and well defined surrounding edges for the watershed void finding. This process makes it possible to find very large scale 'voids' that are below-average in density while having smaller scale substructure.

Examples of such super-voids found using WVF with a filter scale of $R_f = 10h^{-1}$Mpc are shown in figure 5. The WVF method recovers several super-voids that had been previously identified in the sample volume, eg: Tully (1986).†

4. MMF-NEXUS: Multi-scale parameter-free analysis

The idea of doing image segmentation by mimicking the brain goes back the the work of Marr (1982) and his concept of the *primal sketch*. In a 2-dimensional continuous grey-tone image, Marr considered the contours on where the grey-scale level ϕ has $\nabla^2\phi = 0$. By looking at differently (Gaussian) smoothed versions of the same image Marr constructed what he described as a *primal sketch*. Over the past decades this notion has been generalised and refined by a number of authors in diverse fields: one of the main changes has been to study the Hessian of the field ϕ: $\mathcal{H}_{ab} = \partial^2\phi/\partial x_a\partial x_b$. This was further modified and brought to the problem of cosmic structure finding by Aragón-Calvo *et al.* (2007b, and references therein).

4.1. *Halo spins and bulk motions*

One of the more open questions of cosmology has always been the question of whether there are any systematic alignments in the spins of galaxies. The advent of large N-body simulations has allowed this question to be studied within the framework of the dark matter halos that form in these simulations. Use of simulations greatly simplifies the question since there are no projection effects to deal with and the spin axes pf the halos is well defined. Once we can ambiguously identify filaments, the issue is further simplified when fixing attention to looking for any alignment of galaxies in the filaments.

The galaxies in the filaments are moving along the filaments under the influence of the general tidal field of the nearest galaxy cluster into which they are falling. Hence

† The diameters assigned by WVF are somewhat smaller than those given by Tully (renormalized to $H_0 = 100$). Moreover, they are far from being spherical which makes assignment of a size somewhat ambiguous.

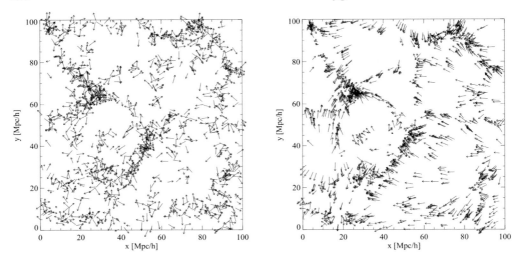

Figure 6. *Left:* Orientation of halo spin directions. *Right:* Halo bulk motions. The spin axes appear random, any systemic orientation is marginal. However, the systemic bulk flows of filaments is manifest. (From numerical simulations reported by Trowland *et al.* (2013)).

one might expected that Filaments are a good place to search for alignments with the structure. Moreover, since these are the same tidal fields that determine the bulk motions within the filaments, we would likewise expect to find correlations of spin axis with various aspects of the bulk flow.

4.2. *Alignments of halo spins and filaments*

The situation is nevertheless not totally straightforward, as can be seen from the left hand panel of Fig. 6 (Trowland *et al.* 2013): the small arrows depict the spin axes of haloes in the simulation and they are not manifestly lined up. However, the right hand panel shows that the filaments do have a unambiguous bulk motions (see also Rieder *et al.* (2013)).

The first report of systemic alignment of spins in N-body filaments and walls was that of Aragón-Calvo *et al.* (2007a) who did find evidence for non-random alignment in the halos, and, interestingly, reported that the low-mass objects tended to have there spin axes aligned within the host structure, while the higher mass objects had their spin axes aligned perpendicular to the host structure. This result that has since been confirmed by a number of groups (see, for example, Forero-Romero *et al.* (2014) who provide a nice overview of previous results). There have also been confirmations that the filaments are correlated with various aspects the underlying velocity field (van de Weygaert & van Kampen 1993; Libeskind *et al.* 2014; Tempel *et al.* 2014). As seen in the right hand panel of Fig. 6, the correlations in the velocity field are rather strong.

4.3. *Filamentary structure and Galaxy alignments in the SDSS*

The situation with real data (2 + 1 data) is less straightforward. Redshift distortions make it more difficult to define filaments that are oriented close to the line of sight, this is evident in Figure 2. Moreover, there is a difficulty in accurately defining the spin axis of real galaxies which are determined from their baryonic component rather than their dark matter component. The selection of edge-on galaxies removes the spin axis orientation issue, and selection of well defined filaments that are almost transverse to the line of sight removes filaments that result from redshift distortions. Using a sample of

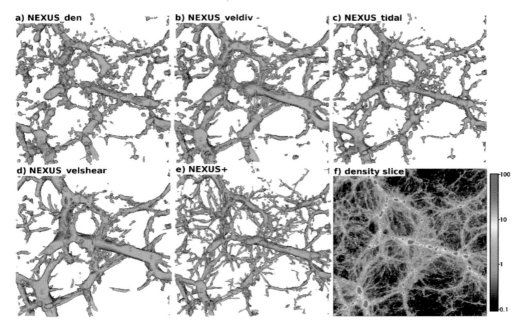

Figure 7. Structures in different fields as revealed by the NEXUS and NEXUS+ techniques. We clearly see the structural differences and inter-relationships between the various field that determine the physical nature of the cosmic web. The density field as delineated by NEXUS+ using the logarithm of the density shows a finer structure than using NEXUS on the density field.

~500 filaments found in the DR5 release of the SDSS using MMF, Jones *et al.* (2010) found evidence of a bimodal distribution of the spins relative to the local filament axis. Tempel *et al.* (2013) analyzed filaments found in the SDSS (DR8) using their BISOUS method and found evidence for alignments in spiral galaxies with the spin axis along the filament, and for E/S0 galaxies, but with their minor axes aligned perpendicular to the host filament.

5. NEXUS and NEXUS+

NEXUS (Cautun *et al.* 2013) is a scale-free and parameter-free structure classifier and extends and improves on the original MMF structure classifier (Aragón-Calvo *et al.* 2007b). Like MMF, NEXUS was designed to handle density fields, but, importantly, it is equally aimed at determining the structural morphology of the divergence and shear of the particle velocity field, as well as the structure in the gravitational tidal field. The NEXUS+ variant of NEXUS deals with the log-density field. Structures found in the various fields are shown in Figure 7.

5.1. *Determining scale-free parameter-free morphology*

The field values are re-sampled onto a grid using DTFE, and then a hierarchy of smoothed fields is generated on a set of scales $\{k\}$ † The local environment is characterized on each

---

† The smoothing kernel used was Gaussian on a set of scales differing by factors of 2. Earlier experiments by BJ (unpublished) had shown that the use of spline or wavelet kernels, or the use of finer hierarchies made relatively little difference when structural morphology was based simply on the eigenvalues of the Hessian of the field values.

scale, k, by an *environment signature*, S_k, that is constructed from the local eigenvalues of the Hessian of the underlying field values smoothed on that scale. The scale free value of the environment signature, S, is then simply the maximum value of S_k.

The map of S-values has a large dynamic range and only the largest value of S are significant, the low ones being noise. A threshold, S_{thresh} must therefore be applied for each of the cluster, filament and wall-like structure maps: the significant structure is set of above-threshold points. The major technical issue is the method whereby the critical threshold is determined, and how to determine that automatically. Details are in Cautun *et al.* (2013, Section 2).

One of the key technical differences between NEXUS and MMF lies in this automated thresholding mechanism. For finding filamentary structure in the density field, MMF calculated a threshold by finding a critical value of a percolation parameter. NEXUS uses a threshold value where the mass fraction, ΔM^2, in the filaments is most rapidly changing. As it turns out, for the density field, the percolation and mass fraction thresholds are very close, and hence the resulting filamentary structure is very similar in both approaches. Using ΔM^2 solves an issue regarding the walls in that were found in MMF, and works particularly well for the structures based on properties of the velocity field or gravitational potential. The details are given in Cautun *et al.* (2013, Appendix A).

5.2. *NEXUS structures in N-body simulations*

Application of NEXUS+ to an N-body model, where the densities and velocities are known, allows structural characterization of the density field, the velocity divergence and shear fields, the tidal field and the log-density field. Figure 7 displays the results.

Although the perceived structure is similar for all tracers, the details of the structures that are found vary perceptibly among the different tracers used in the mapping. For example, the filamentary structure found log-density field, labeled "NEXUS+" in the figure, covers a large range of scales: it is clearly advantageous to use this method for density fields. The velocity field tracers tend to be smoother, but neverthelss show structures of all scales. It is also to be noted that many filaments are often "open-ended" - they disappear into voids. These differences are potentially important since they provide clues as to how the structures evolved.

Galaxy formation and evolution are strongly environment dependent and NEXUS provides a tool for objectively identifying and analyzing the local environment of a galaxy at any stage of its evolution. This is discussed exhaustively in Cautun *et al.* (2014), where the focus in on the halos in the simulation, and their morphological environments. The simulations used are the $z = 0$ volumes of the MS-I (Springel *et al.* 2005) and MS-II (Boylan-Kolchin *et al.* 2009) simulations, which use the canonical WMAP cosmological parameters. We show just one set of results from the NEXUS+ analysis analysis of these simulations in 8. The graph in the left panel shows the cumulative mass function of objects as a function of environment. The right panel shows pie-diagrams of the morphological make-up by volume fraction and by mass fraction: the voids obviously dominate the volume fraction while the filaments dominate the mass fraction.

6. The 6-dimensional and 3-dimensional views

It is clear that 6-dimensional, phase space, structure classifiers should do a better job at assigning cosmic structures to different morphology classes than do the 3-dimensional classifiers or the $2 + 1$-dimensional classifiers. However, the central issue is to understand what each of the classifiers is measuring and to discuss how the findings of these classifiers

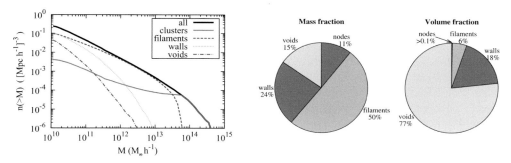

Figure 8. Left: The cumulative halo mass function segmented according to the components of the cosmic web as identified by NEXUS+, normalized according to the volume of the whole simulation box. Right: The mass and volume fractions occupied by cosmic web environments detected by the NEXUS+ method. (Cautun *et al.* 2014, Figures 19, 8).

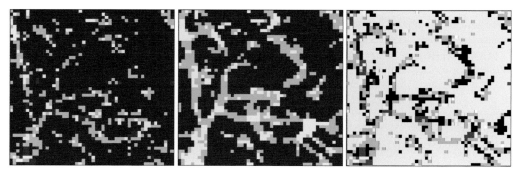

Figure 9. A simple comparison of the spatial structure found by the $6 - D$ Origami and $3D$ Nexus filament finders. The panel on the left is the outcome of the ORIGAMI method, while the center panel is the result from NEXUS. The panel on the right has been constructed by recoloring all the NEXUS structure in black and transparently superposing the ORIGAMI image. Note that, in this, way some of the NEXUS (black) structures are hidden beneath the ORIGAMI structures and the gray area show where both agree there is no structure. (From the Lorentz Center Web Comparison Project, Leiden February 2014.)

relate to one another. As a specific example, consider the ORIGAMI (Neyrinck 2012) and NEXUS methods.

6.1. *ORIGAMI and NEXUS*

The ORIGAMI approach considers the full 6-dimensional phase space structure at the time the initial conditions are set. The inter-penetration of opposingly moving streams of particles in (x, y, z) configuration space appears as a folding of the phase space sheet on which the particle lie in the full 6-dimensional phase space (see also Hidding *et al.* (2014)). The structure is delineated by these folds, hence the name "origami".

NEXUS works entirely on the snapshot of the 3-space particle configuration at any given time, examining the eigenvalues of the scale-independent Hessian of the density or log-density field, and takes no account of how that structure originated. NEXUS is a powerful perceptual model having no dynamical rationale.

6.2. *Why do the ORIGAMI and NEXUS results look so similar?*

The images in Figure 9 show a slice from the results of ORIGAMI and NEXUS analysis of an N-body model that was provided to participants in the Lorentz Center Cosmic Web

Comparison meeting (Leiden, February 2014) for the purpose of structure classification. The spatial structures identified in the slice are highlighted†.

The two left-most panels show the structure assignment by the ORIGAMI and NEXUS classifications schemes. The rightmost panel highlights the commonality between the two approaches by showing the ORIGAMI identified structures superposed on the NEXUS image. The NEXUS structure has been entirely rendered in black to make the comparison clearer. This could be done with far more rigor, but even at this simple level we see the remarkable agreement between the two approaches.

The approaches are totally different, and yet the correspondence between the classifications, at first sight, appears quite remarkable. We see something similar in Figure 7 where NEXUS analysis of an N-body simulation leads to density field maps (a) and (e) that compare in detail with the velocity divergence field and shear field maps (b) and (d). This is, of course, a consequence of the equations of motion of the inhomogeneous cosmic medium: the Euler equation tells us that the rate of change of the velocity is governed by the gravitational potential gradient. The velocity field gradient, ie. the velocity divergence and shear fields, are driven by the gravitational tidal shear, which, via the Poisson equation, is directly proportional to the density.

In the Zel'dovich approximation this is almost trivial since in that approximation the velocity potential and the gravitational potential are proportional to one another.‡

7. Concluding remarks

Zeldovich's simple nonlinear approximation to the evolution of cosmic structure from given initial conditions is a powerful tool for understanding the evolution of the cosmic web. The large scale structures that are the consequence of that approximation have a direct reflection in the web-like structures seen in N-body simulations.

Objectively segmenting cosmic structure into its main component parts is a key factor in understanding the environmental issues which affect our understanding of galaxy evolution and out ability to analyses and interpret velocity fields. There is now a wide variety of techniques available that achieve this segmentation, though most of them are not scale-independent and require tuning of parameters to obtain optimal results. We have focused on our NEXUS/NEXUS+ scheme and shown, in that case, that there is a remarkable level of agreement between the structure found by ORIGAMI analysis of the Zel'dovich caustics in the phase space of the initial conditions and by NEXUS analysis of the final conditions.

Acknowledgements

BJ would like to thank the IAU and the organisers for making it possible to participate in this meeting.

† Since this is a thin slice some of the apparently filamentary structure will in fact be a slice through a wall.
‡ For flow having zero vorticity, the velocity field can be written $\mathbf{v} \propto \mathbf{\nabla}\mathcal{U}$, where $\mathcal{U}$ is the velocity potential and the proportionality factor depends only on time. If ϕ is the gravitational potential, then the Zel'dovich approximation is $\mathcal{U} \propto \phi$ (Gramman 1993; Jones 1999). With this $\partial_i\partial_j\mathcal{U} = \partial_\backslash\partial_|\phi$, where ∂_i denotes $\partial/\partial x_i$. All the derivatives $\partial_i\partial_j\phi$ are determined by the density field (Poisson's equation and the boundary conditions). $\partial_i\partial_j\mathcal{U} = \partial_i v_j$ is the velocity shear tensor.

References

Boylan-Kolchin, M. *et al.* 2009 *MNRAS* 398, 1150

Aragón-Calvo, M. A. *et al.* 2007 *ApJ* 655, L5

Aragón-Calvo M.A. *et al.*(MMF) 2007 *A&A* 474, 315

Aragón-Calvo M.A. *et al.*(Spineweb) 2010 *ApJ* 723, 364

Bardeen, J. M. *et al.* (BBKS) 1986 *ApJ* 304, 15

Cautun, M., van de Weygaert, R. & Jones, B. J. T. 2013 *MNRAS* 429, 1286

Cautun, M., *et al.* 2014 *MNRAS* 441, 2923

Colberg, J. M. *et al.* 2008 *MNRAS* 387, 933

Colless, M. *et al.* (2dF) 2001 *MNRAS* 328, 1039

Davis, M. *et al.* (DEFW) 1985 *ApJ* 292, 371

de Lapparent, V., Geller, M. J. & Huchra, J. P. 1986 *ApJ* 302, L1

Dicke, R. H. *et al.* 1965 *ApJ* 142, 414

Doroshkevich, A. G. & Shandarin, S. F. 1974 *Soviet Ast.* 18, 24

Doroshkevich, A. G. & Shandarin, S. F. 1978 *Soviet Ast.* 22, 653

Forero-Romero, J. E., Contreras, S. & Padilla, N. 2010 *MNRAS*, 2014, 443, 1090

Frenk, C. S. *et al.* 1985 *Nature* 317, 595

Gramman, M. 1993 *ApJ* 405, 449

Hidding, J., Shandarin, S. F. & van de Weygaert, R. 2014 *MNRAS* 437, 3442

Jones, B. J. T., Palmer, P. L. & Wyse, R. F. G. 1981 *MNRAS* 197, 967

Jones, B. J. T. 1999 *MNRAS* 307, 376

Jones, B. J. T., van de Weygaert, R. & Aragón-Calvo, M. A. 2010 *MNRAS* 408, 897

Kaiser, N. 1984 *ApJ* 284, L9

Libeskind, N. I, Hoffman, Y. & Gottlöber, S. 2014 *MNRAS* 441, 1974

Marr, D. 1982 *Vision* W.H. Freeman

Neyrinck, M. C. (ZOBOV) 2008 *MNRAS* 386, 2101

Neyrinck, M. C. (ORIGAMI) 2012 *MNRAS* 427, 494

Peebles, P. J. E. 1965 *ApJ* 142, 1317

Peebles, P. J. E. & Yu, J. T. 1970 *ApJ* 162, 815

Penzias, A. A. & Wilson, R. W. 1965 *ApJ* 42, 419

Platen, E., van de Weygaert, R. & Jones, B. J. T. (WVF) 2007 *MNRAS* 380, 551

Platen, E., van de Weygaert, R. & Jones, B. J. T. 2008 *MNRAS* 387, 128

Rieder, S. *et al.* 2013 *MNRAS* 435, 222

Sachs, R. K. & Wolfe, A. M. 1967 *ApJ* 147, 73

Silk, J. 1967 *Nature* 215, 1155

Springel, V. *et al.* 2005 *Nature* 435, 629

Sullivan, W. 1986 *New York Times* 1986, January 5. http://www.nytimes.com/1986/01/05/science/new-view-of-universe-shows-sea-of-bubbles-to-which-stars-cling.html

Sunyaev, R. A. & Zeldovich, Ya. B. 1970 *Ap&SS* 7, 3

Tempel, E., Stoica, R. S. & Saar, E. 2013 *MNRAS* 428, 1827

Tempel, E. *et al.* 2014 *MNRAS* 437, L11

Trowland, H. E., Lewis, G. F. & Bland-Hawthorn, J. 2013 *ApJ* 762, 72

Tully, R.B. 1986 *ApJ* 303, 25

van de Weygaert, R. & van Kampen, E. 1993 *MNRAS* 263, 481

van de Weygaert, R. & Schaap, W. 2007 in *Data Analysis in Cosmology* Springer-Verlag http://arxiv.org/abs/0708.1441

Way, M. J., Gazis, P. R. & Scargle, J. D. 2014 http://arxiv.org/abs/1406.6111

White, S. D. M. 1984 *ApJ* 286, 38

Zel'dovich, Ya. B. 1970 *A&A* 5, 84

The Zeldovich Universe:
Genesis and Growth of the Cosmic Web
Proceedings IAU Symposium No. 308, 2014
R. van de Weygaert, S. Shandarin, E. Saar & J. Einasto, eds.

© International Astronomical Union 2016
doi:10.1017/S1743921316009935

Filamentary pattern in the cosmic web: galaxy filaments as pearl necklaces

Elmo Tempel[1,2] and Maarja Bussov[1]

[1]Tartu Observatory, Observatooriumi 1, 61602 Tõravere, Estonia
email: elmo.tempel@to.ee

[2]NICPB, Rävala pst 10, 10143 Tallinn, Estonia

Abstract. Galaxies form chains (filaments) that connect groups and clusters of galaxies. The filamentary network includes nearly half of the galaxies and is visually the most striking feature in cosmological maps. We study the distribution of galaxies along such a filamentary network, trying to find specific patterns. Our galaxy filaments are defined using the Bisous process. We use the two-point correlation function and the Rayleigh Z-squared statistic to study how the galaxies are distributed along the filaments. We show that galaxies and galaxy groups are not uniformly distributed along filaments, but tend to form a regular pattern. The characteristic length of the pattern is 7 h^{-1}Mpc. A slightly smaller characteristic length 4 h^{-1}Mpc can also be found, using the Z-squared statistic. One can say that galaxy filaments are like pearl necklaces, where the pearls are galaxy groups distributed more or less regularly along the filaments. We propose that this well defined characteristic scale could be used as a cosmological test.

Keywords. Methods: numerical – methods: observational – large-scale structure of Universe.

1. Introduction

It is well known that galaxy filaments are visually the most dominant structures in the galaxy distribution, being part of the so-called cosmic network. Presumably, nearly half ($\sim 40\%$) of the galaxies (or mass in simulations) are located in filaments (Jasche *et al.* 2010; Cautun *et al.* 2014). Tempel *et al.* (2014a) shows that filaments extracted from the spatial distribution of galaxies/haloes are also dynamical structures that are well connected with the underlying velocity field. Galaxy filaments that connect groups and clusters of galaxies are also affecting the evolution of galaxies (Tempel & Libeskind 2013; Zhang *et al.* 2013).

In this paper we study the distribution of galaxies along galaxy filaments to search for regularities in galaxy and group distributions. Such a clustering pattern exists at least in some filaments as shown decades ago (e.g. Jõeveer *et al.* 1978; Einasto *et al.* 1980). In the current study, we use the two-point correlation function and the Rayleigh Z-squared statistics to look for the regularity in the galaxy distribution. We suggest that the regularities we find in the galaxy distribution could be used as cosmological probes for dark energy and dark matter.

2. Data and methods

The present work is based on the SDSS DR10. We use the galaxy and group samples as compiled in Tempel *et al.* (2014b) that cover the main contiguous area of the survey (the Legacy Survey, approximately 17.5% from the full sky). The flux-limited catalogue extends to the redshift 0.2 (574 h^{-1}Mpc) and includes 588193 galaxies and 82458 groups with two or more members.

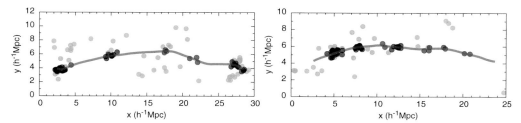

Figure 1. Examples of filaments and their spines. Red points show galaxies in filaments (closer than 0.5 h^{-1} Mpc to the filament axis) that are located in groups with 5 or more members. Blue points show other galaxies in filaments. Grey points are background galaxies that are not located in these filaments. The thick green line shows the spine of a filament.

In Tempel *et al.* (2014b) the redshift-space distortions, the so-called finger-of-god (FoG) effects, are suppressed using the rms sizes of galaxy groups in the plane of the sky and their rms radial velocities as described in Liivamägi *et al.* (2012). We calculate the new radial distances for galaxies in groups in order to make the galaxy distribution in groups approximately spherical. We note that such a compression will remove the artificial line-of-sight filament-like structures.

The detection of filaments is performed by applying an object/marked point process with interactions (the Bisous process) to the distribution of galaxies. This algorithm provides a quantitative classification which agrees with the visual impression of the cosmic web and is based on a robust and well-defined mathematical scheme. A detailed description of the Bisous model is given in Stoica *et al.* (2007, 2010) and Tempel *et al.* (2014c). For reader convenience, a brief and intuitive description is given below.

The marked point process we propose for filament detection is different from the ones already used in cosmology. In fact, we do not model galaxies, but the structure outlined by galaxy positions.

This model approximates the filamentary network by a random configuration of small segments (thin cylinders). We assume that locally galaxies may be grouped together inside a rather small cylinder, and such cylinders may combine to form a filament if neighbouring cylinders are aligned in similar directions. This approach has the advantage that it is using only positions of galaxies and does not require any additional smoothing to create a continuous density field.

The solution provided by our model is stochastic. Therefore, we find some variation in the detected patterns for different MCMC runs of the model. The main advantage of using such a stochastic approach is the ability to give simultaneous morphological and statistical characterisation of the filamentary pattern.

In practice, after fixing the approximate scale of the filaments, the algorithm returns the filament detection probability field together with the filament orientation field. Based on these data, filament spines are extracted and a filament catalogue is built. Every filament in this catalogue is represented as a spine: a set of points that define the axis of the filament.

The spine detection we use is based on two ideas. First, filament spines are located at the highest density regions outlined by the filament probability maps. Second, in these regions of high probability for the filamentary network, the spines are oriented along the orientation field of the filamentary network.

The filaments are extracted from a flux-limited galaxy sample, hence, the completeness of extracted filaments decreases with distance. In Tempel *et al.* (2014c) we showed that the volume filling fraction of filaments is roughly constant with distance if filaments longer than 15 h^{-1}Mpc are considered. Therefore, in the current study we are using only filaments longer than this limit. In addition, this choice is justified, since longer filaments allow us to study the distribution of galaxies along the filaments, which is the purpose of this paper. The remaining incompleteness of filaments in our sample is not a problem, because we are analysing single filaments. The filaments we use are the strongest filaments (or segments of filaments) in the sample.

To suppress the flux-limited sample effect, we volume-limit the galaxy content for single filaments. For that, we find for every filament the maximum distance (from the observer) of its galaxies and the corresponding magnitude limit and use only galaxies brighter than that. Since the majority of filaments extend over a relatively narrow distance interval, the number of excluded galaxies in every filament is small.

To study the galaxy spacing along the filaments, we project every galaxy to the filament spine. Hence, the distance is measured along the filament spine. Figure 1 illustrates extracted galaxy filaments and their spines in the field of galaxies. For our analysis, we use only filaments that contain at least 10 galaxies. When studying galaxy groups, we use only filaments where the number of groups per filament is at least 5.

To study galaxy correlations in filaments, we use the two-point correlation function $\xi(\mathbf{r})$ that measures the excess probability of finding two points separated by a vector $\mathbf{r}$ compared to that probability in a homogeneous Poisson sample. For galaxy filaments, the correlation function along the filament can be expressed in terms of the distances between the galaxies measured along the filament.

We estimate $\xi(r)$ following the Landy-Szalay border-corrected estimator (Landy & Szalay 1993). We generated a random distribution of points for each filament considered, and estimated the correlation function $\xi(r)$:

$$\widehat{\xi}(r) = 1 + \frac{DD(r)}{RR(r)} - 2\frac{DR(r)}{RR(r)}, \tag{2.1}$$

where $DD(r)$, $RR(r)$, and $DR(r)$ are the probability densities for the galaxy-galaxy, random-random, and galaxy-random pairs, respectively, for a pair distance r.

Using Eq. (2.1), we estimate the correlation function for each filament separately. The final pair correlation function, averaged over all filaments, is estimated as a sum over all filaments. We estimate the statistical error on our $\xi^{\mathrm{fil}}(r)$ measurements with the standard jackknife method, where we omit one filament under consideration at a time. In our figures we show the 95% confidence intervals.

To test whether the galaxy distribution might have regularity along the filaments we use the Rayleigh (or Z-squared) statistic. It is an excellent method when the event rate (in our case, the number of galaxies per filament) is low. The method has been used to detect periodicity in time series for the data in the form of discrete events (photon arrival times) and can be applied to detect periodicity in the galaxy distribution along filaments, where the galaxy positions can be considered as events.

The algorithm works as follows. For each filament, we produce a periodogram using the Z_1^2 (Rayleigh statistic),

$$Z_1^2 = \frac{2}{N}\left[\left(\sum_{j=1}^{N}\cos\phi_j\right)^2 + \left(\sum_{j=1}^{N}\sin\phi_j\right)^2\right], \tag{2.2}$$

where N is the number of galaxies in a filament and $\phi_j = 2\pi l_j/d$ is the phase value for a galaxy j for a fixed period d; l_j is a distance of the galaxy j along the filament spine from the beginning of the filament.

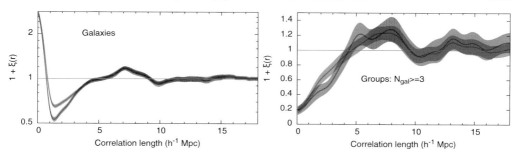

Figure 2. Galaxy (left panel) and group (right panel) correlation function along filaments. Blue and red lines with filled regions show the correlation function together with its 95% confidence limits (based on a jackknife estimate) for galaxies/groups closer than 0.5 and 0.25 h^{-1}Mpc to the filament axis, respectively.

To measure the Z-squared statistic for a period d, we are using only filaments longer than $2d$. This assures that there are at least two periods for each filament. We are deriving a null-hypothesis probability function using Monte Carlo simulations with N (the number of galaxies in a filament) data points assuming an uniform distribution of points along the spine of the filament. This allows us also to estimate the confidence intervals for our measured signal.

We compute the Z_1^2 statistic as a function of a period d for every filament and then we find the average signal using the filament length L_{fil} as the weight.

3. Results and discussion

Figure 2 (left panel) shows the galaxy correlation function along filaments. Three specific features are seen in this correlation function: a maximum near the zero pair distance, a minimum that follows it, and a bump close to 7 h^{-1}Mpc. The first maximum is caused by galaxy groups. It shows that galaxies are not distributed uniformly in the space, they form groups and clusters, as it is well known. The minimum next to the first maximum shows that groups themselves are not distributed uniformly along filaments. It shows that two groups cannot be located directly close to each other (merging groups are exceptions) and there exists some preferred minimum distance between galaxy groups. This is also expected since matter is falling into groups and there is not enough matter in a close-by neighbourhood of groups to form another group. The most interesting feature is the small maximum close to 7 h^{-1}Mpc. It shows that galaxies (and also groups) have a tendency to be separated by this distance along the filaments.

In Fig. 2 (right panel) we show the correlation function for galaxy groups with at least three members. In this figure, the smoothing scale is twice as large as used for the galaxy correlation functions and we analyse only filaments that contain at least five groups. We see that the bump around 7 h^{-1}Mpc is present also for groups, indicating that groups themselves have some preferred distance between them. We analysed groups with different minimum group richness (the number of galaxies in groups). The 7 h^{-1}Mpc bump was higher if isolated galaxies ($N_{\mathrm{gal}} = 1$) were excluded.

The biggest difference when comparing the correlation function for galaxies (left panel in Fig. 2) and for groups (right panel in Fig. 2) is the fact that there is no zero-distance maximum for groups. This shows that groups themselves do not form clusters, which is expected. The minimum close to zero distance in the group correlation function shows that there is a lack of other groups around each group. The radius of the group influence extends to 4 h^{-1}Mpc, after that the correlation function is mostly flat.

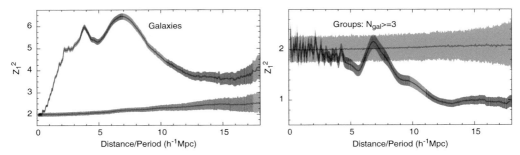

Figure 3. The dependence of the Rayleigh (Z-squared) statistic Z_1^2 on distance (period). The left panel shows the results for galaxies closer than $0.5\ h^{-1}\mathrm{Mpc}$ to the filament axis and right panel shows the results for galaxy groups. The red line shows the Z_1^2 statistic together with the jackknife 95% confidence estimate. The blue line shows the results from Monte Carlo simulations for the null hypothesis together with the 95% confidence limits.

Figure 3 (left panel) shows the Z-squared statistic for all galaxies closer than $0.5\ h^{-1}\mathrm{Mpc}$ to the filament axis. The Z-squared statistic based on galaxies is shown with red lines, where the shaded region shows the 95% confidence limits. The blue line shows the statistic for the null hypothesis using Monte Carlo simulation for a Poisson sample. The shaded region shows the 95% confidence limits for this case. For Monte Carlo simulation, the filaments and numbers of galaxies per filament are the same as for the real sample, but galaxies are Poisson distributed. Since we are averaging filaments with different lengths, the Z-squared statistic for the Poisson sample is not exactly around 2 as predicted by theory. Since the deviation is small, it does not affect our conclusions.

The maximum around $7\ h^{-1}\mathrm{Mpc}$ that was visible in the correlation functions is also visible in Fig. 3, confirming that galaxies are distributed along filaments in some regular pattern. Interestingly, the Z_1^2 statistic shows that there is also a small maximum around $4\ h^{-1}\mathrm{Mpc}$. This indicates that between two groups that are separated by $7\ h^{-1}\mathrm{Mpc}$ there is quite often another group.

Figure 3 (right panel) shows the Z-squared statistic for groups with three or more member galaxies. The blue region shows the Monte Carlo simulation results for a Poisson sample. The Z_1^2 statistic for groups lies considerably below of that for the Poisson sample. This indicates that galaxy groups are distributed along filaments more regularly than in the Poisson case. This is expected, since galaxy groups cannot be located directly close to each other, there is some minimum distance between the groups that is also visible in the two-point correlation functions. We analysed groups with different minimum group richness. The $7\ h^{-1}\mathrm{Mpc}$ scale was the strongest for groups with at least three member galaxies. It shows that the regularity is stronger if the weakest groups are excluded.

In Fig. 3 we see that the maxima around 4 and $7\ h^{-1}\mathrm{Mpc}$ that define the preferred scales for the galaxy distribution, are also visible in the group distributions. This indicates that it is related to galaxy clustering.

We refer to Tempel *et al.* (2014d) for more detailed analysis of the results presented here.

4. Conclusions

Using the Bisous model (marked point process with interactions) we extracted the galaxy filaments from the SDSS spectroscopic galaxy survey. The diameter of the extracted filaments is roughly $1\ h^{-1}\mathrm{Mpc}$ and the catalogue of filament spines is built as described in Tempel *et al.* (2014c). Using the galaxies and groups in filaments (with a distance from the filament axis less than $0.5\ h^{-1}$Mpc) we studied how the galaxies/groups are distributed along the filament axis. The main results of our study can be summarised as following.

• The galaxy and group distributions along filaments show a regular pattern with a preferred scale around $7\ h^{-1}$Mpc. A weaker regularity is also visible at a scale of $4\ h^{-1}$Mpc. The regularity of the distribution of galaxies along filaments is a new result that might help to understand structure formation in the Universe.

• The pair correlation functions of galaxies and groups along filaments show that around each group, there is a region where the number density of galaxies/groups is smaller than on average.

• Galaxy groups in the Universe are more uniformly distributed along filaments than in the Poisson case.

The clustering pattern of galaxies and groups along filaments tells us that galaxy filaments are like pearl necklaces, where the pearls are galaxy groups that are distributed along the filaments in some regular pattern.

We suggest that the measured regularity of the galaxy distribution along filaments could be used as a cosmological probe to discriminate between various dark energy and dark matter cosmological models. Additionally, it can be used to probe environmental effects in the formation and evolution of galaxies. We plan to test these hypothesis in our following analysis using N-body simulations and deep redshift surveys like the Galaxy And Mass Assembly (GAMA) and the VIMOS Public Extragalactic Survey (VIPERS).

Acknowledgements. This work was supported by ESF grants/projects MJD272, IUT40-2, TK120. We thank SDSS for the publicly available data releases.

References

Cautun, M., van de Weygaert, R., Jones, B. J. T., & Frenk, C. S. 2014, *MNRAS*, 441, 2923

Einasto, J., Jõeveer, M., & Saar, E. 1980, *Nature*, 283, 47

Jasche, J., Kitaura, F. S., Li, C., & Enßlin, T. A. 2010, *MNRAS*, 409, 355

Jõeveer, M., Einasto, J., & Tago, E. 1978, *MNRAS*, 185, 357

Landy, S. D. & Szalay, A. S. 1993, *ApJ*, 412, 64

Liivamägi, L. J., Tempel, E., & Saar, E. 2012, *A&A*, 539, A80

Stoica, R. S., Martínez, V. J., & Saar, E. 2007, *Journal of the Royal Statistical Society Series C*, 56, 459

Stoica, R. S., Martínez, V. J., & Saar, E. 2010, *A&A*, 510, A38

Tempel, E. & Libeskind, N. I. 2013, *ApJL*, 775, L42

Tempel, E., Libeskind, N. I., Hoffman, Y., Liivamägi, L. J., & Tamm, A. 2014a, *MNRAS*, 437, L11

Tempel, E., Tamm, A., Gramann, M., Tuvikene, T., Liivamägi, L. J., Suhhonenko, I., Kipper, R., Einasto, M., & Saar, E. 2014b, *A&A*, 566, A1

Tempel, E., Stoica, R. S., Martínez, V. J., Liivamägi, L. J., Castellan, G., & Saar, E. 2014c, *MNRAS*, 438, 3465

Tempel, E., Kipper, R., Saar, E., Bussov, M., Hektor, A., & Pelt, J. 2014d, arXiv:1406.4357

Zhang, Y., Dietrich, J. P., McKay, T. A., Sheldon, E. S., & Nguyen, A. T. Q. 2013, *ApJ*, 773, 115

The Zeldovich Universe:
Genesis and Growth of the Cosmic Web
Proceedings IAU Symposium No. 308, 2014
R. van de Weygaert, S. Shandarin, E. Saar & J. Einasto, eds.

© International Astronomical Union 2016
doi:10.1017/S1743921316009947

Adaptive density estimator
for galaxy surveys

Enn Saar[1,2]

[1]Tartu Observatory, Tõravere, Tartumaa, Estonia
email: saar@aai.ee

[2]Estonian Academy of Sciences, Kohtu 4, Tallinn, Estonia

Abstract. Galaxy number or luminosity density serves as a basis for many structure classification algorithms. Several methods are used to estimate this density. Among them kernel methods have probably the best statistical properties and allow also to estimate the local sample errors of the estimate. We introduce a kernel density estimator with an adaptive data-driven anisotropic kernel, describe its properties and demonstrate the wealth of additional information it gives us about the local properties of the galaxy distribution.

Keywords. Cosmology: large-scale structure of universe, surveys, galaxies: statistics

1. Introduction

Galaxy position (redshift) surveys give us maps of the universe, formed by the mutual positions of many galaxies. Such maps can be studied as they are, but in many cases the first processed product of the survey would be the continuous (matter or luminosity) density map for the survey volume. Such a map gives us an understanding of the cosmography of the survey, forms a basis for finding and classifying the constituents of the large-scale structure, and is probably useful for many other applications.

Density estimation has become an art in itself in recent years, especially with application of Bayesian methods. One of the most impressive cosmological papers in recent years (Jasche & Wandelt (2012)) demonstrated how one can find the density distribution for a galaxy catalog with photometric redshifts. The errors of photometric redshifts are usually so large that they smear up all the large-scale structure. Jasche & Wandelt showed that using an isotropic covariance matrix as a natural prior one can recover the real structure. Think about it – the only additional requirement is a very natural and simple requirement of local statistical isotropy, and it does practically all the work.

Another long-reaching effort along similar lines is the work by Kitaura *et al.* (2012) who reconstruct the density distribution in the local universe by guessing the initial conditions compatible with the present galaxy distribution, taking them to the present by numerical simulations, and comparing them with real galaxy positions. All this work is one step in a MCMC chain, so the total work is enormous. But the result is certainly the best picture of the local universe for that moment.

Both groups are continuing their work and improving their methods, I recommend to check the literature, searching by the authors. But although this approach is solid and the results are impressive, it is very expensive in terms of computer time. The observational data is represented by numbers of galaxies in spatial cells that cover the survey volume, and every number is an independent variable. The typical number of the cells is about 10^6 to 10^7, and this is the dimension of the space where the MCMC has to work. This demands huge computing power, sophisticated methods, and it is difficult to imagine that it would be possible to sample the posterior distribution uniformly.

As future survey volumes are growing fast, we need to also use fast (simple) methods of density estimation, both as an alternative methods or better inputs for the Bayesian methods.

2. Density estimation

The simplest way to estimate the density is to use histograms, that for a 3-D world translate to disjoint volume elements, usually cubic cells, and to count galaxies in these cells. Although such approach is frequently used, it is not the best way to get the density. The most evident drawback is that the galaxy numbers in adjacent cells may crucially depend on the arbitrary location of cell boundaries – we may as well find a whole galaxy cluster in a cell, as to break it into two halves by a happy boundary. Statisticians have long known that there is a much better way, the kernel density estimation (see, e.g., citeSilverman86).

For a 1-D case, the density $\rho(x)$ for a discrete sample of n points with the positions x_i can be found in any point x as

$$\rho(x) = \frac{1}{nh} \sum_i^n K\left(\frac{x - x_i}{h}\right), \qquad (2.1)$$

where x_i are the coordinates of the sample points, summation extends over all these points, $K(x; h)$ is the kernel, and h is the kernel scale. Kernels may be quite arbitrary, there are only four conditions that they must satisfy:

$$K(x) > 0, \qquad (2.2)$$

$$\int K(x)dx = 1, \qquad (2.3)$$

$$\int xK(x)dx = 0, \qquad (2.4)$$

$$\int x^2 K(x)dx < \infty. \qquad (2.5)$$

In other words, the kernel must be a symmetric probability distribution of finite variance.

Practice has shown that the exact functional form of a kernel does not matter much, but the scale does – choosing the right scale we can minimize the goodness measure of our density estimate, its MISE (mean integrated squared error):

$$\text{MISE}(h) = E \int (\hat{\rho}(x; h) - \rho(x))^2 \, dx,$$

where $\hat{\rho}(x; h)$ is the estimate of the density found using the scale h, $\rho(x)$ is the true density, and E denotes the expectation value. The MISE is, in fact, the only number that is used to compare how effective the density estimators are.

The formula 2.1 is referred to as a fixed kernel estimate. In practice, the density distributions are frequently non-uniform, and we could get a better estimate by varying the scale h. These estimates are called adaptive estimates, and there are two kinds of them: the sandbox estimate where the kernel size depends on the data points:

$$\rho(x) = \frac{1}{n} \sum_i^n \frac{1}{h_i} K\left(\frac{x - x_i}{h_i}\right),$$

and the balloon estimate where the kernel size depends on the position where we estimate the density:

$$\rho(x) = \frac{1}{nh(x)} \sum_i^n K\left(\frac{x - x_i}{h(x)}\right)$$

There are several empirical rules and iterative methods to find better h_i or $h(x)$.

In a multidimensional case, the kernel $k(\mathbf{x})$ is also multidimensional:

$$\rho(\mathbf{x}) = \frac{1}{nh} \sum_i^n K\left(\frac{\mathbf{x} - \mathbf{x}_i}{h}\right). \tag{2.6}$$

It is usually constructed as a spherical kernel or a direct product of one-dimensional kernels (for Gaussian kernels, this is the same). As for adaptive kernels, the one-dimensional thinking has carried over to the multidimensional case, and for adaptive density estimates, people usually try to construct adaptive kernels as spherical kernels of different scale. This is not good, as it smears up the local galaxy distribution. Another possibility is to use products of one-dimensional kernels of different scale; these scales have to be prescribed, somehow. But there is more freedom in the multidimensional case. For example, a (balloon) density estimate using a Gaussian kernel is

$$\rho(\mathbf{x}) = \frac{1}{(2\pi|\Sigma|)^{D/2}} \sum_i^n \exp\left[-\frac{1}{2}(\mathbf{x} - \mathbf{x}_i)^T \Sigma^{-1}(\mathbf{x} - \mathbf{x}_i)\right]. \tag{2.7}$$

Here the scaling is fixed by the covariance matrix $\sum(\mathbf{x})$, and this is, in general, not diagonal.

We propose to use a local estimate for this matrix:

$$\Sigma(\mathbf{x}) = \frac{1}{n} \sum_i^n w(\mathbf{x}, \mathbf{x}_i, R)(\mathbf{x}_i - \mathbf{x})(\mathbf{x}_i - \mathbf{x})^T,$$

that is similar to the usual covariance matrix estimate, but is restricted to a region near the point $\mathbf{x}$ by the weight function $w(\mathbf{x}_i, \mathbf{x}_j, R)$. We choose it to be also Gaussian of rms R:

$$w(\mathbf{x}_i, \mathbf{x}_j, R) = \frac{1}{(2\pi)^{D/2} R^D} \exp\left[-\frac{1}{2R^2}(\mathbf{x}_i - \mathbf{x}_j)^2\right].$$

It is easy to see that in the case of a locally constant density, our covariance matrix will be diagonal,

$$\Sigma = R^2 \mathbf{I},$$

where $\mathbf{I}$ is the unit matrix of dimension D. In a general case, the density distribution is described not only by the scalar $\rho(\mathbf{x})$, but, in addition, by the eigenvalues $S_k, k \in [1, D]$ of the covariance matrix, and by the eigenvectors $\mathbf{v}_k$, the axes of the covariance ellipsoid. This describes the local anisotropy of the galaxy distribution.

Solving for eigensystems is usually a delicate iterative process. But for three-dimensional maps there are direct analytical algoritms giving both the eigenvalues and eigenvectors (see, e.g., Kopp (2008)). The necessary libraries are freely available†. And analytical formulae exist also for two-dimensional matrices. So for the usual 3D and 2D maps the algorithm is fast.

† See http://www.mpi-hd.mpg.de/personalhomes/globes/3x3/

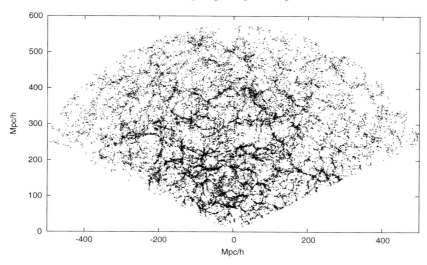

Figure 1. The 2D projection of galaxies for the thin Sloan sample slice used.

3. Examples

For data, we use the SDSS DR8 galaxies, where we have found the groups and spheri-sized their velocity space fingers (Tempel *et al.* (2012)). To see better how the algorithm works, we show a 2-D example, and select for that galaxies from the equatorial slice, where the SDSS survey coordinate $\eta \in [-2, 2]$ degrees. Fig. 1 shows the projected galaxy distribution in this slice. We compare the Gaussian and adaptive densities for a sensible $R = 10\mathrm{Mpc}h^{-1}$. As expected, adaptive kernels restore the density better than the standard Gaussian one. For the 2D case, we get also the axes ratio that describes the local anisotropy of the galaxy distribution.

4. Summary

The main advantage of the present approach is that it allows us to use the local anisotropy on the galaxy distribution. There are many ways to describe the galaxy maps and to classify the elements of the large-scale structure that rely on the local properties of the smooth density field. It is clear that density maps obtained by counting galaxy numbers in cells or by using isotropic kernels smooth out this information to a very large extent.

The local adaptive estimates of the kernel width are also useful, but not to such extent. We know that galaxy distribution is of a multiscale nature – there are clusters, groups, and filaments of different scale. For one smoothing scale, we get filaments, for another scale, these filaments may form a wall. So there probably is no unique density distribution. These advices us that we should not use our algorithm in an iterative way, although it may be tempting.

The computer code for the algorithm can be found on GitHub†. It is written in C, and includes several tricks to speed up the algorithm – arranging the data for a fast neighbour search, using a compact kernel for the final density estimation, etc.

† https://github.com/esaar/andens.

E. Saar

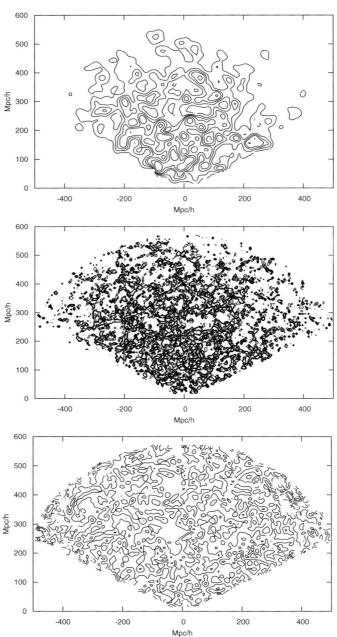

Figure 2. Sloan slice densities for Gaussian smoothing ($R = 10\mathrm{Mpc}h^{-1}$), upper panel, and for adaptive smoothing for the same initial σ, and axes ratios for the adaptive kernels.

References

Jasche, J. & Wandelt, B. D 2012, *Monthly Notices of the Royal Astronomical Society*, 425, 1042
Kitaura, F.-S., Erdoğdu, P., Nuza, S. E., Khalatyan, A., Angulo, R. E, Hoffman, Y., & Gottlöber, S. 2012, *Monthly Notices of the Royal Astronomical Society*, 427, L35

Silverman, B. W. 1986, *Density Estimation for Statistics and Data Analysis. Chapman & Hall, London*

Kopp, J. 2008, *International Journal of Modern Physics C*, 19, 523

Tempel, E., Tago, E.,& Liivamägi, L. J. 2012, *Astronomy and Astrophysics*, 540, A106

The Zeldovich Universe:
Genesis and Growth of the Cosmic Web
Proceedings IAU Symposium No. 308, 2014
R. van de Weygaert, S. Shandarin, E. Saar & J. Einasto, eds.

© International Astronomical Union 2016
doi:10.1017/S1743921316009959

Environmental density of galaxies from SDSS via Voronoi tessellation

Dobrycheva, D.[1], Melnyk, O.[2,3], Elyiv, A.[1,3] and Vavilova, I.[1]

[1] Main Astronomical Observatory of the National Academy of Sciences of Ukraine
27 Akademika Zabolotnoho St., 03680 Kyiv, Ukraine
email: dariadobrycheva@gmail.com, andrii.elyiv@gmail.com, irivav@mao.kiev.ua

[2] Astronomical Observatory, Taras Shevchenko National University of Kyiv
3 Observatorna St., 04053 Kyiv, Ukraine email: melnykol@gmail.com

[3] Dipartimento di Fisica e Astronomia, Universita di Bologna

Abstract. The aim of our work was to determine the environmental density of galaxies from SDSS DR9 using the Voronoi tessellation. We constructed the 3D Voronoi tessellation for the volume-limited galaxy sample within $0.02 < z < 0.1$ and $-24 < M_r < -20.7$ using an inverse volume of Voronoi cell as a parameter describing the local environmental density of a galaxy. It allowed us to inspect the morphology - density relation. We obtained that the early type galaxies prefer to reside in the Voronoi cells of smaller volumes (i.e. dense environments) than the late type galaxies, which are located in the larger Voronoi cells (i.e. sparse environments).

Keywords. methods: data analysis; surveys; galaxies: fundamental parameters, general

1. Introduction

The morphology of galaxies as well as their mass, color, gas content, star formation rate, and metallicity depend on the environment: a fraction of the early type (red) galaxies is higher in regions with elevated concentrations of galaxies, while the late (blue) type galaxies predominate in the general field (Dressler 1980; Einasto *et al.* 2003, Blanton *et al.* 2005). In this work we determined the environmental density of galaxies from SDSS DR9 using the Voronoi tessellation method.

2. The sample and the Voronoi tessellation method

We have compiled the sample of 124,000 galaxies from the spectroscopic SDSS DR9 limiting it by the absolute magnitude $-24 < M_r < -20.7$ and redshift $0.02 < z < 0.1$ (Dobrycheva 2013). We divided this sample on the early type (E+S0) galaxies and late type (S+Irr) galaxies using the color (g-i) and inverse concentration indices R50/R90 according to the criteria proposed by Melnyk *et al.* (2012): $(0.95 < g - i < 1.5$ and $0 < R50/R90 < 0.4)$ for E+S0; $(0 < g - i < 0.95, 0 < R50/R90 < 0.6)$ and $(0.95 < g - i < 1.5, 0.4 < R50/R90 < 0.6)$ for S+Irr.

To determine the environmental density of galaxies we used the Voronoi tessellation method because it detects effectively both spherical and prolongate large-scale structures (Melnyk *et al.* 2006, Elyiv *et al.* 2009, Vavilova *et al.* 2009, Zaninetti 2010, Way *et al.* 2011, Scoville *et al.* 2013). The Voronoi tessellation operates with only the positions and radial velocities of galaxies for dividing the entire 3D space containing galaxies into elementary volumes. Those of the elementary volumes, which are located closer to a given galaxy than to the remaining galaxies, form the volume of the Voronoi cell for this galaxy. The galaxy itself becomes the nucleus of this cell. Thus, after applying the

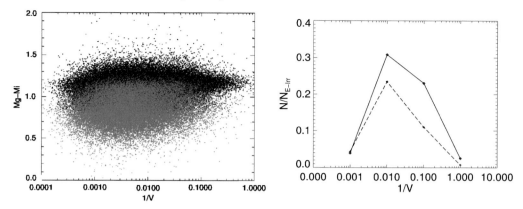

Figure 1. Left: The dependence of galaxy color on the inverse volume of the Voronoi cells for early (black dots) and late (gray dots) type galaxies. Right: The distributions of early (solid line) and late (dashed line) type galaxies by the inverse volume of the Voronoi cells.

Voronoi tessellation we obtain the distribution of volumes of the Voronoi cells (each cell contains one nucleus/galaxy). The local environmental density for given galaxy (or its concentration) is described by the inverse volume of its Voronoi cell: $n = 1/V$.

3. Results

The dependence of galaxy color on the inverse volume of the Voronoi cells $(1/V)$ for E+S0 (black dots) and S+Irr (gray dots) galaxies are presented in Fig. 1 (left). One can see, there is a general tendency that the early type galaxies are located in higher density regions than the late type galaxies. The same conclusion may be obtained from the inspection of Fig. 1 (right), where the distribution of galaxies by the inverse volume $(1/V)$ for the early and late types are shown. The number of galaxies in each bin is normalized by the total number of galaxies. We see also that most of the bright galaxies belongs to the early types. The mean values of $1/V$ for E+S0 and S+Irr types galaxies are 0.02±0.05 and 0.01±0.04, respectively.

We follow this research with the aim to consider the evolution of early and late type galaxies with redshift and luminosity in details. The accentuation will be pointed to the fractions of fainter satellites ($M_r >$-20.7) of galaxies, which are located in the neighborhood of galaxies from the studied sample.

References

Blanton, M. R., Lupton, R. H., Schlegel, D. J., *et al.* 2005, *ApJ*, 631, 208

Dobrycheva, D. V. 2013, *Odessa Astronomical Publications*, 26, 187

Dressler, A. 1980, *AJ*, 236, 351

Einasto, M., Einasto, J., Muller, V. *et al.* 2003, *A&A*, 401, 851

Elyiv, A. A., Melnyk, O. V., & Vavilova, I. B. 2009, *MNRAS*, 394, 1409

Melnyk, O. V., Elyiv, A. A., & Vavilova, I. B. 2006, *Kinemat. Fiz. Nebesn. Tel*, 22, 283

Melnyk, O. V., Dobrycheva, D. V., & Vavilova, I. B. 2012, *Astrophysics*, 55, 293

Scoville, N., Arnouts, S., Aussel, H. *et al.* 2013, *ApJS*, 206, 3

Vavilova, I. B., Melnyk, O. V., & Elyiv, A. A. 2009, *AN*, 330, 1004

Way, M. J., Gazis, P. R., & Scargle, J. D. 2011, *ApJ*, 727, 48

Weinmann, S. M., van den Bosch, F. C., Yang, X., & Mo, H. J. 2006, *MNRAS*, 366, 2

Zaninetti, L., 2010, *Rev. Mexicana AyA*, 46, 115

The Zeldovich Universe:
Genesis and Growth of the Cosmic Web
Proceedings IAU Symposium No. 308, 2014
R. van de Weygaert, S. Shandarin, E. Saar & J. Einasto, eds.
© International Astronomical Union 2016
doi:10.1017/S1743921316009960

Quantifying the Cosmic Web using the Shapefinder diagonistic

Prakash Sarkar

Department of Theoretical Physics, TIFR, Mumbai-400005, India
email: prakash@theory.tifr.res.in

Abstract. One of the most successful method in quantifying the structures in the Cosmic Web is the Minkowski Functionals. In 3D, there are four minkowski Functionals: Area, Volume, Integrated Mean Curvature and the Integrated Gaussian Curvature. For defining the Minkowski Functionals one should define a surface. We have developed a method based on Marching cube 33 algorithm to generate a surface from a discrete data sets. Next we calculate the Minkowski Functionals and Shapefinder from the triangulated polyhedral surface. Applying this methodology to different data sets , we obtain interesting results related to geometry, morphology and topology of the large scale structure

Keywords. methods: data analysis - galaxies: statistics - large-scale structure of universe

1. Introduction

A visual inspection of the galaxy surveys, like SDSS, 2dFGRS etc, reveals that the galaxies are arranged into a magnificent architecture. This architecture, a complex distribution of galaxies in different structural elements (filaments, sheets and clusters), is coined as Cosmic Web. Quantifying the Cosmic Web is a major problem in modern cosmology.

There are many statistical tools to quantify the Cosmic Web. One of the most effective diagnostic to quantify the Cosmic Web is the Minkowski Functionals (henceforth MFs), introduced in Cosmology by Mecke *et al.*(1994). Sahni *et al.* (1998) introduced Shapefinders, using the ratio of MFs, to quantify individual structural elements in the Cosmic Web. The 2D version of Shapefinders was introduced by Bharadwaj *et al.*(2000).

In the present work, we will describe an ansatz to accurately estimate the MFs and hence the shapefinders. This include the construction of smoothed density field from a point distribution and then construct triangulated isodensity surface using Marching cube 33 algorithm (Chernyaev (1995)). We have studied this in details and applied it to SDSS datasets.

2. Minkowski Functionals and Shapefinders

In 3D, there are four MFs, namely the Volume (V), Surface Area(S), Integrated mean curvature (C) and Genus(G). These MFs provide a global characterization of individual structures in the Cosmic Web. Using the above MFs, Sahni *et al.* (1998) devised shapefinders which include those which have dimension of length as well as those that are dimensionless, to classify the large scale structure. The shapefinder that have dimension of length, $\mathcal{H}_1 = 3V/S$, $\mathcal{H}_2 = S/C$, $\mathcal{H}_3 = C/4\pi G$, are used to evaluate the size of the structures. Based on $\mathcal{H}_i$, the useful dimensionless shapefinders $-$ Planarity $(\mathcal{P} = (\mathcal{H}_2 - \mathcal{H}_1)/(\mathcal{H}_2 + \mathcal{H}_1))$ and Filamentarity $(\mathcal{F} = (\mathcal{H}_3 - \mathcal{H}_2)/(\mathcal{H}_3 + \mathcal{H}_2))$ are defined to quantify the shape of 3D objects.

It is interesting to note that $\mathcal{P}, \mathcal{F} \leqslant 1$. The dimensionless shapefinders ($\mathcal{P}, \mathcal{F}$) along with the Genus(G) gives the complete information about the shape and the topology of the structure.

For estimating MFs the first basic step is to construct a isodensity surface. For that, we require to construct smoothed density field defined on a rectangular grid from point distribution. Then given a density threshold (ρ_{TH}), one can construct a triangulated surface using Marching Cube-33 algorithm. For these triangulated surfaces, the MFs can be evaluated by the following equations (also see Sheth (2006) for details):

(a) The **Volume** is given by $V = \sum_{i=1}^{N_T} V_i = \sum_{i=1}^{N_T} \frac{1}{3} S_i (\hat{n} \cdot \vec{P})_i$ where V_i is the volume of the individual tetrahedron, $\hat{n}$ is the normal vector of ith triangle and $\vec{P}$ is the position vector of the centroid of ith triangle. N_T is the total number of triangles and S_i is the area of the ith triangle.

(b) The **Surface Area** is given by $S = \sum_{i=1}^{N_T} S_i$.

(c) The **Genus** is given by $G = 1 - \frac{\chi}{2} = 1 - \frac{N_T - N_E + N_V}{2}$, where χ is the Euler characteristics, N_E and N_V denotes the total number of triangle-edges and triangle-vertices of the triangulated surface respectively.

(d) The **Integrated Mean Curvature** is estimated by calculating the curvature tensor per triangle based on vertex normals on the triangulated surface as describe in Rusinkiewicz (2004)

3. Test with idealised structures

We first check reliability of the above technique with Standard Eikonal Surfaces. The three different structures that we consider correspond to Spherical, Ellipsoidal and Torus. We have generated a spherical surface by assigning density (ρ) on a grid using the following relations,

$$\rho(i, j, k) = \begin{cases} \rho_0/R & \text{if } (i, j, k) \neq (11, 11, 11) \\ \rho_0 & \text{if } (i, j, k) = (11, 11, 11) \end{cases} \tag{3.1}$$

where R is the radius of the sphere. The threshold ρ_{TH} associated with sphere of radius R is ρ/ρ_{TH}. Figure 1(a) shows the percentage error in the MFs as a function of radius of sphere. We note that the percentage errors for all radius ($R > 1$) are less than 0.5 for all MFs. The genus estimated is 0 for sphere of all radii.

We next study the accuracy for triaxial ellipsoid, which depending on the relative scales of the three axes (a, b, c) can be prolate, oblate or spherical. We first study the Prolate deformation. We start with $a >> b = c$, and then increase the b and c-axis, until we get $a = b = c$. Figure 1(b) shows the percentage error of MFs as they evolve with the dimensionless variable c/a. The results clearly demonstrate the matching MFs with the exact value to a remarkable degree of accuracy. At the $c/a \geqslant 0.2$, the percentage error of all MFs is less than 0.7%. Next we study the oblate deformation of the Ellipsoid. We start we $a << b = c$ and increase a until to a value such that $a = b = c$. As a increases, the planarity starts decreasing. Figure 1(c) shows the variation of MFs with dimensionless variable a/c. The figure clearly shows that the MFs obtained using triangulation match the exact value with high degree of accuracy. We find that C is estimate to the worst accuracy with maximum percentage error of 1.2%. The genus estimated is 1 for both types of ellipsoid irrespective of axis length.

We now demonstrate the accuracy of ansatz for Torus deformation. The torus can be describe by three parameters a, b and c. We have considered the deformations of a two kind of torus – Circular Torus ($a = c << b$) having circular cross-sections and elliptical Torus ($a \neq c << b$) having elliptical cross-section. It is an important test, since it contains

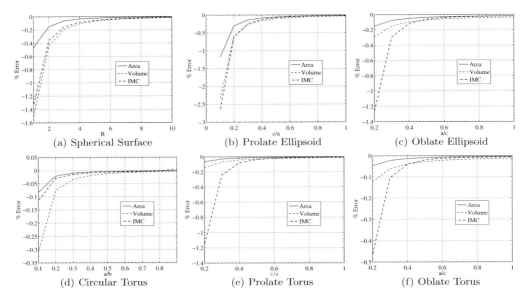

Figure 1. The figures shows the variation of percentage error of different MFs with Radius (for sphere) or the ratio of the axis length.

both convex and concave region. For both deformations we have consider $b = 20$. Figure 1(d) shows the variation of percentage error for different MFs with a. The small error becomes even smaller with increasing a. We next study the accuracy for two different types of elliptical Torus $-$, prolate ($a > c$, $a = 10$) and oblate ($c > a$, $c = 10$). Figure 1(e) and 1(f) shows the variation of percentage error estimated by ansatz with respect to increasing values c/a and a/c respectively. The results again prove the accuracy of the algorithm. Our results clearly demonstrate the excellent agreement of the calculate value of Minkowski Functionals with the exact analytic formulae.

4. Results for the SDSS and Conclusion

We now apply the MFs estimator to quantify the shape of the galaxy distribution in the Sloan Digital Sky Survey Data Release 7. A volume-limited sample of $116,877$ galaxies is constructed by restricting the absolute r-band magnitude brighter than -21.6. The details of the sample have been discussed in Park et $al.$(2012). The analysis was carried out by identifying structures using Friend-of-Friend (fof) algorithm. The critical linking length is determined by choosing that linking length which results in maximum number of structures. In our case the critical linking length is found out to be 5.6 h^{-1} Mpc when the minimum number of member galaxies is set to 20. There are about 873 structures discovered by fof method. For each structures, we apply a Cloud in Cell (CIC) technique to construct a density field on the grid of size 1 h^{-1} Mpc. Next we smooth this density field with a Gaussian kernel of smoothing length 1.9 h^{-1} Mpc. Now considering a density threshold of 4.15 times the mean density to estimate the Minkowski Functionals and Shapefinders. For comparison we have used 200 mock samples extracted from the Horizon Run 2 (Kim et $al.$(2011)) simulation. Each Mock samples are analyse in the same way as the SDSS data. This gives the linking lengths of 5.55 ± 0.13 h^{-1} Mpc from these 200 mock SDSS samples.

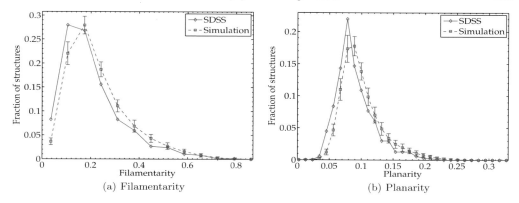

(a) Filamentarity   (b) Planarity

Figure 2. This left and right panel show the fraction of structures with Filamentarity and Planarity respectively. The red curve in the figure shows the results of SDSS while the blue dash curve shows the results of Simulations. 200 Mock samples are used to estimate the $1 - \sigma$ error-bars.

Figure 2(a) and 2(b) shows the fraction of structures with different Filamentarity and Planarity value respective. The red curve shows that of SDSS and the blue dash curve shows the results for the Mock samples. We find that more than 60% structures have Filamentarity values in between 0.1 and 0.3. While more that 80% of structures are found to have planarity values in the range 0.05 to 0.15. We find that the results from the Millennium Simulation shifts toward the right to those obtained for the actual SDSS data. This discrepancy is seen for both figures. The origin of this discrepancy is, at the moment, not clear and is an issue we plan to address in future work.

In summary, we can say that the ansatz discussed above gives an accurately estimate the Minkowski Functionals and shapefinders for any arbitrary shape. When apply to structures identified from SDSS galaxies using Friend-of-Friend algorithm, we observe a clear dominance of Filamentarity and Planarity at value 0.2 and 0.1 respectively. We have seen that the Minkowski Functionals and shapefinder estimated using the above method is a robust method to accurately quantify the individual structures in the galaxy distribution.

Acknowledgements

I am grateful to Prof. Changbom Park for providing the structures identified using Friend-of-Friend algorithm from SDSS and the Simulation data and Prof. Varun Sahni for useful discussions.

References

Bharadwaj, S., Sahni, V., Sathyaprakash, B. S., Shandarin, S. F., & Yess, C. 2000, *ApJ*, 528, 21
Chernyaev, E. V. 1995, *Technical report*, Marching Cubes 33: Construction of Topologically Correct Isosurfaces
Kim, J., Park, C., Rossi, G., Lee, S. M., Gott, J. R., & III, 2011, *Journal of Korean Astronomical Society*, 44, 217
Mecke, K. R., Buchert, T., & Wagner, H. 1994, *A&A*, 288, 697
Park, C., Choi, Y.-Y., Kim, J., *et al.* 2012, *ApJ*(letters), 759, L7
Rusinkiewicz, S. 2004, *Symposium on 3D Data Processing, Visualization, and Transmission*
Sahni, V., Sathyaprakash, B. S., & Shandarin, S. F. 1998, *ApJ*(letters), 495, L5
Sheth, J. V., 2006, *Ph.D. Thesis*

CHAPTER 4B.

Cosmic Web
Reconstruction

*Florent Leclercq working on
the reconstruction of the Universe.*

*Adi Nusser paying close attention,
looking for an opportunity to pose an incisive question.*

The Zeldovich Universe:
Genesis and Growth of the Cosmic Web
Proceedings IAU Symposium No. 308, 2014
R. van de Weygaert, S. Shandarin, E. Saar & J. Einasto, eds.

© International Astronomical Union 2016
doi:10.1017/S1743921316009972

Big Data of the Cosmic Web

Francisco-Shu Kitaura

Leibniz Institute for Astrophysics (AIP),
An der Sternwarte 16, 14482 Potsdam
email: kitaura@aip.de

Abstract. One of the main goals in cosmology is to understand how the Universe evolves, how it forms structures, why it expands, and what is the nature of dark matter and dark energy. Next decade large and expensive observational projects will bring information on the structure and the distribution of many millions of galaxies at different redshifts enabling us to make great progress in answering these questions. However, these data require a very special and complex set of analysis tools to extract the maximum valuable information. Statistical inference techniques are being developed, bridging the gaps between theory, simulations, and observations. In particular, we discuss the efforts to address the question: What is the underlying nonlinear matter distribution and dynamics at any cosmic time corresponding to a set of observed galaxies in redshift space?

An accurate reconstruction of the initial conditions encodes the full phase-space information at any later cosmic time (given a particular structure formation model and a set of cosmological parameters). We present advances to solve this problem in a self-consistent way with Big Data techniques of the Cosmic Web.

Cosmology is experiencing a golden era. A large number of galaxy surveys are planned to produce an enormous avalanche of data. These aim at understanding the nature of dark matter and dark energy, two still unknown components, which make up about 95% of the whole energy budget in the Universe. The scientific goal is to unveil the accelerated expansion of the Universe and the hidden mechanisms of cosmic structure formation, which ultimately led to the place we occupy in the cosmos.

Answering these questions from analyzing vast amounts of data will demand complex data mining techniques able to extract the maximum cosmological information. In particular we aim at doing a global analysis of the data to break all possible degeneracies making the least possible assumptions. This implies using as input data in such a joint analysis the closest form to the raw data. Moreover, one would like to include as many data sets as possible, to combine them in a self-consistent way. These would range from the cosmic microwave background (CMB), over the 21 cm line, the Lyman alpha (and beta) forest, the Lyman alpha emitters, the distribution of quasars, galaxies, and clusters, to the corresponding lensing maps throughout cosmic history. Focusing for instance on just the galaxy distribution, a number of issues needs to be considered. These can be related to observational systematics, such as, the survey geometry; the completeness on the sky; the photometric calibration; the photometric redshift uncertainty (for photometric surveys); the stellar contamination; etc. Other systematic effects can be due to intrinsic physical aspects, such as, the nonlinear, nonlocal, stochastic, and assembly (luminosity dependent) galaxy bias; the coherent and dispersed peculiar velocities; the gravitational mode coupling; the baryonic effects; etc. The complexity of the problem scales dramatically with the volume and resolution we need to achieve, pushed by the requirements of the new generation of surveys covering increasingly larger volumes and fainter objects.

New galaxy surveys aim at going deeper in redshirt by not only using longer exposure times to determine the redshifts with absorption spectra, but also exploiting the characteristic OII doublet seen in emission for strongly star forming galaxies (see, e.g., Dawson *et al.* 2013). Such objects are correspondingly called emission line galaxies (eLGs). As a consequence a large variety of galaxies tracing different density regimes of the large-scale structure will be available. In addition to full gravity calculations, effective theories based on analytical models of structure formation become necessary to shed light on the physical problem and save computational costs inherent to N-body based computations. We will discuss below how such models enable us to make a Big Data analysis of the cosmic web.

0.1. *The legacy of Zel'dovich*

The year 2014 was the commemoration of the 100th birth anniversary of Yakov Borissowitsch Seldowitsch, also known in English as Zel'dovich. While his contributions range from chemistry, over hydrodynamics, atomic nuclei, elementary particles to astrophysics, we want to highlight one of his main contributions to cosmology: the formation of the so-called Zel'dovich pancakes, i.e., the cosmic filamentary network. Zel'dovich proposed in 1970 an elegant solution to cosmological structure formation based on what we nowadays call linear Lagrangian perturbation theory, or Zel'dovich approximation, in which matter tracers move along the paths defined by the initial displacements at early times, or equivalently high redshifts. This approximation is able to describe remarkably well the quasi-nonlinear regime of structure formation and in particular the formation of the cosmic web. However, this picture led to the top-down scenario, where large structures form first and then are fragmented to form smaller ones, which is disfavored by observations. Simulations based on N-body solvers during the 80s and 90s, describing the interaction between matter tracers forming virialised structures, helped to develop the current bottom-up paradigm, in which smaller structures merge to form larger ones. After the accuracy of gravity solvers was found to be crucial to understand structure formation and computational progress made it possible to perform ever larger N-body simulations, the Zel'dovich approximation was confined during some period of time to academic and rather historical studies. Nevertheless, the need to understand the large-scale structure in rapidly increasing volumes pushed by the development of large galaxy redshift surveys starting during the 2000s with the Sloan Digital Sky Survey, has drawn the attention back to approximate gravity solvers, and analytical models. In fact only the order of a dozen large volume N-body simulations have been done, which do not achieve the resolution required for eLGs. Luckily some part of the astrophysics community never ceased investigating Zel'dovich's legacy and set the basis for a whole branch of methods, which are turning out to be very useful to analyze and understand observations of the large-scale structure.

Let us list here some of the methods which rely on the Zel'dovich approximation and find modern applications:

• **setup of initial conditions for N-body simulations**

The Zel'dovich approximation has turned out to be very useful to setup N-body simulations by relating the primordial fluctuations of the Universe to the initial velocities of matter particles (Springel *et al.* 2005). While starting at high enough redshifts ($\gtrsim 100$) the plane Zel'dovich approximation is still being used to setup initial conditions, more sophisticated versions have been developed including second order Lagrangian perturbation theory, i.e., tidal field corrections, which however follow the same idea introduced by Zel'dovich. These have become standard and a number of codes are publicly available (Crocce *et al.* 2006; Jenkins 2009), including primordial non-Gaussianities (Scoccimarro *et al.* 2012).

- **mock galaxy catalogues**

Full N-body simulations are necessary to enable an accurate understanding of structure formation, and, in particular, the distribution of galaxies. However, they are too expensive to be massively produced to compute covariance matrices, and, hence, derive the error bars corresponding to the measured galaxy clustering from observations. Nevertheless, these calculations are necessary to obtain reliable reference halo catalogues. One can aim at including all known physical processes in structure formation simulations, however covering very limited volumes (see, e.g., Illustris simulations, Vogelsberger *et al.* 2014). A number of techniques can convert dark matter halo catalogues into simulated galaxy ones, such as semi-analytic models (e.g., White & Frenk 1991, Kauffmann *et al.* 1993, Cole *et al.* 1994, Somerville & Primack 1999, Cole *et al.* 2000, Croton *et al.* 2006, De Lucia & Blaizot 2006, Benson 2012), halo abundance matching (HAM, e.g., Kravtsov *et al.* 2004, Tasitsiomi *et al.* 2004, Vale & Ostriker 2004, Conroy *et al.* 2006, Kim *et al.* 2008, Guo *et al.* 2010, Wetzel *et al.* 2010, Behroozi *et al.* 2010, Trujillo *et al.* 2011, Leauthaud *et al.* 2011), or halo occupation distribution (HOD, e.g., Berlind *et al.* 2002,Kravtsov *et al.* 2004, Zentner *et al.* 2005, Zehavi *et al.* 2005, Zheng *et al.* 2007, Skibba & Sheth 2009, Ross & Brunner 2009, Zheng *et al.* 2009, White *et al.* 2011) techniques. A number of methods have pioneered the efforts of producing halo catalogues with approximate gravity solvers, such as the peak-patch formalism (Bond & Myers 1996), Pinocchio (Monaco *et al.* 2002), and PThalos (Scoccimarro & Sheth 2002). These approaches were however far from achieving percentage accuracy in reproducing the clustering statistics, and therefore more recent methods have been developed, such as PATCHY (Kitaura *et al.* 2014) and EZmocks (Chuang *et al.* 2015). The latter ones rely on a set of effective bias parameters, which can be calibrated with reference catalogues to obtain the desired high accuracies in the different 1-, 2-, 3- point statistics, aiming at being precise at the 4-point statistics (covariance matrices). All these methods rely in some form or the other on the Zel'dovich approximation.

- **speed-up of N-body solvers**

Another recent idea relies on exploiting the accuracy of Lagrangian perturbation theory to efficiently solve gravity, and compute the remaining mode coupling term with an N-body solver, saving hereby a considerable amount of computational effort (Tassev *et al.* 2013). A similar idea is applied using the spherical collapse model for the short range force to yield approximate extremely efficient structure formation solvers using augmented Lagrangian perturbation theory (ALPT, Kitaura & Heß 2013).

- **modification of cosmological parameters in an N-body simulation**

The cosmology of an N-body simulation can be modified for many purposes to great accuracy by using the Zel'dovich approximation to trace the dark matter particles to higher redshifts and then back to low redshifts, changing the cosmological parameters in this process (Angulo & White 2010).

- **modification of large-scale modes in an N-body simulation**

By constructing a simulation, which has only small-scale power, one can add different large-scale power realizations to it relying on the Zel'dovich approximation. Care must be taken to correct for the mode coupling originated through the large-scale modes, which cause a local modification of the growth. Therefore by looking up the positions of dark matter particles at different redshift snapshots in the N-body simulation, one can mimic that mode coupling. This technique allows to make a reduce number of N-body simulations combined with large numbers of Zel'dovich large-scale calculations to compute covariance matrices (Schneider *et al.* 2011).

Figure 1: Flowchart of the KIGEN-code based on a Bayesian Networks Machine Learning approach. The primordial density fluctuations (initial conditions) are obtained from iteratively sampling Gaussian fields, which lead to cosmic structures compatible with the distribution of galaxies given a particular structure formation model. Starting from some initial guess (steps 0-3) structure formation is simulated forward (steps 4-5), observational effects like redshift-space distortions caused by peculiar motions (steps 6-7) and selection function effects according to the magnitude limited survey (step 8) are taken into account and the resulting mock observations (step 9) are matched with the observations in a likelihood comparison process (steps 0-1). The results are used to improve the initial conditions in the next iteration.

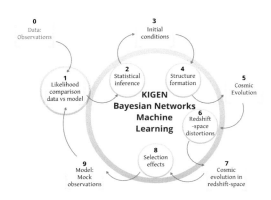

- **modeling the correlation function of halo clustering**

The Zel'dovich approximation has been demonstrated to give a precise description of the gravitational mode coupling introduced in the baryon acoustic oscillations (Tassev & Zaldarriaga 2013). Hence it can be used to model the correlation function after cosmic evolution (McCullagh & Szalay 2012). Including galaxy bias and redshift space distortions originated by the peculiar motions of galaxies requires additional modeling (White 2015).

- **reconstruction of baryon acoustic oscillations**

The baryon acoustic oscillations (BAO) can be used as a standard ruler to measure the scale of the Universe at different epochs, and thereby study dark energy. However, they are distorted by gravitational evolution. To enhance the BAO signal one can undo gravity by moving the galaxies back in time using the Zel'dovich approximation (Eisenstein *et al.* 2007).

- **reconstruction of primordial fluctuations**

Lagrangian perturbation theory can be used to recover the initial conditions of the Universe on scales smaller than the BAO scale, as we will discuss below.

All the above-mentioned techniques enable an efficient analysis of the large-scale structure. We will discuss below how these techniques can be combined with Bayesian techniques to recover the full phase space information of the cosmic web.

1. Big Data of the Cosmic Web

The analysis of the large-scale structure from a galaxy distribution requires the characterization of the underlying dark matter field, which governs the dynamics. Although the primordial fluctuations are closely Gaussian distributed, gravity couples different modes and the formation of the cosmic web introduces an anisotropy in the three-point correlation function. One therefore needs, from a statistical point of view, to jointly constrain all the higher order moments of the dark matter distribution, the galaxy bias, and the peculiar velocity field (Kitaura & Enßlin 2008). This can be a very complex task from the mathematical and computational point of view (Schaap & van de Weygaert 2000, Kitaura *et al.* 2010, Jasche & Kitaura 2010, Jasche *et al.* 2010, Platen *et al.* 2011, Kitaura *et al.* 2012c), especially including higher order correlation functions (e.g., Kitaura 2012).

The initial conditions of the Universe encode the full phase-space (density and peculiar velocity fields) information of any later cosmic time with a given cosmological structure formation model. It is thus tempting to reconstruct the initial conditions to characterize the large-scale structure.

Previous pioneering attempts to recover them have in most of the cases either ignored the relative movement of structures due to gravitation (see, e.g., Weinberg 1992, Kravtsov *et al.* 2002, Klypin *et al.* 03), or relied on linear theory (Nusser 1992, Kolatt *et al.* 1996, Mathis *et al.* 2002, Eisenstein *et al.* 2007, Padmanabhan *et al.* 2012, Doumler *et al.* 2013). Some nonlinear attempts can be found in the literature (see, e.g., Gramann93, Croft & Gaztanaga 1997, Narayanan 1998, Monaco & Efstasthiou 1999, Kitaura & Angulo 2012). Other approaches have aimed at solving the boundary problem of finding the initial positions of a set of matter tracers governed by the Eulerian equation of motion and gravity with the least action principle (see Peebles 1989, Nusser *et al.* 2000, Branchini *et al.* 2002). A similar approach consists on relating the observed positions of galaxies in a geometrical way to a homogeneous distribution by minimizing a cost function (Frisch *et al.* 2002, Brenier *et al.* 2003, Lavaux 2010). All these approaches have one fundamental aspect in common: they aim at finding a single *optimal* solution to the initial conditions boundary problem.

Nevertheless, once shell-crossing starts, two matter tracers can have extremely close positions, but very different peculiar motions. It, thus, becomes impossible to know where the tracers came from in a unique way. Moreover, matter collapses to compact objects, which did not exist in the past, but had an extended Lagrangian region. Therefore, statistical forward approaches have been introduced (see the KIGEN-code: Kitaura 2013, Kitaura *et al.* 2012 and other approaches: Jasche & Wandelt 2013, Wang *et al.* 2013). While KIGEN is a particle-based approach, the other approaches are grid based. Patrick Bos presented in this IAU meeting the first method of the latter kind, which also includes a self-consistent treatment of RSD.

These methods exploit the fact that the statistical description of the density fields at the initial conditions is simple, as it must be closely Gaussian distributed. In particular the KIGEN-code permits one for the first time to deal with any kind of structure formation model in a probabilistic way. It is based on a Bayesian networks machine-learning algorithm, which iteratively samples Gaussian fields, whose phases are constrained by the distribution of observed tracers given a structure formation and a cosmological model. It also includes redshift-space distortions (coherent and virialised peculiar motions) in the likelihood comparison to the observations. In this sense, it is also the first self-consistent phase-space reconstruction method.

One can write down the Hamiltonian equations of motion for any analytical structure formation model, and sample from a posterior defined with such a model within a Bayesian approach. However, every time one changes the structure formation model one needs to rewrite the Hamiltonian sampler. Also it is very difficult to include in a self-consistent way RSD. The approach used in KIGEN splits the sampling problem into two main Gibbs sampling steps. The first step assumes that the positions of dark matter particles at the initial conditions tracing the primordial fluctuations are known. This is a statistical problem known in cosmology as the constrained realization (Hoffman & Ribak 1991, van de Weygaert & Bertschinger 1996), in which the Gaussian field is sampled compatible with a number of constraints. The second step is based on assuming that the

F.-S. Kitaura

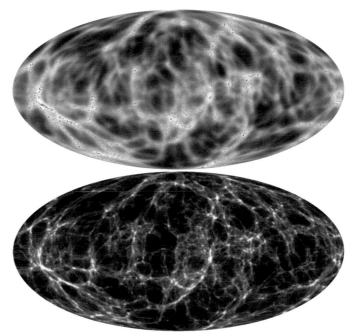

Figure 2: The upper panel shows the sky projection of all galaxies in the 2MRS catalog (red dots) at distances of 170 to 280 million light-years and their exquisite correlation with the mean over 25 reconstructed samples of the nonlinear ALPT cosmic web (grey scale) using the KIGEN code. The lower panel shows the same projection, but for the dark matter field of one constrained N-body simulation.

primordial Gaussian field is known, and given a set of constraints at low redshift, one needs to find the constraints at high redshift. This problem is also a statistical problem which needs to include a structure formation model translating the position of matter tracers at initial times to collapsed objects at late times, which can be in turn compared with the position of observed objects. A likelihood comparison process selects the collapsed objects, which are compatible under certain criteria (some minimum distance with a scatter). Since these collapsed objects have been obtained from a constrained simulation starting at initial times, we have full knowledge of it phase-space at all times. This enables us to look up the linking list of objects and the corresponding initial positions. Once we have them we can go to the first step. One can include RSD in the position of the objects and all kind of systematic effects. We note, that this approach is completely flexible to adopt any kind of structure formation model. As a generalization one can substitute positions of matter tracers with displacement fields (Kitaura in prep.). The method does not change. Here the idea is based on the reverse concept of setting initial conditions for an N-body simulation with the Zel'dovich approximation. In the same way, a Gaussian field determines the initial linear displacement field, the reverse is also true. One has only to consider that the likelihood comparison must be done with a nonlinear displacement, while the constrained realization must be done with the linear one.

The flowchart of the KIGEN Bayesian Networks Machine Learning approach is presented in Fig. 1. We note that the likelihood comparison can be used to improve the

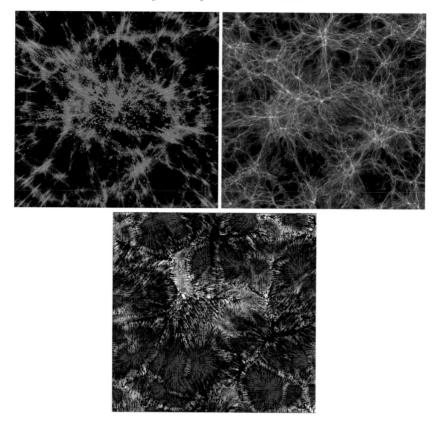

Figure 3: Left panel: slice through the super-galactic plane of the rendering of dark matter particles moved to redshift space from one particular constrained simulation presented in Heß *et al.* (2013), where the plots are instead shown based on the haloes to avoid an excess of virial motions, and the 2MRS galaxies (including <5% mocks in the galactic plane) are represented with red dots. Right panel: same slice, but after applying the phase-space mapping with the method presented in Abel *et al.* (2012). (Credit for the phase-space computation based on the reconstruction performed with KIGEN: Steffen Heß, Devon Powell, Ralf Kaehler & Tom Abel 2012). The lower panel shows the peculiar velocity field obtained with KIGEN using ALPT.

initial conditions since we have the full information available about the trajectories of the matter tracers from some starting high redshift until the redshift of the observations.

The KIGEN-code has been tested with a semi-analytic halo-model based galaxy mock catalog to demonstrate that the recovered initial conditions are closely unbiased with respect to the actual ones from the corresponding N-body simulation (seeKitaura 2013). It has also been applied to the Two-Micron All-Sky Redshift Survey (2MRS: Huchra *et al.* 2012) to perform a cosmography analysis and determine the proper motion of the Local Group finding a close agreement with the direction of the Cosmic Microwave Background (CMB) dipole and explaining about 80 % of its speed (see Figs. 2, 3 and Kitaura *et al.* 2012), and to search for the missing baryons in the warm hot inter-galactic medium (Suarez-Velasquez *et al.* 2013). A thorough analysis of the high performance of the KIGEN-code and its robustness with constrained N-body simulations has also been done (see Heß *et al.* 2013). We have also investigated the cosmic web in the local Universe (Nuza *et al.* 2014). Here it could be shown that the relation between galaxy morphology and environment becomes clearer when properly correcting for RSD. The left panel in Fig. 3 shows that the likelihood comparison is done in redshift space. The knowledge of the primordial fluctuations opens a new possibility to analyze the cosmic web by using the full phase-space (Shandarin *et al.* 2012, Abel *et al.* 2012, Falck *et al.* 2012). We present a first application of such a method on observations in the right panel of Fig. 3, showing the reconstructed real-space cosmic web.

2. Conclusions

The huge amount of data from galaxy surveys will permit us to map the Universe with unprecedented accuracy. We have reached an era in which we need to develop complex data mining techniques to extract the hidden information in the data. We have shown that great advances are been carried in the study and characterization of the large-scale structure. Here we find that the statistics of the cosmic primordial fluctuations are well described by Gaussian distribution functions. Within a Bayesian framework, we can thus use simple priors and encode the physical structure formation models in the likelihood, when comparing to observations. Many developments need still to be done, to sample over the cosmological parameters, over the growth rate, and over the bias model, to include light-cone effects, and to include in the analysis additional cosmological probes, such as CMB, lensing, Lyman-alpha forest, etc. This is an exciting time to work in cosmology.

Acknowledgements

I thank the organizers of the IAU meeting for letting me give my first key note speaker talk at the celebration of Zel'dovich's anniversary in June 2014. I want to specially thank Steffen Heß for a wonderful collaboration. The upper plots in Fig. 3 have been made by him. I also thank Tom Abel and his group, in particular Devon Powell, for providing us the phase-space mapping code, which permitted us to make the first calculation of such kind, based on real observations. A more advanced calculation served for an article in National Geographic `http://ngm.nationalgeographic.com/2015/01/hidden-cosmos/fly-through-video`. All the material presented in this work had been presented in the original talk I gave in Tallin.

References

T. Abel, O. Hahn, & R. Kaehler, *MNRAS*, **427**, 61, (2012)
R. E. Angulo & S. D. M. White, *MNRAS*, **405**, 143, (2010)
P. S. Behroozi, C. Conroy, & R. H. Wechsler, *ApJ*, **717**, 379, (2010)
A. A. Berlind, & D. H. Weinberg, *ApJ*, **575**, 587, (2002)
J. R. Bond & S. T. Myers, *ApJS*, **103**, 1, (1996)
E. Branchini, A. Eldar, & A. Nusser, *MNRAS*, **335**, 53, (2002)
A. J. Benson, *NA*, **17**, 175, (2012)
Y. Brenier, U. Frisch, M. Henon, G. Loeper, S. Matarrese, R. Mohayaee, & A. Sobolevskii, *MNRAS*, **346**, 501, (2003)
C.-H. Chuang, F.-S. Kitaura, F. Prada, C. Zhao, & G. Yepes, *MNRAS*, **446**, 2621, (2015)
S. Cole, C. G. Lacey, C. M. Baugh, & C. S. Frenk, *MNRAS*, **319**, 168, (2000)
S. Cole, A. Aragon-Salamanca, C. S. Frenk, J. F. Navarro, & S. E. Zepf, *MNRAS*, **271**, 781, (1994)
C. Conroy, R. H. Wechsler, & A. V. Kravtsov, *ApJ*, **647**, 201, (2006)

M. Crocce, S. Pueblas, & R. Scoccimarro, *MNRAS*, **373**, 369, (2006)

R. A. C. Croft & E. Gaztanaga, *MNRAS*, **285**, 793, (1997)

D. J. Croton, V. Springel, S. D. M. White, G. De Lucia, C. S. Frenk, L. Gao, A. Jenkins, G. Kauffmann, *et al.* , *MNRAS*, **365**, 11, (2006)

K. S. Dawson, D. J. Schlegel, C. P. Ahn, S. F. Anderson, É. Aubourg, S. Bailey, R. H. Barkhouser, J. E. Bautista, *et al.* , *AJ*, **145**, 10, (2013)

G. De Lucia & J. Blaizot, *MNRAS*, **375**, 2, (2007)

& T. Doumler, Y. Hoffman, H. Courtois, & S. Gottlöber, 2013, *MNRAS*, **430**, 888

D. J. Eisenstein, H.-J. Seo, E. Sirko, & D. N. Spergel, *ApJ*, **664**, 675, (2007)

B. L. Falck, M. C. Neyrinck, & A. S. Szalay, *ApJ*, **323**, 1, (2012)

U. Frisch, S. Matarrese, R. Mohayaee, & A. Sobolevski, *Nature*, **417**, 260, (2002)

Q. Guo, S. White, C. Li, & M. Boylan-Kolchin, *MNRAS*, **404**, 1111, (2010)

M. Gramann, *ApJ*, **405**, 449, (1993)

S. Heß, F.-S. Kitaura, & S. Gottlöber, *MNRAS*, **435**, 3 , 2065, (2013)

Y. Hoffman & E. Ribak, *ApJL*, **380**, L5, (1991)

J. P. Huchra, L. M. Macri, K. L. Masters, T. H. Jarrett, et al., *Rev.Astrn.Astrophys.*, **199**, 26, (2012)

J. Jasche, & F.-S. Kitaura, *MNRAS*, **407**, 29, (2010)

J. Jasche, F.-S. Kitaura, C. Li, & T. A. Enßlin, *MNRAS*, **409**, 355, (2010)

J. Jasche & B. D. Wandelt, *MNRAS*, **425** , 1042, (2013)

A. Jenkins, *MNRAS*, **403** , 1859, (2009)

G. Kauffmann & S. D. M. White, *MNRAS*, **261**, 921, (1993)

J. Kim, C. Park, & Y.-Y. Choi, *ApJ*, **683**, 123, (2008)

F.-S. Kitaura, *MNRAS*, **420** , 2737, (2012)

F.-S. Kitaura, *MNRAS*, **429**, 1, L84, (2013)

F.-S. Kitaura & R. E. Angulo, *MNRAS*, **425**, 4, 2443, (2012)

F.-S. Kitaura, R. E. Angulo, Y. Hoffman & S. Gottlöber, *MNRAS*, **425**, 4, 2422, (2012)

F.-S. Kitaura & T. A. Enßlin, *MNRAS*, **389**, 497, (2008)

F.-S. Kitaura, J. Jasche, & R. B. Metcalf, *MNRAS*, **403**, 589, (2010)

F.-S. Kitaura, P. Erdogdu, S. E. Nuza, A. Khalatyan, R. E. Angulo, Y. Hoffman & S. Gottlöber, *MNRAS*, **427**, 1, L35, (2012)

F.-S. Kitaura, S. Gallerani, & A. Ferrara, *MNRAS*, **420**, 61, (2012)

F.-S. Kitaura & S. Heß, *MNRAS*, **435**, 1, L78, (2013)

F.-S. Kitaura, G. Yepes, & F. Prada, *MNRAS*, **439**, 21, (2014)

A. Klypin, Y. Hoffman, A. V. Kravtsov, & S. Gottlöber, *ApJ*, **596**, 19, (2003)

T. Kolatt, A. Dekel, G. Ganon, & J. A. Willick, *ApJ*, **458**, 419, (1996)

A. V. Kravtsov, A. A. Berlind, R. H. Wechsler, A. A. Klypin, S. Gottlöber, B. Allgood, & J. R. Primack, *ApJ*, **609**, 35, (2004)

A. V. Kravtsov, A. Klypin, & Y. Hoffman, *ApJ*, **571**, 563, (2002)

G. Lavaux, *MNRAS*, **406** , 1007, (2010)

A. Leauthaud, J. Tinker, P. S. Behroozi, M. T. Busha, & R. H. Wechsler, *ApJ*, **738**, 45, (2011)

H. Mathis, G. Lemson, V. Springel, G. Kauffmann, S. D. M. White, A. Eldar, & A. Dekel, *MNRAS*, **333**, 739, (2002)

N. McCullagh, & A. S. Szalay, *MNRAS*, **752**, 21, (2012)

P. Monaco & G. Efstathiou, *MNRAS*, **308**, 763, (1999)

P. Monaco, T. Theuns, G. Taffoni, F. Governato, T. Quinn, & J. Stadel, *ApJ*, **564**, 8, (2002)

V. K. Narayanan & D. H. Weinberg, *ApJ*, **508**, 440, (1998)

A. Nusser, & E. Branchini, *MNRAS*, **313**, 587, (2000)

A. Nusser, & A. Dekel, *ApJ*, **391**, 443, (1992)

S. E. Nuza, F.-S. Kitaura, S. Heß, N. I. Libeskind, & V. Müller, *MNRAS*, **445**, 988, (2014)

N. Padmanabhan, X. Xu, D. J. Eisenstein, R. Scalzo, A. J. Cuesta,K. T. Mehta, & E. Kazin, *MNRAS*, **427**, 2132, (2012)

J. Peebles, *ApJ*, **344**, L53, (1989)

E. Platen, R. van de Weygaert, B. J. T. Jones, G. Vegter, & M. A. A. Calvo, *MNRAS*, **416**, 2494, (2011)

A. J. Ross & R. J. Brunner, *MNRAS*, **399**, 878, (2009)

W. E. Schaap & R. van de Weygaert, *AAP*, **363**, L29, (2000)

M. D. Schneider, S. Cole, C. S. Frenk, & I. Szapudi, *ApJ*, **737**, 11, (2011)

R. Scoccimarro, & R. K. Sheth, *MNRAS*, **329**, 629, (2002)

R. Scoccimarro, L. Hui, M. Manera, & K. C. Chan, *PRD*, **85**, 083002, (2012)

S. Shandarin,S. Habib, & K. Heitmann, *PRD*, **85**, 083005, (2012)

R. A. Skibba & R. K. Sheth, *MNRAS*, **392** , 1080, (2009)

R. S. Somerville & J. R. Primack, *MNRAS*, **310** , 1087, (1999)

V. Springel, S. D. M. White, A. Jenkins, C. S. Frenk, N. Yoshida, L. Gao, J. Navarro, R. Thacker, *et al.* , *Nature*, **435**, 629, (2005)

I. Suarez-Velasquez, F.-S. Kitaura, F. Atrio-Barandela & J. Mucket, *ApJ*, **769**, 1, 7, (2013)

A. Tasitsiomi, A. V. Kravtsov, S. Gottlöber, & A. A. Klypin, *ApJ*, **607**, 125, (2004)

S. Tassev & M. Zaldarriaga, *JCAP*, **4**, 13, (2013)

S. Tassev, M. Zaldarriaga, & D. J. Eisenstein, *JCAP*, **6**, 36, (2013)

S. Trujillo-Gomez, A. Klypin, J. Primack, & A. J. Romanowsky, *ApJ*, **742**, 16, (2011)

A. Vale & J. P. Ostriker, *MNRAS*, **353**, 189, (2004)

M. Vogelsberger, S. Genel, V. Springel, P. Torrey, D. Sijacki, D. Xu, G. Snyder, D. Nelson, & L. Hernquist, *MNRAS*, **444**, 1518, (2014)

H. Wang, H. J. Mo, X. Yang, & F. C. van den Bosch, *ApJ*, **772**, 63, (2013)

D. H. Weinberg, *MNRAS*, **254**, 315, (1992)

A. R. Wetzel & M. White, *MNRAS*, **403** , 1072, (2010)

R. van de Weygaert, & E. Bertschinger, *MNRAS*, **281**, 84, (1996)

M. White, *MNRAS*, **450**, 3822, (2015)

M. White, M. Blanton, A. Bolton, D. Schlegel, J. Tinker, A. Berlind, L. da Costa, E. Kazin, *et al.* , *ApJ*, **728**, 126, (2011)

S. D. M. White, & C. S. Frenk, *ApJ*, **379**, 52, (1991)

I. Zehavi, Z. Zheng, D. H. Weinberg, J. A. Frieman, A. A. Berlind, M. R. Blanton, R. Scoccimarro, R. K. Sheth *et al.* , *ApJ*, **630**, 1, (2005)

Y. B. Zel'dovich, *AAP*, **5**, 84, (1970)

A. R. Zentner, A. A. Berlind, J. S. Bullock, A. V. Kravtsov, & R. H. Wechsler, *ApJ*, **624**, 505, (2005)

Z. Zheng, A. L. Coil & I. Zehavi, *ApJ*, **667**, 760, (2007)

Z. Zheng, I. Zehavi, D. J. Eisenstein, D. H. Weinberg, & Y. P. Jing, *ApJ*, **707**, 554, (2009)

The Zeldovich Universe:
Genesis and Growth of the Cosmic Web
Proceedings IAU Symposium No. 308, 2014
R. van de Weygaert, S. Shandarin, E. Saar & J. Einasto, eds.

© International Astronomical Union 2016
doi:10.1017/S1743921316009984

Bayesian inference of the initial conditions from large-scale structure surveys

Florent Leclercq[1,2,3]

[1]Institut d'Astrophysique de Paris (IAP), UMR 7095, CNRS - UPMC Université Paris 6,
98bis boulevard Arago, F-75014 Paris, France

[2]Institut Lagrange de Paris (ILP), Sorbonne Universités,
98bis boulevard Arago, F-75014 Paris, France

[3]École polytechnique ParisTech,
Route de Saclay, F-91128 Palaiseau, France

email: `florent.leclercq@polytechnique.org`

Abstract. Analysis of three-dimensional cosmological surveys has the potential to answer outstanding questions on the initial conditions from which structure appeared, and therefore on the very high energy physics at play in the early Universe. We report on recently proposed statistical data analysis methods designed to study the primordial large-scale structure via physical inference of the initial conditions in a fully Bayesian framework, and applications to the Sloan Digital Sky Survey data release 7. We illustrate how this approach led to a detailed characterization of the dynamic cosmic web underlying the observed galaxy distribution, based on the tidal environment.

---

1. Introduction

How did the Universe begin? This question has unusual status in physical sciences due to several profound specificities of cosmology. As the Universe is everything that exists in the physical sense, there is no exteriority nor anteriority. The experiment is unique and irreproducible, and the properties of the Universe cannot be determined statistically on a set. The energy scales at stake in the early Universe are orders of magnitude higher than anything we can reach on Earth. Finally, reasoning in cosmology is "bottom-up" in the sense that the final state is known and the initial state has to be inferred. In the context of the cosmic web, we aim at a physical reconstruction of the pattern of initial density fluctuations that gave rise to the present network of clusters, filaments, sheets and voids. Due to the computational challenge and to the lack of detailed physical understanding of the non-Gaussian and non-linear processes that link galaxy formation to the large-scale dark matter distribution, this question has only recently been tackled. Here, we describe progress towards full reconstruction of four-dimensional state of the Universe and illustrate the use of these results for cosmic web classification in the initial and final conditions.

2. Statistical approach: Bayesian inference

Cosmological observations are subject to a variety of intrinsic and experimental uncertainties (incomplete observations – survey geometry and selection effects –, cosmic variance, noise, biases, systematic effects), which make the inference of signals a fundamentally ill-posed problem. For this reason, no unique recovery of the initial conditions from which the present-day cosmic web originates is possible; it is more relevant to quantify a probability distribution for such signals, given the observations. Adopting this point

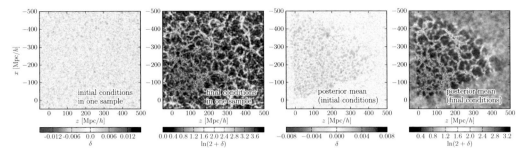

Figure 1. Bayesian large-scale structure inference with BORG in the SDSS DR7. Slices through one sample of the posterior for the initial and final density fields (left) and posterior mean in the initial and final conditions (right). The input galaxies are overplotted on the final conditions as red dots.

of view for large-scale structure surveys, Bayesian probability theory offers a conceptual basis for dealing with the problem of inference in presence of uncertainty.

The introduction of a physical model in the likelihood (gravitational structure formation is the generative model for the complex final state, starting from a simple initial state – Gaussian or nearly-Gaussian initial conditions) generally turns large-scale structure analysis into the task of inferring initial conditions (Jasche & Wandelt 2013a; Kitaura 2013; Wang *et al.* 2013). It is important to notice that this framework requires at no point the inversion of the flow of time, but solely depends on forward evaluations of the dynamical model.

Significant difficulty arises from the very large dimension of the parameter space to be explored (phenomenon usually referred to as the curse of dimensionality, Bellman 1961). However, the problem can still be tractable thanks to powerful sampling techniques such as Hamiltonian Markov Chain Monte Carlo (HMC, Duane *et al.* 1987).

3. Physical reconstructions

The inference code BORG (Bayesian Origin Reconstruction from Galaxies, Jasche & Wandelt 2013a) uses HMC for four-dimensional inference of density fields in the linear and mildly non-linear regime. The physical model for gravitational dynamics included in the likelihood is second-order Lagrangian perturbation theory (2LPT), linking initial density fields (at a scale factor $a = 10^{-3}$) to the presently observed large-scale structure (at $a = 1$). The galaxy distribution is modeled as a Poisson sample from these evolved density fields. The algorithm self-consistently accounts for observational uncertainty such as noise, survey geometry, selection effects and luminosity dependent galaxy biases (Jasche & Wandelt 2013a,b).

In Jasche *et al.* (2014), we apply the BORG code to 372,198 galaxies from the `Sample dr72` of the New York University Value Added Catalogue (NYU-VAGC, Blanton *et al.* 2005), based ot the final data release (DR7) of the Sloan Digital Sky Survey (SDSS, Adelman-McCarthy *et al.* 2008; Padmanabhan *et al.* 2008).

Each inferred sample (Fig. 1, left) is a "possible version of the truth" for the formation history of the Sloan volume, in the form of a full physical realization of dark matter particles. The variation between samples (Fig. 1, right) quantifies joint and correlated uncertainties inherent to any cosmological observation and accounts for all non-linearities and non-Gaussianities involved in the process of structure formation. In particular, it quantifies complex information propagation, translating uncertainties from observations to inferred initial conditions.

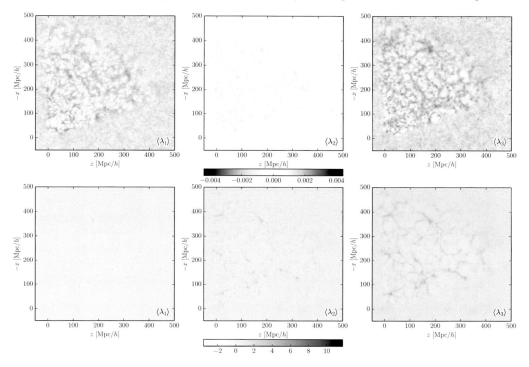

Figure 2. Mean of the posterior pdf for the eigenvalues $\lambda_1 < \lambda_2 < \lambda_3$ of the tidal field tensor in the initial (top) and final (bottom) conditions for the large-scale structure in the Sloan volume.

4. Cosmic web analysis

The results presented in § 3 form the basis of the analysis of Leclercq *et al.* (in prep.), where we classify the cosmic large scale structure into four distinct web-types (voids, sheets, filaments and clusters) and quantify corresponding uncertainties. We follow the dynamic cosmic web classification procedure proposed by Hahn *et al.* (2007) (see also Forero-Romero *et al.* 2009; Hoffman *et al.* 2012), based on the eigenvalues $\lambda_1 < \lambda_2 < \lambda_3$ of the tidal tensor T_{ij}, Hessian of the rescaled gravitational potential: $T_{ij} \equiv \partial^2 \Phi / \partial \mathbf{x}_i \, \partial \mathbf{x}_j$, where Φ follows the Poisson equation ($\nabla^2 \Phi = \delta$). It is important to note, that the tidal tensor, and the rescaled gravitational potential are both physical quantities, and hence their calculation requires the availability of a full physical density field in contrast to a smoothed mean reconstruction of the density field. In figure 2, we show the posterior mean for $\lambda_1, \lambda_2, \lambda_3$ as inferred by BORG is our reconstructions.

A voxel is defined to be in a cluster (resp. in a filament, in a sheet, in a void) if three (resp. two, one, zero) of the λs are positive (Hahn *et al.* 2007). The basic idea of this dynamic classification approach is that the eigenvalues of the tidal tensor characterize the geometrical properties of each point in space.

Our approach propagates uncertainties to structure type classification and yields a full Bayesian description in terms of a probability distribution, indicating the possibility to encounter a specific structure type at a given position in the observed volume. More precisely, by applying the above classification procedure to all density samples, we are able to estimate the posterior of the four different web-types, conditional on the observations. The mean of these pdfs are represented in Fig. 3. There, it is possible to follow the

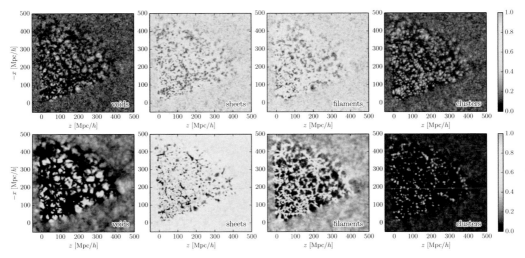

Figure 3. Mean of the posterior pdf for the four different web-types in the initial (top) and final (bottom) conditions for the large-scale structure in the Sloan volume.

dynamic evolution of specific structures. For example, one can observe the voids expand and the clusters shrink, in comoving coordinates.

Acknowledgements

I thank Jacopo Chevallard, Jens Jasche and Benjamin Wandelt for a fruitful collaboration on the projects presented here. I acknowledge funding from an AMX grant (École polytechnique) and Benjamin Wandelt's senior Excellence Chair by the Agence Nationale de la Recherche (ANR-10-CEXC-004-01). This work made in the ILP LABEX (ANR-10-LABX-63) was supported by French state funds managed by the ANR within the Investissements d'Avenir programme (ANR-11-IDEX-0004-02).

References

Adelman-McCarthy, J. K. Agüeros, M. A., *et al.* 2008, *Astrophys. J. Supp.*, 175, 297
Blanton, M. R., Schlegel, D. J., Strauss, M. A., *et al.* 2005, *AJ*, 129, 2562
Bellman, R. E. 1961, *Adaptive Control Processes: A Guided Tour* (Princeton University Press)
Duane, S., Kennedy, A. D., Pendleton, B. J., & Roweth, D. 1987, *Physics Letters B*, 195, 216
Forero-Romero, J. E., Hoffman, Y., Gottlöber, S., Klypin, A., & Yepes, G. 2009, *Mon. Not. R. Astron. Soc.*, 396, 1815
Hahn, O., Porciani, C., Carollo, C. M., & Dekel, A. 2007, *Mon. Not. R. Astron. Soc.*, 375, 489
Hoffman, Y., Metuki, O., Yepes, G., *et al.* 2012, *Mon. Not. R. Astron. Soc.*, 425, 2049
Jasche, J. & Wandelt, B. D. 2013, *Mon. Not. R. Astron. Soc.*, 432, 894
Jasche, J. & Wandelt, B. D. 2013, *ApJ*, 779, 15
Jasche, J., Leclercq, F., & Wandelt, B. D. 2014, arXiv:1409.6308
Kitaura, F.-S. 2013, *Mon. Not. R. Astron. Soc.*, 429, L84
Leclercq, F., Jasche, J., Chevallard, J. & Wandelt, B. D. in prep
Padmanabhan, N., Schlegel, D. J., Finkbeiner, D. P., *et al.* 2008, *ApJ*, 674, 1217
Wang, H., Mo, H. J., Yang, X., & van den Bosch, F. C. 2013, *ApJ*, 772, 63

The Zeldovich Universe:
Genesis and Growth of the Cosmic Web
Proceedings IAU Symposium No. 308, 2014
R. van de Weygaert, S. Shandarin, E. Saar & J. Einasto, eds.
© International Astronomical Union 2016
doi:10.1017/S1743921316009996

Bayesian Cosmic Web Reconstruction: BARCODE for Clusters

E. G. Patrick Bos[1], Rien van de Weygaert[1], Francisco Kitaura[2], and Marius Cautun[3]

[1]Kapteyn Astron. Inst., Univ. of Groningen, Groningen, the Netherlands,
email: pbos@astro.rug.nl

[2]Leibniz-Inst. für Astrophysik Potsdam (AIP), An der Sternwarte 16, 14482 Potsdam, DE

[3]Inst. for Computational Cosmology, Univ. of Durham, Durham DH1 3LE, United Kingdom

Abstract. We describe the Bayesian BARCODE formalism that has been designed towards the reconstruction of the Cosmic Web in a given volume on the basis of the sampled galaxy cluster distribution. Based on the realization that the massive compact clusters are responsible for the major share of the large scale tidal force field shaping the anisotropic and in particular filamentary features in the Cosmic Web. Given the nonlinearity of the constraints imposed by the cluster configurations, we resort to a state-of-the-art constrained reconstruction technique to find a proper statistically sampled realization of the original initial density and velocity field in the same cosmic region. Ultimately, the subsequent gravitational evolution of these initial conditions towards the implied Cosmic Web configuration can be followed on the basis of a proper analytical model or an N-body computer simulation. The BARCODE formalism includes an implicit treatment for redshift space distortions. This enables a direct reconstruction on the basis of observational data, without the need for a correction of redshift space artifacts. In this contribution we provide a general overview of the the Cosmic Web connection with clusters and a description of the Bayesian BARCODE formalism. We conclude with a presentation of its successful workings with respect to test runs based on a simulated large scale matter distribution, in physical space as well as in redshift space.

1. Introduction

On Megaparsec scales, matter and galaxies have aggregated into a complex network of interconnected filaments and walls. This network, which has become known as the *Cosmic Web* (Bond *et al.* 1996), contains structures from a few megaparsecs up to tens and even hundreds of megaparsecs of size. It has organized galaxies and mass into a wispy web-like spatial arrangement, marked by highly elongated filamentary, flattened planar structures and dense compact clusters surrounding large near-empty void regions. Its appearance has been most dramatically illustrated by the maps of the nearby cosmos produced by large galaxy redshift surveys such as the 2dFGRS, the SDSS, and the 2MASS redshift surveys (Colless et al. 2003; Tegmark *et al.* 2004; Huchra *et al.* 2012), as well as by recently produced maps of the galaxy distribution at larger cosmic depths such as VIPERS (Guzzo *et al.* 2013, 2014).

The Cosmic Web is one of the most striking examples of complex geometric patterns found in nature, and certainly the largest in terms of size. Computer simulations suggest that the observed cellular patterns are a prominent and natural aspect of cosmic structure formation through gravitational instability (Peebles 1980). According to the *gravitational instability scenario*, cosmic structure grows from tiny primordial density and velocity perturbations. Once the gravitational clustering process has progressed beyond the initial linear growth phase, we see the emergence of complex patterns and structures in the density field. Within this context, the Cosmic Web is the most prominent manifestation of the Megaparsec scale tidal force field.

The recognition of the *Cosmic Web* as a key aspect in the emergence of structure in the Universe came with early analytical studies and approximations concerning the emergence of structure out of a nearly featureless primordial Universe. In this respect the Zel'dovich formalism (Zeldovich 1970) played a seminal role. It led to view of structure formation in which planar pancakes form first, draining into filaments which in turn drain into clusters, with the entirety forming a cellular network of sheets.

As borne out by a large sequence of N-body computer experiments of cosmic structure formation, web-like patterns in the overall cosmic matter distribution do represent a universal but possibly transient phase in the gravitationally driven emergence and evolution of cosmic structure. N-body calculations have shown that web-like patterns defined by prominent anisotropic filamentary and planar features — and with characteristic large underdense void regions — are a natural manifestation of the gravitational cosmic structure formation process. Within the context of gravitationally driven structure formation, the formation and evolution of anisotropic structures is a direct manifestation of the gravitational tidal forces induced by the inhomogeneous mass distribution. It is the anisotropy of the force field and the resulting deformation of the matter distribution which which are at the heart of the emergence of the web-like structure of the mildly nonlinear mass distribution.

Interestingly, for a considerable amount of time the emphasis on anisotropic collapse as agent for forming and shaping structure was mainly confined the Soviet view of structure formation, Zel'dovich's pancake picture, and was seen as the rival view to the hierarchical clustering picture which dominated the western view. The successful synthesis of both elements culminated in the Cosmic Web theory (Bond *et al.* 1996), which stresses the dominance of filamentary shaped features instead of the dominance of planar pancakes in the pure Zel'dovich theory.

Perhaps even more important is the identification of the intimate dynamical relationship between the filamentary patterns and the compact dense clusters that stand out as the nodes within the cosmic matter distribution: filaments as cluster-cluster bridges (also see van de Weygaert & Bertschinger 1996; Bond *et al.* 1998; Colberg *et al.* 2005; van de Weygaert & Bond 2008). In the overall cosmic mass distribution, clusters — and the density peaks in the primordial density field that are their precursors — stand out as the dominant features for determining and outlining the anisotropic force field that generates the cosmic web.

In this study, we pursue this observation and investigate in how far we should be able to outline in a given volume of the observed Universe the web-like dark matter distribution on the basis of the observed cluster distribution. The challenge of this question is that of translating a highly biased population of compact massive clusters, the result of a highly nonlinear evolution, into the corresponding implied mildly nonlinear Cosmic Web. This inversion problem calls in several state-of-the-art statistical inversion techniques, in combination with advanced analytical prescriptions for the mildly nonlinear gravitational evolution of the cosmic density field. Before describing the BARCODE Bayesian reconstruction formalism, we will first provide the context of the gravitationally induced formation of filaments and walls.

1.1. *Tidal Dynamics of the Cosmic Web*

Within the context of gravitational instability, it are the gravitational tidal forces that establish the relationship between some of the most prominent manifestations of the structure formation process. When describing the dynamical evolution of a region in the density field it is useful to distinguish between large scale "background" fluctuations δ_b and small-scale fluctuations δ_f. Here, we are primarily interested in the influence of the smooth large-scale field.

To a good approximation the smoother background gravitational force $\mathbf{g}_\mathrm{b}(\mathbf{x})$ in and around the mass element includes three components. The *bulk force* $\mathbf{g}_\mathrm{b}(\mathbf{x}_{pk})$ is responsible for the acceleration of the mass element as a whole. Its divergence $(\nabla \cdot \mathbf{g}_\mathrm{b})$ encapsulates the collapse of the overdensity while the tidal tensor T_{ij} quantifies its deformation,

$$g_{\mathrm{b},i}(\mathbf{x}) = g_{\mathrm{b},i}(\mathbf{x}_{pk}) + a \sum_{j=1}^{3} \left\{ \frac{1}{3a}(\nabla \cdot \mathbf{g}_\mathrm{b})(\mathbf{x}_{pk})\,\delta_{\mathrm{ij}} - T_{ij} \right\} (x_j - x_{pk,j}). \qquad (1.1)$$

The tidal shear force acting over the mass element is represented by the (traceless) tidal tensor T_{ij},

$$T_{\mathrm{ij}} \;\equiv\; -\frac{1}{2a}\left\{ \frac{\partial g_{\mathrm{b},i}}{\partial x_i} + \frac{\partial g_{\mathrm{b},j}}{\partial x_j} \right\} + \frac{1}{3a}(\nabla \cdot \mathbf{g}_\mathrm{b})\,\delta_{\mathrm{ij}} \qquad (1.2)$$

in which the trace of the collapsing mass element, proportional to its overdensity δ, dictates its contraction (or expansion). For a cosmological matter distribution the close connection between the local force field and global matter distribution follows from the expression of the tidal tensor in terms of the generating cosmic matter density fluctuation distribution $\delta(\mathbf{r})$ (van de Weygaert & Bertschinger 1996):

$$T_{ij}(\mathbf{r}) = \frac{3\Omega H^2}{8\pi} \int d\mathbf{r}' \,\delta(\mathbf{r}') \left\{ \frac{3(r'_i - r_i)(r'_j - r_j) - |\mathbf{r}' - \mathbf{r}|^2\,\delta_{ij}}{|\mathbf{r}' - \mathbf{r}|^5} \right\} - \frac{1}{2}\Omega H^2\,\delta(\mathbf{r},t)\,\delta_{ij}.$$

The Megaparsec scale tidal shear forces are the main agent for the contraction of matter into the sheets and filaments which trace out the Cosmic Web. The anisotropic contraction of patches of matter depends sensitively on the signature of the tidal shear tensor eigenvalues. With two positive eigenvalues and one negative, $(-++)$, we will see strong collapse along two directions. Dependent on the overall overdensity, along the third axis collapse will be slow or not take place at all. Likewise, a sheetlike membrane will be the product of a $(--+)$ signature, while a $(+++)$ signature inescapably leads to the full collapse of a density peak into a dense cluster.

2. Clusters and the Megaparsec Tidal Force Field

Given our understanding of the tidal force configuration that will induce collapse into a filamentary configuration, we may deduce what a typical mass distribution configuration generating such a force field. Figure 1 shows that in the primordial Gaussian density field a generic configuration generating such a quadrupolar force field is that of two prominent density peaks (van de Weygaert & Bertschinger 1996). In the subsequent nonlinear evolution, these evolve into two massive clusters. Meanwhile, the mass in the intermediate region is gravitationally contracted towards an elongated filamentary configuration.

This demonstrates the intimate relationship of compact highly dense massive cluster peaks as the main source of the Megaparsec tidal force field: filaments should be seen as tidal bridges between cluster peaks. This may be directly understood by realizing that a $(-++)$ tidal shear configuration implies a quadrupolar density distribution (eqn. 1.3). In other words, the cluster peaks induce a compressional tidal shear along two axes and a dilational shear along the third axis. This means that, typically, an evolving filament tends to be accompanied by two massive cluster patches at its tip. These overdense protoclusters are the source of the specified shear.

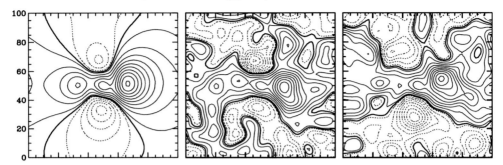

Figure 1. The implied configuration in a Gaussian random field around a location with filamentary tidal compression: clearly visible are two protocluster density peaks that generate the specified tidal force field. This is the typical configuration out of which a filamentary bridge between two massive clusters will evolve. From: van de Weygaert & Bertschinger 1996

The latter explains the canonical *cluster-filament-cluster* configuration so prominently recognizable in the observed Cosmic Web: the nonlinear gravitational evolution results in the typical configuration of filamentary bridges suspended between two or more clusters. Figure 2 illustrates this close connection between cluster nodes and filamentary bridges. It also illustrates the correspondence with the anisotropic force field, as may be inferred from the bars that depict the compressional component of the tidal force field. The typical compressional force directed towards the heartline of the filaments is lined up in a spatially coherent field of tidal bars along the spine of the filaments, suspended in between the typical tidal configuration around cluster peaks. Filaments between two clusters that are mutually aligned show this effect even more strongly. Indeed, these filaments turn out to be the strongest and most prominent ones. Studies have also quantitatively confirmed this expectation in N-body simulations (see Colberg *et al.* 2005).

For a full understanding of the dynamical emergence of the Cosmic Web along the lines sketched above, we need to invoke two important observations stemming from intrinsic correlations in the primordial stochastic cosmic density field. When restricting ourselves to overdense regions in a Gaussian density field we find that mildly overdense regions do mostly correspond to filamentary $(-++)$ tidal signatures (Pogosyan *et al.* 1998). This explains the prominence of filamentary structures in the cosmic Megaparsec matter distribution, as opposed to a more sheetlike appearance predicted by the Zel'dovich theory.

The same considerations lead to the finding that the highest density regions are mainly confined to density peaks and their immediate surroundings. Following the observation that filaments and cluster peaks are intrinsically linked, the Cosmic Web theory of Bond *et al.* (1996) formulates the theoretical framework that describes the development of the Cosmic Web in the context of a hierarchically evolving cosmic mass distribution. A prime consideration is that cluster peaks play a dominant dynamical role in the Megaparsec force field. This is clearly borne out in the map of the tidal force amplitude in fig. 2. This concerns a typical LCDM mass distribution, in this case that of the Millennium simulation (Springel *et al.* 2005).

Bond *et al.* (1996) argued how clusters, or rather peaks in the primordial density field, can be used to define the bulk of the filamentary structure of the corresponding Cosmic Web. They demonstrated how on the basis of a mass ordering of cluster peaks in a particular cosmic volume, one may predict in increasing detail the outline of the filamentary spine of the Cosmic Web in that same volume. For the accuracy of the

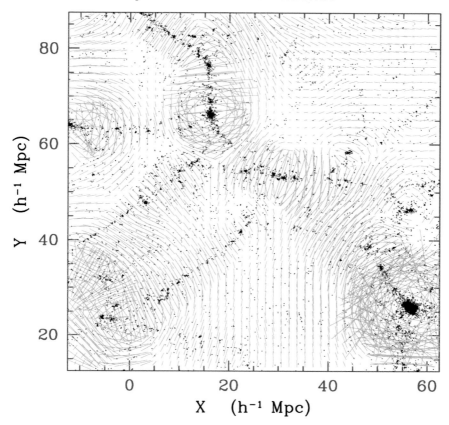

Figure 2. The relation between the *Cosmic Web*, the clusters at the nodes in this network and the corresponding compressional tidal field pattern. It shows the matter distribution at the present cosmic epoch, along with the (compressional component) tidal field bars in a slice through a simulation box containing a realization of cosmic structure formed in an open, $\Omega_\circ = 0.3$, Universe for a CDM structure formation scenario (scale: $R_G = 2h^{-1}\mathrm{Mpc}$). The frame shows structure in a $5h^{-1}\mathrm{Mpc}$ thin central slice, on which the related tidal bar configuration is superimposed. The matter distribution, displaying a pronounced web-like geometry, is clearly intimately linked with a characteristic coherent compressional tidal bar pattern. From: van de Weygaert 2002

predicted web-like structure, it is not sufficient to only take along the location and density excess of the primordial peak, but also its shape and orientation. Anisotropic peaks substantially enhance filament formation through the quadrupolar gravitational field configuration that they induce. When two such anisotropic peaks are mutually aligned, this effect becomes even stronger. In summary, the prominence — i.e. mass surface density and mass — of the implied filamentary bridges is sensitively dependent on a range of aspects of the cluster peaks. The filaments are more prominent and dense when the generating cluster peaks are

1. more massive
2. at a closer distance to each other
3. are strongly anisotropic, i.e. elongated, flattened or triaxial
4. have a favorable mutual orientation, in particular when their long axis is aligned.

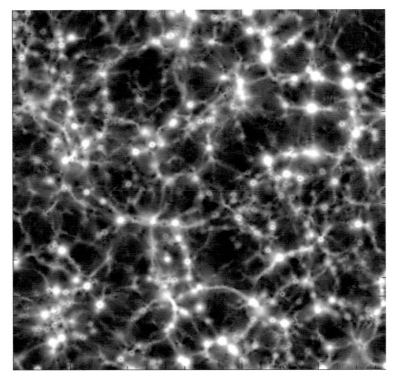

Figure 3. Clusters and Tidal force field. The image shows the magnitude of the cosmic tidal field, filtered on a 1 Mpc/h scale. Clearly visible is the dominance of the clusters in this field, they stand out as nodes in the web-like network of relatively tenuous filamentary connections. The image concerns the mass distribution in the LCDM cosmology based Millennium simulation. Image courtesy: E. Platen.

3. Cluster-based Reconstruction of the Filamentary Spine

Based on the above we arrive at the core question of our program. It involves the question whether on the basis of a complete or well-selected sample of clusters we may predict, or reconstruct, the filamentary Cosmic Web.

This question involves a few aspects. One direct one is the concentration on the filamentary spine of the Cosmic Web. Another integral anisotropic component of the Cosmic Web are also the walls — or membranes — in the Cosmic Web. After all, they occupy — besides void regions — the major share of the volume in the Universe. However, in the observational situation walls are far less outstanding than the filaments. First, filaments represent the major share of the mass content in the universe (Cautun *et al.* 2014). Secondly, as their matter content is distributed over a two-dimensional planar structure, their surface mass density is considerably lower than that of the one-dimensional filamentary arteries. In an observational setting this is augmented by an additional third effect. Cautun *et al.* (2014) showed that the halo mass spectrum is a sensitive function of the Cosmic Web environment and has shifted to considerably lower masses in wall-like environments. This explains why it is so hard to trace walls in magnitude-limited galaxy surveys: because walls are populated but relatively minor galaxies, they tend to be missed in surveys with a magnitude limit that is too high. For practical observational applications, we may therefore be forced to concentrate on the filamentary aspect of the cosmic mass distribution.

Figure 4. Aim of filament reconstruction program. Given a given sample of clusters (center), constrain the outline of filamentary spine of the Cosmic Web (right) and its configuration in the dark matter distribution (left).

Following the considerations outlined above, we arrive at the definition of our program. Given cluster locations and properties, we wish to reconstruct their surroundings and study the Cosmic Web as defined by the clusters. The program entails the following principal target list (see fig.4).

1. Use Cosmic Web theory to predict the location of filaments given cluster observations including shape, size and mutual alignment.

2. Predict the prominence and other intrinsic characteristics of filaments.

3. Investigate the application to the observational reality, taking into account systematic biases and effects, and observational errors.

3.1. *Observational context and potential*

The observational significance of our program is based on the realization that, beyond a rather limited depth, it will be far more challenging to map the filamentary features in a cosmic volume than it is to map its cluster population. While filaments are well-known features in galaxy surveys of the nearby Universe, such as 2dFGRS and SDSS, it is hard to trace them at larger depths. Moreover, mapping the outline of the dark matter Cosmic Web will be almost beyond any practical consideration, although (Dietrich *et al.* 2012) recently managed to trace a strong intercluster filament. On the other hand, the cluster distribution is or will be accessible throughout large parts of the observable universe. Armed with the ability to reconstruct the Cosmic Web given the observed cluster constraints, we would be enabled to explore the possibility to study its characteristics or observe filaments and investigate their properties in targeted campaigns. To this end, it is possible to probe the cluster distribution in different wavelength regimes. Most promising are the cluster samples that will be obtained from upcoming X-ray regime surveys, of which in particular 50k-100k cluster sample of the eROSITA all-sky survey will be an outstanding example (Merloni *et al.* 2012). Given the implicit relation to the tidal force field, major gravitational weak lensing surveys — such at present the KIDS survey an the future Euclid survey — will certainly open the possibility to explore the dark matter Cosmic Web in considerably more detail, and relate its characteristics to the observed galaxy distribution. Perhaps even most promising will be upcoming observations of the Sunyaev-Zel'dovich effect, which will potentially allow a volume-complete sample of clusters throughout the observable universe.

4. BARCODE Reconstruction Formalism

There are several methods that allow the reconstruction of a full density field based on incomplete sampled information. The first systematic technique for allowing the inference on the basis of a restricted set of observational constraints is that of the constrained random field formalism (Bertschinger 1987; Hoffman & Ribak 1991; van de Weygaert & Bertschinger 1996), and its implementation in observational settings involving measurement errors in which the Wiener filter is used to compute the mean field implied by the observational data (Bond 1995; Zaroubi *et al.* 1995). However, strictly speaking it is only suited for Gaussian random fields, so that in a cosmological context it is merely applicable to configurations that are still — nearly — in the linear regime of structure formation (Bardeen *et al.* 1986; Adler 1981; Adler & Taylor 2007).

For the project at hand we seek to translate constraints based on a set of cluster observations into a density field. Clusters, however, are highly nonlinear objects. It is extremely challenging to relate an evolved cluster to the primordial density peak out of which it evolved. This is a consequence of the complex hierarchical buildup of a cluster. A cluster emerges as the result of the assembly of a few major subclumps, augmented by the infall of numerous small clumps and the influx of a continuous stream of mass, along with the gravitational and tidal influence of surrounding matter concentrations (see Ludlow & Porciani 2011, for an insightful study). Any attempt to relate a present-day cluster directly to a unique primordial density peak is doomed.

We therefore follow a "from the ground up" approach, in the context of a Bayesian reconstruction formalism. Besides allowing us to find statistically representative primordial configurations that would relate to the observed cluster configurations that emerged as a result of fully nonlinear evolution, it has the advantage of enabling the self-consistent incorporation of any number of additional physical and data models. In recent years, several groups have worked on the development and elaboration of sophisticated and elaborate Bayesian inference techniques. Original work along these lines by Kitaura & Ensslin (2008) culminated in the development of KIGEN (Kitaura 2012), on which the present BARCODE formalism is based. KIGEN allows the translation of the observed distribution of galaxies, such as e.g. those in the 2MRS survey, into a realization of the implied primordial density and velocity field in the sample volume (Kitaura 2012; Hess *et al.* 2013). Subsequent gravitational evolution reveals the implied nonlinear structure in the dark matter distribution. Along the same lines, elaborate Bayesian reconstruction formalisms have been developed by Jasche, Wandelt, Leclercq and collaborators (Jasche & Wandelt 2013; Leclercq 2015). The latter allowed even the study of the void population in the dark matter distribution implied by the SDSS survey (Leclercq *et al.* 2015).

While the Bayesian formalism is basically oriented towards obtaining constrained samples of the primordial Gaussian density and velocity field underlying the measured observational reality that was imposed as constraint, the intention of our BARCODE program concerns the Cosmic Web that is emerging as a result of the gravitational evolution of structure. To this end, the primordial field sampled from the posterior probability distribution form the initial conditions for the subsequent gravitational evolution. The latter is followed by means of an N-body code or of an analytical structure formation model. In our case, we may analyze the structure of the various structural components of the emerging Cosmic Web.

4.1. *Bayesian formalism*

The basic philosophy of Bayesian reconstructions is to find a set of realizations that form a statistically representative sample of the posterior distribution of allowed models or configurations given the (observed) data sample. In essence, Bayesian inference is about finding the model that best fits the data, while taking into account all prior knowledge. Philosophically, it is based on the interpretation of the concept of probability in terms of the degree of belief in a certain model or hypothesis (Neal 1993; Jaynes 2003).

Central to Bayesian inference is Bayes' theorem, which formalizes the way to update our a priori belief $\mathscr{P}(m) = P(m)$ in a model m into an a posteriori belief $P(m|d)$, by accounting for new evidence from data d,

$$P(m|d) = \frac{P(d|m)P(m)}{P(d)} \equiv \frac{\mathscr{L}(d|m)\mathscr{P}(m)}{P(d)}, \qquad (4.1)$$

in which we recognize the following terms and probabilities,

  (a) the *prior* $\mathscr{P}(m) = P(m)$: the probability distribution that describes our a priori belief in the models m, without regard for the observed data;
  (b) the *likelihood* $\mathscr{L}(d|m) = P(d|m)$: the probability function that describes the odds of observing the data d given our models;
  (c) the *posterior* $P(m|d)$: the updated belief in our models, given the data;
  (d) and the *evidence* $P(d)$: the chance of observing the data without regard for the model. In Bayesian inference this term can be ignored, as we are only interested in the relative probabilities of different models, meaning that the evidence always factors out.

In the context of our Bayesian reconstruction formalism, the data d are the constraints inferred or processed from the raw cluster observations. As we will argue below, in sect. 4.2, there are several means in which galaxy or cluster observations can be imposed as constraints. In order to facilitate the imposition of heterogeneous cluster data, the BARCODE algorithm has chosen for a representation in terms of values on a regular grid in Eulerian space.

The data are specified at Eulerian positions $\{\boldsymbol{x}\}$. Lagrangian coordinates $\{\boldsymbol{q}\}$ and Eulerian coordinates are related via the displacement field $\boldsymbol{s}(\boldsymbol{q})$,

$$\boldsymbol{x}(\boldsymbol{q}) = \boldsymbol{q} + \boldsymbol{s}(\boldsymbol{q}).$$

The displacement field entails the dynamics of the evolving system.

4.2. *the* BARCODE *constraints*

The formalism that we develop with the specific intention of imposing a set of sparse and arbitrarily located cluster constraints is called BARCODE. It is an acronym for BAyesian Reconstruction of COsmic DEnsity fields.

In BARCODE we have translated the observational (cluster) data at a heterogeneous sample of Eulerian locations $\{\boldsymbol{x}_G\}$ into a regular image grid $\{\rho^{\mathrm{obs}}(\boldsymbol{x})\}$, represented on a regular Eulerian grid. A crucial ingredient to facilitate this transformation is the use of a mask. Given the sparseness and heterogeneous nature of the imposed cluster constraints, the formalism includes a mask to indicate the regions that have not, incompletely or selectively covered.

It is in particular on the aspect of using an regular grid $\{\rho^{\mathrm{obs}}(\boldsymbol{x})\}$ to transmit the observational constraints that BARCODE follows a fundamentally different approach from the one followed in the related KIGEN formalism (Kitaura 2012). KIGEN is able to process the galaxy locations in the dataset directly for imposing the data constraints. It involves two stochastic steps instead of one. It first samples a density field given the galaxy distribution. In a second step it then samples a new particle distribution, given the sampled density field and a model of structure formation. It renders KIGEN highly flexible. It does require the posterior probability function to be differentiable, and is able to implicitly take into account the bias description for the galaxy sample.

A grid-based formalism like BARCODE, on the other hand, requires an explicit bias description. Also, it requires the posterior to be differentiable, because of the use of the Hamiltonian Monte Carlo HMC sampling procedure (see sect. 4.6).

4.3. *the signal:* BARCODE *samples*

In the context of the Bayesian formalism, the signals that BARCODE is sampling are the models constrained by the data. In our application they are realizations of the primordial density field $\{\delta(\mathbf{q})\}$. The core task of the reconstruction formalism is therefore the proper sampling of a realization of the primordial density field $\{\delta(\mathbf{q})\}$, given the posterior distribution function for $\{\delta(\mathbf{q})\}$. The posterior distribution function is determined on the basis of the input data and the assumed cosmological model.

The key product of the procedure is therefore a sample $\delta(\mathbf{q})$ of the *primordial density fluctuation field*. For practical computational reasons, the primordial field is specified at the Lagrangian locations $\{\mathbf{q}\}$ on a regular grid. In our test runs, we used a grid with $N_x = 64$ cells on a side for this discrete representation of an intrinsically fully continuous field on a finite regular grid. The presently running implementations have $N_x = 128$ or $N_x = 256$ cells. Each of the density values $\delta(\mathbf{q}_1)$ is one of the stochastic variables to be sampled. The total signal $\mathcal{S}$,

$$\mathcal{S} = \{\delta(\mathbf{q}_1), \delta(\mathbf{q}_2), \dots, \delta(\mathbf{q}_N)\} = \{\delta(\mathbf{q}_i)\}, \tag{4.2}$$

defines a $D = N_x^3$ dimensional probabilistic space. In other words, for $N_x = 64$ we are dealing with the sampling of a full Bayesian signal in a 262144-dimensional probabilistic space.

4.4. *the* BARCODE *Model Prior*

The only thing we know about the primordial density field — without regard for the data — is that it must be a Gaussian random field; this should fully characterize our prior belief in such a field. The prior in our model is therefore a multivariate Gaussian distribution characterized by a zero mean and a correlation matrix determined by the correlation function as measured from the cosmic microwave background:

$$\mathscr{P}(\{\delta(\mathbf{q})\}|P(k)) = \frac{1}{\sqrt{(2\pi)^N \|S\|}} \exp\left(-\frac{1}{2} \sum_{i,j} \delta(\mathbf{q}_i)\xi^{-1}{}_{ij}\delta(\mathbf{q}_j) \right), \tag{4.3}$$

where ξ_{ij}^{-1} is the inverse of the correlation matrix ξ_{kl}, the real-space dual of the Fourier-space power spectrum $P(k)$ with elements

$$\xi_{kl} = \xi(|\mathbf{q}_k - \mathbf{q}_l|), . \tag{4.4}$$

The cosmological model underlying our reconstruction enters via the cosmological power spectrum $P(k)$. It contains implicit information on global cosmological parameters, on the nature of the primordial density fluctuations, and via these the nature of dark matter.

In addition to the primordial fluctuations, we also need to specify the structure formation model, which transforms the Lagrangian to Eulerian space density fields via the transformation $f : \delta(q) \to \rho(x)$ from a Lagrangian space overdensity field $\delta(q)$ to Eulerian space density field $\rho(x)$. In the context of gravitationally driven structure formation, one could employ an N-body simulation to optimally follow the structure formation process, including the complex nonlinear aspects. However, in practice it would be unfeasible to include this in the formalism, as it would take an overpowering computational effort to numerically compute derivatives with respect to the signal $\mathcal{S}$.

For the computation of the displacement, BARCODE therefore prefers to use analytical prescriptions. The most straightforward option is the Zel'dovich formalism. Other options are higher-order Lagrangian perturbation schemes such as 2LPT or ALPT (ALPT, or Augmented Lagrangian Perturbation Theory, is a structure formation model that combines 2LPT on large scales with spherical collapse on small scales).

Closely related to the latter is a model $f : \boldsymbol{x} \rightarrow \boldsymbol{\zeta}$ that subsequently transforms Eulerian coordinates to redshift space coordinates, and hence transforms the modeled density field from Eulerian space to redshift space for the situations in which the observational data are in fact redshift data. This is implicitly included in the structure formation model via its prescription for the corresponding velocity perturbations.

In addition to these cosmological and structure formation models, we also need to take along model descriptions relating the observed galaxy or cluster distribution to the underlying distribution. This includes factors such as a noise model for statistical sources of error, like measurement uncertainties in the observations, and a bias model connecting the galaxy/cluster distribution to the density field.

4.5. *MCMC sampling and the Hamiltonian Monte Carlo Procedure*

On the basis of the set of data constraints, and the model assumptions discussed in the previous section, we obtain the posterior distribution function

$$P(\mathcal{S}) = P(\{\delta(\boldsymbol{q}_1), \delta(\boldsymbol{q}_2), \ldots, \delta(\boldsymbol{q}_N)\}) \,. \tag{4.5}$$

To obtain meaningful constrained realizations for our problem, we sample the posterior distribution. Given the complex nature of the posterior distribution function in an $D = N_x^3$ dimensional probability space, an efficient sampling procedure is crucial in order to guarantee a statistically proper set of samples $\mathcal{S}$.

Markov chain Monte Carlo (MCMC) methods try to alleviate the computational costs of sampling a high-dimensional probability distribution. They do this by constructing a so called Markov chain of samples s_i. In our case, this becomes a chain of density fields, each sampled on a regular grid. Such a chain may be seen as a random walk through the parameter space of the pdf $P(\{\delta(\boldsymbol{q})\})$. Each visited location in the high dimensional space corresponds to one particular sample realization of the density field $\delta(\boldsymbol{q})$. To do so, MCMC chains automatically visit high probability regions in direct proportion to their relative probability, and do not waste computational tie on low probability areas. The performance gain can be orders of magnitude for high-dimensional problems. The resulting sampling ensemble is equivalent to an ensemble from the full posterior $P(\{\delta(\boldsymbol{q})\})$.

Arguably the best known MCMC method is the Metropolis-Hastings sampler. Another well-known algorithm is the closely related concept of Gibbs sampling, basically a multi-dimensional extension of Metropolis-Hastings algorithm. In BARCODE we follow a far more efficient methodology, that of the Hamiltonian Monte Carlo sampling.

4.6. *Hamiltonian Monte Carlo sampling — HMC*

Hamiltonian sampling is based on a physical analogy, and uses concepts from classical mechanics — specifically Hamiltonian dynamics — to solve the statistical sampling problem (Neal (1993, 2011); Jasche & Kitaura (2010), also see Leclercq (2015) for an excellent summary of the ideas involved). Following the concept of Hamiltonian mechanics, a Markov chain is produced in the form of an *orbit* through parameter space, which is found by solving the Hamiltonian equations of motion:

$$\frac{\mathrm{d}q_i}{\mathrm{d}\tau} = \frac{\partial \mathcal{H}}{\partial p_i}$$

$$\frac{\mathrm{d}p_i}{\mathrm{d}\tau} = -\frac{\partial \mathcal{H}}{\partial q_i}\,, \qquad (4.6)$$

where τ is an artificial "time" analogue that we merely use to evolve our Markov chain. Important in this respect is also the time reversibility of Hamiltonian dynamics. The positions q_i, the momenta p_i and the Hamiltonian $\mathcal{H}$, the sum of a potential term $\mathcal{E}$ and a kinetic term $\mathcal{K}$,

$$\mathcal{H} = \mathcal{E} + \mathcal{K}\,, \qquad (4.7)$$

are not real physical quantities, but equivalent stochastic variables. It is straightforward to interpret the meaning of the position vectors q_i. They represent the primordial density field realization $\delta(\boldsymbol{q})$, and the dynamical orbit the chain of density field samplings,

$$q_i = \delta(\boldsymbol{q}_i)\,. \qquad (4.8)$$

In order to avoid confusion, notice that the stochastic coordinate analogues q_i are something completely and fundamentally different than the Lagrangian coordinates $\boldsymbol{q}$. The potential term $\mathcal{E}$ is identified with the probability distribution function P, and is equal to its negative logarithm, $\mathcal{E} = -\log P$. The D-dimensional auxiliary momentum vector p_i is related to the steepness of the change in density field sample between sampling steps. The kinetic energy term $\mathcal{K}$ is the quadratic sum of these auxiliary momenta,

$$\mathcal{K} = \frac{1}{2}\boldsymbol{p}^T \mathcal{M}^{-1} \boldsymbol{p}\,. \qquad (4.9)$$

in which the symmetric positive definite mass matrix term $\mathcal{M}$, called mass in analogy to the situation in the context of classical mechanics, characterizes the inertia of the sampling orbit moving through parameter space. It is of crucial importance for the performance of the formalism: if the mass term is too large it will result in slow exploration efficiency, while masses which are too light will result in large rejection rates.

As in the case of other MCMC algorithms, for the startup of the procedure there is an additional important aspect, the *burn-in* phase. An MCMC chain starts up from an initial guess for the signal S to be sampled, and subsequently evolves a chain from there. As we may not necessarily start from a region of high probability, it may take the MCMC chain a series of iterations before reaching this region. The period up to this point is called the burn-in phase of the chain. The samples obtained in the burn-in phase are discarded from the ensemble, since they are not part of a representative sample.

5. BARCODE validation: Cosmic Web reconstruction

In BARCODE we have implemented an HMC sampler of primordial density fields $\delta(q)$ given density data from observations. The algorithm can deal both with real-space density fields, as well as density fields sampled in redshift space. To incorporate a structure formation model, we can use either the Zel'dovich approximation or an ALPT structure formation model, which combines the higher-order 2LPT Lagrangian model on large scales with spherical collapse on small scales (Hess *et al.* 2013). The implementation may invoke a Gaussian or Poisson error model for the observed densities.

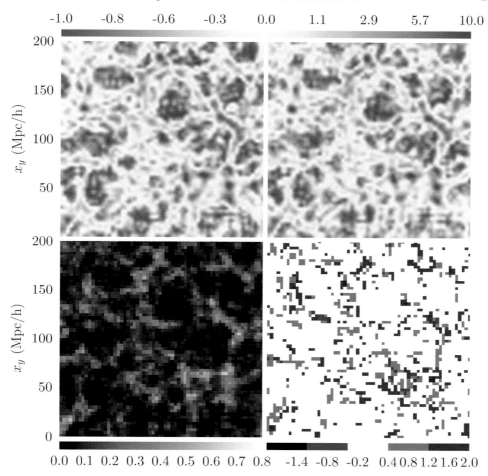

Figure 5. Comparison of true to sampled *Eulerian* density fields. *Top left:* true. *Top right:* mean of 60 samples (iteration 500, 550, ..., 3500) after burn-in phase. *Bottom left:* standard deviation of 60 samples. *Bottom right:* difference of true and mean fields. ¿From E.G.P. Bos 2016

5.1. *Test results*

Here we first present the result of some numerical experiments, demonstrating the potential to reconstruct successfully the web-like outline of the cosmic matter distribution, given an observed sample of the large scale density distribution.

As a test, we have evaluated the performance of our code on a LCDM dark matter simulation, in which the matter distribution in a box with a side of $L = 200 \ h^{-1}$Mpc is represented on a grid of $N_x = 64$ cells. It means the stochastic sampling process takes place in a 262144-dimensional space. We find that the algorithm starts constraining the large scales (low modes) and then increasingly gains power towards small scales (high modes).

Following the burn-in phase, the chain remains in the high probability region around the "true" density. The sampling chain is expected to vary around this true density due to the random kicks from the (statistical, auxiliary) momentum. Figure 5 shows the comparison between the Eulerian space mean density field of 60 samples — obtained

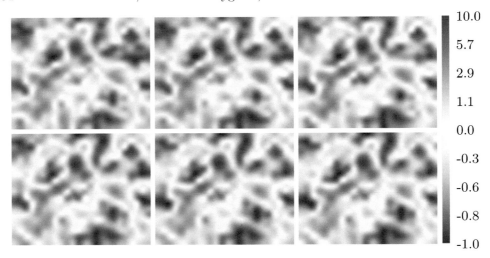

Figure 6. Zoom-ins of Eulerian space density field samples at iterations 500, 1000, 1500, 2000, 2500 and 3000 (left to right). ¿From E.G.P. Bos 2016

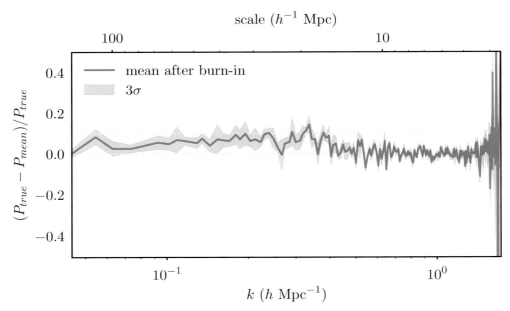

Figure 7. Relative differences between true and mean Lagrangian power spectra for 60 post-burn-in samples. The shaded area represents (three times) the standard deviation of the power spectra of the samples. ¿From E.G.P. Bos 2016

from 3000 iterations — and the true underlying density field. The match between the two is very good. We notice this on account of a visual comparison between the matter distribution in the top frames. This is strongly supported by the quantitative comparison in terms of the standard deviation of the 60 samples (bottom left) and the difference field between the true and mean fields (bottom right-hand plot).

The differences between mean and sampled field turn out to be rather subtle. In Eulerian space they tend to be most prominent in the high-density regions. In part, this can be understood from the realization that there is more variation in those regions in combination with the fact that there are not enough samples to completely average out these variations.

To appreciate the variation between field variations, in figure 6 we zoom in on a couple of post-burn-in iterations. It shows that differences are hard to discern, in particularly at this low resolution. At a careful inspection we may notice some minor differences. For instance, the height of the two large peaks vary significantly. Also, the shape of the "cloud" in the void, just on the lower right from the center, varies from almost spherical with a hole in the middle to an elongated and even somewhat curved configuration.

Yet, overall the reconstructions appear to be remarkably good. This is true over the entire range of scales represented in the density field. The power spectra of the chain samples properly converge to the true power spectrum, as may be inferred from figure 7.

6. Reconstructions in Redshift Space

One important caveat in the basic BARCODE algorithm is that in the observational practice we tend to use redshifts as measures for distance. Because redshifts also include the contribution from peculiar motions, the measured distribution does not concern co-moving space x, but the distribution in redshift space ζ.

The direct implication is that our data will be deformed by so-called redshift space distortions. Clusters will seem radically stretched along the line of sight due to the internal virialized velocities of their galaxies. On the other hand, large scale overdense structures appear squashed along the line of sight, while underdense structures are observed to be larger in the radial direction than in reality. It is one of the principal reasons for the appearance of *Great Walls* perpendicular to the line of sight. All these effects must be taken into account in order to accurately measure volumes, densities, shapes and orientations of the components of the Cosmic Web.

There are several options to accommodate redshift space distortions when processing measurements. Most of these involve a form of a posteriori or a priori "correction" of the redshift space effects. A priori corrections include, for instance, the detection of "Fingers of God", followed by an automatic reshaping into spheres. Possible a posteriori corrections include the reshaping of the power spectrum to account for the "Kaiser effect" of squashing along the line of sight. However, all these methods are approximations and most of them focus on one aspect of the redshift space transformation only.

We opt for a self-consistent redshift space correction in the BARCODE formalism. This concerns a one-step transformation of *data model* itself into redshift space. To the best of our knowledge, this is the first self-consistent redshift space data based reconstruction of the primordial density fluctuations. To this end, we adjust our BARCODE data model to not only transform from Lagrangian coordinates q to Eulerian coordinates x using a structure formation model, but subsequently also from x to the corresponding redshift space.

Since our structure formation formalisms are particle based (using Lagrangian perturbation theory), this can easily be achieved by means of one more coordinate mapping from Eulerian space to redshift space. In this way, we incorporate automatically and directly all possible redshift space effects in our reconstructions, insofar as the structure formation model includes them. The one additional complication for such a strategy in the context of the Hamiltonian Monte Carlo formalism, is that for the redshift space transformation to work it must facilitate analytical derivatives of the posterior distribution with respect to the sampled fields. To this end, we have derived and implemented the required equations in the BARCODE code.

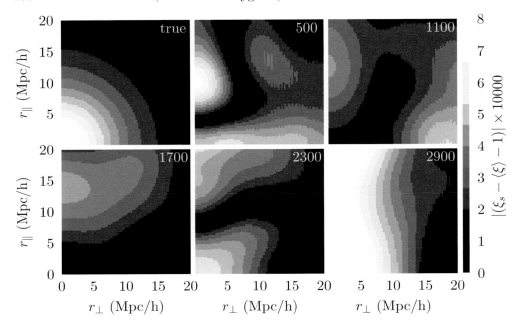

Figure 8. Five sample Eulerian 2D correlation functions minus the mean of all reconstructions, compared to the true field in the top left panel.

6.1. *Mapping from Eulerian to Redshift space*

Redshift space coordinates $\boldsymbol{\zeta}$ can be defined as

$$\boldsymbol{\zeta} \equiv \boldsymbol{x} + \frac{1}{Ha}\left((\boldsymbol{v} - \boldsymbol{v}_{\mathrm{obs}}) \cdot \hat{\boldsymbol{x}}\right)\hat{\boldsymbol{x}} \equiv \boldsymbol{x} + \boldsymbol{s}_\zeta , \qquad (6.1)$$

where $\boldsymbol{\zeta}$ is a particle's location in redshift space, $\boldsymbol{x}$ is the Eulerian comoving coordinate, $\boldsymbol{v}$ is the particle's (peculiar) velocity and $\boldsymbol{v}_{\mathrm{obs}}$ is the velocity of the observer. a is the cosmological expansion factor, which converts from physical to comoving coordinates, and H is the Hubble parameter at that expansion factor, which converts the velocity terms to distances with units $h^{-1}\mathrm{Mpc}$. We call $\boldsymbol{s}_\zeta$ the *redshift space displacement field*, analogously to the displacement field $\boldsymbol{s}$ that transforms coordinates from Lagrangian to Eulerian space in Lagrangian perturbation theory.

An important point to keep in mind is that in equation 6.1, the comoving coordinates are defined such that the observer is at the origin. The unit vector $\hat{\boldsymbol{x}}$, and thus the entire redshift space, changes completely when the origin shifts.

6.2. *Test results: redshift space correlation functions*

Overall, in our 64^3 test-runs based on redshift space renderings of the simulations described in the previous section we found that an excellent performance. It manages to deal with all redshift space artifacts, and differences between the sampled density field and the true density field remains limited to small local variations. This is confirmed by the power spectrum, which seems to match the expected true power spectrum in all necessary detail. Our BARCODE formalism has fully and self-consistently accounted for the Kaiser redshift space effect and even additional mildly nonlinear effects such as in the triple value region corresponding to the infall region around clusters.

Perhaps the most convincing demonstration of this success is that of the 2D correlation function $\xi(\pi,\sigma)$. It measures the correlation function by segregating the contributions along two direction one along the line of sight, the other perpendicular. In the case of an intrinsic anisotropies in the data, we will obtain a deviation from a perfectly circular $\xi(\pi,\sigma)$.

As for the BARCODE reconstructions, the two-dimensional correlation functions of the reconstructions match the true — almost isotropic and perfectly circular — one when using the redshift space model. In other words, it shows that it manages to eliminate large scale redshift space anisotropies. Figure 8 shows the differences of the Eulerian 2D correlation functions $\xi(\pi,\sigma)$ of five BARCODE samples compared to that for the true field. The differences are very small, in the order of $\delta\xi/\xi \approx 10^{-3}$. Nonetheless, we may note some variation in the differences. This is a result of a minor effect, a result of the MCMC sampling process itself introducing additional anisotropies in the samples

In summary, we may conclude that the BARCODE formalism has successfully incorporated the ability to operate directly in redshift space.

References

Adler R.J. 1981, *The Geometry of Random Fields* (Wiley)

Adler R.J. & Taylor J.E. 2007, *Random Fields and Geometry* (Springer)

Bardeen J.M., Bond J.R., Kaiser N. & Szalay A.S. (BBKS) 1986, *ApJ*, 304, 15

Bertschinger E. 1987, *ApJL*, 323, 103

Bond J.R. 1995, *Phys Rev. Lett.*, 74, 4369

Bond J.R., Kofman L. & Pogosyan D. 1996, *Nature*, 6575, 603

Bond J.R., Kofman L., Pogosyan D. & Wadsley J. 1998, in S. Colombi, Y. Mellier eds., *Wide Field Surveys in Cosmology, 14th IAP meeting)* (Editions Frontieres, p. 17

Bos E.G.P. 2016, PhD thesis, Univ. Groningen

Cautun M., van de Weygaert R., Jones B.J.T., & Frenk C.S. 2014, *MNRAS*, 441, 2923

Colberg J.M., Krughoff K.S. & Connolly A.J. 2005, *MNRAS*, 359, 272

Colless M., *et al.* 2003, arXiv:0306581

Dietrich, J. P., Werner, N., Clowe, D., Finoguenov, A., Kitching, T., Miller, L. & Simionescu, A. 2012, *Nature*, 487, 202

Guzzo L. & Teh VIPERS team 2013, *The Messsenger*, 151, 41

Guzzo L., *et al.* 2014, *AA*, 566, 108

Hess S., Kitaura F.-S. & Gottlöber S. 2013, *MNRAS*, 435, 2065

Hoffman Y. & Ribak E. 1991, *ApJL*, 380, 5

Huchra J.P., *et al.* 2012, *ApJS*, 199, 26

Jasche J. & Kitaura F.-S. 2010 *MNRAS*, 407, 29

Jasche J. & Wandelt B.-D. 2013 *MNRAS*, 432, 894

Jaynes E.T. 2003, *Probability Theory: The Logic of Science* (Cambridge Univ. Press)

Kitaura F.-S. & Ensslin T.A. 2008, *MNRAS*, 389, 497

Kitaura F.-S. 2012, *MNRAS*, 429L, 84

Leclercq F., Jasche J., Sutter P.M., Hamaus N. & Wandelt B. 2015, *JCAP*, 03, 047

Leclercq F. 2015, *Bayesian large-scale structure inference and cosmic web analysis*, PhD thesis, Univ. Pierre et Marie Curie, Institut d'Astrophysique de Paris

Ludlow A.D. & Porciani C. 2011, *MNRAS*, 413, 1961

Merloni, A. *et al.* & the German eROSITA Consortium 2012, *ArXiv e-prints*, 1209.3114

Neal R.M. 1993, *Probabilistic Inference Using Markov Chain Monte Carlo Methods*, Technical Report CRG-TR-93-1, Dept. Comp. Science, Univ. Toronto

Neal R.M. 2011 *MCMC using Hamiltonian dynamics*, in S. Brooks, A. Gelman, G. Jones & X-L Meng *Handbook of Markov Chain Monte Carlo*. (Chapman & Hall/CRC Press)

Peebles P.J.E. 1980, *The large-scale structure of the universe* (Princeton Univ. Press)

Pogosyan D., Bond J.R., Kofman L. & Wadsley J. 1998, in S. Colombi, Y. Mellier eds., *Wide Field Surveys in Cosmology, 14th IAP meeting) (Editions Frontieres*, p. 61

Springel, V. *et al.* 2005, *Nature*, 435, 629

Tegmark M., SDSS collaboration 2004 *Ap.J.*, 606, 702

van de Weygaert R. & Bertschinger E. 1996, *MNRAS*, 281, 84

van de Weygaert R. & Bond J.R. 2008, *Clusters and the Theory of the Cosmic Web*, in M. Plionis, O. López-Cruz & D. Hughes eds., *A Pan-Chromatic View of Clusters of Galaxies and the Large-Scale Structure*, LNP 740 (Springer), p. 335

Zaroubi S., Hoffman Y., Fisher K. B. & Lahav O. 1995, *ApJ*, 449, 446

Zeldovich, Ya. B. 1970, *AA*, 5, 84

CHAPTER 4C.

Cosmic Web
Clustering

*Impressions of the IAU308 audience during a session,
at the Conference Hall of Tallinn University.*

The Zeldovich Universe:
Genesis and Growth of the Cosmic Web
Proceedings IAU Symposium No. 308, 2014
R. van de Weygaert, S. Shandarin, E. Saar & J. Einasto, eds.

ⓒ International Astronomical Union 2016
doi:10.1017/S1743921316010000

Evolution of the galaxy correlation function at redshifts $0.2 < z < 3$

Andrzej M. Sołtan

Nicolaus Copernicus Astronomical Center,
Bartycka 18, 00-716 Warsaw, Poland
email: soltan@camk.edu.pl

Abstract. We determine the auto-correlation function (ACF) of galaxies using massive deep galaxy surveys for which distances to individual objects are assessed using photometric redshifts. The method is applied to the 2deg COSMOS survey of ~ 300000 galaxies with $i+ < 25$ and $z_{ph} \lesssim 3$. The distance estimates based on photometric redshifts are not sufficiently accurate to be directly used to determine the ACF. Nevertheless, the photometric redshifts carry statistical information on the data distribution on (very) large scales. The investigation of the surface distribution of galaxies in several redshift (=distance) bins allows us to determine the spatial (3D) ACF over the redshift range of $0.2 - 3.2$ or look back time of $2.4 - 11.5$ Gy.

Keywords. galaxies: distances and redshifts, galaxies: statistics, large-scale structure of universe

1. Observational material

Efficient means to quantify the galaxy clustering at small and medium scales are the auto-correlation functions (ACF). The two-point ACF $\xi(r) = \langle n(\mathbf{x})\, n(\mathbf{x} + \mathbf{r}) \rangle / \langle n \rangle^2 - 1$, where $n(\mathbf{x})$ is the position dependent spatial density of galaxies and $\langle ... \rangle$ denotes averaging over the survey volume. Over a wide range of separations r the ACF is adequately approximated by a power law $\xi(r) = (r/r_o)^\gamma$.

If no information on the galaxy radial distance is available, the amplitude of spatial correlations is derived from the 2D correlation function. Here, we use the photometric redshifts as a tool to assess the number of the galaxy-galaxy pairs resulting from the clustering and due to random coincidences in the celestial sphere. The photometric redshifts (z_{ph}), as compared to spectroscopic ones, are substantially less accurate distance indicators. Nevertheless, they provide a raw estimate of the galaxy position. Thus, photometric redshifts could be used to statistically identify and extract random pairs, and to increase in this way the S/N ratio of the correlation amplitude measurement. The multi-band photometry allows for massive estimates of redshifts down to magnitudes of ~ 25 reaching $z \approx 3$ (e.g. Ilbert *et al.* 2009) In the calculations below we carefully take into account an imprecise nature of photometric redshifts.

The COSMOS galaxies are distributed within a square of 84 arcmin a side centered at $\alpha_c = 150.1°$ and $\delta_c = 2.2°$ (Taniguchi *et al.* 2007). The catalog contains 385065 objects in the deep Subaru Area, of which almost 252000 have been classified as galaxies brighter than $i_{AB}^+ = 25$. For all these galaxies z_{ph} are listed in the COSMOS photometric redshift catalog by Ilbert *et al.* (2009) available through the Web site of IPAC/IRSA. A comparison of z_{ph} data with the available z_{sp} measurements shows that a large majority of z_{ph} is subject to minimal errors. However, below $i_{AB}^+ = 23$ sharply increases number of 'catastrophic' errors, where the $z_{ph} - z_{sp}$ differences are scattered without any characteristic scale. We precisely model the $z_{ph} - z_{sp}$ deviations in the consecutive magnitude bands.

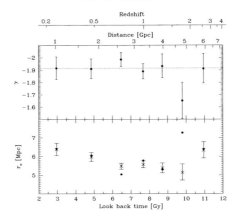

Figure 1. Variations of slope and the normalization of the ACF of the most luminous galaxies in the data as a function of look back time. The simultaneous fitting of γ and r_o (full dots) induces a correlation of both parameters. Crosses show the r_o distribution for the slope fixed at the average value of $\gamma = -1.92$.

2. Results

We take the photometric redshifts, z_{ph}, as a working estimate of the comoving distances for all the galaxies. The whole galaxy sample is divided into 7 distance bins between ~ 650 and $\sim 6550\,\text{Mpc}$. The radial depth of each bin is larger than $600\,\text{Mpc}$. Thus, it is also much larger than the maximum distance at which the ACF differs from 0.

First, we determine the 2D ACF for the each distance bin. Around randomly chosen galaxy from the given bin cluster predominantly galaxies belonging to the same bin. Because of the z_{ph} errors, one can expect also some residual enhancement of galaxies assigned to other bins. Nevertheless, both distributions are distinctly different. A comparison of the surface density profiles formed by galaxies from the same bin and from all the other bins, accompanied by the $z_{ph} - z_{sp}$ statistics, allows us to determine the ACF assuming the power law: $w(p) = w_o\, p^\zeta$, where p is the distance in the plane of the sky.

Exponent ζ in the power law approximation of the 2D ACF is directly related to the slope of the spatial ACF: $\gamma = \zeta - 1$. To retrieve the normalization of the spatial correlations, r_o, from the 2D function one needs the information on the radial distribution of the average galaxy density in the sample. In the present analysis the galaxy concentration varies systematically with the distance. A radial scale of this trend is substantially larger than the clustering scale and it can be appropriately modeled using the photometric redshifts. Our estimates of γ and r_o, shown in Fig. 1, cover a very wide range of redshifts, presently unavailable to the spectroscopic redshift surveys in massive scale. The present measurements of the ACF parameters at moderate and high redshifts essentially do not deviate from those determined at low redshifts. It indicates that both the ACF slope and the correlation length remained stable for at least 11 billion years.

Acknowledgments Based on zCOSMOS observations carried out using the Very Large Telescope at the ESO Paranal Observatory under Programme ID: LP175.A-0839. This work has been partially supported by the Polish NCN grant 2011/01/B/ST9/06023.

References

Ilbert, O., Capak, P., Salvato, M., *et al.* 2009, *ApJ*, 690, 1236
Taniguchi, Y, Scoville, N, Murayama, T., *et al.* 2007, *ApJS*, 172, 9

The Zeldovich Universe:
Genesis and Growth of the Cosmic Web
Proceedings IAU Symposium No. 308, 2014
R. van de Weygaert, S. Shandarin, E. Saar & J. Einasto, eds.

ⓒ International Astronomical Union 2016
doi:10.1017/S1743921316010012

Color and magnitude dependence of galaxy clustering

Volker Müller

Leibniz-Institut für Astrophysik Potsdam
D-14482 Potsdam, Germany
email: vmueller@aip.de

Abstract. A quantitative study of the clustering properties of galaxies in the cosmic web as a function of absolute magnitude and colour is presented using the SDSS Data Release 7 galaxy redshift survey. We compare our results with mock galaxy samples obtained with four different semi-analytical models of galaxy formation imposed on the merger trees of the Millenium simulation.

Keywords. Cosmology: obserations - cosmology: theory - galaxies: statistics

1. Data and mock sample selection

We study galaxy clustering using the SDSS redshift survey DR. We use a large contiguous region of the Northern Galactic cap with 7500 deg^2. Photometric calibration and k-correction to redshift $z = 0$ is done in standard way using the New York University Value-Added Galaxy Catalog by Blanton *et al.* (2005).

Starting from the observed R-band magnitude and redshift distributions, we define volume-limited galaxy samples m1 to m12 (Mueller *et al.* 2011) to investigate the dependence of the correlation function on absolute magnitude and of color taking a separation line $U - R = 1.8 - 0.05 \times (R + 19)$ between red and blue galaxies.

For comparison we use four sets of mock galaxy samples constructed using the Millenium simulation and semi-analytical models of galaxy formation from merger trees of haloes in the simulation due to Croton *et al.* (2006, C06), De Lucia & Blaizot (2007, D07), Font *et al.* (2008, F08), and Guo *et al.* (2011, G11)

2. Correlation analysis

The correlation functions are evaluated using the full selection mask of the survey. Errors are estimated with 10 bootstrap resamplings of the data, cp. Fig. 1. The solid line in the left panel shows the result corresponding to all galaxies for the sample m1. We fit a power law at the correlation length, $\xi(r_0) = 1$, with slope 1.4. The dashed line for red galaxies lie about 0.2 dex above that of the full galaxy sample, the dot-dashed line for blue galaxies are 0.15 dex below. For the other samples we get similar results, only the difference of the clustering between red and blue galaxies gets smaller as magnitudes increase. The right panel shows the ratio between the full correlation functions of the sample m4 and all four mock catalogues. The correlation functions of models C06 (solid line) and G11 (dot-dashed line) reproduce the shape of the observed correlation function over almost all spatial scales. However, the clustering amplitude is underpredicted by about 20 percent. Acceptable results are also obtained for the model D07, while F08 overpredicts the clustering of close pairs by up to a factor of two.

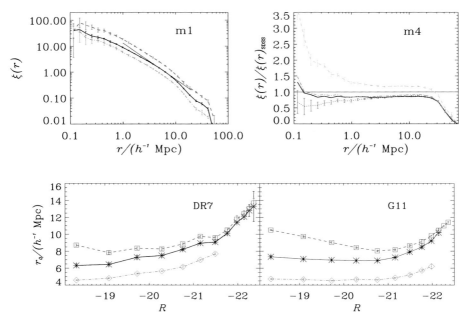

Figure 1. *Upper left:* Two-point correlation function for sample m1 with all galaxies (solid line), red galaxies (dashed line) and blue galaxies (dot-dashed line). *Upper right:* Ratio between model correlation functions and SDSS galaxies for the m4 sample. The mock samples are due to C06 (solid line), D07 (dotted line with error bars), F08 (dashed line with error bars), and G11 (dash-dotted line). *Lower left:* Correlation length as a function of mean R-magnitude for samples from m1 to m12 for all galaxies (stars and solid line), red galaxies (open squares and dashed lines), and blue galaxies (open diamonds and dash-dotted line). *Lower right:* Corresponding mock samples in the G11 model with 2 σ error bars.

The results can be collected in a compact way by the magnitude dependence of the correlation length. The left panel of the lower Fig. 1 shows the correlation length for samples m1 to m12. The solid, dashed and dot-dashed lines correspond to all, red, and blue galaxies, respectively. The right panel shows the results of the G11 model. The correlation lengths of all and blue galaxies stay nearly constant between $R = -18.4$ and $R = -21$, while the correlation length of red galaxies decreases. This is due to the large number of satellites present among faint galaxies. At brighter magnitudes the correlation length increases due to the higher bias of more massive haloes. The remaining semi-analytical models display similar trends. They reproduce qualitatively the clustering dependence as a function of magnitude and colour. However, quantitatively, there exist significant differences, with the F08 model showing the smallest discrepancies for scales above $1\ h^{-1}$ Mpc. Mark correlations complement these results.

References

Blanton, M., Schlegel, D., Strauss, M., *et al.* 2005, *ApJ*, 129, 2562
Croton, D., Springel, V., White, s., *et al.* 2006, *MNRAS*, 365, 11
De Lucia, G. & Blaizot, J. 2007, *MNRAS*, 375, 2
Font, A., Bower, R., McCarchy, I., *et al.* 2008, *MNRAS*, 193, 353
Guo, Q., White, S., Boylan-Kolchin, M., *et al.* 2011, *MNRAS*, 413, 101
Müller, V., Hoffmann, K., & Nuza, S. E. 2011, *Baltic Astronomy*, 20, 259

The Zeldovich Universe:
Genesis and Growth of the Cosmic Web
Proceedings IAU Symposium No. 308, 2014
R. van de Weygaert, S. Shandarin, E. Saar & J. Einasto, eds.
ⓒ International Astronomical Union 2016
doi:10.1017/S1743921316010024

Local Large-Scale Structure and the Assumption of Homogeneity

Ryan C. Keenan[1], Amy J. Barger[2,3,4] and Lennox L. Cowie[3]

[1] Academia Sinica Institute of Astronomy and Astrophysics
P.O. Box 23-141, Taipei 10617, Taiwan
email: rkeenan@asiaa.sinica.edu.tw

[2] Dept. of Astronomy, University of Wisconsin-Madison
475 N. Charter St., Madison, WI 53706, USA

[3] Dept. of Physics and Astronomy, University of Hawaii
2505 Correa Rd., Honolulu, HI 96822, USA

[4] Institute for Astronomy, University of Hawaii, 2680 Woodlawn Dr., Honolulu, HI 96822, USA

Abstract. Our recent estimates of galaxy counts and the luminosity density in the near-infrared (Keenan *et al.* 2010, 2012) indicated that the local universe may be under-dense on radial scales of several hundred megaparsecs. Such a large-scale local under-density could introduce significant biases in the measurement and interpretation of cosmological observables, such as the inferred effects of dark energy on the rate of expansion. In Keenan *et al.* (2013), we measured the $K-$band luminosity density as a function of distance from us to test for such a local under-density. We made this measurement over the redshift range $0.01 < z < 0.2$ (radial distances $D \sim 50 - 800\ h_{70}^{-1}$ Mpc). We found that the shape of the $K-$band luminosity function is relatively constant as a function of distance and environment. We derive a local ($z < 0.07$, $D < 300\ h_{70}^{-1}$ Mpc) $K-$band luminosity density that agrees well with previously published studies. At $z > 0.07$, we measure an increasing luminosity density that by $z \sim 0.1$ rises to a value of ~ 1.5 times higher than that measured locally. This implies that the stellar mass density follows a similar trend. Assuming that the underlying dark matter distribution is traced by this luminous matter, this suggests that the local mass density may be lower than the global mass density of the universe at an amplitude and on a scale that is sufficient to introduce significant biases into the measurement of basic cosmological observables. At least one study has shown that an under-density of roughly this amplitude and scale could resolve the apparent tension between direct local measurements of the Hubble constant and those inferred by Planck team. Other theoretical studies have concluded that such an under-density could account for what looks like an accelerating expansion, even when no dark energy is present.

Keywords. cosmology: observations — cosmology: large-scale structure of universe — galaxies: fundamental parameters — galaxies: luminosity function

1. Introduction

The assumption that the universe is homogeneous on large scales is a fundamental pillar of our current concordance cosmology. However, it is well known that very large-scale inhomogeneity exists in the form of sheets, voids, and filaments of matter . As structure formation proceeds, gravity acts to pile more matter onto the over-dense regions, and simultaneously evacuate the under-dense regions. Models have shown that this phenomenon implies that cosmological observables measured by an observer will vary significantly depending on where that observer is located with respect to these large-scale structures.

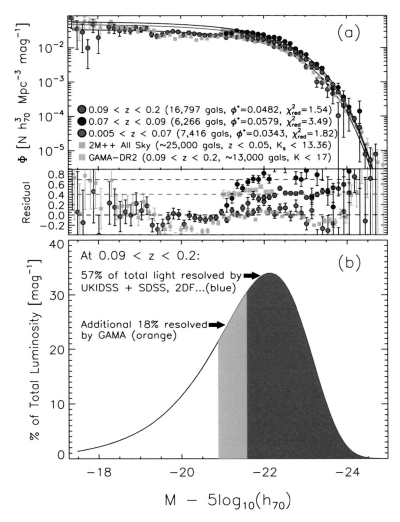

Figure 1. (a) The UKIDSS-LAS $K-$band luminosity function split into three redshift ranges: $0.005 < z < 0.07$ (red), $0.07 < z < 0.09$ (green), and $0.09 < z < 0.2$ (blue). We separate the redshift range $0.07 < z < 0.09$ to demonstrate that the excess at $z > 0.09$ is not due to the Sloan Great Wall or the other over-densities we observe at higher declination in this redshift range. Here we fit ϕ^* in each redshift range with α and M^* fixed (as described in Keenan *et al.* 2013). We list the $\chi^2_{\rm red}$ values for each redshift range in the figure, and note that relatively good fits can be obtained using fixed LF shape parameters and letting the normalization (ϕ^*) vary as a function of redshift. Residuals are shown relative to the low-redshift normalization. Here we also include our own analysis of the 2M++ catalog (all sky, $K_{\rm s,AB} < 13.36$, $\sim 25,000$ galaxies). We have fit the normalization of the LF derived from the 2M++ catalog (LF and residuals shown as light blue squares) with the same shape parameters as for the UKIDSS sample. To extend the faint end of the $z > 0.09$ LF, we use the GAMA redshift sample ($K_{\rm AB} < 17$, shown as orange squares). (b) The relative contribution to the total luminosity density (% per magnitude) as a function of absolute magnitude. The blue shaded region shows the range of magnitudes covered by the UKIDSS sample at $z > 0.09$. In orange we show the extra fraction of light resolved by extending the faint end of the LF with the GAMA sample. This demonstrates that, at $z > 0.09$, we are making a robust measurement of the peak of the luminosity density distribution (occurring at $M \sim M^*$) and resolving $\sim 75\%$ of the total light.

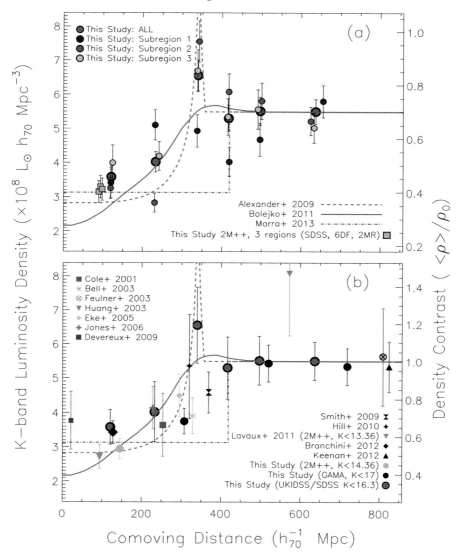

Figure 2. K−band luminosity density as a function of comoving distance. (a) Our measured K−band luminosity density for the full sample (red circles) versus different directions on the sky (green, blue, and orange circles corresponding to different subregions in the UKIDSS sample). Light blue squares indicate the K−band luminosity density we measure in three different directions (SDSS, 6DFGRS, and 2MR regions) using the 2M++ all sky catalog compiled by Lavaux & Hudson (2011). (b) Our measured K−band luminosity density for the full sample (red circles) as a function of comoving distance compared with other studies from the literature. Our estimate of the $K_{s,AB} < 14.36$ luminosity density from the 2M++ catalog (SDSS and 6DFGRS regions only) is shown as a light blue circle. Our estimates in three redshift bins for the GAMA survey only ($K_{AB} < 17$, same methods as for the UKIDSS sample) are shown as black circles. The density contrast, $\langle \rho \rangle / \rho_0$, is displayed on the right-hand vertical axis. The scale of the right-hand axis was established by performing an error-weighted least-squares fit (for the normalization only, not shape) of the radial density profile of Bolejko & Sussman (2011) (gray solid curve) to all the luminosity density data in panel (b). The dashed curve shows the radial density profile of Alexander *et al.* (2009). Both Alexander *et al.* (2009) and Bolejko & Sussman (2011) claimed these density profiles can provide for good fits to the SNIa data without dark energy. The dash-dot curve shows the scale and amplitude of the "Hubble bubble" type perturbation that Marra *et al.* (2013) would require to explain the discrepancy between local measurements of the Hubble constant and those inferred by Planck.

2. Our Study and its Implications

In two previous papers (Keenan *et al.* 2010, 2012), we investigated large-scale inhomogeneity in the local universe. These studies were inspired, in part, by the fact that a number of recent cosmological modeling efforts have shown that the observational phenomena typically used to infer an accelerating expansion, namely the "dimming" of type Ia supernovae, can be equally well fit by invoking a large local under-density instead of dark energy (e.g., Alexander *et al.* 2009; Bolejko & Sussman 2011). Other models (Marra *et al.* 2013) have shown that, even in the context of a universe dominated by dark energy, a large local under-density could explain the apparent discrepancy between local measurements of the Hubble constant and those inferred by the Planck team (Planck Collaboration *et al.* 2013).

In the study presented here, we investigated the $K-$band luminosity density (as a proxy for stellar mass density) as a function of distance from our position in the local universe. To accomplish this we combined photometry from the UKIRT Infrared Deep Sky Large Area Survey (UKIDSS-LAS, Lawrence *et al.* 2007) and the Two Micron All Sky Survey Extended Source Catalog (2MASS-XSC, Skrutskie *et al.* 2006) with redshifts from the the Sloan Digital Sky Survey (SDSS, York *et al.* 2000), the Two-degree Field Galaxy Redshift Survey (2dFGRS, Colless *et al.* 2001), the Galaxy And Mass Assembly Survey (GAMA, Driver *et al.* 2011) the Two Micron Redshift Survey (2MRS, Erdoğdu *et al.* 2006), and the Six Degree Field Galaxy Redshift Survey (6DFGS, Jones *et al.* 2009).

Key results from our study (Keenan *et al.* 2013) are presented in the figures below. In Figure 1a, we show our measurement of the $K-$band luminosity function using our compiled sample in three separate redshift ranges (with residuals showing the normalization offset between these). In Figure 1b, we show the relative contribution to the total light as a function of absolute magnitude to demonstrate that we are resolving $\sim 75\%$ of the total light at $z < 0.2$. In Figure 2a, we show a comparison of our estimate of the total luminosity density in different directions on the sky. In Figure 2b, we show our results compared to other studies from the literature and models. We conclude that the local universe appears under-dense on a scale and amplitude sufficient to introduce significant biases into local measurements of the expansion rate, and, according to some models, to cause what looks like an accelerating expansion even when no dark energy is present.

References

Alexander, S., Biswas, T., Notari, A., & Vaid, D. 2009, *JCAP*, 9, 25
Bolejko, K. & Sussman, R. A. 2011, *Physics Letters B*, 697, 265
Colless, M., *et al.* 2001, *MNRAS*, 328, 1039
Driver, S., *et al.* 2011, *MNRAS*, 413, 971
Erdoğdu, P., *et al.* 2006, *MNRAS*, 373, 45
Jones, D. H., *et al.* 2009, *MNRAS*, 399, 683
Keenan, R. C., Barger, A. J., & Cowie, L. L. 2013, *ApJ*, 775, 62
Keenan, R. C., *et al.* 2012, *ApJ*, 754, 131
Keenan, R. C., *et al.* 2010, *ApJS*, 186, 94
Lavaux, G. & Hudson, M. J. 2011, *MNRAS*, 416, 2840
Lawrence, A., *et al.* 2007, *MNRAS*, 379, 1599
Marra, V., *et al.* 2013, *Physical Review Letters*, 110, 241305
Planck Collaboration *et al.* 2013, ArXiv e-prints
Skrutskie, M. F., *et al.* 2006, *AJ* 131, 1163
York, D. G., *et al.* 2000, *AJ* 120, 1579

The Zeldovich Universe:
Genesis and Growth of the Cosmic Web
Proceedings IAU Symposium No. 308, 2014
R. van de Weygaert, S. Shandarin, E. Saar & J. Einasto, eds.

ⓒ International Astronomical Union 2016
doi:10.1017/S1743921316010036

Identification of
Extremely Large Scale Structures in SDSS-III

Shishir Sankhyayan[1], J. Bagchi[2], P. Sarkar[3], V. Sahni[2] and J. Jacob[4]

[1]Indian Institute of Science Education and Research, Pune, India
email: `shishir.sankhyayan@students.iiserpune.ac.in`

[2]Inter University Center for Astronomy and Astrophysics, Pune, India
[3]Tata Institute of Fundamental Research, Mumbai, India
[4]Newman College, Kerela, India

Abstract. We have initiated the search and detailed study of large scale structures present in the universe using galaxy redshift surveys. In this process, we take the volume-limited sample of galaxies from Sloan Digital Sky Survey III and find very large structures even beyond the redshift of 0.2. One of the structures is even greater than 600 Mpc which raises a question on the homogeneity scale of the universe. The shapes of voids-structures (adjacent to each other) seem to be correlated, which supports the physical existence of the observed structures. The other observational supports include galaxy clusters' and QSO distribution's correlation with the density peaks of the volume limited sample of galaxies.

Keywords. Cosmology: large scale structures

1. Introduction

It is very well known (from observations) that galaxies are distributed in the universe in the form of "cosmic web" (or cosmic network). The cosmic web consists of voids (regions of least density of galaxies), walls (slightly higher density planar regions of galaxies), filaments (regions of cross-sections of walls) and super-clusters (regions of intersections of filaments). Galaxy redshift surveys can be used to construct this three dimensional distribution. Comparison of these observations with the theoretical predictions decides to what extent the theory is correct. In this work, we examines the SDSS-III data and try to find extremely large coherent structures.

2. Data and Methods of Analysis

We made a volume limited sample of a region of space from the spectroscopic galaxy sample of SDSS III. The selected region spans 30° in Right Ascension and 2.5^o in Declination. The redshift is choosen to be between 0.25 and 0.42. The final volume limited sample contains 4156 galaxies.

Since the distribution has a very small extent in declination, we neglect the declination axis and proceed with two dimensional analysis to find the coherent structures in this region. We have used two different methods to identify the clusters within this field. First, smoothed density using cloud in cells method, in which two dimensional galaxy distribution is first converted to density field on a grid. The mean density of the field is $\bar{\rho}$. Then density field is smoothened. Over-dense regions/cells are defined as the cells with densities $\rho > \bar{\rho}$ and voids as cells with densities $\rho \leqslant 0.2\bar{\rho}$. The friend-of-friend (FoF) algorithm is then applied to extract the clusters/groups.

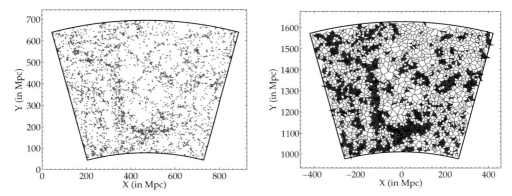

Figure 1. Left Panel: Volume limited sample of 4156 galaxies in comoving coordinates. Right Panel: Voronoi tessellation of the distribution of galaxies. Filled cells have $\rho > \bar{\rho}$. The largest structure's end to end length is $> 600 Mpc$.

The second method uses density distribution using Voronoi Tesselation, in which two dimensional galaxy distribution is first converted to density field using voronoi tessellation. The mean density of the field is $\bar{\rho}$. Over-dense regions/cells are defined as cells having densities, $\rho > \bar{\rho}$. The FoF algorithm is then applied to extract the clusters/groups.

3. Results, GMBCG Clusters and QSOs in the Field

After applying the two methods to extract coherent structures, we get an extremely large and linear structure which has an end to end linear length greater than 600 Mpc. Figure 1 shows the volume limited sample (left) and density field of Voronoi tesselation (right). The right panel shows the largest coherent structure, in an inverted S-shape.

We also compare the positions of the known clusters which have their spectroscopic redshifts available from GMBCG cluster catalog and the QSOs (from SDSS-III) in the field. A strong correlation between density peaks and cluster positions can be easily seen. The QSOs distribution also traces the largest structure very well. These correlations suggest that the largest structure found is composed of smaller clusters and is actually a physical structure.

4. Discussion and Conclusion

Two different methods give the same result - an extremely large structure of galaxies which has a length scale greater than 600 Mpc. The strong correlation of galaxy clusters with high density regions of the largest structure and the distribution of QSOs tracing the structure strongly suggest that this is an extremely dense structure. The size of this structure clearly raises the question on the homogeneity scale of the universe. There are other big structures near this extremely large structure. We will analyze those in our further studies.

References

Hao, J., McKay, T. A., Koester, B. P., *et al.* 2010, *ApJS*, 191, 254

The Zeldovich Universe:
Genesis and Growth of the Cosmic Web
Proceedings IAU Symposium No. 308, 2014
R. van de Weygaert, S. Shandarin, E. Saar & J. Einasto, eds.
© International Astronomical Union 2016
doi:10.1017/S1743921316010048

Constraints on radio source clustering towards galaxy clusters: application for cm-wavelength simulations of blind sky surveys.

Bartosz Lew

Toruń Centre for Astronomy, Nicolaus Copernicus University,
ul. Gagarina 11, 87-100 Toruń, Poland
email: `blew@astro.uni.torun.pl`

Abstract. We derive constraints on radio source clustering towards *Planck*-selected galaxy clusters using the NVSS point source catalog. The constraint can be used for making a more realistic Sunyaev-Zeldovich effect (SZE) mocks, calculating predictions of detectable clusters count and for quantifying source confusion in radio surveys.

Keywords. SZE, cosmological simulations, galaxy clusters, radio surveys

1. Introduction

Galaxy clusters are becoming observationally useful probes of cosmological models. In future thousands galaxy clusters will be detected via the SZE. Radio sources tend to correlate with galaxy cluster directions and constitute an important SZE contamination. The degree of the correlation and its redshift dependence remain uncertain due to poor statistics of faint radio source populations and insofar small SZE galaxy cluster samples. In Coble *et al.* (2007) a constraint on the source clustering was derived using the 28.5-GHz survey yielding the radio source overabundance towards galaxy cluster centres a factor of 8.9 higher as compared to the clusters outskirts.

In preparation for the cm-wavelength sky surveys planned for RT32/OCRA-f (One Centimetre Receiver Array installed on 32-m radio telescope in Toruń, Poland) and possibly for the envisaged Hevelius (a 100-m radio telescope) we calculated a possible scientific output from such blind surveys in terms of the number of detectable point sources and SZE detectable galaxy clusters using hydrodynamic simulations of large scale structure formation (Lew *et al.* (2014)). In that work we used the early Planck cluster sample (henceforth ESZ) (Planck Collaboration *et al.* (2011)) and we have shown that the point source clustering properties may significantly alter SZE counts predictions depending on the telescope beamwidth and observing frequency. In this report we revisit the point source clustering properties towards galaxy clusters using the extended cluster sample (henceforth PSZ) described in Planck Collaboration *et al.* (2014).

2. Point source clustering from NVSS and Planck

In Lew *et al.* (2014) we introduced a simple statistic to quantify the clustering of radio sources towards galaxy clusters as a function of angular distance $\theta_{\max}$ from cluster centres:

$$\rho_N(\theta_{\max}) = \frac{1}{\pi \theta_{\max}^2 N_0} \int_0^{\theta_{\max}} \frac{\partial N(\theta)}{\partial \theta} d\theta \approx \frac{1}{\pi \theta_{\max}^2 N_0} \sum_i A_1(\theta_i) \qquad (2.1)$$

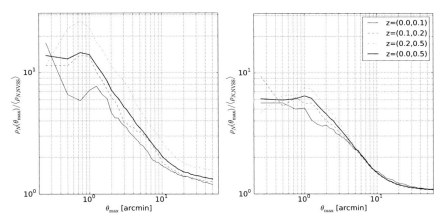

Figure 1. NVSS-normalised point source overdensity in ESZ (*left*) and PSZ (*right*) samples as a function of angular distance from the cluster centres. Individual lines represent overdensities for sub-samples selected according to the clusters redshift ranges.

where $\rho_N(\theta_{\max})$ is the cumulative solid-angle source number density, N_0 is the total number of clusters in the sample and $A_1(\theta_i) = 1$ if the radio source is within the angular distance $\theta_{\max}$ from its associated cluster's centre and $A_1 = 0$ otherwise. The summation extends over all radio sources.

We apply this statistic to the ESZ and PSZ samples which we cross-correlated with the 1.4 GHz NVSS radio source catalogue (Condon *et al.* (1998)) out to $60'$ form the cluster centres. In this cross-correlation study for ESZ (PSZ) sample 142 (993) out of 189 (1227) clusters were used. In Fig. 1 we plot the result relative to the NVSS average value of source density: $\langle \rho_{N,\mathrm{NVSS}} \rangle \approx 0.0135\,\mathrm{arcmin}^{-2}$ and with $N_0 = 1$.

The point source overdensity relative to the cluster outskirts is clearly detected in the PSZ sample as well, but its peak value is systematically lower from the values inferred from ESZ sample in all redshift bins. It ranges between 5 and 10 within the innermost $1'$. These lower values should be expected given that PSZ contains clusters yielding lower SZE flux densities, having lower masses and hence a lower source richness. A more detailed analysis and taking account of redshift space selections of radio sources (with redshifts determined via eg. optical identification) is needed to derive mass-richness scaling relations, while future wide-area cm-wavelength surveys are required to understand a possible spectral dependence of the clustering. The clustering constraints are useful in making mock surveys more realistic especially when deriving survey confusion limits and the expected SZE cluster counts.†

References

Coble, K., *et al.* 2007, *AJ* 897, 134 [astro-ph/0608274]
Condon, J. J., *et al.* 1998, *AJ* 1693-1716, 115
Lew, B., *et al.* 2014, *submitted to JCAP* [arxiv:1410:3660]
Planck Collaboration: Ade, P. A. R., *et al.* 2011, *A&A* A8, 536 [arxiv:1101.2024]
Planck Collaboration: Ade, P. A. R., *et al.* 2014, *A&A accepted* [arxiv:1303.5062]

† This work was financially supported by the Polish National Science Centre through grant DEC-2011/03/D/ST9/03373. A part of this project has made use of "Program Obliczeń WIElkich Wyzwań nauki i techniki" (POWIEW) computational resources (grant 87) at the Poznań Supercomputing and Networking Center (PSNC).

CHAPTER 5.

Megaparsec Velocity Flows

Birdeye's view of the IAU Symposium 308 Banquet.

The White Hall in the 16th century
House of the Brotherhood of the Black Heads.

The Zeldovich Universe:
Genesis and Growth of the Cosmic Web
Proceedings IAU Symposium No. 308, 2014
R. van de Weygaert, S. Shandarin, E. Saar & J. Einasto, eds.

© International Astronomical Union 2016
doi:10.1017/S174392131601005X

Cosmicflows-2

R. Brent Tully[1], Hélène M. Courtois[2], Yehuda Hoffman[3] and Daniel Pomarède[4]

[1]Institute for Astronomy, University of Hawaii,
Honolulu, HI, 96822, USA. email: tully@ifa.hawaii.edu

[2]University of Lyon 1/CNRS/IN2P3, Lyon, France

[2]Hebrew University of Jerusalem, 91904 Jerusalem, Israel

[3]Institut de Recherche sur les Lois Fondamentales de l'Univers, CEA/Saclay, 91191
Gif-sur-Yvette, France

Abstract. A compendium of over 8000 galaxy distances has been accumulated. Distance measurements permit the separation of observed velocities into cosmic expansion and peculiar velocity components. Only the radial component of peculiar velocities can be measured and individual errors are large, but a Wiener Filter procedure permits the reconstruction of three-dimensional motions and the density field that is responsible for these motions. A coherent flow pervades the entire domain of $\pm 15,000$ km/s. Techniques are discussed for the separation of local and tidal components of the flow. Laniakea supercluster is identified as a region of contiguous infalling flows.

Keywords. Galaxies: distances and redshifts; peculiar velocities; dark matter; large scale structure of galaxies.

1. Distances

The overarching goal of the Cosmic Flows program is to map the distribution of matter in the local region of the universe from the departures from cosmic expansion engendered by the matter. The overall program has many elements that have been discussed in detail in other publications. The intent of this brief article is to provide a roadmap through the labyrinth of activities.

A primary product of the program is distances to individual galaxies. If the interest is in maps of peculiar velocities, the radial component of deviations from Hubble expansion, then only relative distances are needed. One needs to cancel the overall expansion with a value of the Hubble Constant appropriate to the distance measurements. A zero-point scale error factors out.

Of course, it makes sense to match as well as possible to the best information available on all scales. The expectation has been raised that the Hubble Constant can be determined with an accuracy of a few percent (Riess *et al.* 2011; Freedman *et al.* 2012). It does not take too great a memory to recall a time when the value of H_0 was very contentious. Figure 1 is offered in the spirit of a warning. The plot shows a recent version of the correlation between galaxy luminosity and rotation rate at H band. Four galaxies are identified by name. Prior to circa 1990, the absolute calibration depended on just these galaxies. All four of these galaxies lie within 1σ of the mean relation defined by current much larger samples. Yet they all happen to lie fainter than the mean correlation at their observed linewidths. But 1σ amounts to a 10% offset, the difference between H_0 in the mid 80's (Pierce & Tully 1988) and H_0 in the mid 70's. The most recent calibration of the luminosity − linewidth correlation combines luminosity information obtained in the satellite infrared with optical photometry, uses 36 absolute calibrators, and is linked to

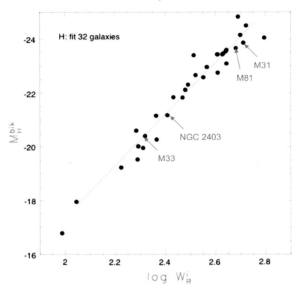

Figure 1. Luminosity-linewidth correlation for galaxies with either Cepheid or tip of the red giant branch distances. Prior to early 1990's, only the 4 identified galaxies had known Cepheid distances.

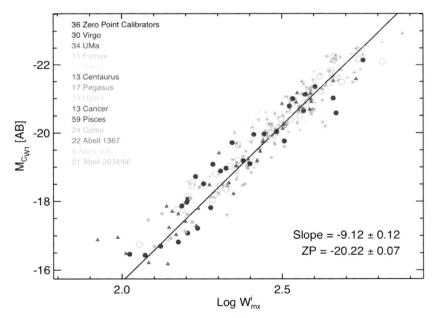

Figure 2. Luminosity–linewidth relation using WISE photometry at 3.4 μm including an adjustment for a color term. A slope template is created from a sample of galaxies in 13 clusters and the zero point is established by 36 nearby galaxies with Cepheid or TRGB distances.

the SNI*a* methodology which allows H_0 to be determined at distances well beyond the domain affected by deviant motions. This procedure leads to the measurement $H_0 = 74.4$ km/s/Mpc with a 68% probability uncertainty of 4% (Neill *et al.* 2014).

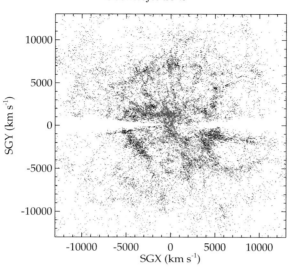

Figure 3. Projection of the distribution of galaxies on the supergalactic equator. All galaxies from the 2MASS $K < 11.75$ redshift survey with SGZ $< \pm 4000$ km/s in black, galaxies with distance measures in Cosmicflows-1 in red, and galaxies with distance measures in Cosmicflows-2 in green. The plane of our Galaxy lies at SGY=0 and causes the wedges of evident incompletion.

The initial release from the program, renamed Cosmicflows-1 (Tully *et al.* 2008), has been superceded by Cosmicflows-2 (Tully *et al.* 2013). Six primary methodologies have been combined for the measurement of distances: one based on the Cepheid period−luminosity relation, a second on the luminosity of the tip of the red giant branch, the third on the properties of surface brightness fluctuations in systems dominated by old populations, a forth on the elliptical galaxy fundamental plane correlation, a fifth on the spiral galaxy luminosity−linewidth relation, and finally on the standard candle nature of type I*a* supernovae. Members of our collaboration have made major contributions with two of the methodologies. Cosmicflows-2 contains almost 300 tip of the red giant branch distance measurements based on imaging using Hubble Space Telescope (Jacobs *et al.* 2009); subsequent observations have increased the number to almost 400. Meanwhile roughly 4000 distances in Cosmicflows-2 are contributed by the luminosity−linewidth correlation, combining optical photometry (Courtois *et al.* 2011a) and neutral Hydrogen global profile linewidths (Courtois *et al.* 2011b). Infrared photometry at 3.6 μm obtained with Spitzer Space Telescope provided an alternative calibration (Sorce *et al.* 2013). In the next release of Cosmic Flows the infrared photometry will make a much larger contribution (Sorce *et al.* 2014; Neill *et al.* 2014). Figure 2 shows the most recent version of the luminosity−linewidth relation based on 3.4 μm photometry with WISE, the Wide-field Infrared Survey Explorer.

In all, Cosmicflows-2 provides distances for over 8000 galaxies, with roughly 6000 contributed by the luminosity−linewidth method, 1500 coming from the fundamental plane, and 1200 given by methods that afford individual distances with accuracies of 10% or better. The projection of these galaxies in supergalactic coordinates is shown in Figure 3.

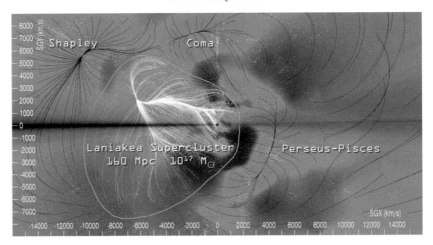

Figure 4. Laniakea supercluster of galaxies. The color palette indicates densities running from low at blue to high at red. Local flows in white lie within the region bounded by the orange outline. External flows to other attractors are in black. Important features are labeled. The horizontal dark wedge identifies the zone of obscuration.

2. Velocities and Densities

The Cosmicflows-2 data set is characterized by high density and accuracy nearby degrading to sparse sampling and large uncertainties at greater distances. The procedure of Wiener Filtering provides a Bayesian probability that the data fits a prior model consistent with a WMAP or Planck power spectrum assuming initial Gaussian fluctuations. Constrained realizations sample the statistical scatter about the mean Wiener Filter field. The reconstruction is only valid in the linear regime but this approximation holds down to scales of a few Mpc in the velocity field. At large scales, coherence in velocity flows carry information on tidal influences significantly beyond the range of distance measurements.

An initial analysis with Cosmicflows-1 data (Courtois *et al.* 2012, 2013) has been followed with a Cosmicflows-2 study (Tully *et al.* 2014). A coherent flow is evident across the entire domain of observations extending $\pm15,000$ km/s, with the Shapley Concentration at the evident terminus. The extensive nature of the flow results in the unsatisfactory situation that there is still no resolution of the extent of the influences responsible for our motion with respect to the cosmic microwave background. In the study reported in Nature the emphasis was on intermediate scale structures. By giving consideration to the reconstructed densities in a restricted region, local flows due to that distribution of matter can be determined. Tidal components from regions external to the restricted region are given by the vector subtraction of the local velocities from the global velocities. This procedure permits the isolation of individual basins of attraction. It is of great interest that the current velocity field information permits a definition of the full extent of the basin of attraction that includes our galaxy. A view in Figure 4 of a slice in the supergalactic equatorial plane illustrates the structure that is identified.

Our galaxy is found to lie in a basin of attraction that extends across roughly 160 Mpc diameter in all three orthogonal directions, a volume enclosing 10^{17} $M_\odot$ and including 100,000 large galaxies. The densest part is in the vicinity of the Norma and Centaurus clusters at roughly the location of what has been called the Great Attractor region. The outer reaches tend to lie in voids but there are interesting locations where apparent filaments run between adjacent basins of attraction. The filaments are shearing at the juncture between the entities. The name Laniakea supercluster has been given to our home large scale structure (in Hawaiian: 'lani' = heaven; 'aiea' = extremely large). Our Milky Way galaxy lies at the periphery of Laniakea, near the boundary with the adjacent Perseus–Pisces structure.

Acknowledgements. Among the many collaborators who have participated, thanks are especially due to Andy Dolphin, Rick Fisher, Stefan Gottlöber, Philippe Héraudeau, Brad Jacobs, Tom Jarrett, Igor Karachentsev, Barry Madore, Dmitry Makarov, Lidia Makarova, Don Neill, Luca Rizzi, Mark Seibert, Ed Shaya, Jenny Source, and Po-Feng Wu. The name Laniakea was suggested by Nawa'a Napoleon. Support has been provided by the US National Science Foundation award AST09-08846, several awards from the Space Telescope Science Institute in connection with Hubble Space Telescope observations, an award from the Jet Propulsion Lab for observations with Spitzer Space Telescope, and NASA award NNX12AE70G for the analysis of data from the Wide-field Infrared Survey Explorer. Additional support was provided by the Israel Science Foundation and the Lyon Institute of Origins and Centre National de la Recherche Scientifique.

References

Courtois, H. M., Tully, R. B., & Héraudeau, P. 2011a, *MNRAS*, 415, 1935
Courtois, H. M. *et al.* (2011b), *MNRAS*, 414, 2005
Courtois, H. M. *et al.* (2012), *ApJ*, 744, 43
Courtois, H. M. *et al.* (2013), *AJ*, 146, 69
Freedman, W. L. *et al.* 2012, *ApJ*, 758, 24
Jacobs, B. A. *et al.* 2009, *AJ*, 138, 332
Neill, J. D. *et al.* 2014, *ApJ*, 792, 129
Pierce, M. J. & Tully, R. B. 1988, *ApJ*, 330, 579
Riess, A. G. *et al.* 2011 *ApJ*, 730, 119
Sorce, J. G. *et al.* 2013 *ApJ*, 765, 94
Sorce, J. G. *et al.* 2014 *MNRAS*, 444, 527
Tully, R. B. *et al.* 2008 *ApJ*, 676, 184
Tully, R. B. *et al.* 2013 *AJ*, 146, 86
Tully, R. B, Courtois, H. M, Hoffman, Y., & Pomarède, D. 2014 *Nature*, 513, 71

The Zeldovich Universe:
Genesis and Growth of the Cosmic Web
Proceedings IAU Symposium No. 308, 2014
R. van de Weygaert, S. Shandarin, E. Saar & J. Einasto, eds.

© International Astronomical Union 2016
doi:10.1017/S1743921316010061

Re-examination of Large Scale Structure & Cosmic Flows

Marc Davis[1] and Adi Nusser[2]

[1]Departments of Astronomy & Physics, University of California at Berkeley, CA 94720
email: *mdavis@berkeley.edu*

[2]Physics Department and the Asher Space Science Institute-Technion,
Haifa 32000, Israel
email: *adi@physics.technion.ac.il*

Abstract. Comparison of galaxy flows with those predicted from the local galaxy distribution ended as an active field after two analyses came to vastly different conclusions 25 years ago, but that was due to faulty data. All the old results are therefore suspect. With new data collected in the last several years, the problem deserves another look. The goal is to explain the 640 km/s dipole anisotropy of the CMBR. For this we analyze the gravity field inferred from the enormous data set derived from the 2MASS collection of galaxies (Huchra *et al.* 2005), and compare it to the velocity field derived from the well calibrated SFI++ Tully-Fisher catalog (Springob *et al.* 2007). Using the "Inverse Method" to minimize Malmquist biases, within 10,000 km/s the gravity field is seen to predict the velocity field (Davis *et al.* 2011) to remarkable consistency. This is a beautiful demonstration of linear perturbation theory and is fully consistent with standard values of the cosmological variables.

1. Comparison of Observed Velocity Field with Gravitational Field

This is a conference proceeding where I summarize several recent publications on peculiar velocities. In particular the brief discussion is based on Davis *et al.* (2011) (hereafter D11), Nusser & Davis (1994) (hereafter ND94), and Davis *et al.* (1996) (hereafter DNW96). Interested parties will find complete references therein.

The analysis of Davis *et al.* (2011) fits the peculiar velocity field given by the SFI++ Tully-Fisher whole sky sample of 2830 galaxies with redshifts $cz < 10,000$ km/s (Springob *et al.* 2007) to a set of orthogonal polynomials by means of an inverse Tully-Fisher (ITF) procedure. The peculiar velocity field derived from this sample is then compared to the gravity field from the largest whole sky redshift survey, the 2MRS survey (Huchra *et al.* 2005). This catalog is K band selected 2MASS galaxies and has been extended to 43,500 galaxies to $K \leqslant 11.75$ and $|b| > 5°$ or $|b| > 10°$ near the galactic center. In our lifetime, the redshift catalog and derived gravity field is unlikely to improve enough to bother, since it is not the limiting noise. For improvements in the future, one should work on enlarging the TF data.

Peculiar velocities are unique in that they provide explicit information on the three dimensional mass distribution, and measure mass on scales of $10 - 60h^{-1}$ Mpc, a scale untouched by alternative methods. Here we will be concerned with a comparison of the observed peculiar velocities on the one hand and the velocities derived from the fluctuations in the galaxy distribution on the other. The basic physical principle behind this comparison is simple. The large scale flows are almost certainly the result of the process of gravitational instability with overdense regions attracting material, and underdense regions repelling material. Initial conditions in the early universe might have been somewhat chaotic, so that the original peculiar velocity field (i.e. deviations from Hubble flow) was uncorrelated with the mass distribution, or even contained vorticity. But those components of the velocity field which are not coherent with the density fluctuations will adiabatically decay as the Universe expands, and so at late times one expects the velocity field to be aligned with the gravity field, at least in the limit of small amplitude fluctuations (Peebles 1980; Nusser *et al.* 1991). In the linear regime, this relation implies a simple proportionality between the gravity field $\mathbf{g}$ and the velocity field $\mathbf{v_g}$, namely $\mathbf{v_g} \propto \mathbf{g}\,t$ where the only possible time t is the Hubble time. The exact expression depends on the mean cosmological density parameter Ω and is given by Peebles (1980),

$$\mathbf{v_g}(r) = \frac{2f(\Omega)}{3H_0\Omega}\mathbf{g}(\mathbf{r}) \ . \tag{1.1}$$

Given complete knowledge of the mass fluctuation field $\delta_\rho(\mathbf{r})$ over all space, the gravity field $\mathbf{g}(\mathbf{r})$ is

$$\mathbf{g}(\mathbf{r}) = G\bar{\rho}\int d^3\mathbf{r}'\delta_\rho(\mathbf{r}')\frac{\mathbf{r}' - \mathbf{r}}{|\mathbf{r}' - \mathbf{r}|^3} \ , \tag{1.2}$$

where $\bar{\rho}$ is the mean mass density of the Universe. If the galaxy distribution at least approximately traces the mass on large scale, with linear bias b between the galaxy fluctuations δ_G and the mass fluctuations (i.e. $\delta_g = b\delta_\rho$), then from (1.1) and (1.2) we have

$$\mathbf{v_g}(r) = \frac{H_0\beta}{4\pi\bar{n}}\sum_i\frac{1}{\phi(r_i)}\frac{\mathbf{r_i} - \mathbf{r}}{|\mathbf{r_i} - \mathbf{r}|^3} + \frac{H_0\beta}{3}\mathbf{r} \ , \tag{1.3}$$

where $\bar{n}$ is the true mean galaxy density in the sample, $\beta \equiv f(\Omega)/b$ with $f \approx \Omega^{0.55}$ the linear growth factor (Linder 2005), and where we have replaced the integral over space with a sum over the galaxies in a catalog, with radial selection function $\phi(r)$†. The second term is for the uniform component of the galaxy distribution and would exactly cancel the first term in the absence of clustering within the survey volume. Note that the result is insensitive to the value of H_0, as the right hand side has units of velocity. We shall henceforth quote all distances in units of $\mathrm{km\,s^{-1}}$. The sum in equation (1.3) is to be computed in real space, whereas the galaxy catalog exists in redshift space. As we shall see in §2, the modified equation, which includes redshift distortions, maintains a dependence on Ω and b through the parameter β. Therefore, a comparison of the measured velocities of galaxies to the predicted velocities, $\mathbf{v_g}(r)$, gives us a measure of β. Further, a detailed comparison of the flow patterns addresses fundamental questions regarding the way galaxies trace mass on large scales and the validity of gravitational instability theory.

2. Methods

In this section we outline our method described in ND94, ND95 and DNW96 for deriving the smooth peculiar velocities of galaxies from an observed distribution of galaxies in redshift space and, independently, from a sample of spiral galaxies with measured circular velocities η and apparent magnitudes m.

Here we restrict ourselves to large scales where linear-theory is applicable. We will use the method of ND94 for reconstructing velocities from the 2MRS. This method is particularly convenient, as it is easy to implement, fast, and requires no iterations. Most

---

† $\phi(r)$ is defined as the fraction of the luminosity distribution function observable at distance r for a given flux limit; see (e.g. Yahil *et al.* 1991).

importantly, this redshift space analysis closely parallels the ITF estimate described below. We next present a very brief summary of the methodology.

We follow the notation of DNW96. The comoving redshift space coordinate and the comoving peculiar velocity relative to the Local Group (LG) are, respectively, denoted by $\boldsymbol{s}$ (i.e. $s = cz/H_0$) and $\boldsymbol{v}(\boldsymbol{s})$. To first order, the peculiar velocity is irrotational in redshift space (Chodorowski & Nusser 1999) and can be expressed as $\boldsymbol{v}_g(\boldsymbol{s}) = -\boldsymbol{\nabla}\Phi(\boldsymbol{s})$ where $\Phi(\boldsymbol{s})$ is a potential function. As an estimate of the fluctuations in the fractional density field $\delta_0(\boldsymbol{s})$ traced by the discrete distribution of galaxies in redshift space we consider,

$$\delta_0(\boldsymbol{s}) = \frac{1}{(2\pi)^{3/2}\bar{n}\sigma^3} \sum_i \frac{w(L_{0i})}{\phi(s_i)} \exp\left[-\frac{(\boldsymbol{s}-\boldsymbol{s}_i)^2}{2\sigma^2}\right] - 1. \tag{2.1}$$

where $\bar{n} = \sum_i w(L_{0i})/\phi(s_i)$ and w weighs each galaxy according to its estimated luminosity, L_{0i}. The 2MRS density field is here smoothed by a gaussian window with a redshift independent width, $\sigma = 350 \text{ km s}^{-1}$. This is in contrast to DNW96 where the *IRAS* density was smoothed with a width proportional to the mean particle separation. The reason for adopting a constant smoothing for 2MRS is its dense sampling which is nearly four time higher than *IRAS* . We emphasize that the coordinates s are in *observed redshift* space, expanded in a galactic reference frame. The only correction from pure redshift space coordinates is the collapse of the fingers of god of the known rich clusters prior to the redshift space smoothing (Yahil et al. 1991). Weighting the galaxies in equation (2.1) by the selection function and luminosities evaluated at their redshifts rather than the actual (unknown) distances yields a biased estimate for the density field. This bias gives rise to Kaiser's rocket effect (Kaiser 1987).

To construct the density field, equation 2.1, we volume limit the 2MRS sample to 3000 km/s, so that $\phi(s < 3000) = 1$, resulting in $\phi(s = 10000) = 0.27$ (Westover 2007). In practice, this means we delete galaxies from the 2MRS sample fainter than $M_* + 2$. Galaxies at 10,000 km/s therefore have $1/\phi = 3.7$ times the weight of foreground galaxies in the generation of the velocity field, v_g.

If we expand the angular dependence of Φ and $\delta_0(\boldsymbol{s})$ redshift space in spherical harmonics in the form,

$$\Phi(\boldsymbol{s}) = \sum_{l=0}^{\infty} \sum_{m=-l}^{l} \Phi_{lm}(s) Y_{lm}(\theta, \varphi) \tag{2.2}$$

and similarly for δ_0, then, to first order, Φ_{lm} and δ_{0lm} satisfy,

$$\frac{1}{s^2}\frac{\mathrm{d}}{\mathrm{d}s}\left(s^2\frac{\mathrm{d}\Phi_{lm}}{\mathrm{d}s}\right) - \frac{1}{1+\beta}\frac{l(l+1)\Phi_{lm}}{s^2} \tag{2.3}$$

$$= \frac{\beta}{1+\beta}\left(\delta_{0lm} - \kappa(s)\frac{\mathrm{d}\Phi_{lm}}{\mathrm{d}s}\right),$$

where

$$\kappa = \frac{\mathrm{d}\ln\phi}{\mathrm{d}s} - \frac{2}{s}\frac{\mathrm{d}\ln w(L_{0i})}{\mathrm{d}\ln L_{0i}} \tag{2.4}$$

represents the correction for the bias introduced by the generalized Kaiser rocket effect. As emphasized by ND94, the solutions to equation (2.3) for the monopole ($l = 0$) and the dipole ($l = 1$) components of the radial peculiar velocity in the LG frame are uniquely determined by specifying vanishing velocity at the origin. That is, the radial velocity field at redshift s, when expanded to harmonic $l \leqslant 1$, is not influenced by material at redshifts greater than s.

In this paper, we shall consider solutions as a function of β. Davis *et al.* (2011) also fit a second parameter, α, defining a power law form $w_i \propto L_i^\alpha$ for the galaxy weights and found $\alpha \approx 0$ was the best fit. The large-scale gravity field is best estimated if all the galaxies are equal-weighted, that is, they all have the same mass. This makes sense if you remember that each point in the 2MRS represents the mass on scales of ~ 4 Mpc.

3. Generating Peculiar Velocities

Given a sample of galaxies with measured circular velocity parameters, $\eta_i \equiv \log w_i$, linewidth w_i, apparent magnitudes m_i, and redshifts z_i, the goal is to derive an estimate for the smooth underlying peculiar velocity field. We assume that the circular velocity parameter, η, of a galaxy is, up to a random scatter, related to its absolute magnitude, M, by means of a linear *inverse* Tully-Fisher (ITF) relation, i.e.,

$$\eta = \gamma M + \eta_0. \tag{3.1}$$

One of the main advantages of inverse TF methods is that samples selected by magnitude, as most are, will be minimally plagued by Malmquist bias effects when analyzed in the inverse direction (Schechter 1980; Aaronson *et al.* 1982). We write the absolute magnitude of a galaxy,

$$M_i = M_{0i} + P_i \tag{3.2}$$

where

$$M_{0i} = m_i + 5\log(z_i) - 15 \tag{3.3}$$

and

$$P_i = 5\log(1 - u_i/z_i) \tag{3.4}$$

where m_i is the apparent magnitude of the galaxy, z_i is its redshift in units of $\mathrm{km\,s^{-1}}$, and u_i its radial peculiar velocity in the LG frame.

4. The Solution in Orthogonal Polynomials

Functions based on Y_{lm} are a poor description.of the complex flows of LSS, giving rise to correlated residuals, but with only ~ 20 numbers to describe the field, we get $\chi^2/\mathrm{dof} = 1$ when we compare the gravity and velocity fields; 25 years ago, the same comparison gave $\chi^2/\mathrm{dof} = 2$ (Davis *et al.* 1996). In the interval, the *IRAS* gravity field has been replaced by the 2MRS, but the two gravity fields are essentially identical. The TF data has been updated to the SFI++ catalogue, which makes all the difference; the old data was constructed of 4 separate catalogues and it was not uniformly calibrated.

The choice of radial basis functions for the expansion of the modes can be made with considerable latitude. The functions should obviously be linearly independent, and close to orthogonal when integrated over volume. They should be smooth and close to a complete set of functions up to a given resolution limit. Spherical harmonics and radial Bessel functions are an obvious choice, but Bessel functions have a constant radial resolution with distance whereas the measured peculiar velocities have velocity error that scales linearly with distance. We deal with this problem by choosing to make the Bessel functions a function of y instead of r by means of the transformation

$$y(r) = (\log(1 + (r/r_s)))^{1/2} \tag{4.1}$$

where $r_s = 5000$ km/s. The resulting radial functions oscillate more rapidly toward the origin then they do toward the outer limit, a physically desirable behavior.

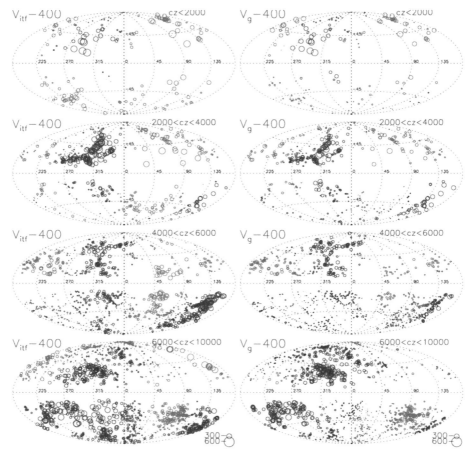

Figure 1. The derived peculiar velocities $\mathbf{v}_{itf}$ and $\boldsymbol{g}$ of SFI++ galaxies on aitoff projections on the sky in galactic coordinates. Red points have positive peculiar velocity and blue points have negative. The rows correspond to galaxies with $cz < 2000$, $2000 < cz < 4000$ $4000 < cz < 6000$ km s^{-1} and $6000 < cz < 10000$ km/s, respectively. The size of the symbols is linearly proportional to the velocity amplitude (see key to the size of the symbols given at the bottom of the figure). In order to better see the differences, a 400 km/s dipole, in the direction of the CMB dipole, has been subtracted from the $\mathbf{v}_{itf}$ and $\boldsymbol{g}$ velocities.

5. The Resulting Velocity and Gravity Fields

In the aitoff projections in Figure 1 we plot the TF peculiar velocities of the SFI++ galaxies, $\mathbf{v}_{itf}$ and the derived gravity modes, $\boldsymbol{g}$, for galaxies in redshift shells, $cz < 2000$, $2000 < cz < 4000$, $4000 < cz < 6000$, and $6000 < cz < 10000$ kms^{-1}. The projections are in galactic coordinates centered on l,b = 0 and with b = 90 at the top. Figure 1 is shown with $\beta = 0.35$; the amplitude of $\boldsymbol{g}$ is almost linear with β, giving a powerful diagnostic. Our best fit is $\beta = 0.33 \pm 0.04$. The key point is to note that the residuals are small for the entire sky and have amplitude that is constant with redshift. The amplitude and coherence of the residuals $\mathbf{v}_{itf} - \boldsymbol{g}$ is the same as for the mock catalogs in figure 2, where for example the lower picture shows $\mathbf{v}_{itf} - \boldsymbol{g}$ for real and mock catalogs. The mocks show the viability of the full procedure (Davis *et al.* 2011).

Figure 1 says it all – the agreement between the inferred velocity field and the gravitational expectations is spectacularly good at all distances. These two fields could have

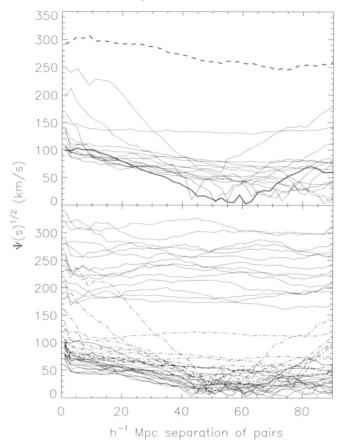

Figure 2. *Top:* The velocity correlation of the real data and 15 mock catalogs. The dashed red and solid red curves curve are, respectively, the correlations of v_{itf} and $\mathbf{v}_{itf} - \mathbf{g}$ in the real data. This plot shows that the 20 mode expansion removes virtually the entire velocity field. The blue lines are each correlations of $v_{itf} - \mathbf{g}$ for the mock catalogs. *Bottom:* Velocity correlations for 15 mock catalogs. The red curves are the velocity of $\mathbf{v}_{itf}$, the dot-dashed curves show the correlation of $(\mathbf{v}_{true} - \mathbf{v}_g)$, and the blue curves correspond to $\mathbf{v}_{true} - \mathbf{v}_{itf})$. Both $\mathbf{v}_{true}$ and $\mathbf{v}_g$ are first smoothed with the 20 mode expansion before the autocovariance is computed. Note that the correlation of $\mathbf{v}_{itf} - \mathbf{g}$ is only slightly worse than the correlation of $\mathbf{v}_{true} - \mathbf{v}_{gs}$, showing that the velocity reconstruction dominates the errors. Note also that we are plotting the square root of the velocity correlation Ψ.

been very discrepant; the only parameter of the fitting is β. The flow field is complex, as galaxies respond to their local gravity field. All the argumentation of 25 years ago is irrelevant. Note that we are only using 20 numbers to describe the local field, thus smoothing out the small scale velocity field.

5.1. *Residual Velocity Correlations*

The residuals, both in the real and mock data, have error fields, $\mathbf{v}_{itf} - \mathbf{g}$, that show large regions of coherence. To address the significance of these errors, we show in figure 2 the velocity correlation function (Gorski *et al.* 1989) defined as

$$\Psi(s; u) = \frac{\sum_{pairs} u_1 u_2 \cos\theta_{12}}{\sum_{pairs} \cos^2 \theta_{12}} \tag{5.1}$$

where the sum is over all pairs, 1 and 2, separated by vector distance $\mathbf{s}_{12}$ (in redshift space), θ_{12} is the angle between points 1 and 2, and u is either v_{itf} (dashed red) or $\mathbf{v}_{itf} - \boldsymbol{g}$ (red for data, blue for 15 mock catalogs), At small lags for the real data, the function $\Psi(r; \mathbf{v}_{itf} - \boldsymbol{g})$ is a factor of 3 less than $\Psi(s; \mathbf{v}_{itf})$, about the same as for the mock catalogs. Note how the large coherence of $\mathbf{v}_{itf}$ is enormously diminished in $\Psi(s > 2000\mathrm{km/s}; \mathbf{v}_{itf} - \boldsymbol{g})$. This shows that the coherence seen in the residual field, figure 2, is expected and is not a problem. The large scale drift of a sample is demonstrated by the persistent amplitude of Ψ beyond $\approx 60 - 80$ Mpc.

The bottom panel of figure 2 shows velocity correlations for 15 mock catalogs where the actual velocity $\mathbf{v}_{\mathbf{true}}$ generated in the nbody code and then smoothed with the 20 mode expansion can be compared to either $\mathbf{v}_{itf}$ or $\boldsymbol{g}$. Note that the raw velocities, $\mathbf{v}_{\mathbf{itf}}$ (red), have enormous correlation that reaches large lag, while the correlations, $(\mathbf{v}_{\mathbf{true}} - \mathbf{v}_{itf})$, (blue) are extremely small. This is because the only difference with v_{true} is the gaussian error in $\Delta\eta = .05$ that affects $\mathbf{v}_{itf}$. The blue curves show this error is not a problem, because the mode expansions are insensitive to gaussian noise in the 2500 galaxies, i.e. they are essentially perfect. This demonstrates that even though the TF noise is as large as for the actual data, the ability to find the correct flow, when characterized by only 20 numbers, is intact.

This demonstrates that the description of the full velocity field by the specification of 20 numbers, specifying the amplitude of the modes, is essentially complete.

6. Summary

- We see no evidence that the dark matter does not follow the galaxy distribution, and it is consistent with constant bias on large scales. There is no evidence for a non-linear bias in the local flows. A smooth component to the universe is not something testable with these methods.
- Linear perturbation theory appears to be adequate for the large scales tested by our method; the comparison of $\mathbf{v_p}$ and $\mathbf{g}$ is so precise as to be a stunning example of the power of linear theory!
- Our estimate of σ_8 gives the most precise value at $z \sim 0$ and is useful for tests of the growth rate and Dark Energy.
- The velocity-gravity comparison measures the acceleration on scales in the range $10 - 60$ Mpc. and since we derived a similar value of β as for clusters of galaxies, we conclude that dark matter appears to fully participate in the clustering on scales of a few Megaparsecs and larger.
- We find no evidence for large-scale flows, and the small residuals are completely consistent with LCDM (Nusser et al. 2014). Note that our analysis has not used the CMBR dipole, but we see a velocity field that is fully consistent with the CMBR dipole radiation. We see no evidence that the dipole in the CMBR is produced by anything other than our motion in the universe.
- The field of Large Scale Flows, apart from going deeper with TF data, appears to this observer to have finally reached its original goal. Remember that 25 years ago, there were no CMBR results measuring Ω_m, and the large scale flows were going to give us the long-sought answer. But the TF data of 25 years ago was not well calibrated and gave inconsistent results, so we lost ground. Now we can state that the LS flows are consistent with standard parameters.
- This finishes the study of the local velocity field, and now I can retire!

References

Aaronson, M., Huchra, J., Mould, J., Schechter, P. L., & Tully, R. B. 1982, ApJ, 258, 64

Chodorowski, M. J. & Nusser, A. 1999, *MNRAS*, 309, L30

Davis, M., Nusser, A., Masters, K. L., Springob, C., Huchra, J. P., & Lemson, G. 2011, *MNRAS*, 413, 2906

Davis, M., Nusser, A., & Willick, J. A. 1996, ApJ, 473, 22

Gorski, K. M., Davis, M., Strauss, M. A., White, S. D. M., & Yahil, A. 1989, ApJ, 344, 1

Huchra, J., Jarrett, T., Skrutskie, M., Cutri, R., Schneider, S., Macri, L., Steining, R., Mader, J., Martimbeau, N., & George, T. 2005, in Astronomical Society of the Pacific Conference Series, Vol. 329, Nearby Large-Scale Structures and the Zone of Avoidance, ed. A. P. Fairall & P. A. Woudt, 135-+

Kaiser, N. 1987, *MNRAS*, 227, 1

Linder, E. V. 2005, Physical Review D., 72, 043529

Nusser, A. & Davis, M. 1994, *ApJL*, 421, L1

Nusser, A., Dekel, A., Bertschinger, E., & Blumenthal, G. R. 1991, ApJ, 379, 6

Nusser, A., Davis, M., & Branchini, E. 2014, ApJ, 788, 157

Peebles, P. J. E. 1980, The large-scale structure of the universe (Princeton University Press)

Schechter, P. L. 1980, Astronomical Journal, 85, 801

Springob, C. M., Masters, K. L., Haynes, M. P., Giovanelli, R., & Marinoni, C. 2007, ApJ. S, 172, 599

Westover, M. 2007, PhD thesis, Harvard University

Yahil, A., Strauss, M. A., Davis, M., & Huchra, J. P. 1991, ApJ, 372, 380

The Zeldovich Universe:
Genesis and Growth of the Cosmic Web
Proceedings IAU Symposium No. 308, 2014
R. van de Weygaert, S. Shandarin, E. Saar & J. Einasto, eds.

© International Astronomical Union 2016
doi:10.1017/S1743921316010073

Cosmological parameters from the comparison of peculiar velocities with predictions from the 2M++ density field

Michael J. Hudson[1,2]**, Jonathan Carrick,**[1] **Stephen J. Turnbull**[1] **and Guilhem Lavaux,**[1,2,3,4,5]

[1] Department of Physics & Astronomy, University of Waterloo, Waterloo, ON, N2L 3G1, Canada
[2] Perimeter Institute for Theoretical Physics, 31 Caroline St. N., Waterloo, ON, N2L 2Y5, Canada
[3] CNRS, UMR7095, Institut d'Astrophysique de Paris, F-75014, Paris, France
[4] Sorbonne Universités, UPMC Univ Paris 06, UMR7095, Institut d'Astrophysique de Paris, F-75014, Paris, France
[5] Canadian Institute for Theoretical Astrophysics, University of Toronto, 60 St. George Street, Toronto, ON M5S 1A7, Canada
email: mjhudson@uwaterloo.ca

Abstract. Using redshifts from the 2M++ redshift compilation, we reconstruct the density of galaxies within 200 h^{-1} Mpc, and compare the predicted peculiar velocities Tully-Fisher and SNe peculiar velocities. The comparison yields a best-fit value of $\beta^* \equiv \Omega_m^{0.55}/b^* = 0.431 \pm 0.021$, suggesting $\Omega_m^{0.55}\sigma_{8,\mathrm{lin}} = 0.401 \pm 0.024$, in good agreement with other probes. The predicted peculiar velocity of the Local Group from sources within the 2M++ volume is 540 ± 40 km s^{-1}, towards $l = 268° \pm 4°$, $b = 38° \pm 6°$, which is misaligned by only $10°$ with the Cosmic Microwave Background dipole. To account for sources outside the 2M++ volume, we fit simultaneously for β^* and an external bulk flow in our analysis. The external bulk flow has a velocity of 159 ± 23 km s^{-1} towards $l = 304° \pm 11°$, $b = 6° \pm 13°$.

Keywords. galaxies: distances and redshifts, large-scale structure of universe, cosmological parameters, dark matter

1. Introduction

The comparison between density and velocity fields allows one to measure two important cosmological parameters. The first is $\beta^* \equiv \Omega_m^{0.55}/b^*$, where b^* is the bias of L^* galaxies. With a measurement of σ_8^*, the L^* galaxy density fluctuation in an 8 h^{-1}Mpc sphere, this can be converted into $\Omega_m^{0.55}\sigma_{8,\mathrm{lin}}$. The second parameter is the contribution to the large-scale flow arising from matter beyond the limits of the density field, V_{ext}. This is sensitive to the growth rate and matter power spectrum and on very large scales. In this contribution, we summarize results from Carrick et al. (2015). We refer the reader to that paper for full technical details.

2. 2M++ Density Field

The reconstructed galaxy density field is based on the 2M++ redshift compilation (Lavaux & Hudson 2011), which in turn is based on the 2MRS redshift survey (Huchra et al. 2012), the 6dF galaxy redshift survey (Jones et al. 2009) and the Sloan Digital Sky Survey (Abazajian et al. 2009). We correct for selection effects using the usual methods. An iterative method is used to obtain the reconstructed real-space positions of galaxies,

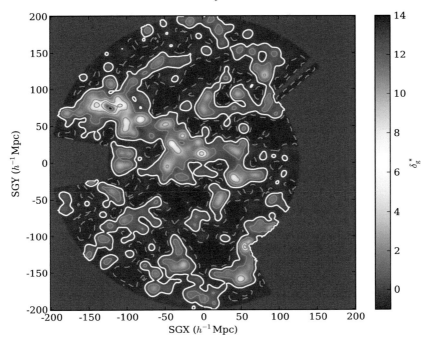

Figure 1. The Supergalactic Plane (SGZ= 0) of the 2M++ luminosity-weighted galaxy density contrast field, reconstructed with $\beta^* = 0.43$ smoothed with a Gaussian kernel of radius 4 h^{-1}Mpc. The dashed contour is $\delta_g^* = -0.5$, the bold white contour is $\delta_g^* = 0$, and successive contours thereafter increase from 1 upwards in steps of 3. The Galactic plane runs roughly along the SGY= 0 axis. The Shapley Concentration is located at (SGX, SGY) $\simeq$ (-125, 75) h^{-1}Mpc, the Virgo Supercluster directly above the LG, the Hydra-Centaurus Supercluster at (-40, 20) h^{-1}Mpc, and the Perseus-Pisces Supercluster is at (40, -30) h^{-1}Mpc. (Reproduced from Fig. 4 of "Cosmological parameters from the comparison of peculiar velocities with predictions from the 2M++ density field," , Carrick et al., MNRAS, 450, 317).

after having smoothed the density field with a Gaussian of width 4 h^{-1}Mpc. In the Appendices of Carrick et al. (2015), we show via N-body simulations that this method yields unbiased predicted peculiar velocities with a scatter of 140 km s^{-1}. Fig. 1 shows the supergalactic plane of the 2M++ galaxy density field.

We find no evidence of a large-scale underdensity within the 2M++, consistent with the results of Böhringer et al. (2015).

3. Comparison with Tully-Fisher and SNe Peculiar Velocity Data

We then compare the predicted peculiar velocities from 2M++ with peculiar velocity data from SFI++ (Springob et al. 2007, Tully-Fisher) and the "First Amendment" supernova sample (Turnbull et al. 2012). We use several methods to make the comparison: a direct method including a correction for inhomogeneous Malmquist bias, and in the case of the TF data, an inverse "VELMOD" method . We find these methods give consistent results, within the uncertainties. The best fitting β^* is 0.431 ± 0.021 with $V_{\text{ext}} = 159 \pm 23$ km s^{-1} towards $l = 304° \pm 11°$, $b = 6° \pm 13°$.

When combined with a measurement of $\sigma_{8,g}^*$, β^* can be used to constrain the degenerate parameter combination $f\sigma_8 = \beta^*\sigma_{8,g}^*$. From 2M++, we use counts in cells within radial shells and obtain the value $\sigma_{8,g}^* = 0.99 \pm 0.04$. The product of the growth factor and

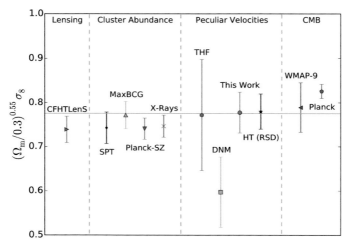

Figure 2. Comparison of $f\sigma_{8,\text{lin}}$ measured results. Values plotted above derived from weak-lensing (Kilbinger et al. 2013, CFHTLenS) and cluster abundances [Reichardt et al. (2013, SPT), Rozo et al. (2010, MaxBCG), Planck Collaboration et al. (2013, Planck-SZ), Vikhlinin et al. (2009, X-rays)] have assumed a value of $\Omega_m = 0.3$ in mapping constraints to $\Omega_m^{0.55}\sigma_8$. Results obtained through previous analyses of measured peculiar velocities are also shown [Turnbull et al. (2012, THF), Davis et al. (2011, DNM)], as well as from redshift space distortions (Hudson & Turnbull 2012, HT). CMB results are from WMAP9 and the Planck Collaboration (2015) . The horizontal line is the error-weighted mean of all values ($f\sigma_8 = 0.400 \pm 0.005$), shown here for reference. (Reproduced from Fig. 8 of "Cosmological parameters from the comparison of peculiar velocities with predictions from the 2M++ density field," , Carrick et al., MNRAS, 450, 317).

non-linear σ_8 is thus $f\sigma_8 = 0.427 \pm 0.026$. We convert our non-linear value of σ_8 to a linearized value and obtain the constraint $f\sigma_{8,\text{lin}} = 0.401 \pm 0.024$.

Our value of $f\sigma_{8,\text{lin}} = 0.40 \pm 0.02$ is in good agreement with those obtained using the same methodology, such as Turnbull et al. (2012) (0.40 ± 0.07), Pike & Hudson (2005) (0.44 ± 0.06). It is, however, in slight tension with the result of Davis et al. (2011) who found 0.31 ± 0.04. We have also compared our value $f\sigma_{8,\text{lin}}$ to constraints placed on a degenerate combination of Ω_m and σ_8 through independent means. In particular, our value is in excellent agreement with a different peculiar velocity probe, namely measurements of $f(z)\sigma_8(z)$ at different redshifts via redshift space distortions, which yield a best-fit value of $f\sigma_8 = 0.40 \pm 0.02$ (Hudson & Turnbull 2012). Fig. 2 shows a comparison between measurements of $f\sigma_{8,\text{lin}}$ by several different techniques. There is some tension between some results *e.g.* Kilbinger et al. (2013) and Planck-SZ (Planck Collaboration et al. 2013) versus Planck CMB temperature (Planck Collaboration et al. 2015). The peculiar velocity result presented here is consistent with all of these values.

4. The Large Scale Velocity Field

The value of V_{ext} is consistent with previous results on a similar scale (Turnbull et al. 2012), who found $150 \pm 43\,\text{km s}^{-1}$ towards $l = 345°$, $b = 8°$ from a comparison of the A1 SNe with the PSCz reconstruction(Branchini et al. 1999).

It is interesting to compare the predicted bulk flow in a $50\ h^{-1}\text{Mpc}$ Gaussian window with observations. The 2M++ velocity model predicts a flow of 227 ± 25 km s^{-1} towards $l = 293°$, $b = 14°$, an amplitude consistent with the cosmic variance expected in ΛCDM. This is smaller than the value of 407 ± 81 km s^{-1} towards $l = 287°$, $b = 8°$ found by Watkins et al. (2009), and 292 ± 27 km s^{-1} towards $l = 297°$, $b = 7°$ by Hong et al. (2014)

but consistent with the 249 ± 76 km s^{-1} towards $l = 319°$, $b = 7°$ found by Turnbull et al. (2012).

5. Conclusions

By comparing the 2M++ density field with observational peculiar velocity data sets, we obtain a value of $f\sigma_{8,\mathrm{lin}}$ is consistent with previous measurements from RSD. It lies between the lower values from small-scale probes such as weak gravitational lensing and the slightly higher values predicted by Planck (Planck Collaboration et al. 2015).

The residual bulk flow, i.e. the contribution to the bulk flow due to sources outside the 2M++ volume, V_{ext}, is significantly different from zero, indicating that we have not yet resolved all of the sources of the LG's motion.

The resulting 2M++ density and peculiar velocity fields obtained from this analysis are made available at `cosmicflows.uwaterloo.ca` and `cosmicflows.iap.fr`

References

Abazajian, K. N. *et al.*, 2009, *ApJS*, 182, 543

Böhringer, H., Chon, G., Bristow, M., & Collins, C. A., 2015, *A&A*, 574, A26

Branchini, E. *et al.*, 1999, *MNRAS*, 308, 1

Carrick, J., Turnbull, S. J., Lavaux, G., & Hudson, M. J., 2015, *MNRAS*, 450, 317

Davis, M., Nusser, A., Masters, K. L., Springob, C., Huchra, J. P., & Lemson, G., 2011, *MNRAS*, 413, 2906

Hong, T. *et al.*, 2014, *MNRAS*, 445, 402

Huchra, J. P. *et al.*, 2012, *ApJS*, 199, 26

Hudson, M. J. & Turnbull S. J., 2012, *ApJL*, 751, L30

Jones, D. H. *et al.*, 2009, *MNRAS*, 399, 683

Kilbinger, M., Fu, L., Heymans, C., Simpson, F., Benjamin, J., Erben, & T., Harnois-Deraps, 2013, *MNRAS*, 430, 735

Lavaux, G. & Hudson, M. J., 2011, *MNRAS*, 416, 2840

Pike, R. W. & Hudson, M. J., 2005, *ApJ*, 635, 11

Planck Collaboration *et al.*, 2013, ArXiv e-prints

Planck Collaboration *et al.*, 2015, ArXiv e-prints

Reichardt, C. L. *et al.*, 2013, *ApJ*, 763, 127

Rozo, E. *et al.*, 2010, *ApJ*, 708, 645

Springob, C. M., Masters, K. L., Haynes, M. P., Giovanelli, R., & Marinoni, C., 2007, *ApJS*, 172, 599

Turnbull, S. J., Hudson, M. J., Feldman, H. A., Hicken, M., Kirshner, R. P., & Watkins, R., 2012, *MNRAS*, 420, 447

Vikhlinin, A. *et al.*, 2009, *ApJ*, 692, 1060

Watkins, R., Feldman, H. A., & Hudson, M. J., 2009, *MNRAS*, 392, 743

The Zeldovich Universe:
Genesis and Growth of the Cosmic Web
Proceedings IAU Symposium No. 308, 2014
R. van de Weygaert, S. Shandarin, E. Saar & J. Einasto, eds.

© International Astronomical Union 2016
doi:10.1017/S1743921316010085

Dynamics of pairwise motions
in the Cosmic Web

Wojciech A. Hellwing[1,2]

[1] Interdisciplinary Center for Mathematical and computational modelling (ICM),
University of Warsaw, ul. Pawińskiego 2a, 02-186 Warsaw, Poland
email: wojciech.hellwing@durham.ac.uk

[2] Institute for Computational Cosmology, University of Durham,
Science Site, South Road, DH1-3LE Durham, UK

Abstract. We present results of analysis of the dark matter (DM) pairwise velocity statistics in different Cosmic Web environments. We use the DM velocity and density field from the Millennium 2 simulation together with the NEXUS+ algorithm to segment the simulation volume into voxels uniquely identifying one of the four possible environments: nodes, filaments, walls or cosmic voids. We show that the PDFs of the mean infall velocities v_{12} as well as its spatial dependence together with the perpendicular and parallel velocity dispersions bear a significant signal of the large-scale structure environment in which DM particle pairs are embedded. The pairwise flows are notably colder and have smaller mean magnitude in wall and voids, when compared to much denser environments of filaments and nodes. We discuss on our results, indicating that they are consistent with a simple theoretical predictions for pairwise motions as induced by gravitational instability mechanism. Our results indicate that the Cosmic Web elements are coherent dynamical entities rather than just temporal geometrical associations. In addition it should be possible to observationally test various Cosmic Web finding algorithms by segmenting available peculiar velocity data and studying resulting pairwise velocity statistics.

Keywords. cosmic web, pairwise velocity, large-scale structure, dark matter

1. Introduction

The Cosmic Web (CW) is the most salient manifestation of the anisotropic nature of gravitational collapse, the motor behind the formation of structure in the cosmos. Structure in the Universe has risen out of tiny primordial (Gaussian) density and velocity perturbations by means of gravitational instability. N-body computer simulations have profusely illustrated how the primordial density field transforms into a pronounced and intricate filigree of filamentary features, dented by dense compact clumps at the nodes of the network (Bond *et al.* 1996). Moreover, because of the hierarchical nature of the cosmic matter distribution, also filaments, sheets and voids emerge by the gradual merging of smaller scale specimen. In the process, small filaments align themselves along the direction of the emerging larger scale filament. The emerging picture is therefore one of a primordially and hierarchically defined skeleton whose weblike topology is imprinted over a wide spectrum of scales. Weblike patterns on ever larger scales get to dominate the density field as cosmic evolution proceeds, and as small scale structures merge into larger ones. On the other hand, within the gradually emptying void regions the topological outline of the early weblike patterns remains largely visible as a faint remnant of past glory (see e.g. Fig. 1).

So far most of the authors have focused on the clustering and density-based statistics analysis of the CW, as it is seen in computer N-body simulations and galaxy catalogues (*e.g.* Springel *et al.* 2006; Aragón-Calvo *et al.* 2010; Cautun *et al.* 2014; Metuki *et al.* 2014;

Falck *et al.* 2014; Nuza *et al.* 2014). In this contribution we want to adopt a different, yet complementary to previous studies, approach. Namely we will study the dynamics of the Cosmic Web elements as reflected by the statistics of pairwise motions observed inside different environments. The dynamical studies of the CW elements are very important, both for providing complementary understanding their internal dynamics and kinematics, and for established whether the observed large-scale structure environments are just reflection of the geometrical and spatial galaxy distribution correlations or consists of a more dynamically coherent objects.

2. The pairwise motions

In this work we will consider lower-moments of the statistics of pairwise Dark Matter (DM) velocities as tracers of the underlying dynamics driven by gravity in different CW environments. The statistic of the mean relative pairwise velocity of galaxies v_{12} – the *streaming velocity* – reflects the "mean tendency of well-separated galaxies to approach each other" (Peebles 1980). This statistic was introduced by Davis&Peebels (Davis & Peebles 1977) in the context of the kinetic BBGKY theory which describes the dynamical evolution of a system of particles interacting via gravity. In the fluid limit, its equivalent is the pair-density weighted relative velocity

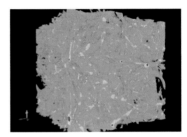

Figure 1. Rendering of the Cosmic Web elements from Millenium Simulation II as identified by the NEXUS algorithm. Different colours mark different elements of the Cosmic Web: dense nodes (clusters) are marked by red, elongated filaments are yellow, the pervading network of walls we depict by green and the remaining empty spaces correspond to cosmic voids

$$\mathbf{v}_{12}(r) = \langle \mathbf{v}_1 - \mathbf{v}_2 \rangle_\rho = \frac{\langle (\mathbf{v}_1 - \mathbf{v}_2)(1 + \delta_1)(1 + \delta_2) \rangle}{1 + \xi(r)} ,$$

$$(2.1)$$

where $\mathbf{v}_1$ and $\delta_1 = \rho_1/\langle\rho\rangle - 1$ stand for the peculiar velocity and fractional matter density contrast taken at point $\mathbf{r}_1$, $r = |\mathbf{r}_1 - \mathbf{r}_2|$, and $\xi(r) = \langle \delta_1 \delta_2 \rangle$ is the 2-point density correlation function. The $\langle \cdots \rangle_\rho$ is the pair-weighted average, which differs from normal spatial averaging by the weighting factor $\mathcal{W} = \rho_1 \rho_2/\langle \rho_1 \rho_2 \rangle$. Note $\mathcal{W}$ is proportional to the number density of pairs. The gravitational instability theory predicts that the $v_{12}(r)$ magnitude is governed by the 2-point correlation function $\xi(r)$ and the growth rate of matter density perturbations $f \equiv d\ln D_+/d\ln a$ (where $D_+(a)$ is the linear growing mode solution, and a is the cosmological scale factor) through the pair conservation equation (Peebles 1980). Juszkiewicz *et al.* (1999) provided a closed-form expression that is a good approximation to the solution of the pair conservation equation for universes with Gaussian initial conditions:

$$v_{12} = -\frac{2}{3} H_0 r f \bar{\bar{\xi}}(r)[1 + \alpha \bar{\bar{\xi}}(r)] ,$$

$$(2.2)$$

where

$$\bar{\xi}(r) = (3/r^3) \int_0^r \xi(x) x^2 dx \equiv \bar{\bar{\xi}}(r)[1 + \xi(r)] .$$

$$(2.3)$$

Here, α is a parameter that depends on the logarithmic slope of $\xi(r)$ and $H_0 = 100\, h$ km s^{-1} Mpc^{-1} is the present day value of the Hubble constant. It is clear that $v_{12}(r)$ is a strong function of $\xi(r)$ and f, which both in general will take different local averages in different CW environments. This motivates us to consider low-order moments of the pairwise velocity distribution as tracers of CW specific local dynamics that should be reflected in motions of galaxies and DM. Specifically, in our analysis we will consider the following quantities:

- the mean radial pairwise velocity, v_{12};
- the dispersion in the (radial) pairwise velocities (not centred), $\sigma_\parallel = \langle v_{12}^2 \rangle^{0.5}$;
- the mean transverse velocity of pairs, $v_\perp$;
- the dispersion of the transverse velocity of pairs, $\sigma_\perp = \langle v_\perp^2 \rangle^{0.5}$.

It is well known that the cosmic velocity field has large coherence length (*e.g.* Chodorowski & Ciecieląg 2002; Ciecieląg *et al.* 2003), much larger than the density field. Thus the contribution of the large-scale velocity modes is always significant, when one concerns the peculiar velocities. However this disadvantage (from the point of view of probing the velocity properties driven by local environment) is not present for the pairwise velocity statistics we have just described above. This is because at a given separation r the velocity difference between a galaxy pair does not receive any net contribution from modes with wavelengths larger than the considered separation, since those give the same contribution for both particles/galaxies. Hence, at larger scales, for which the galaxies in a pair inhabit different haloes, the distribution of v_{12} factorises into two individual peculiar velocity distributions for each galaxy/particle and those are always sensitive to non-linearities driven by virial motions within a given host halo alone (see Scoccimarro (2004) for more details). Thanks to this, the pairwise velocity statistics is automatically free of large-scale velocity modes contributions, and hence is very well suited for the kind of analysis we want to perform.

3. Dynamics of the Cosmic Web

To study the dynamics of pairwise motions in different large-scale structure environments we use the DM density and velocity fields from the Millennium-II simulation (Boylan-Kolchin *et al.* 2009). To segment the simulation volume into voids, walls, filaments and cluster voxels we use the NEXUS+ algorithm (Cautun *et al.* 2013) applied on the DM density field computed on a 256^3 grid (with a grid cell width of $0.39h^{-1}$ Mpc). For such defined setup we obtain a catalogue of voxels fully segmenting the 3D simulation cube into four possible different CW environments (see Fig. 1). The construction of the NEXUS+ algorithm is such that the determination of the local environment is based on spatial variations along three Cartesian directions in the density field. As a consequence our voids correspond to spatial regions that are relatively empty with spherical-like symmetry, wall constitute thin planar structures, finally the filaments are elongated fibres. Our node environment corresponds to regions centred around massive clusters and their local infall regions.

Having assessed the segmentation of our simulation volume into the CW elements we have computed all related pairwise statistics (v_{12}, $\sigma_\parallel$, $v_\perp$ and $\sigma_\perp$) for randomly Poisson sub-sampled DM particles. We use 1/100th of the original number of DM simulation particles. This sub-sampling leave us with a sufficient number of DM particles to study pairwise motions statistics for separations $\geqslant 0.5h^{-1}$ Mpc, since at this separation we have an averaged number of 10 particle pairs per voxel.

We begin by plotting the probability density functions for v_{12} values computed at two different averaged pair separations: 0.5 and $2h^{-1}$ Mpc. The results are shown in the two panels of the Fig. 2. The reference point is marked always by a solid black line with open circle data points and corresponds to a result obtained from the whole uniform volume (*i.e.* without segmenting into different morphology elements). The data corresponding to the following environments we mark by: red (node/cluster), blue (filaments), orange

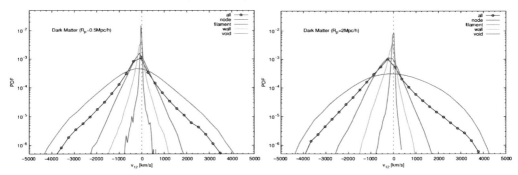

Figure 2. The probability density functions for the mean pairwise streaming velocity v_{12} computed at pair separation of $0.5h^{-1}$ Mpc (left panel) and $2h^{-1}$ Mpc (right panel). The line marking is the following: black with open circles - averaged for all cosmic environments (*i.e.* whole volume), red line - node/cluster environments, blue - filaments, orange - walls and we use green for cosmic voids.

(walls) and green for voids. The both panels illustrate the striking difference of the v_{12} distribution functions among different environments. We denote that the width of the distribution (roughly corresponding here to dispersion) increase significantly as we move from voids and walls towards denser environments of filaments and clusters. This is clearly visible at both considered separations. This indicate that pairwise motions are much hotter in nodes and filaments than in voids and walls. As we have described above, the mean streaming pairwise velocity is very sensitive to perturbation modes whose length is smaller than a considered pair separation. This property of the pairwise statistics is now clearly consistent with our measured v_{12} PDFs. It is evident that in environments like cosmic clusters and filaments the degree of non-linearity (in both the density and velocity field) is much higher compared to relatively quieter cosmic walls and filaments. This difference is mostly driven by violent relaxation processes and thermalised virial motions that tend to dominate the peculiar velocity field inside clusters and dense filaments regions (*e.g.* Cautun *et al.* 2014, 2013). Secondly we can denote that the degree of (a)symmetry of the PDFs are subject to a significant environment-driven variation. This can be especially seen on the left panel of Fig. 2, where we plot the probability density functions for pairs at $R_p = 2h^{-1}$ Mpc separation. The *stable clustering* model (Peebles 1980) predicts that at weakly non-linear scales there should be slight excess of negative pairwise velocities reflecting the gravitational clustering mechanism that still operates at those scales. Namely between two regimes of fully virialised motions at small scales and Hubble expansion dominated at large scales there should be a region where some significant part of the peculiar motions still have potential character. We can observed this by noting that the cluster/node PDF function on the right panel has much higher symmetry then the PDFs of the rest of the environments. Such symmetric v_{12} PDF is a characteristic of a fully randomised motions supported by velocity dispersion. This is the case inside virialised haloes, which clearly dominate the pairwise flows signal in our node/cluster environments.

As a complementary probe of the pairwise motions we also study the minus mean infall velocities ($-v_{12}$) and parallel and perpendicular pair velocities dispersion ($\sigma_\parallel$, $\sigma_\perp$) as a function of a pair separation R_p. We plot the corresponding results in the left and right panel of the Fig. 3. We use the same colours as before to mark results for different CW elements. On the left panel for comparison we also draw a purple line depicting the Hubble expansion velocity. The results shown here are consistent with the previously

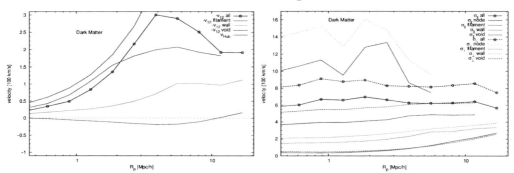

Figure 3. The minus mean streaming velocities - $-v_{12}$ (left panel) and parallel and perpendicular pair velocities dispersions (right panel) - $\sigma_\parallel$, $\sigma_\perp$ - as a function of a pair separation R_p. The lines colouring is consistent with the previous figure.

discussed PDF properties. We note both higher values of the mean infall velocities as well as much higher dispersions of both the infall and perpendicular pair-velocity components for high density environments of nodes and filaments. In contrast the pairwise motions in walls and especially in voids are characterised by very small dispersions (are very cold). We can also denote another interesting observation regarding the dynamics of pairwise motions in cosmic voids. Namely from the shape of the $-v_{12}(R_p)$ curve on the left panel we can infer that, on average, the pairwise motions in voids have *super Hubble* character. This means that inside the cosmic voids the expansion is faster then the background Hubble expansion rate. Although this result was known for some time (*e.g.* Sheth & van de Weygaert 2004), as the inside regions of cosmic voids can be treated locally as approximate Friedman Universes with lower effective Ω_m value then the cosmic mean, it is very interesting to confirm this result in our data for pairwise motions. The last thing we want the emphasise here is the lack of the offset between two different velocity directions dispersions for the cosmic voids. This can be gauged by the offset between solid and dashed lines of the same colour on the right panel of the Fig. 3. The offset between $\sigma_\perp$ and $\sigma_\parallel$ is largest for the cluster environment and decrease as we move to filaments and walls. However even for cosmic walls this offset is still significant (of the order of $\sim 33\%$), while in the case of the cosmic voids we observe that values and shapes of the both dispersions lines are fully consistent with each other. We speculate that this feature is driven by two facts: (i) the structure formation has ceased inside cosmic voids, and (ii) there are no preferred directions for the voids segments reflecting their average close-to spherical symmetry.

4. Conclusions

We have studied the dynamics and statistical properties of dark matter particles pairwise motions in four different CW elements: nodes/clusters, filaments, walls and voids. We have shown that both the probability density functions for the mean infall velocities as well as scale dependence of v_{12} and two connected velocity dispersions are taking significantly different values and shapes in different large-scale environments. The flows in voids and walls are much colder and also have smaller magnitude, when compare to pairwise motions in denser node and filament environments. This indicate that the NEXUS+ algorithm is very robust in determining different CW elements. The algorithm does not use any velocity data whatsoever, and yet the CW elements delineated by this method are characterised by pairwise motions whose statistics are consistent with simple theoretical predictions.

Our finding also indicate that the CW and its elements, as first defined by Bond *et al.* (1996), is not only real in the sense of the DM/galaxy density field morphology, but is also present at the level of peculiar motions of DM. This is clearly seen once the contribution of large-scale velocity modes was removed, which can be obtained for example by considering pairwise statistics as we did in this study. Such a strong correlations between the distribution and dispersion of mean streaming velocities and the CW morphology also indicate that the cosmic velocity-density relation has different limits of validity in different environments. This reflects the different degree of non-linearity in the structure formation in voids, walls, filaments and nodes.

Acknowledgements

This work would not be possible without a significant contribution of Marius Cautun, Rien van de Weygaert and Carlos S. Frenk. I am very grateful for their help and many stimulating discussions. The author acknowledge the support received from Polish National Science Center in grant no. DEC-2011/01/D/ST9/01960 and ERC Advanced Investigator grant of C. S. Frenk, COSMIWAY.

References

Aragón-Calvo, M. A., van de Weygaert, R., & Jones, B. J. T. 2010, *MNRAS*, 408, 2163, 1007.0742

Bond, J. R., Kofman, L., & Pogosyan, D. 1996, *Nature*, 380, 603, astro-ph/9512141

Boylan-Kolchin, M., Springel, V., White, S. D. M., Jenkins, A., & Lemson, G. 2009, *MNRAS*, 398, 1150, 0903.3041

Cautun, M., van de Weygaert, R., & Jones, B. J. T. 2013, *MNRAS*, 429, 1286, 1209.2043

Cautun, M., van de Weygaert, R., Jones, B. J. T., & Frenk, C. S. 2014, *MNRAS*, 441, 2923, 1401.7866

Chodorowski, M. J., & Ciecieląg, P. 2002, *MNRAS*, 331, 133, astro-ph/0109291

Ciecieląg, P., Chodorowski, M. J., Kiraga, M., Strauss, M. A., Kudlicki, A., & Bouchet, F. R. 2003, *MNRAS*, 339, 641, astro-ph/0010364

Davis, M., & Peebles, P. J. E. 1977, *ApJS*, 34, 425

Falck, B., Koyama, K., Zhao, G.-b., & Li, B. 2014, *JCAP*, 7, 58, 1404.2206

Juszkiewicz, R., Springel, V., & Durrer, R. 1999, *ApJL*, 518, L25, astro-ph/9812387

Metuki, O., Libeskind, N. I., Hoffman, Y., Crain, R. A., & Theuns, T. 2014, ArXiv e-prints, 1405.0281

Nuza, S. E., Kitaura, F.-S., Heß, S., Libeskind, N. I., & Müller, V. 2014, *MNRAS*, 445, 988, 1406.1004

Peebles, P. J. E. 1980, The large-scale structure of the universe (Research supported by the National Science Foundation. Princeton, N.J., Princeton University Press, 1980. 435 p.)

Scoccimarro, R. 2004, *Phys. Rev. D*, 70, 083007, astro-ph/0407214

Sheth, R. K., & van de Weygaert, R. 2004, *MNRAS*, 350, 517, astro-ph/0311260

Springel, V., Frenk, C. S., & White, S. D. M. 2006, *Nature*, 440, 1137, astro-ph/0604561

The Zeldovich Universe:
Genesis and Growth of the Cosmic Web
Proceedings IAU Symposium No. 308, 2014
R. van de Weygaert, S. Shandarin, E. Saar & J. Einasto, eds.

ⓒ International Astronomical Union 2016
doi:10.1017/S1743921316010097

Galaxy and Mass Assembly (GAMA): galaxy pairwise velocity dispersion

Jon Loveday[1], Leonidas Christodoulou[1] and the GAMA team

[1]Astronomy Centre, University of Sussex, Falmer, Brighton BN1 9QH, UK
email: J.Loveday@sussex.ac.uk

Abstract. We describe preliminary measurements of the pairwise velocity dispersion (PVD) of galaxies in the Galaxy and Mass Assembly (GAMA) survey as a function of projected separation and galaxy luminosity. Due to the faint magnitude limit ($r < 19.8$) and highly-complete spectroscopic sampling of the GAMA survey, we are able to measure the PVD to smaller scales and for lower-luminosity galaxies than previous SDSS-based work. We see no strong scale-dependence at most luminosities in the quasi-linear regime. We observe an apparent drop in PVD towards very small scales (below $\approx 0.1 h^{-1}$ Mpc), but this could in part be due to a restriction of the streaming model employed. At intermediate scales, the PVD is highest (~ 500 km/s) at intermediate luminosities, dropping at both fainter and brighter luminosities.

Keywords. galaxies: kinematics and dynamics, galaxies: statistics

1. Introduction

The pairwise velocity dispersion of galaxies, PVD, or σ_{12}, is an important quantity for modeling the galaxy redshift space correlation function. It can be used to test predictions of galaxy formation and evolution models and of the cold dark matter paradigm in general. In addition, σ_{12} is a parameter that strongly affects cosmological parameter fits as it correlates with measurements of the growth rate of structure.

The Galaxy and Mass Assembly (GAMA) survey (Driver et al. 2011) is ideal for measuring the PVD due to (i) being two magnitudes fainter than the SDSS main galaxy sample, and (ii) having very high (> 98 per cent) spectroscopic completeness, even in high-density regions. In this contribution, we measure the PVD from the GAMA-II equatorial regions, covering a total area of 180 square degrees. Throughout, we assume a Hubble constant of $H_0 = 100h$ km s^{-1} Mpc^{-1} and an $\Omega_M = 0.3, \Omega_\Lambda = 0.7$ cosmology in calculating distances, co-moving volumes and luminosities.

2. Galaxy clustering measurements

Galaxy clustering is measured as a function of separation parallel ($r_\parallel$) and perpendicular ($r_\perp$) to the line of sight using the Landy & Szalay (1993) estimator,

$$\xi(r_\perp, r_\parallel) = \frac{DD - 2DR + RR}{RR}, \tag{2.1}$$

where DD, DR and RR are the normalised numbers of data-data, data-random and random-random pairs in a given $(r_\perp, r_\parallel)$ bin. The random catalogue is generated within the survey mask and with a radial distribution determined by replicating each galaxy a number of times within its observable redshift range while correcting for radial density fluctuations (Cole 2011; Loveday 2014). Two-dimensional correlation functions are shown for blue and red GAMA-II galaxies in Fig. 1

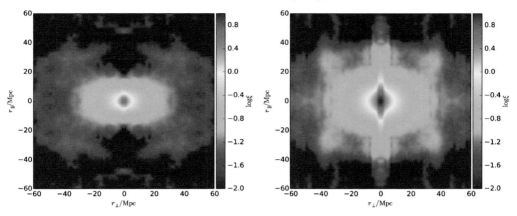

Figure 1. The two-dimensional correlation function $\xi(r_\perp, r_\parallel)$ for blue (left) and red (right) galaxies in GAMA-II. Elongation along the line of sight as small projected separations is clearly visible within the red galaxy sample. Large-scale infall is visible in both samples as a compression of the clustering signal along the line of sight (y-axis) at larger separations.

In the "streaming" model (Peebles 1980; Davis & Peebles 1983; Zehavi et al. 2002), the two-dimensional correlation function $\xi(r_\perp, r_\parallel)$ is given by a convolution of the isotropic real-space correlation function $\xi_r(r)$ with the pairwise velocity distribution $f(v)$:

$$1 + \xi(r_\perp, r_\parallel) = H_0 \int_{-\infty}^{\infty} \left[1 + \xi_r \left(\sqrt{r_\perp^2 + y^2} \right) \right] f(v) dy. \tag{2.2}$$

The pairwise velocity distribution is assumed to follow an exponential distribution

$$f(v) = \frac{1}{\sqrt{2}\sigma_{12}} \exp \left(-\frac{\sqrt{2}|v|}{\sigma_{12}} \right), \tag{2.3}$$

$$v \equiv H_0(r_\parallel - y) + \bar{v}_{12}(r), \tag{2.4}$$

where $\bar{v}_{12}(r)$ is the mean radial pairwise velocity at separation r; we use the expression given by Juszkiewicz et al. (1999, equation 6).

The real-space correlation function $\xi_r(r)$ may be estimated by first integrating $\xi(r_\perp, r_\parallel)$ along the line of sight direction $r_\parallel$ to obtain the projected correlation function

$$w_p(r_\perp) = 2 \int_0^{r_{max}} \xi(r_\perp, r_\parallel) dr_\parallel, \tag{2.5}$$

and then performing the inversion

$$\xi(r) = -\frac{1}{\pi} \int_r^{r_{max}} w_p(r_\perp)(r_\perp^2 - r^2)^{-1/2} dr_\perp. \tag{2.6}$$

This integral is evaluated by linearly interpolating between the binned $w_p(r_\perp)$ values (Saunders et al. 1992).

3. Results

In Fig. 2 we show the PVD σ_{12} as a function of projected separation in bins of absolute magnitude. The data from Li et al. (2006) provide only an approximate comparison with previous results, since Li *et al.* calculate σ_{12} as a function of the amplitude of the total

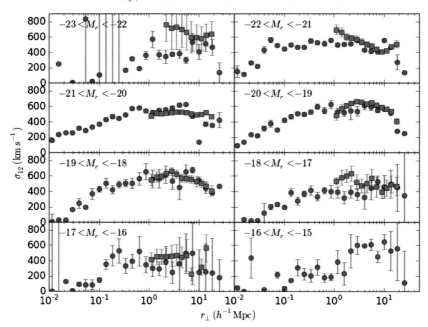

Figure 2. Pairwise velocity dispersion for GAMA-II (this work, blue circles) and SDSS (Li et al. 2006, green squares) as a function of projected separation $r_\perp$, in absolute magnitude bins is labeled.

wave-vector k rather than just the projected component. We assume when plotting their data points that $r_\perp = \sqrt{2}\pi/k$. Given the uncertainties, our results are broadly consistent with those of Li *et al.* where we overlap. Due to the high spectroscopic completeness of GAMA, we are able to estimate σ_{12} to much smaller scales, as small as $10\ h^{-1}$ kpc, with reasonable random errors for most luminosity subsamples. However, we caution that the apparent strong decline of σ_{12} to the smallest scales may be due at least in part to a breakdown of the streaming model on extremely non-linear scales. We plan to test this issue in the near future using N-body simulations.

In Fig. 3 we show the PVD σ_{12} as a function of absolute magnitude in bins of projected separation. At intermediate scales, 0.1–$3.0\ h^{-1}$ Mpc, σ_{12} is highest (around 400–600 km s^{-1}) at intermediate luminosities, $-22 \lesssim M_r \lesssim -18$ mag, decreasing at both faint and bright ends. Unlike Li *et al.*, we see no increase in σ_{12} in the highest luminosity bin. We note that in Fig. 6 of Li *et al.*, the error bars for the most luminous galaxies are very large, and so this apparent increase in σ_{12} is not very significant.

In future, we plan to compare estimates of the PVD obtained in Fourier space using the "damped Kaiser" model (Peacock & Dodds 1993; Li et al. 2006) and to investigate the dependence of the PVD on location within the cosmic web, on stellar mass and on redshift.

JL acknowledges support from the Science and Technology Facilities Council (grant number ST/I000976/1) and is very grateful to the IAU and the symposium organizers for their generous travel support. The GAMA website is: http://www.gama-survey.org/.

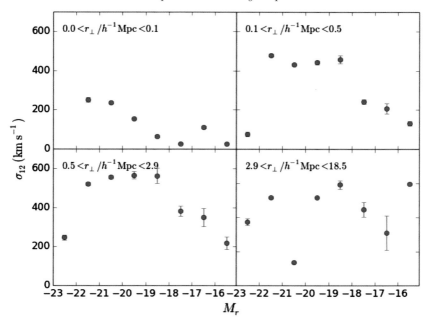

Figure 3. GAMA-II pairwise velocity dispersion as a function of absolute magnitude in bins of projected separation $r_\perp$.

References

Cole, S., 2011, *MNRAS*, 416, 739

Davis, M. & Peebles, P. J. E., 1983, *ApJ*, 267, 465

Driver, S. P. *et al.*, 2011, *MNRAS*, 413, 971

Juszkiewicz, R., Springel, V., & Durrer, R., 1999, *ApJ*, 518, L25

Landy, S. D. & Szalay, A. S., 1993, *ApJ*, 412, 64

Li, C., Jing, Y. P., Kauffmann, G., Borner, G., White, S. D. M., & Cheng, F. Z., 2006, *MNRAS*, 368, 37

Loveday, J., 2014, in *IAU Symp.* 306

Peacock, J. A. & Dodds, S. J., 1993, *MNRAS*, 267, 13

Peebles, P. J. E., 1980, *The large-scale structure of the universe.* Princeton University Press

Saunders, W., Rowan-Robinson, M., & Lawrence, A., 1992, *MNRAS*, 258, 134

Zehavi, I. *et al.*, 2002, *ApJ*, 571, 172

The Zeldovich Universe:
Genesis and Growth of the Cosmic Web
Proceedings IAU Symposium No. 308, 2014
R. van de Weygaert, S. Shandarin, E. Saar & J. Einasto, eds.

© International Astronomical Union 2016
doi:10.1017/S1743921316010103

Large-scale peculiar velocities through the galaxy luminosity function at $z \sim 0.1$

Martin Feix[1], Adi Nusser[1,2] and Enzo Branchini[3,4,5]

[1]Department of Physics, Israel Institute of Technology - Technion, Haifa 32000, Israel
email: mfeix@physics.technion.ac.il

[2]Asher Space Science Institute, Israel Institute of Technology - Technion, Haifa 32000, Israel

[3]Department of Physics, Università Roma Tre, Via della Vasca Navale 84, Rome 00146, Italy

[4]INFN Sezione di Roma 3, Via della Vasca Navale 84, Rome 00146, Italy

[5]INAF, Osservatorio Astronomico di Roma, Monte Porzio Catone, Italy

Abstract. Peculiar motion introduces systematic variations in the observed luminosity distribution of galaxies. This allows one to constrain the cosmic peculiar velocity field from large galaxy redshift surveys. Using around half a million galaxies from the SDSS Data Release 7 at $z \sim 0.1$, we demonstrate the applicability of this approach to large datasets and obtain bounds on peculiar velocity moments and σ_8, the amplitude of the linear matter power spectrum. Our results are in good agreement with the ΛCDM model and consistent with the previously reported $\sim 1\%$ zero-point tilt in the SDSS photometry. Finally, we discuss the prospects of constraining the growth rate of density perturbations by reconstructing the full linear velocity field from the observed galaxy clustering in redshift space.

Keywords. cosmology: theory, large-scale structure of universe, cosmological parameters, cosmology: observations, methods: statistical, galaxies: distances and redshifts

1. Velocities from the variation of observed galaxy luminosities

To linear order in perturbation theory, the observed redshift z of a galaxy typically deviates from its cosmological redshift z_c according to (Sachs & Wolfe 1967)

$$\frac{z - z_c}{1 + z} = \frac{V(t, r)}{c} - \frac{\Phi(t, r)}{c^2} - \frac{2}{c^2} \int_{t(r)}^{t_0} \mathrm{d}t \frac{\partial \Phi \left[\hat{r}r(t), t\right]}{\partial t} \approx \frac{V(t, r)}{c},$$

where V is the (physical) radial peculiar velocity of the galaxy, r is a unit vector along the line of sight to the object, and Φ denotes the usual gravitational potential. Here we explicitly assume low redshifts such that the velocity V is the dominant contribution, and we further consider all fields relative to their present-day values at t_0.

As the shift $z - z_c$ enters the calculation of distance moduli $\mathrm{DM} = 25 + 5 \log_{10}[D_L/\mathrm{Mpc}]$, where D_L is the luminosity distance, observed absolute magnitudes M differ from their true values $M^{(t)}$. We thus have

$$M = m - \mathrm{DM}(z) - K(z) + Q(z) = M^{(t)} + 5 \log_{10} \frac{D_L(z_c)}{D_L(z)},$$

where m is the apparent magnitude, the function $Q(z)$ accounts for luminosity evolution, and $K(z)$ is the K-correction (Blanton & Roweis 2007). On scales where linear theory provides an adequate description, the variation $M - M^{(t)}$ of magnitudes distributed over the sky is systematic, and therefore, contains information on the peculiar velocity field.

Given a suitable parameterized model $V(\hat{r}, z)$ of the radial velocity field, the idea is now to maximize the probability of observing galaxies with magnitudes M_i given only their redshifts and angular positions $\hat{\boldsymbol{r}}_i$ on the sky, i.e.,

$$P_{\text{tot}} = \prod_i P\left(M_i | z_i, V(\hat{\boldsymbol{r}}_i, z_i)\right) = \prod_i \left(\phi(M_i) \Big/ \int_{M_i^+}^{M_i^-} \phi(M)\mathrm{d}M\right),$$

where we assume that redshift errors can be neglected (Nusser *et al.* 2011), $\phi(M)$ denotes the galaxy luminosity function (LF), and the corresponding limiting magnitudes $M^{\pm}$ depend on $V(\hat{\boldsymbol{r}}, z)$ through the cosmological redshift z_c. Here the motivation is to obtain a maximum-likelihood estimate of $V(\hat{\boldsymbol{r}}, z)$ by finding the set of velocity model parameters which minimizes the spread in the observed magnitudes.

Tammann *et al.* (1979) first adopted this approach to estimate the motion of Virgo relative to the local group, and recently, Nusser *et al.* (2011) used it to constrain bulk flows in the local Universe from the 2MASS Redshift Survey (Huchra *et al.* 2012).

2. Constraints on the cosmic peculiar velocity field at $z \sim 0.1$

Galaxies from the Sloan Digital Sky Survey (SDSS) Data Release 7 (Abazajian *et al.* 2009) probe the cosmic velocity field out to $z \sim 0.1$. Here we report results obtained from applying the luminosity method to a subset of roughly half a million galaxies (for additional details, see Feix *et al.* 2014).

Data. In our analysis, we used the latest version of the NYU Value-Added Galaxy Catalog (NYU-VAGC; Blanton *et al.* 2005). Giving the largest spectroscopically complete galaxy sample, we adopted (Petrosian) $^{0.1}r$-band magnitudes, and chose the subsample NYU-VAGC `safe` to minimize incompleteness and systematics. Our final sample contained only galaxies with $14.5 < m_r < 17.6$, $-22.5 < M_r - 5\log_{10} h < -17.0$, and $0.02 < z < 0.22$ (relative to the CMB frame). In addition, we employed a suite of galaxy mock catalogs mimicking the known systematics of the data.

Radial velocity model. We considered a bin-averaged velocity model $\tilde{V}(\hat{\boldsymbol{r}})$ in two redshift bins, $0.02 < z < 0.07$ and $0.07 < z < 0.22$. For each bin, the velocity field was further decomposed into spherical harmonics, i.e.

$$a_{lm} = \int \mathrm{d}\Omega \tilde{V}(\hat{\boldsymbol{r}}) Y_{lm}(\hat{\boldsymbol{r}}), \qquad \tilde{V}(\hat{\boldsymbol{r}}) = \sum_{l,m} a_{lm} Y_{lm}^*(\hat{\boldsymbol{r}}), \qquad l > 0,$$

where the sum over l is cut at some maximum value $l_{\max}$. Because the SDSS data cover only part of the sky, the inferred a_{lm} are not statistically independent. The impact of the angular mask was studied with the help of suitable galaxy mock catalogs. The monopole term ($l = 0$) was not included since it is degenerate with an overall shift of magnitudes.

LF estimators. Reliably measuring the galaxy LF represents a key step in our approach. To assess the robustness of our results with respect to different LF models, we analyzed the data using LF estimators based on a Schechter form and a more flexible spline-based model, together with several combinations and variations thereof. For simplicity, we also assumed a linear dependence of the luminosity evolution with redshift.

Bulk flows and higher-order velocity moments. Accounting for known systematic errors in the SDSS photometry, our "bulk flow" measurements are consistent with a standard ΛCDM cosmology at a 1–2σ confidence level in both redshift bins. A joint analysis of the corresponding three Cartesian components confirmed this result. To characterize higher-order moments as well, we further obtained direct constraints on the angular velocity power spectrum $C_l = \langle |a_{lm}|^2 \rangle$ up to the octupole contribution. The estimated C_l were found compatible to be with the theoretical power spectra of the ΛCDM cosmology.

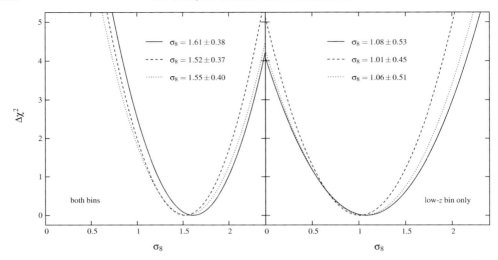

Figure 1. Raw estimates of σ_8 obtained from the NYU-VAGC: shown is the derived $\Delta\chi^2$ as a function of σ_8 for both redshift bins (left panel) and the first redshift bin with $0.02 < z < 0.07$ only (right panel), adopting different estimators of the LF (solid, dashed, and dotted lines).

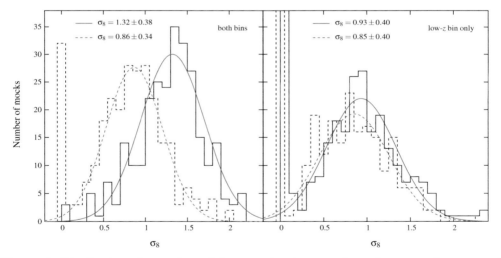

Figure 2. Distribution of σ_8 estimated from mock galaxy catalogs: shown are the recovered histograms (black lines) and respective Gaussian fits with (solid lines) and without (dashed lines) the inclusion of a systematic (randomly oriented) tilt in the galaxy magnitudes, using the information in both redshift bins (left) and the bin with $0.02 < z < 0.07$ only (right).

Constraints on σ_8. Assuming a prior on the C_l as dictated by the ΛCDM model with fixed Hubble constant and density parameters, we independently estimated the parameter σ_8 which determines the amplitude of the velocity field. Due to the presence of a dipole-like tilt in the galaxy magnitudes (Padmanabhan *et al.* 2008), the obtained raw estimates of σ_8 were expected to be biased toward larger values (Fig. 1). After correcting for this magnitude tilt with the help of our mocks (Fig. 2), we eventually found $\sigma_8 \approx 1.1 \pm 0.4$ for the combination of both redshift bins and $\sigma_8 \approx 1.0 \pm 0.5$ for the low-z bin only, where the low accuracy is due to the limited number of galaxies. This confirms our method's validity in view of future datasets with larger sky coverage and better photometric calibration.

3. Toward constraints on the linear growth rate

A very interesting aspect of our luminosity-based approach is the possibility to place bounds on the growth rate of density perturbations, $\beta = f(\Omega)/b$ (where b is the linear galaxy bias), by modeling the large-scale velocity field directly from the observed clustering of galaxies in redshift space (Nusser & Davis 1994). Such bounds are complementary to and — regarding ongoing and future redshift surveys — expected to be competitive with those obtained from redshift-space distortions (Nusser *et al.* 2012).

To get an idea of how well the method could constrain β at $z \sim 0.1$ from SDSS galaxies, we used mocks generated from the Millennium Simulation (Springel *et al.* 2005; Henriques *et al.* 2012) to create full-sky catalogs which otherwise shared all characteristics of the real SDSS data. Adopting a radial velocity model proportional to the true one smoothed over spheres of $10h^{-1}$ Mpc radius, the luminosity method was applied to samples with around 2×10^5 galaxies and correctly recovered the velocity field. The error on the proportionality constant typically yielded ±0.2–0.3 if only the contribution of multipoles with $l > 25$ (an appropriate value for the SDSS geometry) is taken into account. Assuming an accurate velocity reconstruction for these modes, we expect a similar situation for β. A further complication is that the angular mask may introduce bias as a consequence of multipole mixing. This and other technical issues mainly related to the reconstruction of the velocity field are currently under detailed investigation.

4. Outlook

Current and next-generation spectroscopic surveys are designed to reduce data-inherent systematics because of larger sky coverage and improved photometric calibration in ground- and space-based experiments (e.g, Levi *et al.* 2013; Laureijs *et al.* 2011). The method considered here does not require accurate redshifts and can be used with photo-metric redshift surveys such as the 2MASS Photometric Redshift catalog (2MPZ; Bilicki *et al.* 2014) to recover signals on scales larger than the spread of the redshift error.

Together with our results, these observational perspectives give us confidence that the luminosity-based method will be established as a standard cosmological probe, in-dependent from and alternative to the more traditional ones based on galaxy clustering, gravitational lensing and redshift-space distortions.

References

Abazajian, K. N., Adelman-McCarthy, J. K., Agüeros, M. A., *et al.* 2009, *ApJS*, 182, 543
Bilicki, M., Jarrett, T. H., Peacock, J. A., *et al.* 2014, *ApJS*, 210, 9
Blanton, M. R., Schlegel, D. J., Strauss, M. A., *et al.* 2005, *AJ*, 129, 2562
Blanton, M. R. & Roweis, S. 2007, *AJ*, 133, 734
Feix, M., Nusser, A., & Branchini, E. 2014, *JCAP*, 09, 019
Henriques, B. M. B., White, S. D. M., & Lemson, G. 2012, *MNRAS*, 421, 4
Huchra, J. P., Macri, L. M., Masters, K. L., *et al.* 2012, *ApJS*, 199, 26
Laureijs, R., Amiaux, J., Arduini, S., *et al.* 2011, *arXiv:1110.3193*
Levi, M., Bebek, C., Beers, T., *et al.* 2013, *arXiv:1308.0847*
Nusser, A. & Davis, M. 1994, *ApJ (Letters)*, 421, L1
Nusser, A., Branchini, E., & Davis, M. 2011, *ApJ*, 735, 77
Nusser, A., Branchini, E., & Davis, M. 2012, *ApJ*, 744, 193
Padmanabhan, N., Schlegel, D. J., & Finkbeiner, D. P. 2008, *ApJ*, 674, 1217
Sachs, R. K. & Wolfe, A. M. 1967, *ApJ*, 147, 73
Springel, V., White, S. D. M., Jenkins, A., *et al.* 2005, *Nature*, 435, 7042
Tammann, G. A., Yahil, A., & Sandage, A. 1979, *ApJ*, 234, 775

The Zeldovich Universe:
Genesis and Growth of the Cosmic Web
Proceedings IAU Symposium No. 308, 2014
R. van de Weygaert, S. Shandarin, E. Saar & J. Einasto, eds.

© International Astronomical Union 2016
doi:10.1017/S1743921316010115

Measuring the cosmic bulk flow with 6dFGSv

Christina Magoulas[1,2]**, Christopher Springob**[2,3,4]**, Matthew Colless**[5]**,
Jeremy Mould**[4,6]**, John Lucey**[7]**, Pirin Erdoğdu**[8] **and D. Heath Jones**[9]

[1]Department of Astronomy, University of Cape Town, Private Bag X3, Rondebosch 7701, RSA,
email: cmagoulas@ast.uct.ac.za

[2]Australian Astronomical Observatory, PO Box 915, North Ryde, NSW 1670, Australia,
[3]ICRAR, The University of Western Australia, Crawley, WA 6009, Australia,
[4]ARC Centre of Excellence for All-sky Astrophysics (CAASTRO),
[5]RSAA, The Australian National University, Canberra, ACT 2611, Australia,
[6]CAS, Swinburne University, Hawthorn, VIC 3122, Australia,
[7]Department of Physics, University of Durham, Durham, DH1 3LE, UK,
[8]Australian College of Kuwait, PO Box 1411, Safat 13015, Kuwait,
[9]School of Physics, Monash University, Clayton, VIC 3800, Australia.

Abstract. While recent years have seen rapid growth in the number of galaxy peculiar velocity measurements, disagreements remain about the extent to which the peculiar velocity field - a tracer of the large-scale distribution of mass - agrees with both ΛCDM expectations and with velocity field models derived from redshift surveys. The 6dF Galaxy Survey includes peculiar velocities for nearly 9 000 early-type galaxies (6dFGSv), making it the largest and most homogeneous galaxy peculiar velocity sample to date. We have used the 6dFGS velocity field to determine the amplitude and scale of large-scale cosmic flows in the local universe and test standard cosmological models. We also compare the galaxy density and peculiar velocity fields to establish the distribution of dark and luminous matter and better constrain key cosmological parameters such as the redshift-space distortion parameter.

Keywords. galaxies: distances and redshifts, cosmology: observations - distance scale - large-scale structure of universe

---

1. Introduction

Peculiar velocities are a direct, unbiased tracer of the underlying distribution of mass in the universe that are regulated by the scale and amplitude of fluctuations in the density field. The peculiar velocity field is therefore a powerful cosmological probe that can provide independent constraints on the parameters defining models of large-scale structure formation. It is sensitive to mass fluctuations on the largest scales, up to $\sim 100\,h^{-1}$ Mpc, and remains the only such probe in the low-redshift universe.

The dipole moment of the peculiar velocity field, also known as the *bulk flow*, is a measure of the large-scale, coherent motion of matter. The most recent peculiar velocity studies consist of samples containing a large number of measurements (on the order of 5000) to reach a consensus in the scale of these flows and also establish whether they are consistent with the predictions of ΛCDM. When averaged over a large enough volume, cosmological models predict that the bulk flow should approach the Hubble flow, commonly measured as a convergence to the rest frame of the cosmic microwave background (CMB). Whilst there is growing consensus in the *direction* of the bulk flow found by multiple studies, inconsistencies in the observed amplitude and scale still remain.

The distortion of the galaxy distribution in redshift-space by the peculiar velocity field can be characterized by the linear redshift distortion parameter, β. The form of this distortion is related to the growth rate of structure, $f = \Omega_{\mathrm{m}}^{0.55}$, by $\beta = f/b$, under the assumption that the galaxy density (δ_{g}) and matter density (δ_{m}) fluctuations are related by a linear bias parameter, b, such that $b = \delta_{\mathrm{g}}/\delta_{\mathrm{m}}$. By comparing the observed peculiar velocity field to a reconstructed prediction of the velocity field, we can determine the β parameter, linking the total mass density and the bias in the distribution of galaxies relative to the underlying distribution of mass.

Using the Fundamental Plane (FP), we have measured distances and peculiar velocities for nearly 9 000 6dFGS galaxies, as described in Springob *et al.* (2014), to form 6dFGSv - the largest and most homogeneous peculiar velocity sample to date. Using a maximum-likelihood (ML) approach, we measure the overall bulk galaxy motions from the 6dFGS velocity field for the local volume of the universe, finding broad agreement with the predicted density and velocity fields constructed from galaxy redshift surveys.

2. The 6dFGSv peculiar velocity field

In this study we use the 6dFGSv to analyse the peculiar velocity field of the southern hemisphere in the nearby ($z < 0.055$) universe. 6dFGSv provides the largest single sample of galaxy peculiar velocity measurements to date and is also more homogeneous than most previous large peculiar velocity samples. It is drawn from more than 11 000 galaxies in the 6dFGS redshift sample (6dFGSz, Jones *et al.*, 2009) for which we have FP data (see Campbell *et al.*, 2014). The methods for deriving peculiar velocities with a Bayesian approach are detailed in Springob *et al.* (2014), where smoothed maps of the individual velocities are used to perform a cosmographical analysis.

In Magoulas *et al.* (in prep.), we develop a ML model for fitting the parameters defining the peculiar velocity field (such as β and a bulk flow, $\mathbf{u}$), incorporating the velocity model reconstructions from Erdoğdu *et al.* (2006) and Branchini *et al.* (1999) as well as the 3D Gaussian FP model of Magoulas *et al.* (2012). We adopt a forward modelling approach: that is, we specify a velocity field model, apply it to the galaxies in the 6dFGSv sample, and determine whether this leads to a better ML fit of the FP in the observational parameter space, given the characterization of the FP as a 3D Gaussian distribution, and including the uncertainties in the observations and the correlations between them. We prefer the forward approach because in principle it allows simultaneous fitting of the FP and individual peculiar velocities, and in practice it provides better control over the errors and the biases they induce in the fit.

3. Redshift-space distortion parameter

We explore how well the peculiar velocities from 6dFGSv are traced by the reconstructions of both Erdoğdu *et al.* (2006) and Branchini *et al.* (1999). These models reconstruct the 3D density and velocity fields from redshift catalogues, and allow a velocity-velocity comparison with the observed velocity data to estimate the β parameter.

Erdoğdu *et al.* (2006) reconstruct the density and velocity field from the 2MASS redshift survey out to a distance of $200\,h^{-1}$ Mpc (with an $8\,h^{-1}$ Mpc grid) generated assuming a fiducial value of $\beta_{\mathrm{fid}} = 0.4$. The iterative model of Branchini *et al.* (1999) provides a complementary reconstruction of the local density and velocity field from the PSCz redshift survey out to a depth of $180\,h^{-1}$ Mpc with $\beta_{\mathrm{fid}} = 0.5$ (and a $\sim 2.8\,h^{-1}$ Mpc grid).

We compare the 6dFGSv sample with the reconstructed 2MRS velocity field (using the ML method described in Section 2) and obtain a fit of $\beta = 0.30 \pm 0.08$ (see Table 1). This is in good agreement with other studies including Davis *et al.* (2011) who find $\beta = 0.33\pm0.04$ and also Bilicki *et al.* (2011) who find $\beta = 0.38\pm0.04$ (within $\sim 1.5\sigma$ of our

Table 1. Best-fit values (and rms scatter) of β and $\mathbf{u}$ for the 6dFGSv sample ($N_g = 8885$ galaxies), fitting (1) β only, (2) both β and $\mathbf{u_{res}}$ and (3) $\mathbf{u_{tot}}$ only. For the total and residual bulk flow we also include the calculated values for the magnitude of the bulk flow ($|\mathbf{u}|$) in km s^{-1} and the bulk flow direction (l, b) in degrees.

Model	v_n	$\langle\beta\rangle$ [-]	u_x [km s^{-1}]	u_y [km s^{-1}]	u_z [km s^{-1}]	$|\mathbf{u}|$ [km s^{-1}]	l °	b °
(1) β	2MRS	0.30 ± 0.08	-	-	-	-	-	-
(2) β, $\mathbf{u_{res}}$	2MRS	0.20 ± 0.06	-251 ± 70	$+182 \pm 47$	-27 ± 51	324 ± 45	305 ± 23	35 ± 13
(1) β	PSCz	0.57 ± 0.11	-	-	-	-	-	-
(2) β, $\mathbf{u_{res}}$	PSCz	0.33 ± 0.10	-274 ± 75	$+106 \pm 55$	-9 ± 49	308 ± 59	312 ± 23	22 ± 14
(3) $\mathbf{u_{tot}}$	-	-	-358 ± 80	$+118 \pm 53$	-81 ± 56	397 ± 68	303 ± 8	17 ± 11

value) from a measurement of the clustering dipole. However Pike & Hudson (2005) find a substantially larger ($\sim 2.8\sigma$) value for β of 0.49 ± 0.04 from a comparison of their own 2MRS field reconstruction with observed velocities from three surveys. Such discrepancies may result from sensitivity to the assumed value of the linear bias in different samples the size and coverage of the samples, or the uncertainties in the peculiar velocities.

For the comparison with the PSCz reconstruction we find a best-fit β of 0.57 ± 0.11 in agreement with previous PSCz comparisons by Nusser *et al.* (2001) with $\beta = 0.5 \pm 0.1$ and also $\beta = 0.55 \pm 0.06$ found by Radburn-Smith, Lucey & Hudson (2004).

4. The Local Bulk Flow Motion

For the 6dFGSv volume, we measure a total bulk flow amplitude of $|u_{tot}| = 397\pm68$ km s^{-1} in the direction $(l, b) = (303°\pm8°, 17°\pm11°)$, larger but still comparable to previous measurements. In Figure 1 we compare the amplitude of the 6dFGSv bulk flow to other measurements shown at the effective scale of the samples from which they were derived (approximately at the limiting radius of the sample or the error-weighted radius of a Gaussian window; although different volumes sample the local structures influencing the bulk flow in different ways). In general, the measured bulk motions of most studies are within the 90% range of theoretical expectations from ΛCDM; although the 6dFGSv bulk flow amplitude, whilst outside the 90% confidence range, is still consistent with ΛCDM at the 1.5σ level. The previous measurements are also consistent with a general trend of decreasing bulk flow amplitude at large radii.

The residual bulk flow amplitude from 6dFGSv, after accounting for the mass distribution in the 2MRS field, is $|u_{res}| = 324 \pm 45$ km s^{-1} in the direction $(l, b) = (305°\pm23°, 35°\pm13°)$. In comparison, the residual bulk flow from the PSCz velocity field has a slightly smaller amplitude of $|u_{res}| = 308\pm59$ km s^{-1} in the direction $(l, b) = (312°\pm23°, 22°\pm14°)$. Both of these residual flows (and the total bulk flow) tend to point in the direction of the Shapley supercluster at $(312°, 31°)$, suggesting Shapley is playing a dominant role in the motions of the 6dFGSv volume. The 6dFGSv residual bulk flow is also a significant proportion of the total bulk flow, suggesting the supercluster might be underestimated in the model volume, either because it is under-sampled at the edge of the survey or because it extends further out.

The results presented here are preliminary, and the methods used to fit the velocity field are still being improved. In future, we plan to extend the comparison of the 6dFGSv observations to other reconstructions including the predicted 2M++ velocity field from Carrick *et al.* (submitted) and Hudson & Carrick (this volume).

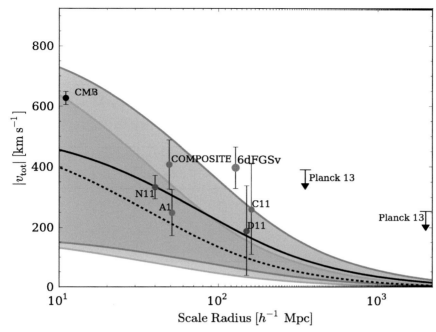

Figure 1. The bulk flow amplitude as a function of scale. The 6dFGSv bulk flow measurement is indicated in red at the radius of a sphere having the same volume as the hemispherical 6dFGSv survey. The predicted rms bulk flow in a flat ΛCDM model ($\Omega_{\mathrm{m}} = 0.274, h = 0.704$ and $\sigma_8 = 0.811$) is shown as the solid (dashed) black line for a top-hat (Gaussian) window function. The light blue and green shadings around these lines are the 90% range of scatter from cosmic variance. Bulk flow measurements from recent studies, coloured according to the most appropriate window function (blue for top-hat, green for Gaussian), are shown for Nusser *et al.* (2011, N11), Watkins *et al.* (2009, COMPOSITE), Turnbull *et al.* (2012, A1), Colin *et al.* (2011, C11), Dai *et al.* (2011, D11) and Planck Collaboration (2013, Planck13) and also the Local Group motion with respect to the CMB (Kogut *et al.*, 1993).

References

Bilicki, M., Chodorowski, M., Jarrett, T., & Mamon, G. A. 2011, *ApJ*, 741, 31
Branchini, E. *et al.* 1999, *MNRAS*, 308, 1
Campbell, L. *et al.* 2014, *MNRAS*, 443, 1231
Colin, J., Mohayaee, R., Sarkar, S., & Shafieloo, A. 2011, *MNRAS*, 414, 264
Dai, D.-C., Kinney, W. H., & Stojkovic, D. 2011, *JCAP*, 4, 15
Davis, M. *et al.* 2011, *MNRAS*, 413, 2906
Erdoğdu, P. *et al.* 2006, *MNRAS*, 373, 45
Jones, D. H. *et al.* 2009, *MNRAS*, 399, 683
Kogut, A. *et al.* 1993, *ApJ*, 419, 1
Magoulas, C. *et al.* 2012, *MNRAS*, 427, 245
Nusser, A. & Davis, M. 2011, *ApJ*, 736, 93
Nusser, A. *et al.* 2001, *MNRAS*, 320, 3
Pike, R. W. & Hudson, M. J. 2005, *ApJ*, 635, 11
The Planck Collaboration 2013, *ArXiv e-prints*, 1303.5090
Radburn-Smith, D. J., Lucey, J. R., & Hudson, M. J. 2004, *MNRAS*, 355, 4
Springob, C. M. *et al.* 2014, *MNRAS*, 445, 3
Turnbull, S. J. *et al.* 2012, *MNRAS*, 420, 447
Watkins, R., Feldman, H. A., & Hudson, M. J. 2009, *MNRAS*, 392, 743

The Zeldovich Universe:
Genesis and Growth of the Cosmic Web
Proceedings IAU Symposium No. 308, 2014
R. van de Weygaert, S. Shandarin, E. Saar & J. Einasto, eds.

© International Astronomical Union 2016
doi:10.1017/S1743921316010127

Towards an accurate model of redshift-space distortions: a bivariate Gaussian description for the galaxy pairwise velocity distributions

Davide Bianchi[1,2], Matteo Chiesa[1,2] and Luigi Guzzo[1]

[1]INAF – Osservatorio Astronomico di Brera, via Emilio Bianchi 46, I-23807 Merate, Italy
[2]Dipartimento di Fisica, Università degli Studi di Milano, via Celoria 16, I-20133 Milano, Italy
email: `davide.bianchi@brera.inaf.it`

Abstract. As a step towards a more accurate modelling of redshift-space distortions (RSD) in galaxy surveys, we develop a general description of the probability distribution function of galaxy pairwise velocities within the framework of the so-called streaming model. For a given galaxy separation $\vec{r}$, such function can be described as a superposition of virtually infinite local distributions. We characterize these in terms of their moments and then consider the specific case in which they are Gaussian functions, each with its own mean μ and variance σ^2. Based on physical considerations, we make the further crucial assumption that these two parameters are in turn distributed according to a bivariate Gaussian, with its own mean and covariance matrix. Tests using numerical simulations explicitly show that with this compact description one can correctly model redshift-space distorsions on all scales, fully capturing the overall linear and nonlinear dynamics of the galaxy flow at different separations. In particular, we naturally obtain Gaussian/exponential, skewed/unskewed distribution functions, depending on separation as observed in simulations and data. Also, the recently proposed single-Gaussian description of redshift-space distortions is included in this model as a limiting case, when the bivariate Gaussian is collapsed to a two-dimensional Dirac delta function. More work is needed, but these results indicate a very promising path to make definitive progress in our program to improve RSD estimators.

Keywords. cosmology: large-scale structure of the Universe, dark energy, theory.

1. Key concepts

The exact relation between real- and redshift-space correlation function is provided by the streaming model (Fisher 1995; Scoccimarro 2004):
$1 + \xi_S(s_\perp, s_\parallel) = \int dr_\parallel \, [1 + \xi_R(r)] \, \mathcal{P}(r_\parallel - s_\parallel | \vec{r})$, where $r^2 = r_\parallel^2 + r_\perp^2$ and $r_\perp = s_\perp$.
$\mathcal{P}(v_\parallel | \vec{r})$ is the line-of-sight pairwise velocity distribution at separation $\vec{r}$.

1.1. Modelling the velocity distribution

At each separation $\vec{r}$ we describe the velocity distribution $\mathcal{P}$ as
$\mathcal{P}(v_\parallel) = \int d\mu d\sigma \, \mathcal{P}_L(v_\parallel | \mu, \sigma) \, \mathcal{F}(\mu, \sigma)$, where $\mathcal{P}_L$ is a generic local velocity distribution parameterizable by its mean μ and standard deviation σ. $\mathcal{F}$ is the joint distribution of μ and σ.

1.2. Local Gaussianity (LG)

We explore the specific case in which the local distribution $\mathcal{P}_L$ is a Gaussian function:
$\mathcal{P}(v_\parallel) = \int d\mu d\sigma \, \mathcal{G}(v_\parallel | \mu, \sigma) \, \mathcal{F}(\mu, \sigma)$, where $\mathcal{G}(v_\parallel | \mu, \sigma) = \frac{1}{\sqrt{2\pi}\sigma} \exp\left[-\frac{(v_\parallel - \mu)^2}{2\sigma^2}\right]$.

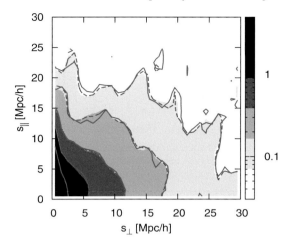

Figure 1. The redshift-space correlation function ξ_S measured from the simulated sample. The grayscale contours correspond to the direct measurement; the blue dashed contours correspond to fitting each local distribution of pairwise velocities $\mathcal{P}_L$ with a Gaussian function and measuring its two moments μ and $\sigma 2$ to empirically build their distribution function $\mathcal{F}$; the red solid curves are instead based on the further assumption that $\mathcal{F}$ is described by bivariate Gaussian. In practice, the contours demonstrate the impact of reducing the degrees of freedom in the form of the distribution function of pairwise velocities. The level of fidelity of the red solid contours when compared to the gray-scale ones shows the goodness of the bivariate Gaussian assumption. Note that the "unsmoothed" appearance of ξ_S is not at all an issue, reflecting the limited number of "local samples" involved in the specific evaluation. The goal of this exercise is to show that the same ξ_S can be obtained when using the directly measured velocity distribution, or its modelization under the increasing assumptions of the LG and GG models.

1.3. Gaussian (local) Gaussianity (GG)

We then make the further assumption that $\mathcal{F}(\mu, \sigma)$ is a bivariate Gaussian:
$\mathcal{P}(v_\parallel) = \int d\mu d\sigma \, \mathcal{G}(v_\parallel | \mu, \sigma) \, \mathcal{B}(\mu, \sigma)$, where $\mathcal{B}(\mu, \sigma) = \frac{1}{2\pi \sqrt{\det(C)}} \exp\left[-\frac{1}{2}\Delta^T C^{-1}\Delta\right]$, with $\Delta_1 = \mu - \bar{\mu}$, $\Delta_2 = \sigma - \bar{\sigma}$ and C representing the μ-σ covariance matrix.

2. Validation of the model

To test the goodness of the LG and GG descriptions we directly measure the local distributions of pairwise velocities from the MultiDark Bolshoi simulation (Riebe *et al.* 2013). The main result of our analysis is reported in Fig. 1 and discussed in the corresponding caption. More details can be found in Bianchi *et al.* (2014) togheter with additional results and a discussion on their potential implications in modelling RSD.

References

Bianchi, D., Chiesa, M., & Guzzo, L. 2014, *arXiv*:1407.4753
Fisher, K., B. 1995, *ApJ*, 448, 494
Riebe, K. *et al.* 2013, *Astronomische Nachrichten*, 334, 691
Scoccimarro, R. 2004, *Phys. Rev D*, 70, 083007

The Zeldovich Universe:
Genesis and Growth of the Cosmic Web
Proceedings IAU Symposium No. 308, 2014
R. van de Weygaert, S. Shandarin, E. Saar & J. Einasto, eds.

© International Astronomical Union 2016
doi:10.1017/S1743921316010139

Redshift-Space Distortions and $f(z)$ from Group-Galaxy Correlations

F. G. Mohammad[1], S. de la Torre[2], L. Guzzo[1], D. Bianchi[1] and J. A. Peacock[3]

[1] INAF-Osservatorio Astronomico di Brera, IT-23807, Merate (LC), Italy
email: faizan.mohammad@brera.inaf.it

[2] LAM - Laboratoire dAstrophysique de Marseille, 13388 Marseille France

[3] Institute for Astronomy-The University of Edinburgh, Edinburgh EH9 3HJ, U.K.

Abstract. We investigate the accuracy achievable on measurements of the the growth rate of structure $f(z)$ using redshift-space distortions (RSD), when (a) these are measured on the group-galaxy cross correlation function; (b) the latter is expanded over a modified version of the conventional spherical armonics, *"truncated multipole moments"*. Simulation results give first indications that this combination can push systematic errors on $f(z)$ below 3%, using scales $r \geqslant 10h^{-1}\mathrm{Mpc}$.

Keywords. Large-scale structure, cosmological parameters, clustering

---

Linear "Redshift-Space Distortions" (RSD) in two-point galaxy correlations (2PCF) measure the growth rate of structure $f(z)$ and represent a powerful test of gravity on cosmological scales. Yet, modelling the observed anisotropy of clustering to extract the linear RSD signal is complicated by non-linear contributions: the perused *Dispersion Model* applied to the galaxy 2PCF introduces systematic effects $\sim 10\%$ on the recovered values of f [e.g. Bianchi *et al.* (2012)]. Together with building more realistic models, improvement could be obtained by (a) choosing specific tracers that are less sensitive to non-linear contributions; (b) using a different two-point statistics. We show here results using the group-galaxy cross correlation (CCF), together with a modified multipole expansion, the *"truncated multipole moments"*, $\hat{\xi}^{(l)}$, which provide the practical advantage of allowing the exclusion of small scales dominated by non-linear distortions.

We first adapt the linear model [Kaiser (1987), Hamilton (1992)] to the group-galaxy CCF

$$\xi^s_{cr,Lin}(r_p, \pi) = \sum_{l=0,2,4} \xi^{s,(l)}_{cr,Lin}(s) L_l(\mu), \qquad (0.1)$$

with $\xi^{s,(l)}_{cr,Lin}(s)$ being the multipoles of the 2PCF. This expression depends directly on the parameter $\beta_{gal}(z) = f(z)/b_{gal}(z)$, the relative bias $b_{12} = b_{gal}/b_{gr}$ and the real-space 2PCF $\xi(r)$. We define the truncated multipoles of the 2PCF as

$$\hat{\xi}^{s,(l)}(s) = \frac{2l+1}{2} \int_{-\bar{\mu}}^{+\bar{\mu}} \xi^s(s, \mu) L_l(\mu) d\mu, \qquad (0.2)$$

where $\bar{\mu} = \sqrt{1 - (\bar{r}_p/s)^2}$, and $\bar{r}_p$ is the minimum transverse scale to be included in the fit. Standard multipoles $\xi^{s,(l)}$ would correspond to $\bar{\mu} = 1$. We fit to either the full 2PCF or the multipoles the standard dispersion model, in which the linear model above is empirically corrected for non-linearities through a scale-independent pairwise velocity dispersion [Peacock & Dodds (1994)].

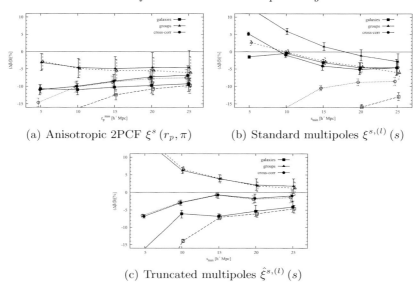

(a) Anisotropic 2PCF $\xi^s\left(r_p, \pi\right)$ (b) Standard multipoles $\xi^{s,(l)}\left(s\right)$

(c) Truncated multipoles $\hat{\xi}^{s,(l)}\left(s\right)$

Figure 1. Results of fitting the 2PCF and CCF of galaxies and groups, using the techniques discussed in the text.

All this is tested over an ensemble of simulated galaxy/halo catalogues obtained from the MultiDark Run1 (MDR1) n-body simulation [Prada *et al.* (2012)], from which we compute both ensemble statistical errors and systematic differences with respect to the known simulation growth rate (measured through β).

Figure 1 compares systematic and statistical errors obtained on the parameter β when using the 2PCF for galaxies (squares), groups (i.e. halos, triangles) and their CCF (circles) and fitting respectively the full two-point functions and the standard or truncated multipoles (panels a, b, c). Continuous lines use the full dispersion model while dashed lines correspond to fits using the pure linear model. The top-left panel simply confirms the results of Bianchi *et al.* (2012) for the auto-correlation function, i.e. a general underestimation of the growth rate, which becomes smaller for groups, i.e. higher-mass halos. Fitting the standard multipoles $\xi^{s,(l)}$, instead, yields systematic errors that are very sensitive to the range of scales included in the fit for all methods, reducing to within $\pm 5\%$ for scales $\gtrsim 15h^{-1}$Mpc. We interpret this as the effect of the projection of the 2PCF on Legendre polynomials which spreads over all scales s, non linearities originally confined to small transverse scales r_p. We defined the truncated multipoles $\hat{\xi}^{s,(l)}$ specifically to avoid this. The third panel in fact shows an even different trend, with the cross correlation providing a stable systematic error on β, confined within a few percent when using scales $\gtrsim 10h^{-1}$Mpc. In all cases the discrepancy between the dispersion model and the linear one decreases moving to larger scales. In Mohammad *et al.* (2014, in preparation) we also discuss the role of properly accounting for the covariance matrix.

References

Bianchi, D., Guzzo, L., Branchini, E., *et al.* 2012, *MNRAS*, 427, 2420
Hamilton, A. J. S. 1992, *APJL*, 385, L5
Kaiser, N. 1987, *MNRAS*, 227, 1
Peacock, J. A. & Dodds, S. J. 1994, *MNRAS*, 267, 1020
Prada, F., *et al.*, *MNRAS*, 423, 3018

The Zeldovich Universe:
Genesis and Growth of the Cosmic Web
Proceedings IAU Symposium No. 308, 2014
R. van de Weygaert, S. Shandarin, E. Saar & J. Einasto, eds.

© International Astronomical Union 2016
doi:10.1017/S1743921316010140

Quasars as tracers of cosmic flows

J. Modzelewska[1], B. Czerny[1], M. Bilicki[2], K. Hryniewicz[1,3], M. Krupa[4], F. Petrogalli[1], W. Pych[1], A. Kurcz[4] and A. Udalski[5]

[1] Nicolaus Copernicus Astronomical Center, Bartycka 18, 00-716 Warsaw, Poland
email: `jmodzel@camk.edu.pl`, `bcz@camk.edu.pl`

[2] Astrophysics, Cosmology and Gravity Centre, Department of Astronomy,
University of Cape Town, Rondebosch, South Africa

[3] ISDC Data Centre for Astrophysics, Observatoire de Geneve,
Universite de Geneve, Chemin d'Ecogia 16, 1290 Versoix, Switzerland

[4] Astronomical Observatory of the Jagiellonian University, Orla 171, 30-244 Cracow, Poland

[5] Warsaw University Observatory, Al. Ujazdowskie 4, 00-478 Warszawa, Poland

Abstract. Quasars, as the most luminous persistent sources in the Universe, have broad applications for cosmological studies. In particular, they can be employed to directly measure the expansion history of the Universe, similarly to SNe Ia. The advantage of quasars is that they are numerous, cover a broad range of redshifts, up to $z = 7$, and do not show significant evolution of metallicity with redshift. The idea is based on the relation between the time delay of an emission line and the continuum, and the absolute monochromatic luminosity of a quasar. For intermediate redshift quasars, the suitable line is Mg II. Between December 2012 and March 2014, we performed five spectroscopic observations of the QSO CTS C30.10 ($z = 0.900$) using the South African Large Telesope (SALT), supplemented with photometric monitoring, with the aim of determining the variability of the line shape, changes in the total line intensity and in the continuum. We show that the method is very promising.

Keywords. accretion disks - black hole physics, emission line, quasar - individual: CTS C30.10

1. Introduction

Quasars represent the high luminosity tail of active galactic nuclei (AGN). Among their multiple applications are probing the intergalactic medium (Borodoi *et al.* 2014) or providing information on massive black hole growth (Kelly *et al.* 2011), but they have been also proposed as promising tracers of the expansion of the Universe. The latter two aspects mostly rely on the presence of the Broad Emission Lines in quasar spectra. The BLR (Broad Line Region) is unresolved but the spectral variability allows to measure the size of the BLR from the time delay between the lines and of the continuum. For sources at redshifts $0.4 < z < 1.5$ the suitable line for such study is Mg II, monitored in the optical range. This delay is then used to determine the absolute quasar luminosity (see Czerny *et al.* 2013) employing the idea of dust origin of the BLR (Czerny & Hryniewicz 2011).

2. Results

Using the South African Large Telescope (SALT), we obtained five spectra of the QSO CTS C30.10, taken in a period of 15 months. All the spectra were analysed separately, in a relatively narrow spectral range of $2700 - 2900$ Å in the rest frame. We used 16 different FeII pseudo-continuum templates and we fit the spectra with the continuum and the Mg II line at the same time. The Mg II line in this source had to be modelled by two separate

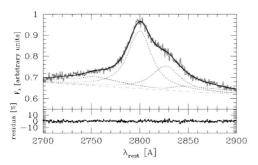

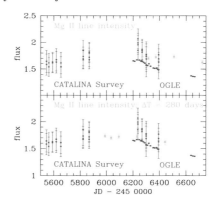

Figure 1. Best fit and residual for the 1st observation of the two kinematic components in emission of the Mg II line (dotted lines) and FeII theoretical templates [Bruhweiler & Verner (2008), d12-m20-20-5]; continuous lines show the model and the data, dashed lines give the underlying power law.

Figure 2. Time evolution of the V-band flux and the MgII line intensity as measured (upper panel), and after a shift by 280 days corresponding to a plausible time delay (lower panel).

kinematic components, meaning that CTS C30.10 is a type B source. We considered two components in emission with a double Lorentzian line shape, which provided the best fit. Using photometry from the Optical Gravitational Lensing Experiment (OGLE), we were able to calibrate the spectra properly and to obtain the calibrated line and continuum luminosity. The time dependence of the SALT Mg II flux, and OGLE and CATALINA Survey continuum luminosity are shown in Fig. 2 (Modzelewska *et al.* 2014). The monitoring of the distant quasar for 15 months has not allowed yet for any firm conclusion on the time delay between the continuum and the Mg II line; however, we can try to make some preliminary estimates based on the fact that the continuum had a clear maximum just at the beginning of our monitoring campaign.

3. Summary

Reverberation studies of quasars can be used as new cosmology probes of the expansion of the Universe. The understanding of the formation of the BLR in AGN, and in particular of the properties of the Mg II line, is also important in a much broader context. The measurement important for cosmological applications is the time delay, and this can be determined well in type B sources. Our monitoring has been too short so far to allow for a detection, but the variability pattern of the line and the continuum seems to suggest a delay of about 300 days.

References

Borodoi, R., Lilly, S. J., Kacprzak, G. G., & Churchill, C. W. 2014, *ApJ*, 784, 108
Bruhweiler, F. & Verner, E. 2008, *ApJ*, 675, 83
Czerny, B., & Hryniewicz, K. 2011, *A&A*, 525, L8
Czerny, B., Hryniewicz, K., Maity, I., *et al.* 2013, *A&A*, 556, A97
Kelly, B. C., Vestergaard, M., *et al.* 2010, *ApJ*, 719, 1315
Modzelewska, J., Czerny, B., Hryniewicz, K. *et al.* 2014, *A&A*, in press, (arXiv1408.1520)

CHAPTER 6

The Gaseous Cosmic Web

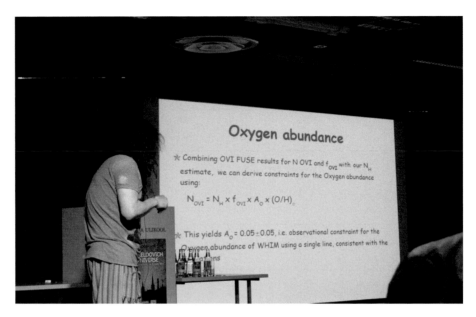

Jukka Nevalainen absorbed by WHIM.

Per Lilje in a serious assessment of presented matters.

The Zeldovich Universe:
Genesis and Growth of the Cosmic Web
Proceedings IAU Symposium No. 308, 2014   © International Astronomical Union 2016
R. van de Weygaert, S. Shandarin, E. Saar & J. Einasto, eds.   doi:10.1017/S1743921316010152

The Lyα forest and the Cosmic Web

Avery Meiksin

Institute for Astronomy, University of Edinburgh,
Edinburgh, Scotland, U.K.
email: a.meiksin@ed.ac.uk

Abstract. The accurate description of the properties of the Lyman-α forest is a spectacular success of the Cold Dark Matter theory of cosmological structure formation. After a brief review of early models, it is shown how numerical simulations have demonstrated the Lyman-α forest emerges from the cosmic web in the quasi-linear regime of overdensity. The quasi-linear nature of the structures allows accurate modeling, providing constraints on cosmological models over a unique range of scales and enabling the Lyman-α forest to serve as a bridge to the more complex problem of galaxy formation.

Keywords. intergalactic medium, quasars: absorption lines, galaxies: formation, large-scale structure of universe, dark matter, cosmological parameters

1. Introduction

One of the great achievements of twentieth century cosmology is the development of the Cold Dark Matter (CDM) paradigm. It provides a framework for describing the growth of large-scale cosmological structure formation from galactic scales to the cosmological horizon. Within the context of Friedmann-Lemaître-Robertson-Walker expanding universe models, the precise predictions of ΛCDM for the angular structure of fluctuations in the Cosmic Microwave Background (CMB) have been confirmed with an accuracy previously unprecedented in cosmology. It provides a broadly successful description of the abundance of galaxies and their clustering, and will be a key ingredient in a complete description of the origin of galaxies.

One of the major successes of the ΛCDM model is its description of the gaseous material between the galaxies, the Intergalactic Medium (IGM). The accuracy of its predictions for the IGM are second only to the CMB, and are of even wider phenomenological scope. With only one principal adjustable parameter, the mean ionization background, the statistics of the neutral hydrogen component are recovered with spectacular success.

The intergalactic medium acts also as the arena for galactic and Active Galactic Nuclei (AGN) feedback and provides a proving ground for theories of galaxy formation. In addition to hydrogen and helium, the IGM contains metals in various ionization stages. These are believed to have been deposited by galaxies through wind ejection during energetic periods of star formation. The enhanced ionization near Quasi-Stellar Objects (QSOs) (the "proximity effect") allows estimates of the Ultra-Violet (UV) metagalactic ionizing background which agree with the values required by simulations. The levels suggest QSOs contribute substantially to the ionizing budget, and may dominate at redshifts $1 < z < 3$, but are not adequate alone at higher redshifts. The most likely candidates for additional ionizing photons are galaxies, yet the direct detection of the required ionizing flux continues to prove elusive. The helium was almost certainly ionized by the hard radiation from QSOs. The ionization scenario itself, however, is still uncertain, and may have been protracted in time. At still earlier times, the ionization impact of the first stars

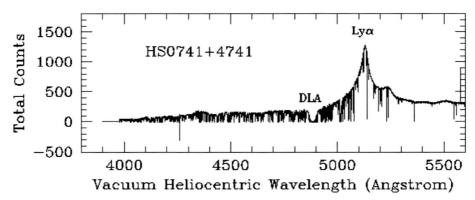

Figure 1. Spectrum of the $z = 3.2$ QSO HS0741+4741. The redshifted Lyα λ1216A emission line of the QSO is indicated. The fluctuations at shorter wavelengths comprise the Lyα forest, including a prominent Damped Lyα Absorption (DLA) system. Absorption features at longer wavelength arise from intervening absorption by metal ions such as C II, C IV, Si IV (Songaila 1996). (Figure courtesy of A. Songaila.)

will produce a distinctive signature in the 21cm absorption from the IGM against the CMB during the Epoch of Reionization.

Most recently, the study of the IGM has been undergoing a renascence with tomographic observations made possible by the huge numbers of bright QSOs delivered through ever deeper sky surveys. No longer are studies restricted to the occasional line-of-sight to a background QSO: multiple QSOs nearby on the sky are making 3D tomographic probing a reality. Motivated by attempts to constrain the evolution of the "dark energy" equation of state, surveys with multiple lines of sight passing near foreground galaxies are also fast presenting a novel means for probing the environment of the galaxies, including the possibility of the direct detection of the impact of galactic winds.

A comprehensive review of the observations and physics of the intergalactic medium is provided in Meiksin (2009).

2. Observational probes of the IGM

The main observational phenomenon for testing model predictions of the IGM is the Lyα forest, the absorption features visible in the spectra of high redshift Quasi-Stellar Objects (QSOs) due to the scattering of Lyα photons by intervening intergalactic neutral hydrogen. A representative spectrum is shown in Fig. 1. The redshifted Lyα λ1216A emission line is clearly visible. The "noise" at shorter wavelengths is not noise at all, but the Lyα forest. For comparison, the noise level is apparent at longer wavelengths, where even there absorption lines are visible. Some of these are also due to intervening intergalactic gas, but from ions of metals, such as carbon and silicon, mixed in the gas.

The flux fluctuations are characterized by their optical depths, which depend on the H I column density N_{HI} and velocity width, or Doppler parameter b, of the system:

$$\tau_0^{\mathrm{HI}} \simeq 0.38 \left(\frac{N_{\mathrm{HI}}}{10^{13}\ \mathrm{cm}^{-2}} \right) \left(\frac{b}{20\ \mathrm{km\ s}^{-1}} \right)^{-1}. \tag{2.1}$$

The absorption systems are classified according to the inferred H I column densities: systems with $\log_{10} N_{\mathrm{HI}} > 17.2$ are Lyman Limit Systems, as they are optically thick at the H I photoelectric edge, while systems with $\log_{10} N_{\mathrm{HI}} > 20.3$ are Damped Lyα absorbers, showing prominent radiation damping wings from predominantly neutral hydrogen. The

Table 1. Summary of absorption line system properties

Absorber class	$N_{\rm HI}$ (cm^{-2})	b^a (km s^{-1})	$n^b_{\rm abs}$ (m^{-3})	T^b (K)	Size (kpc)	$[M/H]^c$	N_0^d	γ^d
Lyα forest	$\lesssim 10^{17}$	15–60	$0.01 - 1000$	$5000 - 50000$	15–1000(?)	-3.5 – -2	6.1	2.47
LLS	$10^{17} - 10^{19}$	∼ 15	$\sim 10^3 - 10^4$	∼ 30000	–	-3 – -2	0.3	1.50
Super LLS	$10^{19} - 2 \times 10^{20}$	∼ 15	$\sim 10^4$	∼ 10000	–	-1 – +0.6	0.03	1.50
DLA	$> 2 \times 10^{20}$	∼ 15	$\sim 10^7$; $\sim 10^4$	∼ 100; ∼ 10000	∼ 10 – 20(?)	-1.5 – -0.8	∼ 0.03	∼ 1.5

^{a} Approximate ranges. Not well determined for most Lyman Limit Systems and super Lyman Limit Systems.
^{b} Values not well constrained by direct observations.
^{c} Approximate metallicity range, expressed as a logarithmic fraction of solar: $[M/H] = \log_{10}(M/H) - \log_{10}(M/H)_\odot$.
^{d} For the following H I column density and redshift ranges. For the Lyα forest: $13.64 < \log_{10} N_{\rm HI} < 17$ and $1.5 < z < 4$; for Lyman Limit Systems: $\log_{10} N_{\rm HI} > 17.2$ and $0.32 < z < 4.11$. The same evolution rate is adopted for super Lyman Limit Systems. The evolution rate of Damped Lyα Absorbers over the range $2 < z < 4$ is consistent with that of Lyman Limit Systems, but poorly constrained by observations.

number of systems per unit redshift interval evolves rapidly, approximately as a power law $dN/dz = N_0(1 + z)^\gamma$. Characteristic values for the absorbers and their inferred physical properties are provided in Table 1.

The absorption systems contribute substantially to the effective optical depth of the IGM, defined through $\exp(-\tau_{\rm eff}) = \langle \exp(-\tau) \rangle$, where the average is carried over a broad region of a spectrum with flux values $\exp(-\tau)$, where τ denotes the optical depth per pixel. For absorption systems optically thin at line centre,

$$\tau_{\rm eff} \simeq \frac{3}{8\pi} \Gamma_\alpha \lambda_\alpha^3 \frac{\langle n_1 \rangle}{H(z)}, \tag{2.2}$$

where Γ_α is the spontaneous decay rate of the Lyα transition, with line-centre wavelength λ_α, $\langle n_1 \rangle$ is the spatially averaged density of neutral hydrogen and $H(z)$ is the Hubble parameter at redshift z. In the limit that the hydrogen is uniform, the expression is known as the Gunn-Peterson optical depth (Gunn & Peterson 1965). For a set of discrete absorption systems of internal hydrogen density $n_{\rm abs}$, the spatially averaged hydrogen density is $\langle n_1 \rangle = Q_{\rm abs} n_{\rm abs}$, where $Q_{\rm abs}$ is the volume filling factor of the clouds.

If the gas density in the absorbers is so high the absorption features become optically thick, then the effective optical depth reduces to

$$\tau_{\rm eff} \simeq 3 Q_{\rm abs} \frac{b}{H(z)L}, \tag{2.3}$$

where b is the velocity width of the absorbers and L is their line-of-sight thickness. For typical values of $b = 25\,{\rm km s}^{-1}$ and $L = 15$ kpc from Table 1, an optical depth of $\tau_{\rm eff} \lesssim 0.3$ at $z = 3$ would correspond to $Q_{\rm abs} \simeq 0.02$, so that optically thick absorbers would occupy only a small volume of the IGM.

The possibility of a dense underbrush of unresolved optically thin Lyα lines frustrated any attempt to secure the discovery of a diffuse homogeneous component of the IGM (e.g., Jenkins & Ostriker 1991). At the same time, size constraints on the absorbers from common features measured in parallel lines of sight gave preference to large systems tens of kiloparsecs across (e.g., Smette *et al.* 1992), although the errors were large. A possible way out of the absence of any Gunn-Peterson signal was that a large fraction, possibly most, of the baryons were contained in the Lyα forest, leaving only a small residual diffuse component (McGill 1990; Meiksin & Madau 1993), although such a conclusion depended on the unknown geometry of the absorbers (Rauch & Haehnelt 1995).

3. Early models

The simplest models for the systems giving rise to the Lyα absorbers were isolated gaseous clouds, probably motivated by the analogy of nebular clouds in the interstellar medium of the Milky Way. If the IGM were neutral, the spectrum of a QSO source shortward of its Lyα emission line would be black, so great is the resonance line scattering cross section for Lyα. The absence of a detectable trough led to the suggestion that the IGM was most likely highly ionized (Gunn & Peterson 1965). A model put forward for the Lyα absorbers was one of clouds pressure-confined by a hot ionized IGM (Sargent *et al.* 1980). The origin of the clouds was unexplained. In succeeding years, suggestions included supernova remnants in galactic haloes, winds from dwarf galaxies and shells of cosmologically expanding shock waves.

The Cold Dark Matter model of structure formation (Peebles 1982; Blumenthal *et al.* 1984; Peebles 1984) provided a natural origin: gas gravitationally bound to small dark matter haloes, so-called "minihaloes" (Rees 1986). A halo with a mass of $\sim 10^9 \, M_\odot$ would retain gas at photoionization temperatures, and copious numbers of such haloes should exist at high redshifts. The Lyα forest would have been a triumphant prediction of the CDM model had it not already been observed.

The model generalizes: minihaloes are only one form of non-linear dark matter structures. Zel'dovich pointed out that gravitational collapse will generally produce 1D sheets (Zel'dovich 1970). Let $\mathbf{x}$ be the comoving position of a particle with initial comoving position $\mathbf{q}$. Its peculiar velocity is then $\mathbf{v} = a\dot{\mathbf{x}}$, where $\Psi(\mathbf{q})$ describes the initial deformation of the density field, and $a(t) = 1/(1+z)$ is the expansion factor of the Universe at the epoch corresponding to redshift z. Zel'dovich showed that the positions of the particle coordinates will evolve approximately according to

$$\mathbf{x}(\mathbf{q},t) = \mathbf{q} - D(t)\nabla\Psi(\mathbf{q}), \quad \mathbf{v}(\mathbf{q},t) = -a\dot{D}(t)\nabla\Psi(\mathbf{q}), \tag{3.1}$$

where the gradient is with respect to $\mathbf{q}$ and $D(t)$ is the growth factor for linear perturbations, $D(t) = (\dot{a}/a)\int^a da/\dot{a}^3$ (Peebles 1993).

The physical radius is related to the comoving radius through $\mathbf{r} = a(t)\mathbf{x}$. Conservation of mass gives for the density, up until the time of caustic formation, $\rho(\mathbf{x},t) = \rho(\mathbf{q})/|d^3\mathbf{x}/d^3\mathbf{q}|$, where the denominator is the determinant of the Jacobian of the coordinate transformation. The solution is exact for a 1D slab. The density prior to caustic formation grows as $\rho(x,t) = \rho_0 a(t)^{-3}/[1 - D(t)d^2\Psi/dq^2]$. A density caustic forms when the denominator vanishes. Gravitational instability ensures that a uniform density ellipsoidal density perturbation will collapse most rapidly along its shortest axis, forming a "Zel'dovich pancake."

Lyα absorption systems would arise naturally as sheet-like caustics in physical space (Miralda-Escudé & Rees 1993; Meiksin 1994) or even redshift space (McGill 1990). A wide range of models could then account for the Lyα forest, from spheres to sheets, from explosion-driven shocks to gravitational collapse.

Gathered in the city of Tallinn, with its medieval old city and thirteenth century monastery, we're reminded of the wisdom of the thirteenth century English monk William of Ockham and the point of logic he famously advocated on the economy of causes, known today as Occam's Razor: "Plurality should not be posited without necessity." The various models for the Lyα absorbers led to much discussion and debate as to which was the correct one. But another medieval doctrine rivaled Occam's Razor, dubbed by the historian A.O. Lovejoy the Principle of Plenitude: "The universe is a *plenum formarum* in which the range of conceivable diversity of *kinds* of . . . things is exhaustively exemplified." In a phenomenologically rich subject like astrophysics, a wide range of causes often produce similar signals. Under the ΛCDM paradigm, the physical systems that give rise to the Lyα absorbers are indeed highly diverse. Occam's Razor instead prevails in another guise: the variation in structure of the absorbers is a consequence of the stochastic

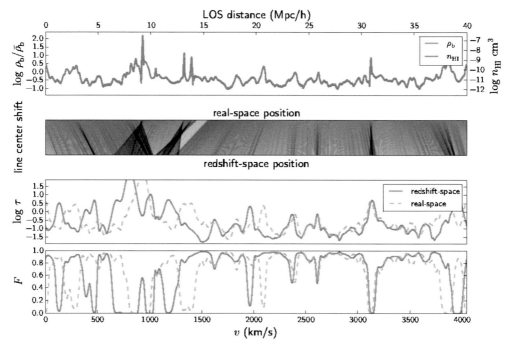

Figure 2. A line-of-sight cut through a ΛCDM simulation with 2048^3 grid zones in a box $40h^{-1}$ Mpc (comoving) on a side of the Lyα forest at $z = 2.5$. The top panel shows the total baryon overdensity and neutral hydrogen density, which are strongly correlated. The second panel shows the mapping between the real space and redshift space of gas parcels, allowing for the peculiar motion of the gas. The bottom two panels show the optical depth and flux, respectively, in both real space and velocity space. Redshift-space distortions both shift the positions of absorption features and alter their blending. From Lukić *et al.* (2014). (Figure reproduced by permission of the AAS.)

geometry of a 3D gaussian random field as realized in the process of large-scale structure formation (Bond *et al.* 1986; Bond, Kofman & Pogosyan 1996).

4. Cosmological simulations

By the late 1980s, a paradigm shift was underway in the description of the origin and structure of the Lyα forest: the absorption features would arise from the filamentary structures seen in cosmological hydrodynamical simulations (Lattanzio, Bond & Monaghan 1989). A flurry of simulations followed, using both grid-based (Cen *et al.* 1994; Zhang et al. 1995; Miralda-Escudé *et al.* 1996) and particle-based (Katz *et al.* 1996; Wadsley and Bond 1997) hydrodynamics codes. The results were sensational. Exquisite agreement was achieved between the simulations and observations of the Lyα forest.

Lines of sight through a simulation volume reveal that features originate from a range of moderate density fluctuations, and from underdense regions as well, as shown in Fig. 2. Underlying thermal motions will broaden spectral features relative to the density fluctuations, while bulk motions shift the features, sometimes sharpening them into structures similar to velocity caustics.

The evolution of the numbers of absorbers is explained predominantly as a consequence of cosmological expansion. As the density of the Universe decreases with time,

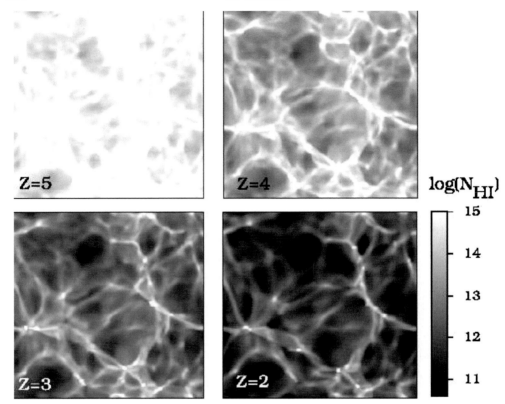

Figure 3. Evolution of the H I column density. The filamentary structure of the IGM is nearly stationary in the comoving frame. By contrast, the decline in the physical gas density results in rapid evolution of the H I column density, and the IGM becomes increasingly transparent with time. From Zhang *et al.* (1998). (Figure reproduced by permission of the AAS.)

the structures along a line of sight that give rise to systems of a given H I column density range become fewer and fewer, as shown in Fig. 3.

Fig. 4 shows that the absorption systems originate from systems with a variety of geometries, depending on column density. The highest column density systems originate in spheroidal haloes, which form at the intersections of elongated filamentary structures that give rise to the most abundant absorbers detected. Lower column density systems originate in sheets extending between the filaments, while the lowest column density systems, too small to detect in H I but visible as He II absorbers, derive from fluctuations in underdense regions.

The simulations produce excellent agreement with the statistical properties of the Lyα forest. One of the most basic is the distribution of flux per pixel. Agreement is acheived at the few percent level, as shown in Fig. 5. The data are sufficiently precise that this statistic has been used to discriminate between rival cosmologies and CDM power spectrum parameters (e.g., Meiksin *et al.* 2001).

The next higher order statistic, the flux power spectrum, shows excellent agreement at small wavenumbers (Croft *et al.* 1998). Following the measurement of the flux power spectrum using 3035 QSO spectra taken as part of the Sloan Digital Sky Survey (SDSS), cosmological simulations of the IGM were used to place constraints on the amplitude

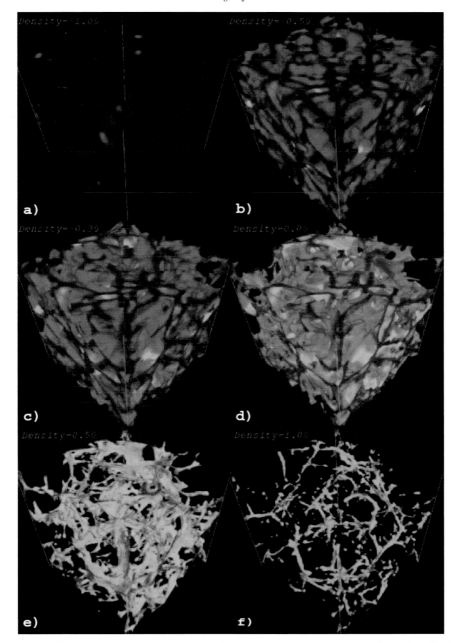

Figure 4. Isodensity contour surfaces of log baryon overdensity at $z = 3$ for a flat Cold Dark Matter dominated universe. The contour levels are $\log_{10}(\rho/\langle\rho\rangle) = -1.0, -0.5, -0.3, 0.0, 0.5$ and 1.0. Low density regions are amorphous structures filling sheets at the mean density which intersect at overdense filaments. The filaments intersect at highly overdense collapsed halos. From Zhang *et al.* (1998). (Figure reproduced by permission of the AAS.)

A. Meiksin

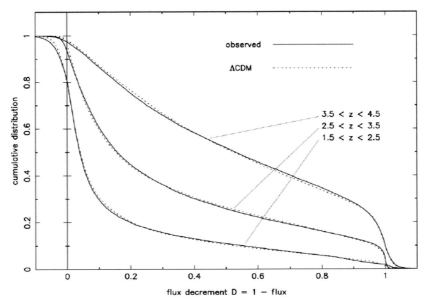

Figure 5. Comparison of the cumulative flux per pixel distribution from a ΛCDM simulation with data based on high resolution Keck spectra of 7 QSOs. The data have been grouped into three redshift intervals and the observations corrected to take into account evolution in the mean transmission within each interval. Noise has been added to the simulated data to match the noise properties of the measured spectra. From Rauch *et al.* (1997). (Figure reproduced by permission of the AAS.)

and tilt of the primordial power spectrum (McDonald *et al.* 2005, Viel & Haehnelt 2006) and on the summed mass of massive neutrino species (Viel, Haehnelt & Springel 2010).

The predicted flux power spectra do not match the observations at high wavenumbers, where line crowding and broadening cut off the power. This is likely in part due to the uncertain temperature structure of the IGM. Early numerical simulations already revealed one important discrepancy between the simulations and the observations: the simulations predict a greater number of narrow lines than measured (Davé *et al.* 1997). The likely culprit is helium and its photoionization, which were not adequately treated.

Since the discovery of intergalactic helium (Jakobsen *et al.* 1994), the observational measurement campaign of intergalactic helium has raised many puzzles. The effective optical depth of singly-ionized helium fluctuates wildly above a redshift of ∼ 2.7 (e.g., Syphers & Shull 2014; Worseck *et al.* 2014). It is unknown why. It may indicate a fluctuating metagalactic UV background, as may be expected from the relatively rare QSOs, as only QSOs have an adequately hard spectrum to photoionize He+; stellar-dominated galaxy spectra are too soft. Possibly the reionization of helium is a long and drawn out process. Since helium reionization will act as an extra source of heating, contributing to the thermal broadening of the lines, a theory of the structure of the IGM will remain incomplete without an understanding of the helium reionization process.

The reionation of helium is one of several outstanding problems of the structure of the IGM. Another is the effect galaxy formation and feedback from forming stars and black holes have on the IGM, especially near galaxies. Indeed, simulations of the IGM offer the possibility of turning the problem around by using them to constrain galaxy formation models.

5. New directions

Observational studies of the IGM continue to flourish, stimulating ever more accurate and realistic simulations. I conclude with some directions the field is taking or may take in the near future.

IGM tomography Motivated by using the Lyα forest to search for evolution in the dark energy equation of state, the BOSS experiment of SDSS-III has substantially increased the sky density of QSO lines of sight. The close proximity of large numbers of lines of sight has enabled 3D correlations in the forest to be measured over large scales for the first time using a sample of some 14000 QSOs. A positive detection of the Baryonic Acoustic Oscillation signature has been made at $z \simeq 2.4$ (Slosar *et al.* 2013), consistent with the standard ΛCDM model. Continuing observations will tighten constraints.

An intriguing spinoff of the sample is the possibility of measuring the coherence length of structure in the IGM using phase angle differences in Fourier modes of close QSO pair spectra (Rorai, Hennawi & White 2013). The technique promises a novel means of placing constraints on the temperature evolution and degree of clumpiness of the IGM.

Helium reionization Early numerical simulations simplified the reionization of hydrogen and helium by treating the gas as optically thin at the photoelectric edges. A correct treatment presents several modeling challenges: (1) a large simulation volume is required to capture a representative sample of QSOs, (2) radiative transfer must be included to produce the correct temperatures, (3) radiative hydrodynamical (RHD) simulations are necessary to track the photoevaporation of gas clumps overrun and heated by the advancing helium ionization fronts and (4) the mean intrinsic shape of the QSO spectra at energies exceeding the He+ photoelectric edge is not well known, and may be both luminosity and redshift dependent. While photoionization simulations in large volumes have been performed with radiative transfer in a post-processing phase (McQuinn *et al.* 2009; Compostella, Cantalupo & Porciani 2014), true RHD simulations have been so far restricted to smaller volumes (Meiksin & Tittley 2012).

Additional heating mechanisms Sources of high energy radiation, such as a hard x-ray background, may inject additional energy into the IGM through Compton heating (Madau & Efstathiou 1999). Additional mechanisms include heating through plasma instabilities triggered by TeV γ-rays from blazars (Chang, Broderick & Pfrommer 2012). No doubt additional heating sources will be suggested in the future which may affect the temperature of the IGM and constraints placed on the properties of QSO sources inferred from their contribution to heating and the shape of the UV metagalactic background radiation field.

Gas around galaxies The gaseous environment of galaxies is expected to be complex. Star formation within forming galaxies is believed to be regulated by the inflow of cold gas and the resulting feedback in the form of supernovae-driven winds or active galactic nuclei (Dekel *et al.* 2009). The large samples of QSOs discovered by the SDSS and the 2dF survey have opened up a new means of probing the gaseous environment of galaxies through the absorption signature against background QSOs (e.g., Crighton *et al.* 2011; Rudie *et al.* 2012). The larger BOSS samples will further enhance these studies. IGM simulations allowing for star formation and feedback offer a new avenue for constraining galaxy formation models (e.g., Kollmeier *et al.* 2006; Rakic *et al.* 2013; Shen *et al.* 2013; Meiksin, Bolton & Tittley 2014).

Metal absorption systems The discovery of metal absorption systems in the IGM predates the identification of intergalactic Lyα absorption. While metals are widely regarded to have been introduced into the IGM by galaxies, possibly in winds as part of the galaxy formation process, it is unknown at what stage this happened. Even the structure of the metal absorbers is unclear. Photoionization modeling suggests they are co-extensive with the H I systems in which they are embedded on the scales of ~ 10 kpc (e.g., Simcoe *et al.* 2006), but they need not be uniformly distributed within them. They could, for instance, be transient nodules only 100 pc in size (Schaye, Carswell & Kim 2007).

Establishing the physical state of the gas and the spectral shape and intensity of the radiation field that ionizes them is essential for inferring the metallicity of the absorbers and its evolution. Doing so would enable the metal absorbers to be used as a means for tracing the history of cosmic star formation. The spectral capacity of the TMT and the E-ELT will permit high precision studies of the metal systems. It will also directly reveal the presence, if any, of metals in the low density diffuse IGM for the first time, eliminating the major modeling uncertainty of radiative transfer through layers of hydrogen and helium optically thick at the photoelectric edges. They raise the possibility as well of detecting signatures of time-dependent metal ionization, placing constraints on the lifetimes or variability of QSO sources (Reynolds 2010; Oppenheimer & Schaye 2013).

Missing physics We do not know if we have all the right physics. For example, intergalactic shocks could produce a pervasive intergalactic magnetic field. Even a weak field could alter the gas fluctuations sufficiently to have implications for the inferred ionization background required to recover the measured mean flux through the IGM (Chongchitnan & Meiksin 2014). The cold dark matter particle may interact with itself, annihilate or decay, altering gas density profiles in collapsed haloes, including minihaloes. Possibly even general relativity must be modified on scales that would affect the large-scale distribution of matter. While current observations do not require any of these more exotic possibilities, as simulations and observations of the IGM continue to increase in precision discrepancies may find the IGM to be the first arena to offer solid evidence for the need of additional physics.

References

Bardeen, J. M., Bond, J. R., Kaiser, N., & Szalay, A. S. 1986, *ApJ*, 304, 15
Blumenthal, G. R., Faber, S. M., Primack, J. R., & Rees, M. J. 1984, *Nature*, 311, 517
Bond, J. R., Kofman, L., & Pogosyan, D. 1996, *Nature*, 380, 603
Cen, R., Miralda-Escudé, J., Ostriker, J. P., & Rauch, M. 1994, *ApJ* (Letters), 437, L9
Chang, P., Broderick, A. E., & Pfrommer, C. 2012, *ApJ*, 752, 23
Chongchitnan, S. & Meiksin, A. 2014, *MNRAS*, 437, 3639
Compostella, M., Cantalupo, S., & Porciani, C. 2014, *arXiv*, 1407.1316
Crighton N. H. M.. *et al.* 2011, *MNRAS*, 414, 28
Croft, R. A. C., Weinberg, D. H., Katz, N., & Hernquist, L. 1998, *ApJ*, 495, 44
Davé, R., Hernquist, L., Weinberg, D. H., & Katz, N. 1997, *ApJ*, 477, 21
Dekel, A. *et al.* 2009, *Nature*, 457, 451
Gunn, J. E. & Peterson, B. A. 1965, *ApJ*, 142, 1633
Jakobsen, P., Boksenberg, A., Deharveng, J. M., Greenfield, P., Jedrzejewski, R., & Paresce, F. 1994, *Nature*, 370, 35
Jenkins, E. B. & Ostriker, J. P. 1991, *ApJ*, 376, 33
Katz, N., Weinberg, D. H., Hernquist, L., & Miralda-Escudé, J. 1996, *ApJ* (Letters), 457, L57
Kollmeier, J. A., Miralda-Escudé, J., Cen, R., & Ostriker, J. P. 2006, *ApJ*, 638, 52
Lattanzio, J. C., Bond, J. R., & Monaghan, J. J. 1989, *BAAS*, 21, 1216
Lukić, Z., Stark, C. W., Nugent, P., White, M., Meiksin, A., & Almgren, A. 2014, *ApJ* (in press)
Madau, P. & Efstathiou, G. 1999, *ApJ* (Letters), 517, L9
McDonald, P. *et al.* 2005, *ApJ*, 635, 761
McGill, C. 1990, *MNRAS*, 242, 544
McQuinn, M., Lidz, A., Zaldarriaga, M., Hernquist, L., Hopkins, P. F., Dutta, S., & Faucher-Giguère, C.-A. 2009, *ApJ*, 694, 842

Meiksin, A. 1994, *ApJ*, 431, 109

Meiksin, A. 2009, *Rev. Mod. Phys.*, 81, 1405

Meiksin, A., Bolton, J. S., & Tittley, E. R. 2014, *MNRAS* (in press)

Meiksin, A., Bryan, G., & Machacek, M. 2001, *MNRAS*, 327, 296

Meiksin, A. & Madau, P. 1993, *ApJ*, 412, 34

Meiksin, A. & Tittley, E. R. 2012, *MNRAS*, 423, 7

Miralda-Escudé, J. & Rees, M. J. 1993, *MNRAS*, 260, 617

Miralda-Escudé, J., Cen, R., Ostriker, J. P., & Rauch, M. 1996, *ApJ*, 471, 582

Oppenheimer, B. D. & Schaye, J. 2013, *MNRAS*, 434, 1063

Peebles, P. J. E.. 1982, *ApJ* (Letters), 263, L1

Peebles, P. J. E.. 1984, *ApJ*, 277, 470

Peebles, P. J. E.. 1993, *Principles of Physical Cosmology* (Princeton, NJ: Princeton University Press)

Rakic, O., Schaye, J., Steidel, C. C., Booth, C. M., Dalla Vecchia, C., & Rudie, G. C. 2013, *MNRAS*, 433, 3103

Rauch, M., Miralda-Escudé, J., Sargent, W. L. W., Barlow, T. A., Weinberg, D. H., Hernquist, L., Katz, N., Cen, R., & Ostriker, J. P. 1997, *ApJ*, 489, 7

Rauch, M. & Haehnelt, M. G. 1995, *MNRAS*, 275, L76

Rees, M. J. 1986, *MNRAS*, 218, 25P

Reynolds, S. 2010, PhD thesis, University of Edinburgh

Rorai, A., Hennawi, J. F., & White, M. 2013, *ApJ*, 775, 81

Rudie, G. C. *et al.* 2012, *ApJ*, 750, 67

Sargent, W. L. W., Young, P. J., Boksenberg, A., & Tytler, D. 1980, *ApJS*, 42, 41

Schaye, J., Carswell, R. F., & Kim, T.-S. 2007, *MNRAS*, 379, 1169

Shen, S., Madau, P., Guedes, J., Mayer, L., Prochaska, J. X., & Wadsley, J. 2013, *ApJ*, 765, 89

Simcoe, R. A., Sargent, W. L. W., Rauch, M., & Becker, G. 2006, *ApJ*, 637, 648

Slosar, A., *et al.* 2013, *JCAP*, 4, 26

Smette, A., Surdej, J., Shaver, P. A., Foltz, C. B., Chaffee, F. H., Weymann, R. J., Williams, R. E., & Magain, P. 1992, *ApJ*, 389, 39

Songaila, A. 2006, *AJ*, 131, 24

Syphers, D. & Shull, J. M. 2014, *ApJ*, 784, 42

Viel, M. & Haehnelt, M. G. 2006, *MNRAS*, 365, 231

Viel, M., Haehnelt, M. G., & Springel, V. 2010, *JCAP*, 6, 15

Wadsley, J. W. & Bond J. R. 1997, *ASP-CS*, 123, 332

Worseck, G., Prochaska, J. X., Hennawi, J. F., & McQuinn, M. 2014, *arXiv*, 1405.7405

Zel'dovich, Ya. B. 1970, *A & A*, 5, 84

Zhang, Y., Meiksin, A., Anninos, P., & Norman M. L. 1998, *ApJ*, 495, 63

Zhang, Y., Anninos, P., & Norman M. L. 1995, *ApJ* (Letters), 453, L7

The Zeldovich Universe:
Genesis and Growth of the Cosmic Web
Proceedings IAU Symposium No. 308, 2014
R. van de Weygaert, S. Shandarin, E. Saar & J. Einasto, eds.

ⓒ International Astronomical Union 2016
doi:10.1017/S1743921316010164

Lyα Forest Tomography of the $z > 2$ Cosmic Web

Khee-Gan Lee

Max-Planck Institut für Astronomie,
Königstuhl 17, Heidelberg, 69117, Germany

Abstract. The hydrogen Lyα forest is an important probe of the $z > 2$ Universe that is otherwise challenging to observe with galaxy redshift surveys, but this technique has traditionally been limited to 1D studies in front of bright quasars. However, by pushing to faint magnitudes ($g > 23$) with 8-10m large telescopes it becomes possible to exploit the high area density of high-redshift star-forming galaxies to create 3D tomographic maps of large-scale structure in the foreground. I describe the first pilot observations using this technique, as well discuss future surveys and the resulting science possibilities for galaxy evolution and cosmology.

Keywords. cosmology: observations — galaxies: high-redshift — intergalactic medium — quasars: absorption lines — surveys — techniques: spectroscopic

1. Introduction

For decades, the hydrogen Lyα forest absorption observed in the spectra of background quasars has been a crucial probe of the $z > 2$ universe. Since the realization in the mid-1990s that residual neutral hydrogen in the photoionized intergalactic medium (IGM) is a non-linear tracer of the overall matter density distribution in the large-scale cosmic web (the 'fluctuating Gunn-Peterson' paradigm Cen *et al.* 1994; Bi *et al.* 1995), the Lyα forest has enabled statistical studies of large-scale structure (LSS) at high redshifts.

The Lyα forest in each quasar spectrum is a one-dimensional probe of the foreground IGM, although a small number of close-pairs exist (e.g., D'Odorico *et al.* 2006). The area density of available quasars increase with limiting magnitude, and at faint magnitudes it becomes possible to study the 3D correlations of the Lyα forest across different lines-of-sight: the BOSS Lyα forest survey (Lee *et al.* 2013) has observed $g \leqslant 21.5$ quasars with an average area density of $15 \deg^{-2}$, enabling measurement of the large-scale two-point Lyα forest correlation function in 3D (Slosar *et al.* 2011) and subsequently the detection of the $\approx 100\,h^{-1}$ Mpc baryon acoustic oscillation scale (Slosar *et al.* 2013; Delubac *et al.* 2014), which is useful as a cosmological distance measure.

Beyond clustering studies, it is also possible to directly interpolate separate Lyα forest sightlines to create a 3D map of the hydrogen absorption. This 'Lyα forest tomography' was conceptually first proposed by Pichon *et al.* (2001) (see also Caucci *et al.* 2008), although the first pilot observations were not conducted until recently (Lee *et al.* 2014b). In this article we will give a brief overview of this exciting new technique in terms of feasibility and initial pilot observations, as well as discuss future prospects.

2. Feasibility

In simplest terms, Lyα forest tomography simply interpolates between a set of Lyα forest spectra to create a map of the foreground IGM absorption. Therefore the typical transverse separation, $\langle d_\perp \rangle$, of the background sources must match the spatial scale of the desired map, ϵ_{3D}. At $z = 2.3$, the typical transverse separation of the BOSS Lyα forest sightlines is $\langle d_\perp \rangle \sim 20\,h^{-1}$ Mpc which is larger than most scales of interest, although there are efforts to create tomographic maps with BOSS (Cisewski *et al.* 2014).

The $\langle d_\perp \rangle$ is reduced by including fainter background sources, but the quasar luminosity function is relatively shallow (Palanque-Delabrouille *et al.* 2013) and even at $g \leqslant 24$, the typical sightline separation probed by quasars is $\langle d_\perp \rangle \sim 15\,h^{-1}$ Mpc. However, at $g \gtrsim 23$, star-forming galaxies (SFGs) begin to dominate the overall UV luminosity function (Reddy *et al.* 2008), allowing very dense grids to sample the foreground forest: at $g \leqslant [24, 25]$ the sightline separations are $\langle d_\perp \rangle \approx [4, 1]\,h^{-1}$ Mpc, respectively.

The question now turns to the requirements in terms of spectral resolution and signal-to-noise (S/N) in the data. If high-resolution ($R \sim 30,000$, $S/N \gtrsim 20$ per pixel) spectra are needed, then Lyα tomography will be a limited 'novelty' technique even with 30m class telescopes, due to the extreme exposure times (> 10hr) needed to obtain such data with $g \sim 24$ sources. Conversely, relatively noisy (S/N of a few) moderate-resolution spectra ($R \sim 200 - 1000$) are already achievable with existing 8-10m telescopes.

In Lee *et al.* (2014a), we studied this issue using both simulations and analytic calculations. In terms of spectral resolution, the desired reconstruction scale, ϵ_{3D}, must be resolved along the line-of-sight (LOS), therefore one needs merely $R \geqslant 1300(1\,h^{-1}\ \mathrm{Mpc}/\epsilon_{3D})$ $[(1+z)/3.25]^{-1/2}$. As for the necessary S/N, the requirements depend on ϵ_{3D} as well as the desired map fidelity. Since there will always be reconstruction errors on scales approaching the skewer separation, $\langle d_\perp \rangle$, due to the finite transverse sampling ('aliasing'), the individual spectra do not need to oversample the absorption along the LOS. In simulated reconstructions using noisy mock data, we found that $S/N = 4$ per Å at a survey limit of $g = 24.2$ is sufficient to give tomographic maps that look similar to the 'true' simulated field on scales of $\epsilon_{3D} \approx 3.5\,h^{-1}$ Mpc. This is achievable with existing telescopes: e.g. with the VLT-VIMOS spectrograph (Le Fèvre *et al.* 2003) such data would require 6hr exposure times per pointing. The requirements do increase if we desire greater spatial resolutions: Lyα forest tomography resolving $\epsilon_{3D} \lesssim 1\,h^{-1}$ Mpc (i.e. $\lesssim 400$kpc physical at $z \sim 2.5$) would require $g > 25$ background galaxies, which are accessible only with future 30m-class telescopes.

3. Pilot Observations

In March 2014, we obtained moderate-resolution ($R \approx 1000$) spectra of 24 SFGs at redshifts $2.3 \lesssim z \lesssim 3$, down to limiting magnitudes of $g \sim 24.7$, using the LRIS spectrograph on the Keck I telescope on Mauna Kea, Hawaii. These SFGs were all within a $5' \times 14'$ area of the COSMOS field (Scoville *et al.* 2007), resulting in an effective area density of ~ 1000 deg^{-2} or a nominal transverse separation of $\langle d_\perp \rangle \approx 2.3\,h^{-1}$ Mpc.

The data was taken with ≈ 2hr exposure times resulting in S/N $\sim 1.5 - 4$ per pixel. It may seem aggressive to use such noisy data, but our tomographic reconstruction uses a Wiener filtering algorithm (Wiener 1942) that incorporates noise-weighting. Another concern is the possible contamination by intrinsic SFG absorption lines, but studies of composite SFG spectra (Shapley *et al.* 2003; Berry *et al.* 2012) have revealed a relative lack of strong intrinsic absorption lines in the relevant $1040 - 1180$Å restframe region once the foreground Lyα forest fluctuations are averaged out by stacking. Moreover, UV spectroscopic observations of a small number of low-z SFGs (e.g. Heckman *et al.* 2011), in which the Lyα forest is negligible, have revealed a relatively flat 'continuum', with only 2-3 strong absorbers that are straightforward to mask. This is supported by the

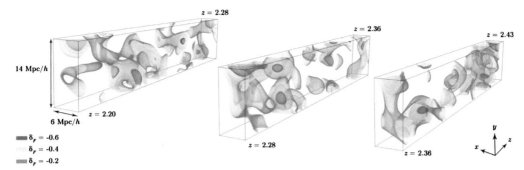

Figure 1. 3D visualization of the Lyα forest tomographic map obtained from the foreground absorption in 24 background SFG spectra, split into 3 different redshift segments for clarity. The color scale represents the relative Lyα absorption, such that lower values correspond to higher densities. Excerpted from Lee *et al.* (2014b).

high-resolution spectrum of the the lensed SFG cB158, which allowed Voigt-profile analysis of the absorbers in its Lyα forest region (Savaglio *et al.* 2002).

In Figure 1 we show the resulting 3D tomographic Lyα forest reconstruction (see Lee *et al.* 2014b), spanning $2.2 \leqslant z \leqslant 2.45$ with a spatial resolution of $\epsilon_{3D} \approx 3.5 \, h^{-1}$ Mpc. The elongated map geometry is because bad weather limited us to a small transverse area, whereas each sightline probes a long pathlength $\sim 300 \, h^{-1}$ Mpc along the LOS.

Despite the narrow geometry, one sees large coherent structures spanning $\gtrsim 10 \, h^{-1}$ Mpc across the entire transverse dimension. These structures are contributed by multiple sightlines and are therefore unlikely to be due to random noise fluctuations nor intrinsic absorption (since the background SFG redshifts are not aligned). Simulated reconstructions on mock data with the exact same spatial sampling and S/N distribution yield a good recovery of LSS features on scales of a few Mpc, although there are distortions on smaller scales. We also compared the map with a small number of coeval galaxies with known spectroscopic redshifts (primarily from zCOSMOS, Lilly *et al.* 2007). In Lee *et al.* (2014b) we have shown that the galaxies live preferentially in lower-flux regions (i.e. overdensities), although this correlation is weakened somewhat by redshift errors on the coeval galaxies and tomographic reconstruction errors.

4. Science Applications & Future Prospects

Our pilot observations have established the feasibility of Lyα forest tomography, and motivates the COSMOS Lyman-Alpha Mapping and Tomography Observations (CLAMATO) survey, which aims to obtain ~ 1000 SFG spectra within $\sim 1 \text{deg}^2$ of the COSMOS field, aimed at creating a tomographic reconstruction of the Lyα forest absorption at $\langle z \rangle \sim 2.3$ over a comoving volume equivalent to $\sim (100 \, h^{-1} \text{ Mpc})^3$.

This survey will require ~ 15 nights of large-telescope time, and will have various science applications: **(I)** The large-scale morphology and topology of the cosmic web has never been studied at $z \gtrsim 1$, and the tomographic map will allow us to study this on scales of $\gtrsim 3-4 \, h^{-1}$ Mpc. **(II)** Using Lyα forest absorption as a proxy for the underlying density field, we will be able to study the properties of $z \sim 2$ coeval galaxies as a function of their large-scale environment, yielding unique insights into this crucial era in galaxy formation and evolution. **(III)** Within the volume covered by the map, we expect to find ~ 10 progenitors of massive $M > 10^{14} M_\odot$ galaxy clusters. At $z \gtrsim 2$, these protoclusters are expected to manifest themselves as overdensities of a few on spanning $\sim 10 \, h^{-1}$ Mpc

True Lyman-alpha Transmission Field (3Mpc/h smoothing)   Reconstructed Lyman-alpha Transmission Field (3 Mpc/h smoothing)

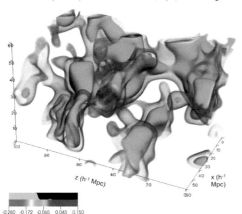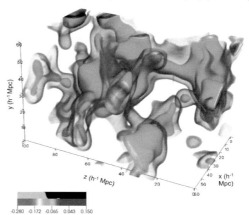

-0.280 -0.172 -0.065 0.043 0.150   -0.280 -0.172 -0.065 0.043 0.150

Figure 2. (Right) Tomographic reconstruction using realistic mock data simulating the $1 \deg^2$ CLAMATO survey, compared with the true 3D absorption field (left). Both fields have an effective smoothing of $\epsilon_{3D} = 3 \, h^{-1}$ Mpc. The simulated survey data clearly recovers the LSS in the volume. Note that the z-dimension here is the LOS dimension, and in the real survey will span $\sim 2 - 3\times$ the distance shown here.

(c.f. Chiang *et al.* 2013), which should be easily detectable with Lyα forest tomography. Finally, **(IV)** the data will allow us to probe 3D small-scale Lyα forest clustering on scales of $\lesssim 10 \, h^{-1}$ Mpc, complementary to the larger scales probed by the BOSS survey (Slosar *et al.* 2011). These measurements will be valuable in arriving at a comprehensive model of the Lyα forest absorption.

References

Berry, M., Gawiser, E., Guaita, L., *et al.* 2012, ApJ, 749, 4
Bi, H., Ge, J., & Fang, L.-Z. 1995, ApJ, 452, 90
Caucci, S., Colombi, S., Pichon, C., *et al.* 2008, MNRAS, 386, 211
Cen, R., Miralda-Escudé, J., Ostriker, J. P., & Rauch, M. 1994, ApJL, 437, L9
Chiang, Y.-K., Overzier, R., & Gebhardt, K. 2013, ApJ, 779, 127
Cisewski, J., Croft, R. A. C., Freeman, P. E., *et al.* 2014, MNRAS, 440, 2599
Delubac, T., Bautista, J. E., Busca, N. G., *et al.* 2014, ArXiv e-prints, arXiv:1404.1801
D'Odorico, V., Viel, M., Saitta, F., *et al.* 2006, MNRAS, 372, 1333
Heckman, T. M., Borthakur, S., Overzier, R., *et al.* 2011, ApJ, 730, 5
Le Fèvre, O., Saisse, M., Mancini, D., *et al.* 2003, in SPIE Conference Series, Vol. 4841, ed.
    M. Iye & A. F. M. Moorwood, 1670–1681
Lee, K.-G., Hennawi, J. F., White, M., Croft, R. A. C., & Ozbek, M. 2014a, ApJ, 788, 49
Lee, K.-G., Bailey, S., Bartsch, L. E., *et al.* 2013, AJ, 145, 69
Lee, K.-G., Hennawi, J. F., Stark, C., *et al.* 2014b, ArXiv e-prints, arXiv:1409.5632
Lilly, S. J., Le Fèvre, O., Renzini, A., *et al.* 2007, ApJS, 172, 70
Palanque-Delabrouille, N., Magneville, C., Yèche, C., *et al.* 2013, A&A, 551, A29
Pichon, C., Vergely, J. L., Rollinde, E., Colombi, S., & Petitjean, P. 2001, MNRAS, 326, 597
Reddy, N. A., Steidel, C. C., Pettini, M., *et al.* 2008, ApJS, 175, 48
Savaglio, S., Panagia, N., & Padovani, P. 2002, ApJ, 567, 702
Scoville, N., Aussel, H., Brusa, M., *et al.* 2007, ApJS, 172, 1
Shapley, A. E., Steidel, C. C., Pettini, M., & Adelberger, K. L. 2003, ApJ, 588, 65
Slosar, A., Font-Ribera, A., Pieri, M. M., *et al.* 2011, JCAP, 9, 1
Slosar, A., Iršič, V., Kirkby, D., *et al.* 2013, JCAP, 4, 26
Wiener, N. 1942, 'Interpolation, extrapolation and smoothing of stationary time series' (MIT)

The Zeldovich Universe:
Genesis and Growth of the Cosmic Web
Proceedings IAU Symposium No. 308, 2014
R. van de Weygaert, S. Shandarin, E. Saar & J. Einasto, eds.
© International Astronomical Union 2016
doi:10.1017/S1743921316010176

The intergalactic medium in the cosmic web

Nicolas Tejos†

Department of Astronomy and Astrophysics,
UCO/Lick Observatory, University of California,
1156 High Street, Santa Cruz, CA 95064, USA
email: `ntejos@ucolick.org`

Abstract. The intergalactic medium (IGM) accounts for $\gtrsim 90\%$ of baryons at all epochs and yet its three dimensional distribution in the cosmic web remains mostly unknown. This is so because the only feasible way to observe the bulk of the IGM is through intervening absorption line systems in the spectra of bright background sources, which limits its characterization to being one-dimensional. Still, an averaged three dimensional picture can be obtained by combining and cross-matching multiple one-dimensional IGM information with three-dimensional galaxy surveys. Here, we present our recent and current efforts to map and characterize the IGM in the cosmic web using galaxies as tracers of the underlying mass distribution. In particular, we summarize our results on: (i) IGM around star-forming and non-star-forming galaxies; (ii) IGM within and around galaxy voids; and (iii) IGM in intercluster filaments. With these datasets, we can directly test the modern paradigm of structure formation and evolution of baryonic matter in the Universe.

Keywords. intergalactic medium; cosmology: large scale structure of the Universe; galaxies: formation; quasars: absorption lines

1. Introduction

The physics of the intergalactic medium (IGM) and its connection with galaxies are key to understanding the evolution of baryonic matter in the Universe. The IGM is the main reservoir of baryons at all epochs (e.g. Fukugita *et al.* 1998; Shull *et al.* 2012), and provides the primordial material for forming galaxies. Once galaxies are formed, supernovae (SNe) and active-galactic nuclei (AGN) feedback inject energy in the interstellar medium, some of which escapes the galaxies as winds, enriching the IGM with metals (e.g. Wiersma *et al.* 2011; Ford *et al.* 2014). Because of the continuous interplay between the IGM and galaxies, it is sensible (if not necessary) to study these two concepts simultaneously (e.g. Morris *et al.* 1993; Lanzetta *et al.* 1995; Tripp *et al.* 1998; Chen & Mulchaey 2009; Prochaska *et al.* 2011; Tumlinson *et al.* 2011; Tejos *et al.* 2014; Werk *et al.* 2014).

The large scale environment in which matter resides also plays an important role. Given that baryonic matter is expected to fall into the considerably deeper gravitational potentials of dark matter, the IGM gas and galaxies should be predominantly found at such locations, forming the so-called 'cosmic web' (Bond *et al.* 1996). Galaxies appear to follow the filamentary structure which simulations predict (e.g. Springel *et al.* 2006), and their properties are partly shaped by environmental effects (e.g. Dressler 1980; Skibba *et al.* 2009). However, much less is known about the *actual* properties and distribution of the IGM in different cosmological environments. This is so because the only feasible way to observe the bulk of the IGM is through intervening absorption line systems in the spectra of bright background sources (e.g. quasi-stellar objects, gamma-ray bursts, galaxies), which limits its characterization to being one-dimensional.

† On behalf of our full collaboration.

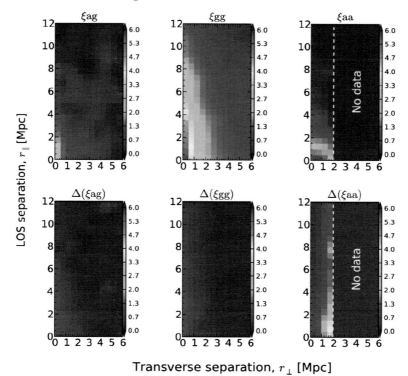

Figure 1. Two-dimensional two-point correlation functions (top panels) at $z \lesssim 1$, and their respective uncertainties (bottom panels). From left to right: H I–galaxy (ξ_{ag}), galaxy–galaxy (ξ_{gg}) and H I–H I (ξ_{aa}) cross-correlations. Figure adapted from Tejos *et al.* (2014).

The advent of big galaxy surveys such as the 2dFGRS (Colless *et al.* 2001) or the SDSS (Abazajian *et al.* 2009), have revolutionized the study of the cosmic web and the large-scale structure (LSS) of the Universe. This is eloquently demonstrated by the plethora of LSS catalogs that are currently available: from galaxy voids (e.g. Pan *et al.* 2012; Sutter *et al.* 2012; Nadathur & Hotchkiss 2014; Way *et al.* 2014), galaxy filaments (e.g. Tempel *et al.* 2014), to galaxy groups and clusters (e.g. Hao *et al.* 2010; Rykoff *et al.* 2014). By combining and cross-matching multiple one-dimensional IGM information with galaxy and LSS surveys, an averaged three dimensional picture can be obtained. Here, we present our recent and current efforts to map and characterize the IGM in the cosmic web using galaxies as tracers of the underlying mass distribution.

2. The IGM-galaxy cross-correlation

The two-point correlation function between neutral hydrogen (H I) and galaxies is a powerful statistical technique to assess the connection between the IGM and galaxies (e.g. Chen *et al.* 2005; Ryan-Weber 2006; Wilman *et al.* 2007; Chen & Mulchaey 2009; Shone *et al.* 2010; Tejos *et al.* 2014).

In Tejos *et al.* (2014), we have recently published observational results on the H I–galaxy two-point cross-correlation at $z \lesssim 1$ (ξ_{ag}; see Fig. 1). These results come from the largest sample ever done for such an analysis, comprising about ~ 700 H I absorption line systems in the UV spectra of 8 background QSOs, in 6 different fields observed with the HST, and about ~ 17000 galaxies with spectroscopic redshifts around these QSOs sightlines, coming from our own spectroscopic surveys, and previously published catalogs by the VVDS (Le Fèvre *et al.* 2013) and GDDS (Abraham *et al.* 2004) galaxy surveys.

Apart from ξ_{ag}, we also measured the H I–H I (ξ_{aa}) and galaxy-galaxy (ξ_{gg}) two-point auto-correlations. Our survey is one of the few in which these three quantities have been measured from the same dataset, and independently from each other. Comparing the results from ξ_{ag}, ξ_{aa} and ξ_{gg}, we constrained the IGM-galaxy statistical connection, as a function of both H I column density and galaxy star formation activity, on $\sim 0.5 - 10\,\mathrm{Mpc}$ scales. Our results are consistent with the following conclusions: (i) the bulk of H I systems on $\sim$ Mpc scales have little velocity dispersion ($\lesssim 120\,\mathrm{km\,s^{-1}}$) with respect to the bulk of galaxies (i.e. no strong galaxy outflow/inflow signal is detected); (ii) the vast majority ($\sim 100\%$) of H I systems with $N_{\mathrm{HI}} > 10^{14}\,\mathrm{cm^{-2}}$ and star-forming galaxies are distributed in the same locations, together with $75\pm15\%$ of non-star forming galaxies; (iii) $25\pm15\%$ of non-star-forming galaxies reside in galaxy clusters and are not correlated with H I systems at scales $\lesssim 2\,\mathrm{Mpc}$; and (iv) $> 50\%$ of H I systems with $N_{\mathrm{HI}} < 10^{14}\,\mathrm{cm^{-2}}$ reside within galaxy voids and hence are not correlated with luminous galaxies.

3. The IGM within and around galaxy voids

In Tejos *et al.* (2012) we have recently measured the properties of H I absorption line systems within and around galaxy voids at $z \leqslant 0.1$, using the galaxy void catalog published by Pan *et al.* (2011) and the low-z H I absorption line catalog published by Danforth & Shull (2008). Our key findings can be summarized as follows: (i) there is a significant excess of IGM gas at the edges of galaxy voids with respect to the random expectation, *consistent* with the overdensity of galaxies defining such voids; and (ii) inside galaxy voids the IGM gas matches the random expectation, *inconsistent* with the underdensity of galaxies defining such voids. In other words, there were no apparent IGM voids detected at the positions of galaxy voids.

We also showed that the column density (N_{HI}) and Doppler parameter (b_{HI}) distributions of H I lines inside and outside galaxy voids were not remarkably different, with only a $\sim 95\%$ and $\sim 90\%$ probability of rejecting the null-hypothesis of both samples coming from the same parent population, respectively. Still, a trend was present, in which galaxy void absorbers have systematically lower values of both N_{HI} and b_{HI} than those found outside galaxy voids. By performing a similar analysis using a state-of-the-art hydrodynamical cosmological simulation (GIMIC; Crain *et al.* 2009), we showed that these observed trends are qualitatively consistent with current theoretical expectations. However, more quantitative comparisons of the gas properties in galaxy voids between hydrodynamical simulations tuned to explore low density environments (e.g. Ricciardelli *et al.* 2013) and observations (e.g. Tejos *et al.* 2012), are required.

4. The IGM in intercluster filaments

Galaxy clusters represent the densest nodes in the cosmic web (with dark matter halo masses of $M \gtrsim 10^{14} M_\odot$) and as such, N-body numerical simulations predict a high probability of finding intercluster filaments between galaxy cluster pairs when separated by $\lesssim 10 - 20\,\mathrm{Mpc}$ (e.g. Colberg *et al.* 2005; González & Padilla 2010). Hydrodynamical simulations predict that an important fraction of baryons at low-z are in a diffuse, shock heated gas phase with $T \sim 10^5 - 10^6\,\mathrm{K}$ in these dense filaments, commonly referred to as the warm-hot intergalactic medium (WHIM; Cen & Ostriker 1999; Davé *et al.* 2001). However, this WHIM has been very elusive and difficult to observe (e.g. Richter *et al.* 2006).

Here, we have presented preliminary results on the properties of the IGM in inter-cluster filaments at $0.1 \leqslant z \leqslant 0.47$ by using a single QSO observed with the HST/COS UV spectrograph, whose sightline intersects 7 independent cluster-pairs at impact parameters $< 5\,\mathrm{Mpc}$ from the intercluster axes. This technique allowed us to perform, for the first time, a systematic and statistical measurement of the incidence of H I and O VI absorption lines associated to intercluster filaments. We constrained the geometry and physical properties of the IGM gas lying between clusters. Our results are consistent with a filamentary geometry for the gas, and the presence of both broad H I ($> 50\,\mathrm{km\ s^{-1}}$) and O VI hint towards the existence of a WHIM (Tejos *et al.* in prep.).

This work was partly funded by CONICYT/PFCHA 72090883 (Chile).

References

Abraham, R. G., Glazebrook, K., McCarthy, P. J., *et al.* 2004, *AJ*, 127, 2455
Abazajian, K. N., Adelman-McCarthy, J. K., Agüeros, M. A., *et al.* 2009, *ApJS*, 182, 543
Bond, J. R., Kofman, L., & Pogosyan, D. 1996, *Nature*, 380, 603
Cen, R. & Ostriker, J. P. 1999, *ApJ*, 514, 1
Chen, H.-W., Lanzetta, K. M., Webb, J. K., & Barcons, X. 1998, *ApJ*, 498, 77
Chen, H.-W., Prochaska, J. X., Weiner, B. J., *et al.* 2005, *ApJ*(Letters), 629, L25
Chen, H.-W. & Mulchaey, J. S. 2009, *ApJ*, 701, 1219
Colberg, J. M., Krughoff, K. S., & Connolly, A. J. 2005, *MNRAS*, 359, 272
Colless, M., Dalton, G., Maddox, S., *et al.* 2001, *MNRAS*, 328, 1039
Crain, R. A., Theuns, T., Dalla Vecchia, C., *et al.* 2009, *MNRAS*, 399, 1773
Danforth, C. W. & Shull, J. M. 2008, *ApJ*, 679, 194
Davé, R., Cen, R., Ostriker, J. P., *et al.* 2001, *ApJ*, 552, 473
Dressler, A. 1980, *ApJ*, 236, 351
Ford, A. B., Davé, R., Oppenheimer, B. D., *et al.* 2014, *MNRAS*, 444, 1260
Fukugita, M., Hogan, C. J., & Peebles, P. J. E. 1998, *ApJ*, 503, 518
González, R. E. & Padilla, N. D. 2010, *MNRAS*, 407, 1449
Hao, J., McKay, T. A., Koester, B. P., *et al.* 2010, *ApJS*, 191, 254
Lanzetta, K. M., Bowen, D. V., Tytler, D., & Webb, J. K. 1995, *ApJ*, 442, 538
Le Fèvre, O., Cassata, P., Cucciati, O., *et al.* 2013, *A&A*, 559, A14
Morris, S. L., Weymann, R. J., Dressler, A., *et al.* 1993, *ApJ*, 419, 524
Nadathur, S. & Hotchkiss, S. 2014, *MNRAS*, 440, 1248
Pan, D. C., Vogeley, M. S., Hoyle, F., Choi, Y.-Y., & Park, C. 2012, *MNRAS*, 421, 926
Prochaska, J. X., Weiner, B., Chen, H.-W., Mulchaey, J., & Cooksey, K. 2011, *ApJ*, 740, 91
Ricciardelli, E., Quilis, V., & Planelles, S. 2013, *MNRAS*, 434, 1192
Richter, P., Savage, B. D., Sembach, K. R., & Tripp, T. M. 2006, *A&A*, 445, 827
Ryan-Weber, E. V. 2006, *MNRAS*, 367, 1251
Rykoff, E. S., Rozo, E., Busha, M. T., *et al.* 2014, *ApJ*, 785, 104
Shone, A. M., Morris, S. L., Crighton, N., & Wilman, R. J. 2010, *MNRAS*, 402, 2520
Shull, J. M., Smith, B. D., & Danforth, C. W. 2012, *ApJ*, 759, 23
Skibba, R. A., Bamford, S. P., Nichol, R. C., *et al.* 2009, *MNRAS*, 399, 966
Springel, V., Frenk, C. S., & White, S. D. M. 2006, *Nature*, 440, 1137
Sutter, P. M., Lavaux, G., Wandelt, B. D., & Weinberg, D. H. 2012, *ApJ*, 761, 44
Tejos, N., Morris, S. L., Crighton, N. H. M., *et al.* 2012, *MNRAS*, 425, 245
Tejos, N., Morris, S. L., Finn, C. W., *et al.* 2014, *MNRAS*, 437, 2017
Tempel, E., Stoica, R. S., Martínez, V. J., *et al.* 2014, *MNRAS*, 438, 3465
Tripp, T. M., Lu, L., & Savage, B. D. 1998, *ApJ*, 508, 200
Tumlinson, J., Thom, C., Werk, J. K., *et al.* 2011, *Science*, 334, 948
Way, M. J., Gazis, P. R., & Scargle, J. D. 2014, arXiv:1406.6111
Werk, J. K., Prochaska, J. X., Tumlinson, J., *et al.* 2014, *ApJ*, 792, 8
Wiersma, R. P. C., Schaye, J., & Theuns, T. 2011, *MNRAS*, 415, 353
Wilman, R. J., Morris, S. L., Jannuzi, B. T., Davé, R., & Shone, A. M. 2007, *MNRAS*, 375, 735

The Zeldovich Universe:
Genesis and Growth of the Cosmic Web
Proceedings IAU Symposium No. 308, 2014
R. van de Weygaert, S. Shandarin, E. Saar & J. Einasto, eds.

© International Astronomical Union 2016
doi:10.1017/S1743921316010188

Finding and characterising WHIM structures using the luminosity density method

Jukka Nevalainen[1], L. J. Liivamägi[1], E. Tempel[1], E. Branchini[2], M. Roncarelli[3], C. Giocoli[3], P. Heinämäki[4], E. Saar[1], M. Bonamente[5], M. Einasto[1], A. Finoguenov[5], J. Kaastra[6], E. Lindfors[4], P. Nurmi[4] and Y. Ueda[7]

[1] Tartu Observatory, Observatooriumi 1, 61602 Tõravere, Estonia
email: jukka@to.ee

[2] University Roma Tre, via della Vasca Navale 84, 00146 Roma, Italy

[3] University of Bologna, viale Berti Pichat 6/2, I-40127 Bologna, Italy

[4] Tuorla Observatory, Väisäläntie 20, FI-21500 Piikkiö, Finland

[5] University of Alabama in Huntsville, Huntsville, AL 35899, USA

[6] SRON Netherlands Institute for Space Research, Sorbonnelaan 2, 3584 CA Utrecht, the Netherlands

[7] Kyoto Observatory, Yoshida-honmachi, Sakyo-ku, Kyoto 606-8501 JAPAN

Abstract. We have developed a new method to approach the missing baryons problem. We assume that the missing baryons reside in a form of Warm Hot Intergalactic Medium, i.e. the WHIM. Our method consists of (a) detecting the coherent large scale structure in the spatial distribution of galaxies that traces the Cosmic Web and that in hydrodynamical simulations is associated to the WHIM, (b) mapping its luminosity into a galaxy luminosity density field, (c) using numerical simulations to relate the luminosity density to the density of the WHIM, (d) applying this relation to real data to trace the WHIM using the observed galaxy luminosities in the Sloan Digital Sky Survey and 2dF redshift surveys. In our application we find evidence for the WHIM along the line of sight to the Sculptor Wall, at redshifts consistent with the recently reported X-ray absorption line detections. Our indirect WHIM detection technique complements the standard method based on the detection of characteristic X-ray absorption lines, showing that the galaxy luminosity density is a reliable signpost for the WHIM. For this reason, our method could be applied to current galaxy surveys to optimise the observational strategies for detecting and studying the WHIM and its properties. Our estimates of the WHIM hydrogen column density N_H in Sculptor agree with those obtained via the X-ray analysis. Due to the additional N_H estimate, our method has potential for improving the constrains of the physical parameters of the WHIM as derived with X-ray absorption, and thus for improving the understanding of the missing baryons problem.

Keywords. cosmology: large-scale structure of universe, galaxies: intergalactic medium

1. Introduction

At low redshifts (z<2) all observations of the visible matter sum up only to $\sim$70% of the expected cosmological mass density of baryons (e.g. Shull *et al.*, 2012). Large scale structure formation simulations suggest that these missing baryons reside in the form of Warm-Hot Intergalactic Matter (WHIM) in the filamentary cosmic web structure connecting the clusters of galaxies and superclusters (e.g. Cen & Ostriker, 1999, 2006). At the predicted temperatures of 10^{5-7} K and densities $10^{-6}-10^{-4}$ cm^{-3} the X-ray emission from single WHIM structures is too faint to be detected with current instrumentation.

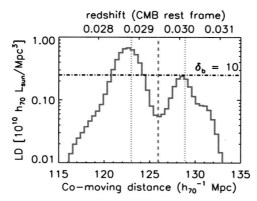

Figure 1. Luminosity density (LD) profile along the Sculptor Wall galaxy structure at z∼0.03 as obtained from the 2dF data is shown with a (blue) solid line. The best-fit value and the 1σ uncertainties of the redshift of the X-ray absorber (Fang *et al.*, 2010) are shown in (red) dashed and dotted lines. Note that the redshifts and distances are reported in CMB rest frame. The luminosity density value corresponding to baryon overdensity $\delta_b = 10$, as estimated with Eq. 2.1, is denoted with horisontal dash-dot line.

However, the column densities of the highly ionised WHIM metals along large–scale filamentary structures can reach a level of 10^{15-16} cm^{-2}, imprinting detectable absorption features on the soft X-ray spectra of background sources (e.g. Nicastro *et al.*, 2010).

While the X-ray absorption measurements are crucial in the analysis of WHIM, they are sparse. To improve this, we focus on *known foreground large scale structures* traced by galaxies, like the cosmic filaments, that most likely contain WHIM. We have applied the Bisous model (Stoica *et al.*, 2005) to trace and extract the filamentary network in the galaxy redshift surveys SDSS DR8 (Tempel *et al.*, 2014) and 2dF. Thus we have an extensive data base of potential WHIM structure locations, sizes and redshifts.

We report distances and lengths in co-moving coordinates (unless stated otherwise), using $\Omega_m = 0.3$, $\Omega_\Lambda = 0.7$ and H = 70 km s^{-1} Mpc^{-1}.

2. The method

2.1. *Luminosity density fields*

We have examined the above galaxy structures using the luminosity density (LD) method, pioneered by our group (e.g. Liivamägi *et al.*, 2012). With this method we quantify the 3-dimensional galaxy distribution using the optical (R - band) light in galaxies. We assume that every galaxy is a visible member of a density enhancement (group or cluster). We place the galaxies at the mean distance of the group or the cluster, to correct for the effect of the dynamical velocities (finger-of-god). We correct the galaxy luminosities for the observational magnitude limited samples by a weighting factor that accounts for the group galaxies outside the visibility window. The galaxy luminosity distribution is then smoothed with a B3-spline kernel function. The smoothing length determines the characteristic scale of the objects under study. We have adopted 1.4 Mpc as the smoothing scale, in order to match the filament widths. We then sample the LD distribution at the points of a uniform grid, encompassing the survey volume, with a sampling scale of 1.4 Mpc, thus creating the LD field. For a given galaxy structure, we use the LD field to evaluate the LD profile (see Fig. 1).

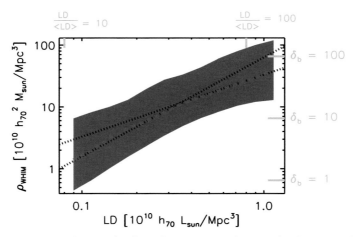

Figure 2. The luminosity density (LD) - WHIM density (ρ_{WHIM}) relation, and the 1σ uncertainties derived from the cosmological hydrodynamical simulations (Cui *et al.*, 2012) are indicated with (red) dashed line and the (blue) shaded region, respectively. The black dotted lines indicate the power-law approximations to the relation (Eq. 2.1). The (green) symbols indicate different levels of overdensity of baryons and luminosity.

2.2. *Large Scale Structure simulations*

If the galaxies follow the underlying dark matter (DM) potential, similarly as the WHIM, then the luminosity density field in filaments can be used to trace the missing baryons. To test the validity of this assumption we have applied our filament–finding and LD field algorithms to a distribution of mock galaxies, DM and the WHIM (i.e. gas at temperatures 10^{5-7} K) at z = 0 in the hydrodynamical simulations (Cui *et al.*, 2012). These simulations make use of smoothed particle hydrodynamics in GADGET-3 code to produce dark matter and diffuse baryonic components within a box with a size of 570 Mpc. The simulations involve radiative cooling, star formation and feedback from supernova remnants. The galaxies are created by populating the DM haloes using a halo occupation distribution constrained with the SDSS data (Zehavi *et al.*, 2011). The simulated data are adequate to follow the different density components with a resolution of ~1 Mpc.

Indeed the LD and WHIM density (ρ_{WHIM}) in the above simulations correlate rather well (Pearson correlation coefficient = 0.85). We used this correlation to derive a quantitative relation between these, i.e. the LD - ρ_{WHIM} relation, within the filaments identified by the mock galaxies (see Eq. 2.1 and Fig. 2).

$$\rho_{WHIM} \approx 63 \times LD^{1.6}, when\ LD < 0.35$$
$$\rho_{WHIM} \approx 33 \times LD^{1.0}, when\ LD > 0.35, \tag{2.1}$$

where LD and ρ_{WHIM} are expressed in units $10^{10}\ h_{70}\ L_{\odot}\ Mpc^{-3}$ and $10^{10}\ h_{70}^{2}\ M_{\odot}\ Mpc^{-3}$.

We use this relation to convert the observational (SDSS and 2dF) LD values into ρ_{WHIM} estimates. We then integrate the ρ_{WHIM} profile of a given filament to obtain the WHIM hydrogen column density N_H. The observed LD profile values typically correspond to baryon overdensities of ~ 10 (see Fig. 1), consistent with those predicted by simulations for the WHIM filaments (e.g. Cen & Ostriker 1999, 2006).

Table 1. Redshifts and WHIM hydrogen column densities $N_{\rm H}$ in Sculptor

system	X-ray[1] redshift[3]	LD[2] redshift[3]	X-ray[1] $\log N_{\rm H}$	LD[2] $\log N_{\rm H}$
Sculptor Wall (SW)	0.030–0.032	0.028–0.033	21.0[20.1–22.3]	20.0[19.6–20.6]
Pisces-Cetus (PC)	0.060–0.063	0.060–0.063	20.1[19.9–20.3]	19.8[19.4–20.3]
Farther Sculptor Wall (FSW)	0.125–0.127	0.128–0.129	20.8[20.0–21.2]	19.9[19.6–20.5]

Notes:
[1] Redshift centroid (Chandra) and WHIM $N_{\rm H}$ estimates based on X-ray absorption measurements (SW: Fang *et al.*, 2010; PC and FSW: Zappacosta *et al.*, 2010). The $\log N_{\rm H}$ values for SW are derived from the reported N_{OVII}, assuming O abundance 0.1 Solar, O/H ratio from Grevesse & Sauval, 1988, and T = 10^6 K.
[2] Our luminosity density - based estimates for the redshift range corresponding to the extent of the given system intersected by the H2356-309 sightline, and WHIM $N_{\rm H}$ of the given system.
[3] In the observational frame.

3. Results

In order to test the feasibility of our WHIM $N_{\rm H}$ estimation method, we extracted the LD profile along the line-of-sight to the blazar H2356-309 behind the Sculptor Wall, where X-ray measurements of WHIM absorption have been obtained (Fang *et al.*, 2010; Buote *et al.*, 2009, Zappacosta *et al.*, 2010). We found ∼10 Mpc long luminosity density structures at redshifts consistent with those measured with X-rays (see Fig. 1). Our LD-WHIM density relation yields WHIM $N_{\rm H}$ values consistent with those measured in X-rays (see Table 1), proving the reliability of our column density estimation method.

4. Discussion

Our plan is to apply our method to current galaxy surveys (and e.g. to Euclid results in near future) to optimise the observational strategies for detecting and studying the WHIM and its properties. In particular, we will cross-correlate the significant WHIM structures found from e.g. SDSS and 2dF with bright background blazars. We aim at obtaining XMM-Newton/RGS and Chandra/LETGS data of such blazars in a high flux state, which are located in the line-of-sight to WHIM structures with the highest WHIM column density estimates. This will be useful also for next generation X-ray telescopes like ATHENA. We will also extend this work to include far-ultraviolet WHIM measurements with e.g. FUSE and HTS of the low temperature WHIM.

References

Buote, D., Zappacosta, L., Fang, T., *et al.*, 2009, *ApJ*, 695, 1351
Cen, R., & Ostriker, J., 1999, *ApJ*, 519, 109
Cen, R., & Ostriker, J., 2006, *ApJ*, 650, 560
Cui, W., Borgani, S., & Dolag, K., *et al.*, 2012, *MNRAS*, 423, 2279
Fang, T., Buote, D., Humphrey, P. *et al.*, 2010, *ApJ*, 714, 1715
Grevesse & Sauval, 1988, *SSR*, 85, 161
Liivamägi, L., Tempel, E., & Saar, E., 2012, *A&A*, 539, A80
Nicastro, F., Krongold, Y., Fields, D., *et al.*, 2010, *ApJ*, 715, 854
Shull, J., Smith, B., Danforth, C., *et al.* 2012 *ApJ*, 759, 23
Stoica, R., Martinez, V., Mateu, J., & Saar, E., 2005, *A&A*, 434, 423
Tempel, E., Stoica, S., & Martinez, V., *et al.* 2014, *MNRAS*, 438, 3465
Zappacosta, L., Nicastro, F., Maiolino, R., *et al.*, 2010, *ApJ*, 717, 74
Zehavi, I., Zheng, Z., Weinberg, D., *et al.* 2011, *ApJ*, 736, 59

The Zeldovich Universe:
Genesis and Growth of the Cosmic Web
Proceedings IAU Symposium No. 308, 2014
R. van de Weygaert, S. Shandarin, E. Saar & J. Einasto, eds.

© International Astronomical Union 2016
doi:10.1017/S174392131601019X

Radiative Feedback Effects during Cosmic Reionization

David Sullivan[1] and Ilian T. Iliev[1,2]

[1] Astronomy Centre, Department of Physics & Astronomy, Pevensey II Building, University of Sussex, Falmer, Brighton BN1 9QH, United Kingdom
email: D.Sullivan@sussex.ac.uk

[2] Speaker, email: I.T.Iliev@sussex.ac.uk

Abstract. We present coupled radiation hydrodynamical simulations of the epoch of reionization, aimed at probing self-feedback on galactic scales. Unlike previous works, which assume a (quasi) homogeneous UV background, we self-consistently evolve both the radiation field and the gas to model the impact of previously unresolved processes such as spectral hardening and self-shielding. We find that the characteristic halo mass with a gas fraction half the cosmic mean, $M_c(z)$, a quantity frequently used in semi-analytical models of galaxy formation, is significantly larger than previously assumed. While this results in an increased suppression of star formation in the early Universe, our results are consistent with the extrapolated stellar abundance matching models from Moster *et al.* 2013.

Keywords. Radiative transfer, methods: numerical, stars: formation, galaxies: formation, galaxies: high-redshift, intergalactic medium, large-scale structure of universe

1. Introduction

During the first billion years after the big bang, the large-scale cosmic web of structures we see today began to form. This was followed by the first stars and galaxies, which brought an end to the Dark Ages (Rees, 1999). These first luminous sources are thought to be the prime candidates which fuelled cosmic reionization, the last major phase transition of the Universe, from a neutral Inter-Galactic Medium (IGM) following recombination to the ionized state it remains in today. The physical processes which drive reionization encapsulate several areas of research, from cosmology and galaxy formation to radiative transfer and atomic physics. Even with the wealth of present-day observational information at our disposal, these processes are still not fully understood. Therefore we cannot model reionization analytically, instead turning to numerical simulations using observations to constrain our models.

Existing semi-analytical models of galaxy formation rely on Cosmological hydrodynamics simulations to calibrate their gas accretion recipes throughout cosmic time. To date, all such simulations have incorporated a (quasi) homogeneous, and instantaneous, UV background (e.g. Haardt and Madau, 2001; Faucher-Giguére *et al.*, 2009) as a cheaper alternative to full radiation hydrodynamics (Shapiro *et al.*, 1994; Gnedin, 2000a; Hoeft *et al.*, 2006; Okamoto *et al.*, 2008). Such models are however a crude approximation, neglecting much of the relevant physics during reionization (i.e self-shielding gas and spectral hardening). Photo-heating raises the temperature of the IGM to $T_{\rm IGM} \sim 10^4 K$, therefore the intergalactic Jeans mass increases substantially, raising the minimum mass of galaxies (Rees, 1986; Efstathiou, 1992; Gnedin and Ostriker, 1997; Miralda-Escude & Rees, 1997). Halos whose virial temperature $T_{vir} \lesssim T_{\rm IGM}$ undergo photo evaporation due to their small potential wells, returning their gas to the IGM. This suppression at early

Run	Box Size (Mpc/h)	Cells/Particles	Resolution (kpc/h)	Radiative Transfer
ATON	4	256^3	15.6	Yes
Fiducial	4	256^3	15.6	No

Table 1. Summary of simulations. Radiative transfer/feedback was neglected in the fiducial model, allowing us to isolate the impact on gas during reionization.

times could be key to rectifying simulations with observations, addressing issues such as the missing satellites problem.

The primary motivation of this work is to improve existing semi-analytical models, by consistently following the evolution of radiation, gas, star formation and dark matter to probe this suppression. In §2 we briefly discuss the methods and codes to be used, followed by a description of the simulation parameters used. Finally, we present our results in §3 and offer some early conclusions and future work in §4.

2. Codes and summary of simulations

Despite the dawn of Petascale computing and large-scale parallel algorithms, modeling the complex interplay of both radiative and hydrodynamic feedback on the formation of the first stars and galaxies remains computationally challenging. Relevant time-scales differ by orders of magnitude between the gas and radiation, as the speed of light is much greater than the local sound speed, therefore fully-coupled radiation hydrodynamics (RHD) simulations require very short time steps to follow their evolution accurately. Initial attempts in the literature were limited to post-processing radiative transfer, however state-of-the-art algorithms have began to surface in recent years which perform this step in situ (Reynolds *et al.*, 2009; Aubert and Teyssier, 2010; Wise & Abel, 2010; Rosdahl *et al.*, 2013 to list a few). We make use of the Eulerian Adaptive Mesh Refinement (AMR) code *RAMSES* (Teyssier, 2002) coupled to the *ATON* code (Aubert and Teyssier, 2008), which uses a moment-based radiative transfer scheme to follow the non-equilibrium thermochemistry of hydrogen.

RAMSES employs a second-order accurate hydrodynamics solver based on Godunov's method, a modern shock capturing scheme known for it's accuracy, and the N-body (dark matter and stars) using a Cloud In Cell (CIC) interpolation scheme on the mesh (Particle-Mesh, or PM). To accelerate the radiative transfer step the *ATON* code utilises graphical processing units (GPUs) to retain the full speed of light (up to 80x speed up over CPUs), however the complex coupling between GPUs and the conventional CPUs, which *RAMSES* uses, results in a loss of the AMR making the code uni-grid. To achieve galactic scale resolutions we have focused on small volume simulations here, however larger volume AMR simulations are currently underway (left to future discussion).

A brief summary of the setup for all simulations is presented in Table 1. Two simulations were carried out, with/without the radiative transfer of ionizing UV photons, initialised from redshift $z = 100$ with identical initial conditions. A geometrically flat ΛCDM Universe is assumed with cosmological parameters $\Omega_m = 0.276$, $\Omega_\Lambda = 0.724$, $\Omega_b = 0.045$, $h \equiv H_0/100$ km s^{-1} $= 0.703$ and $\sigma_8 = 0.811$ where the symbols take their usual meanings. The star forming interstellar medium (ISM) is defined as all gas with over-densities $\Delta \equiv \rho/\langle\rho\rangle > 50$. Both runs include supernovae feedback and were terminated at $z \sim 3$ due to time constraints.

Dark matter (sub) halos are catalogued using the *Rockstar* phase-space halo finder (Behroozi *et al.*, 2013), adapted to work within the yt project (Turk *et al.*, 2011). Particles are first divided into 3D Friends-of-Friends (FOF) groups which are each analysed in 6D phase space to give robust, grid and shape independent results. Those which are

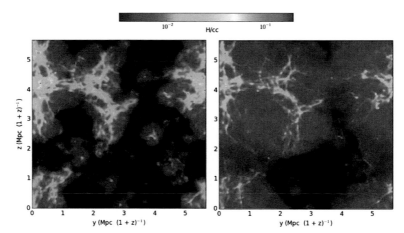

Figure 1. Full-box projections of the 3D volume for our ATON simulation. The colour bar denotes gas density at redshift 8.9 (left) and 7.5 (right) in units of hydrogen atoms per cubic cm. Dark/light regions correspond to neutral/ionized hydrogen, respectively. The box is completely reionized by $z \sim 6.5$, and demonstrates an inside-out reionization.

gravitationally unbound are removed and halo quantities are computed using the definition of the virial over-density from Bryan and Norman, 1998. Finally, merger trees are generated using the consistent-trees algorithm (Behroozi *et al.*, 2013).

3. Results

The cosmic reionization of hydrogen can be characterised by two key stages, following the terminology of Gnedin, 2000b. During the initial "pre-overlap" stage, the Universe is predominately neutral with exception for small HII regions surrounding the first luminous sources. These ionization fronts initially propagate out slowly from high density filaments, until they reach the diffuse IGM. This occurs in approximately 10% of the Hubble time, leading to a sharp peak in the ionizing background and triggering an end to the "overlap" stage. There have been several studies which have investigated the progress and geometry of reionization (Ciardi *et al.*, 2003; Iliev *et al.*, 2006; Mellema *et al.*, 2006; Zahn *et al.*, 2007; McQuinn *et al.*, 2007; Iliev *et al.*, 2007; Alvarez and Abel, 2007; Mesinger & Furlanetto, 2007; Geil and Wyithe, 2008; Choudhury *et al.*, 2009), most of which are consistent with an inside-out progression, i.e where high density regions become ionized earlier, on average (Fig. 1).

Following the completion of the overlap stage, the gas must cool to enable collapse to high densities. In this regime atomic cooling is extremely efficient, as shown in Fig. 2 at $10^4 < T < 10^5$ K in both panels. Regions which were ionized later, predominately diffuse gas in the Circum-Galactic Medium (CGM) or voids, remain hotter on average due to their later, on average, reionization and longer cooling time ($t_{\rm cool} \sim 1/n_H$), as well as some additional local heating due to shocks from supernovae and structure formation. In the fiducial case, most of the gas remains on the adiabat $T \propto \rho^{\gamma-1}$, where γ is the ratio of specific heats, therefore the gas is cold enough to accrete onto low mass halos which are gas poor in our ATON simulation.

The temperature floor in the left panel of Fig. 2 acts to suppress gas accretion onto halos whose virial temperature is below that of the IGM. Pre-reionization, the gas fraction of halos scatters around the cosmic mean $\langle f_b \rangle \equiv \Omega_b/\Omega_m$ as a result of tidal interactions. Photo-heating leads to a sharp drop in the gas fraction of halos below a characteristic

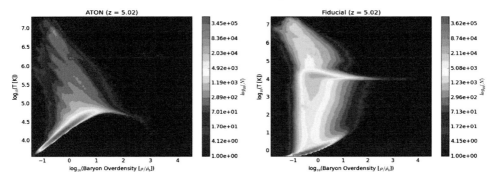

Figure 2. Phase diagrams in temperature-density space for our models. Gas at temperatures of $\sim 10^4$K achieve equilibrium between photo-heating and atomic cooling, collapsing to high densities. Supernovae shock heat the gas to temperatures $\gg 10^4$K, which cannot be maintained by photo-heating alone. The color bar represents the number of cells occupying a given state.

mass scale $M_c(z)$, which sets the halo mass at which the gas fraction is half the cosmic mean. Low mass (mini) halos, whose potential wells are too small to retain photo-heated gas, are photo-evaporated resulting in a baryon mass of the order a fer percent of their total mass. Many semi-analytical models of galaxy formation implement the analytical formula proposed by Gnedin, 2000a to determine the gas fractions of halos:

$$f_b(M, z) = \langle f_b \rangle \left\{ 1 + (2^{\alpha/3} - 1) \left(\frac{M}{M_c(z)} \right)^{-\alpha} \right\}^{-\frac{3}{\alpha}} \tag{3.1}$$

where M is the halo mass, $M_c(z)$ is the characteristic mass and α is an exponent which controls how steep the transition is from baryon poor to rich. For $M \gg M_c(z)$ the first term dominates in the parenthesis, therefore the gas fraction of massive halos approaches the cosmic mean. Both Hoeft *et al.*, 2006 and Okamoto *et al.*, 2008 have tested this and found $\alpha = 2$ gives good agreement with their simulations, albeit with varying conclusions on what sets M_c. Fig. 3 shows the redshift evolution of both the characteristic mass (right vertical axis) and α (left vertical axis), from Okamoto *et al.* Hoeft *et al.* and our ATON simulation. Radiative transfer leads to a significantly larger value for M_c at all times in comparison to the two previous models, while α scatters between ~ 1 at early times to ~ 2 following reionization. Only distinct halos, identified from merger trees, are used here to remove contamination by tidal stripping.

Finally, we show stellar abundance matching models for both simulations listed in Table 1, with comparison to those of Moster *et al.*, 2013, in Fig. 4. The shaded regions show the 1σ deviation from the Moster *et al.* model, for which our ATON simulation shows very good agreement. Our fiducial model suffers from a high star forming efficiency from as early as $z = 9$, suggesting the local reionization history is important prior to global reionization. Importantly, the agreement with Moster *et al.* is consistent at masses near the characteristic mass scale shown in Fig. 3, suggesting we are neither over or under suppressing star formation in these halos.

4. Conclusions

Using fully-coupled radiation hydrodynamics, we have probed the accuracy of semi-analytical models of gas accretion during reionization commonly used in galaxy formation models. Radiative transfer is largely considered to be a second order component in most astrophysical applications, however the significance of the self-feedback as a result of such

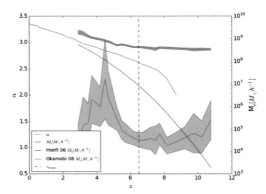

Figure 3. Evolution of the characteristic mass, M_c (red) and the exponent α (blue) from Eqn. 3.1. The green and purple lines denote the values given by Okamoto *et al.*, 2008 and Hoeft *et al.*, 2006, respectively. Shaded regions show 1σ standard deviations, and the dashed line denotes the redshift of reionization for our ATON simulation.

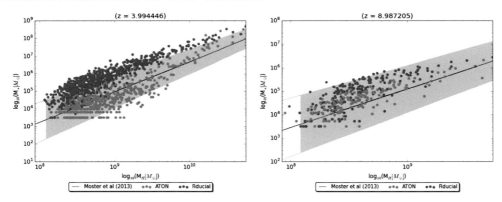

Figure 4. Stellar mass against halo mass with the 1σ standard deviation (shaded) abundance matching model of Moster *et al.*, 2013. The ATON simulation (red) shows strong agreement, suggesting that suppression during reionization is able to regulate early star formation in comparison to our Fiducial model (blue). It should however be noted that Moster *et al.* computed the relative abundance of stars from the Millennium simulation Springel *et al.*, 2005 , which do not resolve halos below $\sim 10^{10}$ M$_\odot$, therefore the above is an extrapolation.

processes has been highlighted here. The increase in the characteristic mass shown in Fig. 3 is largely due to the inside-out progression of reionization in our simulations. Stellar sources of ionizing UV radiation form out of dense clumps, from which the first HII regions expand from. As the recombination time is large in such clumps, the majority of the soft-UV is absorbed, preferentially heating over-dense gas, with only hard-UV photons able to penetrate extending out to the surrounding diffuse gas. This effect is known as spectral hardening, which is absent from previous simulations (Hoeft *et al.*, 2006; Okamoto *et al.*, 2008) due to their (quasi) homogeneous UV background, hence these models under-predict the heating rate in over-dense regions. This was initially exposed by Okamoto *et al.*, 2008, who artificially increased the heating rate as a function of density to test this hypothesis, arriving at the same conclusion as this work.

Our strong agreement with the stellar abundance match models of Moster *et al.*, 2013 suggests that radiative feedback is very efficient at suppressing star formation during cosmic reionization, allowing baryons to remain in the IGM to fuel later periods of star formation. This is consistent with observations, which have long conflicted with numerical simulations which form stars too efficiently in the early Universe (Trujillo-Gomez *et al.*, 2013 and references there-in). It is however not clear whether abundance matching models are accurate at the halo masses we have shown, and the results here are just an extrapolation. To probe whether our combination of radiative and supernovae feedback is feasible, larger simulation volumes must be explored to populate the rare high mass end of the halo mass function (we leave this to future work). Furthermore, our spatial resolution (dx $\sim$ 15 kpc/h comoving) is still quite coarse which may wash out much of the impact of self-shielding (see Fig. 2, no gas is resolved with densities $\rho/\bar{\rho}_b \gtrsim 10^3$). We will explore this with a series of upcoming simulations, using the AMR (RHD) code *RAMSES-RT* (Rosdahl *et al.*, 2013).

References

Alvarez, M. a. & Abel, T. (2007). *MNRAS*, 380(654).

Aubert, D. & Teyssier, R. (2008). *MNRAS*, 387(1).

Aubert, D. & Teyssier, R. (2010). *ApJ*, 724(1).

Behroozi, S., Wechsler, R., H., & Wu, H. (2013). *ApJ*, 762(2).

Behroozi *et al.* (2013). *ApJ*, 763(1).

Bryan, G. L. & Norman, M. L. (1998). *ApJ*, 495(1).

Choudhury, T. R., Haehnelt, M. G., & Regan, J. (2009). *MNRAS*, 394(2).

Ciardi, B., Ferrara, A., & White, S. D. M. (2003). *MNRAS*, 344(1).

Efstathiou, G. (1992). *MNRAS*, 256(2).

Faucher-Giguére, C.-A., Lidz, A., Zaldarriaga, M., & Hernquist, L. (2009). *ApJ*, 703(2).

Geil, P. M. & Wyithe, S. (2008). *MNRAS*, 386(3).

Gnedin, N. Y. (2000a). *ApJ*, 542(2).

Gnedin, N. Y. (2000b). *ApJ*, 535(2).

Gnedin, N. Y. & Ostriker, J. P. (1997). *ApJ*, 486(2).

Haardt, F. & Madau, P. (2001). *Clust. Galaxies High Redshift Universe Obs. X-Rays, ed. D. M. Neumann J. T. T. Van (CEA Saclay)*.

Hoeft, M., Yepes, G., Gottl, S., & Ring, C. (2006). *MNRAS*, 371(1).

Iliev, I. T., Mellema, G., Pen, U., Merz, H., Shapiro, P. R., & Alvarez, M. A. (2006). *MNRAS*, 369(4).

Iliev, I. T., Mellema, G., Shapiro, P. R., & Pen, U.-l. (2007). *MNRAS*, 376(2).

McQuinn, M., Lidz, A., Zahn, O., Dutta, S., & Hernquist, L. (2007). *MNRAS*, 377(3).

Mellema, G., Iliev, I. T., Pen, U.-L., & Shapiro, P. R. (2006). *MNRAS*, 372(2).

Mesinger, A. & Furlanetto, S. (2007). *ApJ*, 669(2).

Miralda-Escude, J. & Rees, M. J. (1997). *ApJ*, 497(1).

Moster, B. P., Naab, T., & White, S. D. M. (2013). 20.

Okamoto, T., Gao, L., & Theuns, T. (2008). 10.

Rees, M. J. (1986). *MNRAS*, 218(1).

Rees, M. J. (1999). *Phys. Rep.*, 333.

Reynolds, D. R., Hayes, J. C., Paschos, P., & Norman, M. L. (2009). *J. Comput. Phys.*, 228(18).

Rosdahl, J., Blaizot, J., Aubert, D., Stranex, T., & Teyssier, R. (2013). *MNRAS*, 436(3).

Shapiro, P. R., Giroux, M. L., & Babul, A. (1994). *ApJ*, 427(1).

Springel *et al.* (2005). *Nature*, 435(7042).

Teyssier, R. (2002). *A&A*, 385.

Trujillo-Gomez, S., Klypin, A., Col, P., Ceverino, D., Arraki, S., & Primack, J. (2013). *pre-print arXiv:1311.2910*.

Turk, M. J., Smith, B. D., Oishi, J. S., Skory, S., Skillman, S. W., Abel, T., & Norman, M. L. (2011). *ApJ*, 192(1).

Wise, J. H. & Abel, T. (2010). 37.

Zahn, O., Lidz, A., McQuinn, M., Dutta, S., Hernquist, L., Zaldarriaga, M., & Furlanetto, S. R. (2007). *ApJ*, 654(1).

The Zeldovich Universe:
Genesis and Growth of the Cosmic Web
Proceedings IAU Symposium No. 308, 2014
R. van de Weygaert, S. Shandarin, E. Saar & J. Einasto, eds.

© International Astronomical Union 2016
doi:10.1017/S1743921316010206

Magnetogenesis at Cosmic Dawn

J-B. Durrive[1,2] and M. Langer[1,2]

[1]Institut d'Astrophysique Spatiale, Bâtiment 121, Univ. Paris-Sud, UMR8617, Orsay, F-91405
[2]CNRS, Orsay, F-91405. email: `jdurrive@ias.u-psud.fr`

Abstract. We present a mechanism for generating cosmological magnetic fields during the Epoch of Reionization, based on the photoionization of intergalactic hydrogen. A general formula is presented, together with an example numerical application which yields magnetic field strengths between 10^{-23} to 10^{-19} G on intersource scales. This mechanism, which operates all along Reionization around any ionizing source, participates to the premagnetization of the whole intergalactic medium. Also, the spatial configuration of these fields may help discriminate them from those produced by other mechanisms in future observations.

Keywords. Cosmology:theory, magnetic fields, large-scale structure of universe

Revisiting Langer *et al.* (2005), we present an analytical model of cosmological magnetogenesis based on plasma-radiation interactions which occur during the Epoch of Reionization. As the first luminous objects formed, they emitted ionizing radiation which photoionized the neutral intergalactic medium (IGM), thus generating currents and inducing magnetic fields. The reionization of the Universe was accomplished by Population III stars, first galaxies and first quasars. Each source formed a fully ionized area (Strömgren sphere) around itself but long mean free path photons (UV and X) escaped into the IGM. Those photons knocked out free a fraction of bound electrons and transferred to them their momentum, generating radial currents and induced magnetic fields. But magnetic fields generated by adjacent currents compensate each other, unless these currents have different intensities. This condition is actually satisfied thanks to inhomogeneities in the IG M. Therefore magnetic fields are generated *where the IGM is inhomogeneous*. More precisely, we show in detail in Durrive & Langer (2014) that sources generate outside their Strömgren spheres the following magnetic field:

$$\vec{B}(t,\vec{r}) = \frac{1}{ex_e}\left(N_1 \frac{\vec{\nabla}x_e}{x_e} + N_2\vec{\nabla}\int_{r_s}^{r} n_{HI}dr\right) \times \frac{\hat{r}}{4\pi r^2}\, t$$

where, for $i = 1, 2$, $N_i(t,\vec{r}) = \int_{\nu_0}^{\infty} f_{\mathrm{mt}}\sigma_{\nu}^i L_{\nu}e^{-\tau_{\nu}}\, d\nu$. In this expression, the first term is local while the second is global. They correspond to the necessity that two adjacent currents have different intensities. This is satisfied either when the matter configuration differs in two adjacent volume elements, or when the intensity of the ionizing radiation incident on two adjacent volume elements differs. This is reflected in the equation above, where the local term corresponds to local inhomogeneities in the electron fraction, and the global term corresponds to the transverse variation of photon absorption along adjacent lines-of-sight. In addition, we recover naturally the geometric dilution of photons, the strength being proportional to r^{-2}. Note also that the strength is linearly growing with time, essentially because we assumed a constant luminosity, and the mechanism operates until the source dies. N_1 and N_2 characterize the impact that the source has at distance $\vec{r}$ at time t. Indeed, they contain the fraction of momentum transferred from photons to electrons f_{mt}, the photoionization cross section σ_{ν}, the spectrum of the source L_{ν} and the optical depth τ_{ν}.

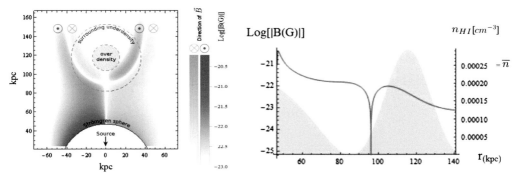

Figure 1. Left: Spatial distribution around an IGM inhomogeneity of the B-field (in Gauss) with a $z = 10$ galaxy source. The symbols $\odot$ and $\otimes$ indicate that the field points towards and away from the reader, respectively. Right: Strength of the B-field along a path through the inhomogeneity (sketched in grey).

Redshift	Source	$Log[B(G)]$	scale (kpc)	$\frac{1}{2}d_{intersource}$ (kpc)
30	*Pop III*	$-19/-21$	0.3/1	10
20		$-19/-21$	0.5/1	
20	*First galaxy*	$-20/-22$	10/15	25
10		$-21/-22$	30/100	
10	*Quasar*	$-21/-22$	300/1000	1000
6		$-22/-23$	500/1500	

Figure 2. Typical values of the resulting fields and scales.

This analytical formula is valid for any ionizing source during Reionization. As an illustrative numerical application, we considered the case of Population III clusters and first galaxies with spectra computed with the Yggdrasil model (Zackrisson *et al.*, 2011), as well as quasars with spectra fitted from observations (Shang *et al.*, 2011), with a mildly non-linear inhomogeneity outside the Strömgren sphere of the source. The results are that, depending on their spectrum, luminosity, lifetime and epoch, different sources generate different magnetic fields (cf. figures 1 and 2). Namely, Pop III clusters generate stronger fields on short scales, while quasars magnetize less but over huge distances. First galaxies combine high amplitudes and large scales (intersource scales). Also, the geometrical configuration of the fields relfects the axisymmetry of the matter distribution with respect to the ionizing source, and fields generated in overdensities have opposite orientation to those generated in underdensities (cf. figure 1).

Conclusion. The mechanism presented here operates with any ionizing source, at any time during the Epoch of Reionization. The resulting magnetic strengths are comparable to those generated by other astrophysical mechanisms, but they appear on entire intersource distances. Therefore it contributes to the premagnetization of the whole Universe. Also, the specific spatial configuration of the generated fields might help discriminate from other cosmological magnetogenesis mechanisms with future observations (e.g. SKA) of the evolved magnetic fields.

References

Langer, M., Aghanim, N., & Puget, J-L. 2005, *A& A*, 443, 367-372

Durrive, J-B., & Langer, M. 2014 (to be submitted)

Zackrisson, E., Rydberg, C.-E., Schaerer, D., Östlin, G., & Tuli, M. 2003, *ApJ*, 740, 13

Shang, Z., Brotherton, M., Beverley, J., Wills, D. *et al.* 2011, *ApJS*, 196, 2

The Zeldovich Universe:
Genesis and Growth of the Cosmic Web
Proceedings IAU Symposium No. 308, 2014
R. van de Weygaert, S. Shandarin, E. Saar & J. Einasto, eds.

© International Astronomical Union 2016
doi:10.1017/S1743921316010218

Zeldovich and the Missing Baryons, Results from Gravitational Lensing

Rudolph E. Schild

Harvard-Smithsonian Center for Astrophysics,
60 Garden St, Cambridge, MA 02138, U.S.A.

Abstract. Central to Zeldovich's attempts to understand the origin of cosmological structure was his exploration of the fluid dynamical effects in the primordial gas, and how the baryonic dark matter formed. Unfortunately microlensing searches for condensed objects in the foreground of the Magellanic Clouds were flawed by the assumption that the objects would be uniformly (Gaussian) distributed, and because the cadence of daily observations strongly disfavored detection of planet mass microlenses. But quasar microlensing showed them to exist at planetary mass at the same time that a hydro-gravitational theory predicted the planet-mass population as fossils of turbulence at the time of recombination ($z = 1100$; Gibson 1996, 2001). Where the population has now been detected from MACHO searches to the LMC (Sumi *et al.* 2011) we compare the quasar microlensing results to the recent determination of the mass distribution function measured for the planetary mass function, and show that the population can account for the baryonic dark matter.

An important component of the Zeldovich sky is the baryonic dark matter, also called the missing baryons. Direct searches for them in the direction of the Galactic Bulge are hampered by the fact that objects smaller than Jupiter have Einstein rings that resolve the illuminated stellar disc of the background star being microlensed, producing the low efficiency of microlensing illustrated for Galactic Bulge backgrounds in Fig. 5 of Sumi *et al.* (2013). However, this is not a problem for searches with gravitational microlensing of quasars, since as discussed by Schild *et al.* (2012) earth mass and even lunar mass microlensing is observable for ordinary microlenses in the foreground galaxy G1 in the the double-image Q0957 gravitational lens system (The First Lens; Schild 1996).

A statistical analysis by Schild *et al.* (2012) of the microlensing statistics for Q0957+561 has shown that the mass function for planetary mass microlensing (nano-lensing) has a measured value of -2.98, where the frequently observed Salpeter exponent is -2.35. In other words, over the observed mass range in Schild (1999) the number of observed objects increases as approximately the third power of the reciprocal mass. Herein we will adopt a value of -3 for the exponent, which was estimated to be -2.98 with an error of +/- 0.5.

These statistics will allow us to formulate a simple estimate of the amount of baryonic dark matter contained in the observed microlensing population, with just the assumption that the planetary range mass function in our Galaxy is the same value in quasar microlensing lens galaxy G1. In other words, since the population condensed as a primordial fog with droplet size approximately $1 M_{Earth}$ the primordial population was reasonably assumed to be a function of the baryonic density at time of recombination, and thus universal.

We begin our calculation by adopting the conclusion of Sumi *et al.* (2011) that in the solar neighborhood the number of Jupiter mass rogue planets equals approximately the number of stars.

Adopting the approximation that the mass of Jupiter M_J is 0.003 times the mass of the sun, $M_\odot$, we easily compute that the number of Earth–mass planets per star is a billion ($\times 30$ the estimate of Gibson 1996), and the mass of each is a millionth of the solar mass. Thus the baryonic mass in a population of rogue planets is 1000 solar masses, which is sufficient to account for the missing baryonic matter, (which has always been assumed to be approximately a factor 100 more than the luminous stellar matter). Thus the population of planetary mass objects seen in quasar microlensing can account for the entire missing baryonic matter.

Surveys looking for microlensing to stars in the Magellanic clouds were undertaken with a low (semi-weekly) cadence and have assumed that the particles are uniformly (Gaussian) distributed. This does not take into account the prediction from the hydro-dynamics community that the objects will be clumped in proto-globular clusters (Jean's clusters, containing $10^6 M_\odot$ and the higher cadence survey of Renault *et al.* (1998) did not use a standard color system to convert from measured colors on the instrumental system,and so do not convert directly to effective temperatures. Evidence for the population seen in microlensing of several other quasars has been given by Nieuwenhuizen *et al.* (2010), who also discuss radio and other observational detections.

References

Gibson, Carl H. 1996, *Applied Mechanics Review*, 49, 299. See also ArXiv:9904.260

Gibson, Carl H. 2001, *arXiv:*astro-ph/0110248

Nieuwenhuizen, T. , Schild, R., & Gibson, Carl H. 2010, *arXiv*, 1011, 2530

Schild, R. 1996, *ApJ*, 464, 125

Schild, R. 1999, *ApJ*, 514, 598

Schild, R., Nieuwenhuizen, T., & Gibson, Carl H. 2012, *Physica Scripta*, 151, 014082

Sumi, T. *et al.* 2011, *Nature*, 473, 349

Sumi, T. *et al.* 2013, *ApJ*, 778, 150

CHAPTER 7.

Galaxy Formation & Evolution in the Cosmic Web

Traditional Estonian Culture and Music.
Performance of the Estonian TV girl's choir,
directed by Arne Saluveer.

CHAPTER 7A.

Galaxy Formation & Evolution

from first galaxies to lonely galaxies:
Saleem Zaroubi and Kathryn Kreckel
pondering where to find them

Photo courtesy: Steven Rieder

Dick Bond and Carlos Frenk
contemplating the wonders of the Universe.

The Zeldovich Universe:
Genesis and Growth of the Cosmic Web
Proceedings IAU Symposium No. 308, 2014
R. van de Weygaert, S. Shandarin, E. Saar & J. Einasto, eds.
© International Astronomical Union 2016
doi:10.1017/S174392131601022X

The origin of the galaxy color bimodality

M. A. Aragón-Calvo[1,2], Mark C. Neyrinck,[2] and Joseph Silk[2,3]

[1] Dept. Physics and Astronomy, Univ. California, Riverside, CA, USA.
email: maragon@ucr.edu

[2] Dept. Physics and Astronomy, Johns Hopkins University.,Baltimore, MD 21218, USA.
[3] Institut d Astrophysique de Paris, Univ. Paris VI, 98 bis boulevard Arago, 75014 Paris, France

Abstract. The star formation history of galaxies is a complex process usually considered to be stochastic in nature, for which we can only give average descriptions such as the color-density relation. In this work we follow star-forming gas particles in a hydrodynamical N-body simulation back in time in order to study their initial spatial configuration. By keeping record of the time when a gas particle started forming stars we can produce Lagrangian *gas-star isochrone surfaces* delineating the surfaces of accreting gas that begin producing stars at different times. These surfaces form a complex a network of filaments in Eulerian space from which galaxies accrete cold gas. Lagrangian accretion surfaces are *closely packed* inside dense regions, intersecting each other, and as a result galaxies inside proto-clusters stop accreting gas early, naturally explaining the color dependence on density. The process described here has a purely gravitational / geometrical origin, arguably operating at a more fundamental level than complex processes such as AGN and supernovae, and providing a conceptual origin for the color-density relation.

Keywords. Large-scale structure of Universe; Galaxy formation, N-body simulations

1. Introduction

The observed properties of galaxies are the combined result of complex internal mechanisms (secular evolution) such as supernovae, AGN feedback, etc. (Powell *et al.* 2011; Larson *et al.* 1980), and ii) external environmental mechanisms such as galaxy interactions and mergers, harassment, etc. (Gunn & Gott 1972; Moore *et al.* 1996; Kawata & Mulchaey 2008). The role of cosmic environment on star formation is evident in processes such as the morphology-density relation (Dressler 1980) and the related color-density relation, which encode the effect of environment (density) on star formation history (color) (Bell *et al.* 2004; Blanton *et al.* 2005). Several mechanisms are assumed to contribute to the observed bimodality in the color distribution and the decreasing fraction of blue galaxies with increasing density, such as galaxy mergers and harassment (Gunn & Gott 1972; Larson *et al.* 1980; Moore *et al.* 1996). These mechanisms however, do not offer a direct link between star formation and environment (Blanton *et al.* 2005; Skibba *et al.* 2009).

Galaxies accrete cold gas via a network of narrow filamentary streams that penetrate deep into the galaxy (Kereš *et al.* 2005; Dekel & Birnboim 2006; Dekel *et al.* 2009; van de Voort *et al.* 2011), fueling star formation shortly after accretion (Bauermeister *et al.* 2010). The gravitational collapse of matter into the galaxy sets a natural order in the accretion of gas, i.e. nearby gas is accreted first while gas in distant reservoirs is accreted later. If star formation closely follows gas accretion, we should then expect a simple relation between star formation time and the original distance between the gas cloud that formed the stars and the galaxy, at least approximately, providing a link between the stellar populations of galaxies, encoded in their color, and their initial spatial configuration.

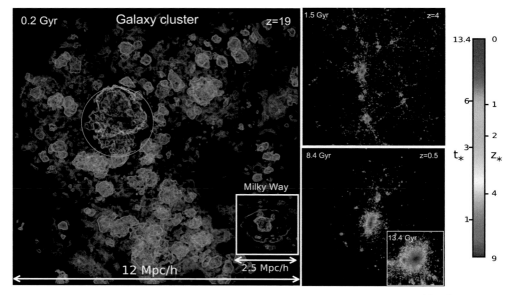

Figure 1. Gas-star conversion times in a galaxy cluster with mass of $\sim 3 \times 10^{14} h^{-1} M_\odot$. Star–forming gas is accreted from increasingly distant isochrone surfaces, centered in the progenitors of massive galaxies (red blobs). Proto-cluster galaxies, being closely packed, are surrounded by a layer of galaxies and are effectively isolated from late star-forming gas (light blue isochrone surfaces). The white circle in the top panel shows the central proto-galaxy in a window carved through the gas cloud. For comparison, we show the Milky Way galaxy (small white square) on the left panel.

2. Simulations and results

We ran a full hydrodynamic zoom resimulation of a galaxy cluster with a present-time mass of $3 \times 10^{14} h^{-1} M_{\odot(z=0)}$. selected from a 64 h^{-1}Mpc box. The high-resolution region was populated with $2.2 \times 10^6 h^{-1} M_\odot$ particles and run using the Gadget-3 code, which implements simple recipes for hydrodynamics and chemical enrichment including stochastic star formation, SN feedback and winds (Springel & Hernquist 2003). From the simulation we identified and followed star-forming gas particles, i.e. gas particles that at some point during the simulation's history produced stars, from the present time back to the initial conditions. For each gas particle we also stored the time when it started producing stars, here referred to as the gas→star conversion time, t_*. By doing so, we were able not only to follow stars after they are formed, but to trace their "progenitor" gas particles back to their Lagrangian positions and study their initial spatial arrangement. Having identified star coming particles and their t_* we produced *gas→star isochrone surfaces*, S_{t_*} define regions of gas that, after being accreted into galaxies, started forming stars at the same time.

Figure 1 shows the *star formation isochrone surfaces* for a galaxy cluster. The central proto-galaxy is the prominent structure near the center of the proto-cluster. The isochrone surfaces are remarkably regular and there is a clear relation between gas-star conversion time and radial distance from centers of proto-galaxies even for this complex cluster formed by a triple major merger. Early star-forming gas is accreted first and so $S_{z_*=9}$ (red surfaces) mark the centers of galaxies. The $S_{z_*=3}$ surfaces (yellow) surrounding satellite galaxies are relatively isolated compared to the $S_{z_*=3}$ surfaces around the central galaxy where surfaces from adjacent galaxies are intersecting. The central galaxy, being surrounded by a compact shell of adjacent satellite galaxies is geometrically constrained to accrete star-forming gas beyond the $S_{z_*=3}$ surface. This can be seen in the almost total lack of $S_{z_*=1}$ surfaces (light blue) around the central galaxy. Satellite galaxies on the other hand are still able to accrete gas at this time although there is no noticeable accretion of star-forming gas after $z \sim 1$.

Figure 1 shows that as a galaxy grows, it carves out increasingly large surfaces (in Lagrangian coordinates) of star-forming gas from it surroundings. These surfaces extend from the center of the proto-galaxy until the maximum radius of influence from which a galaxy can gravitationally accrete mass. Galaxies in dense environments can be seen as a *closely packed system* of S_{z_*} spheres where adjacent galaxies compete for the available gas as they "carve" the proto-cluster's volume. The intersection of isochrone surfaces from adjacent galaxies marks the time where no more gas is available for accretion and star formation. This occurs roughly at half the mean inter-galaxy separation, imposing a fundamental geometric limit to gas accretion in dense environments. On the other hand, galaxies in the outskirts of proto-clusters can have an extended star formation history due to their access to gas in the vicinity of the proto-cluster (Papadopoulos *et al.* 2001; Wolfe *et al.* 2013).

3. Discusion and conclusions

The most straightforward consequence of "galaxy close packing" is that the massive galaxies which are the progenitors of present-time groups and clusters, being surrounded by gas-competing galaxies, become cut off from their gas supply, becoming "quenched" already at early times. The mechanism described here offers a conceptual origin for the observed color-density relation by limiting gas accretion and star formation in dense environments. Galaxies in low-density regions, on the other hand, are not geometrically constrained and can, in principle, freely continue to accrete gas. In addition to this, other processes such as AGN and SN feedback play an important role.

References

Bauermeister, A., Blitz, L., & Ma, C.-P. 2010, *ApJ*, 717, 323
Bell, E. F., Wolf, C., Meisenheimer, K., *et al.* 2004, *ApJ*, 608, 752
Blanton, M. R., Eisenstein, D., Hogg, D. W., Schlegel, D. J., & Brinkmann, J. 2005, *ApJ*, 629, 143
Dekel, A., & Birnboim, Y. 2006, *MNRAS*, 368, 2
Dekel, A., Birnboim, Y., Engel, G., *et al.* 2009, *Nature*, 457, 451
Dressler, A. 1980, *ApJ*, 236, 351
Gunn, J. E., & Gott, J. R., III 1972, *ApJ*, 176, 1
Kawata, D., & Mulchaey, J. S. 2008, ApJLett, 672, L103
Kereš, D., Katz, N., Weinberg, D. H., & Davé, R. 2005, *MNRAS*, 363, 2
Larson, R. B., Tinsley, B. M., & Caldwell, C. N. 1980, *ApJ*, 237, 692
Moore, B., Katz, N., Lake, G., Dressler, A., & Oemler, A. 1996, *Nature*, 379, 613
Papadopoulos, P., Ivison, R., Carilli, C., & Lewis, G. 2001, *Nature*, 409, 58
Powell, L. C., Slyz, A., & Devriendt, J. 2011, *MNRAS*, 414, 3671
Skibba, R. A., Bamford, S. P., Nichol, R. C., *et al.* 2009, *MNRAS*, 399, 966
Springel, V., & Hernquist, L. 2003, *MNRAS*, 339, 289
van de Voort, F., Schaye, J., Booth, C. M., Haas, M. R., & Dalla Vecchia, C. 2011, *MNRAS*, 414, 2458
Wolfe, S. A., Pisano, D. J., Lockman, F. J., McGaugh, S. S., & Shaya, E. J. 2013, *Nature*, 497, 224

The Zeldovich Universe:
Genesis and Growth of the Cosmic Web
Proceedings IAU Symposium No. 308, 2014
R. van de Weygaert, S. Shandarin, E. Saar & J. Einasto, eds.

© International Astronomical Union 2016
doi:10.1017/S1743921316010231

Gas accretion from the cosmic web in the local Universe

J. Sánchez Almeida[1,2], B. G. Elmegreen[3], C. Muñoz-Tuñón[1,2] and D. M. Elmegreen[4]

[1]Instituto de Astrofísica de Canarias, E-38205 La Laguna, Tenerife, Spain
email: jos@iac.es, cmt@iac.es

[2]Departamento de Astrofísica, Universidad de La Laguna, Tenerife, Spain

[3]IBM Research Division, T.J. Watson Research Center, Yorktown Heights, NY 10598, USA
email: bge@us.ibm.com

[4]Department of Physics and Astronomy, Vassar College, Poughkeepsie, NY 12604, USA
email: elmegreen@vassar.edu

Abstract. Numerical simulations predict that gas accretion from the cosmic web drives star formation in disks galaxies. The process is important in low mass haloes ($< 10^{12}$ M$_\odot$), therefore, in the early universe when galaxies were low mass, but also in dwarf galaxies of the local universe. The gas that falls in is predicted to be tenuous, patchy, partly ionized, multi-temperature, and large-scale; therefore, hard to show in a single observation. One of the most compelling cases for gas accretion at work in the local universe comes from the extremely metal poor (XMP) galaxies. They show metallicity inhomogeneities associated with star-forming regions, so that large starbursts have lower metallicity than the underlying galaxy. Here we put forward the case for gas accretion from the web posed by XMP galaxies. Two other observational results are discussed too, namely, the fact that the gas consumption time-scale is shorter than most stellar ages, and the systematic morphological distortions of the HI around galaxies.

Keywords. Galaxies: evolution **and** Galaxies: formation **and** Galaxies: general **and** Galaxies: high-redshift **and** Galaxies: star formation **and** large-scale structure of Universe

1. Star-formation and gas accretion from the cosmic web

Galaxies are not just test particles tracing the cosmic web. Galaxies consume cosmic gas by accretion, and also modify the chemical composition of the web through winds that spread out debris from stellar evolution. Numerical simulations predict that accretion of metal-poor gas from the cosmic web fuels the formation of disk galaxies (e.g., Dekel *et al.* 2009, Genel *et al.* 2012). This cosmological gas supply has a strong dependence on halo mass, and so, on redshift. When the gas encounters a massive halo ($> 10^{12}$M$_\odot$), it becomes shock heated and requires a long time to cool and settle into the galaxy disk. For less massive haloes, cool gas streams can reach the inner halo or disk directly through the so-called cold-flow accretion. Since high redshift haloes tend to be low in mass, cold-flow accretion is predicted to be the main mode of galaxy growth in early times. Ultimately, the galaxies evolve into a quasi-stationary state, where inflows and outflows balance the star formation rate (SFR), a phase that still goes on for most of them.

Although the theoretical predictions are clear, we only have indirect evidence for gas infall feeding star-formation. These evidences are reviewed by Sánchez Almeida *et al.* (2014a), and we refer to this work for full details. Here we illustrate the type of evidence (Sect. 1), emphasizing our contribution to the issue, which is related to inhomogeneities in the distribution of metallicity found in local XMP galaxies (Sect. 2). Other recent reviews covering cosmic gas accretion from different perspectives have been written by

Sancisi *et al.* (2008), Silk & Mamon (2012), Combes (2014), Fraternali (2014), Conselice *et al.* (2013), Benson (2010), and Madau & Dickinson (2014).

2. Observational evidence for accretion in the local Universe

Two examples are discussed, chosen to be representative of a long list of indirect evidence put forward by Sánchez Almeida *et al.* (2014a).

Gas consumption time-scale shorter than stellar ages. The Kennicutt-Schmidt (KS) law (Kennicutt 1998) provides a time-scale to consume the gas available to form stars τ_g. According to the KS law, the SFR scales as a power of the gas mass M_g, with power index close to one, i.e.,

$$\text{SFR} \simeq \frac{M_g}{\tau_g}. \tag{2.1}$$

If a galaxy evolves as a closed box, τ_g gives the time-scale to exhaust the gas available to form stars; τ_g is measured to change from 0.5 to 2 Gyr for galaxies in the redshift range between 2 and 0 (e.g., Genzel *et al.* 2010; Gnedin *et al.* 2014). When outflows are important, as expected in dwarf galaxies (e.g., Peeples & Shankar 2011), the gas consumption time-scale can be much shorter than τ_g (e.g., Sánchez Almeida *et al.* 2014a, Sect. 2.1). Despite this very short gas-consumption time-scale, star-forming galaxies are observed to have stars of all ages from the origin of the Universe to the present (e.g., Heavens *et al.* 2004; Sánchez Almeida *et al.* 2012). The only way to reconcile this short gas-consumption time-scale with the ages of the stars is assuming a continuous gas infall able to replace the gas lost through star-formation and winds.

Morphological distortions of the HI distribution. Neutral gas is present around almost all galaxies. Often the HI maps present non-axisymmetric distortions showing that the gas is not contained in a disk or a spheroid and so, suggesting the gas distribution to be transient. Spirals are known to have extended warped HI envelopes that can be sustained by gas infall (e.g., Bournaud *et al.* 2005). Extreme distortions are actually common among galaxies with large specific SFR in the local universe (Lelli *et al.* 2012; Nidever *et al.* 2013), with kinematically separate components, streamers extending far beyond the optical size, kinematic HI axes offset from the optical axes, and clouds associated with recent starbursts (Johnson *et al.* 2012; Ashley *et al.* 2013). A particularly telling deviation from axi-symmetry has been found recently by Kacprzak *et al.* (2012). They report a bimodality in the azimuthal angle distribution of low ionization gas around galaxies as traced by MgII absorption along QSO lines-of-sight. The circum-galactic gas prefers to lie near the projected galaxy major and minor axes. The bimodality is clear in blue star-forming galaxies whereas red passive galaxies exhibit an excess of absorption along their major axis. These results suggest the bimodality to be driven by gas accretion along the galaxy major axis and gas outflows along the minor axis, with no other clear alternative explanation existing yet.

3. Evidence from extremely metal poor galaxies

The secular evolution of disk galaxies produces a regular pattern with the metallicity decreasing inside out, i.e., having a negative gradient with galactocentric distance (Vilchez *et al.* 1988). The time-scale for gas mixing in a disk is fairly short, on the order of a rotational period or a few hundred Myr. Deviations from negative metallicity gradients are attributed to the recent arrival of cosmic gas that feeds the star formation. If the gas is accreted through the cold-flow mode, it is expected to reach the disks in clumps

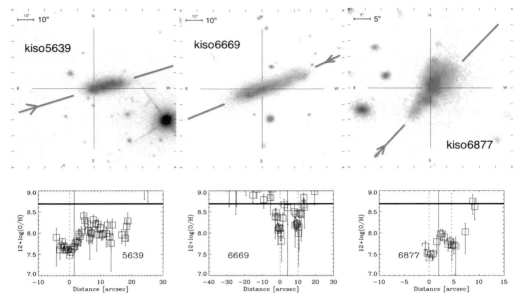

Figure 1. Top: Images of three tadpole galaxies characterized by having a bright peripheral clump on a faint tail. The images have been taken from SDSS, and are displayed with an inverted color palette so that the background sky looks white, and the intrinsically blue galaxies appear reddish. The red line shows the orientation of the spectrograph slit used to measure metallicity variations, and the horizontal scales on the upper left corner of the panels correspond to 5 or 10 arcsec as indicated. Bottom: Oxygen abundance variation across the galaxies on top. The vertical solid line represents the center of rotation, whereas the vertical dotted lines indicate the location of maxima in SFR. Note the existence of abundance variations, with the minima coinciding with the regions of largest SFRs. The thick horizontal solid line indicates the solar metallicity.

often forming stars already (e.g., Dekel *et al.* 2009; Ceverino *et al.* 2010; Genel *et al.* 2012). Alternatively, the external gas streams may fuel the disks with metal-poor gas, so that gas mass builds up developing starbursts through internal gravitational instabilities (e.g., Noguchi 1999; Bournaud & Elmegreen 2009). In any case, the cold-flow accretion is bound to induce metal-poor starbursts.

Metallicity drops associated with intense starbursts may reflect cold-flow accretion. Cresci *et al.* (2010) were the first to identify this pattern in high redshift galaxies. The same kind of metallicity drops appear in local tadpole galaxies, which are often XMPs (Morales-Luis *et al.* 2011). The bright star-forming head of the XMP tadpole has lower metallicity than the underlying galaxy (see Fig. 1). Sánchez Almeida *et al.* (2013) interpret this observation as an episode of gas accretion onto the tadpole head. Localized metallicity drops associated with star-forming regions have also been observed in other objects, including gamma ray burst host galaxies (Levesque *et al.* 2011; Thöne *et al.* 2014), BCD galaxies (Izotov *et al.* 2009; Sánchez Almeida *et al.* 2014b) and dIrr galaxies (Haurberg *et al.* 2013). Variations of metallicity among HII regions located at the same galactocentric distance are not unusual even in large nearby spirals (e.g., Li *et al.* 2013); some of these variations could come from localized accretion events.

4. Take-home message

Numerical simulations tell us that most star-formation in spirals is driven by gas accretion from the cosmic web. It is a process happening at all redshifts. Although this is a solid theoretical prediction, only indirect evidence exist so far. The case of the XMP galaxies results particularly compelling. XMP galaxies seems to be primitive disks in the process of assembling in the nearby universe, where a major cold-flow accretion episode is producing the current starburst. They are extreme cases showing a physical process that affects many other disk galaxies.

References

Ashley, T., Simpson, C. E., & Elmegreen, B. G. 2013, *AJ*, 146, 42

Benson, A. J. 2010, *Physics Reports*, 495, 33

Bournaud, F., Combes, F., Jog, C. J., & Puerari, I. 2005, *A&A*, 438, 507

Bournaud, F. & Elmegreen, B. G. 2009, *ApJ*, 694, L158

Ceverino, D., Dekel, A., & Bournaud, F. 2010, *MNRAS*, 404, 2151

Combes, F. 2014, in *Astronomical Society of the Pacific Conference Series*, Vol. 480, Structure and Dynamics of Disk Galaxies, ed. M. S. Seigar & P. Treuthardt, 211

Conselice, C. J., Mortlock, A., Bluck, A. F. L., Grützbauch, R., & Duncan, K. 2013, *MNRAS*, 430, 1051

Cresci, G., Mannucci, F., Maiolino, R., *et al.* 2010, *Nat*, 467, 811

Dekel, A., Birnboim, Y., Engel, G., *et al.* 2009, *Nat*, 457, 451

Fraternali, F. 2014, in IAU Symposium, Vol. 298, IAU Symposium, ed. S. Feltzing, G. Zhao, N. A. Walton, & P. Whitelock, 228–239

Genel, S., Naab, T., Genzel, R., *et al.* 2012, *ApJ*, 745, 11

Genzel, R., Tacconi, L. J., Gracia-Carpio, J., *et al.* 2010, *MNRAS*, 407, 2091

Gnedin, N. Y., Tasker, E. J., & Fujimoto, Y. 2014, *ApJ*, 787, L7

Haurberg, N. C., Rosenberg, J., & Salzer, J. J. 2013, *ApJ*, 765, 66

Heavens, A., Panter, B., Jimenez, R., & Dunlop, J. 2004, *Nat*, 428, 625

Izotov, Y. I., Guseva, N. G., Fricke, K. J., & Papaderos, P. 2009, *A&A*, 503, 61

Johnson, M., Hunter, D. A., Oh, S.-H., *et al.* 2012, *AJ*, 144, 152

Kacprzak, G. G., Churchill, C. W., & Nielsen, N. M. 2012, *ApJ*, 760, L7

Kennicutt, Jr., R. C. 1998, *ApJ*, 498, 541

Lelli, F., Verheijen, M., Fraternali, F., & Sancisi, R. 2012, *A&A*, 544, A145

Levesque, E. M., Berger, E., Soderberg, A. M., & Chornock, R. 2011, *ApJ*, 739, 23

Li, Y., Bresolin, F., & Kennicutt, Jr., R. C. 2013, *ApJ*, 766, 17

Madau, P. & Dickinson, M. 2014, *ARA&A*, 52, 415

Morales-Luis, A. B., Sánchez Almeida, J., Aguerri, J. A. L., & Muñoz-Tuñón, C. 2011, *ApJ*, 743, 77

Nidever, D. L., Ashley, T., Slater, C. T., *et al.* 2013, *ApJ*, 779, L15

Noguchi, M. 1999, *ApJ*, 514, 77

Peeples, M. S. & Shankar, F. 2011, *MNRAS*, 417, 2962

Sánchez Almeida, J., Elmegreen, B. G., Muñoz-Tuñón, C., & Elmegreen, D. M. 2014a, *A&ARev*, 22, 71

Sánchez Almeida, J., Morales-Luis, A. B., Muñoz-Tuñón, C., *et al.* 2014b, *ApJ*, 783, 45

Sánchez Almeida, J., Muñoz-Tuñón, C., Elmegreen, D. M., Elmegreen, B. G., & Méndez-Abreu, J. 2013, *ApJ*, 767, 74

Sánchez Almeida, J., Terlevich, R., Terlevich, E., Cid Fernandes, R., & Morales-Luis, A. B. 2012, *ApJ*, 756, 163

Sancisi, R., Fraternali, F., Oosterloo, T., & van der Hulst, T. 2008, *A&ARev*, 15, 189

Silk, J. & Mamon, G. A. 2012, Research in Astronomy and Astrophysics, 12, 917

Thöne, C. C., Christensen, L., Prochaska, J. X., *et al.* 2014, *MNRAS*, 441, 2034

Vilchez, J. M., Pagel, B. E. J., Diaz, A. I., Terlevich, E., & Edmunds, M. G. 1988, *MNRAS*, 235, 633

The Zeldovich Universe:
Genesis and Growth of the Cosmic Web
Proceedings IAU Symposium No. 308, 2014   © International Astronomical Union 2016
R. van de Weygaert, S. Shandarin, E. Saar & J. Einasto, eds.   doi:10.1017/S1743921316010243

The role of cold and hot gas flows in feeding early-type galaxy formation

Peter H. Johansson

Department of Physics, University of Helsinki,
Gustaf Hällströmin katu 2a, FI-00014 Helsinki, Finland
email: `Peter.Johansson@helsinki.fi`

Abstract. We study the evolution of the gaseous components in massive simulated galaxies and show that their early formation is fuelled by cold, low entropy gas streams. At lower redshifts of $z \lesssim 3$ the simulated galaxies are massive enough to support stable virial shocks resulting in a transition from cold to hot gas accretion. The gas accretion history of early-type galaxies is directly linked to the formation of their stellar component in the two phased formation scenario, in which the central parts of the galaxy assemble rapidly through in situ star formation and the later assembly is dominated primarily by minor stellar mergers.

Keywords. galaxies: elliptical and lenticular, galaxies: formation, galaxies: evolution

1. Introduction

Recently there has been growing evidence both observationally (e.g. Bezanson *et al.* 2009; Toft *et al.* 2014) and theoretically (e.g. Naab *et al.* 2007, 2009) that massive early-type galaxies assemble in two phases. At high redshifts of $z \sim 3 - 6$ the central parts of the galaxies assemble rapidly through compact in situ star formation fuelled by cold, low entropy gas streams penetrating deep within the galactic halo. At lower redshifts the in situ growth of the stellar component is stifled by the lack of cold gas and instead the late assembly proceeds predominantly through the accretion of stars formed in subunits outside the main galaxy (Oser *et al.* 2010, 2012; Lackner *et al.* 2012).

Here we study in more detail the evolution of the gaseous component in simulated early-type galaxies and show that the two phased formation of the stellar component is driven by a bimodal evolution in the temperatures of the gas accreting onto the forming galaxies. The results are based on a subsample of four galaxies (A2, C2, E2 and U) extracted from Johansson *et al.* (2012) and simulated at high spatial ($\epsilon_\star = 0.125 - 0.25$ kpc) and mass resolution ($m_\star \sim 10^5 - 10^6 M_\odot$). All of the simulations were run using the Gadget-2 smoothed particle hydrodynamics (SPH) code (Springel 2005) and include cooling for a primordial composition, star formation and self-regulating feedback from type II supernovae (Springel & Hernquist 2003).

2. The gas assembly of early-type galaxies

In Fig. 1 we depict the evolution of the gas surface densities, together with the mass-weighted temperatures and entropies for halo A2, which is representative of our simulation sample. At the high redshift of $z = 5$ the forming galaxy is found sitting at the intersection of several gas filaments, which are feeding cold gas directly onto the galaxy as can be seen in the blue filaments in the entropy panel. At the lower redshifts of $z = 3$ and especially at $z = 1$ the halo has grown sufficiently in mass to be able to support strong virial shocks resulting in a majority of hot gas. However, at both of these redshifts there

$$\log(\Sigma_{\text{gas}}) \ [M_\odot/\text{pc}^2] \qquad \log(T) \ [K] \qquad \log(S) \ [\text{kev cm}^2]$$

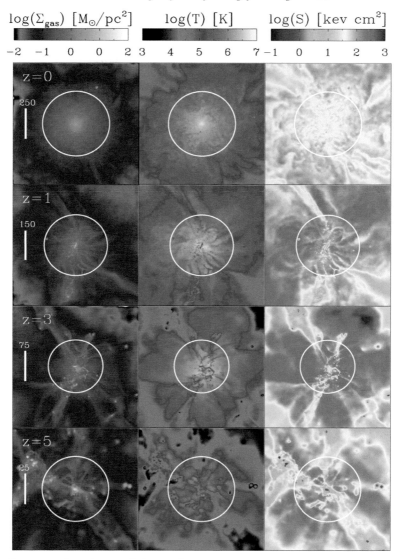

Figure 1. The gas surface density (left panel), the mass-weighted mean gas temperature (middle panel) and mass-weighted mean gas entropy ($S = kTn^{-2/3}$, right panel) are shown for halo A2 at redshifts of $z = 0$, $z = 1$, $z = 3$ and $z = 5$ from top to bottom. The length scale in kpc is indicated by the white bar and the white circles show the corresponding virial radii.

are still some remaining cold gas filaments that are able to penetrate into the central gaseous structure and supply some fuel for star formation. In the final snapshot at $z = 0$ almost all of the gas can be found in a hot diffuse state, with virtually no remaining cold clumpy gas.

3. The temperature structure of the gas

A more detailed view of the evolution of the gas properties of the galaxies can be gained from Fig. 2, where we plot the temperature profiles, entropy distributions and

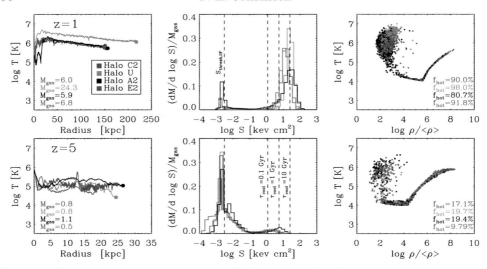

Figure 2. The temperature profile (left panel), the entropy distribution (middle panel) and the phase-space diagram (right panel) for gas within the virial radius for four simulated galaxies at redshifts of $z = 1$ (top panels) and $z = 5$ (bottom panels). The gas mass ($M_{\rm gas}$) in units of $10^{10} M_{\odot}$ together with the fraction ($f_{\rm hot}$) of hot diffuse gas is also indicated.

phase-space diagrams for our simulation sample of four galaxies at redshifts of $z = 5$ and $z = 1$. The mean temperature of the gas is increasing in all of the simulations from $T \sim 10^5$ K at $z = 5$ to temperatures in excess of a million degrees ($T \sim 10^6$ K) by $z = 1$.

In the middle panel the gas entropy distributions defined as $S = kTn^{-2/3}$ are shown, where k is the Boltzmann constant, T is the temperature and n is the number density. The dashed vertical lines indicate the threshold entropy for star formation (neutral gas at $T \sim 10^4$ K) and the entropy values corresponding to the minimum cooling times of 0.1, 1 and 10 Gyr. The entropy distributions are bimodal, at high redshifts the majority of the gas is cold, high-density, star-forming gas that forms a low-entropy peak. By redshift $z = 1$ most of the cold gas has formed stars and the remaining gas is dilute shock-heated gas with long cooling times forming a high-entropy peak. Correspondingly the fraction of hot gas (defined as $T > 2.5 \times 10^5$ K and $n < n_{\rm thresh,SF} = 0.205$ cm^{-3}) increases from $f_{\rm hot} \lesssim 20\%$ at $z = 5$ to $f_{\rm hot} \gtrsim 90\%$ by $z = 1$. At $z = 5$ star-forming gas can clearly be seen in the phase-diagram plot as most of the gas has a density of $\rho > \rho_{\rm thresh,SF}$ and can be found on an equilibrium curve in the $\rho - T$ plane dictated by the self-regulated feedback model (Springel & Hernquist 2003) employed in this study. At $z = 1$, although the majority of the gas is in the hot component, some residual star-forming gas can also be found. This is in contrast with the no feedback simulations of Johansson *et al.* (2009) in which virtually no cold star-forming gas remained at redshifts below $z \lesssim 2$.

4. Cold and hot gas flows

In Fig. 3 we study following Kereš *et al.* (2005) the thermal properties of the accreted gas particles prior to their accretion onto the haloes. The temperature evolution of the accreted gas particles is traced back through the simulation and the maximum temperature is recorded excluding the snapshots when the gas particles were star-forming as in these cases the high temperature was due to supernova feedback.

From Fig. 3 we see that at $z = 5$ almost all of the gas is accreted with a temperature that is below one tenth of the virial temperature, which is typically a few times 10^5 K at

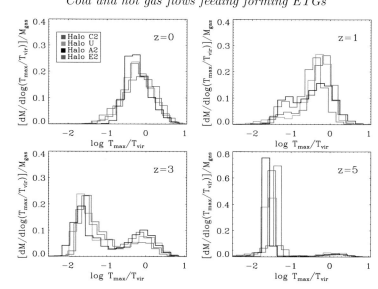

Figure 3. The distribution of the maximum gas temperatures normalized to the virial temperature of gas accreting onto the four simulated galaxies as a function of redshift. At high redshifts the gas predominantly flows in cold, whereas at lower redshifts the gas is accreted in a hot mode.

$z \sim 5$. This indicates that the gas is accreted cold at a typical temperature of a few times 10^4 K at this redshift. At $z = 3$ the distribution is more bimodal with the majority of accreted gas still being cold, however a substantial component of hot gas being accreted at $T \sim T_{\rm vir}$ can also be seen. The transition from predominantly cold accretion to hot accretion occurs at $z \sim 2-3$ when the galaxies reach masses of $M_{\rm halo} = 5 \times 10^{11} - 10^{12} M_\odot$ in good agreement with the predictions of Dekel & Birnboim (2006). At lower redshifts the haloes are sufficiently massive to support stable virial shocks and the vast majority of the gas is accreted in the hot phase.

The transition from a gas accretion model based on cold gas flows at $z \gtrsim 3$ to a hot accretion mode at lower redshifts ($z \lesssim 3$) is directly connected to the two phased picture of early-type galaxy formation. In this picture the early formation is dominated by in situ star formation fuelled by cold gas flows, whereas the later assemble history is dominated by stellar accretion primarily through minor mergers as the source of internal cold gas is exhausted by the transition from cold gas accretion to a hot accretion mode.

References

Bezanson, R., van Dokkum, P. G., Tal, T., *et al.* 2009, *ApJ*, 697, 1290
Dekel, A. & Birnboim, Y. 2006, *MNRAS*, 368, 2
Johansson, P. H., Naab, T., & Ostriker, J. P. 2009, *ApJL*, 697, L38
Johansson, P. H., Naab, T., & Ostriker, J. P. 2012, *ApJ*, 754, 115
Kereš, D., Katz, N., Weinberg, D. H., & Davé, R., 2005, *MNRAS*, 363, 2
Lackner, C. N., Cen, R., Ostriker, J. P., & Joung, M. R., 2012 *MNRAS*, 425, 641
Naab, T., Johansson, P. H., Ostriker, J. P., & Efstathiou, G. 2007, *ApJ*, 658, 710
Naab, T., Johansson, P. H., & Ostriker, J. P. 2009, *ApJL*, 699, L178
Oser, L., Ostriker, J. P., Naab, T., Johansson, P. H., & Burkert, A. 2010, *ApJ*, 725, 2312
Oser, L., Naab, T., Ostriker, J. P., & Johansson, P. H. 2012, *ApJ*, 744, 63
Springel, V. 2005, *MNRAS*, 364, 1105
Springel, V. & Hernquist, L. 2003, *MNRAS*, 339, 389
Toft, S., Smolčić, V., Magnelli, B., *et al.* 2014, *ApJ*, 782, 68

The Zeldovich Universe:
Genesis and Growth of the Cosmic Web
Proceedings IAU Symposium No. 308, 2014
R. van de Weygaert, S. Shandarin, E. Saar & J. Einasto, eds.

© International Astronomical Union 2016
doi:10.1017/S1743921316010255

The disks and spheroid of LTGs in the light of their early web-like organization

R. Domínguez-Tenreiro[1], A. Obreja[1], C. Brook[1], F. J. Martínez-Serrano[2] and A. Serna[2]

[1]Dept. de Física Teórica, Univ. Autónoma de Madrid, E-28049 Cantoblanco Madrid, Spain
[2]Dept. de Física y A.C., Universidad Miguel Hernández, E-03202 Elche, Spain

Abstract. Cosmological hydrodynamical simulations show that the baryonic elements that at $z = 0$ form the stellar populations of late-type galaxies (LTGs), display, at high z, a gaseous web-like organization, where different singular structures (walls, filaments, nodes) show up. The analysis also shows that the spheroid-to-be elements are the first to be involved in the singular structures, while thick and thin disk-to-be elements chronologically follow them. We discuss how these differences at high z can explain the differences among these three components at $z = 0$.

Keywords. cosmology: theory, galaxies: formation, methods: numerical

1. Introduction

The so-called "cold mode" for galaxy mass assembly was first formulated by Binney (1977) and (re)discovered in the last few years (e.g., Birnboim & Dekel 2003; Kereš et al. 2003: Ocvirk et al. 2008; Brooks et al. 2009; Voort et al. 2011). This scenario has recently received much attention because it easily explains halo and galaxy angular momentum acquisition through filaments, (see e.g. Pichon et al. 2011 and this volume; Kimm et al. 2011; Codis et al. 2012; Tillson et al. 2012; Stewart et al. 2013; Dubois et al. 2014; Danovich et al. 2014) and also for its possible implications in the star formation history of galaxies (Kereš et al. 2009; Brooks et al. 2009; Obreja et al. 2013).

On general grounds, cold streams feeding clumps is a general prediction of the Zeldovich Approximation (hereafter ZA, Zeldovich 1970) extended to the Adhesion Model (AM, see e.g. Gurbatov et al. 1989; 2012; Kofman et al. 1990), now confirmed in detail by numerical simulations (e.g., Cautun et al. 2014). Indeed, the AM predicts that, at a given scale, walls surrounding voids, filaments and nodes (i.e., the Cosmic Web, CW, elements, see Bond et al. 1996) are successively formed. Then they vanish due to mass elements flow through the voids towards the walls, through the walls towards the filaments, and finally from these to nodes, where mass elements pile up and virialize after turn-around and collapse (see e.g., Domínguez-Tenreiro et al. 2011). Meanwhile, the same CW elements emerge at larger and larger scales, disappearing later on in favour of voids and nodes.

We can conclude from these considerations that the stellar populations of current galaxies displayed at high z a web-like structure. The question then arises whether or not current stellar populations in galaxies keep memory of their former lives as mass elements involved in the CW dynamics. More specifically, in this talk we focus on the origin of the so-called fine structure of late-type galaxies (hereafter, LTGs). Indeed, stellar populations of LTGs come into three different categories, those forming the spheroid, the thick disk and the thin disk, see Ivezić et al. (2012) and references therein. The differences among the components involve morphologies, stellar ages, velocity fields and chemical compositions, and are thought to be driven by their different specific angular momentum

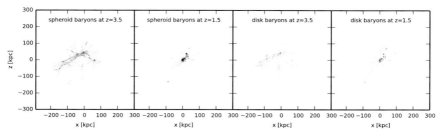

Figure 1. For the HD-5004A galaxy, a projection of the positions of its stellar particles identified at $z = 0$ when traced back to $z_{high} = 3.5$ and 1.5. Left: spheroid-to-be particles. Right: thin disk-to-be particles. Grey (black): gaseous (stellar) particles at the respective z_{high}s.

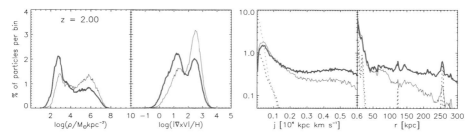

Figure 2. Histograms for different quantities measured on g1536-L* galaxy -to-be particles. Solid red (thick blue) lines stand for gaseous spheroid (disk) precursors, while dashed lines refer to particles that are already stellar at $z = 2$.

j content, decreasing as we go from the thin disk to the spheroid components, with the thick disk in between.

Peebles (1969) and and Doroshkevich (1970) first pointed out that the j_{LV} enclosed within a Lagrangian Volume grows as $a(t)^2$ untill turn around (i.e., the so-called Tidal Torques Theory, based on the ZA). Later on, j_{LV} is roughly conserved, but redistributed within the LV from voids to filaments also as a consequence of the CW dynamics. Indeed, as first shown by Pichon el al. (2011), at the boundaries between voids, matter flows wind up coherently as they meet to form the filaments, causing whirls with their spins roughly parallel to their axes. At the same time, mass elements acquire a net transverse motion (i.e., an orbital j relative to the patch center), due to the asymmetric forces acting on the forming filaments from the voids they divide. Once the filaments form, they collimate longitudinal motions towards the growing central mass accumulation, advecting j in the process and coherently transporting it towards the inner central concentration, either as clumpy or diffuse material. Pichon *et al.* (2011) also show that, at a given time, the j content of infalling matter grows with its distance to the patch center (lever effect). The implications this ordering has onto the j acquisition by gas elements at the scales of *galaxy halos* have been analyzed recently by Stewart *et al.* (2013) and Danovich *et al.* (2014), who in particular have analyzed the gravitational torques responsible for gas and DM j exchanges at the inner halo.

Hydrodynamical simulations in a cosmological context are an adequate tool to study the emergence of the fine-structure of LTGs. Comparisons among the results of different codes are more than advisable, because we are looking for effects coming from a generic and fundamental level of physical description, beyond the astrophysical scale. In view of these considerations, in this talk we present results of LTGs simulations run with two different SPH codes. In designing P-DEVA (Serna *et al.* 2003; Martínez-Serrano *et al.* 2008) the main concern was that j conservation holds as accurately as possible.

In GASOLINE (Wadsley *et al.* 2004), SNe feedback is implemented using the blastwave-formalism (Stinson *et al.* 2006), with a low effective coupling with gas. P-DEVA galaxies have been studied by Doménech *et al.* 2012 at $z = 0$, who analyze some aspects of their fine structure, GASOLINE ones by Brook *et al.* (2012) and Obreja *et al.* (2014). In addition, Obreja *et al.* (2013) and Domínguez-Tenreiro *et al.* (2014) included galaxies of both sets. In all these cases, the consistency with observational data is very satisfactory.

To disentangle the fine structure of a given simulated LTG, Doménech *et al.* (2012) use the kk-means method, an unsupervised clustering algorithm looking for clusters in a 3-dimensional space of kinetic variables for each stellar particle of the LTG. By applying this method, with no priors assumed, the fine structure of LTGs emerges naturally (see Figure 7 in the former paper). Once classified, the systematics of spheroid, thick disk and thin disk observational properties is recovered (see also Obreja *et al.* 2013).

2. Results and Conclusions

To decipher the possible imprints left by the CW dynamics onto the current stellar populations of local LTGs, we have traced back these populations to their progenitors at high z. By plotting the positions of the progenitors of the stars that at $z = 0$ form a LTG, we confirm that indeed they display a clear CW-like structure before collapse, whose evolution is as predicted by the AM. We also note that a main filament outstands in this structure (Dubois *et al.* 2014; Danovich *et al.* 2014) and that the disk is initially normal to this filament, as expected. Moreover, the systematics of j acquisition by disks at the halo scale is recovered as well (see §1).

By seperately plotting the positions of the progenitors of each LTG component, we still found CW-like configurations, but with different characteristics. To illustrate these results, in Fig. 1 we plot, for a simulated LTG in the P-DEVA set, the projections on a given plane of the progenitor positions at $z_{\text{high}} = 3.5$ and 1.5 (collapse). The left panels correspond to the spheroid progenitors, while the right panels correspond to the thin disk antecessors. The visual impression is that spheroid progenitors have been involved in strong shell crossings from earlier on than thin disk progenitors, in such a way that at given zs, they form a CW-like structure whose dynamical state is more advanced than that of the CW corresponding to the thin disk, with the thick disk in between.

To go a step further, the visual differences in Fig. 1 have been quantified by using the V-Web method (Hoffman *et al.* 2012). For a given simulated LTG and different z_{high}s, and for each of its gaseous or stellar particles, the densities, vorticities, j_i, and distances to the particles center-of-mass have been calculated, as well as the respective probability distributions for these quantities. The results we obtain confirm the visual impression that the spheroid-to-be gas particles are the first to be involved in shell-crossing events. To illustrate these results, in Fig. 2 we plot the respective histograms at $z_{\text{high}} = 2$ for a GASOLINE galaxy. The histograms clearly show that, at given times, the spheroid-to-be gas particles statistically have a higher probability to sample the highest densities and vorticities (marking the CW element locations). Moreover, the j histograms confirm that the gas j content increases as time goes by, caused by the more delayed *gaseous* particles having a higher probability of a bigger j (lever effect, see §1). On the contrary, spheroid-to-be particles that by $z_{\text{high}} = 2$ have already turned into stars are j-poorer. In addition, we also see that disk-to-be gaseous particles come on average from further away than spheroid-to-be ones, therefore the former are delayed relative to the later.

We conclude that these results obtained with two codes that are different in their design and subgrid modelling (see §1), strongly suggest that indeed the fine structure in LTGs expresses information imprinted in their baryonic mass elements at high z, when they display a CW configuration, and that this information comes from a fundamental level of physical description such as, e.g., the kinematics involved in the ZA or AM, see §1. In addition, our result that the local LTG fine structure (at disk scale) keeps memory of the CW dynamics at high z suggests that, in the long run, the ordering it imprints in the cosmic inflow somehow overcomes the complex astrophysical processes occurring in the circumgalactic environment of forming disks.

We thank MICINN and MINECO (Spain) for financial support through grants AYA2009-12792-C03-02, -03 and AYA2012-31101 from the PNAyA.

References

Binney, J. 1977, *ApJ*, 215, 492
Birnboim, Y., Dekel A. 2003, *MNRAS*, 345, 349
Bond, J. R., Kofman, L., & Pogosyan, D. 1996, *Nature*, 380, 603
Brook, C. B., Stinson, G. S., Gibson, B. K., Kawata, D., *et al.* 2012, *MNRAS*, 426, 690
Brooks, A. M., Governato, F., Quinn, T., Brook, C. B., & Wadsley, J. 2009, *ApJ*, 694, 396
Cautun, M., van de Weygaert, R., Jones, B. J. T., & Frenk, C. S. 2014, ArXiv e-prints
Codis, S., Pichon, C., Devriendt, J., Slyz, A., Pogosyan, D., Dubois, Y., & Sousbie, T. 2012, *MNRAS*, 427, 3320
Danovich, M., Dekel, A., Hahn, O., Ceverino, D., & Primack, J. 2014, ArXiv e-prints
Doménech-Moral, M., Martínez-Serrano, F. J., Domínguez-Tenreiro, R., & Serna, A. 2012, *MNRAS*, 421, 2510
Domínguez-Tenreiro, R., Oñorbe, J., Martínez-Serrano, F., & Serna, A. 2011, *MNRAS*, 413, 3022
Domínguez-Tenreiro, R., Obreja, A., Granato, G. L., Schurer, A., *et al.* 2014, *MNRAS* 439, 3868
Doroshkevich, A. G. 1970, *Afz*, 6, 320
Dubois, Y., Pichon, C., Welker, C., Le Borgne, D., *et al.* 2014, *MNRAS*, 444, 1453
Gurbatov, S. N., Saichev, A. I., & Shandarin, S. F. 1989, *MNRAS*, 236, 385 XYZ
Gurbatov, S. N., Saichev, A. I., & Shandarin, S. F. 2012, *Physics Uspekhi*, 55, 223
Hoffman, Y., Metuki, O., Yepes, G., Gottlöber, S., Forero-Romero, J. E., Libeskind, N. I., & Knebe, A. 2012, *MNRAS*, 425, 2049
Ivezić, Ž., Beers, T. C., & Jurić, M. 2012, *ARAA*, 50, 251
Kereš, D., Katz, N., Fardal, M., Davé, R., Weinberg, D. H. 2009, *MNRAS*, 395, 160
Kereš, D., Katz, N., Weinberg, D. H., & Davé, R. 2005, *MNRAS*, 363, 2
Kimm, T., Devriendt, J., Slyz, A., Pichon, C., Kassin, S. A., & Dubois, Y. 2011, ArXiv e-prints
Kofman, L., Pogosian, D., & Shandarin, S. 1990, *MNRAS*, 242, 200
Martínez-Serrano, F. J., Serna, A., Domínguez-Tenreiro, R., & Mollá, M. 2008, *MNRAS*, 388, 39
Obreja, A., Brook, C. B., Stinson, G., Domínguez-Tenreiro, R., Gibson, B. K., Silva, L., & Granato, G. L. 2014, *MNRAS*, 442, 1794
Obreja, A., Domínguez-Tenreiro, R., Brook, C., Martínez-Serrano, F. J., *et al.* 2013, *ApJ*, 763, 26
Ocvirk, P., Pichon, C., & Teyssier, R. 2008, *MNRAS*, 390, 1326
Peebles, P. J. E. 1969, *ApJ*, 155, 393
Pichon, C., Pogosyan, D., Kimm, T., Slyz, A., Devriendt, J., & Dubois, Y. 2011 *MNRAS*, 418, 2493
Serna, A., Domínguez-Tenreiro, R., & Sáiz, A., 2003 *ApJ*, 597, 878
Stewart, K. R., Brooks, A. M., Bullock, J. S., Maller, A. H., Diemand, J., Wadsley, J., & Moustakas, L. A. 2013, *ApJ*, 769, 74
Stinson, G., Seth, A., Katz, N., Wadsley, J., Governato, F., & Quinn, T. 2006, *MNRAS*, 373, 1074
Tillson, H., Devriendt, J., Slyz, A., Miller, L., & Pichon, C. 2012, ArXiv e-prints
van de Voort, F., Schaye, J., Booth, C., Haas, M., & DallaVecchia, C. 2011, *MNRAS*, 414, 2458
Wadsley, J. W., Stadel, J., & Quinn, T. 2004, *New Astron.*, 9, 137
Zel'dovich, Y. B. 1970, *A&A*, 5, 84

The Zeldovich Universe:
Genesis and Growth of the Cosmic Web
Proceedings IAU Symposium No. 308, 2014
R. van de Weygaert, S. Shandarin, E. Saar & J. Einasto, eds.

© International Astronomical Union 2016
doi:10.1017/S1743921316010267

Satellites are the main drivers of environmental effects at least to z = 0.7

Katarina Kovač[1] and the zCOSMOS† team

[1]Institute for Astronomy, ETH Zurich, Zurich 8093, Switzerland
email: kovac@phys.astro.ch

Abstract. We study the role of environment in the evolution of galaxies up to z = 0.7 using the final zCOSMOS-bright data set. We use the colour as a proxy for the quenched population, and measure the dependence of the red fraction of galaxies on stellar mass and two environmental indicators: the local overdensity of galaxies δ and a demarcation of galaxies to centrals and satellites. The analysis is carried out by quantifying the role of different quenching processes. We find that the measured dependence of the red fraction of galaxies on stellar mass and environment can be well described by two quenching processes: one related only to stellar mass (mass quenching) and the other related to the local environment (environment quenching). Within the errors, these processes are independent of each other, and consistent with the $z \sim 0$ measurement. Moreover, the red fraction of centrals $f_{r,cen}$ (both singleton centrals and centrals in the groups) does not show any trend with δ and more than 95% of $f_{r,cen}$ is consistent with being produced through the mass quenching alone. The satellite galaxies are redder than the centrals at the same stellar mass and δ, requiring additional environment quenching. Given the observed fractional distribution of satellites at different overdensities, the normalized excess in the red fraction of satellites with respect to the red fraction of centrals is consistent with a scenario in which the satellites account for most of the δ-dependences observed in the overall population of galaxies covering $0.1 < z < 0.7$.

Keywords. galaxies: evolution, cosmology: large-scale structure of universe

1. Introduction

A correlation between various galaxy properties and their environment has been measured up to redshifts of about $z = 1 - 1.5$(e.g. Dressler 1980; Cooper *et al.* 2010; Kovač *et al.* 2014). Galaxies which reside in the regions of higher density are more massive, redder, they are forming less stars, and morphologically they are of an earlier class than their counterparts in the less dense environments (e.g. Kauffmann *et al.* 2004; Cucciati *et al.* 2010; Kovač *et al.* 2010a).

There is a range of physical processes suggested to exist only in dense environments (see e.g. Boselli & Gavazzi 2006), where the majority if not all of the suggested mechanisms are related to the transformation of galaxies after they become satellites. With the recent advances in the spectroscopic surveys, there is growing observational evidence that satellite galaxies are indeed affected by some additional environment specific processes. At the same stellar mass, the fraction of quenched satellites is systematically above those of the centrals, consistent with a scenario in which 40-50% of star-forming galaxies transition to quenched galaxies after they infall into a more massive halo (van den Bosch *et al.* 2008; Peng *et al.* 2012). This fraction stays remarkably constant with stellar mass and redshift up to z = 0.8 (Knobel *et al.* 2013).

† Based on observations undertaken at the European Southern Observatory (ESO) Very Large Telescope (VLT) under Large Program 175.A-0839.

The fraction of quenched satellites also increases with the environmental overdensity, and locally, satellite quenching can account for most of the observed environmental dependence in the quenched fraction of all galaxies (Peng *et al.* 2012). We showed in Kovač *et al.* (2014) that to a good degree this holds to $z = 0.7$ and in this proceeding, we will highlight the key points from that work. Our recent and ongoing more detailed analysis of the observational data shows that the central/satellite paradigm is more complicated: e.g. for massive centrals in the rich groups at $z \sim 0$ the quenched fractions of centrals and satellites are the same (Knobel *et al.* 2014) and the fraction of quenched satellites is correlated with the properties of their central over the whole $0 < z < 0.7$ interval (i.e. galactic conformity, Knobel *et al.* 2014, Kovač *et al.* in prep).

2. Data sample

Our analysis is based on the final zCOSMOS-bright data set of $\sim 17,000$ galaxies with reliable redshift at $z < 1.2$ (Lilly *et al.* 2007, 2009). This sample is limited by a magnitude $i_{ACS} = 22.5$ and it is confined to 1.4 deg^2 within the COSMOS field (Scoville *et al.* 2007). The properties of galaxies (such as luminosities and stellar masses) are based on the the spectral energy distribution (SED) fitting utilising a subset of the 30-band COSMOS photometry (e.g. Capak *et al.* 2007; Sanders *et al.* 2007).

We use two indicators of environment: local overdensity of galaxies and the identification of a galaxy as a central or a satellite. The apertures for the overdensity reconstruction are obtained by measuring distances to the fifth nearest neighbour in the $M_B < -19.3 - z$ sample of galaxies which includes both galaxies with the reliable redshifts and the ZADE-modified photometric probability distribution functions (see Kovač *et al.* 2010b). The central/satellite identification of galaxies is based on the group catalogue of Knobel *et al.* (2012). By definition, the most massive galaxy in a group is termed central, and other galaxies in a given group are marked as satellites. We use the mocks (Henriques *et al.* 2012), modified to match the zCOSMOS selections, to quantify the imperfect separation of galaxies to centrals and satellites and use these corrections to obtain the statistically corrected, true quantities.

The samples which we use in the subsequent analysis contain 2340, 1730, and 610 all, central, and satellite galaxies above the adopted mass limit in $0.1 < z < 0.4$. In the higher redshift bin $0.4 < z < 0.7$, these numbers are 2448, 2062, and 386 respectively. The mass limits are $\log(M_*/M_\odot) = 9.82$ and 10.29 in the lower and higher redshift bins.

3. Colour-density relation for the overall sample

We start by quantifying the relation between the red fraction f_{red} of all zCOSMOS galaxies above the mass completeness limit and their local overdensity and stellar mass. The measured red fractions as a function of overdensity are shown in Fig. 1 for a range of independent stellar mass bins. Even though some of the measured points are discrepant, there is an obvious trend that at a given stellar mass the red fraction of galaxies increases with the local overdensity. Also, at a given value of overdensity, the red fraction increases with the stellar mass. These relations agree (at least qualitatively) with the previously published similar analyses based on the smaller 10k zCOSMOS data set (Cucciati *et al.* 2010; Tasca *et al.* 2009) and a broad range of studies from other surveys up to $z = 1.5$ (e.g. Baldry *et al.* 2006; Peng *et al.* 2010; Cooper *et al.* 2010).

It was shown by Baldry *et al.* (2006) that at $z \sim 0$ (see also Peng *et al.* 2010 for $0.3 < z < 0.6$) the dependence of the red fraction on both stellar mass and overdensity can be well quantified by the following function:

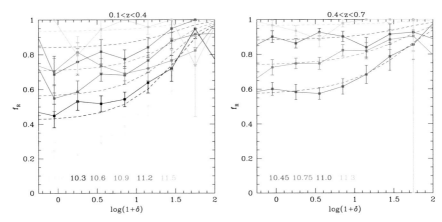

Figure 1. Red fraction of all zCOSMOS galaxies as a function of overdensity in different bins of stellar mass. The left- and right-hand panels correspond to $0.1 < z < 0.4$ and $0.4 < z < 0.7$, respectively. The symbols, connected by solid lines, mark the measured fractions. Going from the bottom to the top, the stellar mass for each curve increases, and the median mass in an increasing order is given at the bottom of each of the panels. The dashed lines correspond to the best-fit model given by equation 3.1. The error bars correspond to half of the 16-84% interval from 100 bootstrapped samples. The figure is taken from Kovač *et al.* (2014, Fig. 3).

$$f_{red}(\delta, M_*) = 1 - exp(-(\delta/p_1)^{p2} - (M_*/p_3)^{p4}) \qquad (3.1)$$

where the $p1 - p4$ parameters are simply the fitting parameters. We explore here if this function can still describe our final zCOSMOS data set. The red fraction from the best-fit model given by Eq. 3.1 is also shown in Fig. 1 (dashed lines). We conclude that the proposed model provides a good description of the zCOSMOS data, as most of the measured and fitted values agree within the 2σ interval. Our additional error analysis does not indicate any systematic effect neither with the mass nor the density.

It is important to stress that one of the properties of the function given by equation 3.1 is that it is separable in its dependency on the mass and environment and it can be rewritten as:

$$f_{red}(\delta, M_*) = \epsilon_\rho(\delta, \delta_0) + \epsilon_m(M_*, M_{*0}) - \epsilon_\rho(\delta, \delta_0)\epsilon_m(M_*, M_{*0}) \qquad (3.2)$$

where $\epsilon_\rho(\delta, \delta_0) = 1 - exp(-(\delta/p_1)^{p2})$ and $\epsilon_m(M_*, M_{*0}) = 1 - exp(-(M_*/p_3)^{p4})$ (Peng *et al.* 2010). Given the definition above, the ϵ_ρ function depends only on environment and the ϵ_m functions depends only on mass. Following Peng *et al.* (2010) we will refer to these two functions as the relative environment and mass quenching efficiencies, respectively. This therefore means that the red fraction of galaxies can be considered to be build through two different quenching processes, one of them dependent only on local density, and the other one dependent only on stellar mass or a closely related property. It is also plausible that the 5-15% difference between the measurement and the best-fit model reflects some quenching processes not related to mass or environment or depending on both of them simultaneously which are not included in this simple f_{red} model.

A comparison of quenching efficiencies obtained from the zCOSMOS data in two red-shift bins and from the $z \sim 0$ data (Peng *et al.* 2010) shows that, within the errors, there is no significant redshift evolution in the ϵ_ρ and ϵ_m functions. This however does not correspond to the static situation in the overall transition of galaxy population from blue to red. This simply means that the physical processes which are responsible for the mass and enviornment quenching act in the same way on galaxies of the same stellar mass and which reside in the quantitatively same environment at different epochs.

Moreover, the relative roles of the environment and mass quenching are expected to change with cosmic time (Fig. 15 in Peng *et al.* 2010), such that the role of environment increases with decreasing stellar mass and decreasing redshift. Qualitatively, this is consistent with the idea that the environment quenching affects the galaxies infalling into the larger haloes, building the large scale structure as cosmic time passes.

4. Red fraction of centrals and satellites

In this section we investigate the role of satellites in producing the observed colour-density relation in the overall population of zCOSMOS galaxies. Given the relatively small number of satellites in our mass complete samples, we proceed by measuring the red fractions of centrals $f_{r,cen}$ and satellites $f_{r,sat}$ in a range of overdensity bins in the carefully constructed mass-matched samples. Centrals and satellites are divided in four quartiles of overdensity (defined by centrals and satellites, respectively, and for each redshift bin) and all samples are matched to the mass distribution of satellites in the reference sample. This was chosen to be the 37.5-62.5% interval in the sorted satellite overdensitites. The median stellar mass is $\log(M_*/M_\odot) = 10.40$ and 10.59 in the lower and higher redshift bin, respectively.

The resulting $f_{r,cen}$, statistically corrected for the impurities in the central/satellite demarcation, are shown in the left-hand panels in Fig. 2 in the four $\delta-$quartiles. Broadly, we do not detect any strong trend between the red fraction of central galaxies and the overdensity in either of the redshift bins. The average red fractions of centrals (dotted lines in the left-hand panels in Fig. 2) are a few percent higher than or consistent with the red fractions expected from the mass quenching alone, estimated to be 0.48 and 0.66 at the median stellar mass in the reference samples in the lower and higher redshift bins, respectively (dashed lines in the left-hand panels in Fig. 2). As the environment quenching accounts for producing less than 5% of red centrals, we adopt in what follows that $f_{r,cen}$ is independent of environment at all overdensities and in both redshift intervals. We take the average red fraction to be a proxy for $f_{r,cen}$. As we do not have a sufficiently large sample of centrals at the highest overdensities probed by the satellites, this statement remains to be tested by future observations.

The purity corrected red fraction of satellites are shown as squares in the left-hand panels in Fig. 2. It is clear that the red fractions of satellites are systematically above the red fractions of centrals at all overdensities over $0.1 < z < 0.7$, requiring some additional quenching channel on top of the mass quenching. We quantify the excess of red satellites with respect to the centrals at a given overdensity δ and mass M_* by defining the satellite quenching efficiency $\epsilon_{sat}(\delta, M_*)$ as

$$\epsilon_{sat}(\delta, M_*) = \frac{f_{r,sat}(\delta, M_*) - f_{r,cen}(M_*)}{f_{b,cen}(M_*)} \quad (4.1)$$

in which $f_{b,cen}(M_*)$ is the blue fraction of centrals. Within the hierarchical scenario, all satellites were centrals in the past, and the $\epsilon_{sat}(\delta, M_*)$ quantity corresponds to the excess fraction of satellites which are environmentally quenched after the infall into a larger halo with respect to the population of star-forming centrals of the same mass.

The $\epsilon_{sat}(\delta, M_*)$ values computed from the zCOSMOS data are shown in the right-hand panels in Fig. 2 as squares. Given the broadness of our $\delta-$bins we also measure $\epsilon_{sat}(\delta, M_*)$

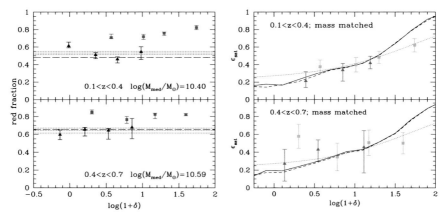

Figure 2. Left: Red fractions of centrals (triangles) and satellites (squares), matched in stellar mass, as a function of overdensity. The fractions measured in $0.1 < z < 0.4$ and $0.4 < z < 0.7$ are shown in the top and bottom panels, respectively. The errors encompass the 16-84% interval of the red fractions in the 20 realisations of the mass-matching. The dotted lines mark the average $f_{r,cen}$ and its $\pm 1\sigma$ values. The dashed lines mark the model $f_{r,cen}$ values produced through the mass quenching alone. Right: Satellite quenching efficiency ϵ_{sat} as a function of overdensity. The squared and triangular symbols correspond to the measurements in the $\delta-$quartiles and in the narrower bins of overdensity, respectively. The error bars encompass the 16-84% uncertainty interval based on the 20 (squares) or 50 (triangles) realisations of the mass-matching. The continuous and dashed lines are the model predicted $\epsilon_\rho(\delta)/f_{sat}(\delta, M_*)$ functions when using the satellite fraction measured in the $\Delta \log(M_*/M_\odot) = 0.5$ bin centered at $\log(M_*/M_\odot) = 10.40$ (top) and 10.59 (bottom), and 10.24 (top) and 10.49 (bottom), respectively. The dotted line is the equivalent $z \sim 0$ $\epsilon_{sat}(\delta)$ measurement from Peng *et al.* (2012). Figures adapted from Kovač *et al.* (2014).

in a narrower range of overdensities where both populations are numerous, still matching the mass distribution of centrals to the satellites in a given $\delta-$bin. These $\epsilon_{sat}(\delta, M_*)$ estimates are shown as triangles in the right-hand panels in Fig. 2. The median stellar mass, when going from the lower to the higher narrower overdensity bins, takes values $\log(M_*/M_\odot) = 10.24$, 10.32 and 10.38 and $\log(M_*/M_\odot) = 10.49$, 10.52, and 10.60 in $0.1 < z < 0.4$ and $0.4 < z < 0.7$, respectively.

Within our mass and environment quenching model, in the situation when the quenched fraction of centrals is produced via the mass quenching process ϵ_m alone, the overall environment quenching ϵ_ρ must come from the satellite population. Then, for the fraction of satellites described by $f_{sat}(\delta, M_*)$, the equality between $\epsilon_{sat}(\delta, M_*)$ and $\epsilon_\rho(\delta)/f_{sat}(\delta, M_*)$ must hold.

The $\epsilon_\rho(\delta)/f_{sat}(\delta, M_*)$ functions obtained from the zCOSMOS data are shown as the black continuous and dashed lines in the right-hand panels in Fig. 2. Different curves are for the fraction of satellites measured at the median stellar mass in the reference overdensity bin and in the lowest median stellar mass in the narrower overdensity bins.

There is an excellent agreement between the purity-corrected $\epsilon_{sat}(\delta, M_*)$ measurements in the narrower $\delta-$bins and $\epsilon_\rho(\delta)/f_{sat}(\delta, M_*)$ in both redshifts. The $\epsilon_{sat}(\delta, M_*)$ values from the quartiles in the overdensity are somewhat discrepant with the prediction, particularly in the lowest and highest overdensities. However, these points cover the broadest range in overdensity and the satellite red fractions measured in these intervals are rather smoothed. Considering the associated uncertainties, the agreement between the measured $\epsilon_{sat}(\delta, M_*)$ values and the $\epsilon_\rho(\delta)/f_{sat}(\delta, M_*)$ function is remarkably good.

Our results constrain that the satellite galaxies must be the dominant population of galaxies driving the overall environmental trends at least up to $z < 0.7$.

5. Conclusions

We have explored the colour-density relation in the final sample of galaxies in $0.1 < z < 0.7$ obtained in the zCOSMOS-bright survey. Using the two environmental indicators: local overdensity of galaxies and the dichotomous sample of centrals and satellites, we reach the following conclusions: 1) red fraction of galaxies f_{red} increases with both stellar mass and overdensity in $0.1 < z < 0.7$. Within the errors, f_{red} is consistent to be separable in stellar mass and environment, indicating the existence of the two independent quenching mechanisms: mass quenching and environment quenching; 2) the differential effect of these two mechanisms does not change with cosmic time and they seem to be the same at $z = 0.7$ as locally; 3) red fraction of centrals is consistent with being independent of overdensity and more than 95% of $f_{r,cen}$ can be produced through the mass quenching alone; 4) the red fraction of satellites requires an additional quenching mechanism with respect to the mass quenching: at the same stellar mass and overdensity, satellites are redder; 5) in the framework of the model of mass and environment quenching efficiencies, our data support the scenario where the satellite quenching efficiency can explain the majority of the overall environmental effects at least up to $z = 0.7$; 6) given the associated uncertainties in our analysis, all these statements should be understood only as approximations to a complex physical reality. In our ongoing work, we are focussing on furthering the understanding of the physical processes governing the observed relations.

References

Baldry, I. K., Balogh, M. L., Bower, R. G., Glazebrook, K., Nichol, R. C., Bamford, S. P., & Budavari, T., 2006, *MNRAS*, 373, 469

Boselli, A. & Gavazzi, G., 2006, *PASP*, 118, 517

Capak, P. *et al.*, 2007, *ApJS*, 172, 99

Cooper, M. C. *et al.*, 2010, *MNRAS*, 409, 337

Cucciati, O. *et al.*, 2010, *A&A*, 524, A2

Dressler, A., 1980, *ApJ*, 236, 351

Henriques, B. M. B., White, S. D. M., Lemson, G., Thomas, P. A., Guo, Q., Marleau, G.-D., & Overzier, R. A., 2012, *MNRAS*, 421, 2904

Kauffmann, G., White, S. D. M., Heckman, T. M., Ménard B., Brinchmann, J., Charlot, S., Tremonti, C., & Brinkmann, J., 2004, *MNRAS*, 353, 713

Knobel, C. *et al.*, 2012, *ApJ*, 753, 121

Knobel, C. *et al.*, 2013, *ApJ*, 769, 24

Knobel, C., Lilly, S. J., Woo, J., & Kovač K., 2014, arXiv1408.2553

Kovač K. *et al.*, 2010b, *ApJ*, 708, 505

Kovač K. *et al.*, 2010a, *ApJ*, 718, 86

Kovač K. *et al.*, 2014, *MNRAS*, 438, 717

Lilly, S. J. *et al.*, 2007, *ApJS*, 172, 70

Lilly, S. J. *et al.*, 2009, *ApJS*, 184, 218

Peng, Y.-j. *et al.*, 2010, *ApJ*, 721, 193

Peng, Y.-j., Lilly, S. J., Renzini, A., & Carollo, M., 2012, *ApJ*, 757, 4

Sanders, D. B. *et al.*, 2007, *ApJS*, 172, 86

Scoville, N. *et al.*, 2007, *ApJS*, 172, 38

Tasca, L. A. M. *et al.*, 2009, *A&A*, 503, 379

van den Bosch F. C., Aquino, D., Yang, X., Mo, H. J., Pasquali, A., McIntosh, D. H., Weinmann, S. M., & Kang, X., 2008, *MNRAS*, 387, 79

The Zeldovich Universe:
Genesis and Growth of the Cosmic Web
Proceedings IAU Symposium No. 308, 2014
R. van de Weygaert, S. Shandarin, E. Saar & J. Einasto, eds.

© International Astronomical Union 2016
doi:10.1017/S1743921316010279

The different lives of galaxies at different environment density levels

Antti Tamm[1], Lauri Juhan Liivamägi[1] and Elmo Tempel[1,2]

[1] Tartu Observatory, Tõravere, 61602 Tartumaa, Estonia
email: antti.tamm@to.ee

[2] National Institute of Chemical Physics and Biophysics, Rävala pst 10, 10143, Tallinn, Estonia

Abstract. We take a closer look at the dependence of the galactic colour histogram on the environment density using a volume-limited sample of SDSS galaxies. We find that the strongest changes with environment are taking place with spiral galaxies. In dense environment, discs become considerably redder, apparently due to the shortage of gas, and less concentrated. Contrary to expectation, the mean Sérsic index of luminous elliptical galaxies decreases in denser environments.

Keywords. galaxies: photometry – galaxies: structure – galaxies: spiral – galaxies: elliptical

1. Introduction

The dependence of galaxy properties on the environment has been known since the papers by Einasto *et al.* (1974), Oemler (1974), Davis & Geller (1976), Dressler (1980), Postman & Geller (1984). The most pronounced effect is the morphology-density relation, which states that the fraction of elliptical galaxies increases at the cost of spiral galaxies in cluster environment with respect to field environment. Exploiting data from extensive redshift surveys, more recent studies have revealed several additional effects of the environment. It has been shown that in addition to the local (i.e. group and cluster) neighbourhood, also the larger-scale environment density (e.g. void vs superclulster) has an impact on galaxy populations (Tempel *et al.* 2011, Lietzen *et al.* 2012, Einasto *et al.* 2014). For example, the luminosity function of elliptical galaxies depends on the large-scale environment density, while spiral galaxies maintain their luminosity function shape.

The transformation of blue star-forming spiral galaxies into red and dead ellipticals and S0-s in dense environments apparently takes place via two major mechanisms: galaxy-galaxy interactions and gas-gas interactions. While the former transforms the structure of galaxies via matter redistribution, the latter does not. Instead, removal of gas suppresses star formation, but has no immediate effect on the galactic structure. As a result, the bimodal colour distribution, corresponding to a bimodal star formation rate distribution, depends heavily of the environmental density (Kauffmann *et al.* 2004).

In this contribution we take a closer look at the impact of environmental desity on galaxy properties by investigating the colours and structural properties of galaxies within high- and low-density large-scale environments.

2. Input

The galaxy sample is based on a volume-limited sample of the Sloan Digital Sky Survey (SDSS; York *et al.* 2000) data release 10 (Ahn *et al.* 2014), taken from the galaxy group catalogue (Tempel *et al.* 2014). The whole sample consists of 105,041 galaxies

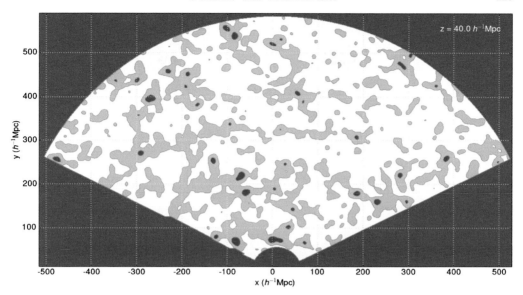

Figure 1. An illustrative slice of the SDSS luminosity density field, smoothed with 8 Mpc B3 spline kernel. Void regions are marked with light gray (yellow), superclusters with dark gray (red) in the paper (electronic) version, respectively. The rest of the field has intermediate luminosity density. See text for the density level definitions.

and is complete down to $M_r = -18$ mag. Further on, the sample is split into different subsamples with correspondingly lower membership.

For characterising the environment, we have applied the large scale luminosity density, derived by smoothing with a B3 spline kernel within 8 Mpc radius (Tempel *et al.* 2014). In addition to specifically probing the large-scale environment, the large smoothing radius minimises the contamination of the density field by the luminosity of the galaxy under consideration and its nearby neighbours. We compare galaxies within two extremes: void environment (density below the mean value of the Universe) and supercluster environment (density more than five times the mean value of the Universe). A slice of the SDSS luminosity density field with the corresponding density levels marked is shown in Fig. 1.

The morphological types of the galaxies (E or S) are taken from two independent sources: the visual classifications from the Galaxy Zoo Project (Lintott *et al.* 2008) and parameter-based determinations from (Tempel *et al.* 2011). A closer inspection reveals that both sources still contain obvious misclassifications, thus for security reasons, we have considered only those galaxies for which the two classifications overlap.

The photometric properties of the galaxies used in this study are taken from the SDSS. In addition, we use structural parameters from Simard *et al.* (2011), where bulge+disc decomposition of the SDSS galaxies was performed.

3. Output

Colour histograms of elliptical and spiral galaxies are shown in Fig. 2. As expected, the colour distribution of elliptical galaxies is relatively concentrated, with a single peak. The large-scale environment has only minor effects, narrowing the distribution slightly and shifting the maximum of $(g-i)$ by about 0.05 mag redward. On the other hand, the colour distribution of spirals is strongly bimodal in both environments. A clear density evolution is present: while blue spirals vastly dominates in void environmets, the numbers of

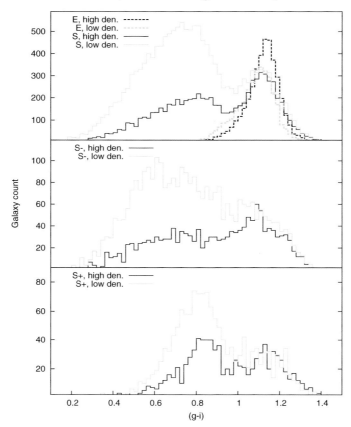

Figure 2. Colour histrogram of the volume-limited sample of SDSS galaxies in voids and in superclusters. Top panel – all ellipticals and spirals; middle panel – low luminosity spirals; bottom panel – high luminosity spirals. The strongest evolution with the large scale environment occurs for low luminosity spirals.

blue and red spirals become roughly balanced in supercluster environments. Apparently, the effect results from suppressed starformation, caused by the deficit of gas in denser environments. For investigating this effect further, we have split the spiral sample into lower (r-band absolute luminosity $L_r < 0.5 \cdot 10^{10} h^{-2} L_\odot$) and higher ($L_r > 1 \cdot 10^{10} h^{-2} L_\odot$) luminosity bins. A comparison of the middle and bottom panels of Fig. 2 shows that the loss of the blue peak is more pronounced for lower luminosity galaxies. In dense environments, low luminosity spirals achieve a nearly flat colour distribution, with the small red peak possibly caused by still inevitably present misclassifications.

Interesting aspects emerge also from correlating the structural parameters derived by Simard *et al.* (2011) with the environment density. For the spiral population of galaxies, the mean Sérsic index n, describing the shape of the surface brightness distribution, increases notably from 1.1 to 1.3. Galaxy-galaxy interactions and accretion of satellites are the potential culprits in this affair. A similar trend can be observed for low luminosity ellipticals (increase from 3.8 to 4), but the opposite takes place with luminous ellipticals (decrease from 3.5 in voids vs 3.0 in superclusters). One might suspect that the decrease of n with environment density is related to the central ellipticals of groups and clusters, but after segregating ellipticals ranking first in luminosity within their groups and clusters from the others (according to Tempel *et al.* 2014), we find that the effect concerns both subsamples of luminous ellipticals.

4. Conclusions

The properties of both elliptical and spiral galaxies change with the large-scale environment density. Void spirals tend to be bluer and more compact than spirals in superclusters. Low- and high-luminosity ellipticals react differently to environment density increase: while both become slightly redder, the former become more centrally concentrated while the latter do not.

Acknowledgements

We acknowledge the financial support from the Estonian Research Council and the Centre of Excellence of Dark Matter in (Astro)particle Physics and Cosmology. This research is based on the SDSS datasets and the authors feel indebted to the SDSS team and their funding organisations; unfortunately, the page limit does not enable us to include the whole officially requested acknowledgment here.

References

Ahn, C. P., Alexandroff, R., Allende Prieto, C., *et al.* 2014, *ApJ*, 211, 17
Davis, M. & Geller, M. J. 1976, *ApJ*, 208, 13
Dressler, A. 1980, *ApJ*, 236, 351
Einasto, J., Saar, E., Kaasik, A., & Chernin, A. D. 1974, *Nature*, 252, 111
Einasto, M., Lietzen, H., Tempel, E., Gramann, M., Liivamägi, L. J., & Einasto, J. 2014, *A&A*, 562, 87
Kauffmann, G., White, S. D. M., Heckman, T. M., Ménard, B., Brinchmann, *et al.*, 2004, *MNRAS*, 353, 713
Lietzen, H., Tempel, E., Heinämäki, P., Nurmi, P., Einasto, M., & Saar, E. 2012, *A&A*, 545, 104L
Lintott, C. J., Schawinski, K., Slosar, A., *et al.* 2008, *MNRAS*, 389, 1179
Oemler, Jr., A. 1974, *ApJ*, 194, 1
Postman, M. & Geller, M. J. 1984, *ApJ*, 281, 95
Simard, L., Mendel, J. T., Patton, D. R., Ellison, S. L., & McConnachie, A. W. 2011, *ApJS*, 196, 11
Tempel, E., Saar, E., Liivamägi, L. J., Tamm, A., Einasto, J., Einasto, M., & Müller, V. 2011, *A&A*, 529, 53
Tempel, E., Tamm, A., Gramann, M., *et al.* 2014, *A&A*, 566, A1
York, D. G., Adelman, J., Anderson, Jr., J. E., *et al.* 2000, *AJ*, 120, 1579

The Zeldovich Universe:
Genesis and Growth of the Cosmic Web
Proceedings IAU Symposium No. 308, 2014
R. van de Weygaert, S. Shandarin, E. Saar & J. Einasto, eds.

© International Astronomical Union 2016
doi:10.1017/S1743921316010280

It takes a supercluster to raise a galaxy

Heidi Lietzen[1,2] and Maret Einasto[3]

[1]Instituto de Astrofísica de Canarias, E-38205 La Laguna, Tenerife, Spain

[2]Universidad de La Laguna, Dept. Astrofísica, E-38206 La Laguna, Tenerife, Spain
email: `hlietzen@iac.es`

[3]Tartu Observatory, 61602 Tõravere, Estonia

Abstract. The properties of galaxies depend on their environment: red, passive elliptical galaxies are usually located in denser environments than blue, star-forming spiral galaxies. This difference in galaxy populations can be detected at all scales from groups of galaxies to superclusters. In this paper, we will discuss the effect of the large-scale environment on galaxies. Our results suggest that galaxies in superclusters are more likely to be passive than galaxies in voids even when they belong to groups with the same richness. In addition, the galaxies in superclusters are also affected by the morphology of the supercluster: filament-type superclusters contain relatively more red, passive galaxies than spider-type superclusters. These results suggest that the evolution of a galaxy is not determined by its local environment alone, but the large-scale environment also affects.

Keywords. Large-scale structure of universe, galaxies: statistics, galaxies: evolution

1. Introduction

Properties of galaxies depend on their environment. In dense environments, massive galaxies are more likely to be red and passive ellipticals than in less dense regions. This color-density or morphology-density relation can be found on many scales of density from the distance to the neighboring galaxies to the location of the galaxy in the cosmic web (Dressler 1980, Postman & Geller 1984, Gómez *et al.* 2003, Einasto & Einasto 1987, Balogh *et al.* 2004, Porter *et al.* 2008, Skibba *et al.* 2009).

The dependence of galaxy properties on the environment can be explained with an evolutionary scenario, where the star formation in a galaxy is quenched by some interaction with the environment. There are several physical processes that can cause the transformation of galaxies from star forming to passive. These include interaction with other galaxies such as mergers (Barnes & Hernquist 1992) and galaxy harassment (Richstone 1976, Moore *et al.* 1996), and interaction with the group or the cluster, such as tidal stripping (Gnedin 2003), ram pressure stripping (Gunn & Gott 1972), or strangulation (Larson *et al.* 1980).

We analyse the galaxy populations in different large-scale environments, concentrating especially on galaxies in small groups. The majority of galaxies are located in groups of a few galaxies or in the field, i.e. not belonging to any group. Figure 1 shows the total number of galaxies in different large-scale environments as a function of group richness. The data is from the Sloan Digital Sky Survey 10th Data Release (SDSS DR10). The group catalog was constructed by Tempel *et al.* (2014) and it contains 588 193 galaxies that form 82 458 groups. The sample is magnitude limited with $r < 17.77$. The large-scale environment was determined using a luminosity-density field with smoothing scale of 8 Mpc (Tempel *et al.* (2014), Liivamägi *et al.* (2012)). The unit for the large-scale density D is the mean density of the whole field. Regions with less than the mean density

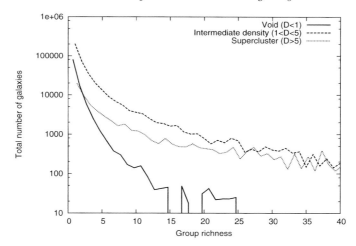

Figure 1. Distribution of galaxies in voids (solid line), intermediate densities (dashed line), and superclusters (dotted line).

can be defined as voids, while densities with more than five times the mean density are superclusters.

The high fraction of galaxies in groups suggests that the groups are important environments in galaxy evolution. Small groups can also be found in any large-scale environment from voids to superclusters, which makes it possible to compare the different large-scale environments. Our aim is to study how the large-scale environment affects the galaxies. We compare the properties of galaxies in high and low density large-scale environments (superclusters and voids), concentrating especially on galaxies in small groups.

2. Results

In Lietzen *et al.* (2012) we used the galaxy catalogs of the eigth data release (DR8) of the SDSS to compare galaxy populations in different large-scale environments. We found that galaxies in superclusters are more likely to be red and passive than galaxies in voids even when they belong to groups with the same richness. This difference was found by comparing the fractions of star-forming and passive galaxies, and by calculating the average colors of galaxies in different environments. Morphology on its own does not depend on the environment: environments of non-star-forming spiral galaxies are similar to those of elliptical galaxies.

As an example of the large-scale effect, we analyse in Fig. 2 the color distribution of galaxies that do not belong to any group (field galaxies) from the SDSS tenth data release (DR10) data. These galaxies represent the poorest group-scale environments in the universe. The distribution is shown separately for supercluster, intermediate density, and void galaxies. The figure shows that field galaxies in superclusters are more likely red than the field galaxies in voids.

Besides the differences between superclusters and low-density environments, there are also differences in the galaxy populations between different superclusters. Superclusters can be divided into two morphological types: spider-type superclusters, which consist of high-density clumps with several outgoing filaments, and filament-type superclusters, in which high-density clumps are connected by only a few galaxy filaments. The galaxy populations in different superclusters were studied in Einasto *et al.* (2014) by

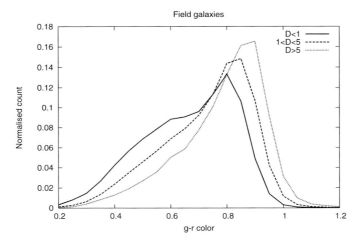

Figure 2. Color distribution of galaxies in voids (solid line), intermediate densities (dashed line), and superclusters (dotted line).

comparing galaxy contents of spider and filament type superclusters. Distributions of color, morphology, stellar mass, and star-formation rate of galaxies in the two types of superclusters and in lower-density environments are shown in Fig. 3. Superclusters of filament type contain a higher fraction of early-type, low star-formation rate galaxies than the spider-type superclusters. There are also significant differences between the galaxy populations of individual superclusters.

3. Discussion

Our results suggest that galaxies in superclusters are different from galaxies in voids, and that different superclusters have different galaxy populations. These results imply that the evolution of galaxies may not be determined by the group-scale environment alone, but the large-scale environment also has an effect.

The physical processes that drive galaxy evolution happen on smaller scales. Star formation in a galaxy may be ended by an interaction with a neighboring galaxy or with the surrounding group of galaxies. Typical sizes of superclusters are of the order of ten megaparsec or more. Therefore, the large-scale effect must be less direct. One possible explanation may be the "speed" of evolution: according to Einasto et al. (2005), the dynamical evolution in high-density global environments the dynamical evolution starts earlies and is more rapid than in low-density environments.

The higher fraction of mature galaxies in denser environments may be interpreted as a product of mass assembly bias. Gao et al. (2005) found in the simulations that the oldest 10 % of halos are more than five times more strongly correlated than the youngest 10 % with the same mass. If old halos also host old galaxies, this could explain the denser environments of the red, passive galaxies.

The morphology of superclusters may affect galaxies in superclusters through different dynamical structure. According to Einasto et al. (2012) superclusters with spider morphology have richer inner structure than those with filament morphology. This means that mergers of clusters may occur more often in the spider superclusters, making them dynamically younger.

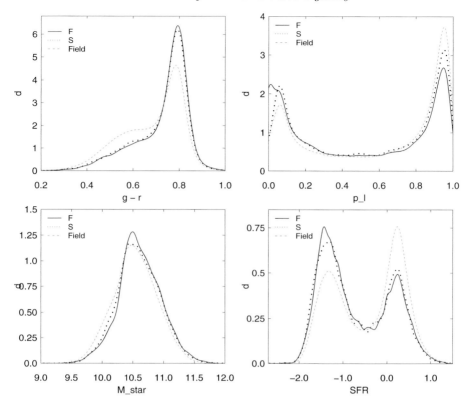

Figure 3. Distribution of $g-r$ color (top left), probability for late-type morphology (top right), stellar mass (bottom left), and star-formation rate (bottom right) for galaxies in filament (solid line) and spider (dotted line) type superclusters and outside superclusters (dashed line).

References

Balogh, M. L., Baldry, I. K., Nichol, R., Miller, C., Bower, R. & Glazebrook, K. 2004 *ApJ*, 615, L101

Barnes, J. E. & Hernquist, L. 1992, *ARA&A*, 30, 705

Dressler, A. 1980, *ApJ*, 236, 351

Einasto, J., Tago, E., Einasto, M., *et al.* 2005, *A&A*, 439, 45

Einasto, M. & Einasto, J. 1987, *MNRAS*, 266, 543

Einasto, M., Lietzen, H., Tempel, E., *et al.* 2014, *A&A*, 562, A87

Einasto, M., Liivamägi, L. J., Tempel, E., *et al.* 2012, *A&A*, 542, A36

Gao, L., Springel, V., & White, S. D. M. 2005, *MNRAS*, 363, L66

Gómez, P. L., Nichol, R. C., Miller, C. J., *et al.* 2003, *ApJ*, 584, 210

Gnedin, O. Y. 2003, *ApJ*, 582, 141

Gunn, J. E. & Gott, III, J. R. 1972, *ApJ*, 176, 1

Larson, R. B., Tinsley, B. M., & Caldwell, C. N., 1980, *ApJ*, 237, 692

Lietzen, H., Tempel, E., Heinämäki, P., *et al.* 2012, *A&A*, 545, A104

Liivamägi, L. J., Tempel, E., & Saar, E. 2012, *A&A*, 539, A80

Moore, B., Katz, N., Lake, G., Dressler, A., & Oemler, A. 1996, *Nature*, 379, 613

Porter, S. C., Raychaudhury, S., Pimbblet, K. A., & Drinkwater, M. J. 2008, *MNRAS*, 388, 1152

Postman, M. & Geller, M. J. 1984, *ApJ*, 281, 95

Richstone, D. O. 1976, *ApJ*, 204, 642

Skibba, R. A., Bamford, S. P., Nichol, R. C., *et al.* 2009, *MNRAS*, 399, 966

Tempel, E., Tamm, A., Gramann, M. *et al.* 2014. *A&A*, 566, A1

The Zeldovich Universe:
Genesis and Growth of the Cosmic Web
Proceedings IAU Symposium No. 308, 2014
R. van de Weygaert, S. Shandarin, E. Saar & J. Einasto, eds.

© International Astronomical Union 2016
doi:10.1017/S1743921316010292

Hierarchical formation of Dark Matter Halos near the Free Streaming Scale, and Their Implications on Indirect Dark Matter Search

Tomoaki Ishiyama

Center for Computational Science, University of Tsukuba, 1-1-1, Tennodai, Tsukuba, Ibaraki,
305-8577, Japan
email: ishiyama@ccs.tsukuba.ac.jp

Abstract. The smallest dark matter halos are formed first in the early universe. According to recent studies, the central density cusp is much steeper in these halos than in larger halos and scales as $\rho \propto r^{-(1.5-1.3)}$. We present results of very large cosmological N-body simulations of the hierarchical formation and evolution of halos over a wide mass range, beginning from the formation of the smallest halos. We confirmed early studies that the inner density cusps are steeper in halos at the free streaming scale. The cusp slope gradually becomes shallower as the halo mass increases. The slope of halos 50 times more massive than the smallest halo is approximately -1.3. The concentration parameter is nearly independent of halo mass, and ruling out simple power law mass-concentration relations. The steeper inner cusps of halos near the free streaming scale enhance the annihilation luminosity of a Milky Way sized halo between 12 to 67%.

Keywords. dark matter, halo, numerical

1. Introduction

The size of the smallest halo is determined by the free streaming scale of the dark matter particles. If dark matter comprises the lightest supersymmetric particle (the neutralino of mass approximately 100 GeV), the estimated corresponding mass of the smallest microhalos is approximately earth-mass (e.g. Hofmann *et al.* 2001; Berezinsky *et al.* 2003).

If many earth-mass microhalos exist at present universe, they could significantly enhance gamma-ray signals by neutralino self-annihilation. Since the gamma-ray flux is proportional to the square of the local dark matter density, the annihilation signal from the Milky Way halo should largely depend on the halo's fine structure. According to recent studies, the central density cusp is much steeper in the smallest halos than in larger halos (Ishiyama *et al.* 2010). Such steep cusps may largely impact on indirect experimental searches for dark matter.

In Ishiyama *et al.* (2010), only three microhalos were simulated. Statistical study, such as the distribution of microhalo density profiles, requires a more extensive dataset. Ishiyama *et al.*(2010) analytically estimated that the formation epoch of microhalos is affected by larger scale density fluctuations. They predicted that the formation epoch of their simulated microhalos was later than the average value. Since halo concentration reflects the cosmic density at which halos collapse, this suggests that their microhalos were less concentrated than indicated by the average. Therefore, to precisely predict the gamma-ray flux, the distribution of both the microhalo density profile shapes and of microhalo concentrations must be elucidated.

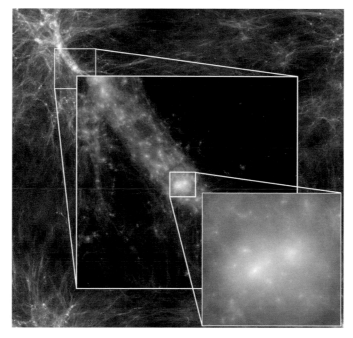

Figure 1. The distribution of dark matter in the A_N4096L400 simulation at $z = 32$. Central and bottom right panels show the enlargement of the largest halo.

We address these questions by large and high resolution cosmological N-body simulations. The detailed explanation of this work can be found in Ishiyama (2014).

2. Numerical Methods

We performed three large cosmological N-body simulations by massively parallel TreePM code, GreeM (Ishiyama *et al.* 2012). Two different initial matter power spectra were used. In two of the simulations, the power spectrum included the sharp cutoff imposed by the free streaming damping of dark matter particles with a mass of 100GeV (Green *et al.* 2004). The third simulation ignored the effect of free streaming damping. The initial conditions were generated by a first-order Zeldovich approximation at $z = 400$.

In the simulations with the cutoff imposed, the motions of 4096^3 particles in comoving boxes of side lengths 400 pc and 200 pc were followed (these simulations are denoted A_N4096L400 and A_N4096L200, respectively). The particle masses were $3.4 \times 10^{-11} M_\odot$ and $4.3 \times 10^{-12} M_\odot$, ensuring that halos at the free streaming scale were represented by $\sim 30,000$ and $\sim 230,000$ particles, respectively. The simulation with no cutoff followed the motions of 2048^3 particles in comoving boxes of side length 200 pc (this simulation is denoted B_N2048L200). Figure 1 shows the dark matter distribution in the A_N4096L400 simulation at $z = 32$.

3. Results

We calculated the spherically averaged radial density profile of each halo within the range $0.02 \leqslant r/r_{\rm vir} \leqslant 1.0$, divided into 32 logarithmically equal intervals. Each density profile deviates to varying extent from the average density profile, mainly because sub-halos exist in the halos. To minimize this effect and obtain proper average radial density

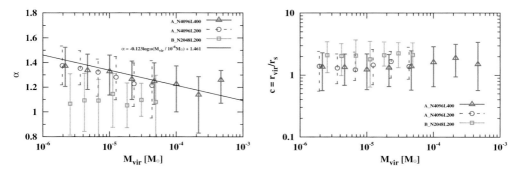

Figure 2. Distributions of slope α and concentration $c = r_{\rm vir}/r_{\rm s}$ of the density profile, plotted against the halo virial mass $M_{\rm vir}$. Circles, triangles and squares show the median value in each mass bin. Whiskers are the first and third quantiles. Black solid line is the best fit power law function [Equation (3.2) in the text].

profiles of halos with a wide mass distribution, we stacked the profiles of similar-mass halos.

To quantify the simulated density structures, we fitted the stacked density profiles to a double power law function, given by

$$\rho(r) = \frac{\rho_0}{(r/r_{\rm s})^\alpha \, (1 + r/r_{\rm s})^{(3-\alpha)}}. \tag{3.1}$$

The power law functions that best fits the relation between mass and α obtained from our simulations is

$$\alpha = -0.123 \log(M_{\rm vir}/10^{-6} M_\odot) + 1.461. \tag{3.2}$$

To visualize the scatter in the density profiles, we fitted the profile of each halo to Equation (3.1) and calculated the median and scatter in each mass bin. These results are shown in Figure 2. The two simulations with different resolutions give similar results (A_N4096L400and A_N4096L200). The median accurately matches the fitting function derived from the stacked density profile. Regardless of halo mass, the first and third quantiles deviate by less than 20% in the A_N4096L400 simulation. Clearly, the B_N2048L200 simulation generates more scatter than the A_N4096L400 simulation.

Note that the concentration parameter is defined differently than the NFW profile, since another fitting function is used. Remarkably, the concentration parameter in both models is nearly independent of halo mass over the range shown in Figure 2. The median concentration in the cutoff model is 1.2–1.7, increasing to 1.8–2.3 without the cutoff.

These results differ from what we see in larger halos (dwarf-galaxy-sized to cluster-sized halos) at lower redshifts (typically less than $z = 5$). The concentrations of these halos weakly depend on the halo mass, and have been fitted to many simple single power law functions. Our results rule out single power law mass-concentration relations.

4. Discussions

The gamma-ray luminosity of a halo by neutralino self-annihilation seen from a distant observer is calculated by the volume integral of the density squared. The steeper cusps obtained in our simulations could significantly enhance the signals. In this section, we evaluate boost factors using the density profiles obtained in our simulations.

Figure 3 shows the annihilation boost factor as a function of the halo mass. We considered three models. In the first model, we used the profile of Equation (3.1) with the

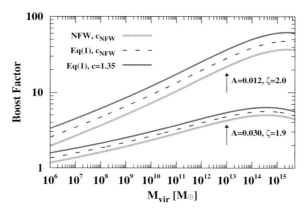

Figure 3. Annihilation boost factor as a function of the halo mass. The results of three models of halos near the free streaming scale are shown. The subhalo mass function $dn/dm = 0.012/M(m/M)^{-2.0}$ gives upper three curves. The subhalo mass function $dn/dm = 0.030/M(m/M)^{-1.9}$ gives lower three curves.

mass-shape relation of Equation (3.2), and the constant concentration $c = 1.35$ at $z = 32$, consistent with our cutoff simulation. In the second model, we used the profile of Equation (3.1) with the mass-shape relation of Equation (3.2), and the mass-concentration relation proposed by Sánchez-Conde and Prada (2014). In the third model, we used the NFW profile and the mass-concentration relation proposed by Sánchez-Conde and Prada (2014). The models with steeper inner cusps raise the boost factor moderately. The boost factors of a Milky Way sized halo ($M = 2.0 \times 10^{12} M_\odot$) with the $\zeta = 2.0$ subhalo mass function are $\sim 17, 22$ and 29 for three models, respectively. Those with the $\zeta = 1.9$ subhalo mass function are $\sim 3.7, 4.2$ and 4.8. Strongly depending on the subhalo mass function and the adopted concentration model, the steeper inner cusps of halos near the free streaming scale enhance the annihilation luminosity of a Milky Way sized halo between 12 to 67%.

Acknowledgement

Numerical computations were partially carried out on Aterui supercomputer at Center for Computational Astrophysics, CfCA, of National Astronomical Observatory of Japan, and the K computer at the RIKEN Advanced Institute for Computational Science (Proposal numbers hp120286 and hp130026). This work has been funded by MEXT HPCI STRATEGIC PROGRAM and MEXT/JSPS KAKENHI Grant Number 24740115.

References

Hofmann, S., Schwarz, D. J., & Stöcker, H. 2001, *PRD*, 64, 083507
Berezinsky, V., Dokuchaev, V., & Eroshenko, Y. 2003, *PRD*, 68, 103003
Ishiyama, T., Makino, J., & Ebisuzaki, T. 2010, *ApJL*, 723, L195
Ishiyama, T. 2014, *ApJ*, 788, 27
Green, A. M., Hofmann, S., & Schwarz, D. J. 2004, *MNRAS*, 353, L23
Ishiyama, T., Nitadori, K., & Makino, J. 2012, in Proc. Int. Conf. High Performance Computing, Networking, Storage and Analysis, SC 12 (Los Alamitos, CA: IEEE Computer Society Press), 5:, (arXiv:1211.4406)
Sánchez-Conde, M. A. & Prada, F. 2014, *MNRAS*, 442, 2271

CHAPTER 7B.

Galaxy Alignments

Church of the Holy Ghost, old town Tallinn.
Clock on the facade has been made by Christian Ackermann.
Photo: Rien van de Weijgaert.

The Zeldovich Universe:
Genesis and Growth of the Cosmic Web
Proceedings IAU Symposium No. 308, 2014
R. van de Weygaert, S. Shandarin, E. Saar & J. Einasto, eds.

© International Astronomical Union 2016
doi:10.1017/S1743921316010309

Why do galactic spins flip in the cosmic web? A Theory of Tidal Torques near saddles.

Christophe Pichon[1]†, Sandrine Codis[1], Dmitry Pogosyan[2], Yohan Dubois[1], Vincent Desjacques[3] and Julien Devriendt[4]

[1] Institut d'Astrophysique de Paris & UPMC, 98 bis Boulevard Arago, 75014, Paris, France
[2] University of Alberta, 11322-89 Avenue, Edmonton, Alberta, T6G 2G7, Canada
[3] Université de Genève 24, quai Ernest Ansermet. 1211, Genève, Switzerland
[4] Sub-department of Astrophysics, University of Oxford, Keble Road, Oxford OX1 3RH

Abstract. Filaments of the cosmic web drive spin acquisition of disc galaxies. The point process of filament-type saddle represent best this environment and can be used to revisit the Tidal Torque Theory in the context of an anisotropic peak (saddle) background split. The constrained misalignment between the tidal tensor and the Hessian of the density field generated in the vicinity of filament saddle points simply explains the corresponding transverse and longitudinal point-reflection symmetric geometry of spin distribution. It predicts in particular an *azimuthal* orientation of the spins of more massive galaxies and spin *alignment* with the filament for less massive galaxies. Its scale dependence also allows us to relate the transition mass corresponding to the alignment of dark matter halos' spin relative to the direction of their neighboring filament to this geometry, and to predict accordingly it's scaling with the mass of non linearity, as was measured in simulations.

Keywords. large-scale structure of universe, gravitational lensing, galaxies: statistics.

1. Introduction

Modern simulations based on a well-established paradigm of cosmological structure formation predict a significant connection between the geometry and dynamics of the large-scale structure on the one hand, and the evolution of the physical properties of forming galaxies on the other. Pichon *et al.* (2011) have suggested that the large-scale coherence, inherited from the low-density cosmic web, explains why cold flows are so efficient at producing thin high-redshift discs from the inside out. They also predicted that the distribution of the properties of galaxies measured relative to their cosmic web environment should reflect such a process. Both numerical (e.g. Hahn *et al.* (2007), Codis *et al.* (2012), Fig 1, Libeskind *et al.* (2013)), and observational evidence (e.g. Tempel & Libeskind (2013)) have recently supported this scenario. In parallel, much observational effort has been invested to control the level of intrinsic alignments of galaxies as a potential source of systematic errors in weak gravitational lensing measurements (e.g. Heavens *et al.* (2000)). It is therefore of interest to explain from first principles why such intrinsic alignments arise, so as to possibly temper their effects (see also Codis *et al.* (2014)).

Yet, understanding the effect of this cosmic anisotropy on galactic morphology is a challenging task. The difficulty is two-fold: i) the geometry of the flow within filaments is complex: the spin distribution is intrinsically point-reflection symmetric relative to saddles, and confined to filaments ii) the cosmic web itself is strongly anisotropic and multiscale. In this paper, we will try and address these challenges and formalize the

† Email: pichon@iap.fr

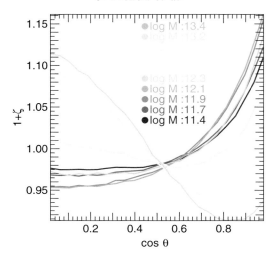

Figure 1. The probability distribution of the cosine of the angle between the spin of dark haloes and the direction of the closest filament as a function of mass in the Horizon simulation. The probability to have a small angle between the halo's spin and the filament's direction first *increases* as mass grows. At larger masses the spin-filament alignment first decays (at $\mathcal{M}_{\rm tr}^{\rm 2D}$), and then flips (at $\mathcal{M}_{\rm tr}^{\rm 3D}$) to predominately orthogonal orientations (from Laigle *et al.* (2014)).

corresponding theory of anisotropic secondary infall. Specifically, we will model the intrinsically 3D geometry of galactic accretion while taking into account the geometry of the tidal and density field near a *typical* saddle point. Indeed, saddle points define an *point process* which accounts for the presence of filaments embedded in walls, two critical ingredient in shaping the spins of galaxies. A proper account of the anisotropy of the environment in this context will allow us to demonstrate why, as measured in simulations, the spin of the forming (*low mass*) galaxies are first aligned with the filaments direction with a quadratric point symmetric geometry (Fig 1 and Laigle *et al.* (2014)). While relying on a straightforward extension of Press Schechter's theory, we will also demonstrate that *massive* galaxies will have their spin preferentially along the azimuthal direction, and predict the corresponding scaling of the spin-flip transition mass with the (redshift dependent) mass of non-linearity, on the basis of the so-called cloud-in-cloud problem, applied at the peak (filamentary) background split level.

Qualitatively, the idea is the following: given a triaxial saddle constraint, the misalignment between the tidal tensor and the Hessian of the density field simply explains the transverse and longitudinal point-reflection symmetric geometry of spin distribution in their vicinity. It arises because the two tensors probe different scales: given their relative correlation lengths, the Hessian probes more directly its closest neighborhood, while the tidal field, somewhat larger scales, see Fig. 2. Within the plane perpendicular to the filament axis at the saddle point, the dominant wall (corresponding to the longer axis of the cross section of the saddle point) will re-orient more the Hessian than the tidal field, which also feels the denser, but typically further away saddle point. This net misalignment will induces spin perpendicular to that plane i.e along the filament. This effect will produce a quadru-polar, point reflection symmetric distribution of the polar component of the spin which will be strongest at some four points, not far off axis. Beyond a couple of correlation lengths away from those four points, the effect of the tidal field induced by the saddle point will subside, as both tensor become more spherical. Conversely, in planes perpendicular to that plane, e.g. containing the dominant wall and the filament, a similar

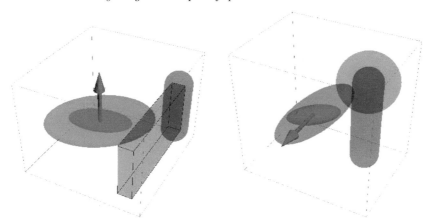

Figure 2. sketch of main differential alignment between hessian and tidal responsible for $\mathbf{e}_z$ and $\mathbf{e}_\phi$ component of spin. *Left:* the two tensors in light and dark red, are misaligned as they feel differently the neighboring wall (blue) and filament (purple), inducing a spin parallel to the filament (red arrow). *Right:* correspondingly, the differential pull from the filament (purple) and the density gradient towards the peak (blue) generates a spin (red arrow) along the azimuthal direction.

process will misalign both tensors. This time, the two anisotropic features differentially pulling the tensors are the filament on the one hand, and the density gradient towards the peak on the other. The net effect of the corresponding misalignment will be to also spin up halos perpendicular to that plane, along the azimuthal direction. By symmetry, an anti-clockwise tidal spin will be generated on the other side of the saddle point.

Hence, as the theory developed below will allow us to predict, the geometry of spin near filament-saddle points is the following: it is aligned with the filament in the median plane (within four anti-symmetric quadrants), and aligned with the azimuthal direction away from that plane, see Fig. 3. The stronger the triaxiality the stronger the amplitude. Conversely, if the saddle point becomes degenerate in one or two directions, the component of the spin in the corresponding direction will vanish. For instance, a saddle point in the middle of a very long filament will only display alignment with that filament axis, with no azimuthal component.

The paper is organised as follows. Section 2 presents the expected Lagrangian spin distribution near filaments, assuming cylindrical symmetry, while Section 3 revisits this distribution in three dimensions for realistic typical 3D saddle points.

2. Spin in cylindrical symmetry

Let us start while assuming that the filament is of infinite extent, so that we can restrict ourselves to cylindrical symmetry in two dimensions. This is of interest as the spin is then along the filament axis by symmetry and its derivation in the context of Tidal Torque theory (TTT) is much simpler. It captures already in part the mass transition, as we can define the mean extension of a given quadrant of spin with a given polarity.

Under the assumption that the *direction* of the spin along the z direction is well represented by the anti-symmetric (Levi Civita) contraction of the tidal field and hessian (e.g. Schäfer & Merkel (2012)), it becomes a quadratic function of the second and fourth derivatives of the potential. As such, it becomes possible to compute expectations of it subject to its relative position to a peak with a given geometry (which would correspond to the cross section of the filament in that plane). In contrast, standard TTT relies, more

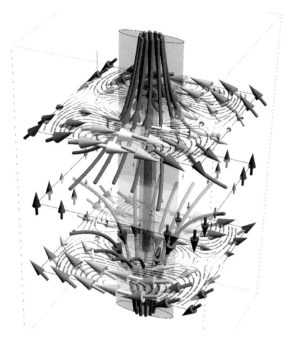

Figure 3. The velocity and Spin flow near a vertical filament (in red) embedded in a (purple) wall. The purple and green flow lines trace the (Lagrangian) 3D velocities (upwards and downwards respectively). The red and blue arrows show the spin 3D distribution, while the three horizontal cross sections show spin flow lines in the corresponding plane. Note that the spin is along $\mathbf{e}_z$ in the mid plane and along $\mathbf{e}_\phi$ away from it, and that it rotates in opposite direction above and below the mid-plane. See also `http://www.iap.fr/users/pichon/AM-near-saddle.html`

correctly, on the inertia tensor in place of the Hessian. Even though they have inverse curvature of each other, their set of eigen-directions are locally the same, so we expect the induced spin *direction*– which is the focus of this paper, to be the same, so long as the inertia tensor is well described by its local Taylor expansion.

Any matrix of second derivatives f_{ij} – rescaled so that $\langle (\Delta f)^2 \rangle = 1$– can be decomposed into its trace Δf, and its detraced components in the frame of the separation $f^+ = (f_{11} - f_{22})/2$ and $f^\times = f_{12}$. Then all the correlations between two such matrices, f_{ij} and g_{ij} can be decomposed irreducibly as follows. Let us call $\xi_{fg}^{\Delta\Delta}$, $\xi_{fg}^{\Delta+}$ and $\xi_{fg}^{\times\times}$ the correlation functions in the frame of the separation (which is the first coordinate here) between the second derivatives of the field f and g separated by a distance r:

$$\xi_{fg}^{\Delta\Delta}(r) = \langle \Delta f \Delta g \rangle, \quad \xi_{fg}^{\Delta+}(r) = \langle \Delta f g^+ \rangle, \quad \xi_{fg}^{\times\times}(r) = \langle f^\times g^\times \rangle.$$

All other correlations are trivially expressed in terms of the above as $\langle f^\times \Delta g \rangle = 0$, $\langle f^+ g^\times \rangle = 0$, $\langle f^+ g^+ \rangle = \frac{1}{4}\xi_{fg}^{\Delta\Delta}(r) - \xi_{fg}^{\times\times}(r)$. Here, we consider two such fields, namely the gravitational potential ϕ and the density contrast δ. In the following these two fields and their first and second derivatives are assumed to be rescaled by their variance $\sigma_0^2 = \langle \phi^2 \rangle$, $\sigma_1^2 = \langle (\nabla\phi)^2 \rangle$, $\sigma_2^2 = \langle (\delta = \Delta\phi)^2 \rangle$, $\sigma_3^2 = \langle (\nabla\delta)^2 \rangle$ and $\sigma_4^2 = \langle (\Delta\delta)^2 \rangle$. The shape parameter is defined as $\gamma = \sigma_3^2/(\sigma_2\sigma_4)$.

The Gaussian joint PDF of the gravitational field, its first and second derivatives and the first and second derivatives of the density is sufficient to compute the expectation of any quantity involving derivatives of the potential and the density up to second order. The two-point covariance matrix can be derived from the power spectrum of the potential, the result being a function of the above defined nine functions (for $fg = \phi\phi, \phi\delta, \delta\delta$). Once the joint PDF is known, it is straightforward to compute conditional PDFs. Simple algebra yield the conditional density and spin as a function of separation and geometry of the saddle. In details, given a contrast ν and a geometry for the saddle defined by $\kappa = \lambda_1 - \lambda_2, I_1 = \lambda_1 + \lambda_2$ (where $\lambda_1 > \lambda_2$ are the two eigenvalues of the Hessian of the density field $\mathbf{H}$ – both negative for a peak), the mean density contrast, $\langle \delta | \text{ext} \rangle$ (in units of σ_2) around the corresponding extremum can be computed as

$$\delta(\mathbf{r},\kappa,I_1,\nu|\text{ext}) = \frac{I_1(\xi^{\Delta\Delta}_{\phi\delta} + \gamma\xi^{\Delta\Delta}_{\phi\phi}) + \nu(\xi^{\Delta\Delta}_{\phi\phi} + \gamma\xi^{\Delta\Delta}_{\phi\delta})}{1-\gamma^2} + 4\left(\hat{\mathbf{r}}^{\mathrm{T}}\cdot\overline{\mathbf{H}}\cdot\hat{\mathbf{r}}\right)\xi^{\Delta+}_{\phi\delta}, \quad (2.1)$$

where $\overline{\mathbf{H}}$ is the detraced Hessian of the density and $\hat{\mathbf{r}} = \mathbf{r}/r$ so that $\hat{\mathbf{r}}^{\mathrm{T}}\cdot\overline{\mathbf{H}}\cdot\hat{\mathbf{r}} = \kappa\cos(2\theta)/2$, with r is the distance to the extremum and θ the angle from the eigen-direction corresponding to the first eigenvalue λ_1 of the extremum. When r goes to zero, given the properties of the ξ functions, the density trivially converges to the constraint ν.

In 2D, the (rescaled) spin is a scalar given by $L_z(\mathbf{r}) = \varepsilon_{ij}\phi_{il}x_{jl}$, where ϵ is the rank 2 Levi-Civita tensor. Hence the spin generated by TTT as a function of the polar position, (r,θ) subject to the same extrema constraint at the origin with contrast, ν, and principal curvatures (λ_1,λ_2) is given by the sum of a quadrupole ($\propto \sin 2\theta$) and an octupole ($\propto \sin 4\theta$, since $\hat{\mathbf{r}}^{\mathrm{T}}\cdot\epsilon\cdot\overline{\mathbf{H}}\cdot\hat{\mathbf{r}} = -\kappa\sin(2\theta)/2$):

$$\langle L_z|\text{ext}\rangle = L_z(\mathbf{r},\kappa,I_1,\nu|\text{ext}) = -16(\hat{\mathbf{r}}^{\mathrm{T}}\cdot\epsilon\cdot\overline{\mathbf{H}}\cdot\hat{\mathbf{r}})\left(L_z^{(1)}(r) + 2(\hat{\mathbf{r}}^{\mathrm{T}}\cdot\overline{\mathbf{H}}\cdot\hat{\mathbf{r}})L_z^{(2)}(r)\right), \quad (2.2)$$

where the octupolar component $L_z^{(2)}$ can be written as $L_z^{(2)}(r) = (\xi^{\Delta\Delta}_{\phi x}\xi^{\times\times}_{\delta\delta} - \xi^{\times\times}_{\phi\delta}\xi^{\Delta\Delta}_{\delta\delta})$, and the quadrupolar coefficient $L_z^{(1)}(r)$ reads

$$L_z^{(1)}(r) = \frac{\nu}{1-\gamma^2}\left[(\xi^{\Delta+}_{\phi\phi} + \gamma\xi^{\Delta+}_{\phi\delta})\xi^{\times\times}_{\delta\delta} - (\xi^{\Delta+}_{\phi\delta} + \gamma\xi^{\Delta+}_{\delta\delta})\xi^{\times\times}_{\phi\delta}\right]$$
$$+ \frac{I_1}{1-\gamma^2}\left[(\xi^{\Delta+}_{\phi\delta} + \gamma\xi^{\Delta+}_{\phi\phi})\xi^{\times\times}_{\delta\delta} - (\xi^{\Delta+}_{\delta\delta} + \gamma\xi^{\Delta+}_{\phi\delta})\xi^{\times\times}_{\phi\delta}\right].$$

Eq. (2.2) is remarkably simple. As expected, the spin, L_z, is identically null if the filament is axially symmetric ($\kappa = 0$). It is zero along the principal axis of the Hessian (where $\theta = 0$ mod $\pi/2$ for which $\hat{\mathbf{r}}^{\mathrm{T}}\cdot\epsilon\cdot\hat{\mathbf{r}} = 0$). Near the peak, the anti-symmetric, $\sin(2\theta)$, component dominates, and the spin distribution is quadripolar (see the midplane of Fig 3).

Let us now understand how much spin is contained within spheres of increasing radius that would feed the forming object at different stage of its evolution. For instance let us assume there is a small-scale overdensity at (one of the four) location of maximum spin (denoted $r_\star$ hereafter) and let us filter the spin field with a top-hat window function centered on $r_\star$ and of radius R_{TH}. The resulting amount of spin as a function of this top-hat scale is displayed in Fig. 4. During the first stage of evolution, the central object will acquire spin constructively until it reaches a Lagrangian size of radius $R_{\mathrm{TH}} = r_\star$ and feels the two neighbouring quadrants of opposite spin direction. The spin amplitude then decreases and becomes even negative before it is fed by the last quadrant of positive spin. The minimum is reached for radius around $2.4r_\star$. This result does not change much with the contrast and the geometry of the peak constraint. Let us now predict the mass that corresponds to maximum spin i.e. the mass contained in a sphere of radius $R_\star$. First, let us compute $r_\star$, as the radius for which $L_z(r,\theta = \pi/4)$ is maximal as a function of r. Indeed, for small enough κ, the quadruple dominates, and the extremum is along $\theta = \pi/4$. The area of a typical quadrant, in which the spin has the same polarity, can then simply be expressed as $\mathcal{A} = \pi\lambda_2/\lambda_1(2r_\star)^2/4$, where $\lambda_1 < \lambda_2 < 0$ are the two eigenvalues of the Hessian and $r_\star = r_\star(\nu,\kappa)$ is the position of a maximum of spin from

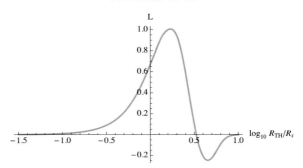

Figure 4. Evolution of the amount of algebraic 2D spin in sphere of radius R_{TH} centered on $r_\star$. The density power spectrum index is $n = -3/2$, the height of the peak in $(0,0)$ is $\nu = 1$ and principal curvatures $\lambda_1 = -1, \lambda_2 = -2$. The amplitude of the spin is normalised by its maximum value around $R_{\mathrm{TH}} = r_\star$.

the peak. Because of the quadrupolar anti-symmetric geometry of the spin distribution near the saddle point, it is typically twice as small (in units of the smoothing length) as one would naively expect.

With prior knowledge of the distribution of the shape, κ and height, ν for 2D peaks and of the maximal area, $\mathcal{A}(\kappa, \nu)$, corresponding to spins with the same polarity, we may *define* the transition corresponding mass as (with Σ_0 the cosmic mean surface density)

$$\mathcal{M}_{\mathrm{tr}}^{2\mathrm{D}}(L_s) = \Sigma_0 \int d\nu \, d\kappa \mathcal{A}(\nu, \kappa) \mathcal{P}(\nu, \kappa | \mathrm{pk}) \,, \tag{2.3}$$

Following Pogosyan *et al.* (2009), it is straightforward to derive this PDF, $\mathcal{P}$, for a *peak* to have height ν and geometry κ, I_1 so that

$$\mathcal{P}(\nu, \kappa, I_1 | \mathrm{pk}) = \frac{\sqrt{3}\kappa |(I_1 - \kappa)(I_1 + \kappa)|}{2\pi\sqrt{1 - \gamma^2}} \Theta(-\kappa - I_1) \exp\left(-\frac{1}{2}\left(\frac{\nu + \gamma I_1}{\sqrt{1 - \gamma^2}}\right)^2 - \frac{1}{2}I_1^2 - \kappa^2\right).$$

Given the shape of $\mathcal{P}$ near it maximum, we can approximate $\mathcal{M}_{\mathrm{tr}}^{2\mathrm{D}}$ in Eq. (2.3) as

$$\mathcal{M}_{\mathrm{tr}}^{2\mathrm{D}}(z) = \Delta N \frac{\lambda_{2,\star}}{\lambda_{1,\star}}\left(\frac{r_\star}{L_s}\right)^2 M_s(z) \equiv \alpha M_s \,, \tag{2.4}$$

where $\Delta N = \mathcal{P}(\nu_\star, \kappa_\star | \mathrm{pk})\Delta\nu\Delta\kappa$, and $M_s(z) \equiv \pi L_s^2(z)\Sigma_0$. Here the λ's and ν are evaluated at the maximum of $\mathcal{P}$ and the Δ's represent the local inverse curvature at the peak of that distribution. For scale invariant power spectra, the calculation shows that $\alpha(n = -1) \sim 1/11$. This is one of the main results of this investigation. It states that, in the framework of anisotropic peak background split of TTT near a typical saddle point (for a GRF of density index ~ -1), the transition mass is *predicted* to be smaller than the scaling mass, M_s, by an order of magnitude. This is what Codis *et al.* (2012) found while analyzing the scaling of the transition mass with the mass of non-linearity (see Fig 1).

3. The 3D spin near and along filaments

Let us now turn to the truly three dimensional theory of tidal torques in the vicinity of a typical filament saddle point, see Fig. 3. The main motivation is that the 3D saddle geometry captures fully the second (spin flip) mass transition. In three dimensions, we must consider two competing processes. If we vary the radius corresponding to the

Lagrangian patch centered on the running point, we have a spin-up (along $\mathbf{e}_z$) arising from the running point to wall-running point to saddle tidal misalignment and a second spin-up (along $\mathbf{e}_\phi$) arising from running point to filament- running point to peak tidal misalignment.

In order to compute the spin distribution, the formalism developed in Section 2 is easily extended to 3D. A critical (including saddle condition) point constraint is imposed. The resulting mean density field subject to that constraint becomes (in units of σ_2):

$$\delta(\mathbf{r}, \kappa, I_1, \nu | \text{ext}) = \frac{I_1(\xi_{\phi\delta}^{\Delta\Delta} + \gamma\xi_{\phi\phi}^{\Delta\Delta})}{1 - \gamma^2} + \frac{\nu(\xi_{\phi\phi}^{\Delta\Delta} + \gamma\xi_{\phi\delta}^{\Delta\Delta})}{1 - \gamma^2} + \frac{15}{2}\left(\hat{\mathbf{r}}^{\mathrm{T}} \cdot \overline{\mathbf{H}} \cdot \hat{\mathbf{r}}\right)\xi_{\phi\delta}^{\Delta+}, \quad (3.1)$$

where again $\overline{\mathbf{H}}$ is the *detraced* Hessian of the density and $\hat{\mathbf{r}} = \mathbf{r}/r$ and we define in 3D $\xi_{\phi x}^{\Delta+}$ as $\xi_{\phi\delta}^{\Delta+} = \langle\Delta\delta, \phi^+\rangle$, with $\phi^+ = \phi_{11} - (\phi_{22} + \phi_{33})/2$. Note that $\hat{\mathbf{r}}^{\mathrm{T}} \cdot \overline{\mathbf{H}} \cdot \hat{\mathbf{r}}$ is a scalar quantity defined explicitly as $\hat{r}_i \overline{H}_{ij} \hat{r}_j$. As in 2D, the expected spin can also be computed. In 3D, the spin is a vector, which components are given by $L_i = \varepsilon_{ijk}\delta_{kl}\phi_{lj}$, with $\boldsymbol{\epsilon}$ the rank 3 Levi Civita tensor. It is found to be orthogonal to the separation and can be written as the sum of two terms

$$\mathbf{L}(\mathbf{r}, \kappa, I_1, \nu | \text{ext}) = -15\left(\mathbf{L}^{(1)}(r) + \mathbf{L}^{(2)}(r)\right) \cdot \left(\hat{\mathbf{r}}^{\mathrm{T}} \cdot \boldsymbol{\epsilon} \cdot \overline{\mathbf{H}} \cdot \hat{\mathbf{r}}\right), \quad (3.2)$$

where $\mathbf{L}^{(1)}$ depends on height, ν, and on the trace of the Hessian I_1 but not on orientation

$$\mathbf{L}^{(1)}(r) = \left(\frac{\nu}{1 - \gamma^2}\left[(\xi_{\phi\phi}^{\Delta+} + \gamma\xi_{\phi\delta}^{\Delta+})\xi_{\delta\delta}^{\times\times} - (\xi_{\phi\delta}^{\Delta+} + \gamma\xi_{\delta\delta}^{\Delta+})\xi_{\phi\delta}^{\times\times}\right]\right.$$
$$\left. + \frac{I_1}{1 - \gamma^2}\left[(\xi_{\phi\delta}^{\Delta+} + \gamma\xi_{\phi\phi}^{\Delta+})\xi_{\delta\delta}^{\times\times} - (\xi_{\delta\delta}^{\Delta+} + \gamma\xi_{\phi\delta}^{\Delta+})\xi_{\phi\delta}^{\times\times}\right]\right)\mathbb{I}_3,$$

and $\mathbf{L}^{(2)}(\mathbf{r})$ now depends on $\overline{\mathbf{H}}$ and on orientation:

$$\mathbf{L}^{(2)}(\mathbf{r}) = -\frac{5}{8}\left[2((\xi_{\phi\delta}^{\Delta+} - \xi_{\phi\delta}^{\Delta\Delta})\xi_{\delta\delta}^{\times\times} - (\xi_{\delta\delta}^{\Delta+} - \xi_{\delta\delta}^{\Delta\Delta})\xi_{\phi\delta}^{\times\times})\overline{\mathbf{H}}\right.$$
$$\left. + ((7\xi_{\delta\delta}^{\Delta\Delta} + 5\xi_{\delta\delta}^{\Delta+})\xi_{\phi\delta}^{\times\times} - (7\xi_{\phi\delta}^{\Delta\Delta} + 5\xi_{\phi\delta}^{\Delta+})\xi_{\delta\delta}^{\times\times})(\hat{\mathbf{r}}^{\mathrm{T}} \cdot \overline{\mathbf{H}} \cdot \hat{\mathbf{r}})\mathbb{I}_3\right],$$

(with $\mathbb{I}_3$ the identity matrix) operating on the *vector* $\left(\hat{\mathbf{r}}^{\mathrm{T}} \cdot \boldsymbol{\epsilon} \cdot \overline{\mathbf{H}} \cdot \hat{\mathbf{r}}\right)_j = \hat{r}_i \epsilon_{ijk}\overline{\mathbf{H}}_{kl}\hat{r}_l$.

Note that all the dependence with the distance r is encoded in the ξ functions, while the geometry of the critical point is encoded in the terms corresponding to the peak height, trace and detraced part of the Hessian, while the orientation of the separation is encoded in $\hat{\mathbf{r}}$. Eq. (3.2) is also remarkably simple: as expected the symmetry of the model induces zero spin along the principal directions of the Hessian (where $\hat{\mathbf{r}}^{\mathrm{T}} \cdot \boldsymbol{\epsilon} \cdot \overline{\mathbf{H}} \cdot \hat{\mathbf{r}} = 0$) and a point reflection symmetry ($\hat{\mathbf{r}} \to -\hat{\mathbf{r}}$), see Fig. 3.

Let us now compute the mean values of ν, $\lambda_1 < \lambda_2 < 0 < \lambda_3$ of a typical filament-type saddle-point. Starting from the so-called Doroskevich formula for the PDF:

$$\mathcal{P}(\nu, \lambda_i) = \frac{135\,(5/2\pi)^{3/2}}{4\sqrt{1 - \gamma^2}}\exp\left[-\frac{1}{2}\zeta^2 - 3I_1^2 + \frac{15}{2}I_2\right] \times (\lambda_3 - \lambda_1)(\lambda_3 - \lambda_2)(\lambda_2 - \lambda_1),$$

where $\zeta = (\nu + \gamma I_1)/\sqrt{1 - \gamma^2}$, $I_1 = \lambda_1 + \lambda_2 + \lambda_3$, $I_2 = \lambda_1\lambda_2 + \lambda_2\lambda_3 + \lambda_1\lambda_3$ and $I_3 = \lambda_1\lambda_2\lambda_3$, subject to the constraint, this PDF becomes

$$\mathcal{P}(\nu, \lambda_i | \text{skl}) = \frac{26460\sqrt{5\pi}\mathcal{P}(\nu, \lambda_i)I_3\Theta(\lambda_3)}{1421\sqrt{2} - 735\sqrt{3} + 66\sqrt{42}}\Theta(-\lambda_2 - \lambda_3), \quad (3.3)$$

after imposing the condition of saddle point $|\det x_{ij}|\delta_D(x_i)\Theta(\lambda_3)\Theta(-\lambda_2)$ and the additional constraint of a skeleton-like saddle, which is $\lambda_2 + \lambda_3 < 0$. The expected value of the density and the eigenvalues at a skeleton saddle position reads $\langle\nu\rangle \approx 1.25\gamma$, $\langle\lambda_1\rangle \approx -1.0$, $\langle\lambda_2\rangle \approx -0.56$ and $\langle\lambda_3\rangle \approx 0.31$.

The transition mass, $\mathcal{M}_{\mathrm{tr}}^{\mathrm{3D}}$ may then be defined as follows. The geometry of the spin distribution near a saddle point allows us to compute the mean orientation of the spin around the saddle point. Let us define $\hat{\theta}$ the flip angle so that

$$\cos\hat{\theta}(\mathbf{r}) = \frac{\mathbf{L}(\mathbf{r}).\mathbf{e}_z}{||\mathbf{L}(\mathbf{r})||}. \tag{3.4}$$

In turn, the shape of the density profile in the vicinity of the most likely skeleton-like saddle point (as defined by equation 3.3), together with an extension of the Press-Schecher theory involving the peak background split allows us to estimate the typical mass of dark halos around the same saddle point.

Indeed, the local mass distribution of halos is expected to vary along the large scale filament due to changes in the underlying long-wave density. In the linear regime, the typical density near the end points of the filament, where it joins the protoclusters, exceeds the typical density near the saddle point by a factor of two (Pogosyan *et al.* (1998)). At epochs before the whole filamentary structure has collapsed, this leads to a shift in hierarchy of the forming halos towards larger masses near filaments end points (the clusters) relative to the filament middle point (the saddle). This can be easily understood using the formalism of barrier crossing (e.g. Bond *et al.* (1991)), which associates the density of objects of a given mass to the statistics for the random walk of halo density, as the field is smoothed with decreasing filter sizes.

Given the Peacock-Heavens (Peacock & Heavens (1990)) approximation, the number density of dark halos in the interval $[M, M + \mathrm{d}M]$ is

$$\frac{dn(M)}{dM}dM = \frac{\rho}{M}f(\sigma^2, \delta_c)d\ln\sigma^2, \tag{3.5}$$

where $f(\sigma^2, \delta_c)$ is given by the function

$$f(\sigma^2, \delta_c) = \exp\left(\frac{1}{\Gamma}\int_0^{\sigma^2}\frac{ds'}{s'}\ln p(s', \delta_c)\right)\left(-\sigma^2\frac{dp(\sigma^2, \delta_c)}{d\sigma^2} - \frac{1}{\Gamma}p(\sigma^2, \delta_c)\ln p(\sigma^2, \delta_c)\right).$$

Here σ^2 is the variance of the density fluctuations smoothed at the scale corresponding to M and $p(\sigma^2, \delta_c) \equiv 1/2\left(1 + \mathrm{erf}(\delta_c/\sqrt{2}\sigma)\right)$ is the probability of a Gaussian process with variance σ^2 to yield value below some critical threshold δ_c. Here $\Gamma \approx 4$. The overall mass distribution of halos is well described by the choice $\delta_{c,0} = 3/5\,(3\pi/2)^{2/3} = 1.686$, motivated by the spherical collapse model. When halos form on top of a large scale structure background, however, the long-wave over-density $\bar{\delta}(z)$ adds to the over-density in the proto-halo peaks. The effect on halo mass distribution, in this peak-background split approach, can be approximated as a *shifted* threshold $\delta_c(z, Z) = 1.686 - \bar{\delta}(z, Z)$ for halo formation as a function of the curvilinear coordinate z along the filament and redshift Z. The corresponding shift in mass can be characterized by the dependence on the threshold of $M_*(\delta_c)$, defined as $\sigma_*(M_*) = \delta_c$, or of the mass $M_p(\delta_c)$ that corresponds to the peak of $f(\sigma^2, \delta_c)$, i.e. the variance $\sigma_p^2(z) \equiv \underset{\sigma^2}{\mathrm{argmax}}(f(\sigma^2, \delta_c(z)))$. When large scale structures are considered as fixed background, the variance of the relevant small scale density fluctuations that are responsible for object formation is reduced, approximately as $\sigma^2 \approx \sigma^2(M) - \sigma^2(M_{\mathrm{LSS}})$ where $\sigma^2(M_{\mathrm{LSS}})$ is the unconstrained large-scale density variance. This correction becomes important, truncating the mass hierarchy at M_{LSS},

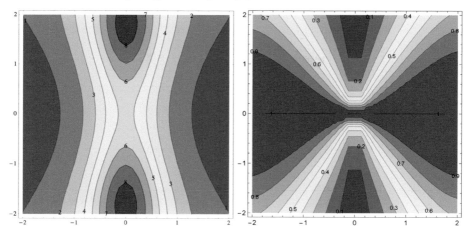

Figure 5. *Left:* logarithmic cross section of $M_p(r,z)$ along the most likely (vertical) filament (in units of $10^{12} M_\odot$). *Right:* corresponding cross section of $\langle \cos \hat\theta \rangle (r,z)$. The mass of halos increases towards the nodes, while the spin flips.

whenever large scale structures are themselves non-linear. Here we choose $\sigma_8 = 0.8$, redshift zero, use the value for mass in a $8h^{-1}$ Mpc comoving sphere for the best-fit cosmological mass density, and approximate the spectrum with a power law of index $n = -2$, which allows to solve for $M(\sigma)$ as

$$M(\sigma, Z) = 2.6 \times 10^{14} M_\odot \left(\frac{\sigma^2 + \sigma^2(M_{\mathrm{LSS}})}{\sigma_8^2 D(Z)^2} \right)^{-\frac{3}{n+3}}. \tag{3.6}$$

We consider filaments defined with $R = 5h^{-1}$Mpc Gaussian smoothing. Then, in addition to a spin orientation map around the saddle point, one can establish a mass map directly from the density map by means of the $M_p(\delta)$ relation. A cut of those two maps is displayed in Fig. 5. The spin flips towards the nodes, while mass increases. In each point of the vicinity of the saddle point, the mass and spin orientation are known so that one can do an histogram and plot the mean orientation as a function of the mass, see Fig. 6. The 3D transition mass for spin flip (i.e. $\cos \hat\theta = 0.5$) is clearly of the order $\mathcal{M}_{\mathrm{tr}}^{\mathrm{3D}} \approx 5\,10^{12} M_\odot$ This mass is in good agreement with the transition mass found in Codis *et al.* (2012).

4. Discussion

Tidal torque theory was revisited while focussing on an anisotropic peak background split in the vicinity of a saddle point. Such point process captures the point-symmetric multipolar geometry of a typical filament embedded in a given wall (Pogosyan *et al.* (1998)). The induced misalignment between the tidal tensor and the Hessian simply explains the surrounding transverse and longitudinal point reflection-symmetric geometry of spin distribution near filaments. It predicts in particular that less massive galaxies have their spin parallel to the filament, while more massive ones have their spin in the azimuthal direction. The corresponding transition masses ($\mathcal{M}_{\mathrm{tr}}^{2/3\mathrm{D}}$, corresponding resp. to maximal alignment and flip, see Fig 1) follows from this geometry, together with their scaling with the mass of non linearity, as observed in simulation. The neighborhood of a *given unique* typical saddle point was considered as a proxy for the behaviour within a Gaussian random field. It is shown elsewhere (Pichon *et al. in prep.*) that it holds statistically.

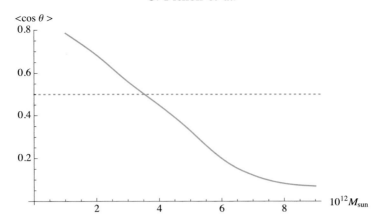

Figure 6. Mean alignment between spin and filament as a function of mass for a filament smoothing scale of 5 Mpc/h. The spin flip transition mass is around $4\,10^{12}\,M_\odot$.

One of the interesting feature of this *Lagrangian* framework is that it captures naturally the arguably non linear *Eulerian* process of spin flip via mergers. Recently, Laigle *et al.* (2014) showed that angular momentum generation of halos is captured via the secondary advection of vorticity which was generated by the formation of filaments. These two (Eulerian versus Lagrangian) descriptions are the two sides of the same coin. The mapping between the two descriptions requires a reversible time integrator, such as the Zeldovitch approximation. In effect, the geometry of the saddle provides a natural 'metric' (the local frame as defined by the Hessian at that saddle point) relative to which the dynamical evolution of dark halos along filaments can be predicted. For instance, from Eq. (3.2) we can compute the loci, along the filament, of maximum angular momentum advection. They characterize the most active regions in the cosmic web for galactic spin up. The argument sketched in Section 3 allows us to assign the corresponding redshift dependent spin-up *mass*, and its evolution with redshift. It should have an observational signature in terms of the cosmic evolution of the SFR, as it corresponds to efficient pristine cold and dense gas accretion, which in turn induces steady star formation.

This work is partially supported by grant ANR-13-BS05-0005 of the french ANR. CP thanks D. Lynden-Bell for encouragement.

References

Bond, J. R., Cole, S., Efstathiou, G. & Nick, K. 1991 *ApJ*, 379, 440
Codis, S., Pichon C., Devriendt, J., Slyz, A. *et al.* 2012 *MNRAS*, 427, 3320
Codis, S., Gavazzi, R., Dubois, Y., Pichon, C. *et al.* 2014 *ArXiv e-prints*
Hahn, O., Porciani, C., Carollo, C. M., & Dekel, A. 2007, *MNRAS*, 375, 489
Heavens, A., Refregier, A. & Heymans, C. 2000 *MNRAS*, 319, 649
Laigle, C., Pichon C., Codis, S., Dubois, Y. *et al.* 2014, *ArXiv e-prints*
Libeskind, N. I., *et al.* 2013, *MNRAS*, 428, 2489
Peacock, J. A. & Heavens, F. 1990, *MNRAS*, 243, 133
Pichon, C. *et al.* 2011, *MNRAS*, 418, 2493
Pogosyan, D., Bond, J. R., & Kofman, L. 1998, *JRASC*, 92, 313
Pogosyan, D., *et al.* 2009, *MNRAS*, 396, 635
Schäfer, B. M. & Merkel, P. M 2012, *MNRAS*, 421, 2751
Tempel, E. & Libeskind, N. I. 2013, *ApJL*, 775, L42

The Zeldovich Universe:
Genesis and Growth of the Cosmic Web
Proceedings IAU Symposium No. 308, 2014
R. van de Weygaert, S. Shandarin, E. Saar & J. Einasto, eds.

© International Astronomical Union 2016
doi:10.1017/S1743921316010310

How do galaxies build up their spin in the cosmic web?

Charlotte Welker[1,2] Yohan Dubois[1,2] Christophe Pichon[1] Julien Devriendt[2] and Sebastien Peirani[1]

[1] Institut d'Astrophysique de Paris,
98bis boulevard Arago, 75014 Paris, France
email: welker@iap.fr

[2] Sub-department of Astrophysics, University of Oxford,
Keble Road, Oxford OX1 3RH, United Kingdom.

Abstract. Using the Horizon-AGN simulation we find a mass dependent spin orientation trend for galaxies: the spin of low-mass, rotation-dominated, blue, star-forming galaxies are preferentially aligned with their closest filament, whereas high-mass, velocity dispersion- supported, red quiescent galaxies tend to possess a spin perpendicular to these filaments. We explore the physical mechanisms driving galactic spin swings and quantify how much mergers and smooth accretion re-orient them relative to their host filaments.

Keywords. large-scale structure, galaxy formation, galaxy evolution, merger.

---

1. Introduction: from haloes to galaxies

Over the past ten years, numerous simulations have reported the influence of the cosmic web on the direction of the angular momentum (AM) of halos. Several numerical investigations (see Dubois *et al.* (2014) for references) argued that the spin of high-mass halos tends to lie perpendicular to their host filament, whereas low-mass halos have a spin preferentially aligned with it. This was confirmed with high degree of accuracy by Codis *et al.* (2012) which quantified a redshift-dependent mass transition $M_{tr,h}$ separating aligned and perpendicular halos. They interpreted the origin of the transition in terms of large-scale cosmic flows: high-mass haloes would have spins perpendicular to the filament because they are the results of major mergers while low-mass haloes acquire their mass by smooth accretion from the vorticity quadrant they are embedded in, resulting in spins parallel to the filament (see also Laigle *et al.* 2014).

In order to make predictions for observational estimators of such correlations, it is important to identify the expected corresponding correlations between *galaxies* and the cosmic web. In this work, we use the `Horizon-AGN` simulation to focus on the influence of the cosmic web as an anisotropic vector of the gas mass and angular momentum which ultimately shapes galaxies. Our purpose is to determine if the mass-dependent halo spin-filament correlations of Codis *et al.* (2012) can be traced through the morphology and physical properties of simulated galaxies.

We first present the method used to identify the effect of the environmentally-driven spin acquisition on morphology, and probe the tendency for galaxies to align or misalign with the cosmic filaments as a function of galactic properties. We then focus on explaining *why* galactic spins swing and investigate the competitive effects of mergers and smooth accretion on alignment.

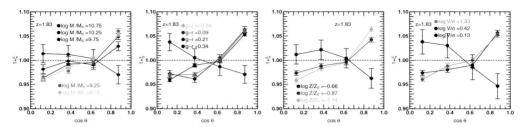

Figure 1. Excess probability, ξ, of the alignment between the spin of galaxies and their closest filament as a function of galactic properties at $z = 1.83$: (from left to right) M_s, $g-r$, metallicity Z and V/σ. Half sigma error bars are shown for readability. Dashed line corresponds to a uniform PDF (excess probability $\xi = 0$).

2. Tracing galactic alignments

Horizon-AGN

Horizon-AGN is a state-of-the-art hydrodynamical cosmological simulation run on a $L_{box} = 100Mpc.h - 1$ box down to z = 1.2 with the AMR code RAMSES (Teyssier (2002), an initial resolution of 1024^3 dark matter particles and 7 levels of refinement. It assumes a ΛCDM cosmology and includes various physical processes like feedback from supernovae and AGN, star formation and cooling from H and He with a contribution from metals. (see details in Dubois *et al.* (2014)) With more than 150 000 galaxies and 300 000 haloes per snapshot, it displays a vast diversity of galaxies, and spans a wide range of galactic morphologies.

Spin orientation distribution for galaxies

As can be seen in Fig. 1, we recover the spin alignment for small galaxies (and perpendicular spin orientation for massive galaxies) consistent with what is already known for dark haloes. Moreover, we find that various galactic properties traced this transition such as colour, age, specific stellar formation rate or metallicity. In a nutshell, low-mass, young, centrifugally supported, metal-poor, bluer galaxies tend to have their spin aligned to the closest filament, while massive, high velocity dispersion, red, metal-rich, old galaxies are more likely to have a spin perpendicular to it (see Dubois *et al.* 2014). The mass-transition, confidently bracketed between $log(M_\mathrm{s}/M_\odot) = 10.25$ and $log(M_\mathrm{s}/M_\odot) = 10.75$, was also found to be consistent with the transition mass for dark haloes estimated in Codis *et al.* (2012).

Morphology tracers and observations

Let us consider in particular the V/σ kinematic parameter, with V the luminosity weighted rotation component of the stellar speeds and σ the dispersion speed of the stars (see Fig. 1). Using the corresponding projected quantities, this pseudo-observable can be used as a morphology tracer. Indeed, $V/\sigma < 1$ corresponds to dispersion supported ellipticals while $V/\sigma > 1$ traces rotation supported spirals. As expected, we find that spirals tend to have spin parallel to the closest filament while the spin of ellipticals tend to be perpendicular. This result allows for direct comparison with recent observations in the SDSS. Indeed Tempel and Libeskind (2013) found a similar trend, giving this numerical distribution some observational support.

The underlying explanation for these trends in various galactic features lies in the fact that they are highly correlated to the corresponding stellar mass and with the AM distribution within the galaxy. This calls for a dynamical scenario, which we develop now.

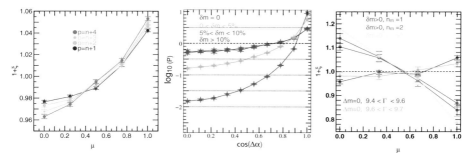

Figure 2. Panel (a) Excess probability, ξ, of μ, the cosine of the angle between the spin of galaxies and their closest filament is shown for 4 consecutive outputs for galaxies which do not merge. Panel (b) the log(PDF) of α, the flip angle of the galactic angular momentum between two consecutive outputs is shown as a function of the merger ratio. Panel (c) same as (a) but as a function of the number of mergers n_m the galaxy has undergone. $\Delta m = 0$, absence of mergers across the history of the galaxy. $\Gamma = \log(M/M_\odot)$. Half sigma error bars are shown for readability. Dashed line is uniform PDF (excess probability $\xi = 0$)

3. Spin swing dynamics

Merger trees

To understand this morphological segregation, we perform a statistical analysis on 22 outputs equally spaced in redshift between $z = 5.2$ and $z = 1.2$, which loosely corresponds to an average time step of 250 Myr. A candidate for merger being defined as any structure with $M_s > 10^8 M_\odot$, we then build merger trees to track back in redshift the most massive progenitor of each galaxy at any time step. We define the merger mass ratio $\delta m = \Delta m_{\rm mer}(z_n)/M_s(z_n)$ with $M_s(z_n)$ the total stellar mass of a galaxy at redshift z_n and $\Delta m_{\rm mer}(z_n)$ the stellar mass accreted through mergers between redshifts z_{n-1} and z_n. In a similar spirit, we also quantify the evolution of the specific angular momentum evolution.

Spin alignments and smooth accretion

We argue that small galaxies with $M_s < M_{tr,s}$ grow their spin through smooth accretion (possibly from cold gas inflows) from the vorticity quadrant they are caught in. In the process, vorticity is transferred to the intrinsic angular momentum of the galaxy. This results in the (re)-alignment of the spin of those galaxies with their surrounding filament over time.

This can be seen on Fig. 2(a) which displays the PDF of μ -the cosine of the spin-filament angle- for galaxies which do not merge over four consecutive outputs (with an average time step of 250 Myr). This shows that the excess probability ξ increases for values close to unity over cosmic time, indicating a tendency for those galaxies to align to their closest filament.

Similarly, we find an average increase in galactic angular momentum amplitude over cosmic time for non-mergers. This confirms that smooth accretion tends to build the galactic spin over time.

Finally, computing the inertia tensor of each galaxy from stellar kinematics and assuming an ellipsoidal shape, we recover the axis lengths and analyze the cumulative probabilities of the axis ratios over cosmic time. We find clear indication that smooth accretion flattens galaxies along this axis over cosmic time.

All findings sustain the above detailed scenario. While it is easy to understand why the aligned component of the spin might not persist when galaxies grow larger than their embedding vorticity quadrant, this does not explain why a clear perpendicular orientation is found for older massive galaxies. This leads us to investigate the effect of mergers.

Spin "anti"-alignments and mergers

Galaxies merge preferentially along filaments. A pair of galaxies about to merge – which are catching up relative to each other – are more likely to display a relative orbital momentum perpendicular to the surrounding filament. As this momentum is converted into intrinsic angular momentum during the merger, the remnant is more likely to display a spin perpendicular to that filament. Moreover, galaxies grow in mass from successive mergers. This was confirmed in the Horizon-AGN simulation where we found a clear correlation between merger fraction and stellar mass. Indeed, massive ellipticals are much more likely to be the result of mergers, which motivate their tendency to grow an orthogonal spin that will persist, and will not re-align from accretion providing that the galaxy is large enough.

To support this proposition, Fig. 2(b) shows the log(PDF) of the spin flip angle between successive outputs for different merger ratios. If all the spins in the sample maintain a fixed orientation, we would expect a Dirac function centered on one, while if all the spins get randomly reinitialized at each time step, we expect the uniform PDF (the dashed line). On this figure, we see clearly that non-mergers maintain their spin orientation (91 percent within a 25 degrees cone) while mergers trigger important swings, with a stronger effect for most massive mergers. This strongly supports the merger-driven swings scenario.

Fig. 2(c) confirms this and displays the PDF of μ (as defined above) for different merger histories: we identify an important excess probability ξ of perpendicular orientation (μ close to 0) for mergers, with a increased amplitude for higher number of mergers across the history of the galaxy. Moreover the magnitude of this signal is at least comparable to that found for dark haloes in Codis *et al.* and three times stronger than the signals obtained for tracers in Dubois *et al.* Consistently, non-mergers tend to remain aligned to their closest filament.

4. Conclusion

• Smooth accretion from the surrounding vorticity quadrant builds up the spin of young galaxies parallel to the filament

• Mergers along the cosmic web flip spins perpendicular to the filament through an orbital-to-intrinsic angular momentum transfer.

• Galactic properties correlated to merger rate trace the galactic flips and allow for detection.

• The scenario detailed here is the eulerian counterpart of the lagrangian theory (anisotropic TTT) presented by Christophe Pichon in the same proceedings, and shows how their predictions pervade down to small scales and low redshifts where non-linear baryonic physics become important.

References

Dubois, Y., Pichon, C., Welker, C. *et al.* 2014, *MNRAS*, in press
Welker, C., Pichon, C., Dubois, Y., Devriendt, J., & Peirani, S. 2014, *MNRAS Letters*, 445,L46-50
Codis, S. *et al.* 2012, *MNRAS*, 427,3320
Laigle, *et al.* 2014, *MNRAS*, arxiv 1310.3801
Tempel, E. & Libeskind 2013, *ApJ*, 42, 775
Teyssier, R. *et al.* 2002, *aap*, 385

This work is partially supported by the grant ANR-13-BS05-0005.

The Zeldovich Universe:
Genesis and Growth of the Cosmic Web
Proceedings IAU Symposium No. 308, 2014
R. van de Weygaert, S. Shandarin, E. Saar & J. Einasto, eds.

© International Astronomical Union 2016
doi:10.1017/S1743921316010322

How the cosmic web induces
intrinsic alignments of galaxies

S. Codis[1]†, Y. Dubois[1], C. Pichon[1], J. Devriendt[2] and A. Slyz[2]

[1]Sorbonne Universités, UPMC Univ. Paris 06 & CNRS, UMR7095,
Institut d'Astrophysique de Paris, 98 bis Boulevard Arago, 75014, Paris, France

[2]Sub-department of Astrophysics, University of Oxford, Keble Road, Oxford OX1 3RH

Abstract. Intrinsic alignments are believed to be a major source of systematics for future generation of weak gravitational lensing surveys like Euclid or LSST. Direct measurements of the alignment of the projected light distribution of galaxies in wide field imaging data seem to agree on a contamination at a level of a few per cent of the shear correlation functions, although the amplitude of the effect depends on the population of galaxies considered. Given this dependency, it is difficult to use dark matter-only simulations as the sole resource to predict and control intrinsic alignments. We report here estimates on the level of intrinsic alignment in the cosmological hydrodynamical simulation HORIZON-AGN that could be a major source of systematic errors in weak gravitational lensing measurements. In particular, assuming that the spin of galaxies is a good proxy for their ellipticity, we show how those spins are spatially correlated and how they couple to the tidal field in which they are embedded. We will also present theoretical calculations that illustrate and qualitatively explain the observed signals.

Keywords. large-scale structure of universe, gravitational lensing, method: numerical

---

1. Introduction

Weak lensing is often presented as a potential powerful probe of cosmology for the coming years with large surveys like DES,
Euclid
or LSST. It relies on the fact that the observed shape of galaxies is distorted because the light path from background sources towards us is bent by the gravitational potential well along the line of sight. Therefore, measuring these distortions directly probes cosmology (cosmological model, dark matter distribution, etc). The idea behind weak lensing cosmic probes is thus to try and detect coherent distortions of the shapes of galaxies, e.g using the two-point correlation function of the ellipticities of galaxies. Note that the apparent ellipticity of a galaxy is induced by the cosmic shear γ (which is related to the projected gravitational potential along the line of sight) but also encompasses the intrinsic ellipticity of that galaxy $e = e_s + \gamma$, where e is the apparent ellipticity and e_s the intrinsic source ellipticity (that would have been observed without lensing). Therefore the (projected) ellipticity-ellipticity two-point correlation function can be written as the sum of a shear-shear term, intrinsic-intrinsic and intrinsic-shear correlations

$$\langle e(\vartheta)e(\vartheta + \theta)\rangle_\vartheta = \langle \gamma\gamma'\rangle + \langle e_s e_s'\rangle + 2\langle e_s\gamma'\rangle , \qquad (1.1)$$

where, for compactness, the prime means at an angular distance θ from the first location. These last two contributions that contaminate the shear signal are the two kinds of intrinsic alignments (IA hereafter), one term being the so-called "II" term $\langle e_s e_s'\rangle$ induced by the intrinsic correlation of the shape of galaxies in the source plane (Heavens *et al.*

† Email: codis@iap.fr

437

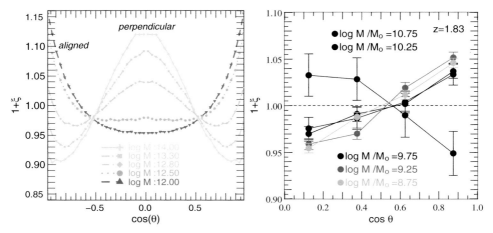

Figure 1. *Left:* excess probability of alignment between the spin and the direction of the closest filament as measured from the 43 millions haloes of the Horizon 4π simulation (Teyssier (2002))) at redshift zero. Different colours correspond to different mass bins from 10^{12} (red) to 10^{14} $M_\odot$ (blue) as labeled. A transition mass is detected at $M_0^s = M_{\rm crit}^s(z=0) \simeq 5(\pm 1) \times 10^{12} M_\odot$: for haloes with $M > M_0^s$, the spin is more likely to be perpendicular to their host filament, whereas for haloes with $M < M_0^s$, the spin tends to be aligned with the closest filament. This figure is from Codis *et al.* (2012). Right-hand panel: same as left panel for the 160 000 galaxies of the Horizon-AGN hydrodynamical simulation at $z = 1.8$ (Dubois *et al.* (2014)).

(2000), Croft & Metzler (2000), Catelan *et al.* (2001)) and the other one is the so-called "GI" term $\langle e_s \gamma' \rangle$ coming from correlations between the intrinsic ellipticity of a galaxy and the induced ellipticity (or shear) of a source at higher redshift (Hirata & Seljak (2004)). Much effort has thus been made to control the level of IA of galaxies as a potential source of systematic errors in weak gravitational lensing measurements although some techniques have been proposed to mitigate their nuisance by making extensive use of photometric redshifts (e.g. Blazek *et al.* (2012)). Direct measurements of the alignment of the projected light distribution of galaxies in wide field imaging data seem to agree on a contamination at a level of a few percents in the shear correlation functions, although the amplitude of the effect depends on the depth of observations, the amount of redshift information and the population of galaxies considered in the sense that red galaxies seem to show a strong intrinsic projected shape alignment signal whereas observations only place upper limits in the amplitude of the signal for blue galaxies (e.g. Joachimi *et al.* (2013a)).

From a theoretical point of view, it has been shown that dark halos (Aragòn-Calvo *et al.* (2007), Paz *et al.* (2008), Codis *et al.* (2012) among many others) and galaxies (Hahn *et al.* (2010), Dubois *et al.* (2014)) are correlated with the cosmic web. Fig. 1 shows that the spin of dark halos (left panel) and galaxies (right panel) is correlated to the direction of the closest filament. Consequently, this large-scale coherence of galaxies could then contaminate significantly the weak lensing observables. Given the inherently anisotropic nature of the large-scale structure and its complex imprint on the shapes and spins of galaxies together with the dependency on the physical properties of the galaxies seen in the observation, it is probably difficult to rely on isotropic linear theory (e.g. Lee & Pen (2001)) or dark matter-only numerical simulations as the sole resort to predict and control IA for weak lensing applications. With the advent of cosmological hydrodynamical simulations, we are now in a position to try and measure IA directly into those simulations instead of relying on pure N-body simulations and semi-analytical

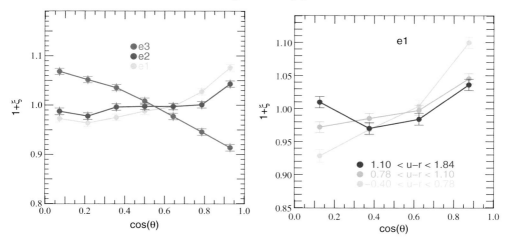

Figure 2. *Left:* PDF of the cosine of the angle between the spin of galaxies and the minor (cyan), intermediate (purple) and major (magenta) eigen-direction of the tidal tensor in the HORIZON-AGN simulation. *Right:* PDF of the cosine of the angle between the spin of galaxies and the minor eigen-direction for different colours as labeled.

models (Schneider & Bridle (2010), Joachimi *et al.* (2013b)). We report on the recent findings of Codis *et al.* (2014) who uses the HORIZON-AGN simulation presented in Dubois *et al.* (2014) at redshift $z = 1.2$ to measure the level of IA taking the spin as a proxy for the shape of galaxies. Section 2 will be devoted to the measured correlations between galaxy shapes and tidal field (related to the "GI" term) and section 3 to the auto-correlation of the intrinsic ellipticities (related to the "II" term).

2. Gravitational-intrinsic correlations

In order to study the correlations between the spin direction and the surrounding tidal field, the traceless tidal shear tensor is computed in the HORIZON-AGN simulation $T_{ij} = \partial_{ij}\Phi - \Delta\Phi\,\delta_{ij}/3$, Φ being the gravitational potential and δ_{ij} the Kronecker delta function. The minor, intermediate and major eigen-directions of the tidal tensor T_{ij} are called $\mathbf{e}_1$, $\mathbf{e}_2$ and $\mathbf{e}_3$ corresponding to the ordered eigenvalues $\lambda_1 \leqslant \lambda_2 \leqslant \lambda_3$ of the Hessian of the gravitational potential, $\partial_{ij}\Phi$. In the filamentary regions, $\mathbf{e}_1$ gives the direction of the filament, while the walls are collapsing along $\mathbf{e}_3$ and extend, locally, in the plane spanned by $\mathbf{e}_1$ and $\mathbf{e}_2$ (Pogosyan *et al.* (1998)).

2.1. *One-point correlation between spin and tidal tensor*

The cosine of the angle between the spin of the galaxies and the three local eigen-directions of the tidal tensor is then measured. An histogram of these values is computed and rescaled so as to give the probability distribution function (PDF) displayed in Fig. 2 (left panel). The spins clearly exhibits a tendency to be aligned with the minor eigen-direction (i.e the filaments) and in a weaker way with the intermediate axis (i.e the wall). The same analysis can be done for different mass and colour samples (see Fig. 2, right panel) : small-mass galaxies tend to have a spin aligned with the minor eigen-direction while more massive galaxies tend to have their spin perpendicular to that direction. The transition seems to occur around $4 \times 10^{10}\,M_\odot$. With regards to colours, the bluest galaxies (defined here by $u - r < 0.78$) are more correlated with the tidal eigen-directions than the red galaxies ($u - r > 1.1$). This can be easily understood as red galaxies are typically

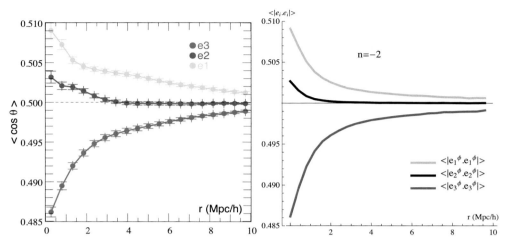

Figure 3. *Left:* Mean cosine of the angle between the spin of galaxies and the minor (cyan), intermediate (purple) and major (magenta) eigen-direction of the tidal tensor as a function of the separation. *Right:* Same as left panel for a Gaussian random field with power-law power spectrum once rescaled so as to match the measured value at zero separation.

massive, while blue galaxies are often small-mass galaxies. At that redshift ($z \sim 1.2$), this implies that red galaxies correspond to objects around the transition mass, whereas blue galaxies are mostly aligned with $\mathbf{e}_1$. At lower redshift, we expect the population of massive galaxies perpendicular to $\mathbf{e}_1$ to increase, so that red galaxies become more correlated. Obviously, we should also keep in mind that applying additional selection cuts on the galaxy samples (mass, luminosity, etc) would change the level of correlation.

2.2. *Two-point correlation between spin and tidal tensor*

Beyond one-point statistics, it is also of interest in the context of weak lensing studies to quantify how this signal pervades when the tidal field at a distance r from the galaxy is considered. Since the tidal field in the vicinity of a galaxy contributes also to the lensing signal carried by more distant galaxies, it is clear that the spin – tidal tensor cross-correlation is closely related to the GI term. In order to o address that question, the correlations between the spins and the eigen-directions of the tidal tensor at comoving distance r are measured. Fig. 3 (left panel) shows the mean angle between the spins and the three eigen-directions of the tidal field $\mathbf{e}_1$ (cyan), $\mathbf{e}_2$ (purple) and $\mathbf{e}_3$ (magenta) as a function of the separation r. As expected, the spin and the tidal eigen-directions de-correlate with increasing separation. However, whereas the signal vanishes on scales $r > 3 \ h^{-1}$ Mpc for the spin to intermediate tidal eigen-direction correlation, it persists on distances as large as $\sim 10 \ h^{-1}$ Mpc for the minor and major eigen-directions of the tidal tensor. This behaviour can be theoretically understood (Codis *et al. in prep*) using a Gaussian random field δ (here we use a power-law power spectrum with spectral index $n = -2$) for which we compute the joint PDF of the second derivatives of its corresponding potential (ϕ_{ij}, ϕ being related to δ by the Poisson equation). Then the mean angle between the eigen-directions of ϕ_{ij} in two locations separated by r can be computed. Once rescaled so as to match the one-point statistics measured in Fig. 2 (here we want to study the evolution of the two-point function with the separation, not its absolute value), we find the function plotted on the right panel of Fig. 3 which interestingly shows the same qualitative behaviour as what is measured in the simulation (left panel).

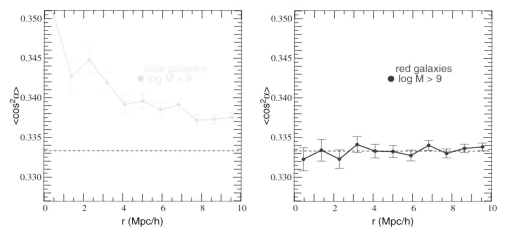

Figure 4. *Left:* mean cosine square of the angle between the spin of blue galaxies separated by r. *Right:* same as left panel for red galaxies. Important correlations on cosmological scales are detected for the blue sample, not for red galaxies.

3. Intrinsic-intrinsic correlations

3.1. *3D spin-spin correlations*

The IA contamination coming from the auto-correlations of the intrinsic ellipticities is now investigated by means of the spin-spin correlation function. The angle between the spins of each pair of galaxies separated by a distance r is computed and the resulting histogram of the square of the cosine of those angles (the polarity is not relevant for weak lensing studies) is shown in Fig. 4 for two different colour samples. A significant spin correlation is detected for blue galaxies out to at least 10 Mpc/h. Conversely, we detect no significant correlations for red galaxies at that redshift ($z = 1.2$).

3.2. *Projected spin-spin correlations*

In order to get closer to weak lensing observables, one question arises: what fraction of the spin-spin correlations remains after projection on the sky? To address this issue, the spins are projected along a given line-of-sight direction in the box and the apparent axis ratio is assumed to be well-approximated by $q = |L_z|/|\mathbf{L}|$, where z is the line of sight direction. The orientation of the major axis of the projected ellipse is $\psi = \pi/2 - \arctan(L_y/L_x)$ so that the complex ellipticity can be written $e = (1-q)/(1+q)\exp(2i\psi)$ in cartesian coordinates. The projected ellipticities can easily be mapped from cartesian (x, y) coordinates to the $(+, \times)$ frame attached to the separation of a given galaxy pair according to the geometric transformation $e_+ = -e_x \cos(2\beta) - e_y \sin(2\beta)$, $e_\times = e_x \sin(2\beta) - e_y \cos(2\beta)$, where β is the angle between the separation and the first cartesian coordinate (x). With those prescriptions, we can estimate the projected correlation functions for a given projected separation θ. For the II component (dropping the subscript s), this reads

$$\xi_+^{\mathrm{II}}(\theta) = \langle e_+ e'_+ + e_\times e'_\times \rangle . \qquad (3.1)$$

This correlation function of the projected spins is displayed on Fig. 5 for different samples of galaxies. The spins of blue and intermediate-mass galaxies are shown to be correlated on scales about 10 arcminutes while (as expected from the 3D study) the signal for red galaxies is compatible with zero. Note that this signal is not contradictory with current observations as it is at a larger redshift.

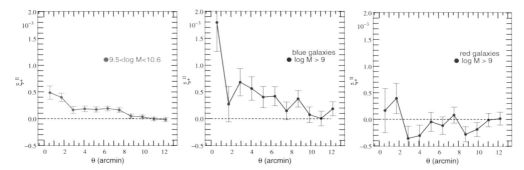

Figure 5. Two-point correlation function of the projected ellipticities of galaxies as a function of angular separation for intermediate mass (left), blue (middle) and red (right) galaxies.

4. Conclusion

In the context of high-precision cosmology (Euclid, DES, LSST, etc), it is crucial to study systematic effects like IA that could significantly contaminate weak lensing observables. Codis *et al.* (2014) found that at redshift $z = 1.2$ in the Horizon-AGN hydrodynamical simulation, galaxy ellipticities are correlated with the tidal field and with themselves on cosmological scales with a level of correlation that depends on mass and colour. After projection, these correlation pervades in particular for blue and intermediate-mass galaxies and could be a major source of contamination for cosmic shear studies.

The post-processing of hydrodynamical simulations represent a novel approach to deal with IA which, unlike semi-analytical modeling or linear theory, takes into account baryonic physics. Mass, colour-dependence and any other selection effects can be modeled accordingly. The analysis presented here (see also Codis *et al.* 2014) is a first step in the accurate modeling of IA effects and paves the way to future more realistic studies (light-weighted measurements on the light cone, etc.).

This work is partially supported by grant ANR-13-BS05-0005 of the french ANR. SC thanks Raphael Gavazzi and Karim Benabed for fruitful comments.

References

Aragòn-Calvo, M. A. *et al.* 2007, *ApJ Let.*, 655, L5
Blazek, J., Mandelbaum, R., Seljak,U., & Nakajima, R. 2012, *JCAP*, 5, 41
Catelan, P., Kamionkowski, M., & Blandford, R. D. 2001, *MNRAS*, 320, L7
Codis, S. *et al.* 2012 *MNRAS*, 427, 3320
Codis, S. *et al.* 2014, *ArXiv e-prints*
Croft, R. A. C. & Metzler, C. A. 2000, *ApJ*, 545, 561
Dubois, Y. *et al.* 2014, *ArXiv e-prints*
Hahn, O., Teyssier, R., & Carollo, C. M. 2010, *MNRAS*, 405, 274
Heavens, A., Refregier, A., & Heymans, C. 2000, *MNRAS*, 319, 649
Hirata, C. M. & Seljak, U. 2004, *Phys. Rev. D*, 70, 063526
Joachimi, B. *et al.* 2013, *MNRAS*, 431, 477
Joachimi, B. *et al.* 2013, *MNRAS*, 436, 819
Lee, J. & Pen, U.-L. 2001, *ApJ*, 555, 106
Paz, D. J., Stasyszyn, F., & Padilla, N. D. 2008, *MNRAS*, 389, 1127P
Pogosyan, D., Bond, J. R., & Kofman, L. 1998, *JRASC*, 92, 313
Schneider, M. D. & Bridle, S. 2010, *MNRAS*, 402, 2127
Teyssier, R. 2002, *MNRAS*, 402, 2127

The Zeldovich Universe:
Genesis and Growth of the Cosmic Web
Proceedings IAU Symposium No. 308, 2014
R. van de Weygaert, S. Shandarin, E. Saar & J. Einasto, eds.
© International Astronomical Union 2016
doi:10.1017/S1743921316010334

Spin Alignment
in Analogues of The Local Sheet

George J. Conidis

Department of Physics and Astronomy, York University, Toronto, Ontario, M3J 1P3, Canada
email: gconidis@yorku.ca

Abstract. Tidal torque theory and simulations of large scale structure predict spin vectors of massive galaxies should be coplanar with sheets in the cosmic web. Recently demonstrated, the giants ($K_s \leqslant$ -22.5 mag) in the Local Volume beyond the Local Sheet have spin vectors directed close to the plane of the Local Supercluster, supporting the predictions of Tidal Torque Theory. However, the giants in the Local Sheet encircling the Local Group display a distinctly different arrangement, suggesting that the mass asymmetry of the Local Group or its progenitor torqued them from their primordial spin directions. To investigate the origin of the spin alignment of giants locally, analogues of the Local Sheet were identified in the SDSS DR9. Similar to the Local Sheet, analogues have an interacting pair of disk galaxies isolated from the remaining sheet members. Modified sheets in which there is no interacting pair of disk galaxies were identified as a control sample.

Galaxies in face-on control sheets do not display axis ratios predominantly weighted toward low values, contrary to the expectation of tidal torque theory. For face-on and edge-on sheets, the distribution of axis ratios for galaxies in analogues is distinct from that in controls with a confidence of 97.6 % & 96.9%, respectively. This corroborates the hypothesis that an interacting pair can affect spin directions of neighbouring galaxies.

Keywords. cosmology: large-scale structure of universe, galaxies: spiral, galaxies: evolution, galaxies: interactions, galaxies: fundamental parameters

1. Introduction & Overview

The relationship between a galaxy's spin (angular momentum) vector and its surrounding environment is of substantial importance to the understanding of galaxy formation and evolution. In the last century, simulations of large scale structure have produced a better understanding of spin alignment in the cosmic web (Faltenbacher *et al.* 2002; Cuesta *et al.* 2008; Paz *et al.* 2008; Zhang *et al.* 2009; Wang *et al.* 2011; Codis *et al.* 2012; Trowland *et al.* 2013; Zhang *et al.* 2013; Tempel *et al.* 2013; Tempel & Libeskind 2013; Cen 2014; Dubois *et al.* 2014). Investigations of alignment in specific environments (voids, sheets, filaments, and nodes) have been carried out using the well established tidal torque theory (Hoyle, 1949; Peebles, P. J. E., 1969; Efstathiou, G., *et al.* 1979; White, S. D. M. 1984; Porciani, *et al.* 2002; Schafer, B. M. 2009). Within a sheet, simulations predict that dark matter haloes above 10^{12} M$_\odot$ should have spin vectors preferentially directed along the mid-plane of the sheet (Aragón-Calvo *et al.* 2007; Hahn *et al.* 2007; Codis *et al.* 2012; Trowland *et al.* 2013). However, such a prediction is untestable unless the assumption is made that spin vectors for dark and baryonic matter are aligned (van den Bosch *et al.* 2002; Yoshida *et al.* 2003; Chen, Jing & Yoshikawa 2003; Kazantzidis *et al.* 2004; Springel, White, & Hernquist 2004; Bailin *et al.* 2005; Berentzen & Shlosman 2006; Gustafsson *et al.* 2006; Adabi *et al.* 2009; Croft *et al.* 2009; Romano-Diaz *et al.* 2009; Bett *et al.* 2010; Sales *et al.* 2012). There have been many observational studies to look for alignment of baryonic spin with cosmic structure, but results have been contradictory. While some researchers have claimed to see alignment (Trujillo *et al.* 2006; Lee & Erdogdu 2007; Paz *et al.* 2008; Jones *et al.* 2010), others have found that spin directions are random (Slosar & White 2009; Cervantes-Sodi *et al.* 2010). In those studies which find evidence for alignment, there is a preference for a disk galaxy to have its spin vector directed along the midplane of its host sheet (Trujillo *et al.* 2006; Varela *et al.* 2012; Tempel *et al.* 2013; Tempel & Libeskind 2013). However, it is important to realize that observational studies constrain spin directions using axis ratios and position angles only. For face-on and edge-on galaxies, there are two possible solutions, and for inclined galaxies there are four. The ambiguity reduces the likelihood of a positive detection of alignment.

Galaxies within a distance of 6 Mpc from the Milky Way are localized in a highly flattened configuration known as the Local Sheet which is distinct from the Local Super-cluster (Peebles *et al.* 2001; Tully *et al.* 2008; Peebles & Nusser 2010; McCall 2014). The extent (diameter) of the Sheet is 10.4 Mpc and the thickness ($2\times$ vertical dispersion) is 0.47 Mpc. The Sheet hosts 14 giants ($K_s \leqslant$ -22.5 mag), of which 2 are in the Local Group and 12 beyond. Those beyond are distributed around a ring with a radius of 3.75 Mpc (the Council of Giants) whose centre is only 1.1 Mpc from the centroid of the Local Group.

McCall (2014) analyzed the distribution of spin vectors of the Local Sheet giants with respect to the giants beyond the Local Sheet in the Local Volume (< 10 Mpc). The Local Sheet giants have a distinct spin distribution from the giants beyond, with vectors arranged around a small circle on the celestial sphere. However, the spin vectors for giants beyond the Local Sheet are arranged around a great circle on the sky which is closely aligned with the supergalactic plane (see Figure 1). The pole of this spin distribution, excluding the three outliers near the poles, is $23° \pm 13°$ from the north galactic pole resulting in a sinusoidal trend in latitude. The mean supergalactic latitude of spin directions is only $-1.8° \pm 20.6°$. Indeed, coplanar spin vectors for galaxies in the luminosity range spanned is predicted by tidal torque theory.

The distinct spin distribution observed for Local Sheet giants has been hypothesized to be the result of tidal torquing induced by the Local Group (McCall 2014). First of all, the Local Group contains 27% of the mass of the Sheet. By virtue of the pair of interacting giants which dominate it, its mass distribution is highly asymmetrical. Also, its location is such that surrounding giants are at comparable distances; the radial dispersion is only ± 0.8 Mpc. Thus, the Local Group is the most likely source of torquing.

2. Do interacting Pairs Influence Spin alignment in Local Sheet Analogues?

To further our understanding of the effect of an interacting pair of galaxies on the spin alignments of galaxies in sheets, analogues of the Local Sheet were identified within the Sloan Digital Sky Survey Data Release 9 (Conidis & McCall, in preparation). As controls, sheets were also identified in which the interacting pair is replaced with a single disk galaxy. It was expected that if an interacting pair torqued fellow sheet members, the two samples would show a distinct difference.

An assessment of spin distributions was carried out by examining axis ratios for disk galaxies in both analogue and control sheets. Axis ratio distributions were derived for edge-on sheets (tilt $> 70°$) and face-on sheets (tilt $< 44°$). The control sheets were predicted to preserve their primordial spin organization, with spin vectors for members pointing along the midplanes of the sheets. This prediction implied that axis ratios for galaxies in a face-on sheet would be clustered about low values. Figure 2a (grey bars)

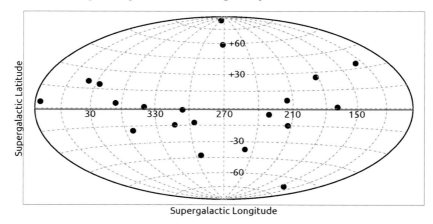

Supergalactic Longitude

Figure 1: A Hammer projection (area preserving) of spin vector directions of the giant galaxies in the Local Volume beyond the Local Sheet (black filled circles). The mean supergalactic latitude ($-1.8°$) is shown as a horizontal blurred black line. The alignment of the giants with the supergalactic plane is in accordance with the prediction of tidal torque theory.

is the observed axis ratio distribution for the disk galaxies in face-on control sheets. It does not cluster at low axis ratios as expected. It is found that the analogues disk galaxies hosted in a face-on sheet configuration tend to favour a low b/a value (Figure 2a, black bars). In face-on analogue and control sheets, 56% and 46%, respectively, of the galaxy population displays b/a $\leqslant$ 0.5. Furthermore the axis ratio distribution for face-on analogues and control have a 0.0029% and 10.1% chance, respectively, of being consistent with a random distribution. Clearly indicating the analogue and control distributions are distinct from each other. As well, the control's face-on distribution cannot be confidently distinguished from random. For edge-on analogue and control sheets (Figure 2b), 54% and 45%, respectively, of the galaxy population displays b/a $\leqslant$ 0.5. The axis ratio distributions for edge-on analogues and controls have a 0.025% and 0.57% chance, respectively, of being consistent with a random distribution. Thus, both distributions are found to be distinct from random.

Cumulative distributions of the axis ratios of galaxies in analogue and control sheets are shown in Figure 3. The likelihood that the spin distributions for analogues and controls are drawn from the same probability distribution are 2.4% and 3.1%, respectively, for face-on and edge-on orientations. This clearly shows that an interacting pair of disk galaxies embedded near the centre of a sheet torques the spins of surrounding members.

3. Implications and Future Work

The distinct difference between the axis ratio distributions for analogue and control sheets, be they face-on or edge-on, validates the hypothesis that an interacting pair of galaxies changes the spin distribution of other members. This supports the conjecture that the Local Group was responsible for torquing the spins of galaxies in the Local Sheet out of alignment with the plane of the Local Supercluster.

Axis ratio distributions for galaxies in control sheets disagree with the predictions of tidal torque theory. Especially, in face-on sheets, axis ratios do not cluster around low values and the distribution can be considered random within errors. Surprisingly, the axis

G. J. Conidis

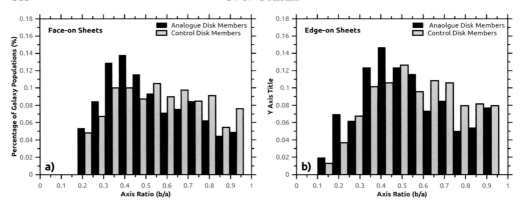

Figure 2: Histograms of axis ratios of disks in analogue (black) and control (grey) sheets. a) face-on sheets (tilt $< 44°$). b) edge-on sheets (tilt $> 70°$).

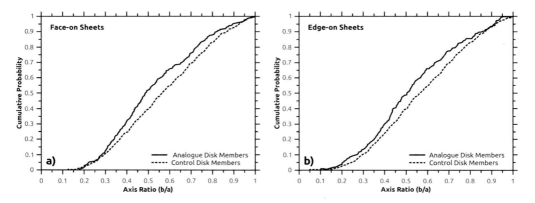

Figure 3: Cumulative distributions (CDs) of the axis ratios of disk galaxies in analogue (solid curve) and control (dashed curve) sheets. a) CDs for face-on sheets (tilt $< 44°$). The Kuiper probability that these CDs were drawn from the same distribution is 2.4%. b) Edge-on sheets (tilt $> 70°$). The Kuiper probability that these CDs were drawn from the same distribution is 3.1%.

ratio distribution for face-on and edge-on analogues show a stronger tendency toward low axis ratios. As well, both these distributions are distinct from random providing additional evidence that a galaxy's orientation is environmentally dependant.

Analyses here are limited by the use of axis ratios as proxies for spin vectors. For a given axis ratio, there are up to four spin directions possible. Currently, the author is engaged in determining unambiguously three-dimensional spin vectors for the disk galaxies in sheets, which will make possible more definitive statements about alignments (or lack thereof) with structure.

Acknowledgements

I would like to thank Prof. Marshall L. McCall for his insightful discussions and helpful comments while conducting this research and producing this document.

References

Abadi, M. G., Navarro, J. F., Fardal, M., Babul, A., & Steinmetz, M., 2009, preprint (arXiv:0902.2477)

Aragón-Calvo, M. A., M. A., van de Weygaert, R., Jones, B. J. T., & van der Hulst, J. M. 2007, *ApJ*, 655, L5

Bailin, J., Kawata, D., Gibson, B. K., Steinmetz, M., Navarro, J. F., Brook, C. B., Gill, S. P. D., Ibata, R. A., Knebe, A., Lewis, G. F., & Okamoto, T. 2005, textit*ApJ*, 627, 17B

Berentzen, I. & Shlosman, I., 2006, *ApJ*, 648, 807

Philip, B., Vincent, E., Carlos, S. F., Adrian, J., & Takashi, O., 2010, *MNRAS*, 404, 1137

Cen, R., 2014, *ApJ*, 785, 15

Cervantes-Sodi, B., Hernandez, X., & Park, C., 2010, *MNRAS*, 402, 1807

Codis, S., Pichon, C., Devriendt, J., *et al.* 2012, *MNRAS*, 427, 3320

Croft, R. A. C., Di Matteo, T., Springel, V., & Hernquist, L., 2009, *MNRAS*, 400, 43

Cuesta, A. J., Betancort-Rijo, J. E., Gottlöber, S., *et al.* 2008, *MNRAS*, 385, 867

Dubois, Y., Pichon, C., Welker, C., *et al.* 2014, *MNRAS*, 444, 1453

Efstathiou, G. & Jones, B. J. T., 1979, *MNRAS*, 186, 133

Faltenbacher, A., Gottlöber, S., Kerscher, M., & Müller, V. 2002, *A&A*, 395, 1

Gustafsson, M., Fairbairn, M., & Sommer-Larsen, J., 2006, *Phys. Rev. D*, 74, 123522

Hahn, O., Carollo, C. M., Porciani, C., & Dekel, A., 2007, *MNRAS*, 381, 41

Hoyle, F., 1949, *MNRAS*, 109, 365

Jones, B. J. T., van de Weygaert, R., & Arag on-Calvo, M. A., 2010, *MNRAS*, 408, 897

Kazantzidis, S., Kravtsov, A. V., Zentner, A. R., Allgood, B., Nagai, D., & Moore, B., 2004, *ApJ*, 611, L73

Lee, J. & Erdogdu, P., 2007, *ApJ*, 671, 1248

McCall, M. L., 2014, *MNRAS*, 440, 405

Paz, D. J., Stasyszyn, F., & Padilla, N. D., 2008, *MNRAS*, 389, 1127

Peebles, P. J. E., 1969, *ApJ*, 155, 393

Peebles, P. J. E. & Nusser, A., 2010, *Nature*, 465, 565

Peebles, P. J. E., Phelps, S. D., Shaya, E. J., & Tully, R. B., 2001, *ApJ*, 554, 104

Porciani, C., Dekel, A., & Hoffman, Y., 2002, *MNRAS*, 332, 325

Romano-Díaz, E., Shlosman, I., Heller, C., & Hoffman, Y., 2009, *ApJ*, 702, 1250

Sales, L. V., Navarro, J. F., Theuns, T., Schaye, J., White, S. D. M., Frenk, C. S., Crain, R. A., & Vecchia, C. D., 2012, 423, 1544

Schäfer, B. M., 2009, *IJMPD*, 18, 173

Slosar, A. & White, M., 2009, *JCAP*, 6, 9

Springel, V., White, S. D. M., & Hernquist, L., 2004, in Ryder S., Pisano D., Walker M., Freeman K., eds, Proc. IAU Symp. 220, Dark Matter in Galaxies. Astron. Soc. Pac., San Francisco, p. 421

Tempel, E., Stoica, R. S., & Saar, E., 2013, *MNRAS*, 428, 1827

Tempel, E. & Libeskind, N. I., 2013, *ApJ*, 775, L42

Trowland, H. E., Lewis, G. F., & Bland-Hawthorn, J., 2013, *ApJ*, 762, 72

Trujillo, I., Carretero, C., & Patiri, S. G., 2006, *ApJ*, 640, L111

Tully, R. B., Shaya, E. J., Karachentsev, I. D., Courtois, H. M., Kocevski, D. D., Rizzi, L., & Peel, A., 2008, *ApJ*, 676, 184

van den Bosch, F. C., Abel, T., Croft, R. A. C., Hernquist, L., & White, S. D. M.,, 2002, *ApJ*, 576, 21

Varela, J., Betancort-Rijo, J., Trujillo, I., & Ricciardelli, E., 2012, *ApJ*, 744, 82

Wang, H., Mo, H. J., Jing, Y. P., Yang, X., & Wang, Y., 2011, *MNRAS*, 413, 1973

White, S. D. M., 1984, *ApJ*, 286, 38

Yoshida, N., Abel, T., Hernquist, L., & Sugiyama, N., 2003, *ApJ*, 592, 645

Zhang, Y., Yang, X., Faltenbacher, A., *et al.*, 2009, *ApJ*, 706, 747

Zhang, Y., Yang, X., Wang, H., *et al.* 2013, *ApJ*, 779, 160

The Zeldovich Universe:
Genesis and Growth of the Cosmic Web
Proceedings IAU Symposium No. 308, 2014
R. van de Weygaert, S. Shandarin, E. Saar & J. Einasto, eds.

© International Astronomical Union 2016
doi:10.1017/S1743921316010346

Galaxy alignment on large and small scales

X. Kang[1]†, W.P. Lin[2], X. Dong[2], Y.O. Wang[2], A. Dutton[3] and A. Macciò[3]

[1]Purple Mountain Observatory, the Partner Group of MPI für Astronomie, 2 West Beijing Road, Nanjing 210008, China
[2]Key Laboratory for Research in Galaxies and Cosmology, Shanghai Astronomical Observatory, Chinese Academy of Science, 80 Nandan Road, Shanghai 200030, China
[3]Max Planck Institut für Astronomie, Königstuhl 17, D-69117 Heidelberg, Germany

Abstract. Galaxies are not randomly distributed across the universe but showing different kinds of alignment on different scales. On small scales satellite galaxies have a tendency to distribute along the major axis of the central galaxy, with dependence on galaxy properties that both red satellites and centrals have stronger alignment than their blue counterparts. On large scales, it is found that the major axes of Luminous Red Galaxies (LRGs) have correlation up to 30Mpc/h. Using hydro-dynamical simulation with star formation, we investigate the origin of galaxy alignment on different scales. It is found that most red satellite galaxies stay in the inner region of dark matter halo inside which the shape of central galaxy is well aligned with the dark matter distribution. Red centrals have stronger alignment than blue ones as they live in massive haloes and the central galaxy-halo alignment increases with halo mass. On large scales, the alignment of LRGs is also from the galaxy-halo shape correlation, but with some extent of mis-alignment. The massive haloes have stronger alignment than haloes in filament which connect massive haloes. This is contrary to the naive expectation that cosmic filament is the cause of halo alignment.

Keywords. large-scale structure of universe, dark matter halo, numerical simulation

---

1. Introduction

The spatial and orientation of galaxies are not randomly distributed in the universe. On small scales the distribution of satellite galaxies is correlated with the central galaxy. For example, in our Milky Way Galaxy and neighboring M31, the satellites are found to be highly anisotropic that most of them are distributed in a great thin plane with a common rotation (e.g., Kroupa *et al.* 2005; Ibata *et al.* 2013). This raises questions if our Milky Way or M31 is anomaly or the formation and accretion of its satellites is along some special direction, or even the Cold Dark Matter theory is wrong (e.g., Kang *et al.* 2005; Pawlowski *et al.* 2014). It is highly possible that the Milky Way or M31 could be an outlier, thus its unique may not violate the general predictions from CDM model. However, the large sky surveys, such as 2dFGRS and SDSS, have found that satellites are still not randomly distributed around central galaxies, but are preferentially distributed along the major axes of centrals. This phenomenon is know as galaxy alignment. (e.g., Yang *et al.* 2006). On large scales, it is found that the shape of Luminous Red Galaxies (LRGs) are correlated up to scales of 30∼70Mpc/h (Okumura *et al.* 2009; Li *et al.* 2013).

Most studies using N-body simulations are devoted to the origin of galaxy alignment, and they have found that the observed alignment can be reproduced if the shape of central galaxies are correlated with that of the dark matter halo (e.g., Kang *et al.* 2007). However, to match the color dependence of galaxy alignment, one has to make different

† Email: kangxi@pmo.ac.cn

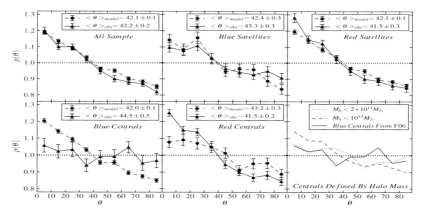

Figure 1. Predicted galaxy alignment (circles connected by dashed lines). The solid lines with triangles show the observational results of Yang *et al.* (2006) from SDSS. Each panel shows different samples.

assumption of how blue and red centrals follow the dark matter haloes (Agustsson & Brainerd 2010). Thus no consistent conclusions about the color dependence of galaxy alignment have been drawn. To study the galaxy alignment in detail and avoid making assumption about how the shape of central galaxy follow the dark matter halo in N-body simulation, we make use of hydro-dynamical simulation which includes important physics governing galaxy formation, such as gas cooling, metal recycling, star formation and supernova feedback. As the simulation includes star formation, the shape of central galaxy and the properties of satellites are self-consistently included in the simulation, thus the predictions can be directly tested against the data.

2. Simulation and Methods

The hydro-dynamical simulation we use is run using the Gadget-2 code (Springel 2005), and it follows 2×512^3 dark matter particles and gas particles in a cube box with each side of $100Mpc/h$. The cosmological parameters are selected as $\Omega_m = 0.268$, $\Omega_\Lambda = 0.732$, $\sigma_8 = 0.85$ and $h = 0.71$. The dark matter haloes and galaxies are both found using the standard friends-of-friends (FoF) algorithm. In each FoF group, the most massive galaxy is defined as the central galaxy, and all others are defined as satellites. The reduced inertia tensor of central galaxy determines its shape and major axis. The distribution of satellites around central is described by the angle between the position of satellite galaxy respect to the major axis of central galaxy. An alignment is obtained if the average angle is less than $45°$. For more details, see Dong *et al.* (2014).

3. Results

Fig. 1 shows the predicted alignment with comparison to the data of Yang *et al.* (2006). The upper left panel shows that the predicted alignment agrees well with the data for all galaxies. Also the simulation reproduces the dependence on color of satellites, as shown in upper middle and right panels that red satellites have stronger alignment. The lower left and middle panels show the alignment for blue and red centrals, and it is found that the predictions are inconsistent with the data. The predicted signal for blue centrals is too high and the one for red centrals is too low. We found that this is due to the neglect of effective feedback in our simulation, such as AGN feedback, which results in too many

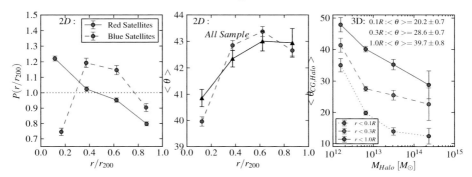

Figure 2. Left panel: satellites radial distribution. Middle: alignment of satellite in different radii. Right panel: central galaxy-halo shape alignment, with color lines for halo shape measured inside different radii.

massive blue central galaxies (see Fig. 1 in Dong *et al.* 2014). Those blue centrals are living in massive haloes where the alignment is larger (see right panel of Fig. 2).

It is known observationally that the color of central galaxy is strongly correlated with the host halo mass (e.g., Yang *et al.* 2008), and our simulation fails to reproduce that correlation. To mimic this effect and see if the dependence on color of central is from the halo mass dependence, in the lower right panel of Fig. 1 we show the prediction for centrals in host halo mass lower than some critical mass (see caption in plot). It is found that the alignment of centrals is a function of halo mass, and if blue centrals are living in haloes with mass lower than $2 \times 10^{11} M_{\odot}$, the predicted alignment is close to the data. Such a critical mass is also obtained from hydro-dynamical simulations (e.g., Kereš *et al.* 2005).

Fig. 2 explains the origin of alignment on galaxy color. The right panel shows the alignment between the shape of central galaxy and dark matter halo inside different radii. It is found that central galaxy traces better the shape of dark matter in inner region (red dotted line) and this alignment increases with halo mass. The left panel plots the radial distribution of red/blue satellites (with red solid and blue dashed lines). It is found that most red satellites stay in the inner halo region and thus trace the shape of dark matter halo there. As central galaxy follows better the shape of halo in inner region, it naturally predicts that red satellites are stronger aligned with central than blue satellites. The middle panel shows the alignment for satellites at different distance to central galaxy and good agreement with data is seen. The halo mass dependence seen in the right panel also explains the observed dependence on color of centrals as most red centrals live in massive haloes.

On large scales, the galaxy alignment is often referred to the shape correlation of central galaxies. It is found that the major axes of LRGs at z~ 0.3 are correlated up to 30 ~70Mpc/h (Okumura *et al.* 2009; Li *et al.* 2013), as shown by the points in upper left panel of Fig. 3. Okumura *et al.* also found that if the major axis of LRG follows that of the dark matter perfectly, the predicted alignment from N-body simulation is higher than the data (red dashed line), but if there is an average misalignment about 35°, the prediction will match well with the data (red solid line). As the LRGs often live in massive haloes connect by cosmic filament, it is natural to ask if the halo shape correlation is related to haloes in filament. The right panel of Fig. 3 shows that shape correlation of haloes in different environment (Zhang *et al.* 2008). It is found that haloes in dense region have stronger correlation than haloes in filament. Thus the observed alignment of LRGs seems to be fixed before the formation of the filaments connecting them.

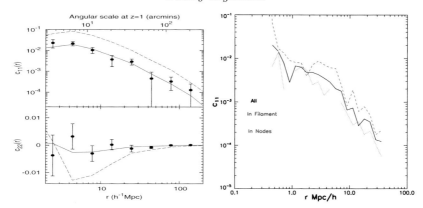

Figure 3. Left panel: the shape correlation of LRGs. Upper panel shows that LRGs are correlated along their major axes, but not along minor axes (lower panel). Right panel: halo shape correlation in different environment. Stronger alignment is found for haloes in dense environment.

4. Conclusion

Galaxy alignment effects are seen on both large and small scales. On small scales, it is mainly caused by the non-spherical nature of dark matter formed in CDM scenario (Jing & Suto 2002). The stronger alignment of red satellites is because they prefer halo inner region inside which the shape of central galaxy and dark matter distribution is well aligned, thus leading to stronger alignment between red satellites and central galaxy. The dependence on color of central is from the combined effects that red centrals live in massive haloes and the alignment between central galaxy and halo shape increases with halo mass. On large scales, the shape correlation of LRGs is determined by the halo-halo shape correlation, but with an average misalignment about 35°. The shape correlation of haloes in dense region is larger than those in filament which connect the massive haloes.

This work is partially supported by the NSFC project No.11333008 and the Strategic Priority Research Program "The Emergence of Cosmological Structures" of the Chinese Academy of Science Grant No. XDB09000000.

References

Agustsson, I. & Brainerd, T. G., 2010, *ApJ*, 709, 1321
Dong, X. C., Lin, W. P., Kang, X., Wang, Y., Dutton, A., & Macciò, A., 2014, *ApJ*, 791, L33
Ibata, R., Lewis, G. F., Conn, A. R., *et al.*, 2013, *Nature*, 493, 62
Jing, Y. P. & Suto, Y., 2002, *ApJ*, 574, 538
Kang, X., Mao, S., Gao, L., & Jing, Y. P., 2005, *A&A*, 437, 383
Kereš, D., Katz, N., Weinberg, D. H., & Dave, R., 2005, *MNRAS*, 363, 2
Kroupa, P., Theis, C. & Boily, C. M., 2005, *MNRAS*, 431, 507
Li, C., Jing, Y. P., Faltenbacher, A., & Wang, J., 2013, *ApJ*, 770, L12
Okumura, T., Jing, Y. P., & Li, C., 2009, *ApJ*, 694, 214
Pawlowski, M. S., *et al.*, 2014, *MNRAS*, 442, 2362
Springel, V., 2005, *MNRAS*, 364, 1105
Yang, X. H., van den Bosch, F. C., Mo, H. J., *et al.*, 2006, *MNRAS*, 369, 1293
Yang, X. H., Mo, H. J., & van den Bosch, F. C., 2008, *ApJ*, 676, 248
Zhang, Y., Yang, X., Faltenbacher, A., Springel, V., Lin, W. P., & Wang, H., 2009, *ApJ*, 706, 747

The Zeldovich Universe:
Genesis and Growth of the Cosmic Web
Proceedings IAU Symposium No. 308, 2014
R. van de Weygaert, S. Shandarin, E. Saar & J. Einasto, eds.

© International Astronomical Union 2016
doi:10.1017/S1743921316010358

Large-scale structure and the intrinsic alignment of galaxies

Jonathan Blazek[1,†], Uroš Seljak[2] and Rachel Mandelbaum[3]

[1] Center for Cosmology and AstroParticle Physics, Department of Physics,
Ohio State University, Columbus, USA

[2] Departments of Physics and Astronomy and Lawrence Berkeley National Laboratory,
University of California, Berkeley, USA

[3] McWilliams Center for Cosmology, Department of Physics, Carnegie Mellon University,
Pittsburgh, PA, USA

Abstract. Coherent alignments of galaxy shapes, often called "intrinsic alignments" (IA), are the most significant source of astrophysical uncertainty in weak lensing measurements. We develop the tidal alignment model of IA and demonstrate its success in describing observational data. We also describe a technique to separate IA from galaxy-galaxy lensing measurements. Applying this technique to luminous red galaxy lenses in the Sloan Digital Sky Survey, we constrain potential IA contamination from associated sources to be below a few percent.

Keywords. Gravitational lensing; large-scale structure of universe; cosmological parameters; galaxies: formation, halos, evolution

1. Introduction

Coherent, large-scale correlations of the intrinsic shapes and orientations of galaxies are a potentially significant source of systematic error in gravitational lensing studies, with the corresponding potential to bias or degrade lensing science results. This intrinsic alignment (IA) of galaxies has been examined through observations (e.g., Hirata *et al.* 2007, Joachimi *et al.* 2011), analytic modeling (e.g., Catelan *et al.* 2001, Hirata & Seljak 2004), and simulations (e.g., Schneider *et al.* 2012, Tenneti *et al.* 2014) - see Troxel & Ishak 2014 for a recent review. As our understanding has improved, IA has also emerged as a potential probe of large-scale structure as well as halo and galaxy formation and evolution (e.g. Chisari & Dvorkin 2013).

We present recent developments in understanding IA. We first describe analytic modeling of IA, focusing on the tidal alignment model (sometimes called the "linear alignment model"). This model is tested and developed to consistently include nonlinear effects (Blazek *et al.* 2011, Blazek *et al.* in prep.). We then present a technique to separate the IA and weak lensing signals in a galaxy-galaxy lensing measurement (Blazek *et al.* 2012).

2. Tidal alignment

Model predictions and comparison to data. The tidal alignment model (Catelan *et al.* 2001, Hirata & Seljak 2004) posits that the intrinsic ellipticity components of a galaxy, denoted $\gamma_{(+,\times)}$ are proportional to the tidal field:

$$\gamma^I_{(+,\times)} = -\frac{C_1}{4\pi G}(\nabla^2_x - \nabla^2_y, 2\nabla_x \nabla_y)S[\Psi(z_{\rm IA})], \qquad (2.1)$$

† email: blazek@berkeley.edu

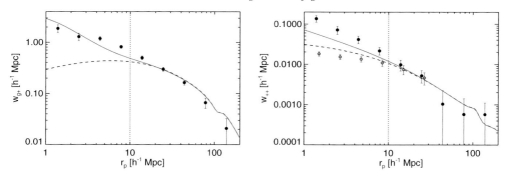

Figure 1. The projected correlation function between galaxy position and shape w_{g+} (*left panel*) and projected shape autocorrelation w_{++} (*right panel*) are shown as a function of projected separation, r_p. Data points show the measurements of Okumura *et al.* 2009, Okumura & Jing 2009. The w_{++} measurements have been projected along the line-of-sight, and open circles indicate the original measurement without a correction required to account for source clustering - see Blazek *et al.* 2011 for more information. Dashed lines indicate the linear order prediction of the tidal alignment model, while solid lines include the Halofit correction for nonlinear dark matter clustering (Smith *et al.* 2003).

where C_1 parameterizes the strength of the alignment, Ψ is the gravitational potential, and S is a filter that smooths fluctuations on halo or galactic scales.† This type of alignment is most likely to arise in elliptical galaxies for which angular momentum does not play a significant role in determining shape and orientation. For spiral galaxies, the acquisition of angular momentum, for instance through "tidal torquing," generally leads to quadratic dependence on the tidal field (e.g., Hirata & Seljak 2004). The tidal alignment contribution is the lowest-order function of the gravitational potential with the necessary symmetry and is thus expected to dominate IA correlations on sufficiently large scales. The redshift at which the alignment is set, z_{IA}, is determined by the astrophysical processes involved in galaxy formation and evolution. It is sometimes assumed that z_{IA} is during matter domination when a halo first forms. However, late-time accretion and mergers could have a significant impact on IA, in which case the relevant z_{IA} could be closer to the observed redshift, allowing for significant nonlinear evolution of the tidal field (Blazek *et al.* in prep.).

To test the predictions of the tidal alignment model, we consider the results of Okumura *et al.* 2009 and Okumura & Jing 2009, who measure the auto- and cross-correlations between the shapes and positions of luminous red galaxies (LRGs) in the Sloan Digital Sky Survey (SDSS). Since they are both highly biased and elliptical, LRGs should exhibit strong alignment in agreement with the tidal alignment model. Indeed, as seen in Fig. 1, the tidal alignment model provides a good description of correlations at large projected separations ($r_p \gtrsim 10\ h^{-1}\mathrm{Mpc}$). Fitting the model on large scales to both auto- and cross-correlations yields a consistent value of $C_1 \rho_{\mathrm{crit}} \approx 0.12 \pm 0.01$ (Blazek *et al.* 2011). Agreement with measurements on smaller scales is improved when the nonlinear evolution of dark matter clustering is included, producing what is sometimes called the "nonlinear alignment" (NLA) model (Bridle & King 2007).

Consistent treatment of nonlinear effects. Four effects can produce nonlinearities in the intrinsic shape correlations: (1) nonlinear dependence of intrinsic galaxy shape on the tidal field (e.g., quadratic tidal torquing); (2) nonlinear evolution of the dark matter

---

† The proper form and scale of this smoothing can have an important impact on model predictions on small scales. It is the subject of ongoing work (Chisari & Dvorkin 2013, Blazek *et al.* in prep.).

density field, leading to nonlinear evolution in the tidal field; (3) a nonlinear bias relationship between the galaxy and dark matter density fields; (4) the IA field actually observed is weighted by the local density of galaxies used to trace the shapes. In the pure tidal alignment model, intrinsic galaxy shapes depend linearly on the tidal field, even on small scales. However, nonlinearities from the other three effects must still be considered. The NLA model includes the nonlinear evolution of the dark matter density but does not consider other nonlinear contributions. Thus, although the NLA approach improves the model fit to data, it is not fully consistent and omits important astrophysical effects. Instead, we examine all nonlinear contributions in the context of tidal alignment (Blazek *et al.* in prep.), including terms that contribute at next-to-leading order while simultaneously smoothing the tidal field (e.g. at the Lagrangian radius of the host halo). We find that the effects of weighting by the local galaxy density can be larger than the correction from the nonlinear evolution of dark matter density, especially in the case of highly biased tracers, since the leading effect from weighting scales with the linear galaxy bias. Contributions from nonlinear galaxy bias are also appreciable, although subdominant.

3. Separating IA from galaxy-galaxy lensing

In galaxy-galaxy lensing, the shapes of background galaxies are used to probe the dark matter distribution around lens objects, often by stacking images at lens centers in order to increase the signal. Scatter in the photometric redshifts for lensing sources (e.g, Nakajima *et al.* 2012) can cause objects that are physically associated with the lens to be assigned a location significantly behind the lens, contaminating the lensing signal with shape correlations between physically associated sources and lens positions. Since the lensing signal measures the projected density profile of the lens objects, it should not depend on the redshift distribution of the background sources, assuming that the distribution is well known. However, as more distant sources (as defined by photometric redshift) are used, the fraction of physically associated objects and the corresponding IA contamination will decrease. Thus, by dividing the lens-source pairs into tomographic bins (by redshift separation), we are able to isolate the lensing signal from the IA signal. We apply this method to LRG lenses (Kazin *et al.* 2010) using the SDSS lensing catalog described in Reyes *et al.* 2012 to place constraints on the average IA contamination per physically associated source (Blazek *et al.* 2012). Results are shown in Fig. 2 and compared with IA measured directly in spectroscopic samples. These constraints correspond to a maximum IA contamination of a few percent at projected separations of $0.1 < r_p < 10 \ h^{-1}\mathrm{Mpc}$. The tighter constraints for red sources is due to their stronger clustering rather than an underlying trend in IA amplitude.

4. Conclusions

Continued progress in understanding IA is critical for current and upcoming lensing experiments. We have presented important developments in modeling IA through the tidal alignment model, including a consistent treatment of nonlinear effects. Combined with results from simulations and techniques for predicting IA in the deeply nonlinear regime (e.g., a halo model approach), this improved tidal alignment model will allow more effective mitigation of IA contamination. We have also developed and applied a method to separate IA from the galaxy-galaxy lensing signal, providing both a clean lensing measurement and a probe of IA in galaxy samples without large numbers of spectroscopic redshifts. This technique should be an important tool in future measurements.

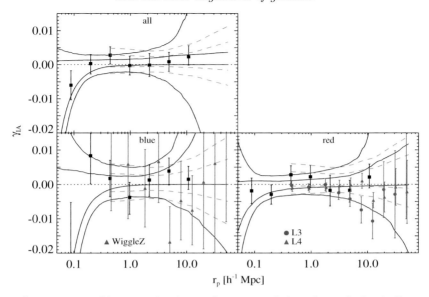

Figure 2. Constraints on IA contamination to the measured shear for each physically associated source (all, blue, and red) are shown. Data points with errors are model-independent constraints. Solid (dashed) lines denote 68% and 95% confidence intervals assuming a power-law (LRG observational) model. Previous spectroscopic results for SDSS red galaxies from Hirata *et al.* 2007 and for WiggleZ galaxies from Mandelbaum *et al.* 2011 are shown for comparison.

References

Blazek, J., Mandelbaum, R., Seljak, U., & Nakajima, R. 2012, *JCAP*, 5, 41
Blazek, J., McQuinn, M., & Seljak, U. 2011, *JCAP*, 5, 10
Blazek, J., Vlah, Z., & Seljak, U. in preparation
Bridle, S. & King, L. 2007, *New Journal of Physics*, 9, 444
Catelan, P., Kamionkowski, M., & Blandford, R. D. 2001, *MNRAS*, 320, L7
Chisari, N. E. & Dvorkin, C. 2013, *JCAP*, 12, 29
Hirata, C. M. *et al.* 2007, *MNRAS*, 381, 1197
Hirata, C. M. & Seljak, U. 2004, *Phys. Rev. D*, 70, 6
Joachimi, B., Mandelbaum, R., Abdalla, F. B., & Bridle, S. L. 2011, *A&A*, 527, A26
Kazin, E. A. *et al.* 2010, *ApJ*, 710, 1444
Mandelbaum, R., *et al.* 2011, *MNRAS*, 410, 844
Nakajima, R., *et al.* 2012, *MNRAS*, 420, 3240
Okumura, T. & Jing, Y. P. 2009, *ApJL*, 694, L83
Okumura, T., Jing, Y. P., & Li, C. 2009, *ApJ*, 694, 214
Reyes, R., *et al.* 2012, *MNRAS*, 425, 2610
Schneider, M. D., Frenk, C. S., & Cole, S. 2012, *JCAP*, 5, 30
Smith, R. E. & *et al.* 2003, *MNRAS*, 341, 1311
Tenneti, A., Mandelbaum, R., Di Matteo, T., Feng, Y., & Khandai, N. 2014, *MNRAS*, 441, 470
Troxel, M. A. & Ishak, M. 2014, arXiv:1407.6990

The Zeldovich Universe:
Genesis and Growth of the Cosmic Web
Proceedings IAU Symposium No. 308, 2014   © International Astronomical Union 2016
R. van de Weygaert, S. Shandarin, E. Saar & J. Einasto, eds.   doi:10.1017/S174392131601036X

The beaming of subhalo accretion

Noam I. Libeskind

Leibniz-Institute für Astrophysik Potsdam (AIP),
An der Sternwarte 16,
D-14482 Potsdam, Germany
email: nlibeskind@aip.de

Abstract. We examine the infall pattern of subhaloes onto hosts in the context of the large-scale structure. We find that the infall pattern is essentially driven by the shear tensor of the ambient velocity field. Dark matter subhaloes are preferentially accreted along the principal axis of the shear tensor which corresponds to the direction of weakest collapse. We examine the dependence of this preferential infall on subhalo mass, host halo mass and redshift. Although strongest for the most massive hosts and the most massive subhaloes at high redshift, the preferential infall of subhaloes is effectively universal in the sense that its always aligned with the axis of weakest collapse of the velocity shear tensor. It is the same shear tensor that dictates the structure of the cosmic web and hence the shear field emerges as the key factor that governs the local anisotropic pattern of structure formation. Since the small (sub-Mpc) scale is strongly correlated with the mid-range ($\sim$ 10 Mpc) scale - a scale accessible by current surveys of peculiar velocities - it follows that findings presented here open a new window into the relation between the observed large scale structure unveiled by current surveys of peculiar velocities and the preferential infall direction of the Local Group. This may shed light on the unexpected alignments of dwarf galaxies seen in the Local Group.

1. Introduction

Zeldovich (1970) first introduced the concept of anisotropy in to the way we think about structure formation in cosmology. Numerical N-body simulations have been, arguably, the main driving force of research on structure formation (e.g. Springel *et al.* 2006, among others). Even a causal visual inspection of cosmological simulations reveals the anisotropic nature of the growth of structure, and evokes the "pancake" theory of Zeldovich and his school of thoght (Zeldovich *et al.* 1982; Doroshkevich *et al.* 1980). The analytical approach to galaxy formation, namely the cooling and fragmentation of gas in dark matter (DM) halos, relies even more heavily on the top-hat model and hence the assumption of spherical spherical symmetry was integrated into the fabric of the theory of galaxy formation (Rees & Ostriker 1977; White & Rees 1978). These studies envisaged galaxy formation proceeding from the heating of gas accreted onto DM halos to virial temperatures and subsequently cooling and fragmenting to form stars.

The notion of the cosmic web provides a very tempting framework for describing the anisotropic mass assembly of halos and galaxies. Subhaloes shape the the halo they inhabit (Faltenbacher *et al.* 2008; Hoffmann *et al.* 2014) and a number of studies have shown how the orientation of galactic spin is tied to large scale structure. Libeskind *et al.* (2013) found that halo spin aligns itself with the cosmic vortical field, while a number of related studies found weaker alignments with the "cosmic web". These numerical approaches have been complimented by a number of recent observational studies that have found similar trends in redshift surveys (e.g. Tempel *et al.* 2013).

In this work the LSS is defined using the eigenvectors of the velocity-shear field. Such a definition is "democratic" in the sense that each point in space has an equally well-defined

LSS irrespective of other environmental factors such as density. Using this definition we show that the accretion of subhaloes onto host haloes is universally reflective of the shear field.

2. Method

In order to examine the anisotropy of the angular infall pattern of subhaloes crossing the virial sphere of their host haloes, we use a DM-only N-body simulation of 1024^3 particles in a $64h^{-1}$ Mpc box. Such a simulation achieves a mass resolution of $1.89 \times 10^7 h^{-1} M_\odot$ per particle and a spatial softening length of $1\ h^{-1}$kpc. A standard WMAP5 (Komatsu *et al.* 2009) ΛCDM cosmology is assumed: $\Omega_\Lambda = 0.72$, $\Omega_\mathrm{m} = 0.28$, $\sigma_8 = 0.817$ and $H_0 = 70$km/s/Mpc. The publicly available GADGET2 code is used and 190 snapshots (equally spaced in expansion factor) are stored from $z = 20$ to $z = 0$.

The velocity shear field is defined by the symmetric tensor: $\Sigma_{ij} = -\frac{1}{2H(z)}\left(\frac{\partial v_i}{\partial r_j} + \frac{\partial v_j}{\partial r_i}\right)$ where $i, j = x, y, z$. The $H(z)$ normalization is used to make the tensor dimensionless and the minus sign is introduced to make positive eigenvalues correspond to a converging flow. As dictated by convention, the eigenvalues are sorted in increasing order ($\lambda_1 > \lambda_2 > \lambda_3$), and the associated eigenvectors are termed $\mathbf{e}_1$, $\mathbf{e}_2$, and $\mathbf{e}_3$.

In order to compute the shear tensor at each point in the simulation, the velocity field is gridded according to a Clouds-In-Cell (CIC) scheme. This is then smoothed with a gaussian kernel in Fourier space. The size of the CIC used here is 256^3, chosen such that every mesh cell contains at least one particle at $z = 0$ .The width of the gaussian smoothing we apply is adaptive and depends on the mass of the halo we wish to examine (see below).

Following Knollman *et al.* (2008) host halo mass is scaled by $M_\star(z)$, the mass of a typically collapsing object at a given redshift, namely $\widetilde{M} = M_\mathrm{halo}/M_\star$. $M_\star(z)$ is defined by requiring that the variance σ^2, of the linear over-density field within a sphere of radius $R(z) = (3M_\star(z)/4\pi\rho_\mathrm{crit})^{1/3}$, should equal to δ_c^2, the square of the critical density threshold for spherical collapse. $M_\star(z)$ is calculated using the cosmological parameters adopted here: at the present epoch $M_\star = 3.6 \times 10^{12} h^{-1} M_\odot$. At $z = 5$, $M_\star(z = 5) \approx 10^8 h^{-1} M_\odot$.

3. Results

The eigenvectors of the shear tensor, evaluated at the position of each host halo, provides the principal orthonormal vectors within which the anisotropy of mass aggregation onto halos can be naturally examined. Because the eigenvectors are orthonormal, they define an "eigen-frame". Each halo has its own eigen-frame, defined by the ambient shear field. Given that the eigenvectors are non directional lines, this corresponds to a single octant of the 3D cartesian coordinate system. The location of where subhaloes cross the halo virial radius ("entry points") is plotted in this eigen-frame. We stack the accretion events onto all host haloes at all redshifts. Fig. 1 shows the entry points in an Aitoff projection for all accretion events. Fig. 2 shows the entry points in an Aitoff projection for "major mergers" where the mass ratio of the accretion event is greater than 1:10.

In each of these figures we show the entry points in the eigenframe defined by the shear computed with three different smoothings that correspond to 4 (upper left), 8 (upper right) and 16 (bottom) virial radii. Host haloes are divided into four mass bins according to $\widetilde{M}$. Starting at "noon" and going clockwise, these are where $\widetilde{M} < 0.1$; $0.1 < \widetilde{M} < 1$; $1 < \widetilde{M} < 10$; and $10 < \widetilde{M}$. In order to quantify the statistical significance

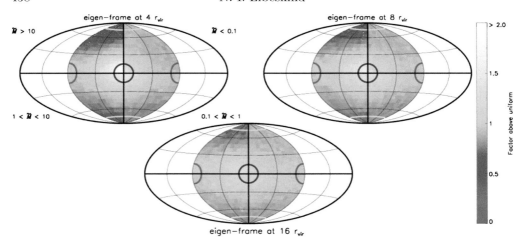

Figure 1. The location of subhalo entry points is shown in an Aitoff projection of the virial sphere. The density of subhalo entry points is shown for eigen-frames smoothed on 4 (upper left), 8 (upper right) and 16 (bottom) virial radii. Starting from "noon" and going clockwise, we show these entry points for accretion events occurring on to host haloes in four different mass ranges $\widetilde{M} < 0.1; 0.1 < \widetilde{M} < 1; 1 < \widetilde{M} < 10$ and $\widetilde{M} > 10$ and at all redshifts below $z \sim 5$. $\widetilde{M}$ is a measure of the halo mass in units of the mass of a collapsing object at each redshift. The density of entry points is normalized to that expected from a uniform distribution, and contoured accordingly. The "north" and "south" pole correspond to $\mathbf{e}_1$; the two mid points on the horizontal axis at $\pm 180°$ to correspond to $\mathbf{e}_2$, while the midpoint corresponds to $\mathbf{e}_1$. The yellow, red and blue circles define areas within 15 degrees of the eigen-frame axes, $\mathbf{e}_1$, $\mathbf{e}_2$, and $\mathbf{e}_3$, respectively.

The distribution of entry points is never consistent with uniform. Instead it universally (irrespective of host halo mass, scale on which the shear is computed or redshift) peaks close to $\mathbf{e}_3$: on large scales the shear tensor dictates the shape of cosmic web and on small scales it determines the infall pattern of satellites

of any anisotropy in the angular entry-point distribution, we divide the number of entry points in a given area on the virial sphere by that expected from a uniform distribution. The variance due to Poisson statistics of a uniform distribution of the same number of points, is small.

There is a strong tendency for the accretion to occur along $\mathbf{e}_3$. Regardless of the host halo mass, the merger ratio or the smoothing used, there is a statistically significant tendency for subhaloes to be accreted closer to $\mathbf{e}_3$ than to either other of the eigenvectors. Recall that $\mathbf{e}_3$ corresponds to the direction of slowest collapse. This is the main result of this paper: *subhaloes are preferentially accreted along the direction that corresponds to slowest collapse.* Note that this effect is greatest for the most massive host haloes and becomes progressively weaker as halo mass decreases. Also, as the gaussian smoothing kernel is increased the effect also weakens. This is expected: large smoothing kernels effectively homogenize the LSS, randomizing the principal direction of the shear tensor. Finally the tendency to be accreted along $\mathbf{e}_3$ is largest for "massive" subhaloes that are greater than 10% of their hosts. In the "best" case, where the smoothing is confined to $4r_{\mathrm{vir}}$, where only the most massive host haloes and the greatest merger events are considered, the mergers are more than 4 times as likely to come along $\mathbf{e}_3$, than expected from a uniform distribution.

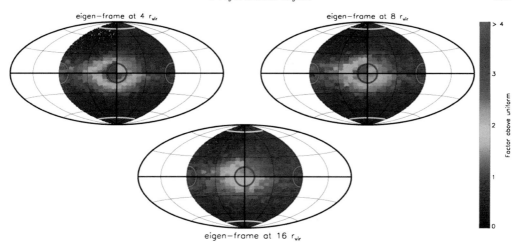

Figure 2. Same as Fig. 1 but only considering subhaloes whose mass is greater than 10% of their host.

By stacking our results in the manner shown, we have explicitly omitted any dependence of subhalo accretion on redshift or absolute halo mass. Below we examine how the funneling of accretion events changes with redshift and host halo mass.

In Fig. 3 we show the probability distribution of the cosine of the angle formed between the subhalo entry point ($\mathbf{r}_{\mathrm{acc}}$) and the eigenvectors of the shear $\mathbf{e}_1$ (black), $\mathbf{e}_2$ (green) and $\mathbf{e}_3$ (magenta), namely $\cos\theta = \mathbf{r}_{\mathrm{acc}} \cdot \mathbf{e}_i$, where $i = 1, 2, 3$. In what follows only this angle is considered. The probability distributions in Fig. 3 are valid for all redshifts but are split by mass (first column: $\widetilde{M} < 0.1$; second column: $0.1 < \widetilde{M} < 1$; third column: $1 < \widetilde{M} < 10$ and fourth column $\widetilde{M} > 10$). Additionally the eigenvectors are computed on three scales corresponding to smoothing of 4 (top), 8 (middle) and 16 (bottom row) times the halo's virial radius. Uniform distributions would be represented by a solid flat line at unity.

4. Summary and Discussion

It has long been realized that DM halos grow in an anisotropic fashion. This has often been claimed to be related to the cosmic web which constitutes the scaffolding for the building of halos and the galaxies within them. The accretion of matter onto halos generally proceeds in two modes: the accretion of clumps and the smooth accretion of diffuse material. In this paper, we have focused on the mergers of small halos with massive ones and analyzed their anisotropic infall pattern with respect to the eigen-frame of each individual host halo. The eigen-frame is defined by the three eigenvectors of the velocity shear tensor evaluated at the position of each halo. Our main finding is that, across a range of halo masses and redshifts, the infall direction of subhaloes is preferentially confined to the plane orthogonal to $\mathbf{e}_1$ (the direction of fastest collapse), within which it is aligned with $\mathbf{e}_3$ (the axis of slowest collapse).

Some of the main characteristics of the infall of subhalos onto more massive halos are listed here:

• Subhaloes tend to be accreted onto hosts from a specific direction with respect to the large scale structure.

• In the case of filaments, the $\mathbf{e}_3$ direction coincides with the "spine" of the filament (Tempel *et al.* 2014). Hence, the well known phenomenon of halos being fed by substructures funneled by filaments is recovered here.

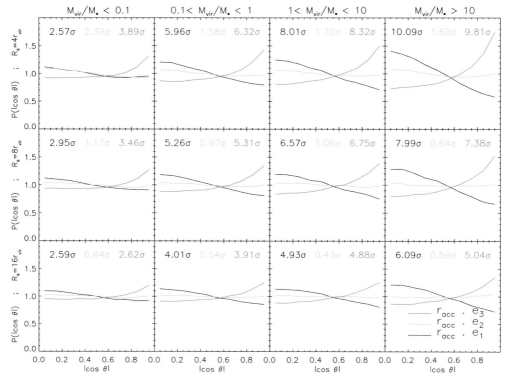

Figure 3. The anisotropic accretion shown in Fig. 1 is quantified by means of a probability distribution, $P(|\cos\theta|)$ of the cosine of the angle made between a subhalo's entry point ($\mathbf{r}_{\rm acc}$) and the eigenvectors $\mathbf{e}_1$ (black), $\mathbf{e}_2$ (green) and $\mathbf{e}_3$ (magenta). The top, middle and bottom rows show the probability distribution when the shear has been smoothed on 4, 8 and $16 r_{\rm vir}$. The probability distributions are split according to value of $\widetilde{M}$, denoted on top of each column. The statistical significance of each probability distribution is characterized by the average offset between it and a random distribution in units of the Poisson error and is indicated by the corresponding colored number in each panel. Distributions that are consistent with random have values $< 1\sigma$.

- The strength of the beaming effect depends somewhat on the length scale used to compute the velocity-shear eigenframe: the smaller the scale, the stronger the beaming.
- More massive subhaloes are more anisotropically accreted onto host haloes than smaller subhaloes.
- Similarly, more massive host haloes accrete subhaloes more anisotropically than smaller ones.
- Accretion at high redshift is more anisotropic than accretion at low redshift, for all masses of hosts and subhaloes.

A somewhat naive reasoning might suggest that halos should be nourished along $\mathbf{e}_1$, the axis of the fastest collapse. However, this reasoning is flawed: halos grows by accreting material from their surrounding, and this occurs most rapidly in the direction of $\mathbf{e}_1$. It follows that at the time a given halo is inspected, the mere existence of that halo implies that much of the surrounding material has already been consumed by the halo along $\mathbf{e}_1$. This leaves the material along $\mathbf{e}_3$ as the main supply of fresh material that feeds the halo.

The beaming of subhalo accretion onto halos in a given bin of $\widetilde{M}$ (namely M_{vir} scaled by the redshift dependent $M_\star(z)$), narrows with increasing redshift (Fig. 4). In a scale-free universe, i.e. an Einstein-de Sitter cosmology with a power law power spectrum, the angular dependence of the accretion is expected to be completely epoch independent. This does not hold for the ΛCDM cosmology assumed here. As the universe evolves it becomes more dominated by the Λ term and consequently the role of gravity (via subhalo dynamics) diminishes with time. This is manifested in the accretion and merger rate: accretion onto halos decreases and the funneling of matter along $\mathbf{e}_3$ gets weaker.

Arguably, the most important ramification of this paper is that anisotropic nature of the mass growth of halos is dictated by the velocity shear tensor and not by cosmic web (Hoffman *et al.* 2012). That is to say, the anisotropic nature of subhalo accretion *does not* depend on the magnitude of the shear tensor's eigenvalues, nor does it depend on the "web environment". The beaming of subhaloes along $\mathbf{e}_3$ occurs equally in knots, filaments, sheets and voids. Rather, the shear tensor is the one that characterizes, shapes and dictates the directions of the cosmic web. This provides further support to earlier claims regarding the dominance of the shear tensor in shaping the large scale structure (Libeskind *et al.* 2012, 2013)

Libeskind, Hoffman & Gottlöber (2014) have recently shown that the principal directions of the shear tensor remain coherent over a wide range of redshifts and spatial scales. This opens interesting possibilities for relating the observed large scale velocity field with the properties of halos, and hence of galaxies and groups of galaxies. The work presented here, combined with observations of the local velocity field, will allow us to thus identify the direction along which most accretion onto the Local Group occurred. Such findings can have important implications on the peculiar geometric set up of dwarf galaxies in the local group.

References

Doroshkevich, A. G., Kotok, E. V., Poliudov, A. N., Shan- darin, S. F., Sigov, I. S., & Novikov, I. D., 1980, *MNRAS*, 192, 321

Faltenbacher, A., Jing, Y. P., Li, C., Mao, S., Mo, H. J., Pasquali, A., & van den Bosch, F. C., 2008, *ApJ*, 675, 146

Hoffmann, K., *et al.* 2014, ArXiv e-prints

Rees, M. J. & Ostriker, J. P., 1977, *MNRAS*, 179, 541

Springel, V., Frenk, C. S., & White, S. D. M., 2006, *Nature,*

White, S. D. M. & Rees, M. J., 1978, *MNRAS*, 183, 341

Zeldovich, Y. B., 1970, *A&A*, 5, 84 440, 1137

Zeldovich, I. B., Einasto, J., & Shandarin, S. F., 1982, *Nature,* 300, 407

Libeskind, N. I., Hoffman, Y., Forero-Romero, J., Gottlöber, S., Knebe, A., Steinmetz, M., & Klypin, A., 2013, *MNRAS,* 428, 2489

Tempel, E., Libeskind, N. I., Hoffman, Y., Liivamägi, L. J., & Tamm, A., 2014, *MNRAS*, 437, L11

Komatsu, E., Dunkley, J., Nolta, M. R., Bennett, C. L., Gold, B., Hinshaw, G., et. al. 2009, *ApJS*, 180, 330

Knollmann, S. R., Power, C., & Knebe, A., 2008, *MNRAS*, 385, 545

Libeskind, N. I., Hoffman, Y., Knebe, A., Steinmetz, M., Gottlöber, S., Metuki, O., & Yepes, G., 2012, *MNRAS*, 421, L137

Libeskind, N. I., Hoffman, Y., Steinmetz, M., Gottlöber, S., Knebe, A., & Hess, S., 2013, *ApJL*, 766, L15

Libeskind, N. I., Hoffman, Y., & Gottlöber, S., 2014,*MNRAS,* 441, 1974

Hoffman, Y., Metuki, O., Yepes, G., Gottlöber, S., Forero-Romero, J. E., Libeskind, N. I., & Knebe, A., 2012, *MNRAS*, 425, 2049

CHAPTER 7C.

Galaxy Formation & Evolution:
Poster papers

The Zeldovich Universe:
Genesis and Growth of the Cosmic Web
Proceedings IAU Symposium No. 308, 2014
R. van de Weygaert, S. Shandarin, E. Saar & J. Einasto, eds.
© International Astronomical Union 2016
doi:10.1017/S1743921316010371

The morphological types of galaxies in the Local Supercluster

K. Bajan[1] P. Flin[2] and W. Godłowski[3]

[1] Mt. Suhora Observatory, Cracow Pedagogical University, Krakow, Poland,
email: kbajan@up.krakow.pl

[2] Institute of Physics, Jan Kochanowski University, Kielce, Poland
email: sfflin@cyf-kr.edu.pl

[3] Institute of Physics, Opole University, Opole, Poland
email: godlowski@uni.opole.pl

Abstract. On the basis of the Hyper – Leda Catalogue (HyperLeda) 8293 galaxies with helio-centric radial velocities below 2500 km s^{-1} were selected; 4570 had known morphological types (4366 had calculated b/a ratio). We checked the frequency of the distribution of various types in the LSC, finding spirals and irregulars most numerous, in accordance with expectations. The axial ratio of galaxy diameters of various types was studied, and the dependence of this parameter on the morphological type was noted.

1. Introduction

Galaxy parameters are used in many astrophysical investigations. Usually, it is accepted that in any larger structure the value of the parameters for all galaxies are equal and mean values are adopted. It is interesting to look in more detail at galaxies in such a structure. In this paper we are trying to look for morphological types of galaxies in the Local Supercluster (LSC), as well as the dependence of the galaxy inclination angle on the morphological type. Assuming an oblate galaxy, the value of the inclination angle i (that is, the angle between the galactic plane and line of sight) is usually determined from the formula (Holmberg 1946):

$$\cos(i) = \left[\frac{\left(\frac{b}{a}\right)^2 - \left(\frac{b_0}{a_0}\right)^2}{1 - \left(\frac{b_0}{a_0}\right)^2} \right]^{-1} . \tag{1.1}$$

where a, b are the observed lengths of the major and minor axes of the galaxy image, while a_0 and b_0 are the theoretical values of these parameters. The ratio $q_0 = b_0/a_0$ depends on the morphological type of the galaxy (Heidmann *et al.* 1972). (The value $q_0 = 0.2$ is usually adopted when the morphological type is unknown.) Current huge databases permit the estimation of this relation based on a much greater sample than previously.

2. Observational data

The Hyper – Leda Catalogue of galaxies was used as our observational basis. From the Catalogue, we extracted galaxies with heliocentric radial velocities $V_r < 2500$ km s^{-1}. We obtained the sample of 8293 galaxies which are presumed to be members of the LSC. The Catalogue listed morphological types for 4570 of these galaxies. This set formed the basic sample for our studies.

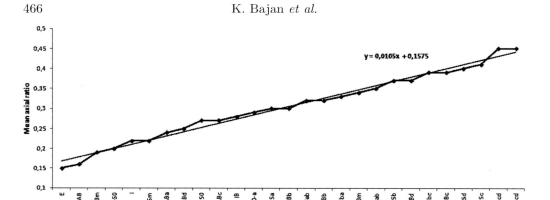

Figure 1. The dependence of mean axial ratio on morphological types, with a fitted line.

3. Results

In the huge volume of the LSC analysed by us, the majority of galaxies are spirals and irregulars. Ellipticals and lenticulars constitute less than 5 % of the population. This finding is in agreement with Dressler's relation (Dressler 1980) that ellipticals are more frequent in the dense region of a structure, while spirals dominate in less populated regions.

We studied the distribution of the galaxy axial ratio for each morphological type in a sample containing only galaxies of known morphological types and given b/a ratios (4366 objects). We considered the value of $\log R_{25}$ (Paturel *et al.* 1991), where R_{25} is the photometric axial ratio of the major to minor diameters (a/b) of the galaxy image at the 25 mag arcsec^{-2} isophote. The existence of a dependence of axial ratio on morphological types is clear: fig. 1 presents the dependence of the mean axial ratio on morphological types with fitted line.

4. Conclusions

Our investigations are based on a large sample of galaxies.
- We confirm the Dressler relation that, in a structure, the galaxy morphological type depends on the space density of galaxies.
- The axial ratio $\log R_{25}$ is function of morphological type. It is greater for late types than early types.
- Spirals with well-established type have the value of this parameter greater than suspected spirals. This is probably due to erroneous identification of these objects as spirals.
- The real ratio q_0 used for calculation of the inclination angle i also depends on the morphological type. This relation is rather weak within 1σ.

References

Hyper Leda *http://leda.univ-lyon1.fr*
Dressler, A. 1980, *ApJ* 236, 351
Heidmann, J. & Heidmann, N., & de Vaucouleurs. G. 1972, *Mem. R. astr. Soc* 75, 85
Holmberg, E. 1996, *Medd. Lund. Obs.* 11, 117
Paturel, G., Fouque, P., Buta, R., & Garcia,A. M. 1991, *A& A*, 243, 319

The Zeldovich Universe:
Genesis and Growth of the Cosmic Web
Proceedings IAU Symposium No. 308, 2014
R. van de Weygaert, S. Shandarin, E. Saar & J. Einasto, eds.

© International Astronomical Union 2016
doi:10.1017/S1743921316010383

Structural decomposition of galaxies in the CALIFA survey

Teet Kuutma[1,2], Antti Tamm[1] and Elmo Tempel[1]

[1]Tartu Observatory, Observatooriumi 1, 61602 Tõravere, Estonia

[2]Institute of Physics, University of Tartu, Riia 142, 51014 Tartu, Estonia

Abstract. Several clues for understanding the nature and evolution of galaxies can be gained by studying galactic structures and their evolution with time and environment. However, even for nearby galaxies, detailed structural decomposition is not a straightforward task. Choosing the number of structural components and the limits placed on their parameters can have a large effect on the derived characteristics of galaxies. For distant galaxies, structural analysis is further hampered by the spatial resolution limits of the imaging. However, by using a relatively robust two-component bulge+disk modelling, galaxies in the nearby Universe can be compared to distant galaxies for tracing signs of evolution in the extracted structures. We start such a study by analysing first a well observed nearby sample of galaxies, using ∼600 targets from the CALIFA survey. We show that even in this small sample of nearby galaxies, the effects of environmental density are already well apparent.

Keywords. galaxies: photometry, galaxies: structure, galaxies: spiral, galaxies: elliptical

1. Data and methodology

We use the Sloan Digital Sky Survey (SDSS; York *et al.* 2000) Data Release 10 images of galaxies, which are selected based on the Calar Alto Legacy Integral Field Area Survey (CALIFA) sample, described by Sanchez *et al.* (2012). For the CALIFA survey, targets have been selected to have an apparent diameter which fits within the FoV of the integral field spectrophotometer on the Calar Alto 3.5 m telescope. An additional selection criterion in redshift range of 0.005 to 0.03 is made. Thus the CALIFA selection criteria create a comprehensive sample of nearby Universe galaxies. The available kinematical data can later be used to verify the meaningfulness of the decomposition.

Structural decomposition was performed on post-stamp cutouts from the corrected SDSS DR10 frames. Galfit (Peng *et al.* 2010) was used to fit galaxy models with two Sersic components. Initial guess parameters for the fitting were derived with Source-Extractor (Bertin & Arnouts 1996).

2. Results

The galaxy sample is characterised with the *g-i* colour-magnitude diagramme on Fig. 1. Spiral and elliptical morphologies, as determined within the Galaxy Zoo project (Lintott *et al.* 2008), are shown. As expected, elliptical galaxies are generally redder, while spirals have a much broader colour span.

The mean values of the main structural parameters of the extracted bulges and discs are given in Table 1. In the table, we compare two subsamples: isolated galaxies vs rich groups (with 10 or more members); the group membership has been taken from the galaxy group catalogue Tempel *et al.* (2014). Both bins contain 44 galaxies. Even for these small subsamples, several expected differences between sparser and denser environments are

	Isolated	Rich groups
B/T$_i$	0.34(±24)	0.43(±26)
N$_{Bulge}$	2.5(±1.3)	3.2(±1.2)
r$_{bulge}$ (kpc)	1.4(±1.0)	1.9(±1.3)
N$_{disk}$	0.6(±0.4)	0.7(±0.4)
r$_{disk}$ (kpc)	4.5(±2.0)	6.0(±3.2)
L$_r$/L$_{r,Sol}$ (10^{10})	1.0(±0.7)	2.2(±1.7)
g-i	1.0	1.2

Table 1. Mean values of some structural characteristics of galaxies in different environments; the uncertainties are shown in brackets. From top to bottom: ratio of bulge to total luminosity in the i band, bulge Sersic index, bulge effective radius in kpc, disk Sersic index, disk effective radius in kpc, total r band luminosity in 10^{10} Solar values, the median of g-i colours.

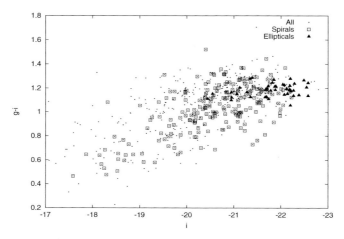

Figure 1. Color-magnitude diagramme of the galaxy sample. The complete sample is shown with dots, morphological determination from Galaxy Zoo is given with an empty box around a dot for spirals and filled triangles for ellipticals.

apparent: galaxies tend to become more luminous, redder, and with higher bulge-to-total ratios in denser environments. Interestingly, the Sersic indices and effective radii increase simultaneously with environmental density, indicating that stellar matter in galaxies becomes less concentrated in cluster environments. This effect is more prominent for bulges; we speculate that it is a natural result of galaxy-galaxy interactions.

The presented results show that even a simple bulge+disc decomposition can be a powerful tool for analysing the structure and evolution of galaxies.

References

Bertin, E. & Arnouts, S. 1996, *A&AS*, 317, 393
Lintott, C. J., Schawinski, K., Slosar, A., *et al.* 2008, *MNRAS*, 389, 1179
Peng, C. Y., Ho, L. C., Impey, C. D., & Rix, H.-W. 2010, *AJ*, 139, 2097
Sanchez, S. F., Kennicutt, R. C., Gil de Paz, A., *et al.* 2012, *A&A*, 538, A8
Tempel, E., Tamm, A., Gramann, M., *et al.* 2014, *A&A*, 566, A1
York, D. G., Adelman, J., Anderson, Jr., J. E., *et al.* 2000, *AJ*, 120, 1579

The Zeldovich Universe:
Genesis and Growth of the Cosmic Web
Proceedings IAU Symposium No. 308, 2014
R. van de Weygaert, S. Shandarin, E. Saar & J. Einasto, eds.
© International Astronomical Union 2016
doi:10.1017/S1743921316010395

Gaseous discs at intermediate redshifts from kinematic data modelling

R. Kipper[1,2], A. Tamm[1], P. Tenjes[1,2] and E. Tempel[1]

[1]Tartu Observatory, Observatooriumi 1, 61602 Tõravere, Estonia

[2]Institute of Physics, University of Tartu, 51010 Tartu, Estonia

Abstract. Our purpose is to measure thickness of gaseous discs in $0 < z < 1.2$ galaxies. As gas dispersions are sensitive to scale height of gaseous discs, we model the kinematics of galaxies using Jeans equations. The resulting thicknesses of gaseous discs at higher redshifts are more thicker (and arbitrary) while nearby ones are thinner. We also found that clumpiness of galaxy is a possible indicator of the gas disc thickness.

Keywords. galaxies: bulges - galaxies: kinematics and dynamics - galaxies: structure

1. Introduction and methods

Thickness of gaseous discs of galaxies is important to study in order to narrow down secular evolution and minor mergers: if we know how thin is gaseous disc, we know the initial thickness of stellar disc. Comparing it with observed thickness of disc it is possible to estimate the effect of secular evolution and possibly minor mergers.

We model the kinematics, measured by Weiner *et al.* (2006), of galaxies from GOODS-N field, by estimating density distribution and using Jeans equations we calculate observable kinematics. Density estimation is based on superposition of four components: stellar disc, bulge, gaseous disc and dark matter halo. The distribution of stellar component comes from photometric modelling while its mass comes from spectral energy distribution fitting. The acquired mass is the sum of bulge and disc mass: the division between them was done manually. We assume that most of gaseous disc parameters are proportional to stellar ones, except mass, which comes from spectral energy distribution fitting and thickness, which is our aim to study and therefore free parameter.

2. Results

An example of our modelling results can be found in Fig. 1. Thickness of gaseous disc is difficult to measure, especially when galaxy is not edge-on. To overcome this difficulty we tried to find more simple estimate for that and our best finding is correlation with galaxy clumpiness. We calculated clumpiness using similar method by Conselice *et al.* (2003) where they subtracted blurred and observed image, summed the result and normalised with galaxy luminosity. Main difference compared with their approach is subtracting clumpiness of model image from observed one. Fig. 2 shows that this can only be rough estimator.

We also see from Fig. 2 that the scale height of gaseous disc is small in nearby galaxies while it has more arbitrary and bigger thicknesses at higher redshifts. This trend is compatible with nearby galaxies where thickness is estimated from dust distribution by Xilouris *et al.* (1999) and Bianchi (2007). Thinning of discs can be explained by "settling down" of gas and having fewer interactions between galaxies.

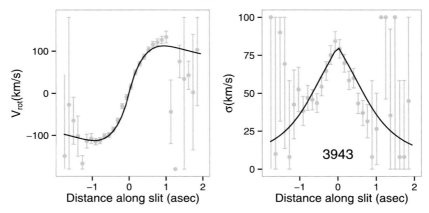

Figure 1. An example of kinematic modelling for galaxy TKRS 3943. Left panel shows rotation curve and right velocity dispersion distribution.

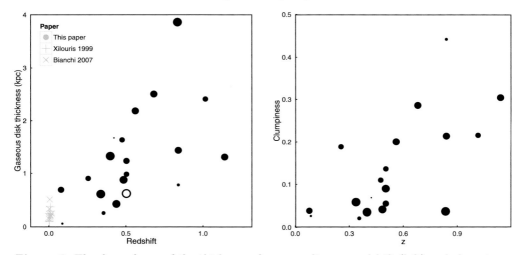

Figure 2. The dependence of the thickness of gaseous discs on redshift (left) and clumpiness (right). The size of points indicate the quality of our modelling.

References

Weiner, B. J., Willmer, C. N. A., Faber, S. M., Melbourne, J., Kassin, S. A., Phillips, A. C., Harker, J., Metevier, A. J., Bogt, N. P., & Koo, D. C. 2006, *A&A*, 653, 1027

Conselice, C. J. 2003, *ApJS*, 147, 1

Bianchi, S. 2007, *A&A*, 471, 765

Xilouris, E. M., Byun, Y. I., Kylafis, N. D., Paleologou, E. V., & Papamastorakis, J. 1999, *A&A*, 344, 868

The Zeldovich Universe:
Genesis and Growth of the Cosmic Web
Proceedings IAU Symposium No. 308, 2014
R. van de Weygaert, S. Shandarin, E. Saar & J. Einasto, eds.

© International Astronomical Union 2016
doi:10.1017/S1743921316010401

The effect of environment on the fundamental plane of elliptical galaxies

R. Kipper[1,2], A. Tamm[1], P. Tenjes[1,2] and E. Tempel[1]

[1]Tartu Observatory, Observatooriumi 1, 61602 Tõravere, Estonia
[2]Institute of Physics, University of Tartu, 51010 Tartu, Estonia

Abstract. We study the effect of environment to fundamental relation of elliptical galaxies. We find that superclusters, filaments and groups give noticeable effect to slope of velocity dispersions while little to luminosity slope.

Keywords. galaxies: elliptical and lenticular, cD - galaxies: clusters: general - galaxies: fundamental parameters

1. Introduction and method

Fundamental plane (FP) is a plane that ties independent parameters of elliptical galaxies: surface brightness inside half-light radius (μ), half light radius (R) and central velocity dispersions (σ):

$$R \sim \sigma^a \mu^b. \tag{1.1}$$

The parameters of FP can be derived from virial theorem, from observations and from simulations. Each of these methods give slightly different values. Taranu *et al.* (2014) simulated spiral galaxy mergers and found $a = 1.7$ and $b = 0.3$. Virial theorem gives $a = 2$ and $b = 0.4$. Observations indicate that a and b values are 1...1.5, 0.25...0.35 respectively depending on observed filter, radius definition and fiber size.

We study FP dependence on three environment classes: superclusters (as large scale luminosity density field), filaments and groups/clusters. Supercluster environment is derived by kernel density distribution (see appendix of Tempel *et al.*2014b). Data is smoothed with kernel size 8 Mpc/h with the idea it is being close to the average size of supercluster. Filaments are derived using Bisous point process method. This method is implemented on SDSS galaxy structure by Tempel *et al.*(2014a). The volume limited group environment dependence is found with friend of friends method in Tempel *et al.*(2014b). We cleaned the sample by selecting only galaxies that have high probability of being elliptical and can be described by one component.

All the input data can be downloaded from Tartu Observatory cosmology department database: http://cosmodb.to.ee.

2. Results

From the Fig. 1 can be seen that the effect of different scale environment (group, filament and large scale density) on the fundamental plane. Dependence is more clear when looking at a parameter, the velocity dispersion slope. In figure is also seen that the effect is similar, therefore it is possible that they describe hierarchical effect: dispersions slope is higher in larger groups and groups are mostly in supercluster or filament regions.

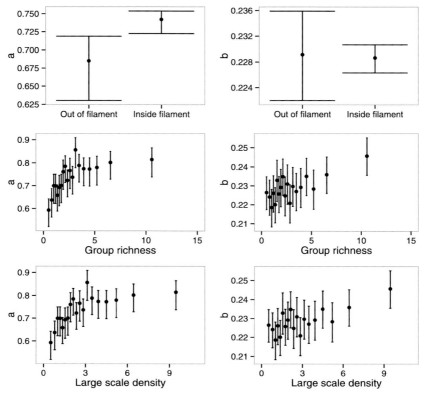

Figure 1. Fundamental plane parameters in filament environment (top panel), group environment (middle panel) and supercluster environment (bottom panel). The errors are 95% confidence intervals.

References

Taranu, D. S., Dubinski, J., & Yee, H. K. C. 2014, arXiv:1406.2693

Tempel, E., Stoica, R. S., Martnez, V. J., Liivamägi, L. J., Castellan, G., & Saar, E. 2014a, *MNRAS*, 438, 3465

Tempel, E., Tamm, A., Gramann, M., Tuvikene, T., Liivamägi, L. J., Suhhonenko, I., Kipper, R., Einasto, M., & Saar, E. 2014b, *A&A*, 566, 16

The Zeldovich Universe:
Genesis and Growth of the Cosmic Web
Proceedings IAU Symposium No. 308, 2014
R. van de Weygaert, S. Shandarin, E. Saar & J. Einasto, eds.

© International Astronomical Union 2016
doi:10.1017/S1743921316010413

Kinematics, structure and environment of three dwarf spheroidal galaxies

M. E. Sharina[1], I. D. Karachentsev[1] and V. E. Karachentseva[2]

[1]Special Astrophysical Observatory, Russian Academy of Sciences, N.Arkhyz, KChR, 369167, Russia
email: sme@sao.ru, ikar@sao.ru

[2]Main Astronomical Observatory, National Academy of Sciences of Ukraine, 27 Akademika Zabolotnoho St., 03680 Kyiv, Ukraine
email: valkarach@gmail.com

Abstract. We explore the environmental status of three low surface brightness dwarf spheroidal galaxies (dSphs) KKH65, KK180 and KK227 using the results of our long slit spectroscopic observations at the 6m telescope of the Russian Academy of Sciences and surface photometry on the Sloan Digital Sky Survey (SDSS) images. The objects were selected by Karachentseva in 2010 as presumably isolated galaxies. The obtained surface brightness profiles demonstrate that our sample dSphs are less centrally concentrated than the objects of the same morphological type in the Virgo cluster (VC). Using the derived kinematic data we searched for possible neighbours of the dSphs within the projected distances from them $R_{proj} < 500$ kpc and with the differences in radial velocities $|\Delta V| < 500$ kms^{-1}. We applied the group finding algorithm by Makarov and Karachentsev to the selected sample. Our analysis shows that the dwarf galaxies of our study are not isolated. KKH65 and KK227 belong to the groups NGC3414 and NGC5371, respectively. KK180 is in the VC infall region. We conclude that it is not possible at the moment to justify the existence of isolated dSphs outside the Local Volume. The searches are complicated due to the lack of the accurate distances to the galaxies farther than 10 Mpc.

Keywords. galaxies: dwarf, galaxies: kinematics and dynamics, galaxies: fundamental parameters

1. Introduction

In the course of systematic spectroscopic and photometric studies of dwarf galaxies in the Local Universe we examine the properties of three objects from the list of 10 candidate isolated early-type dwarf galaxies by Karachentseva *et al.* (2010).

Very few isolated dSphs were found up to date (Karachentseva *et al.* 1999, Karachentsev *et al.* 2001, Makarov *et al.* 2012). It has been proved observationally that most dSphs are located within ∼2 virial radii from a massive neighbour (Karachentsev *et al.* 2005). The evolution of dwarf galaxies is influenced by the environmental factors, and by internal starbursts (e.g. Grebel 2005, Kormendy & Bender 2012 and references therein).

The origin of isolated dSphs is still not fully understood. Ricotti & Gnedin (2005) suggested that these objects may originate in low-mass halos ($< 2 \cdot 10^8 M_\odot$) before the reionization epoch.

2. Results and conclusions

Our spectroscopic observations were carried out with the SCORPIO spectrograph (Afanasiev & Moiseev 2005) equipped with the CCD detector EEV 42-40 and the grism VPHG1200B. The reduction of the photometric and spectroscopic data was performed using the MIDAS (Banse *et al.* 1983) and IRAF (Tody 1993) software systems. The radial velocities were derived using the ULySS program (Koleva *et al.* 2008, 2009).

Photometry on SDSS images and fitting of the surface brightness profiles showed that $g - r$, $r - i$ colours of the three galaxies are typical for dSphs (Sharina *et al.* 2008, 2013 and references therein). The Sersic indices are in the range $0.9 \div 1.1$. Our sample objects are less centrally concentrated than dSphs in the VC (Kormendy *et al.* 2009).

We used the derived radial velocities and integrated magnitudes to identify possible neighbours of KKH65, KK180 and KK227. First, we searched for galaxies in the SDSS and LEDA databases within the projected radii $R_{proj} < 500$ kpc around dSphs and with the differences in radial velocities $|\Delta V| < 500$ kms^{-1}. Then we applied the group finding algorithm by Makarov & Karachentsev (2011) to the selected sample. We considered the studied galaxies to be located at their Hubble distances.

It appears that KKH65 ($V_{LG} \sim 1300$ kms^{-1}, $V_t = 16.9$) is one of the members of the NGC3414 group. The projected separation of KKH65 from the massive lenticular galaxy is ~ 130 kpc. KK180 ($V_{LG} \sim 609$ kms^{-1}, $V_t = 16.1$) has no close bright massive neighbours within $R_{proj} < 500$ kpc and with the differences in radial velocities $|\Delta V| < 500$ kms^{-1}. However, the VC is so massive that KK180 is surely under its gravitational influence. KK227 ($V_{LG} \sim 1900$ kms^{-1}, $V_t = 17.1$) belongs to the group of ~ 50 galaxies around NGC5371.

The exact environmental status of the three dSphs will be established when the accurate distances are known for them and for their neighbouring galaxies.

Acknowledgements. This work was performed under support of the Russian Science Foundation grant N 14-12-00965. IDK thanks a grant RFBR 13-02-90407 Ukr f a. VEK acknowledges a Ukrainian-Russian grant F 53.2/015. We are grateful to Dr. S.N. Dodonov for the technical supervision of our observations. We acknowledge the usage of the HyperLeda (http://leda.univ-lyon1.fr) and the SDSS (http://www.sdss.org) databases.

References

Afanasiev, V. L. & Moiseev, A. V. 2005, *Astron. Lett.*, 31, 194

Banse, K., Crane, P., Grosbol, P., Middleburg, F., Ounnas, C., Ponz, D., & Waldthausen, H. 1983, *The Messenger*, 31, 26

Grebel, E. K. 2005 in Jerjen H., Binggeli B. (eds.), *Near-fields cosmology with dwarf elliptical galaxies*, Proc. IAU Colloq. 198 (Cambridge Univ. Press), p. 1

Karachentsev, I. D., Karachentseva, V. E., & Sharina, M. E. 2005, in Jerjen H., Binggeli B. (eds.), *Near-fields cosmology with dwarf elliptical galaxies*, Proc. IAU Colloq. 198 (Cambridge Univ. Press), p. 295

Karachentsev, I. D., *et al.* 2001, *A&A*, 379, 407

Karachentseva, V. E., Karachentsev, I. D., & Sharina, M. E. 2010, *Astrophysics*, 53, 462

Karachentseva, V. E., Karachentsev, I. D., & Richter, G. M. 1999, *A&AS*, 135, 221

Koleva, M., Prugniel, P., Bouchard, A., & Wu, Y. 2009, *A&A*, 501, 1269

Koleva, M., Prugniel, P., Ocvirk, P., Le Borgne, D., & Soubiran, C. 2008, *MNRAS*, 385, 1998

Kormendy, J. & Bender, R. 2012, *ApJS*, 198, 2

Kormendy, J., Fisher, D. B., Cornell, M. E., & Bender, R. 2009, *ApJS*, 182, 216

Makarov, D., Makarova, L., Sharina, M., Uklein, R., Tikhonov, A., Guhathakurta, P., Kirby, E., & Terekhova, N. 2012, *MNRAS*, 425, 709

Makarov, D. I. & Karachentsev, I. D. 2011, *MNRAS*, 412, 2498

Ricotti, M. & Gnedin, N. Y. 2005, *ApJ*, 629, 259

Tody, D. 1993, *ASP-CS*, 52, 173

Sharina, M. E., Karachentseva, V. E., & Makarov, D. I. 2013, *Advancing the Physics of Cosmic Distances*, Proc. IAU Symp.289 (Cambridge Univ. Press), p. 236

Sharina, *et al.* 2008, *MNRAS*, 384, 1544

The Zeldovich Universe:
Genesis and Growth of the Cosmic Web
Proceedings IAU Symposium No. 308, 2014
R. van de Weygaert, S. Shandarin, E. Saar & J. Einasto, eds.
© International Astronomical Union 2016
doi:10.1017/S1743921316010425

Properties of satellite galaxies
in nearby groups

Jaan Vennik

Tartu Observatory, 61602 Tõravere, Tartumaa, Estonia
email: `vennik@to.ee`

Abstract. We studied the variation of stellar mass and various star-formation characteristics of satellite galaxies in a volume limited sample of nearby groups as a function of their group-centric distance and of their relative line-of-sight velocity in the group rest frame. We found clear radial dependencies, e.g. massive, red and passive satellites being distributed predominantly near the center of composite group. We also found some evidence of velocity modulation of star-forming properties of satellite galaxies near the group virial radius. We conclude that using kinematical data, it should be feasible to separate dynamical classes of bound, in-falling and 'backsplash' satellite galaxies.

Keywords. galaxies: clusters: general, galaxies: statistics, galaxies: evolution

1. Introduction

Satellite galaxies are proper test bodies when studying the physical mechanisms that drive the galaxy evolution in groups and clusters. The star-forming in-falling galaxies are expected to have distinctive properties compared to the passively evolving partly virialised population near the cluster/group center. An intermediate population of 'backsplash' galaxies may have lost a significant fraction of their mass and velocity by crossing the central region and hence they may contribute to the radial velocity segregation in galaxy properties on the outskirts of galaxy groups and clusters. Earlier studies have found statistically significant evidence of velocity segregation of galaxy properties in clusters (e.g. Mahajan *et al.* 2011). According to the group pre-processing scenario of cluster galaxies, similar processes can take place also in groups of galaxies.

2. Data and analysis

Our target groups are extracted from the parent catalogue of groups and clusters of galaxies, based on the spectroscopic sample of the SDSS DR10 (Tempel *et al.* 2014). We selected a sample of 2112 medium richness ($5 \leqslant n_{gal} < 50$), relatively nearby ($0.013 < z < 0.05$) groups, for which are available re-processed photometric data (Blanton *et al.* 2011) in the NASA-Sloan Atlas (http://nsatlas.org). We use also the galaxy stellar masses (M^*), specific star-formation rates (sSFR) and various spectral characteristics (EW(Hα), EW(Hδ), Dn4000 break), obtained from the SDSS CAS.

Our aim is to address the radial trends of various physical and star-formation characteristics of satellite galaxies in groups, and the possible velocity modulations of these trends, conditioned by dynamical evolution in the group core and by occurence of in-falling and 'backsplash' galaxies in radial range of about 1-3 virial radii. For this purpose we stacked together the complete sample of 16284 satellite (rank > 1) galaxies, with limiting absolute magnitudes of $M_r < -18$, using the group-centric distances, normalized by the virial radius (R_{vir}) and relative line-of-sight (*los*) velocities scaled by the

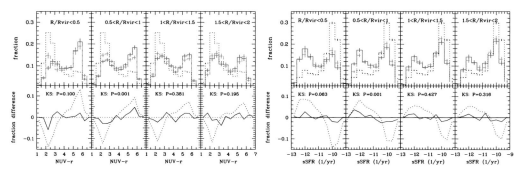

Figure 1. The fraction distributions of the (NUV-r) colours (left panel) and the specific SFRs (right panel) of low- and high-velocity satellite galaxies in four scaled radial bins are compared to one another and to the properties of field galaxies. *Upper section:* Solid (blue) histogram represents satellites with low velocities, dashed (mageneta) histogram is for high velocity satellites; dotted (black) histogram denotes the field galaxies. Error bars are Poisson. *Lower section:* shows the fraction difference between two velocity bins ($frac(< 0.75) - frac(> 0.75)$ - solid line) and the difference between all satellites and field galaxies ($frac(sat.) - frac(field)$ - dotted line). The Kolmogorov-Smirnov (KS) test probabilities (P) for the null hypothesis is indicated for the distribution of satellites in two velocity bins. Fraction of red/passive and blue/star-forming satellite galaxies is changing with group-centric radius. There is an excess of blue high velocity satellites near the center and an excess of red low velocity satellites near the group virial radius. The fraction distributions of properties of field and satellite galaxies are different, however the difference decreases for satellites, which are located at the group outskirts.

velocity dispersion (σ_v) of each group. For comparison purposes we selected a similar sample of field (rank = 0) galaxies from the same parent catalogue. We analyse the fraction distribution for various properties of low-velocity ($|v- < v > |/\sigma_v < 0.75$) and high-velocity ($|v- < v > |/\sigma_v > 0.75$) satellite galaxies in four scaled radial bins ($R/R_{vir} = 0 - 0.5, 0.5 - 1.0, 1.0 - 1.5, 1.5 - 2$).

3. Results

We found statistically significant radial dependencies of satellite properties, e.g. massive, red and passive satellites being distributed predominantly near the center of the composite group. Our study also highlights some velocity modulation of these radial trends, e.g. (i) a mild excess of massive satellites in the group core, and (ii) an excess of red and passive satellites near the group virial radius, amongst the low velocity satellite galaxies (Fig. 1). We conclude that the galaxy properties in and around groups are shaped not only by their stellar mass and environment but also by absolute *los* velocity. Using kinematical data, it should be statistically feasible to separate dynamical classes of bound, in-falling and 'backsplash' satellite galaxies also in poor groups of galaxies.

References

Blanton, M. R., Kazin, E., Muna, D., Weaver, B. A., & Price-Whelan, A. 2011,
Mahajan, S., Mamon, G. A., & Raychaudhury, S. 2011, *MNRAS*, 416, 2882
Tempel, E., Tamm, A., Gramann, M., *et al.* 2014, *A&A*, 566, A1

The Zeldovich Universe:
Genesis and Growth of the Cosmic Web
Proceedings IAU Symposium No. 308, 2014
R. van de Weygaert, S. Shandarin, E. Saar & J. Einasto, eds.
© International Astronomical Union 2016
doi:10.1017/S1743921316010437

Alignments of galaxies and halos in hydrodynamical simulations

Isha Pahwa,[1,2,3]† and Noam I. Libeskind[3]

[1]Inter University Centre for Astronomy and Astrophysics, Post Bag 4,
Pune University Campus, Ganeshkhind, Pune 411007, India

[2]Department of Physics and Astrophysics, University of Delhi, Delhi-110076

[3] Leibniz-Institut für Astrophysik Potsdam (AIP), An der Sternwarte 16,
D-14482 Potsdam, Germany

Abstract. We use a $200\,h^{-1}$Mpc cosmological hydrodynamical simulation to examine the alignments of galaxies with respect to the host halo. We do separate study for the different components of the halo, such as stars, gas and dark matter. We show that angular momentum of gas is more aligned with the angular momentum of host halo compared with the stellar component.

Keywords. Galaxies: halos, Cosmology: theory, dark matter, large-scale structure of universe

---

1. Introduction

The origin of angular momentum is well known problem in the standard picture of cosmic structure formation. Current ideas of how dark matter halos get their spin, rely on torques imparted due to the misalignment of a collapsing Lagrangian region's inertia tensor and the tidal field (Peebles 1969; White 1984). Such torques are only effective in the linear regime, so long as the collapsing region has a large enough "lever arm". Halos (and the galaxies within them) acquire significant angular momentum in the non-linear stage of their evolution. Numerous numerical studies have dealt with the issues related to the non-linear angular momentum acquisition of dark matter halos, its evolution and its relation to the large-scale structure (LSS, Bett *et al.* 2010; Libeskind *et al.* 2012 among others). However, in the non-linear regime, baryons become important. Unlike dark matter, baryons are affected not only by gravity but also by hydrodynamic processes. These processes can dominate galaxy spin and hence, become important to study.

2. Simulation

We carry out a large cosmological hydrodynamical simulation CURIE, using the PMTree-SPH code GADGET-2 (Springel 2005), assuming the standard ΛCDM cosmology (WMAP5). The cosmological parameters are these: $\Omega_m = 0.28$, $\Omega_\Lambda = 0.72$, $h = 0.7 = H_0/100$ kms^{-1} Mpc^{-1}, $\sigma_8 = 0.817$ and $n_s = 0.96$. The simulation box size is $200h^{-1}$Mpc with 2×1024^3 particles. The masses of dark matter, gas and star particles are $4.6 \times 10^8 h^{-1}$M$_\odot$, $9.7 \times 10^7 h^{-1}$M$_\odot$ and $4.85 \times 10^7 h^{-1}$M$_\odot$ respectively. The simulation uses the feedback and star formation rules of Springel & Hernquist (2003). A halo catalogue is obtained by running the publicly available halo finder AHF (Knollmann & knebe 2009) on the particle distribution. We use this simulation at $z = 0$.

† Email: `ipahwa@iucaa.ernet.in`

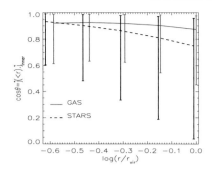

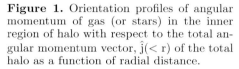

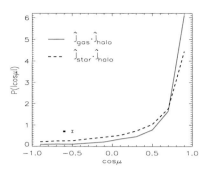

Figure 1. Orientation profiles of angular momentum of gas (or stars) in the inner region of halo with respect to the total angular momentum vector, $\hat{j}(<r)$ of the total halo as a function of radial distance.

Figure 2. Probability distribution of the cosine of angle formed between the angular momentum vector of gas (or stars) in inner region of halos with total angular momentum of halo computed at virial radius.

3. Results

The study of orientations of the angular momentum of gas and star particles with respect to the total angular momentum of the halo is an important input in the formulation of models for galaxy formation. We define a spherical 'inner' region at $0.25\,r_{\rm vir}$ where $r_{\rm vir}$ is the virial radius of the halo. Generally speaking, baryonic effects are dominant over dark matter in this region.

Around 25,000 halos in the mass range $10^{11.5} - 10^{13}h^{-1}{\rm M}_\odot$ are selected for analysis. We ensure that each halo has at least 1000 baryonic particles within $0.25r_{\rm vir}$, such that the computed angular momentum directions are not affected by poor resolution (see Sales *et al.* 2012). Fig. 1 shows the (cosine of the) angle formed between the angular momentum of the gas or stellar component (red solid or black dashed line, respectively) computed at $0.25r_{\rm vir}$ with the angular momentum of the all material (i.e including dark matter) as a function of radius. At each radius, the median and 1σ deviation are plotted. The angular momentum of the gas component in the inner parts of the halo shows a fairly strong alignment with the angular momentum of outer parts of the halo. The alignment of the stellar component with the outer parts of the halo, is not as strong, and the spread in alignment is larger, indicated by the increasing variance around medians.

In Fig. 2 we show the probability distribution of the $\cos\mu$, the cosine of the angle formed between the angular momentum of gas, $\hat{j}_{\rm gas}$ (red, solid line) or stars, $\hat{j}_{\rm star}$ (black, dash line) within the inner part of the halo with the total angular momentum of halo, $\hat{j}_{\rm halo}$ computed at $r_{\rm vir}$. The small error bars in the lower left of the plot indicate the (Poissonian) scatter that one would expect from a uniform distribution of that size. As it is clear from this figure, $\hat{j}_{\rm gas}$ is more aligned with $\hat{j}_{\rm halo}$ than $\hat{j}_{\rm star}$ with $\hat{j}_{\rm halo}$.

IP is supported by the Deutsche Forschungs Gemeinschaft and the hospitality of Leibniz-Institut für Astrophysik Potsdam.

References

Bett, P., Eke, V., Frenk, C. S., Jenkins, A., & Okamoto, T. 2010, *MNRAS* 404, 1137
Knollmann, S. R. & Knebe, A. 2009, *ApJS* 182, 608
Libeskind, N. I., Hoffman, Y., Knebe, A., *et. al.* 2012, *MNRAS* 421, L137
Peebles, P. J. E. 1969, *ApJ* 155, 393
Sales, L. V., Navarro, J. F., Theuns, T., *et. al.* 2012, *MNRAS* 423, 1544
Springel, V. 2005, *MNRAS* 364, 1105
Springel, V. & Hernquist, L. 2003, *MNRAS* 339, 289
White, S. D. M. 1984, *ApJ* 286, 38

The Zeldovich Universe:
Genesis and Growth of the Cosmic Web
Proceedings IAU Symposium No. 308, 2014
R. van de Weygaert, S. Shandarin, E. Saar & J. Einasto, eds.

© International Astronomical Union 2016
doi:10.1017/S1743921316010449

The Binggeli effect

M. Biernacka[1], E. Panko[2], W. Godłowski[3], K. Bajan[4] and P. Flin[1]

[1] Institute of Physics, Jan Kochanowski University, Kielce, Poland
email: `bmonika@ujk.edu.pl`

[2] Kalinenkov Observatory, Nikolaev National University, Nikolaev, Ukraine

[3] Institute of Physics, Opole University, Opole, Poland

[4] Mt. Suhora Observatory, Cracow Pedagogical University, Krakow, Poland

Abstract. We found the alignement of elongated clusters of BM type I and III (the excess of small values of the $\Delta\theta$ angles is observed), having range till about $60h^{-1}Mpc$. The first one is probably connected with the origin of supergiant galaxy, while the second one with environmental effects in clusters, originated on the long filament or plane.

Keywords. galaxy cluster, alignment

1. Introduction

Binggeli (1982) was the first who found that galaxy clusters tend to be aligned pointing each other. Later on the existence of this effect was discussed by several authors and usually the significant alignment was reported. The distance between clusters for which the effect was detected changed from $10h^{-1}Mpc$ till $150h^{-1}Mpc$ and the strength of the effect diminished with distance (Struble & Peebles (1985), Flin (1987), Rhee & Katgert (1987), Ulmer et al. (1989), Plionis (1994), Chambers et al. (2002)). These investigation employed both optical and X-ray data, as well as clusters assigned or not to superclusters. Nowadays it is accepted that the effect is real, not due to selection effects, and it distance scale is between $(10-60)h^{-1}Mpc$. Better understanding of physical processes leading to cluster alignment revealed numerical simulations. This was done in the framework of cold dark matter (CDM) model regarded now as being correct for large scale structure fomation (Onuora & Thomas (2000)) using large - scale simulation found that in LCDM cosmological model for distance up to $30h^{-1}Mpc$, while in tCDM models the range of effects is twice smaller. In SCDM and OCDM models, where smaller scale simuations were performed some alignment effect could be noted.

2. Observational data

The catalogue of 6188 galaxy clusters (hereafter PF, Panko & Flin (2006)) served as observational basis of our studies. It is prepared applying Voronoy tessellation technique (Ramella et al. 1999, 2001) to Muenster Red Sky Survey (MRSS, Ungruhe et al. (2003)). The PF catalogue is statistically complete till $r_F = 18.3^m$. Our cluster contains at least 10 galaxies in the brightnest range m_3, $m_3 + 3^m$ (m_3 is the magnitude of the third brightest galaxy). The distance to clusters were determined using the dependence between the tenth brightest galaxy m_{10} and redshift (Panko et al. (2009)). Clusters morphological types for 1056 PF objects were taken from ACO (Abell et al. (1989)).

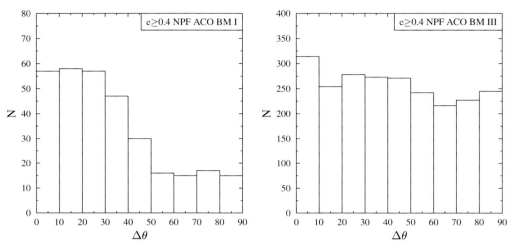

Figure 1. Distribution of the angle $\Delta\theta$ for BM type I and III with galaxy clusters ellipticities $e > 0.4$.

3. Method and results

Method used to determine galaxy clusters shape (ellipticity and position angle) was covariance ellipse method (Carter & Metcalfe (1980)). The method is based on the first five moments of the observed distribution of galaxy coordinates x_i, y_i. The existence of the Binggeli effect was checked studing the angle $\Delta\theta$ between the direction toward neighbouring clusters and the cluster position angle. The coordinate distances between clusters were calculated assuming $h = 0.75$, $q_o = 0.5$.

4. Conclusions

- We found the alignment of elongated clusters of BM type I and III (the excess of small values of the $\Delta\theta$ angles is observed), having range till about $45h^{-1}Mpc$. The BMI was expected due to special role of CD galaxies during formation. The existence of the Binggeli effect in BM III is surprising. It can be due to incorporation of RS types L and F in which alignment is frequently observed.
- The effect is observed only in the sample containing elongated clusters. Cluster ellipticity is not well defined parameter, which was noted by several investigators. The problem with cluster ellipticity determination is not only the problem of the applied method.
- The alignment in BMI is probably connected with the origin of supergiant galaxy, while in the case of BMIII with environmental effects in clusters originated on the one long filament or plane.
- It is quite possible that the origin of galaxy clusters was not a unique, homogeneous process and various factors influenced this process.

References

Abell, G. O., Corwin, H. G., & Olowin, R. P. 1989, *Astrophysical J Sup. Ser.* 70, 1
Binggeli, B. 1982, *AA* 107, 338
Carter, D. & Metcalfe, N. 1980, *MNRAS* 191, 325
Chambers, S. W., Melott, A. L., & Miller, C. J. 2002, *ApJ* 565, 849
Flin, P. 1987, *MNRAS* 228, 941

Onuora, L. I. & Thomas, P. A. 2000, *MNRAS* 319, 614

Panko, E. & Flin, P. 2006, *JAD* 12, 1

Panko, E., Juszczyk, T., Biernacka, M., & Flin, P. 2009, *ApJ* 700, 168

Plionis, M. 1994, *ApJS* 95, 401

Ramella M., Nonino M., Boschin W., & Fadda D. 1999, *in: ASP Conf. Ser. 176, Observational Cosmology: The Development of Galaxy Systems, ed. G. Giuricin,M. Mezzetti, & P. Salucci (San Francisco: ASP)*, s.108

Ramella,M., Boschin,W., Fadda, D. & Nonino, M. 2001, *A & A* 368, 776

Rhee, G. F. R. N. & Katgert, P. 1987, *AA* 183, 217

Struble, M. F. & Peebles, P. J. E.. 1985, *AJ* 90, 582

Ulmer, M. P., McMillan, S. L. W., & Kowalski, M. P. 1989, *ApJ* 338, 711

Ungruhe, R., Seitter, W. C., & Duerbeck, H. W. 2003, *JAD* 9,1

The Zeldovich Universe:
Genesis and Growth of the Cosmic Web
Proceedings IAU Symposium No. 308, 2014
R. van de Weygaert, S. Shandarin, E. Saar & J. Einasto, eds.

© International Astronomical Union 2016
doi:10.1017/S1743921316010450

Luminosity function for galaxy clusters

K. Bajan[1], M. Biernacka[2], P. Flin[2], W. Godłowski[3], E. Panko[4] and J. Popiela[3]

[1] Mt. Suhora Observatory, Cracow Pedagogical University, Krakow, Poland,
email: kbajan@up.krakow.pl

[2] Institute of Physics, Jan Kochanowski University, Kielce, Poland
[3] Institute of Physics, Opole University, Opole, Poland
[4] Kalinenkov Observatory, Nikolaev National University, Nikolaev, Ukraine

Abstract. We constructed and studied the luminosity function of 6188 galaxy clusters. This was performed by counting brightness of galaxies belonging to clusters in the PF catalogue, taking galaxy data from MRSS. Our result shows that the investigated structures are characterized by a luminosity function different from that of optical galaxies and radiogalaxies (Machalski & Godłowski 2000). The implications of this result for theories of galaxy formation are briefly discussed.

1. Introduction

For galaxies, the luminosity function is usually described by the Schechter function (Schechter 1976). This function presents the space density of objects as function of their luminosities. The first parameter has units of number density and provides the normalization. The galaxy luminosity function may have different parameters for different populations and environments; it is not a universal function. One measurement based on field galaxies is (Machalski & Godłowski 2000):

$$\alpha = -1.25, \quad \phi^* = 1.2 \times 10^{-3} h^3 Mpc^{-3}. \tag{1.1}$$

It is often convenient to rewrite the Schechter luminosity function in terms of magnitude, rather than luminosities. In this case the function becomes:

$$n(M)dM = 0.4 \ln 10 \phi^* \left[10^{-0.4(M-M^*)} \right]^{\alpha+1} \exp \left[-10^{-0.4(M-M^*)} \right] dM. \tag{1.2}$$

2. Observational data

The catalogue of 6188 galaxy clusters by Panko & Flin (2006) (hereafter PF) served as the observational basis for our studies. It was prepared by applying Voronoi tessellation (Ramella *et al.* 1999, 2001) to the Muenster Red Sky Survey (MRSS, Ungruhe *et al.* 2003). The PF catalogue is statistically complete up to $r_F = 18.3^m$. Each cluster contains at least 10 galaxies in the brightness range m_3, $m_3 + 3^m$, where m_3 is the magnitude of the third brightest galaxy in the cluster. The distances to the clusters were determined using the relation between the tenth brightest galaxy m_{10} and redshift (Panko *et al.* 2009).

3. Results

In the case where m_3 is good indicator of distance, Fig. 1a,b represents the luminosity function. Fig.1c,d presents the distribution of absolute magnitudes of m_3. The dispersion

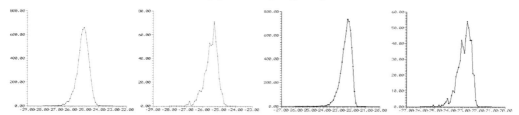

Figure 1. Histograms: number of clusters versus absolute magnitude (First left (1a) – for all clusters; second left (1b) – for the statistically complete sample, $m_3 < 15.3$). Histograms of the absolute magnitude of the m_3 galaxy in each cluster for all clusters and for the statistically complete sample – 1c, d.

is clearly observed. In order to correct for this effect we applied Condon's (1989) method, which consists of weighting each cluster by a weight reversely proportional to the cluster volume V_{max}, where V_{max} is connected with the cluster redshift.

The Condon-method corrected histograms were quite similar to those presented in Fig. 1a,b. The calculated values of the parameters are: $M^* = -23.1$, $\alpha = 4.26$. These values are drastically different from the corresponding values for galaxies see eq. (1.1). On the other hand they are quite close to the values obtained without taking into account the distribution of m_3. These values were also counted using a different approach for the Schechter luminosity function determination, namely that of Efstathiou *et al.* (1988) and Lin & Kirshner (1996), where the maximum likelihood method is applied. Their values was quite similar to those given above.

4. Conclusions

- The third-brightest galaxies in clusters have different absolute magnitudes.
- The cluster luminosity function is significantly different than that obtained for both optical and radiogalaxies (Machalski & Godłowski 2000).
- The shape of the cluster luminosity function, as obtained through this method, is a Gaussian, not a Schechter function.
- The application of Condon method does not change drastically the obtained result.
- A possible explanation of the difference in cluster and galaxy luminosity functions are the different scenarios leading to the formation of these structures (galaxies and clusters).

References

Condon, J. J. 1989, *ApJ* 338, 13
Efstathiou, G., Ellis, R. S., & Peterson, B. A. 1988, *MNRAS* 232, 431
Lin, H. & Kirshner, R. P. 1996, *ApJ* 464, 60
Machalski, J. & Godłowski, W. 2000, *AA* 360, 463
Panko, E. & Flin, P. 2006, *Journal Astron. Data*, 12, 1
Panko, E., Juszczyk, T., Biernacka, M., & Flin, P. 2009, *ApJ* 700, 168
Ramella, M., Nonino, M., Boschin, W., & Fadda, D. 1999, *in: ASP Conf. Ser. 176, Observational Cosmology: The Development of Galaxy Systems, ed. G. Giuricin,M. Mezzetti, & P. Salucci (San Francisco: ASP)*, s.108
Ramella, M., Boschin, W., Fadda, D., & Nonino, M. 2001, *A& A*, 368, 776
Schechter, P. 1976, *ApJ*, 203, 297
Ungruhe, R., Seitter, W. C., & Durbeck, H. W. 2003, *Journal Astronomical Data* 9, 1

The Zeldovich Universe:
Genesis and Growth of the Cosmic Web
Proceedings IAU Symposium No. 308, 2014
R. van de Weygaert, S. Shandarin, E. Saar & J. Einasto, eds.

© International Astronomical Union 2016
doi:10.1017/S1743921316010462

Giant radio galaxies and cosmic web

Pekka Heinämäki

Tuorla Observatory, Department of physics and astronomy, University of Turku
Väisäläntie 20, 21500 Piikkiö, Finland
email: pekheina@utu.fi

Abstract. Giant radio galaxies create the well distinguishable class of sources. These sources are characterized with edge-brightened radio lobes with highly collimated radio jets and large linear sizes which make them the largest individual structures in the Universe. They are also known to be hosted by elliptical/disturbed host galaxies and avoid clusters and high galaxy density regions. Because of GRG, large linear sizes lobes extend well beyond the interstellar media and host galaxy halo the evolution of the radio lobes may depend on interaction with this environment. Using our method to extract filamentary structure of the galaxies in our local universe we study whether radio lobe properties in some giant radio galaxies are determined on an interaction of this filament ambient.

Keywords. Cosmology: large-scale structure of the Universe; galaxies: structure; radio continuum: galaxies

1. Introduction

Among FR II radio galaxies, the largest sources of their projected linear sizes > 0.7 Mpc are called giant radio galaxies (GRG). Because of their large size it is probable that the orientations of these sources are near the plane of the sky and radio lobes extend well beyond the interstellar media and host galaxy halo. As a result, we may expect that plasma of the lobes interacts with intergalactic medium (IGM) modifying morphology and their asymmetry in the lobe lengths. While GRG avoid high density regions they are assumed to be associated with large scale filamentary structures (Subrahmanyan *et al.* 2008) and the evolution of the radio lobes may depend on interaction with ambient gas (Safouris *et al.* 2009).

Relatively little is known about gas associated with galaxy filaments, but according to simulations (Cen&Ostriker 2006) the majority of the missing baryons are hidden filaments in low-density plasma (WHIM). Low density of WHIM gas renders it very difficult to observe. While GRG location and orientation relative to filamentary structure can modify their morphology and radio properties of the radio lobes it allows the use of GRGs as intergalactic barometers (Malarecki *et al.* 2013).

2. Data and Methods

Koziel-Wierzbowska & Stasinska (2011) analyzed FR II radio sources (the Cambridge Catalogues of Radio sources: 3C-9C) and their counterparts in SDSS DR7, finding in total 401 FR II sources. Among these sources we found 22 GRGs. As a first step in this project we study where these GRGs are located in the filament-void network. For that purpose we use Tempel *et al.* (2014) filament catalogue based on Bisous model (Stoica *et al.* 2005). This catalogue is extracted from the SDSS-DR8 and it contains 3D mapping of 15421 filaments including their properties (e.g. galaxy richness, luminosity, length). The radius of the filaments in this catalogue is fixed at 0.5 Mpc/h.

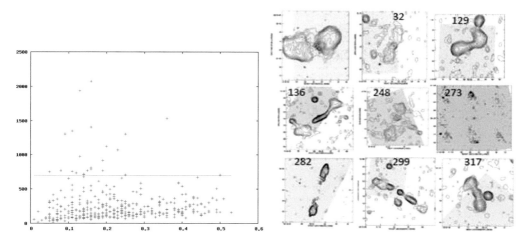

Figure 1. Linear sizes of the FRII sources in the SDSS survey region as a function of redshift. Green line indicates GRG size limit (left panel). Nine radiomaps of the GRGs that have redshifts less than z = 0.155(right panel). Data and figure by Koziel-Wierzbowska & Stasinska(2011)

Filament data extends up to z = 0.155 which reduce the number of GRGs to 9. Our preliminary analysis shows that only 1 case out of these 9 GRGs has host galaxy(obj273) inside the filament (note, despite that its morphology seems to be quite symmetric). In two cases the host galaxy is in 1 Mpc/h radius from the filament axis (obj136 and obj248). We will also analyze whether the asymmetry in these sources is related to the distribution of the galaxies in the local galaxy group and what is the orientation of the nearby filament. In other six cases host galaxies and radio lobes are too far away from the filaments to have interaction with filament ambient. Note that even obj129 seems to be in a void region, it is clearly asymmetric.

3. Summary

The current project is at preliminary stage and the aim of this poster is to describe the idea and the starting point for more detailed forthcoming analysis of the relations between GRG and filaments. The next step is to study how radio lobes are aligned with filaments and to investigate the distribution and physical state of the gas in these regions. Moreover, we are planning to add more GRGs to our analysis.

References

Cen, R. & Ostriker, J. 2006, *APJ*, 650, 60

Fanaroff, B. L. & Riley, J. M. 1974, *MNRAS*, 167, 31

Koziel-Wierzbowska, D. & Stasinska, G. 2011, *MNRAS*, 415, 013

Malarecki, J. M., Staveley-Smith, L., Saripalli, L., Subrahmanyan, R., Jones, D. H., Duffy, A. R., & Rioja, M. 2013, *MNRAS*, 432, 200

Safouris, V., Subrahmanyan, R., Bicknell, G. V., & Saripalli, L. 2009, *MNRAS*, 393, 2

Stoica, R. S, Gregori, P., & Mateu, J. 2005, *Stoch.Process.Appl.*, 115, 1860

Subrahmanyan, R., Saripalli, L.,Safouriys, V., & Hunstead, R. 2008, *ApJ*, 677, 63

Tempel, E., Stoica, R. S., Martinez, V. J., Liivamägi, L. J., Castellan, G., & Saar, E. 2014, *MNRAS*, 438, 3465

The Zeldovich Universe:
Genesis and Growth of the Cosmic Web
Proceedings IAU Symposium No. 308, 2014
R. van de Weygaert, S. Shandarin, E. Saar & J. Einasto, eds.

© International Astronomical Union 2016
doi:10.1017/S1743921316010474

The Direct Collapse of Supermassive Black Hole Seeds

John A. Regan[1], Peter H. Johansson[1] and John H. Wise[2]

[1] Department of Physics, University of Helsinki, Finland
email: john.regan@helsinki.fi

[2] Center for Relativistic Astrophysics, Georgia Institute of Technology, USA

Abstract. The direct collapse model of supermassive black hole seed formation requires that the gas cools predominantly via atomic hydrogen. To this end we simulate the effect of an *anisotropic* radiation source on the collapse of a halo at high redshift. The radiation source is placed at a distance of 3 kpc (physical) from the collapsing object and is set to emit monochromatically in the center of the Lyman-Werner (LW) band. The LW radiation emitted from the high redshift source is followed self-consistently using ray tracing techniques. Due to self-shielding, a small amount of H_2 is able to form at the very center of the collapsing halo even under very strong LW radiation. Furthermore, we find that a radiation source, emitting $> 10^{54}$ ($\sim 10^3$ J_{21}) photons per second is required to cause the collapse of a clump of M $\sim 10^5$ M$_\odot$. The resulting accretion rate onto the collapsing object is ~ 0.25 M$_\odot$ yr^{-1}. Our results display significant differences, compared to the isotropic radiation field case, in terms of H_2 fraction at an equivalent radius. These differences will significantly effect the dynamics of the collapse. With the inclusion of a strong anisotropic radiation source, the final mass of the collapsing object is found to be M $\sim 10^5$ M$_\odot$. This is consistent with predictions for the formation of a supermassive star or quasi-star leading to a supermassive black hole.

Keywords. Cosmology: theory – large-scale structure – black holes physics – methods: numerical – radiative transfer

1. Introduction

The discovery of Super Massive Black Holes (SMBHs) (e.g. Mortlock *et al.* 2011) at very high redshift (z > 6) has led to a difficulty in explaining how black hole seeds formed so early in the Universe. The seed (progenitor for the SMBH) must grow from its initial mass early in the Universe to a mass $M \gtrsim 10^9 M_\odot$ by a redshift of z $\sim$ 6. The direct collapse model of SMBH growth provides a compelling solution. In this model a collapsing baryonic mass, which can cool only through Lyman-α radiation, provides an initial seed mass black hole of $M_{BH} \sim 10^4$ M$_\odot$ (e.g. Regan & Haehnelt. 2009a, Regan *et al.* 2014a). In order for the collapsing gas cloud to cool only through atomic transitions H_2 must be effectively dissociated and the cloud must be deficient in metals. H_2 readily forms in the early Universe and so in order for the halo to cool through atomic transitions only H_2 must be dissociation by a radiative background or else collisionally dissociated though collisions with other primordial elements. The assumption of very low metallicity within the halo is easily satisfied at high redshifts due to the limited time for metals to pollute all collapsing halos.

2. Anisotropic Radiation Sources

In order to dissociate H_2, radiation in the Lyman-Werner (LW) bands is required. The LW band occupies the frequency spectrum between $\sim 11.2 \, - -13.6$ eV. Furthermore, the global LW background at high-z is likely to be relatively low and so a close-by, high

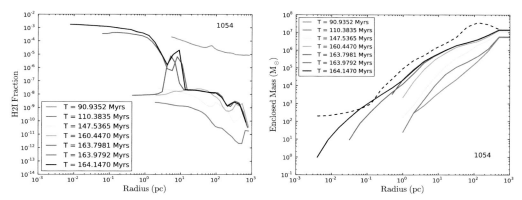

Figure 1. In both panels we show a radial profile centred on the point of maximum density in the simulation i.e. centred on the collapsing halo. In both cases the source is exposed to a radiation source emitting 10^{54} photons per second in the Lyman-Werner band corresponding to $\sim 10^3 \; J_{21}$. In the left hand panel we show the H_2 fraction and in the right hand panel we show the enclosed mass. The dashed line in the right hand panel is the Jeans mass at the last output time.

intensity, LW flux will be required to reach sufficient intensities (e.g. Dijkstra *et al.* 2008). This anisotropic source will be able to effectively dissociate H_2 as long as it is within ~ 10 kpc and has a flux of $\gtrsim 10^3 \; J_{21}$, where $J_{21} = 10^{-21}$ erg cm^{-2} s^{-1} Hz^{-1} sr^{-1}.

3. Simulations & Results

To model the direct collapse of SMBH seeds at high redshift we use hydrodynamical simulations including self-consistent radiative transfer to study the collapse. We use the publicly available Enzo code (Byran *et al.* 2014). We simulate a single halo using a series of fluxes from a source located at a distance of 3 kpc (physical)(Regan *et al.* 2014b). In Figure 1 we show the results from the simulation with a source flux of 1×10^{54} photons per second for clarity ($\sim 10^3 J_{21}$). We found that this was close to the minimum flux required in order that H_2 self-shielding would be overcome.

4. Conclusions

A flux in the LW band of $J \sim$ a few times J_{21} delays the collapse of the gas cloud by up to 70 Myrs as the cooling due to H_2 is inhibited. Fluxes in the LW band of $J \gtrsim 10^3 J_{21}$ cause the formation of a collapsing cloud of $M \sim 10^5$ M$_\odot$ suggesting a SMBH seed of this mass will form if fragmentation can be prevented.

Furthermore, our simulations suggest that an anisotropic source is required and will significantly effect the dynamics of the collapse compared to the isotropic radiation case. Otherwise identical simulations run with isotropic radiation and anisotropic radiation showed differences of up to two orders of magnitude in the H_2 abundances at a given radius which will strongly effect the collapse.

References

G. L. Bryan and The Enzo Collaboration. *AstroPhysical Journal*, 211:19, April 2014.

M. Dijkstra *et al. Monthly Notices*, 391:1961–1972, December 2008.

D. J. Mortlock *et al. Nature*, 474:616–619, June 2011.

J. A. Regan & M. G. Haehnelt. *Monthly Notices*, 393:858–871, March 2009.

J. A. Regan & M. G. Haehnelt. *Monthly Notices*, 396:343–353, June 2009.

J. A. Regan *et al. Monthly Notices*, 439:1160–1175, March 2014.

J. A. Regan, P. H. Johansson, & J. H. Wise. *ArXiv e-prints:1407.4472*, July 2014.

The Zeldovich Universe:
Genesis and Growth of the Cosmic Web
Proceedings IAU Symposium No. 308, 2014
R. van de Weygaert, S. Shandarin, E. Saar & J. Einasto, eds.

© International Astronomical Union 2016
doi:10.1017/S1743921316010486

Co–evolution of black holes and galaxies: the role of selection biases

Laura Portinari

Tuorla Observatory, University of Turku, Väisäläntie 20, FIN-21500 Piikkiö, Finland

Abstract. Quasars are tracers of the cosmological evolution of the Black Hole mass – host galaxy relation, and indicate that the formation of BH anticipated that of the host galaxies. We find that selection effects and statistical biases dominate the interpretation of the observational results; and co-evolution (= constant BH/galaxy mass ratio) is still compatible with observations.

Keywords. Galaxies: active, high-redshift, formation, evolution; quasars: general

In the local Universe, the mass of the central SuperMassive Black Hole (SMBH) correlates with that of the host galaxy with a typical mass ratio $\Gamma \equiv M_{BH}/M_\star \simeq 0.002$. At high redshifts the Γ ratio can be estimated in QSOs, by measuring the SMBH mass via the "virial technique", and determining the host galaxy luminosity (and stellar mass) from deep imaging, after subtracting the (dominant) QSO point–source.

QSO host studies indicate that Γ increases with redshift, and was 5–10 times larger than now by $z = 3$: SMBH appear to have formed before the bulk of their host galaxies (Peng *et al.* 2006; Decarli *et al.* 2010; Merloni *et al.* 2010). However, a proper interpretation of the data must consider a number of selection biases and observational errors.

1. Lauer bias The intrinsic relation between SMBH and host mass has a scatter (about 0.3 dex) and is weighted by the galaxy luminosity function: a very massive BH is more likely to be an outlier in the relation, hosted in a relatively low-mass galaxy and with high Γ, than belong to a very massive host, as these are rare. A QSO host is a typical median host galaxy behind a selected quasar/SMBH mass; as high z QSOs are very massive BH ($10^9 - 10^{10}$ M$_\odot$), they will tendentially trace higher–than–average Γ values creating an apparent evolution (Lauer *et al.* 2007); see Fig. 1a,b.

2. Overluminosity bias of QSO hosts. Nuclear activity at high z is supposed to be associated with recent mergers and starbursts, making QSO hosts overluminous with respect to the general galaxy population for the same stellar (and BH) mass (Fig. 1c). This induces an overestimate of the stellar mass $M_\star$, and an underestimate of the Γ ratio. This bias acts in the opposite direction to the Lauer bias.

3. Shen–Kelly bias is a Malmquist bias on BH masses, as high redshift QSOs trace the most massive, rare BH. The typical error on measured BH masses of 0.4 dex artificially boosts them, by statistically overestimating the mass of smaller, more common BH (Shen & Kelly 2010); see Fig. 1d.

In Portinari *et al.* (2012), we implemented all these biases when comparing QSO host data to the publicly available semi–analytical model (SAM) of De Lucia & Blaizot (2007). Their model predicts a negligible evolution of the "intrinsic" BH–host relation (i.e. Γ varies little with redshift). Yet, once all the biases are included, the model can account for the observed strong variation of Γ, and co–evolution (constant intrinsic Γ) is still compatible with the data (Fig. 1). The model however has difficulty in predicting the number of the most massive BH ($> 10^9$ M$_\odot$) at high z.

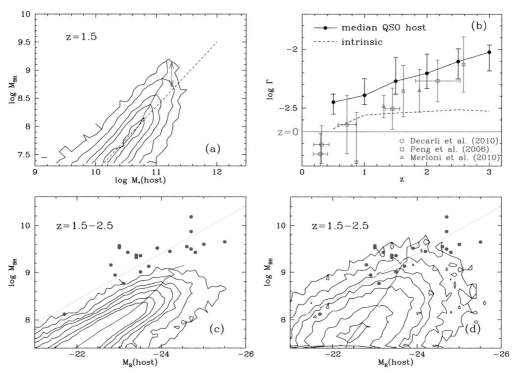

Figure 1. (a) Contour plot of the $(M_{BH}, M_\star)$ distribution of high z galaxies from the SAM of De Lucia & Blaizot (2007). The dashed line marks the "intrinsic" $M_{BH} - M_\star$ relation, the solid line traces the "median" QSO host at given M_{BH}. The difference (updown arrow) translates into a spurious Γ evolution (Lauer bias). **(b)** While the intrinsic Γ ratio evolves little (dashed line) the median QSO hosts (full dots connected by solid line) trace a strong apparent evolution, comparable to the observed one (open symbols). **(c)** In the observational $(M_{BH},$ luminosity) plane, model QSO hosts are overluminous and offset from the observed data (full dots). **(d)** After convolving with the observational error M_{BH}, the theoretical distribution is in fair agreement with observations (Shen–Kelly bias); although it is missing the most massive BH.

The aforementioned biases are due to the intrinsic scatter in the BH–host relations, combined with the fact that QSOs sample the massive end of the BH and galaxy mass function. These biases dominate the interpretation of the observational results, as they mimic, and are hardly distinguishable from, evolution. We should turn to relatively low redshift QSO hosts, where evolution is negligible, to better understand the scatter and biases in the real Universe, disentangled from evolution. The Stripe 82 sample of almost 400 QSO hosts at $z < 0.5$ by Falomo *et al.* (2014) will be useful to that purpose.

References

De Lucia, G. & Blaizot, J., 2007, *MNRAS* 375, 1
Decarli, R., Falomo, R., Treves, A., Labita, M., Kotilainen, J., & Scarpa, R., 2010, *MNRAS* 402, 2453
Falomo, R., Bettoni, D., Karhunen, K., Kotilainen, J. K., & Uslenghi, M., 2014, *MNRAS* 440, 476
Lauer, T. R., Tremain, S., Richstone, D., & Faber, S. M., 2007, *ApJ* 670, 249
Merloni, A., Bongiorno, A., Bolzonella, M., *et al.* 2010, *ApJ* 708, 37
Peng, C. Y., Impey, C. D., Rix, H.-W., *et al.* 2006, *ApJ* 649, 616
Portinari, L., Kotilainen, J., Falomo, R., & Decarli, R., 2012, *MNRAS* 420, 732
Shen, Y. & Kelly, B. C., 2010, *ApJ* 713, 41

The Zeldovich Universe:
Genesis and Growth of the Cosmic Web
Proceedings IAU Symposium No. 308, 2014
R. van de Weygaert, S. Shandarin, E. Saar & J. Einasto, eds.

© International Astronomical Union 2016
doi:10.1017/S1743921316010498

Metallicity evolution in mergers of disk galaxies with black holes

Antti Rantala and Peter H. Johansson

Department of Physics, University of Helsinki,
Gustaf Hällströmin katu 2a, 00560 Helsinki, Finland
`antti.rantala@helsinki.fi`
`peter.johansson@helsinki.fi`

Abstract. We use the TreeSPH simulation code Gadget-3 including a recently improved smoothed particle hydrodynamics (SPH) module, a detailed metallicity evolution model and sophisticated subresolution feedback models for supernovae and supermassive black holes in order to study the metallicity evolution in disk galaxy mergers. In addition, we examine the simulated morphology, star formation histories, metallicity gradients and kinematic properties of merging galaxies and merger remnants. We will compare our simulation results with observations of the early-type Centaurus A galaxy and the currently colliding Antennae galaxies.

Keywords. galaxies: abundances, galaxies: evolution, galaxies: formation, galaxies: interactions, methods: n-body simulations.

1. Introduction

A scenario has been put forward in which intermediate-mass, fast-rotating elliptical galaxies have been formed in mergers of disk galaxies (e.g. Naab *et al.* 2007, Johansson *et al.* 2009). The first galaxy interaction simulations by Toomre & Toomre (1972) contained only collisionless N-body particles. The development of SPH (see Monaghan 1992 for a review and e.g. Hu *et al.* 2014 for recent improvements) and computationally inexpensive gravitational force algorithms (e.g. the tree code) led eventually to modern galaxy merger codes such as Gadget (Springel 2005). Present-day simulations containing subresolution astrophysical models for gas and stellar metallicity, star formation, supernova feedback (Scannapieco *et al.* 2006) and supermassive black hole (SMBH) feedback (Springel *et al.* 2005, Choi *et al.* 2014) enable detailed comparisons with observations.

2. Metals in Merger Simulations

Following the approach of Johansson *et al.* (2009), we construct two disk galaxies consisting of a dark matter halo, a rotationally supported disk of gas and stars, and a central bulge. Both the bulge and halo component have Hernquist-like density profiles while the disk component is of an exponential form. We employ both random orbital geometries for large simulation samples and specific orbital geometries set to match observed galaxy mergers, such as the Antennae galaxies (Karl *et al.* 2010). We set the initial fractional abundances of 10 metal species (C, N, O, Ne, Mg, Si, S, Ca and Fe) in the following way. Observations indicate that a radial metallicity gradient exists in disk galaxies. We use dex units (dex $= 12 + \log 10(X/H)$) in which the radial abundance gradient is close to a linear form: $\mathrm{dex}(r) = \mathrm{dex}(r_0) + k(r - r_0)$, where r_0 is a scale radius and k is the negative metallicity gradient (Zaritsky *et al.* 1994, Kilian *et al.* 1994). The vertical abundance gradient is set in a similar way, motivated by observations from Ivezić *et al.* (2008). The ages of disk and bulge stars are initialized at $t = 0$ from an exponentially decaying star formation rate (SFR $\propto \exp(-t/\tau)$) with $\tau = 1$ Gyr for bulge stars and $\tau = 4$ Gyr for disk stars. The interstellar medium is enriched during the simulation by Type Ia and II supernovae and winds of AGB stars with metal yields calculated by Iwamoto *et al.* (1999), Woosley & Weaver (1995) and Karakas (2010), respectively. Gas metallicity is smoothed using a metal diffusion technique between neighboring gas particles (Aumer *et al.* 2013).

3. Ongoing and Future Work

Currently (September 2014), the initial conditions of the merger simulations have been prepared and high-resolution simulations will be run in late 2014. We study especially the impact of SMBH feedback on the metallicity evolution in galaxy mergers, since previous studies have not considered both detailed metallicity evolution models and black holes at the same time. We also plan to study the triggering mechanisms of active galactic nuclei (AGN) in the aftermath of galaxy mergers as these systems with high BH accretion rates at high redshifts are possible quasar host galaxy candidates. We will produce simulated spectral energy distributions of quasar host galaxies using stellar population modelling techniques and compare the results with the observed properties of quasar host galaxies (Kotilainen *et al.* 2009).

References

Aumer, M., White, S. D. M., Naab, T., & Scannapieco, C. 2013, *MNRAS*, 434, 3142
Choi, E., Naab, T., Ostriker, J. P., Johansson, P. H., & Moster, B. P. 2014, *MNRAS*, 442, 440
Hu, C.-Y., Naab, T., Walch, S., Moster, B. P., & Oser, L. 2014, *MNRAS*, 443, 1173
Johansson, P. H., Naab, T., & Burkert, A. 2009a, *ApJ*, 690, 802
Ivezić, Ž., Sesar, B., Jurić, M., *et al.* 2008, *ApJ*, 684, 287
Iwamoto, K., Brachwitz, F., Nomoto, K., *et al.* 1999, *ApJS*, 125, 439
Karakas, A. I. 2010, *MNRAS*, 403, 1413
Karl, S. J., Naab, T., Johansson, P. H., *et al.* 2010, *ApJ*, 715, L88
Kilian, J., Montenbruck, O., & Nissen, P. E. 1994, *A&A*, 284, 437
Kotilainen, J. K., Falomo, R., Decarli, R., *et al.* 2009, *ApJ*, 703, 1663
Monaghan, J. J. 1992, *ARA&A*, 30, 543
Naab, T., Johansson, P. H., Ostriker, J. P., & Efstathiou, G. 2007, *ApJ*, 658, 710
Scannapieco, C., Tissera, P. B., White, S. D. M., & Springel, V. 2006, *MNRAS*, 371, 1125
Springel, V. 2005, *MNRAS*, 364, 1105
Springel, V., Di Matteo, T., & Hernquist, L. 2005, *MNRAS*, 361, 776
Toomre, A. & Toomre, J. 1972, *ApJ*, 178, 623
Woosley, S. E. & Weaver, T. A. 1995, *ApJS*, 101, 181
Zaritsky, D., Kennicutt, R. C., Jr., & Huchra, J. P. 1994, *ApJ*, 420, 87

CHAPTER 8.

Cosmic Voids

Jaanilaupäev - St. John's Eve,
the most important day and evening in the Estonian calendar.

The Midsummer Night's sun setting over the Baltic Sea,
and casting its last rays over traditional Estonian farms.

Photo courtesy (top): Steven Rieder.
Photo (bottom): Rien van de Weijgaert

CHAPTER 8A.

Cosmic Voids
Structure, Dynamics and Cosmology

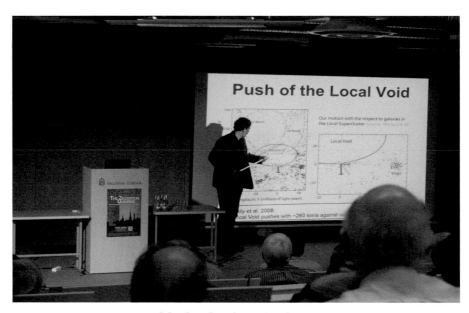

Much ado about Nothing:
Rien van de Weygaert reviewing cosmic voids.

On the stairs, behind the panels,
participants were to be found all over the conference hall.

The Zeldovich Universe:
Genesis and Growth of the Cosmic Web
Proceedings IAU Symposium No. 308, 2014   © International Astronomical Union 2016
R. van de Weygaert, S. Shandarin, E. Saar & J. Einasto, eds.   doi:10.1017/S1743921316010504

Voids and the Cosmic Web: cosmic depression & spatial complexity

Rien van de Weygaert

Kapteyn Astronomical Institute, University of Groningen,
Postbus 800, NL-9700AD, Groningen, the Netherlands
email: weygaert@astro.rug.nl

Abstract. Voids form a prominent aspect of the Megaparsec distribution of galaxies and matter. Not only do they represent a key constituent of the Cosmic Web, they also are one of the cleanest probes and measures of global cosmological parameters. The shape and evolution of voids are highly sensitive to the nature of dark energy, while their substructure and galaxy population provides a direct key to the nature of dark matter. Also, the pristine environment of void interiors is an important testing ground for our understanding of environmental influences on galaxy formation and evolution. In this paper, we review the key aspects of the structure and dynamics of voids, with a particular focus on the hierarchical evolution of the void population. We demonstrate how the rich structural pattern of the Cosmic Web is related to the complex evolution and buildup of voids.

Keywords. Cosmology, large-scale structure, voids, dark energy, modified gravity

---

1. Introduction

Voids form a prominent aspect of the Megaparsec distribution of galaxies and matter (Chincarini & Rood 1975; Gregory & Thompson 1978; Einasto, Joeveer & Saar 1980; Kirshner *et al.* 1981, 1987; de Lapparent, Geller & Huchra 1986; Colless *et al.* 2003; Tegmark *et al.* 2004; Guzzo *et al.* 2013, 2014). They are enormous regions with sizes in the range of $20 - 50h^{-1}$ Mpc that are practically devoid of any galaxy, usually roundish in shape and occupying the major share of space in the Universe (see fig. 1 and van de Weygaert & Platen (2011) for a recent review). Forming an essential and prominent aspect of the *Cosmic Web* (Bond, Kofman & Pogosyan 1996), they are instrumental in the spatial organization of the Cosmic Web (Icke 1984; Sahni, Sathyaprakash & Shandarin 1994; Sheth & van de Weygaert 2004; Aragon-Calvo *et al.* 2010; Einasto *et al.* 2011). Surrounded by elongated filaments, sheetlike walls and dense compact clusters, they weave the salient weblike pattern of galaxies and matter pervading the observable Universe.

Several recent studies came to the realization that voids not only represent a key constituent of the cosmic mass distribution, but that they are also one of the cleanest probes and measures of the global cosmology. Particularly interesting is the realization that their structure, morphology and dynamics reflects the nature of dark energy, dark matter and that of the possibly non-Gaussian nature of the primordial perturbation field. Another major aspect of voids is that their pristine environment represents an ideal and pure setting for the study of galaxy formation and the influence of cosmic environment on the evolution of galaxies. In addition, voids play a prominent role in the reionization process of the universe, forming the principal regions along which the ionizing radiation produced by the first stars in the Universe propages.

In a void-based description of the evolution of the cosmic matter distribution, voids mark the transition scale at which density perturbations have decoupled from the Hubble

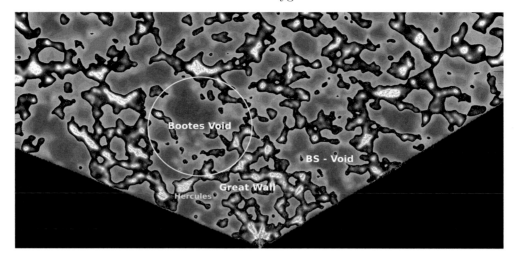

Figure 1. SDSS density map and galaxies in the a region of the SDSS galaxy redshift survey region containing the canonical Boötes void. The DTFE computed galaxy density map, Gaussian smoothed on a scale of $R_f = 1h^{-1}Mpc$, is represented by the color scale map. The galaxies in the SDSS survey are superimposed as dark dots. The underdense voids are clearly outlined as the lighter region outside the high-density weblike filamentary and wall-like features. We have indicated the location of the Hercules supercluster, the CfA Great Wall, and the Boötes void.

flow and contracted into recognizable structural features. At any cosmic epoch the voids that dominate the spatial matter distribution are a manifestation of the cosmic structure formation process reaching a non-linear stage of evolution. On the basis of theoretical models of void formation one might infer that voids may act as the key organizing element for arranging matter concentrations into an all-pervasive cosmic network (Icke 1984; Regős & Geller 1991; van de Weygaert 1991; Sheth & van de Weygaert 2004; Aragon-Calvo et al. 2010; Aragon-Calvo & Szalay 2013). As voids expand, matter is squeezed in between them, and sheets and filaments form the void boundaries. This view is supported by numerical studies and computer simulations of the gravitational evolution of voids in more complex and realistic configurations (Martel & Wasserman 1990; Regős & Geller 1991; Dubinski et al. 1993; van de Weygaert & van Kampen 1993; Goldberg & Vogeley 2004; Colberg et al. 2005; Padilla, Ceccarelli & Lambas 2005; Aragon-Calvo et al. 2010; Aragon-Calvo & Szalay 2013; Sutter et al. 2014b; Wojtak et al. 2016). A marked example of the evolution of a typical large and deep void in a ΛCDM scenarios is given by the time sequence of six frames in fig. 2.2.

The relatively simple structure and dynamical evolution of voids remains strongly influenced by the evolving large scale environment, through dominant tidal influences and direct contact between neighbouring voids. It is this aspect that has been recognized in a surge of recent studies for its considerable potential for measuring the value of cosmological parameters. The evolution of their structure and shape appear to be a direct manifestation of the nature of dark energy, while their dynamics reflects the nature of dark matter or that of the possibly nonstandard nature of gravity.

2. Depressions in the Universe

Voids are an outstanding aspect of the weblike cosmic mass distribution (de Lapparent, Geller & Huchra 1986; Colless et al. 2003; Huchra et al. 2012; Pan et al. 2012; Sutter

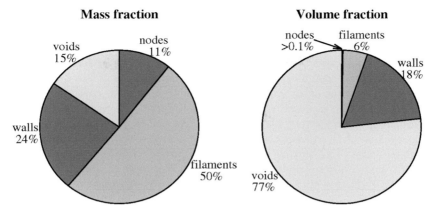

Figure 2. The mass and volume fractions occupied by cosmic web environments detected by the NEXUS+ method. From Cautun *et al.* 2014

et al. 2012). They have been known as a feature of galaxy surveys since the first surveys were compiled (Chincarini & Rood 1975; Gregory & Thompson 1978; Einasto, Joeveer & Saar 1980; Zeldovich *et al.* 1982). Following the discovery by Kirshner *et al.* (1981) and Kirshner *et al.* (1987) of the most dramatic specimen, the Boötes void, a hint of their central position within a weblike arrangement came with the first CfA redshift slice (de Lapparent, Geller & Huchra 1986). This view has been dramatically endorsed and expanded by the redshift maps of the 2dFGRS and SDSS surveys (Colless *et al.* 2003; Tegmark *et al.* 2004). They have established voids as an integral component of the Cosmic Web. The 2dFGRS maps and SDSS maps (see e.g. fig. 1) are telling illustrations of the ubiquity and prominence of voids in the cosmic galaxy distribution. Recent studies have also revealed the prominence of voids in the distant and early U niverse. A beautiful example is that of the deep probe of the VIPERS galaxy redshift survey, which meticulously outlined the Cosmic Web up to redshifts $z \approx 0.5$ (Guzzo *et al.* 2013, 2014). Perhaps even more impressive is the identification of the considerably more intricate void population in the dark matter distribution, as revealed by its reconstruction in the 2MRS survey volume (Kitaura 2012; Hess *et al.* 2013) and the SDSS volume (Leclercq *et al.* 2015; Leclercq 2015).

A relatively crude analysis of voids in computer simulations provides an impression of the status of voids in the Megaparsec universe. They clearly occupy a major share of the volume of the Universe. This has been confirmed in a recent systematic analysis by Cautun *et al.* (2014), which included an inventory of morphological components of the cosmic web with respect in the dark matter distribution in the LCDM Millennium simulation (see fig. 2). Around 77% of the cosmic volume should be identified as a void region. Nonetheless, it represents less than 15% of the mass content of the Universe. This implies that the average density in voids is $\approx 20\%$ of the average cosmic density, which is indeed reasonably close to what is expected on simple theoretical grounds (see below).

2.1. *Voids and the Cosmos*

With voids being recognized as prominent aspects of the Megaparsec galaxy and matter distribution, we should expect them to be a rich source of information on a range of cosmological questions. We may identify at least four:

• They are a prominent aspect of the Megaparsec Universe, instrumental in the spatial organization of the Cosmic Web (Icke & van de Weygaert 1987; van de Weygaert & van

Kampen 1993; Sheth & van de Weygaert 2004; Aragon-Calvo & Szalay 2013). Their effective repulsive influence over their surroundings has been recognized in surveys of the Local Universe (Courtois *et al.* 2012; Tully *et al.* 2014).

• Voids contain a considerable amount of information on the underlying cosmological scenario and on global cosmological parameters. They are one of the cleanest cosmological probes and measures of dark energy, dark matter and tests with respect to possible modifications of gravity as described by General Relativity. This realization is based on the fact that it is relatively straightforward to relate their dynamics to the underlying cosmology, because they represent a relatively modest density perturbation. Their structure and shape, as well as mutual alignment, are direct reflections of dark energy (Park & Lee 2007; Lee & Park 2009; Platen, van de Weygaert & Jones 2008; Lavaux & Wandelt 2010, 2012; Bos *et al.* 2012; Pisani *et al.* 2015). Notable cosmological imprints are also found in the outflow velocities and accompanying redshift distortions (Dekel & Rees 1994; Martel & Wasserman 1990; Ryden & Melott 1996). In particular interesting is the realization that the dynamics and infrastructure of voids are manifestations of the nature of dark matter or of modified gravity (Li 2011; Peebles & Nusser 2010; Clampitt *et al.* 2013; Cai *et al.* 2015; Cautun *et al.* 2016). The cosmological ramifications of the reality of a supersized voids akin to the identified ones by Rudnick, Brown & Williams (2007), Granett, Neyrinck & Szapudi (2009) and Szapudi *et al.* (2015) would obviously be far-reaching.

• The pristine low-density environment of voids represents an ideal and pure setting for the study of galaxy formation and the influence of cosmic environment on the formation of galaxies (e.g. Kreckel *et al.* 2011, 2012). Voids are in particular interesing following the observation by Peebles that the dearth of low luminosity objects in voids is hard to understand within the ΛCDM cosmology (Peebles 2001).

• Voids are prominent in the key reionization transition in the early universe, key targets of LOFAR and SKA (e.g. Furlanetto *et al.* 2006; Morales & Wyithe 2010).

2.2. *Voids in the Local Universe*

The most detailed view of the structure and galaxy population of voids is offered by voids in the local Universe. A particularly important source of information is the Karachentsev LV catalog of galaxies in the Local Volume, a volume-limited sample of galaxies within a radius of 11 Mpc around the Milky Way (Karachentsev *et al.* 2004). It provides a meticulously detailed view of the Local Void that appears to dominate a major fraction of space in our immediate cosmic neighbourhood. The desolate emptiness of this vast volume is most strikingly borne out by the adhesion reconstruction of the Local Void by Hidding *et al.* (2016) and Hidding (2016).

Moving to a slightly larger volume of the surrounding universe, we obtain a more representative impression of the prominence and role of voids in the overall large scale mass distribution. The 2MRS survey (Huchra *et al.* 2012) provides a uniquely complete census of the galaxy distribution out to distances of 100-150 Mpc. It entails the entire cosmic environment out to that distance, and has enabled remarkably precise and detailed reconstructions of the underlying matter distribution in our cosmic neighbourhood.

Figure 2.2 provides a remarkably detailed reconstruction of the cosmic web in the 2MRS volume. It shows the (surface) density of the weblike structures in the Local Universe. These are the result of adhesion simulations by Hidding *et al.* (2016), based on the the constrained Bayesian KIGEN reconstruction by Kitaura (2012) of the initial conditions in the local volume traced by the 2MRS redshift survey. For a given Gaussian primordial field, the adhesion formalism allows the accurate reconstruction of the rich pattern of weblike features that emerge in the same region as a result of gravitational evolution.

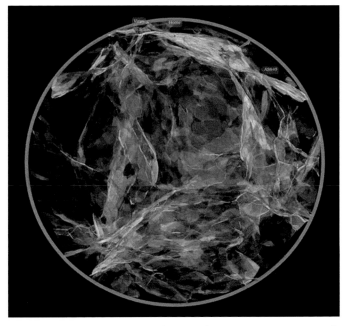

Figure 3. The Local Void. 3D rendering of the adhesion reconstruction of the Local Void. The reconstruction is based on the Bayesian KIGEN reconstruction by Kitaura *et al.* (2012) of the initial conditions in the local volume traced by the 2MRS redshift survey. The Local Supercluster, including the Virgo cluster and the Local Group, are located at the top of the image. Particularly striking is the precipitous emptiness of the Local Void. Image courtesy J. Hidding (see Hidding *et al.* 2016).

The adhesion formalism was applied to 25 constrained realizations of the 2MRS based primordial density field (Hidding *et al.* 2012, 2016). The mean of these realizations gives a reasonably accurate representation of the significant filamentary and wall-like features in the Local Universe. Most outstanding is the clear outline of the void population in the local Universe. The reconstruction also includes the velocity flow in the same cosmic region. It reveals the prominent nature of the outflow from the underdense voids, clearly forming a key aspect of the dynamics of the Megaparsec scale universe.

The Local Universe structure in figure 2.2 presents a telling image of a void dominated large scale Universe. Many of the voids in the adhesion reconstruction can be identified with the void nomenclature proposed by Fairall (Fairall 1998), who mainly identified these voids by eye from the 6dFGRS survey. It is interesting to see that the socalled Tully void appears to be a richly structured underdense region, containing at least the Microscopium Void, the Local Void and the "Trans Tully Void".

3. Formation and Evolution of Voids

Voids emerge out of the density troughs in the primordial Gaussian field of density fluctuations. Early theoretical models of void formation concentrated on the evolution of isolated voids (Hoffman & Shaham 1982; Icke 1984; Bertschinger 1985; Blumenthal *et al.* 1992). Nearly without exception they were limited to spherically symmetric configurations (but see Icke 1984). Neither of these simplifications appears to be close to what happens in reality. Nonetheless, the spherical model of isolated void evolution

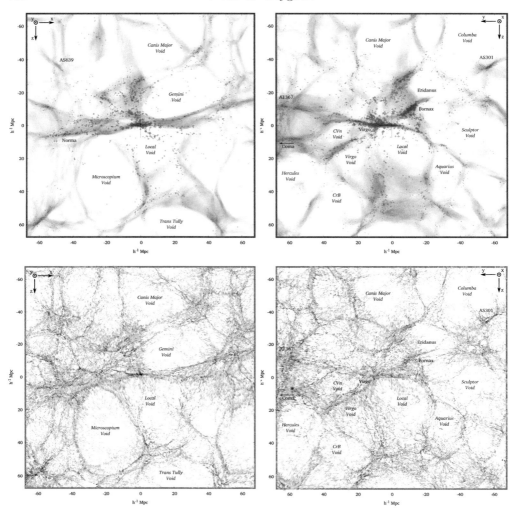

Figure 4. Voids in the Local Universe. The reconstruction of the weblike structure in the local Universe, sampled by the 2MRS survey, has been obtained on the basis of the adhesion formalism applied to a set of 25 constrained Bayesian KIGEN realizations of the primordial density and velocity field in the Local Universe. Top frames: (surface) density field along two perpendicular slices perpendicular to the plane of the Local Supercluster. Note that the density field concerns the dark matter distribution. The red dots are the 2MRS galaxies in the same volume. Bottom frame: the corresponding implied velocity flow in the same slices. Image courtesy J. Hidding (see Hidding *et al.* 2016).

appears to provide us with valuable insights in major physical void characteristics and, as important, with sometimes surprisingly accurate quantitative reference benchmarks (see below).

In the next section (sect.4.2), we will see that a major and dominant aspect of voids is the fact that they are the opposite of isolated objects. To understand the dynamical evolution of voids we need to appreciate two major aspects in which the dynamics of voids differs fundamentally from that off dark halos. Because voids expand and increase in size, they will naturally run up against their expanding peers. Their spatial distribution and organization will be substantially influenced by the way in which the mutually competing

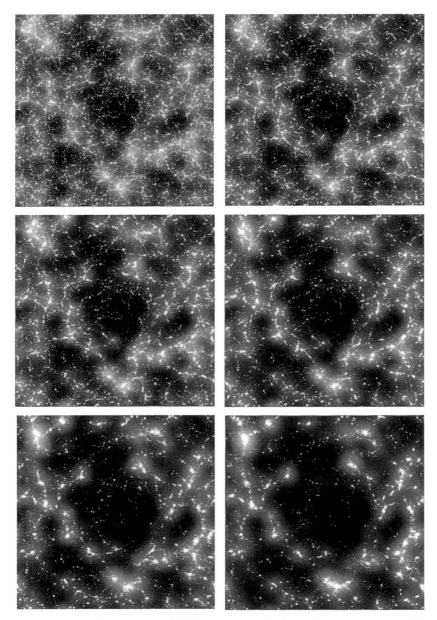

Figure 5. Simulation of evolving void (LCDM scenario). A void in a $n = 0$ power-law power spectrum model. The slice is $50h^{-1}$ Mpc wide and $10h^{-1}$ Mpc thick. Shown are the particles and smoothed density field (smoothed on a scale of $4h^{-1}$ Mpc) at six different timesteps: a=0.05, 0.15, 0.35, 0.55, 0.75 and 1.0. Image courtesy of Erwin Platen

voids distribute their share of space. Equally important is the fact that voids will always be limited to a rather modest density deficit: they cannot become more empty than empty. An immediate consequence of this is that the dynamical influence of the external mass inhomogeneities retains its dominant role in the evolution of a void. Voids will never and cannot decouple from their surroundings!

3.1. *Expansion, Evacuation, Dilution*

The essence of void evolution stems from the fact that they are underdensities in the mass distribution. As a result, they represent regions of weaker gravity. It translates in an effective repulsive peculiar gravitational influence. Most of the principal characteristics of void evolution can be recognized in the illustration of a typical evolving void in a ΛCDM Universe in fig. 2.2 (also see van de Weygaert & van Kampen 1993). The sequence of 6 timestepe of the simulation reveals at least three of the principal aspects of void evolution. Because of the effective repulsive and thus outward peculiar gravitational accleration, initially underdense regions expand faster than the Hubble flow: they expand with respect to the background Universe. Also we see that mass is streaming out of the underdensity. As a result of this evacuation, the density within voids continuously decrease. A clear census of the continuously decreasing mass content in voids can be found in the study of cosmic web evolution by Cautun *et al.* (2014). Isolated voids would asympotically evolve towards an underdensity $\delta = -1$, pure emptiness.

Also observable in figure 2.2 is how the mass distribution inside the void appears to become increasingly uniform. At the same time we see an accumulation of mass around its boundary. This is a direct consequence of the differential outward peculiar gravitational acceleration in voids. Because the density within underdense regions gradually increases outward, we see a decrease of the corresponding peculiar (outward) gravitational acceleration. It means that matter at the centre of voids moves outward faster than matter at the boundary regions. As a result, matter tends to accumulate in - filamentary or planar - ridges surrounding the void, while the interior evolves into a uniform low-density region resembling a low-density homogeneous FRW universe (Goldberg & Vogeley 2004).

Another key feature of void evolution is the diminishing prominence of substructure in its interior. Figure 2.2 clearly shows how the internal structure gradually disappears as mass moves out of the void. This is a direct manifestation of the complex hierarchical evolution of voids (Sheth & van de Weygaert 2004). A direct implication of this is that massive features and objects cannot form in voids, which e.g. manifests itself in a strong shift of the mass spectrum of dark halos - and hence galaxies - towards small masses (Cautun *et al.* 2014). Goldberg & Vogeley (2004) showed how this can be rather accurately modelled by means of a modification of the spectrum of density fluctuations towards one more appropriate for a low-density FRW universe.

Note that while by definition voids correspond to density perturbations of at most unity, $|\delta_v| \leqslant 1$, mature voids in the nonlinear matter distribution do represent *highly nonlinear* features. This may be best understood within the context of Lagrangian perturbation theory (Sahni & Shandarin 1996). Overdense fluctuations may be described as a converging series of higher order perturbations. The equivalent perturbation series is less well behaved for voids: successive higher order terms of both density deficit and corresponding velocity divergence alternate between negative and positive.

3.2. *Spherical Voids*

While in reality voids will never be isolated, nor spherical, we may obtain a basic understanding of the quantitative aspects of void evolution from the simple model of an evolving isolated spherical void. Figure 6 illustrates the evolution of spherical isolated voids. The void in the lefthand is a typical pure and uncompensated "bucket" void†. We notice the principal characteristics of void formation that we discussed in the previous section. The expansion of the void is evident as we see the boundary edge, and the ridge

† This is more commonly known as *tophat*, but given the configuration, *bucket* seems a more appropriate description.

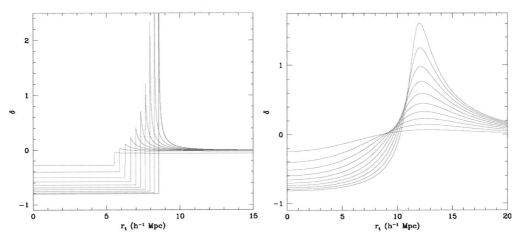

Figure 6. Spherical model for the evolution of voids. Left: a pure (uncompensated) "bucket" void evolving up to the epoch of shell-crossing. Initial (linearly extrapolated) density deficit was $\Delta_{lin,0} = -10.0$, the initial (comoving) radius $\widetilde{R}_{i,0} = 5.0h^{-1}\mathrm{Mpc}$. Righthand: a void with an angular averaged SCDM profile. Initial density deficit and characteristic radius are same as for the tophat void (left). The tendency of this void to evolve into a tophat configuration by the time of shell crossing is clear. Shell-crossing, and the formation of a ridge, happens only if the initial profile is sufficiently steep.

at its boundary, move outward. Meanwhile, matter moves out of the void, leading to an increasingly empty void.

The spherical void model also clarifies an additional major aspect, the formation of a ridge around the expanding void. This is the result of the differential outward expansion of the mass in and around the void. While the uniform underdensity profile implies a uniform expansion within the void, near the boundary the inner layers move outward faster than the more outward layers. This is the result of the latter feeling a more moderate interior underdensity. In all, this ultimately leads to the interior mass shells taking over the initially exterior shells. This leads to a *fundamental* evolutionary timescale for void evolution, that of *shellcrossing*.

Bertschinger (1985) demonstrated that once voids have passed the stage of shellcrossing they enter a phase of self-similar expansion. Subsequently, their expansion will slow down with respect to the earlier linear expansion. This impelled Blumenthal *et al.* (1992) to identify voids in the present-day galaxy distribution with voids that have just reached the stage of shell-crossing. It happens when a primordial density depression attains a linearly extrapolated underdensity $\delta_v = f_v = -2.81$ (strictly speaking for an EdS universe). That happens when a perfectly spherical "bucket" void will have expanded by a factor of 1.72 at shellcrossing, and therefore have evolved into an underdensity of $\sim 20\%$ of the global cosmological density, ie. $\delta_{v,nl} = -0.8$. In other words, the voids that we see nowadays in the galaxy distribution do probably correspond to regions whose density is $\sim 20\%$ of the mean cosmic density (note that it may be different for underdensity in the galaxy distribution).

Interestingly, for a wide range of initial radial profiles, voids will evolve into a bucket shaped void profile. The righthand frame of fig. 6 shows the evolution of a spherical void that develops out of an initial underensity that is an angular averaged density profile of an underdensity in a CDM Gaussian random field (see Bardeen *et al.* 1986; van de Weygaert & Bertschinger 1996). Due to the differential expansion of the interior mass shells, we get an acccumulation of mass near the exterior and boundary of the void, meanwhile evening out the density distribution in the interior. We also recognized this behaviour in the more complex circumstances of the LCDM void in fig. 2.2. To a considerable extent this is determined by the steepness of the density profile of the protovoid depression (Palmer & Voglis 1983). In nearly all conceivable situations the void therefore appears to assume a *bucket shape*, with a uniform interior density depression and a steep outer boundary (fig. 6, righthand frame). The development of a *bucket* shaped density profile may therefore be considered a generic property of cosmic voids.

Recently, there have been a range of studies on the issue of void density profiles, and the question whether they display universal behaviour (see e.g. Hamaus *et al.* 20144; Cautun *et al.* 2016). In a range of studies, Hamaus *et al.* (e.g Hamaus *et al.* 20144) concluded that spherically averaged density profiles of voids do indeed imply a universal density profile, that could be parameterized by 2 factors. Interestingly, these density profiles have a less prominent bucket shaped interior profile than those seen for the spherical voids in fig. 6. This may be understood from the fact that voids in general are not spherical (see sec. 4.2.2), so that spherical averaging will lead to the mixing of different layers in the void's interior. The recent study by Cautun *et al.* (2016) confirms this: when taking into account the shape of voids, a remarkably strong bucket void density profile appears to surface.

The corresponding void expansion velocity profile confirms the above. A uniform density distribution within a void's interior directly translates into a superHubble velocity outflow, ie. an outflow in which the expansion velocity scales linear with the distance to the void's center. In the next section we will discuss this within the context of the dynamics of voids.

4. Void Dynamics

Soon after their discovery, various studies pointed out their essential role in the organization of the cosmic matter distribution (Icke 1984; Regős & Geller 1991), based on the realization that they have a substantial dynamical influence on their surroundings. Voids have an effective repulsive influence over their surroundings (see below). Even though they represent relatively limited density deficits - voids will never exceed $|\delta| > 1$ - the fact that they occupy a major share of cosmic volume translates into a dominant factor in the dynamical interplay of forces on Megaparsec scales. The flow in and around the void is dominated by the outflow of matter from the void, culminating into the void's own expansion near the outer edge. This has indeed been recognized in various galaxy surveys and surveys of galaxy peculiar motions in our Local Universe. A telling illustration is the map of velocity flows implied by the PSCz galaxy redshift survey (see fig. 7). The gravitational impact of the Sculptor Void on our immediate cosmic vicinity is directly visible. The influence of voids on the peculiar velocities of galaxies have been even recognized at an individual level (Tully *et al.* 2008).

To fully to understand the dynamical influences on the evolution of voids, and as well to appreciate the dynamical impact of voids on their environment, we may obtain substantial insight from the idealized configurations of a spherically isolated void, and that of the force field on a homogeneous ellipsoidal void. The spherical model allows us to understand the principal aspects of the flow field in the interior of voids. The ellipsoidal model allows us to evaluate the importance of external forces on the dynamical evolution of voids.

In terms of their dynamics, the key observation is that external mass inhomogeneities retain a dominant role in the evolution of void. Key to this realization is the fact that voids are always be limited to a rather modest density deficit of not more than $\delta = -1$.

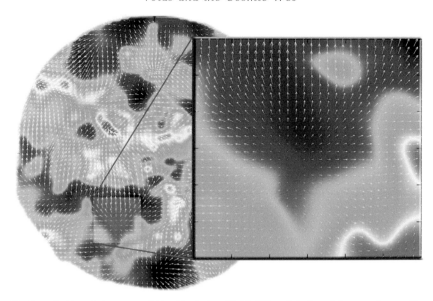

Figure 7. Gravitational impact of the Sculptor Void. The righthand frame shows the inferred velocity field in and around the Sculptor void near the Local Supercluster. The colour map represents the density values, with dark blue at $\delta \sim -0.75$ and cyan near $\delta \sim 0.0$. The vectors show the implied velocity flow around the void, with a distinct nearly spherically symmetric outflow. It is a zoom-in onto the indicated region in the density and velocity map in the Local Universe (lefthand) determined on the basis of the PSCz galaxy redshift survey. From: Romano-Díaz & van de Weygaert 2007.

This will strongly limit their own local gravitational action. It also means that the implicit tidal influences induces by the large scale environment of voids represent a major factor in their dynamical evolution. Their shape, mutual alignment and overall expansion are aspects that are strongly dependent on these external tidal influences.

It also means that global cosmological influences retain a strong impact on the void's evolution. While highly nonlinear overdensities internally hardly notice the presence of the repulsive force induced by the presence of dark energy, in the desolate interiors of voids dark energy plays an even considerably more prominent role than on global cosmological scales (Park & Lee 2007; Lee & Park 2009; Lavaux & Wandelt 2010, 2012). Along the same line, it is in the diluted density field of a void that the imprint of possible modifications of the force of gravity will be most noticeable (Li 2011; Clampitt *et al.* 2013). While this direct dynamical influence of dark energy, and gravity modifications, in voids may be one factor of importance, there is an equally and possibly even more sizeable secondary influence. The dominant role of external tidal influences on the void's dynamics depends on the amplitude of the inhomogeneities in the large scale environment of voids. The growth rate of mass fluctuations is a sensitive function of factors such as the nature of dark energy, and the nature of gravity on large scales, as these influence the dynamical timescales involved in the formation of structure.

To fully appreciate the forces at work in voids, we first looks at their own principal imprint, that of the void's expansion and the corresponding outflow velocities. Subsequently, we will investigate the external influences on the basis of the ellipsoidal model.

4.1. *Superhubble Bubbles*

Because voids are emptier than the rest of the universe, they have an outward directed velocity flow: mass is flowing out of the void. Evidently, the resulting void velocity field profile is intimately coupled to that of its interior density field profile.

In the situation of a mature, evolved void, the velocity field of a void resembles that of a Hubble flow, in which the outflow velocity increases linearly with distance to the void center. In other words, voids are *super-Hubble bubbles* (Icke 1984)! The linear velocity increase is a manifestation of its general expansion. The near constant velocity divergence within the void conforms to the *super-Hubble flow* expected for the near uniform interior density distribution that voids attain at more advanced stages (see fig. 6).

It is straightforward to appeciate this from the *continuity equation*. For a uniform density field, it tells us that the velocity divergence in the void will be uniform, corresponding to a Hubble-like outflow. Because voids are emptier than the rest of the universe they will expand faster than the rest of the universe, with a net velocity divergence equal to

$$\theta \; = \; \frac{\nabla \cdot \mathbf{v}}{H} = 3(\alpha - 1)\,, \qquad \alpha = H_{\mathrm{void}}/H\,, \tag{4.1}$$

where α is defined to be the ratio of the super-Hubble expansion rate of the void and the Hubble expansion of the universe. Inspection of the velocity flow in the Sculptor void (see fig. 7) shows that this is indeed a good description of the observed reality.

van de Weygaert & van Kampen (1993) confirmed that the velocity outflow field in viable cosmological scenarios does indeed resemble that of a superHubble expanding bubble (see fig. 8b). They also established that the superHubble expansion rate is directly proportional to the nonlinear void density $\Delta(t)$ (see fig. 8a),

$$H_{\mathrm{void}}/H \; = \; -\frac{1}{3} f(\Omega)\, \Delta(t)\,. \tag{4.2}$$

This relation, known within the context of a linearly evolving spherical density perturbation, in the case of fully evolved voids appears to be valid on the basis of the *nonlinear* void density deficit. As van de Weygaert & van Kampen (1993) illustrated, voids should therefore be considered as distinctly nonlinear objects. Several recent studies (e.g. Hamaus *et al.* 20144) have confirmed this finding for voids in a range of high resolution cosmological simulations.

4.2. *Tidal Forces - Ellipsoidal Voids*

Of decisive importance for understanding the dynamics and evolution of voids is the realization that voids will never and cannot decouple from their surroundings. One aspect of this is that voids expand and increase in size, they will naturally run up against their expanding peers. Their spatial distribution and organization will be substantially influenced by the way in which the mutually competing voids distribute their share of space.

The second strong environmental influence on the evolution of voids is that of the tidal influences induced by the large scale environment of voids. Their shape, mutual alignment and overall expansion are aspects that are strongly dependent on these external tidal influences.

4.2.1. *Homogeneous Ellipsoidal Model*

Arguably one of the most direct and transparent means of developing our insight into the tidally affected dynamics of voids is to look at the evolution of homogeneous underdense ellipsoids. This idea was forwarded by Icke (1984). In particular the dominant

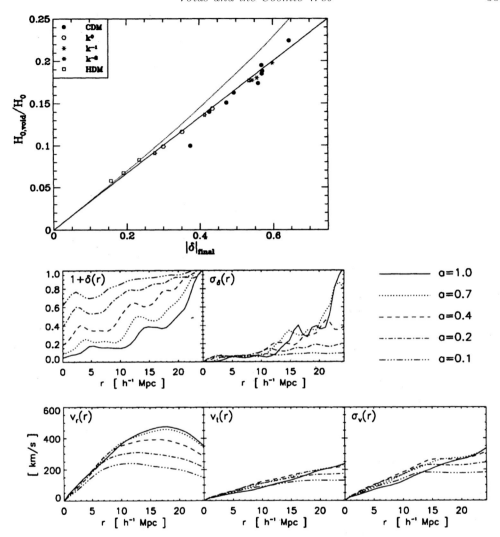

Figure 8. Superhubble Void expansion. Bottom rows: the density profiles and velocity profiles of an evolving void in a constrained simulation of a void in a CDM cosmology. Clearly visible is the evolution of the void towards a bucket shaped density profile, and the corresponding development of a linear superHubble flow field in the interior of the void. Top: the superHubble void expansion rate, in units of the global Hubble parameter $H(t)$ for a set of simulated voids in various cosmologies, as a function of the underdensity $\Delta(t)$ of a void (here called δ_{final}. From: van de Weygaert & van Kampen 1993.

role of external tidal forces on the evolution of voids can be understood by assessing the evolution of homogeneous underdense ellipsoids, within the context of *Homogeneous Ellipsoidal Model* (Icke 1973; White & Silk 1979; Eisenstein & Loeb 1995; Bond & Myers 1996; Desjacques 2008).

It is interesting to realize that the ellipsoidal model, even while idealized, is a rather good approximation for the main aspects of a void's evolution. In many respects the homogeneous model is far more suitable as an approximation for underdense regions than it is for overdense ones: voids expand and evolve towards an increasingly uniform interior density field over a vast range of their volume (see fig. 6): while they expand their interior gets drained of matter and develops a flat "bucket-shaped" density profile. Overdense regions contract into more compact and hence steeper density peaks, so that the area in which the ellipsoidal model represents a reasonable approximation will continuously shrink. Evidently, the approximation is restricted to the interior and fails at the void's outer fringes. The latter is a result of the accumulation of material near the void's edge, and the encounter with neighbouring features of the cosmic web and surrounding voids.

The homogeneous ellipsoidal model assumes an object to be a region with a triaxially symmetric ellipsoidal geometry and a homogeneous interior density, embedded within a uniform background density ρ_u. Consider the simple situation of the external tidal shear directed along the principal axes of the ellipsoid. The gravitational acceleration along the principal axes of an ellipsoid with over/underdensity δ can be evaluated from the expression for the corresponding scale factors $\mathcal{R}_i$,

$$\frac{d^2\mathcal{R}_m}{dt^2} = -4\pi G \rho_u(t) \left[\frac{1+\delta}{3} + \frac{1}{2}\left(\alpha_m - \frac{2}{3}\right)\delta\right]\mathcal{R}_m - \tau_m\,\mathcal{R}_m + \Lambda\mathcal{R}_m\,,\quad (4.3)$$

where we have also taken into account the influence of the cosmological constant Λ. The factors $\alpha_m(t)$ are the ellipsoidal coefficients specified by the integral equation,

$$\alpha_m(t) = \mathcal{R}_1(t)\mathcal{R}_2(t)\mathcal{R}_3(t)\int_0^\infty \frac{d\lambda}{(\mathcal{R}_m^2(t)+\lambda)\prod_{n=1}^3(\mathcal{R}_n^2(t)+\lambda)^{1/2}}\,.\quad (4.4)$$

The influence of the external (large-scale) tidal shear tensor $T_{mn}^{(ext)}$ enters via the eigenvalue τ_m. From eqn. 4.3, it is straightforward to appreciate that as δ grows strongly nonlinear, the relative influence of the large-scale (near-)linear tidal field will decline. However, the density deficit δ of voids will never exceed unity, $|\delta| < 1$, so that the importance of the factor τ_m remains relatively large.

The impact of the external tidal forces can be so strong they not only effect an anisotropic expansion of the void, but even may manage to make an initially spherically expanding void to collapse. The latter can be seen in fig. 9, where we compare the evolution of the two initially spherical and equivalent void regions. The isolated one assumes the regular evolution of a spherical isolated void, while its peer develops into an increasingly anisotropic configuration. At some point the external tides go as far to squeeze the void into contraction and ultimately towards collapse. This is the situation that we find in the hierarchical buildup of voids, to be described in the next section 5. As illustrated in the accompanying illustration of the hierarchically evolving void population, small voids near the boundary of large void bubbles get squeezed into collapse as a result of the tidal impact of their overdense surroundings.

4.2.2. *Void Shapes and Tides*

Icke (1984) pointed out that any (isolated) aspherical underdensity will become more spherical as it expands. The effective gravitational acceleration is stronger along the short axis than along the longer axes. For overdensities this results in a stronger inward acceleration and infall, producing increasingly flattened and elongated features. By contrast, for voids this translates into a larger *outward* acceleration along the shortest axis so that asphericities will tend to diminish. For the interior of voids this tendency has been confirmed by N-body simulations (van de Weygaert & van Kampen 1993).

In reality, voids will never reach sphericity. Even though voids tend to be less flattened or elongated than the halos in the dark matter distribution, they are nevertheless quite nonspherical: Platen, van de Weygaert & Jones (2008) find that they are slightly prolate

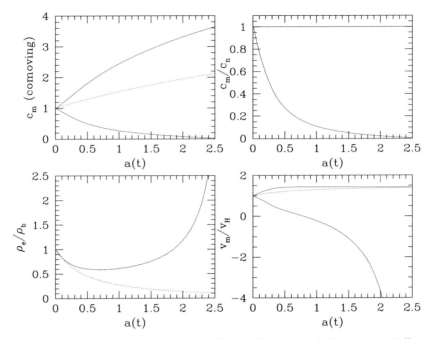

Figure 9. Anisotropic Void Expansion and Collapse. Evolution of the same initially spherical void. Dashed line: without external influences Solid line: under the influence of an axisymmetric and linearly evolving external tidal force field $(E_{mm,0} = (-E, -E, 2E))$. *Topleft:* (comoving) void axis, $c'_m = c_m/a(t)$. *Topright:* axis ratios c_2/c_1 and c_3/c_1. *Bottom Left:* Internal density ρ_e of the void, in units of the cosmic density $\rho_u(t)$. *Bottom Right:* The velocity v_m along the axes of the voids, in units of the Hubble velocity v_H. Note the collapse along axis 1 and 2.

with axis ratios of the order of $c : b : a \approx 0.5 : 0.7 : 1$. This agrees with the statement by Shandarin *et al.* (2006) and Park & Lee (2007) that in realistic cosmological circumstances voids will be nonspherical. This is also quite apparent in images of simulations, such as those in figure 5.

The flattening is a result of large scale dynamical and environmental factors, amongst which we can identify at least two important factors (Platen, van de Weygaert & Jones 2008). Even while their internal dynamics pushes them to a more spherical shape they will never be able to reach perfect sphericity before encountering surrounding structures such as overdense filaments or planar walls. Even more important is the fact that, for voids, the external tidal influences remain important (see discussion above). These external tidal forces are responsible for a significant anisotropic effect in the development of the voids. In extreme cases they may even induce fullscale collapse and demolition of the void.

4.2.3. *Void Shape Measurement: the WVF Watershed Void Finder*

Figure 4.2.4 illustrates a typical example of the shape distribution of voids in a ΛCDM cosmology. The identification of voids by means of the watershed transform, which delineates the region of a void independent of its scale and shape, a direct objective measurement of the volume, shape and orientation of the void population is obtained. This may be directly inferred from the comparison between the bottom panels with the dark matter distribution in the top panel of fig. 4.2.4. Introduced and proposed by Platen, van de Weygaert & Jones (2007) (also see contribution by Jones & van de Weygaert in this volume), the Watershed Void Finder (WVF) identifies voids via a watershed transform applied to the DTFE (Delaunay Tessellation Field Estimator) density field reconstruction (Schaap & van de Weygaert 2000; van de Weygaert & Schaap 2009). The latter exploits the scale and shape sensitivity of Voronoi and Delaunay tessellations to retain the intricate multiscale and weblike nature of the mass distribution probed by a discrete particle or galaxy distribution. The idea of the use of the watershed transform for the objective identification of voids, given the close relation to the topology of the weblike Megaparsec matter distribution (see Aragon-Calvo et al. 2010), has in the meantime been recognized as a true watershed with respect to setting a standard definition for what should be considered as a void (Colberg et al. 2008; Neyrinck 2008; Sutter et al. 2015; Nadathur & Hotchskiss 2015).

4.2.4. Void Alignments

Large scale tidal influences not only manifest themselves in the shaping of individual voids. They are also responsible for a distinct alignment of substructures along a preferred direction, while they are also instrumental in their mutual arrangement and organization. Locally, the orientation of a void turns out to be strongly aligned with the tidal force field generated by structures on scales up to at least $20 - 30h^{-1}$Mpc. This goes along with a similar mutual alignment amongst voids themselves. They have strongly correlated orientations over distances $> 30h^{-1}$Mpc (Platen, van de Weygaert & Jones 2008), a scale considerably exceeding the typical void size. It forms a strong confirmation of the large scale tidal force field as the dominant agent for the evolution and spatial organization of the Megaparsec Universe, as emphasized by the Cosmic Web theory (Bond, Kofman & Pogosyan 1996; van de Weygaert & Bond 2008).

4.2.5. Void Shapes and Dark Energy

A third interesting influence on the shape of vois is the possible impact of dark energy on the dynamics and evolution of voids. This influence is a result of its direct repulsive effect on the force field of the void, as well as indirectly via the external tidal force field induced by the surrounding large scale inhomogeneities.

Following the earlier suggestions by Park & Lee (2007); Lee & Park (2009), studies by Wandelt and collaborators (Lavaux & Wandelt 2010; Biswas et al. 2010) showed that void shapes may be used as precision probes of dark energy. Biswas et al. (2010) even quoted the possibility of improving the figure of the Dark Energy Task Force figure of merit by an order of hundred for future experiments like Euclid. An elaborate study by Bos et al. (2012) of void shape evolution in simulations within a range of different dark energy cosmologies confirmed the high level of sensitivity of the shapes of dark matter voids to the underlying dark energy (see fig. 4.2.4). However, it is less straightforward to reach similar conclusions of the shapes of voids in the observed galaxy distribution (Bos et al. 2012).

Lavaux & Wandelt (2012) forwarded the suggestion of using the Alcock-Paczysnki test on the stacking of voids in a sufficiently large cosmic volume. Assuming that the orientation of voids is random, the shapes of properly scaled voids would average out to purely spherical. This comes along with the significant advantage of substantially increasing the signal-to-noise of the resulting void stack. Given the observation of the voids in a galaxy redshift survey, as a result of the cosmic expansion the resulting void stack in redshift space will be stretched differently in the radial direction than in the transverse direction. The ratio between the transverse and radial stretching of an intrinsically spherical feature yields direct information on the angular diameter distance and Hubble parameter at a redshift z. Since the suggestion by Lavaux & Wandelt (2012), several groups realized

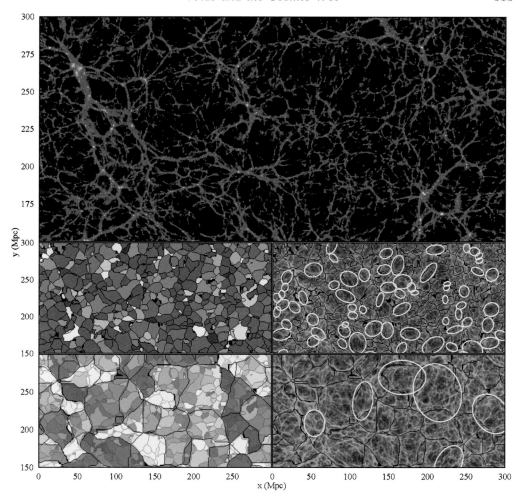

Figure 10. Shapes of voids in a ΛCDM simulation. Top: a density field slice of thickness $0.25h^{-1}$Mpc, and size 300 by 150 h^{-1}Mpc. *Bottom left two*: the corresponding distribution of voids. The voids have been identified with the Watershed Void Finder (Platen *et al.* 2007), for the top panel a Gaussian filter with radius $R_f = 1.5h^{-1}$Mpc, for the bottom panel $R_f = 6.0h^{-1}$Mpc (the $1.5h^{-1}$Mpc ones are transparently inset). *Bottom right two*: the shapes of the watershed identified voids have been determined via the inertia tensor of the corresponding region. A random selection of the ellipsoid fits (yellow) is overlaid on the density field (now in grayscale), again at two radii 1.5 and $6.0h^{-1}$Mpc. From Bos *et al.* 2012.

the large cosmological potential of the application of the Alcock-Paczysnki test to voids in large surveys. It has also resulted in a range of interesting studies (e.g. Sutter *et al.* 2012b, 2014; Pisani *et al.* 2014; Pisani 2014), and is generating substantial interest for exploiting voids in upcoming large galaxy surveys (e.g. Nadathur 2016).

Of key importance for a succesful application of the Alcock-Paczysnki test to samples of voids is to correct for some systematic effects. One concerns the correction for the peculiar velocity outflow of the void itself (see Pisani 2014, for a clear discussion). Another effect that should be accounted for is the alignment of voids on large scales, as discussed in the previous subsection 4.2.4 (also see fig. 4.2.4). Clearly, if the region from which the void sample is extracted is not much larger than the scale on which tidally induced alignment are expected, the stacked voids may not define an intrinsically spherical object and hence influence the outcome of the determined cosmological parameters.

4.3. *Dynamical Influence of Voids*

Various studies have found strong indications for the imprint of voids in the peculiar velocity flows of galaxies in the Local Universe. Bothun *et al.* (1992) made the first claim of seeing pushing influence of voids when assessing the stronger velocity flows of galaxies along a filament in the first CfA slice. Stronger evidence came from the extensive and systematic POTENT analysis of Mark III peculiar galaxy velocities (Willick *et al.* 1997) in the Local Universe (Dekel, Bertschinger & Faber 1990; Bertschinger *et al.* 1990). POTENT found that for a fully selfconsistent reconstruction of the dynamics in the Local Universe, it was inescapable to include the dynamical influence of voids (Dekel 1994).

With the arrival of new and considerably improved data samples the dynamical influence of voids in the Local Universe has been investigated and understood in greater detail. The reconstruction of the density and velocity field in our local cosmos on the basis of the 2MASS redshift survey has indeed resulted in a very interesting and complete view of the dynamics on Megaparsec scales. This conclusion agree with that reached on the basis of an analysis of the peculiar velocity of the Local Group by Tully *et al.* (2008). Their claim is that the Local Void is responsible for a considerable repulsive influence, accounting for ~ 259 km s^{-1} of the ~ 631 km s^{-1} Local Group motion with respect to the CMB (also see fig. 2.2).

The substantial dynamical role of voids in the large scale Universe has been most convincingly demonstrated in the recent advances in the study of cosmic flows enabled by the completion of the Cosmicflows2 and Cosmicflows3 surveys (Tully *et al.* 2008; Courtois *et al.* 2012). The implied velocity-based reconstructions of the Local Universe, culminating in the suggestion of the hypercluster Laniakea as local dynamical entity (Tully *et al.* 2014), clearly reveal the substantial repulsive influence of voids in the Local Universe.

5. Void Sociology & Void Hierarchy: Bubbles in Soapsuds

Computer simulations of the gravitational evolution of voids in realistic cosmological environments do show a considerably more complex situation than that described by idealized spherical or ellipsoidal models (Martel & Wasserman 1990; Regős & Geller 1991; Dubinski *et al.* 1993; van de Weygaert & van Kampen 1993; Goldberg & Vogeley 2004; Colberg *et al.* 2005; Padilla, Ceccarelli & Lambas 2005; Ceccarelli *et al.* 2006; Bos *et al.* 2012; Aragon-Calvo & Szalay 2013; Sutter *et al.* 2014b; Wojtak *et al.* 2016). In recent years the huge increase in computational resources has enabled N-body simulations to resolve in detail the intricate substructure of voids within the context of hierarchical cosmological structure formation scenarios (Mathis & White 2002; Gottlöber *et al.* 2003; Goldberg & Vogeley 2004; Colberg *et al.* 2005; Padilla, Ceccarelli & Lambas 2005; Ceccarelli *et al.* 2006; Bos *et al.* 2012; Aragon-Calvo & Szalay 2013; Sutter *et al.* 2014b; Wojtak *et al.* 2016). They confirm the theoretical expectation of voids having a rich substructure as a result of their hierarchical buildup (see e.g. fig. 5).

Sheth & van de Weygaert (2004) treated the emergence and evolution of voids within the context of *hierarchical* gravitational scenarios. It leads to a considerably richer view of the evolution of voids. The role of substructure within their interior and the interaction with their surroundings turn out to be essential aspects of the *hierarchical* evolution of the void population in the Universe. An important guideline are the heuristic void

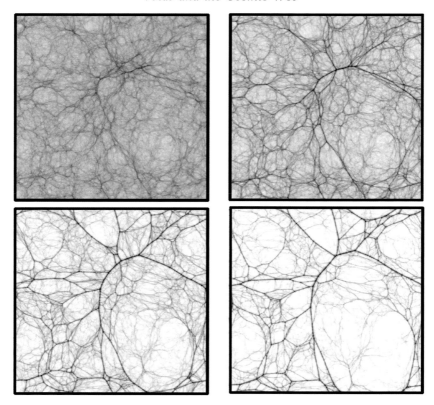

Figure 11. The hierarchical evolution of a void. The evolving mass distribution in and around a large void in an LCDM cosmology. The evolution runs from top lefthand to the bottom righthand frame. The structure depicted shows the result of an adhesion calculation of the evolving weblike network of walls, filaments and nodes in the cosmic volume. It shows in detail and with considerable contrast the evolution of the void population. From Hidding *et al.* 2016.

model simulations by Dubinski *et al.* (1993), and the theoretical void study by Sahni, Sathyaprakash & Shandarin (1994) within the context of a Lagrangian adhesion model description.

In some sense voids have a considerably more complex evolutionary path than overdense halos. Their evolution is dictate by two processes: their *merging* into ever larger voids as well as the *collapse* and disappearance of small ones embedded in overdense regions (see fig. 12). As argued by Sheth & van de Weygaert (2004), the implied hierarchical development of voids, akin to the evolution of overdense halos, may be described by an *excursion set* formulation (Press & Schechter 1974; Bond *et al.* 1991; Sheth 1998). To take account of the more complex evolutionary history, the evolution of voids needs to be described by a *two-barrier* excursion set formalism.

The resulting hierarchical buildup of the void population resembles that of a gradually diluting soapsud. At early time, emerging from a primordial Gaussian random field, the first bubbles to appear are small matured voids. The ones coexisting within the realm of a larger underdense region, will gradually merge into a larger void. This is the typical fate of a void in a *void-in-void* configuration. Meanwhile, small voids that find themselves in overdense regions or near the boundary of larger underdensities get squeezed out of existence, the fate for a void in a *void-in-cloud* configuration. What remains is a sud of larger void bubbles. This sequence proceeds continuously. Given the almost volume-filling nature of voids, this void hierarchical process conjures up the impression of a weblike pattern evolving as the result of continuously merging bubbles, with filaments, walls and cluster nodes at the interstices of the network (see Icke & van de Weygaert 1987; Aragon-Calvo 2014, for a geometric model along these lines).

5.0.1. *Void Merging*

First, consider a small region which was less dense than the critical void density value. It may be that this region is embedded in a significantly larger underdense region which is also less dense than the critical density. Many small primordial density troughs may exist within the larger void region. Once small density depressions located within a larger embedding underdensity have emerged as true voids at some earlier epoch, their expansion tends to slow down.

When the adjacent subvoids meet up, the matter in between is squeezed in thin walls and filaments. The peculiar velocities perpendicular to the void walls are mostly suppressed so that the flow of matter is mostly confined to tangential motions. The subsequent merging of the subvoids is marked by the gradual fading of these structures while matter evacuates along the walls and filaments towards the enclosing boundary of the emerging void (Dubinski *et al.* 1993) (also see top row fig. 12). The timescale on which the internal substructure of the encompassing void is erased is approximately the same as that on which it reaches maturity.

The final result is the merging and absorption of the subvoids in the larger void emerging from the embedding underdensity. Hence, as far as the void population is concerned only the large void counts, while the smaller subvoids should be discarded as such. Only a faint and gradually fading imprint of their original outline remains as a reminder of the initial internal substructure.

5.0.2. *Void Collapse*

A *second* void process is responsible for the radical dissimilarity between void and halo populations. If a small scale minimum is embedded in a sufficiently high large scale density maximum, then the collapse of the larger surrounding region will eventually squeeze the underdense region it surrounds: the small-scale void will vanish when the region around it has collapsed completely. Alternatively, though usually coupled, they may collapse as a result of the tidal force field in which they find themselves. If the void within the contracting overdensity has been squeezed to vanishingly small size it should no longer be counted as a void (see fig. 12, bottom row).

When inspecting the evolving matter distribution in and around voids, such as in the high resolution configuration of fig. 5, we find that most of the collapsing or squeezed voids are small voids to be found in and near the boundary regions of large underdense void regions. The small voids in these regions are strained by the high density boundary regions, mostly filaments and clusters, and in many situations squeezed out of existence. On the other hand, subvoids in the interior of a larger void tend to merge with surrounding peers. A detailed assessment of the evolution of these regions reveals that the void collapse process is an important aspect of the evolution and buildup of the cosmic web. Interestingly, recent observational studies have indeed identified the existence of collapsing void regions by studying the redshift structure around voids in the SDSS galaxy redshift survey (Paz 2013).

The collapse of small voids is an important aspect of the symmetry breaking between underdensities and overdensities. In the primorial Universe, Gaussian primordial conditions involve a perfect symmetry between under- and overdense. Any inspection of a galaxy redshift map or an N-body simulation shows that there is a marked difference

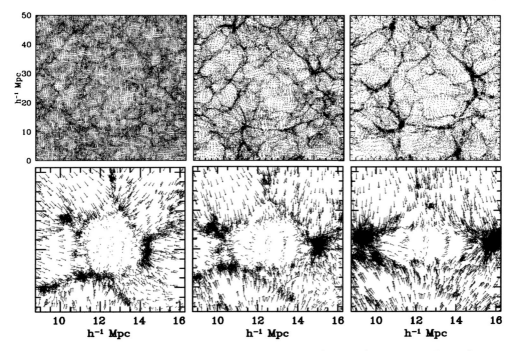

Figure 12. The two modes of void evolution: void merging (top row) and void collapse (bottom row). Top: three timesteps of evolving void structure in a 128^3 particle N-body simulation of structure formation in an SCDM model ($a_{\mathrm{exp}} = 0.1, 0.3, 0.5$). The sequence shows the gradual development of a large void of diameter $\approx 25 h^{-1}\mathrm{Mpc}$ as the complex pattern of smaller voids and structures which had emerged within it at an earlier time, merge with one another. It illustrates the *void-in-void* process of the evolving void hierarchy. Bottom: a choice of three collapsing voids in a constrained N-body simulation, each embedded within an environment of different tidal shear strength. The arrows indicate the velocity vectors, showing the infall of outer regions onto the void region. As a result the voids will be crushed as the surrounding matter rains down on them.

between matter clumps and voids. While the number density of halos is dominated by small objects, void collapse is responsible for the lack of small voids.

5.1. *Void Excursions*

Marked by the two processes of merging and collapse, the complex hierarchical buildup of the void population may be modelled by a two-barrier excursion set formalism (Sheth & van de Weygaert 2004). The barriers refer to the critical (linear) density thresholds involved with the *merging* and *collapse* of voids. Whenever the linearly extrapolated (primordial) $\delta_L(\vec{r}, t|R)$ on a scale R,

$$\delta_L(r, t|R) \;=\; \frac{D(t)}{D(t_i)}\, \delta_L(r, t_i|R)\,, \tag{5.1}$$

exceeds a density threshold f_c it will collapse. For an Einstein-de Sitter $\Omega_m = 1$ Universe the critical value has the well-known value $f_c \simeq 1.686$ (Gunn & Gott 1972). A void will form when an underdensity reaches the critical density threshold of *shell-crossing*, corresponding to a value of $f_v \simeq -2.81$ for spherical voids in an Einstein-de Sitter Universe. The linear density growth factor $D(t)$, normalized to unity at the present epoch, follows from the integral (Heath 1977; Peebles 1980; Hamilton 2001; Lahav &

Suto 2004),

$$D(t) = D(t, \Omega_{m,0}, \Omega_{\Lambda,0}) = \frac{5\,\Omega_{m,0}\,H_0^2}{2}\, H(a) \int_0^a \frac{da'}{a'^3 H^3(a')}. \tag{5.2}$$

Emerging from a primordial Gaussian random field, many small voids may coexist within one larger void. Small voids from an early epoch will merge with one another to form a larger void at a later epoch. The excursion set formalism takes account of this *void-in-void* configuration by discarding these small voids from the void count. To account for the impact of voids disappearing when embedded in collapsing regions, the two-barrier formalism also deals with the *void-in-cloud* problem.

By contrast, the evolution of overdensities is governed only by the *cloud-in-cloud* process: the *cloud-in-void* process is much less important, because clouds which condense in a large scale void are not torn apart as their parent void expands around them. This asymmetry between how the surrounding environment affects halo and void formation is incorporated into the *excursion set approach* by using one barrier to model halo formation and a second barrier to model void formation. Only the first barrier matters for halo formation, but both barriers play a role in determining the expected abundance of voids.

5.2. *Void Population Statistics.*

The analytical evaluation of the two-barrier random walk problem in the extended Press-Schechter approach leads directly to a prediction of the distribution function $n_v(M)$ for voids on a mass scale M (or corresponding void size R†). The resulting void spectrum‡. is peaked, with a sharp cutoff at both small and large values of the peak mass $M_{v,*}$ (fig. 5.2, bottom lefthand frame),

$$n_v(M)\,dM \quad \approx \tag{5.3}$$

$$\sqrt{\frac{2}{\pi}}\,\frac{\rho_u}{M^2}\, \nu_v(M)\, \exp\left(-\frac{\nu_v(M)^2}{2}\right)\left|\frac{d\ln\sigma(M)}{d\ln M}\right| \exp\left\{-\frac{|f_v|}{f_c}\frac{\mathcal{D}^2}{4\nu_v^2} - 2\frac{\mathcal{D}^4}{\nu_v^4}\right\},$$

where $\sqrt{\nu_v(M)}$ is the fractional relative underdensity,

$$\nu_v(M) \equiv \frac{|f_v|}{\sigma(M)}, \tag{5.4}$$

in which with the dependence on the mass scale M entering via the rms density fluctuation on that scale, $\sigma(M)$. The quantity $\mathcal{D}$ is the *"void-and-cloud parameter"*, $\mathcal{D} \equiv |f_v|/(f_c + |f_v|)$. It parameterizes the impact of halo evolution on the evolving population of voids: the likelihood of smaller voids being crushed through the *void-in-cloud* process decreases as the relative value of the collapse barrier f_c with respect to the void barrier f_v becomes larger.

Assessment of the relation above shows that the population of large voids is insensitive to the *void-in-cloud* process. The large mass cutoff of the void spectrum is similar to the ones for clusters and reflects the Gaussian nature of the fluctuation field from which the objects have condensed. The *characteristic void size* increases with time: the gradual

---

† The conversion of the void mass scale to equivalent void radius R is done by assuming the simplest approximation, that of the spherical tophat model. According to this model a void has expanded by a factor of 1.7 by the time it has mature, so that $V_v = (M/\rho_u) * 1.7^3$.

‡ Note that for near-empty voids the mass scale M is nearly equal to the corresponding void mass deficit.

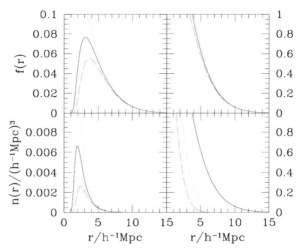

Figure 13. Distribution of void radii predicted by the two-barrier extended PS formalism, in an Einstein de-Sitter model with $P(k) \propto k^{-1.5}$, normalized to $\sigma_8 = 0.9$ at $z = 0$. Top left panel shows the mass fraction in voids of radius r. Bottom left panel shows the number density of voids of radius r. Note that the void-size distribution is well peaked about a characteristic size provided one accounts for the void-in-cloud process. Top right panel shows the cumulative distribution of the void volume fraction. Dashed and solid curves in the top panels and bottom left panel show the two natural choices for the importance of the void-in-cloud process discussed in the text: $\delta_c = 1.06$ and 1.686, with $\delta_v = -2.81$. Dotted curve shows the result of ignoring the *void-in-cloud* process entirely. Clearly, the number of small voids decreases as the ratio of $\delta_c/|\delta_v|$ decreases. Bottom right panel shows the evolution of the cumulative void volume fraction distribution. The three curves in this panel are for $\delta_c = 1.686(1 + z)$, where $z = 0$ (solid), 0.5 (dotted) and 1 (dashed). From Sheth & van de Weygaert 2004.

merging of voids into ever larger ones is embodied in a self-similar shift of the peak of the void spectrum.

When evaluating the corresponding fraction of contained in voids on mass scale M, $f(M) = M n_v(M)/\rho_u$, we find that this also peaks at the characteristic void scale. It implies a mass fraction in voids of approximately thirty percent of the mass in the Universe, with most of the void mass to be found in voids of this characteristic mass.

While the two-barrier excursion set formalism offers an attractive theoretical explanation for the distinct asymmetry between clumps and voids and for the peaked void size distribution, realistic cosmological simulations are needed to identify where the disappearing small-scale voids are to be found in a genuine evolving cosmic matter distribution.

An important aspect of the implied void population is that it is approximately *space-filling*. It underlines the adagio that the large scale distribution of matter may be compared to a *soapsud of expanding bubbles*. This follows from evaluation of the cumulative integral

$$\mathcal{F}_V(M) \equiv \int_M^\infty (1.7)^3 \frac{M' n_v(M')}{\rho_u} dM' . \qquad (5.5)$$

where the factor 1.7 is an estimate of the excess expansion of the void based upon the spherical model for void evolution (see footnote). The top righthand panel of fig. 5.2 shows the resulting (current) cumulative void volume distribution: for a finite value of void radius R the whole of space indeed appears to be occupied by voids, while there is the gradual shift of the cumulative volume distribution towards larger voids. In other words, the correct image appears to be that of a gradually unfolding bubbly universe in which the average size of the voids grows as small voids merge into ever larger ones.

While the SvdWmodel has proven to provide a good model for the evolving void population, it should be considered as a basic framework. Substantial finetuning to realistic circumstances, devoid of spherical symmetry and other implicit simplifications, will likely improve the model. One major issue concerns the value of the typical void formation threshold or density barrier. The simple choice for the shellcrossing of a hypothetical spherical configuration certainly deserves scrutiny, and considerable improvements to this have been forwarded (Paranjape *et al.* 2012; Jennings *et al.* 2013; Pontzen *et al.* 2016).

Also, to connect the SvdW formalism to the voids in the observed galaxy distribution involves a range of factors. Galaxies are diluted and biased tracers of the underlying dark matter distribution. While the dark matter voids define a rich and complex structural pattern (see e.g. Leclercq *et al.* 2015; Leclercq 2015), voids in the galaxy distribution will only represent a limited aspect of the void populations. Furlanetto & Piran (2006) elaborated on the SvdW formalism to describe what it would imply for voids in the galaxy distribution.

The advantage of having an analytical formalism for describing the void population has recently been recognized in the context of casting constraints on the nature of dark energy. By evaluating the implied void population in a range of dark energy scenarios, Pisani *et al.* (2015) demonstrated that in upcoming surveys such as Euclid it will be possible to infer competitive estimates of the equation of state of dark energy, as well as on its time derivative. The use of the analytical description of the two-barrier excursion set model provides highly versatile and flexible framework for readily identifying the most sensitive factors and situations. The same philosophy was followed in the elaboration by Li & Efstathiou (2012), who applied the excursion set formalism towards a general description of the evolving void population in modified gravity scenarios.

5.3. *Void Substructure*

An important issue within the hierarchically proceeding evolution of voids and the Cosmic Web is the fate of its substructure. In voids the diluted and diminished infrastructure remains visible, at ever decreasing density contrast, as cinders of the earlier phases of the *void hierarchy* in which the substructure stood out more prominently.

N-body simulations show that voids do retain a rich infrastructure. Examples such as the fig. 5, and images of large cosmological simulations such as the Millennium and the Illustris simulation (Springel *et al.* 2005) show that while void substructure does fade, it does not disappear. We may find structures ranging from filamentary and sheetlike structures to a population of low mass dark matter halos and galaxies (see e.g. Rieder *et al.* 2013). Although challenging, these may also be seen in the observational reality. The galaxies that populate the voids do currently attract quite some attention (see next section). Also, the SDSS galaxy survey has uncovered a substantial level of substructure within voids, such as a the VGS31 filament by Beygu *et al.* (2013).

The gradual fading of the initially rich substructure as a void expands, and its matter Rcontent evacuates its interior, is clearly visible in the development of the void region in fig. 2.2. This may be seen as the slowing of structure growth in the low-density environment of voids. The resulting observation is one in which the density distribution in a void region undergoes a manifest and continuous transformation towards an ever larger dominating scale, while it mostly retains the topological features of the initial density field. In a sense, it reflects of the buildup of the Cosmic Web itself, its basic pattern already recognizable in the primordial matter distribution (Bond, Kofman & Pogosyan 1996; van de Weygaert & Bond 2008).

Acknowledgements

I wish to thank the many collaborators and students that over many years have accompanied me in my explorations of the desolate but wildly interesting void regions in our universe. In particular I wish to thank Ravi Sheth, Johan Hidding, Marius Cautun, Patrick Bos, Bernard Jones, Erwin Platen, Mathijs Dries and Vincent Icke for major contributions to the work described in this paper. RvdW is particularly grateful to J. Hidding for permission to use the images in figures 3, 4 and 9 in advance of publication.

References

Aragon-Calvo, M. A., Platen, E., van de Weygaert, R. & Szalay, A. S. 2010, *ApJ* 723, 364
Aragon-Calvo, M. A., van de Weygaert, R., Araya-Melo, P. A., Platen, E. & Szalay, A. S. 2010, *MNRAS* 404, 89
Aragon-Calvo, M. A. & Szalay, A. S. 2013, *MNRAS* 28, 3409
Aragon-Calvo, M. A. 2014, arXiv:14409.8661
Bardeen, J. M., Bond, J. R., Kaiser, N. & Szalay, A. S. (BBKS) 1986, *ApJ*, 304, 15
Bertschinger, E. 1985, *ApJ* 295, 1
Bertschinger, E., Dekel, A, Faber, S. M., Dressler, A. & Burstein, D. 1990, *ApJ* 364, 370
Beygu, B., Kreckel, K., van de Weygaert, R., van der Hulst, J. M. & van Gorkom, J. H. 2013, *AJ* 145, 120
Biswas, R., Alizadeh, E., & Wandelt, B. 2010, *PhRvD* 82, 3002
Blumenthal, G. R., Da Costa, L., Goldwirth, D. S., Lecar, M. & Piran, T., 1992, *ApJ* 388, 324
Bond, J. R., Cole, S., Efstathiou, G., & Kaiser, N. 1991, *ApJ* 379, 4440
Bond, J. R. & Myers, S. T. 1996 *ApJS*, 103, 1
Bond, J. R., Kofman, L. & Pogosyan, D. Yu. 1996 *Nature* 380, 603
Bos, E. G. P., van de Weygaert, R., Dolag, K. & Pettorino, V. 2012, *MNRAS* 426, 440
Bothun, G. D., Geller, M. J., Kurtz, M. J., Huchra, J. P. & Schild, R. E. 1992 *ApJ* 395, 347
Cai, Y-C., Padilla, N. & Li, B. 2015, *MNRAS* 451, 1036
Cautun, M., van de Weygaert, R., Jones, B. J. T., & Frenk, C. S. 2014, *MNRAS*, 441, 2923
Cautun, M., Cai, Y-H., & Frenk, C. S. 2016, *MNRAS* 457, 2540
Ceccarelli, L., Padilla, N. D., Valotto, C., & Lambas, D. G 2006 *MNRAS* 373, 1440
Chincarini, G. & Rood, H. J. 1975 *Nature* 257, 294
Clampitt, J., Cai, Y-C., & Li, B. 2013, *MNRAS* 431, 749
Colberg, J. M., Sheth, R. K., Diaferio, A., Gao, L. & Yoshida, N. 2005, *MNRAS* 360, 216
Colberg, J. M. *et al.* 2008, *MNRAS* 387, 933
Colless, M., *et al.* 2003, ArXiv:0306581
Courtois, H., Hoffman, Y., Tully, B., & Gottlöber, S. 2012, *ApJ* 744, 43
Dekel, A., Bertschinger, E., & Faber, S. M. 1990, *ApJ* 364, 349
Dekel, A. 1994, *ARAA* 32, 371
Dekel, A. & Rees, M. J. 1994, *ApJL* 443, L1
Desjacques, V. 2008, *MNRAS* 388, 638
Dubinski, J., da Costa, L. N., Goldwirth, D. S., Lecar, M., & Piran, T. 1993, *ApJ* 410, 458
Einasto, J., Joeveer, M., & Saar, E. 1980, *Nature* 283, 47
Einasto, J., *et al.* 2011, *A&A* 534, 128
Eisenstein, D. J. & Loeb, A. 1995, *ApJ* 439, 520
Fairall, A. P. 1998, *Large-scale structures in the Universe* (Wiley)
Furlanetto, S. R. & Piran, T. 2006, *MNRAS* 366, 467
Furlanetto, S., Peng, Oh. S., & Briggs, F. 2006, *Phys. Rep.* 433, 181
Goldberg, D. M. & Vogeley, M. S. 2004, *ApJ* 605, 1
Gottlöber S., Lokas E. L., Klypin, A., & Hoffman, Y. 2003, *MNRAS* 344, 715
Granett, B. R., Neyrinck, M. C., & Szapudi, I. 2008, *ApJ* 701, 414
Gregory, S. A. & Thompson, L. A. 1978, *ApJ* 222, 784
Gunn, J. E. & Gott, J. R. 1972, *ApJ* 176, 1

Guzzo, L. & Teh VIPERS team 2013, *The Messsenger*, 151, 41

Guzzo, L., *et al.* 2014, *AA*, 566, 108

Hamaus, N., Sutter, P. M., & Wandelt, B. D. 2014, *PhRvL* 112, 1302

Hamilton, A. J. S. 2001, *MNRAS* 322, 419

Heath, D. J. 1977, *MNRAS* 179, 351

Hess, S., Kitaura, F.-S., & Gottlöber, S. 2013, *MNRAS*, 435, 2065

Hidding, J., van de Weygaert, R., Vegter, G., Jones, B. J. T., & Teillaud, M. 2012, Video publication SOCG12, Symposium on Computational Geometry, arXiv1205.1669

Hidding, J., van de Weygaert, R., Kitaura, F.-S., & Hess, S. 2016 *MNRAS*, to be subm.

Hidding, J. 2016, *PhD thesis*, Univ. Groningen

Hoffman, Y. & Shaham, J. 1982, *ApJL* 262, L23

Huchra, J. P. *et al.* 2012, *ApJS*, 199, 26

Icke, V. 1973, *A&A*, 27, 1

Icke, V. 1984, *MNRAS* 206, 1P

Icke, V. & van de Weygaert, R. 1987, *A&A* 18, 16

Jennings, E., Li, Y., & Hu, W. 2013, *MNRAS* 434, 2167

Karachentsev, I. D., Karachentseva, V. E., Huchtmeier, W. K., & Makarov, D. I. 2004, *AJ* 127, 2031

Kirshner, R. P., Oemler, A., Schechter, P. L., & Shectman, S. A. 1981, *ApJL* 2448, L57

Kirshner, R. P., Oemler, A., Schechter, P. L., & Shectman, S. A. 1987, *ApJ* 314, 493

Kitaura, F.-S. 2012, *MNRAS*, 429L, 84

Kreckel, K., Platen, E., Aragon-Calvo, M. A., van Gorkom, J. H., van de Weygaert, R., van der Hulst, J. M., Kovac, K., Yip, C.-W., & Peebles, P. J. E. 2011, *AJ* 141, 4

Kreckel, K., Platen, E., Aragon-Calvo, M. A., van Gorkom, J. H., van de Weygaert, R., van der Hulst, J. M., & Beygu, B. 2012, *AJ* 144, 16

Lahav, O. & Suto, Y. 2004, *Living Reviews in Relativity* 7, 82pp

Lavaux, G. & Wandelt, B. 2010, *MNRAS* 403, 1391

Lavaux, G. & Wandelt, B. 2012, *ApJ* 754, 109

de Lapparent, V., Geller, M. J., & Huchra, J. P. 1986, *ApJL* 302, L1

Leclercq, F., Jasche, J., Sutter, P. M., Hamaus, N., & Wandelt, B. 2015, *JCAP*, 03, 047

Leclercq, F. 2015, *Bayesian large-scale structure inference and cosmic web analysis*, PhD thesis, Univ. Pierre et Marie Curie, Institut d'Astrophysique de Paris

Lee, J. & Park, D. 2009, *ApJ*, 696, 10

Li, B. 2011, *MNRAS* 411, 2615

Li, B. & Efstathiou, G. 2012, *MNRAS* 421, 1431

Martel, H. & Wasserman, I. 1990, *ApJ* 348, 1

Mathis, H. & White, S. D. M. 2002, *MNRAS* 337, 1193

Morales, M. F. & Wyithe, J. S. B. 2010, *ARAA* 48, 127

Nadathur, S. & Hotchkiss, S. 2015, *MNRAS* 454, 2228

Nadathur, S. 2016, *MNRAS* in press, arXiv:1602.04752

Neyrinck, M. 2008, *MNRAS* 386, 2101

Padilla, N. D., Ceccarelli, L., & Lambas, D. G. 2005, *MNRAS* 363, 977

Palmer, P. L. & Voglis, N. 1983, *MNRAS* 205, 543

Pan, D., Vogeley, M. S., Hoyle, F., Choi, Y.-Y., & Park, C. 2012 *MNRAS* 421, 926

Paranjape, A., Lam, T. Y., & Sheth, R. K. 2012, *MNRAS* 420, 1648

Park, D. & Lee, J. 2007, *Phys. Rev. Lett.* 98, 081301

Paz, D., Lares, M., Ceccarelli, L., Padilla, N., & Lambas, D. G. 2013, *MNRAS* 436, 3480

Peebles, P. J. E. 1980, *The large-scale structure of the universe* (Princeton University Press)

Peebles, P. J. E. 2001, *ApJ* 557, 495

Peebles, P. J. E. & Nusser, A. 2010, *Nature* 465, 565

Pisani, A., Lavaux, G., Sutter, P. M., & Wandelt, B. D. 2014, *MNRAS* 443, 3238

Pisani, A. 2014, *Cosmology with Cosmic Voids*, PhD thesis, Univ. Pierre et Marie Curie, Institut d'Astrophysique de Paris

Pisani, A., Sutter, P. M., Hamaus, N., Alizadeh, E., Biswas, R., Wandelt, B. D., & Hirata, C. M. 2015, *PhRvD* 92, 3531

Platen, E., van de Weygaert, R., & Jones, B. J. T. 2007, *MNRAS* 380, 551

Platen, E., van de Weygaert, R., & Jones, B. J. T. 2008, *MNRAS* 387, 128

Pontzen, A., Slosar, A., Roth, N., & Peiris, H. V. 2016, *PhRvD* 93, 3519

Press, W. H. & Schechter, P. 1974, *ApJ* 187, 425

Regős E. & Geller, M. J. 1991, *ApJ* 373, 14

Rieder, S., van de Weygaert, R., Cautun, M., Beygu, B., & Portegies-Zwart, S. 2013 *MNRAS* 435, 222

Rudnick, L., Brown, S., & Williams, L. R. 2007, *ApJ* 671, 40

Ryden, B. S. & Melott, A. L. 1996, *ApJ* 470, 160

Sahni, V., Sathyaprakash, B. S., & Shandarin, S. F. 1994, *ApJ* 431, 20

Sahni, V. & Shandarin, S. F. 1996, *MNRAS* 282, 641

Schaap, W. E. & van de Weygaert, R. 2000, *A&A* 363, L29

Shandarin, S., Feldman, H.A., Heitmann, K., & Habib, S. 2006 *MNRAS* 376, 1629

Sheth, R. K. 1998, *MNRAS* 300, 1057

Sheth, R. K. & van de Weygaert, R. 2004, *MNRAS* 350, 517

Springel, V. *et al.* 2005, *Nature* 435, 629

Sutter P., Lavaux, G., Wandelt, B. D., & Weinberg, D. H. 2012, *ApJ* 761, 44

Sutter, P., Lavaux, G., Wandelt, B. D., & Weinberg, D. H. 2012b, *ApJ* 761, 187

Sutter, P., Pisani, A., Wandelt, B. D., & Weinberg, D. H. 2014, *443*, 2983

Sutter, P. M., Elahi, P., Falck, B., Onions, J., Hamaus, N., Knebe, A., Srisawat, C., & Schneider, A. 2014b, *MNRAS* 445, 1235

Sutter, P.M., *et al.* 2015 *A & C* 9, 1

Szapudi, I., *et al.* 2015, *MNRAS* 450, 288

Tegmark, M. & SDSS collaboration 2004, *ApJ* 606, 702

Tully, R. B., Shaya, E. J., Karachentsev, I. D., Courtois, H., Kocevski, D. D., Rizzi, L., & Peel, A. 2008, *ApJ* 676, 184

Tully, R. B., Courtois, H., Hoffman, Y., & Pomarede, D. 2014, *Nature* 513, 71

van de Weygaert, R. 1991, *Voids and the geometry of large scale structure*, PhD thesis, Leiden University

van de Weygaert, R. & van Kampen E. 1993, *MNRAS* 263, 481

van de Weygaert, R. & Bertschinger, E. 1996, *MNRAS*, 281, 84

van de Weygaert, R. & Bond, J. R. 2008, *Clusters and the Theory of the Cosmic Web*, in M. Plionis, O. López-Cruz & D. Hughes eds., *A Pan-Chromatic View of Clusters of Galaxies and the Large-Scale Structure*, LNP 740 (Springer), p. 335

van de Weygaert, R. & Platen, E. 2011, *IJMPS* 1, 41

van de Weygaert, R. & Schaap, W. E. 2009, in V. Martínez *et al.* eds., *Data Analysis in Cosmology*, LNP 665, 291

White, S. D. M. & Silk, J. 1979, *ApJ* 231, 1

Willick, J. A., Courteau, S., Faber, S. M., Burstein, D., Dekel, A., & Strauss, M. A. 1997, *ApJS* 109, 333

Wojtak, R., Powell, D., & Abel, T. 2016, *MNRAS* 458, 4431

Zeldovich, Ya. B., Einasto, J., & Shandarin, S. F. 1982, *Nature* 300, 407

The Zeldovich Universe:
Genesis and Growth of the Cosmic Web
Proceedings IAU Symposium No. 308, 2014
R. van de Weygaert, S. Shandarin, E. Saar & J. Einasto, eds.

© International Astronomical Union 2016
doi:10.1017/S1743921316010516

Answers from the Void:
VIDE and its Applications

P. M. Sutter[1,2,3]†, N. Hamaus[1,2], A. Pisani[1,2], G. Lavaux[1,2] and B. D. Wandelt[1,2,4,5]

[1]Sorbonne Universités, UPMC Univ Paris 06, UMR7095, F-75014, Paris, France
[2]CNRS, UMR7095, Institut d'Astrophysique de Paris, F-75014, Paris, France
[3]Center for Cosmology and AstroParticle Physics, Ohio State University, Columbus, USA
[4]Department of Physics, University of Illinois at Urbana-Champaign, Urbana, USA
[5]Department of Astronomy, University of Illinois at Urbana-Champaign, Urbana, USA

Abstract. We discuss various applications of VIDE, the Void IDentification and Examination toolkit, an open-source Python/C++ code for finding cosmic voids in galaxy redshift surveys and N-body simulations. Based on a substantially enhanced version of ZOBOV, VIDE not only finds voids, but also summarizes their properties, extracts statistical information, and provides a Python-based platform for more detailed analysis, such as manipulating void catalogs and particle members, filtering, plotting, computing clustering statistics, stacking, comparing catalogs, and fitting density profiles. VIDE also provides significant additional functionality for pre-processing inputs: for example, VIDE can work with volume- or magnitude-limited galaxy samples with arbitrary survey geometries, or dark matter particles or halo catalogs in a variety of common formats. It can also randomly subsample inputs and includes a Halo Occupation Distribution model for constructing mock galaxy populations. VIDE has been used for a wide variety of applications, from discovering a universal density profile to estimating primordial magnetic fields, and is publicly available at http://bitbucket.org/cosmicvoids/vide_public and http://www.cosmicvoids.net.

Keywords. cosmology: large-scale structure of universe, methods: data analysis, methods: n-body simulations, dark matter

1. Introduction

First discovered over thirty years ago (e.g., Gregory & Thompson 1978), cosmic voids are emerging as a novel and innovative probe of both cosmology and astrophysics through their intrinsic properties (e.g., Bos *et al.* 2012), exploitation of their statistical isotropy via the Alcock-Paczyński test (Sutter et al. 2014), or cross-correlation with the cosmic microwave background (Ilić et al. 2013; Planck Collaboration 2013; Cai *et al.* 2014). Additionally, fifth forces and modified gravity are unscreened in void environments, making them singular probes of exotic physics (e.g., Li et al. 2012; Clampitt et al. 2013; Carlesi *et al.* 2014).

Voids are also fascinating objects to study by themselves. For example, there appears to be a self-similar relationship between voids in dark matter distributions and voids in galaxies (Sutter *et al.* 2014) via a universal density profile (Hamaus et al. 2014) (hereafter HSW). Observationally, the anti-lensing shear signal (Melchior *et al.* 2014; Clampitt & Jain 2014) and Ly-alpha absorption measurements (Tejos *et al.* 2012) have illuminated the dark matter properties in voids.

However, there are only a few public catalogs of voids identified in galaxy redshift surveys, primarily the SDSS (e.g., Pan *et al.* 2012; Sutter *et al.* 2012). And while there are many published methods for finding voids based on a variety of techniques, most codes remain private. In order

† email: `sutter@iap.fr`

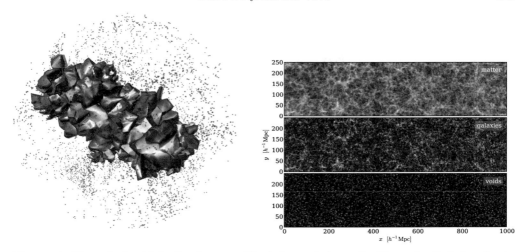

Figure 1. Watershed void finding in VIDE. The left panel shows the Voronoi cells belonging to a void embedded in an observed galaxy population. *Reproduced from Sutter et al. (2012).* The right panel shows projections of dark matter, mock galaxies, and void positions in a simulation. *Reproduced from Hamaus et al. (2014).*

to accelerate the application of voids, it is essential to provide easy-to-use, flexible, and strongly supported void-finding codes.

In this work we discuss VIDE, for Void IDentification and Examination, a toolkit based on the publicly-available watershed code ZOBOV (Neyrinck 2008) for finding voids but considerably enhanced and extended to handle a variety of simulations and observations (Sutter et al. 2014). VIDE also provides an extensive interface for analyzing void catalogs.

2. Void Finding

VIDE is able to identify voids in N-body simulation snapshots produced by GADGET (Springel 2005), FLASH (Dubey et al. 2008), RAMSES (Teyssier 2002) 2HOT (Warren 2013), and generic ASCII files. VIDE can either find voids directly in the full simulation, or randomly subsampled populations, as well as halo catalogs and mock galaxy samples using the Halo Occupation Distribution formalism (Berlind & Weinberg 2002). For observational datasets, to find voids in galaxy surveys the user must provide a list of galaxy positions and a pixelization of the survey mask using HEALPIX (Gorski et al. 2005)†.

The core of our void finding algorithm is ZOBOV (Neyrinck 2008), which creates a Voronoi tessellation of the tracer particle population and uses the watershed transform to group Voronoi cells into zones and subsequently voids (Platen et al. 2007). The Voronoi tessellation provides a density field estimator based on the underlying particle positions.

ZOBOV first groups nearby Voronoi cells into zones, then merges adjacent zones into voids by finding minimum-density barriers between them. We impose a density-based threshold within ZOBOV where adjacent zones are only added to a void if the density of the wall between them is less than 0.2 times the mean particle density $\bar{n}$. This density criterion prevents voids from expanding deeply into overdense structures and limits the depth of the void hierarchy (Neyrinck 2008).

In this picture, a void is simply a basin in the Voronoi density field bounded by a common set of higher-density ridgelines, as demonstrated by Figure 1, which shows a typical 20 h^{-1}Mpc void identified in the SDSS DR7 galaxy survey (Sutter *et al.* 2012). This also means that voids may

† http://healpix.jpl.nasa.gov

have any *mean* density, since the watershed includes in the void definition all wall particles all the way up to the very highest-density separating ridgeline.

We have made several modifications and improvements to the original ZOBOV algorithm. First, we have strengthened the algorithm with respect to numerical precision by rewriting large portions in a templated C++ framework. We enforce bijectivity in the Voronoi graph, so that the tessellation is self-consistent, and use an improved volume-splitting technique to minimize the number of difference operators at the edge of the box when joining subregions. Finally, we have optimized several portions of the central watershed algorithm, enabling the identification of voids in simulations with up to 1024^3 particles in ~ 10 hours using 16 cores.

To prevent the growth of voids outside survey boundaries in observational datasets, we place a large number of mock particles along any identified edge and along the redshift caps. We assign essentially infinite density to these mock particles, preventing the watershed from including zones external to the survey. Since the local volumes of the edge galaxies are arbitrary, we prevent these mock particles from participating in any void by disconnecting their adjacency links in the Voronoi graph.

We use the mean particle spacing to set a lower size limit for voids because of Poisson shot noise. VIDE does not include any void with effective radius smaller than $\bar{n}^{-1/3}$, where $\bar{n}$ is the mean number density of the sample. We define the effective radius as as the radius of a sphere with the same total volume of all the Voronoi cells that make up the void.

Figure 1 shows an example void population with VIDE, taken from the analysis of Hamaus *et al.* (2014). In this figure we show a slice from an N-body simulation, a set of mock galaxies painted onto the simulation using the HOD formalism discussed above, and the distribution of voids identified in the mock galaxies.

VIDE automatically provides some basic derived void information, such as the *macrocenter*, or volume-weighted center of all the Voronoi cells in the void, and the ellipticity.

3. Post-Processing & Analysis

VIDE provides a Python-based application programming interface (API) for loading and manipulating the void catalog and performing analysis and plotting.

Via the API the user has immediate access to all basic and derived void properties (ID, macrocenter, radius, density contrast, RA, Dec, hierarchy level, ellipticity, etc.) as well as the positions (x, y, z, RA, Dec, redshift), velocities (if known), and local Voronoi volumes of all void member particles. The user can also access all particle and galaxy sample information, such as redshift extents, the mask, simulation extents and cosmological parameters. Upon request the user can also load all particles in the simulation or observation.

VIDE includes several built-in plotting routines. First, users may plot cumulative number functions of multiple catalogs on a logarithmic scale. Volume normalizations are handled automatically, and 1σ Poisson uncertainties are shown as shaded regions, as demonstrated in Figure 3. Secondly, users may plot a slice from a single void and its surrounding environment. In these plots we bin the background particles onto a two-dimensional grid and plot the density on a logarithmic scale. We draw the void member galaxies as small semi-transparent disks with radii equal to the effective radii of their corresponding Voronoi cells.

The user can directly compare two void catalogs by using a built-in function for computing the overlap. This function attempts to find a one-to-one match between voids in one catalog and another and reports the relative properties (size, ellipticity, etc.) between them. It also records which voids cannot find a reliable match and how many potential matches a void may have. A more detailed discussion of the matching process can be found in Sutter *et al.* (2014).

VIDE allows the user to compute simple two-point clustering statistics, i.e. power spectra and correlation functions.

Three-dimensional stacks of voids are useful in a variety of scenarios, and the user may construct these with built-in functions. The stacked void may optionally contain only void member particles or all particles within a certain radius. With this stacked profile VIDE can build a spherically averaged radial density profile and fit the universal HSW void profile to it.

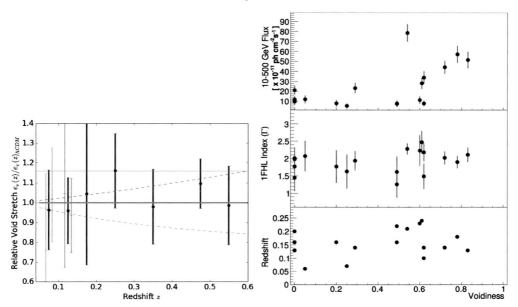

Figure 2. Left panel: a detection of the Alcock-Paczyński effect using stacked cosmic voids in the SDSS DR7 and DR10. The slight deviation from unity in the expected void stretch can be used to extract cosmological parameters. *Reproduced from Sutter et al. (2014)*. Right panel: GeV luminosity versus void fraction along the line of sight for Fermi-detected blazars. *Reproduced from Furniss et al. (2014)*.

All proper normalizations are handled internally. Figure 3 shows an example of VIDE-produced density profiles and best-fit HSW profile curves.

4. Applications

Throughout its development VIDE has been used in a variety of applications. Initially, the enhancements to ZOBOV were used to create a set of public catalogs from both observations (Sutter *et al.* 2012, 2014) and simulations (Sutter *et al.* 2014). The toolkit presented in this work is fully compatible with those public releases.

The publication of simple void properties, such as sky position and effective radius, enabled direct cross correlations with other datasets leading to several interesting results. Examples of such correlations include: galaxy shear maps leading to a detection of a weak lensing effect inside voids (Melchior *et al.* 2014), the cosmic microwave background giving rise to a detection of the integrated Sachs-Wolfe effect (Planck Collaboration 2013; Ilić et al. 2013), and positions of high-energy blazars putting limits on the intergalactic magnetic field (Furniss *et al.* 2014). The latter correlation is shown in the right panel of Figure 2. Additionally, cross-correlations of void centers with galaxy positions have recently been demonstrated to allow the construction of a static ruler (Hamaus *et al.* 2014), and the three-dimensional arrangement of galaxies around stacked voids has been exploited for an Alcock-Paczyński test (Sutter et al. 2014), as shown in the left-hand panel of Figure 2.

Derived void statistical information, such as radial density profiles (as shown in Figure 3), have led to the discovery of a universal density profile (Hamaus et al. 2014; Sutter *et al.* 2014) and a method for constructing real-space profiles in a parameter-independent manner (Pisani *et al.* 2013). The abundance of voids is potentially a very powerful probe of cosmology, as shown in Sutter *et al.* (2014) and highlighted in Figure 3.

Lastly, the ability to match and compare void properties from one catalog to another (a built-in feature of VIDE) has been used to examine the relationship of voids identified in sparse,

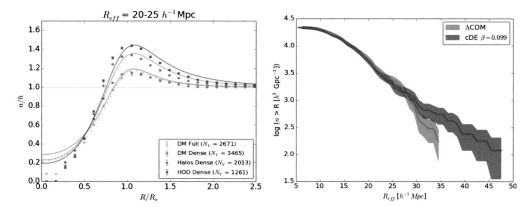

Figure 3. Left panel: example one-dimensional radial density profiles of stacked voids (points with error bars) and best-fit curves (thin lines) using the HSW profile. *Reproduced from Sutter et al. (2014).* Right panel: example cumulative void number functions from simulations. *Reproduced from Sutter et al. (2014).*

biased galaxy samples to dark matter underdensities (Sutter *et al.* 2014) and to reconstruct the life history of voids using merger trees (Sutter et al. 2014).

5. Conclusions

We have presented and discussed the capabilities of VIDE, a new Python/C++ toolkit for identifying cosmic voids in cosmological simulations and galaxy redshift surveys, as well as a brief overview of its current applications. VIDE performs void identification using a substantially modified and enhanced version of the watershed code ZOBOV. Furthermore, VIDE is able to support a variety of mock and real datasets, and provides extensive and flexible tools for loading and analyzing void catalogs. We have highlighted these analysis tools (e.g., filtering, plotting, clustering statistics, stacking, profile fitting) using examples from previous and current research using VIDE.

The analysis toolkit bundled in VIDE enables a wide variety of both theoretical and observational void research. As its most basic, the void properties made available to the user, such as sky positions, shapes, and sizes, permit simple explorations of void properties and cross-correlation with other datasets. The user may also use void member particles and their associated volumes for examining galaxy properties in low-density environments. Cross-matching by overlapping catalogs can be useful for understanding the impacts of peculiar velocities or galaxy bias, as well as providing a platform for studying the effects of modified gravity or fifth forces on a void-by-void basis. Void power spectra, shape distributions, number functions, and density profiles, easily accessible via VIDE, are sensitive probes of cosmology. Users may also access HSW density profiles, enabling theoretical predictions of the ISW or gravitational lensing signals.

The past few years have seen an enormous growth in void interest and research. This research includes new void-finding algorithms, studies of void properties, investigations and forecasts of cosmological probes, and explorations into the nature of void themselves from theoretical and numerical viewpoints. Put simply, VIDE is designed to meet the growing demand of next-generation void science.

The VIDE code and documentation is currently hosted at http://bitbucket.org/cosmicvoids/vide_public, with links to numbered versions at http://www.cosmicvoids.net.

Acknowledgments

The authors acknowledge support from NSF Grant NSF AST 09-08693 ARRA. BDW acknowledges funding from an ANR Chaire d'Excellence (ANR-10-CEXC-004-01), the UPMC Chaire Internationale in Theoretical Cosmology, and NSF grants AST-0908 902 and AST-0708849. This work made in the ILP LABEX (under reference ANR-10-LABX-63) was supported by French state funds managed by the ANR within the Investissements d'Avenir programme under reference ANR-11-IDEX-0004-02.

References

Berlind, A. A. & Weinberg, D. H., 2002, ApJ, 575, 587

Bos, E. G. P., van de Weygaert, R., Dolag, K., & Pettorino, V., 2012, Mon. Not. R. Astron. Soc., 426, 440

Cai, Y.-C., Neyrinck, M. C., Szapudi, I., Cole, S., & Frenk, C. S., 2014, ApJ, 786, 110

Carlesi, E., Knebe, A., Lewis, G. F., Wales, S., & Yepes, G., 2014, Mon. Not. R. Astron. Soc., 439, 2943

Clampitt, J., Cai, Y.-C., & Li, B., 2013, Mon. Not. R. Astron. Soc., 431, 749

Clampitt, J. & Jain, B., 2014, ArXiv e-prints: 1404.1834

Dubey A., Reid L. B., & Fisher R., 2008, *Physica Scripta*, T132, 014046

Furniss, A., Sutter, P. M., Primack, J. R., & Dominguez, A., 2014, ArXiv e-prints: 1407.6307

Gorski K. M., Hivon E., Banday A. J., Wandelt B. D., Hansen F. K., Reinecke M., & Bartelmann M., 2005, ApJ, 622, 759

Gregory, S. A. & Thompson, L. A., 1978, ApJ, 222, 784

Hamaus, N., Sutter, P. M., & Wandelt, B. D., 2014, *Physical Review Letters*, 112, 251302

Hamaus, N., Wandelt, B. D., Sutter, P. M., Lavaux, G., & Warren M. S., 2014, *Physical Review Letters*, 112, 041304

Ilić S., Langer, M., & Douspis, M., 2013, Astron. & Astrophys., 556, A51

Li, B., Zhao, G.-B., & Koyama, K., 2012, Mon. Not. R. Astron. Soc., 421, 3481

Melchior, P., Sutter, P. M., Sheldon, E. S., Krause, E., & Wandelt B. D., 2014, Mon. Not. R. Astron. Soc., 440, 2922

Neyrinck, M. C., 2008, Mon. Not. R. Astron. Soc., 386, 2101

Pan, D. C., Vogeley, M. S., Hoyle, F., Choi, Y.-Y., & Park, C., 2012, Mon. Not. R. Astron. Soc., 421, 926

Pisani, A., Lavaux, G., Sutter, P. M., & Wandelt, B. D., 2013, ArXiv e-prints: 1306.3052

Planck Collaboration 2013, ArXiv e-prints: 1303.5079

Platen E., van de Weygaert R., & Jones B. J. T., 2007, Mon. Not. R. Astron. Soc., 380, 551

Springel, V., 2005, Mon. Not. R. Astron. Soc., 364, 1105

Sutter, P. M., Carlesi, E., Wandelt, B. D., & Knebe, A., 2014, ArXiv e-prints

Sutter, P. M., Elahi, P., Falck, B., Onions, J., Hamaus, N., Knebe A., Srisawat, C., & Schneider, A., 2014, ArXiv e-prints: 1403.7525

Sutter, P. M., *et al.*, 2014, ArXiv e-prints: 1406.1191

Sutter, P. M., Lavaux, G., Hamaus, N., Wandelt, B. D., Weinberg D. H., & Warren, M. S., 2014, Mon. Not. R. Astron. Soc., 442, 462

Sutter, P. M., Lavaux, G., Wandelt, B. D., & Weinberg, D. H., 2012, ApJ, 761, 44

Sutter, P. M., Lavaux, G., Wandelt, B. D., Weinberg, D. H., & Warren M. S., 2014, Mon. Not. R. Astron. Soc., 438, 3177

Sutter, P. M., Lavaux, G., Wandelt, B. D., Weinberg, D. H., Warren M. S., & Pisani, A., 2014, Mon. Not. R. Astron. Soc., 442, 3127

Sutter, P. M., Pisani, A., & Wandelt, B. D., 2014, ArXiv e-prints: 1404.5618

Tejos, N., Morris, S. L., Crighton, N. H. M., Theuns, T., Altay, G., & Finn, C. W., 2012, Mon. Not. R. Astron. Soc., 425, 245

Teyssier, R., 2002, Astron. & Astrophys., 385, 337

Warren, M. S., 2013, in Proceedings of SC13: International Conference for High Performance Computing, Networking, Storage and Analysis 2HOT: An Improved Parallel Hashed Oct-Tree N-Body Algorithm for Cosmological Simulation. p. 72

The Zeldovich Universe:
Genesis and Growth of the Cosmic Web
Proceedings IAU Symposium No. 308, 2014
R. van de Weygaert, S. Shandarin, E. Saar & J. Einasto, eds.

© International Astronomical Union 2016
doi:10.1017/S1743921316010528

Void Dynamics

Nelson D. Padilla[1,2], Dante Paz[3], Marcelo Lares[3], Laura Ceccarelli[3], Diego García Lambas[3], Yan-Chuan Cai[4] and Baojiu Li[4]

[1] Instituto de Astrofísica,
Universidad Católica de Chile,
Vicuña Mackenna 4860, Santiago, Chile
email: npadilla@astro.puc.cl

[2] Centro de Astro-Ingeniería,
Universidad Católica de Chile,
[3] Instituto de Astronomía Teórica y Experimental (IATE),
Laprida 922, Córdoba, Argentina
[4] Institute for Computational Cosmology
Durham University
South Road, Durham, DH1 3LE, UK

Abstract. Cosmic voids are becoming key players in testing the physics of our Universe. Here we concentrate on the abundances and the dynamics of voids as these are among the best candidates to provide information on cosmological parameters. Cai, Padilla & Li (2014) use the abundance of voids to tell apart Hu & Sawicki $f(R)$ models from General Relativity. An interesting result is that even though, as expected, voids in the dark matter field are emptier in $f(R)$ gravity due to the fifth force expelling away from the void centres, this result is reversed when haloes are used to find voids. The abundance of voids in this case becomes even lower in $f(R)$ compared to GR for large voids. Still, the differences are significant and this provides a way to tell apart these models. The velocity field differences between $f(R)$ and GR, on the other hand, are the same for halo voids and for dark matter voids. Paz *et al.* (2013), concentrate on the velocity profiles around voids. First they show the necessity of four parameters to describe the density profiles around voids given two distinct void populations, voids-in-voids and voids-in-clouds. This profile is used to predict peculiar velocities around voids, and the combination of the latter with void density profiles allows the construction of model void-galaxy cross-correlation functions with redshift space distortions. When these models are tuned to fit the measured correlation functions for voids and galaxies in the Sloan Digital Sky Survey, small voids are found to be of the void-in-cloud type, whereas larger ones are consistent with being void-in-void. This is a novel result that is obtained directly from redshift space data around voids. These profiles can be used to remove systematics on void-galaxy Alcock-Pacinsky tests coming from redshift-space distortions.

Keywords. Cosmic Voids, Cosmology, Large Scale Structure of the Universe, Peculiar velocities

1. Introduction

Cosmic voids are underdense regions in the universe which occupy a significant fraction of the total volume, and as such are potentially powerful tools for statistical studies of the Universe. They are the result of the history of growth of perturbations since, the faster the virialised structures gain mass, the faster other parts of the Universe must empty of material to feed this mass increase. Therefore, statistics of the abundance and the rate of growth of voids are related to the growth factor in the universe, which in turn is related to cosmology. Regarding the rate of growth of voids, it is possible to study it via the evolution of the abundance of voids, or directly measuring velocity profiles around voids.

The abundance of voids has been proposed as a way to test cosmology. However, this has proved to be challenging given the difference in abundance obtained using different void finders (e.g. Colberg *et al.* 2005). Jennings *et al.* (2013) have recently obtained a ~ 16 percent agreement between simulations and excursion set predictions (see for instance Sheth & van de Weygaert, 2004), but in their case they use the dark matter particles to do this comparison.

The dynamics around voids has been studied in simulations and data (e.g., Padilla *et al.* 2005, Ceccarelli *et al.*, 2006, respectively) where the amplitude of peculiar velocities was shown to reach a maximum of a few hundreds of kms^{-1}, and has now been included in studies of density profiles of voids in the directions parallel and perpendicular to the line of sight; if the voids are spherical and the effects from peculiar velocities are either consistently similar for different types of voids, or can be modeled, these can be used to disaffect the redshift space distortions on the profile along the line of sight. Once this is done, the remaining differences in the two directions can be used to study cosmology using the Alcock-Pacinsky test (e.g. Lavaux & Wandelt 2012, Sutter *et al.* 2014).

However, there is still debate on whether the profiles (density and velocity) of voids are self-similar, of if they separate into (at least) two different families of voids. Sheth & van de Weygaert (2004) propose the existence of two distinct types of voids. The ones called void-in-void which are underdense up to very large distances from the void centre (several void radii), and those that are embedded in a larger overdensity, called void-in-cloud. It is clear that the density profiles of these two types will be different. Ceccarelli *et al.* (2013) selected voids that lie in an overdensity at tree times the void radius as voids-in-clouds, and voids that do not lie in an overdensity at this same radius as voids-in-voids and showed that the velocity profiles of these two families of voids are different. Void-in-clouds show an infall of matter beyond three times the void radius. Void-in-voids only show outflows. Therefore, in order to be able to do high precision Alcock-Pacinsky tests it will be necessary to take into account these differences fully.

In this proceedings we will first briefly discuss the void finding algorithm adopted for the works whose results will be presented here; then we will show very recent results on the abundance of voids near the present epoch of the universe for the ΛCDM universe but also for alternative gravity $f(R)$ models. We will also show the first measurements of void expansion using real data by means of fitting the void-galaxy cross-correlation function in the spectroscopic Sloan Digital Sky Survey (SDSS). The works we concentrate on are Cai, Padilla & Li (2014) and Paz *et al.* (2013), respectively.

2. Finding voids

The works we refer to in this proceeding use voids identified using modified versions of the Padilla *et al.* (2005, P05 from now on) finder. This finder is explained in detail in P05, so we only include a brief description of the algorithm. The steps that are followed are first, to take a number of prospective void centres (either by making a full spatial search or starting from low density regions), and grow spheres around these until the density within them reaches a threshold value. For Paz *et al.* (2013, P13 from this point on) and Cai, Padilla & Li (2014, C14) these thresholds were 0.1 and 0.2 times the average density of the mass or tracers, respectively. The size of the sphere at this point is defined as the void radius.

Depending on whether the tracers are sparse or not, a minimum number of tracers (DM particles or galaxies in C14 and P13, respectively) is required for a void to be accepted into the final sample. In the case of using massive haloes in C14, no minimum number of haloes is required to lie within the voids.

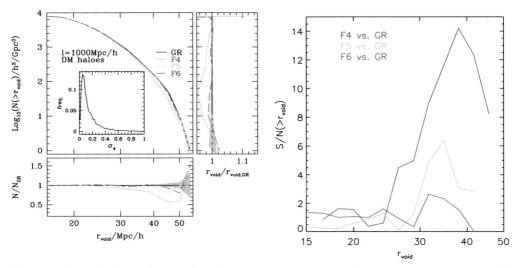

Figure 1. Left: Void abundances when dark matter haloes are used as tracers of the density field. The main panels shows the abundances of voids above a minimum void radius, with differently line types and colours for different gravity models as indicated in the figure key. The right panel shows the ratio of abundances at fixed cumulative density. The bottom panel shows the ratio of abundances at a fixed void radius. The inset in the main panel shows the distribution function of σ_4, the relative dispersion in the distance to the first four haloes above the void density threshold, for GR only; a small value of σ_4 indicates a well centred void. Right: The corresponding S/N from void abundances.

The modifications to the P05 finder in each case, C14 and P13, are described in detail in each paper. In C14 the modifications improve the convergence with numerical resolution and is adapted to be applied to massive haloes, which suffer from being sparse samplers of the density field. In particular, since no minimum number of haloes within voids is required when haloes are used to detect voids, a new parameter, σ_4 is defined,

$$\sigma_4 \equiv \sqrt{\left\langle \left(\frac{d_i - \langle d_i \rangle_4}{\langle d_i \rangle_4} \right)^2 \right\rangle_4},$$

that measures the dispersion of the distance between the void centre and the nearest four haloes. The smaller this parameter is the better defined the void centre is, and the more spherical its shape is. In P13 the modifications concentrate on adapting the finder to work in redshift space in a survey affected by a complicated angular mask (though it is applied to volume limited samples). The space density of galaxies used by P13 makes the sparseness not as important as in C14.

3. Void abundance

If the interest is to compare model predictions with different cosmologies or different recipes for gravity, it is valid to use a single void finding algorithm. This is the spirit with which C14 search for ways to tell apart modified $f(R)$ gravity models from General Relativity. By using the P05 with the modifications we mentioned earlier, they are able to find voids in simulations with GR and with Hu & Sawicki (2007) $f(R)$ models with different amplitudes of the scalar field strength of $|f_{R0}| = 10^{-6}$, 10^{-5} and 10^{-4} (referred to as F6, F5 and F4 from this point on). The simulations cover a volume of 1 cubic

Gigaparsec, and all share the same initial conditions, and were run with the ECOSMOG code (Li *et al.*, 2012). This way all the differences found in the resulting voids arise solely from the different recipes for gravity.

C14 first identified voids in the matter density field traced by dark matter particles, and they were able to confirm analytic expectations that in $f(R)$ models, voids tend to be emptier due to the fifth force acting in the opposite direction to that of Newtonian gravity (Clampitt *et al.* 2013). The differences in abundance between F4 and GR are above 100 percent with extremely high significances.

However, when using haloes with masses $\gtrsim 10^{13} h^{-1} M_\odot$ (the lower limit in mass is slightly different in each model so that the number density of haloes is the same), they found the result shown in the left panels of Figure 1. In the figure it is clear that the voids found in $f(R)$ (large ones in particular) are about as abundant as in GR. C14 interpret this result as a subtle effect of the fifth force that could only be fully realised with the non-linear evolution of the density field as it evolved in the numerical simulation. The fifth force, locally, induces the formation of haloes. The structures embedded in voids are therefore located in a region where this force is increased, which causes more massive haloes to form within voids in $f(R)$ in comparison to GR. This makes voids even smaller than in GR when haloes are used to identify them.

The small differences are still significant, to the levels shown in the right main panel in Figure 1, where the peak in signal-to-noise ratio, S/N, is found for haloes of around $30 - 40 h^{-1}$Mpc, and is $S/N \sim 2$ even for F6. Therefore, the abundance of voids can be used to tell apart very mild fifth force models such as F6, although with a rather small significance. We refer the reader to C14 for a full discussion of the implications of this result, as well as for conclusions regarding the density and velocity profiles of voids in GR and $f(R)$ models.

4. Void dynamics

Once the abundance of voids is determined, one direct application that can be followed is to study the dynamics of voids. As shown by C14, when moving from the density field of the mass, to that traced by haloes, the differences between $f(R)$ and GR become smaller. This, however, does not happen when studying velocity profiles.

In a previous paper, P13 studied the dynamics in detail for GR voids. We will concentrate on these results in what follows.

Before going to the actual dynamics, it should be noticed that the source of peculiar velocities is the distribution of mass. Therefore, it is necessary to briefly discuss some aspects of density profiles around voids. Several recent works (e.g. Nadathur *et al.*, 2014) argue that the density profiles around voids are self similar. If this is the case, then one can adopt self-similar density profiles to model the void velocities. However, Sheth & van de Weygaert (2004) showed that there are probably voids of the void-in-cloud and void-in-void types. Ceccarelli *et al.* (2013, C13) adopted a practical definition to separate voids into these two classes. If the integrated overdensity in spheres of 3 times the void radius is positive, then the void is of the void-in-cloud family. If it is negative, it is a void-in-void. Using simulations C13 then showed that indeed, profiles are different for these two families. However, C13 also find that the largest voids, those with radius $> 10 h^{-1}$Mpc, are mostly of the void-in-void type. Therefore, for sparse samples where voids are rather large, the approximation of a self-similar profile is adequate. P13 study the dynamics for voids of sizes going down to $6 h^{-1}$Mpc and the distinction between void-in-cloud and void-in-void is unavoidable.

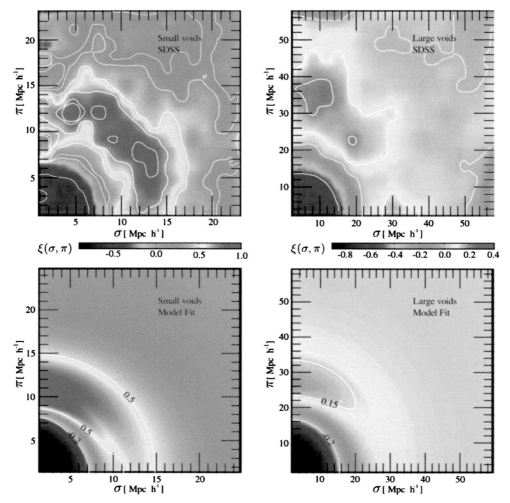

Figure 2. Redshift space distortions of galaxies in the SDSS (upper left panel) and the best fit models (bottom left panels) for small voids-in-clouds ($6 < R_{\mathrm{void}}/\mathrm{Mpc}h^{-1} < 8$). Larger voids ($10 < R_{\mathrm{void}}/\mathrm{Mpc}h^{-1} < 20$) are shown correspondingly on the right panels.

In order to model the integrated density profiles of voids, P13 introduce a simple empirical model that contains the necessary features of the density profiles of the two types. The voids-in-voids have the simplest profile shapes, a continuously rising function to the mean density. The error function $\mathrm{erf}(x)$ behaves as needed,

$$\Delta_R(r) = \frac{1}{2}\left[\mathrm{erf}\left(S \log(r/R)\right) - 1\right]. \tag{4.1}$$

This model depends on two parameters, the void radius R and a *steepness* coefficient S. On the other hand, the profiles of voids-in-clouds are more complex and require two additional parameters in order to account for the overdensity shell surrounding the voids. To the rising term in Eq. 4.1, P13 add a peak in the density profile,

$$\Delta_S(r) = \frac{1}{2}\left[\mathrm{erf}\left(S \log(r/R)\right) - 1\right] + P \exp\left(-\frac{\log^2(r/R)}{2\Theta^2(r)}\right) \tag{4.2}$$

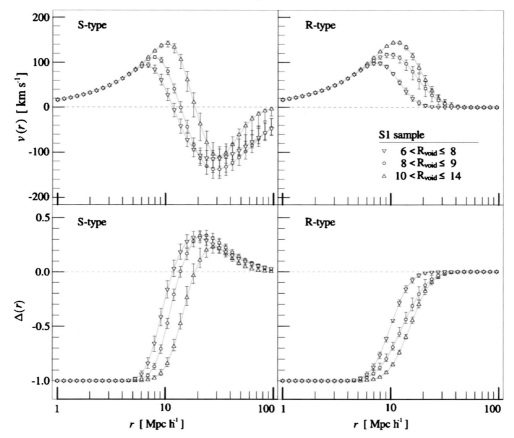

Figure 3. Void–centric radial galaxy density profiles $\Delta(r)$ (lower panels) and void–centric radial galaxy velocity profiles (upper panels) for voids-in-clouds (left) and voids-in-voids (right). Different void radii ranges are indicated with different symbols as shown in the key. Error bars indicate the region enclosing all curves within 68.3% uncertainty in the parameter space. The sample of voids in this case corresponds to sample S1 in P13 which is volume limited out to a redshift $z = 0.08$.

where the Gaussian peak has an asymmetric width,

$$\Theta(r) = \begin{cases} 1/\sqrt{2\,S} & r < R \\ 1/\sqrt{2\,W} & r > R \end{cases} \tag{4.3}$$

This allows them to modify the size of the shell through the W parameter, without changing the inner shape of the profile, related to the parameter S. In this way, a cumulative profile for a void-in-cloud requires four parameters, R, S, P and W.

P13 use these profiles to obtain the average radial velocity profiles around voids using the formalism by Peebles (1979). To fit the measured cross-correlations it is then necessary to convert the density profile into a correlation function. This correlation function is then calculated as a function of two separations, one in the direction of the line of sight, and in the one perpendicular to it. Before adding any velocity information, the correlation function is identical in these two directions. Then the correlation function in the direction of the line of sight is convolved with a peculiar velocity model that responds to the density of matter out to a given separation, and with a gaussian distribution of velocities to take into account random motions within groups of galaxies in overdensities around the voids. For the full details of the calculations we refer the reader to P13.

P13 apply their method to volume limited samples of SDSS, where they identify voids using the modified P05 finder (modifications explained in detail in C13), and measure the void-galaxy cross-correlation functions for the resulting samples. The galaxies used in this measurement are obtained from the full flux limited sample. The results for two sets of voids are shown in the top panels of Figure 2. As can be seen, the correlation function is negative close to the void centres and at the void radius it quickly rises to reach positive values (red regions). These regions are slightly elongated along the line of sight. This shows that there is indeed a velocity dispersion in the void walls, otherwise this ridge would look just as wide in the two directions (parallel and perpendicular to the line of sight). Also, larger voids show a less pronounced ridge (right panels) as these are mostly of the void-in-void type.

P13 study samples of small and large voids which effectively separates the problem into void-in-cloud and void-in-void centres, respectively; note that as the real space profiles are unknown for redshift space data, this is only an expectation that can be confirmed (or not) by looking at the resulting profiles after fitting the observed correlations. The bottom left panel show the result of fitting for the profile parameters, such that the model cross-correlation functions match as closely as possible the measured ones. As can be seen, the model cross-correlation function captures most of the phenomenology seen in the observations.

The profiles corresponding to the best fit parameters are shown in Figure 3. As can be seen the measured cross-correlations are clearly consistent with a picture where small voids have an overdense ridge, and a velocity profile that shows outflow out to the void radius, and then infall at larger separations. This is a clear signature of a void-in-cloud centre. On the other hand, larger voids point to profiles consistent with that of void-in-void centres, with a smoothly rising profile, and outflow velocities at all radii; we show the results of the fit for larger voids in the right panels. This constitutes the first detection of the two types of voids in actual data, although in an indirect way that uses a linear theory model for the velocity profiles. This, however, implies that in order to use void-galaxy correlations to perform Alcock-Pacinsky type tests, one needs to be careful to take into account the proper peculiar velocity field around voids to remove systematics from redshift-space distortions.

5. Conclusions and outlook

Voids are promising structures with a great potential to provide important constraints on cosmology and gravity. Even though there is no clear convergence on the abundance of voids (e.g. Colberg *et al.*, 2005) and therefore this statistics is not suitable for cosmological constraints yet, there are other applications that show excellent capabilities in this sense.

In Cai, Padilla & Li (2014), a single void finder is used, that of Padilla *et al.* (2005) with modifications to allow it to run in parallel and also to provide stable results for sparse samples of haloes. They use this finder to search for voids in simulations with different gravity recipes, corresponding to General Relativity, and to three Hu & Sawicki $f(R)$ models with different strengths of the scalar field parameter $|f_{R0}| = 10^{-6}$, 10^{-5} and 10^{-4} (F6, F5 and F4). All the simulations follow a volume of 1 cubic Gigaparsec to $z = 0$ and start from the same initial conditions. The differences between the populations of voids respond solely to the different strengths of the fifth force (absent in GR). They find that, when the dark matter field (particles) are used to detect voids, those in $f(R)$ models are larger as expected (Clampitt *et al.*, 2013). However, when using massive haloes of $\gtrsim 10^{13} h^{-1} M_\odot$ as tracers, a surprising result arises. Voids are just as abundant in $f(R)$; furthermore, when restricting the comparison to only the largest voids with radii $> 30 - 40 h^{-1}$ Mpc, $f(R)$ voids are less abundant. This difference, although not as large as for voids found from the dark matter field, is still enough for a significant detection of the effect of the fifth force in F5 and F4, and marginally for F6, for a 1 cubic Gigaparsec volume (the significance would be larger for volumes such as those of LSST or EUCLID).

It is important to understand why the abundance of voids found using haloes is lower in $f(R)$. The analysis by C14 indicates that this is due to the fact that the fifth force is stronger in voids (low densities) and this increases the growth of haloes in these regions. Therefore, haloes in voids tend to be more massive in $f(R)$.

In order to tell appart $f(R)$ from GR, a single finder can be used, but it would need to be tested on detailed mock catalogues with the different gravity recipes. Then, it would be possible to search for the gravity that best matches the observed abundances which would be found using the same void finder in actual data.

C14 also point out that the different results found when using the dark matter field compared to using tracer haloes are not present when studying velocity profiles.

To obtain velocity profiles, though, it is of great importance to determine the overdensities surrounding the voids, as these are the source of peculiar velocities. In Paz *et al.* (2013) they set out to use redshift space distortions around voids to learn about the density profiles around voids of the void-in-cloud and void-in-void types (Sheth & van de Weygaert, 2004). P13 propose a four parameter function that describes the full range of void profiles, for these two void types. They use this functional form to obtain redshift space distortions around voids, by calculating the cross-correlation function between voids and galaxies as a function of the separations perpendicular and parallel to the line of sight. When they fit these parameters to obtain models that best reproduce the measured cross-correlations from voids and galaxies in the Sloan Digital Sky Survey, they find that small voids are best fit by void-in-cloud type profiles, whereas large voids are consistent with the void-in-void type. This represents the first indirect detection of these two types of voids using redshift space data.

An interesting application of these predictions for redshift space distortions around voids is to use them to perform Alcock-Pacinsky effect tests (Lavaux & Wandelt 2012, Sutter *et al.* 2014), but removing the distortions coming from peculiar velocities, which could otherwise systematically change the shapes of the voids in the direction parallel to the line of sight.

References

Cai, Y.-C., Padilla, N., & Li, B., 2014, submitted to MNRAS, arXiv:1410.1510
Ceccarelli, L., Padilla, N., Valotto, C., & Lambas, D. G., 2006, *MNRAS*, 373, 1440
Ceccarelli, L., Paz, D., Lares, M., Padilla, N., Lambas, & D. Garca, 2013, *MNRAS*, 434, 1435
Clampitt, J., Cai, Y.-C., & Li, B., 2013, *MNRAS*, 431, 749
Colberg, J. M., *et al.* 2008, *MNRAS*, 387, 933
Hu, W. & Sawicki, I., 2007, *Phys. Rev.*, D76, 064004
Jennings, E., Baugh, C. M., Li, B., Zhao, G.-B., & Koyama, K., 2012, *MNRAS*, 425, 2128
Li, B., Zhao, G., Teyssier, R., & Koyama, K., 2012, *J. Cosmo. As- tropart. Phys.*, 01, 051
Nadathur, S., *et al.*, 2014, ArXiv e-prints
Padilla, N. D., Ceccarelli, L., & Lambas, D., 2005, *MNRAS*, 363, 977
Paz, D., Lares, M., Ceccarelli, L., Padilla, N., & Lambas, D. G., 2013, *MNRAS*, 436, 3480
Peebles, P. J. E., 1979, *AJ*, 84, 730
Sheth, R. K. & van de Weygaert, R., 2004, *MNRAS*, 350, 517

The Zeldovich Universe:
Genesis and Growth of the Cosmic Web
Proceedings IAU Symposium No. 308, 2014
R. van de Weygaert, S. Shandarin, E. Saar & J. Einasto, eds.

© International Astronomical Union 2016
doi:10.1017/S174392131601053X

Modeling cosmic void statistics

Nico Hamaus[1,2,†], P. M. Sutter[1,2,3] and Benjamin D. Wandelt[1,2,4]

[1]Sorbonne Universités, UPMC Univ Paris 06, Paris, France

[2]CNRS, UMR 7095, Institut d'Astrophysique de Paris, Paris, France

[3]Center for Cosmology and AstroParticle Physics, Ohio State University, Columbus, USA

[4]Departments of Physics and Astronomy, University of Illinois at Urbana-Champaign, USA

Abstract. Understanding the internal structure and spatial distribution of cosmic voids is crucial when considering them as probes of cosmology. We present recent advances in modeling void density- and velocity-profiles in real space, as well as void two-point statistics in redshift space, by examining voids identified via the watershed transform in state-of-the-art ΛCDM n-body simulations and mock galaxy catalogs. The simple and universal characteristics that emerge from these statistics indicate the self-similarity of large-scale structure and suggest cosmic voids to be among the most pristine objects to consider for future studies on the nature of dark energy, dark matter and modified gravity.

Keywords. cosmology: large-scale structure of universe, dark matter, cosmological parameters, gravitation, methods: n-body simulations

1. Introduction

A fundamental quantity to describe the structure of cosmic voids in a statistical sense is their spherically averaged density profile. In contrast to the well-known formulas parametrizing density profiles of simulated dark matter halos, rather few models for void density profiles have been developed, mainly focusing on their central regions (e.g., Colberg *et al.* 2005, Padilla *et al.* 2005, Ricciardelli *et al.* 2014), rarely taking into account the compensation walls around voids (Paz *et al.* 2013). We present a simple formula able to accurately describe the real-space density and velocity profile of voids and subvoids of any size and redshift out to large distances from their center (Hamaus *et al.* 2014b).

We further propose a new statistic to be adequate for cosmological applications: void auto-correlations. In contrast to void-galaxy cross- (Padilla *et al.* 2005, Paz *et al.* 2013) and galaxy auto-correlations, this statistic is not directly affected by redshift-space distortions, since the galaxy density does not enter in it. Peculiar motions of galaxies only affect this estimator indirectly when voids are identified in redshift space. However, the difference in void positions from real space is expected to be small, as many galaxies are needed to define a single void, which diminishes the net displacement of its center. This property makes the void auto-correlation a very pristine and model-independent statistic to conduct an Alcock-Paczynski (AP) test (Hamaus *et al.* 2014c).

In this study we apply the publicly available Void IDentification and Examination toolkit (VIDE, Sutter *et al.* 2014b), which is based on the watershed algorithm ZOBOV (Neyrinck 2008), to n-body simulations (Warren 2013) and derived mock galaxy catalogs. The code accounts for the self-similar nature of the cosmic web by arranging a nested hierarchy of voids and subvoids into a tree-like structure.

† email: hamaus@iap.fr

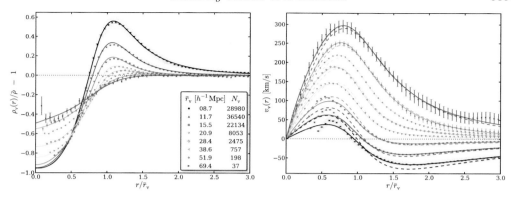

Figure 1. Stacked real-space density (left) and velocity (right) profiles of voids at $z = 0$ with mean effective radii and void counts indicated in the inset. Shaded regions depict the standard deviation σ within each of the stacks (scaled down by 20 for visibility), while error bars show standard errors on the mean profile $\sigma/\sqrt{N_v}$. Solid lines represent individual best-fit solutions and dashed lines show linear theory predictions for velocity obtained from the density stacks.

2. Results

Density structure. We define the void density profile as the spherically averaged relative deviation of mass density around a void center from the mean value $\bar{\rho}$ across the Universe, $\rho_v(r)/\bar{\rho} - 1$. Note that we consider the full void hierarchy including multiple levels of subvoids here, while Nadathur *et al.* 2014 restrict their analysis to only top-level voids in moderate sampling densities, and apply further restrictive cuts. The results from shell-averaging particles around voids in bins (stacks) of effective radius are shown in the left-hand panel of Fig. 1. We have conducted a resolution study in order to ensure our density estimation to be unbiased for all the data points shown.

As expected, stacked voids are deeply underdense inside, with their central density increasing with void size. In addition, the variance of underdense regions is suppressed compared to overdense ones, yielding the smallest error bars in the centers of the emptiest voids. However, note that the void-to-void scatter in the profile decreases towards the largest voids, as can be seen from the shaded regions in Fig. 1. The profiles all exhibit overdense compensation walls with a maximum located slightly outside their mean effective void radius $\bar{r}_v$, shifting outwards for larger voids. The height of the compensation wall decreases with void size, causing the inner profile slope to become shallower. This trend divides all voids into being either over- or undercompensated, depending on whether the total mass within their compensation wall exceeds or falls behind their missing mass in the center, respectively. We propose a simple empirical formula that accurately captures the properties described above:

$$\frac{\rho_v(r)}{\bar{\rho}} - 1 = \delta_c \frac{1 - (r/r_s)^\alpha}{1 + (r/r_v)^\beta} , \qquad (2.1)$$

where δ_c is the central density contrast, r_s a scale radius at which $\rho_v = \bar{\rho}$, and α and β determine the inner and outer slope of the void's compensation wall, respectively. The best fits of this model to the data are shown as solid lines in the left-hand panel of Fig. 1.

Velocity structure. The right-hand panel of Fig. 1 depicts the resulting velocity stacks using the same void radius bins. Note that a positive velocity implies outflow of tracer particles from the void center, while a negative one denotes infall. As the largest voids are undercompensated (void-in-void process in Sheth & van de Weygaert 2004), they are characterized by outflow in the entire distance range. Small voids exhibit infall velocities, as they are overcompensated (void-in-cloud process). This causes a sign change in their velocity profile around the void's effective radius beyond which matter is flowing onto its compensation wall, ultimately leading to a collapse of the void.

The distinction between over- and undercompensation can directly be inferred from velocities, since only overcompensated voids feature a sign change in their velocity profile, while undercompensated ones do not. Consequently, the flow of tracer particles around precisely compensated voids vanishes already at a finite distance to the void center and remains zero outwards. We denote the effective radius of such voids the *compensation scale*. It can also be inferred via clustering analysis in Fourier space, as compensated structures do not generate any large-scale power (Hamaus *et al.* 2014a).

In linear theory the velocity profile can be related to the density using $v_v(r) = -\frac{1}{3}\Omega_m^\gamma H r \Delta(r)$, where Ω_m is the relative matter content in the Universe, $\gamma \simeq 0.55$ the growth index of matter perturbations, H the Hubble constant, and $\Delta(r)$ the integrated density contrast defined as $\Delta(r) = \frac{3}{r^3} \int_0^r [\rho_v(q)/\bar\rho - 1] q^2 \mathrm{d}q$. With Eq. (2.1), this yields

$$\Delta(r) = \delta_c \, {}_2F_1\left[1, \frac{3}{\beta}, \frac{3}{\beta}+1, -(r/r_v)^\beta\right] - \frac{3\delta_c(r/r_s)^\alpha}{\alpha+3} \, {}_2F_1\left[1, \frac{\alpha+3}{\beta}, \frac{\alpha+3}{\beta}+1, -(r/r_v)^\beta\right],$$
(2.2)

where ${}_2F_1$ is the Gauss hypergeometric function. We can use this analytic formula to fit the velocity profiles obtained from our simulations; the results are shown as solid lines in the right-hand panel of Fig. 1. As for the density profiles, the quality of the fits is remarkable, especially for large voids. Only the interiors of smaller voids show stronger discrepancy, which is mainly due to the decreasing validity of the linear theory relation between density and velocity. We obtain best-fit parameter values that are very similar to the ones resulting from the density stacks above. In fact, evaluating the velocity profile at the best-fit parameters obtained from the density stacks yields almost identical results, as indicated by the dashed lines in Fig. 1.

Geometry. Although individual voids exhibit arbitrary shapes and internal structures, their ensemble average must obey statistical isotropy if the cosmological principle holds. In observations, void density profiles are constructed by alignment of the barycenters of each individual void identified in a galaxy survey, and by histogramming the distribution of galaxies around this center. This is equivalent to the void-galaxy cross-correlation function and hence a two-point statistic like the galaxy correlation function, which can be exploited for the inference of cosmological parameters. However, these stacked voids are subject to redshift-space distortions, which are inherent to the galaxies used to construct their profiles (Padilla *et al.* 2005, Sutter *et al.* 2012, Paz *et al.* 2013, Hamaus *et al.* 2014c).

The two-point statistics (auto-power spectrum and auto-correlation function) for voids found in mock galaxies in redshift space are shown in Fig. 2. Dynamic distortions evidently play a minor role here, as the contour lines appear to be fairly circular. Residual anisotropies are consistent with being random fluctuations due to the lower number count of voids compared to galaxies in a given survey volume, but no systematic trends are manifest. The compensation wall of the void auto-correlation function extends to twice the effective void radius $\bar r_v$, because of mutual void exclusion. This feature manifests itself as a ring of suppressed power in Fourier space at a scale $k \simeq \pi/2\bar r_v$ (Hamaus *et al.* 2014c). Existing studies so far have only considered galaxy auto- and void-galaxy cross-correlations for applications of the AP test (Sutter *et al.* 2012), but void auto-correlations can be readily obtained from existing data as well and provide additional information on the inferred geometry of large-scale structure. Moreover, systematic uncertainties and biases from redshift-space distortions appear to be negligible in void auto-correlations, which potentially makes this statistic the cleanest one for detecting geometric distortions.

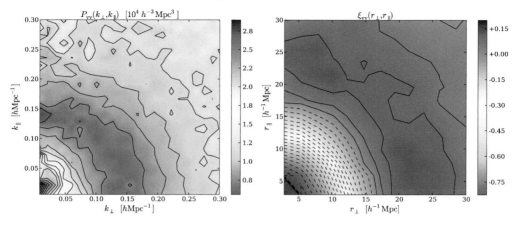

Figure 2. Two-dimensional void power spectrum (left) and void correlation function (right) from galaxies in redshift space at $z = 0$. Solid lines show positive, dashed lines negative contours.

3. Conclusions

There are a number of cosmological applications to make use of the presented functional form of the average void density profile, for example, studies of gravitational (anti)lensing that directly probe the projected mass distribution around voids, which in turn may serve as a tool for constraining models of dark matter, dark energy, and modified gravity. This is thanks to the universal nature of Eq. (2.1), which even describes voids in the distribution of galaxies remarkably well, as demonstrated in Sutter *et al.* 2014a. That paper further pointed out that the impact of tracer sparsity and bias on the definition of voids can be accounted for by simple rescalings of voids.

Moreover, void auto-correlations are a promising statistic for applications of the AP test and constraining the expansion history of the Universe. In addition to galaxy auto- and void-galaxy cross-correlations, they provide complementary information on cosmological parameters, while at the same time being least affected by systematic effects from redshift-space distortions. Our findings corroborate other indications that cosmic voids may indeed offer new and complementary approaches to modeling fundamental aspects of the large-scale structure of our Universe.

References

Colberg, J. M., Sheth, R. K., Diaferio, A., Gao, L., & Yoshida, N. 2005, *MNRAS*, 360, 216
Hamaus, N., Wandelt, B. D., Sutter, *et al.* 2014, *Phys. Rev. Lett.*, 112, 041304
Hamaus, N., Sutter, P. M., & Wandelt, B. D. 2014, *Phys. Rev. Lett.*, 112, 251302
Hamaus, N., Sutter, P. M., Lavaux, G., & Wandelt, B. D. 2014, arXiv:1409.3580
Nadathur, S., Hotchkiss, S., Diego, J. M., *et al.* 2014, arXiv:1407.1295
Neyrinck, M. C. 2008, *MNRAS*, 386, 2101
Padilla, N. D., Ceccarelli, L., & Lambas, D. G. 2005, *MNRAS*, 363, 977
Paz, D., Lares, M., Ceccarelli, L., Padilla, N., & Lambas, D. G. 2013, *MNRAS*, 436, 3480
Ricciardelli, E., Quilis, V., & Varela, J. 2014, *MNRAS*, 440, 601
Sheth, R. K. & van de Weygaert, R. 2004, *MNRAS*, 350, 517
Sutter, P. M., Lavaux, G., Wandelt, B. D., & Weinberg, D. H. 2012 *ApJ*, 761, 187
Sutter, P. M., Lavaux, G., Hamaus, N., *et al.* 2014, *MNRAS*, 442, 462
Sutter, P. M., Lavaux, G., Hamaus, N., *et al.* 2014, arXiv:1406.1191
Warren, M. S. 2013, in Proceedings of SC '13 (ACM, New York, USA), *2HOT: An Improved Parallel Hashed Oct-tree N-body Algorithm for Cosmological Simulation*, pp. 72:1–72:12

The Zeldovich Universe:
Genesis and Growth of the Cosmic Web
Proceedings IAU Symposium No. 308, 2014
R. van de Weygaert, S. Shandarin, E. Saar & J. Einasto, eds.

© International Astronomical Union 2016
doi:10.1017/S1743921316010541

Universal void density profiles from simulation and SDSS

S. Nadathur[1],† S. Hotchkiss[2], J. M. Diego[3], I. T. Iliev[2], S. Gottlöber[4], W. A. Watson[2] and G. Yepes[5]

[1]Department of Physics, University of Helsinki and Helsinki Institute of Physics, P.O. Box 64, FIN-00014, University of Helsinki, Finland

[2]Department of Physics and Astronomy, University of Sussex, Falmer, Brighton, BN1 9QH, UK

[3]IFCA, Instituto de Fisica de Cantabria (UC-CSIC), Avda Los Castros s/n. E-39005 Santander, Spain

[4]Leibniz-Institute for Astrophysics, An der Sternwarte 16, D-14482 Potsdam, Germany

[5]Departamento de Física Teórica, Modulo C-XI, Facultad de Ciencias, Universidad Autónoma de Madrid, 28049 Cantoblanco, Madrid, Spain

Abstract. We discuss the universality and self-similarity of void density profiles, for voids in realistic mock luminous red galaxy (LRG) catalogues from the Jubilee simulation, as well as in void catalogues constructed from the SDSS LRG and Main Galaxy samples. Voids are identified using a modified version of the ZOBOV watershed transform algorithm, with additional selection cuts. We find that voids in simulation are *self-similar*, meaning that their average rescaled profile does not depend on the void size, or – within the range of the simulated catalogue – on the redshift. Comparison of the profiles obtained from simulated and real voids shows an excellent match. The profiles of real voids also show a *universal* behaviour over a wide range of galaxy luminosities, number densities and redshifts. This points to a fundamental property of the voids found by the watershed algorithm, which can be exploited in future studies of voids.

Keywords. catalogues – cosmology: observations – large-scale structure of Universe – methods: numerical – methods: data analysis

---

1. Introduction

Voids are recognised as particularly interesting objects for cosmology for many reasons. Of particular interest recently has been their use as probes of the expansion history via the Alcock-Paczynski test (e.g. Lavaux & Wandelt 2012, Hamaus *et al.* 2014c), void-galaxy correlations (Hamaus *et al.* 2014a, Paz *et al.* 2013) and the weak lensing signal of stacked voids (e.g. Krause *et al.* 2013, Clampitt & Jain 2014). It has even been suggested that the integrated Sachs-Wolfe effect of voids on the CMB can be measured (Granett *et al.* 2008), though theoretical expectations and more recent observational results (e.g. Nadathur *et al.* 2012, Flender *et al.* 2013, Cai *et al.* 2013, Hotchkiss *et al.* 2014) do not support this.

Many of these studies have assumed that voids are *self-similar* objects, in particular that the density distribution in each void can be simply rescaled depending on the size of the void, and sometimes that this distribution is *universal*—that is, that the rescaled void properties are independent of the properties of the tracer population in which the voids were identified or the survey redshift. The form of the density profile itself has also been the subject of study (e.g. Colberg *et al.* 2005, Ceccarelli *et al.* 2013, Nadathur &

† seshadri.nadathur@helsinki.fi

Hotchkiss 2014, Sutter *et al.* 2014, Ricciardelli *et al.* 2014, Hamaus *et al.* 2014b, Nadathur *et al.* 2014), but there is a lack of consensus on the functional form of the profile as well as on the questions of self-similarity and universality.

We make use of data from the Jubilee *N*-body simulation (Watson *et al.*, 2014) and SDSS DR7 galaxy catalogues to further investigate these issues. The voids are identified using a modified version of the ZOBOV watershed transform void finder (Neyrinck 2008) with further selection criteria imposed to avoid spurious detections arising from Poisson noise or survey boundary contamination effects. The void catalogues from SDSS data are taken from Nadathur & Hotchkiss 2014, and cover voids identified in six spectro- scopic volume-limited galaxy samples (*dim1*, *dim2*, *bright1*, *bright2*, *lrgdim* and *lrgbright*; Blanton *et al.* 2005, Kazin *et al.* 2010), with widely varying luminosity cuts and galaxy densities. For Jubilee data we use an HOD model applied to halos on the light cone (Watson *et al.*, 2014, Nadathur *et al.* 2014) to obtain two mock LRG samples, referred to as JDim and JBright, whose properties match those of the *lrgdim* and *lrgbright*, and extract voids from those.

2. Method

To identify voids we use a modification of the ZOBOV algorithm (Neyrinck 2008), which uses a Voronoi tessellation field estimator (VTFE) to reconstruct the galaxy density field from discrete point distribution, and then joins local minima of this field together to form voids according to the watershed algorithm. We account for the finite redshift extents of the samples and the irregular SDSS survey mask through the use of buffer particles at all survey boundaries. We restrict the merging of zones of density minima beyond linking densities $\rho_{\rm link} = 0.3\bar{\rho}$ and apply a strict selection criterion on the minimum VTFE density in the void, $\rho_{\rm min} \leqslant 0.3\bar{\rho}$. This last criterion is introduced to counter shot noise effects. Even in a uniform Poisson distribution of points, ZOBOV will always find spurious 'voids' with $\rho_{\rm min} < \bar{\rho}$; however, fewer than 1% of these spurious voids have $\rho_{\rm min} \leqslant 0.3\bar{\rho}$ (Nadathur & Hotchkiss 2014). We have also examined stricter criteria $\rho_{\rm link} \leqslant 0.2\bar{\rho}$ and $\rho_{\rm min} \leqslant 0.2\bar{\rho}$, which do not materially affect our conclusions.

For each void, we define its centre to be the volume-weighted barycentre of the member galaxies of the void as identified by ZOBOV, $\mathbf{X} = \frac{1}{\sum_i V_i} \sum_i \mathbf{x}_i V_i$, where V_i is the volume of the Voronoi cell of the ith galaxy, and the void effective radius R_v to be the volume of a sphere occupying a volume equal to the sum of the Voronoi volumes of the void member galaxies.

To measure the average (stacked) galaxy density profile, we first assume self-similarity to rescale all distances from the void centre in units of the radius R_v^i and construct a series of spherically symmetric radial shells each of width Δ in rescaled units. Then if mean density in the jth shell is estimated using the VTFE itself

$$\bar{\rho}^j = \frac{\sum_{i=1}^{N_v} \sum_{k=1}^{N_i^j} \rho_k V_k}{\sum_{i=1}^{N_v} \sum_{k=1}^{N_i^j} V_k}, \qquad (2.1)$$

where V_k is the volume of the Voronoi cell of the galaxy k, ρ_k is its density inferred from the inverse of the Voronoi volume; the sum over k runs over all galaxies in the jth shell of void i (not only void member galaxies); and the sum over i includes all voids in the stack. The error in $\bar{\rho}^j$ is estimated from jacknife samples excluding all galaxies from each of the N_v voids in turn.

Nadathur *et al.* 2014 also examined the performance of two other density estimators. The commonly used 'naive' estimator, $\bar{\rho}^j = \left(\sum_{i=1}^{N_v} N_i^j / V_i^j \right) / N_v$, where V_i^j is the volume of the jth radial shell of the ith void and N_i^j is the number of galaxies contained within it,

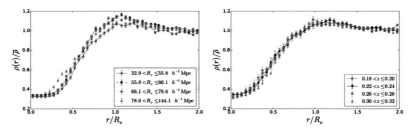

Figure 1. *Left panel:* The average stacked density profiles for voids from the JDim mock catalogue, split into different quartiles of void radius R_v. *Right:* Stacked profiles for voids in four representative redshift bins within the JDim catalogue.

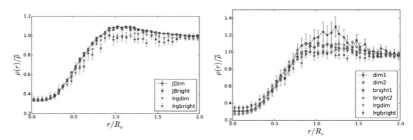

Figure 2. *Left panel:* Stacked density profiles for voids from the SDSS catalogues *lrgdim* and *lrgbright*, compared with those from the corresponding mock catalogues JDim and JBright from Jubilee. *Right:* Stacked profiles for voids from all six SDSS galaxy catalogues.

was shown to be systematically biased low in low density regions such as void centres. A better estimator based on Poisson statistics is $\overline{\rho}^j = \left(\sum_{i=1}^{N_v} N_i^j + 1 \right) / \sum_{i=1}^{N_v} V_i^j$; however, while this is optimal for simulations in a cubic box with periodic boundary conditions, it suffers from volume-leakage effects when the galaxy positions are restricted to a finite (and highly irregular) survey window, which cause it to underestimate the density at distances $r \gtrsim R_v$ for voids close to the survey edge.

3. Results

Self-similarity. Although the stacked density estimator of eq.(2.1) assumes self-similarity, we can also directly test this assumption for consistency by splitting the full stack of voids into different subsets. We find that a small subset ($\lesssim 5\%$) of voids in simulation consist of more than 5 different density minima amalgamated together by the watershed algorithm, and these voids shows a different density profile to the others. This is due to averaging over the internal substructure of these voids, and is also manifest in the greater distance of the barycentre from the minimum density centre. These voids can be removed from the full sample by hand, or they can be eliminated by a stricter (but somewhat arbitrary) choice of the zone merging criterion, $\rho_{\text{link}} \leqslant 0.2\overline{\rho}$ as used by Hamaus *et al.* 2014b. When this is done, the remaining voids show no dependence of the mean stacked profile on the void radius or redshift (Fig. 1). We find that a simple fitting formula of the form $\frac{\rho(r)}{\overline{\rho}} = 1 + \delta \left(\frac{1-(r/r_s)^\alpha}{1+(r/r_s)^\beta} \right)$, with $\delta = -0.69$, $r_s = 0.81 R_v$, $\alpha = 1.57$ and $\beta = 5.72$, provides a reasonable fit to the simulation data. This describes a one-parameter family of curves over the void size parameter R_v.

Agreement with SDSS. Fig. 2 shows that up to distances $r \sim R_v$ there is very good agreement between the density profiles from Jubilee and SDSS data. There is however a residual small difference at $r \sim R_v$, where in simulation we see an overdense compensating wall at the void edge. This may reflect some small inadequacies of the HOD modelling of

LRGs in these regions, but a more likely explanation is that these are residual artefacts of the SDSS survey mask. Further detailed study of the effect of the survey mask in simulations will clarify this issue.

Universality. In Fig. 2, we show the dependence of the average stacked profile on the properties of the tracer galaxies using SDSS data. These samples span a wider range of redshifts (from $z < 0.05$ for *dim1* to $0.16 < z < 0.44$ for *lrgbright*), absolute magnitudes (from $M_r < -18.9$ for *dim1* to $M_g < -21.8$ for *lrgbright*) and mean void sizes (from $\overline{R_v} = 9.6\ h^{-1}$Mpc for *dim1* to $\overline{R_v} = 92.8\ h^{-1}$Mpc for *lrgbright*) than is available in our mock catalogues. Nevertheless, the results indicate a remarkable degree of universality in the stacked void profile across all galaxy samples. This universality is most pronounced close to the void centres; small differences are seen at the void edge. The trend seen in this edge region is consistent with the expectation that samples at lower redshift should show higher densities in the void walls simply due to greater growth of structure at late times.

4. Conclusions

The density profile of voids is a subject of great interest for cosmology. Several studies have implicitly assumed the self-similarity and universality of this profile; our aim in this work has been to examine the validity of these assumptions. We find that voids in the Jubilee simulation are indeed self-similar, and that the measured profile matches that seen for voids in the SDSS DR7. In addition we have shown that void profiles from SDSS galaxy samples covering a wide range of galaxy magnitudes and number densities are also universal, being essentially indistinguishable from each other within the void interior. This significantly extends the results found from simulation. It provides a reference point for comparisons with theoretical models (Sheth & van de Weygaert 2004) and may prove to be a useful observational tool to constrain cosmology.

References

Blanton, M. R. *et al.*, 2005, *AJ*, 129, 2562
Cai, Y.-C., Neyrinck, M. C., Szapudi, I., Cole, S., & Frenk, C. S., 2013, ArXiv e-prints, 1301.6136
Ceccarelli, L., Paz, D., Lares, M., Padilla, N., & Lambas, D. G., 2013, *MNRAS*, 434, 1435
Clampitt, J., Cai, Y.-C., & Li, B., 2013, *MNRAS*, 431, 749
Clampitt, J. & Jain, B., 2014, ArXiv e-prints, arXiv:1404.1834
Colberg, J. M., Sheth, R. K., Diaferio, A., Gao, L., & Yoshida, N., 2005, *MNRAS*, 360, 216
Flender, S., Hotchkiss, S., & Nadathur, S., 2013, *JCAP*, 1302, 013
Granett, B. R., Neyrinck, M. C., & Szapudi, I., 2008, *ApJ*, 683, L99
Hamaus, N., Wandelt, B. D., Sutter, P. M., Lavaux, G., & Warren, M. S., 2014, *Phys. Rev. Lett.*, 112, 041304
Hamaus, N., Sutter, P. M., & Wandelt, B. D., 2014, *Phys. Rev. Lett.*, 112, 251302
Hamaus, N., Sutter, P. M., Lavaux, G., & Wandelt, B. D., 2014, ArXiv e-prints, arXiv:1409.3580
Hotchkiss, S., Nadathur, S., Gottlöber S., *et al.*, 2014, ArXiv e-prints, arXiv:1405.3552
Kazin, E. A. *et al.*, 2010, *ApJ*, 710, 1444
Krause, E., Chang, T.-C., Doré O., & Umetsu, K., 2013, *ApJ*, 762, L20
Lavaux, G. & Wandelt, B. D., 2012, *ApJ*, 754, 109
Nadathur, S. & Hotchkiss, S., 2014, *MNRAS*, 440, 1248
Nadathur, S., Hotchkiss, S., & Sarkar, S., 2012, *JCAP*, 1206, 042
Nadathur, S., Hotchkiss, S., Diego, J. M., *et al.*, 2014, ArXiv e-prints, arXiv:1407.1729
Neyrinck, M. C., 2008, *MNRAS*, 386, 2101
Paz, D., Lares, M., Ceccarelli, L., Padilla, N., & Lambas, D. G., 2013, *MNRAS*, 436, 3480
Ricciardelli, E., Quilis, V., & Varela, J., 2014, *MNRAS*, 440, 601
Sheth, R. K. & van de Weygaert, R., 2004, *MNRAS*, 350, 517
Sutter, P. M., Lavaux, G., Hamaus, N., Wandelt, B. D., Weinberg, D. H., & Warren, M. S., 2014, *MNRAS*, 442, 462
Watson, W. A. *et al.*, 2014, *MNRAS*, 438, 412

The Zeldovich Universe:
Genesis and Growth of the Cosmic Web
Proceedings IAU Symposium No. 308, 2014
R. van de Weygaert, S. Shandarin, E. Saar & J. Einasto, eds.

© International Astronomical Union 2016
doi:10.1017/S1743921316010553

Real-space density profile reconstruction of stacked voids

Alice Pisani[1,2], **P. Sutter**[1,2,3], **G. Lavaux**[1,2] and **B. Wandelt**[1,2,4,5]

[1]Sorbonne Universités, UPMC Univ Paris 06, UMR7095, F-75014, Paris, France

[2]CNRS, UMR7095, Institut d'Astrophysique de Paris, F-75014, Paris, France

[3]Center for Cosmology and AstroParticle Physics, Ohio State University, Columbus, USA

[4]Department of Physics, University of Illinois at Urbana-Champaign, Urbana, USA

[5]Department of Astronomy, University of Illinois at Urbana-Champaign, Urbana, USA

Abstract. Modern surveys allow us to access to high quality large scale structure measurements. In this framework, cosmic voids appear as a new potential probe of Cosmology. We discuss the use of cosmic voids as standard spheres and their capacity to constrain new physics, dark energy and cosmological models. We introduce the Alcock-Paczyński test and its use with voids. We discuss the main difficulties in treating with cosmic voids: redshift-space distortions, the sparsity of data, and peculiar velocities. We present a method to reconstruct the spherical density profiles of void stacks in real space, without redshift-space distortions. We show its application to a toy model and a dark matter simulation; as well as a first application to reconstruct real cosmic void stacks density profiles in real space from the Sloan Digital Sky Survey.

Keywords. cosmology: large-scale structure of universe, methods: dark energy, Abel inverse

1. Introduction

The large scale structure of the Universe is an incredibly powerful tool to constrain cosmic evolution. At large scales, clusters of galaxies, voids and filaments shape the Universe — it is the Cosmic Web. Cosmic voids, discovered in 1978 (Gregory & Thompson, 1978; Jõeveer et al., 1978), are under-dense regions in the Universe with sizes from ten to hundreds of Mpcs. Until very recently, due to the difficulty of extracting data from low density zones, the potential of voids has been under-explored.

With the increasing precision and quality of modern surveys we now have access to high quality large scale structure measurements: the appeal of cosmic voids becomes thus considerable. Being devoid of matter, cosmic voids might be mainly composed of dark energy (Bos et al., 2012) — which strongly justifies their importance for Cosmology, as dark energy is believed to be ∼70% of the Universe and we are still not able to understand it. The efforts of cosmologists seem to converge to a model (called ΛCDM) that leaves many unknowns. The nature — and we could say even more, the existence — of dark energy remains a mystery; and so does the nature of dark matter. In this framework cosmic voids appear as a new tool to use in our research of a correct cosmological model.

Cosmic voids have a much simpler evolution compared to high density zones of the Universe and, as such, constitute a promising laboratory to test dark energy, constrain cosmic expansion and discriminate between cosmological models such as modified gravity models (e. g. Spolyar et al., 2013). The difficulties in extracting information from voids are: the irregular shape of voids, the sparsity of galaxies and the presence of redshift-space distortions (RSD).

In the first part we discuss the use of voids for cosmology, the second part presents the key idea to overcome RSD and its test on a toy model. In the third part we illustrate the application of the algorithm to a simulation and we present the first ever real-space density profiles of stacked voids from SDSS DR7 data, reconstructed without any assumption about RSD. Finally we conclude discussing future applications of the algorithm.

2. Cosmic voids for Cosmology

The shape of cosmic voids is not regular. Cosmic voids are then difficult objects to study because of their irregular shape and of the sparsity of data: galaxies are mostly concentrated in filaments or clusters of galaxies. To overcome both the irregularity of the shape and the small amount of data, Lavaux and Wandelt (2012) suggested a new approach consisting in the stacking of voids. In a homogeneous and isotropic Universe, the average real-space shape of cosmic voids — obtained by stacking — is spherical. Because of its spherical shape, the stacked void can be used to extract cosmological information. To reach such goal, we use void catalogues to obtain stacked voids at different redshift and of different sizes. Then, the spherical profiles of stacked voids can be used as standard spheres, necessary tools for the so-called Alcock-Paczyński test.

The Alcock-Paczyński test (Alcock and Paczyński, 1979), proposed in 1978 by Charles Alcock and Bohdan Paczyński, compares the shape of a standard sphere in real and redshift space to constrain the expansion of the Universe. When applied to voids (Sutter *et al.*, 2014d), the test compares the shape of the stacked distorted void in redshift space and of the stacked spherical void in real space to obtain information about the expansion of the universe. It uses the void as a standard sphere.

The term *standard sphere* is an analogy with the idea of standard candle or standard ruler. A standard candle (ruler) is an object of known luminosity (length) used to measure the expansion of the universe through the difference between the observed and the expected luminosity (length). For a standard sphere the known quantity is the diameter of the sphere, that we measure along and perpendicularly to the line of sight direction. Unfortunately, although void stacks would be perfect as standard spheres, we cannot perform a precise Alcock-Paczyński test without the real space profile of voids (Pisani *et al.* (2014a)): observations will always give us the shape in redshift space.

Indeed, surveys measure the position of galaxies *in redshift space*. Since our universe is expanding, all galaxies are redshifted due to the expansion of space. To this is added the redshift caused by the peculiar motion of the galaxy. Only the line of sight component of velocity affects the galaxy redshift (Hamilton, 1998). In the framework of voids, it means that the real-space spherical shape of voids is distorted in redshift space: we do not measure a spherical shape of the stacked void, we measure a distorted void in redshift space. In the next part we describe the key idea to obtain the shape of voids in real space.

3. Overcoming RSD: real-space density reconstruction

To extract cosmological information from real stacked voids, we need their shape in real space. It is fundamental to recover the real-space density without any assumption about RSD, otherwise the cosmological results would not be independent from the model used for RSD.

The key idea is that RSD only affect the line of sight component of the galaxy positions. If we were able to reconstruct the shape of the stacked void without using this component — that is from the projection on the x-y plane — we could obtain the void in real space (see figure 1, left).

We consider the redshift-space density of the void (number of galaxies per volume element), we then project it by counting all the galaxies in bins in the x-y plane and

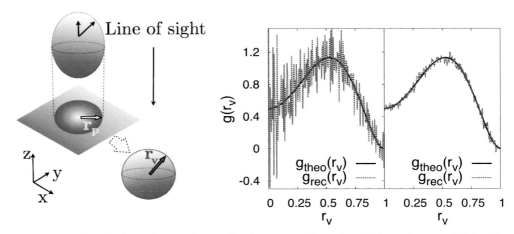

Figure 1. Left: The key idea to obtain the density profiles of voids in real space. Right: Reconstruction of the toy model density profile, without (first plot) and with (second plot) the regularisation algorithm for the inversion. Reproduced from Pisani *et al.* (2014a)

obtain the projected density. To obtain the spherical density profile of the void from the projection, we use Abel inverse transform (Abel, 1842):

$$g(r_{\rm v}) = -\frac{1}{\pi} \int_{r_{\rm v}}^{1} \frac{I'(r_{\rm p})}{\sqrt{r_{\rm p}^2 - r_{\rm v}^2}} dr_{\rm p} \qquad (3.1)$$

where $g(r_{\rm v})$ is the spherical density profile of galaxies in real space we aim to reconstruct, $r_{\rm v} = \sqrt{x^2 + y^2 + z^2}$ is the void's radius, $r_{\rm p} = \sqrt{x^2 + y^2}$ is the radius of the projection onto a plane perpendicular to the line of sight and $I(r_{\rm p})$ is the projected density.

The problem is that the Abel inverse transform, although well mathematically defined (see equation 3.1), is strongly *ill-conditioned*: if there is some noise in the input function $I(r_{\rm p})$ (of which $I'(r_{\rm p})$ is the derivative with respect to $r_{\rm p}$), the reconstruction will be dominated by noise. To overcome the problem of ill-conditioning we have implemented for the case of voids a polynomial regularization of the inversion (Li *et al.*, 2007), allowing to obtain the spherical density profiles of voids in real space.

We test the reconstruction with a toy model: we artificially distort a void stack with a simple model of RSD and we reconstruct the profile in real space. As a pedagogical illustration, we show the reconstruction of the density without any regularization algorithm for the Abel inverse transform and we compare it with the reconstruction that uses our algorithm (see figure 1, right; note that the void is a toy model and the density profile does not reflect the real shape of a void). The ill-conditioning of the reconstruction is well controlled. The next section presents the results with a dark matter particle simulation and real data from SDSS.

4. Results: simulation and SDSS data

In this part, we show the result of the reconstruction (figure 2) of a stacked void from a dark matter simulation. More details can be found in Pisani *et al.* (2014a). As a first application of the algorithm to real data, we reconstruct the density profiles of stacked voids from the void catalogue presented in Sutter *et al.* (2012). We show in figure 2 a reconstructed profile for a real stacked void. Despite the noise present in real data,

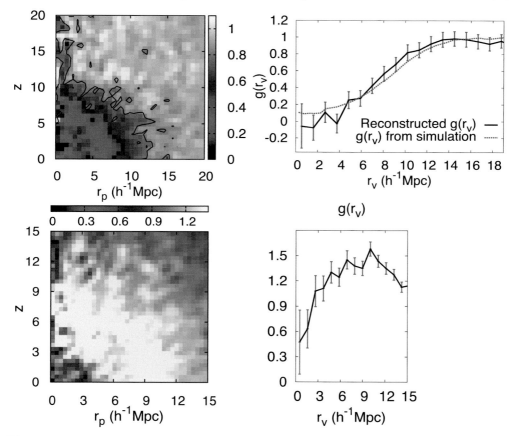

Figure 2. From left to right: density in redshift space from the simulation; comparison of the reconstructed density profile with the profile in real space from the simulation; density in redshift space for a 5-15 $h^{-1} Mpc$ stacked void from SDSS and reconstructed density profile in real space. Reproduced from Pisani *et al.* (2014a).

the algorithm is capable to obtain the density profile of the void, overcoming the ill-conditioning of the inverse. To conclude, in the next section we briefly discuss future applications for the reconstruction of the density profiles.

Conclusion: modelling systematics for the future

We developed an algorithm to reconstruct density profiles of voids in real space without any assumption about RSD. The knowledge of the real space shape of the void is instrumental to improve the application of the Alcock-Paczyński test, since the method for the test is based on the real-space profile of void stacks. Additionally to the real-space modelling, information about peculiar velocity effects on cosmic voids would also enhance the understanding of systematics effects in their use for Cosmology and in the study of void evolution. An analysis of the effects of velocities promises to improve the precision of cosmological constraints from voids. The comparison of mock catalogues of galaxies with and without peculiar velocities allows to obtain guidelines to master such effects (see Pisani *et al.* (2014b)).

The density profiles of stacked voids in real space open the way to better constrain the value of the Hubble constant, cosmological models and new physics on current and future data sets.

Acknowledgements

AP would like to express her gratitude to the organizers for the opportunity to deliver this talk. The authors acknowledge support from BDW's Chaire Internationale at UPMC and ANR-10-CEXC-004- 01).

References

Gregory S. A. & Thompson L. A., 1978, *ApJ*, 222, 784
Jõeveer, M., Einasto, J., & Tago, E., *MNRAS*, 185:357–370, November 1978.
Bos, E., van de Weygaert, R., Dolag, K., & Pettorino, V., *MNRAS*, 426(1)440–461, 2012.
Spolyar, D., Sahlén, M., & Silk, J., *Physical Review Letters*, 111(24):241103, Dec., 2013.
Lavaux, G. & Wandelt, B. D. *ApJ*, 754:109, Aug. 2012.
Alcock, C. & Paczyński, B. *Nature*, 281:358, Oct.1979.
Sutter, P., Pisani, A., & Wandelt, B. D. *MNRAS*, Vol. 443, Issue 4, p. 2983-2990, 2014d
Hamilton, A. J. S. vol. 231 of *Astroph. and Space Science Library*, p.185, 1998
Abel, N. H. *Oeuvres Completes*, 1842, Ed. L. Sylow & S. Lie, Johnson Reprint Corp. 1988
Li, X.-F., Huang, L., & Huang, Y. *Journal of Physics A Math. Gen.*, 40:347–360, Jan. 2007
Pisani, A., Lavaux, G., Sutter, P. M., & Wandelt, B. D. *MNRAS*, 443:3238–3250, Oct. 2014a
Sutter, P. M., Lavaux, G., Wandelt, B. D., & Weinberg, D. H. *ApJ*, 761:44, Dec. 2012.
Pisani, A., Sutter, P. M., & Wandelt, B. D. *in prep.*, 2014b.

The Zeldovich Universe:
Genesis and Growth of the Cosmic Web
Proceedings IAU Symposium No. 308, 2014
R. van de Weygaert, S. Shandarin, E. Saar & J. Einasto, eds.

© International Astronomical Union 2016
doi:10.1017/S1743921316010565

On the universality of void density profiles

E. Ricciardelli[1], V. Quilis[1,2] and J. Varela[3]

[1] Departament d'Astronomia i Astrofísica, Universitat de Valencia, c/ Dr. Moliner 50, E-46100
- Burjassot, València, Spain
email: **elena.ricciardelli@gmail.com**
[2] Centro de Estudios de Física del Cosmos de Aragón (CEFCA), Plaza San Juan 1, 44001
Teruel, Spain
[3] Observatori Astronòmic, Universitat de València, E-46980 Paterna, València, Spain

Abstract. The massive exploitation of cosmic voids for precision cosmology in the upcoming dark energy experiments, requires a robust understanding of their internal structure, particularly of their density profile. We show that the void density profile is insensitive to the void radius both in a catalogue of observed voids and in voids from a large cosmological simulation. However, the observed and simulated voids display remarkably different profile shapes, with the former having much steeper profiles than the latter. We ascribe such difference to the dependence of the observed profiles on the galaxy sample used to trace the matter distribution. Samples including low-mass galaxies lead to shallower profiles with respect to the samples where only massive galaxies are used, as faint galaxies live closer to the void centre. We argue that galaxies are biased tracers when used to probe the matter distribution within voids.

Keywords. cosmology: large-scale structure of universe, cosmology: observations, methods: numerical

1. Introduction

Cosmic voids are recently attracting growing interest thanks to their potential in probing cosmological parameters. In particular, voids are the ideal candidate for probing the expansion history of the Universe through the Alcock-Paczynski test, using the average shape of stacked voids (e.g., Lavaux & Wandelt 2012, Sutter *et al.* 2012). The huge potentiality of voids for precision cosmology requires a robust knowledge of their internal structure, particularly of the density profiles. Recent works have studied the void density profiles without reaching a consensus on the functional form that can reproduce such profiles and on their dependence on void size (e.g., Ricciardelli *et al.* 2013, Ricciardelli *et al.* 2014, Hamaus *et al.* 2014, Nadathur *et al.* 2014). In the present work, we study the density profiles of voids in a catalogue of observed voids and in voids from a large cosmological simulation and assess their universality, that we intend as their insensitivity on void sizes. In doing so, we also provide a robust determination of the systematic effects arising when using different mass tracers.

The catalogue of observed voids is drawn from SDSS-DR7 with the same method adopted by Varela *et al.* 2012. Voids are identified as empty spheres, devoid of galaxies brighter than $M_r - 5 log h = -20.17$. For the present analysis, only voids with radius larger $7 \, h^{-1} \, Mpc$ are considered, resulting in 4453 voids. As for the simulations, we use a version of the MASCLET code (Quilis 2004), that has been designed to follow the formation and evolution of low density regions, to simulate a box of comoving size length $512 \, h^{-1} \, Mpc$. In the simulation, we identify more than 3000 voids larger than $7 \, h^{-1} \, Mpc$.

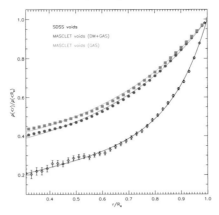

Figure 1. Stacked density profiles for all voids larger than $7\,h^{-1}\,Mpc$ identified in the SDSS database (black diamonds) and the best-fit model (black line). The coloured symbols indicate the stacked profiles of the simulated voids larger than $7\,h^{-1}\,Mpc$, computed using the total density field (blue circles) and the gas density field (orange squares). The solid coloured lines indicate the best-fits for each curve.

2. Results

In Fig. 1 we show the comparison between simulated and observed stacked profiles. The simulated profiles are computed using the density field to trace the matter distribution, whereas in the observed voids the density is traced by the luminous galaxies. The stacked profile is fitted by the two-parameters law proposed in Ricciardelli *et al.* (2013):

$$\frac{\rho(<r)}{\rho_e} = \left(\frac{r}{R_e}\right)^\alpha \exp\left[\left(\frac{r}{R_e}\right)^\beta - 1\right] \tag{2.1}$$

where $\rho(<r)$ is the density enclosed within the void-centric distance r, ρ_e is the density enclosed within the void effective radius R_e and α and β are the best-fit parameters to be obtained from the fit. The functional form expressed in Eq. 2.1 turns out to be adequate in reproducing with good accuracy both the observed and simulated profiles, although the free parameters used in the two cases are significantly different, as the observed profile appears much steeper than the simulated one.

A possible reason for the steepness of the observed density profile could lie in the different tracers used to measure the void density profiles. To assess the impact of the mass tracers, we have built volume limited samples of galaxies up to a given redshift and complete down to the corresponding threshold mass limit (see Ricciardelli *et al.* 2014). It is worth to emphasize that the choice of the galaxy sample used only affects the recovered density profiles, leaving the sample of voids unchanged. In Fig. 2 we show the recovered density profiles using the different galaxy samples. The profiles appear to steepen as galaxies at higher redshift and higher stellar mass are used. Interestingly, the profiles traced with the faintest galaxy samples approach the simulated profiles shown in Fig. 1. The steepening is particularly evident in the evolution of α, that becomes progressively higher as more massive galaxies are concerned. On the other hand, β does not show any clear dependence on the tracers, as the β beta values are just scattered around the reference values. We argue that the steepness of the observed profile, with respect to the simulated ones, can be explained by the absence of tracers in the innermost regions of the observed voids. It is not clear however whether such absence could be solved by using deeper data or it is just a consequence of the galaxy bias. We notice that hat a similar comparison of density profiles measured with different samples of galaxies has

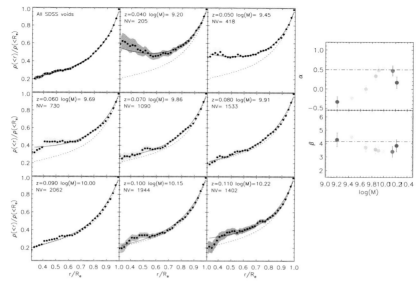

Figure 2. Dependence of the observational void density profiles on the choice of the density tracers. Different panels show the density profiles for samples of voids lying at redshift below that indicated and using, as mass tracers, the galaxies more massive than the threshold mass at that redshift. The black points indicate the radial value of the stacked void, colored shaded regions show the confidence regions determined by means of a bootstrapping, and the black solid line is the best fit. The dotted black line reported in all the panels is the best-fit density profiles of the stack drawn from the parent sample (first panel). The dependence of the best-fit parameters on the threshold mass of the galaxies used is shown in the smaller panels on the right.

been shown by Nadathur *et al.* (2013, 2014), but they do not find any dependence of the profile on the magnitude of the tracers. However, their galaxy samples are relatively bright ($Mr < -18.16 + 5log(h)$). Void galaxies in our SDSS catalogue, are, by definition, fainter than $Mr = -20.17 + 5log(h)$, hence allowing us to probe the profiles using also galaxies with very low mass. In fact, if only the highest mass bins were concerned, we would not observe such dependence of the profile on the galaxy mass. Furthermore, we refer to mass density, whereas the above authors only consider number densities.

To assess the dependence of the void density profiles on the void radius, we refer to the observed voids and divide the void sample in equi-populated subsamples, having ~ 300 voids each. The profiles for different void radii are shown in Fig. 3. All the best-fit parameters, α and β, fall within 1-2 σ of the reference values, derived by fitting the profile of the parent sample at $z < 0.08$, without any dependence on the radius. Indeed, the best-fit profile derived for the parent sample (dotted line) is compatible with the profile shape in all the size bins. Such insensitivity of the void density profile on the void radius is obtained also when using the simulated voids (Ricciardelli *et al.* 2013, Ricciardelli *et al.* 2014).

3. Conclusions

We have shown that the void density profiles recovered by means of the observed and simulated voids share the same qualitative shape, showing a significant underdensity in the centre and a sharp density increase approaching the void edges. Both profiles can be well described by the functional form proposed in Ricciardelli *et al.* (2013). Such profile,

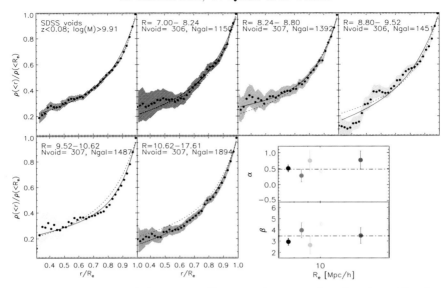

Figure 3. Observational void density profiles as a function of void radius. The first panel shows the density profile measured from an homogenous sample of galaxies, with voids lying at $z < 0.08$ and the mass tracers having stellar mass above $10^{9.9}\,M_\odot$. In the other panels, we show the density profiles of voids within different size bins. The black points indicate the radial value of the stacked void, colored shaded regions show the confidence regions determined by means of a bootstrapping, and the black solid line is the best fit. The dotted black line reported in all the panels is the best-fit density profile shown in the first panel. The lower-right panels show the dependence of the best-fit parameters α and β on void radius.

if properly rescaled, does not depend on the void size, hence we refer to it as universal density profile. It nevertheless has a strong dependence on the mass of the tracers used to constrain the profiles. Within the observed voids, the density profiles recovered by means of faint samples of galaxies are shallower than those determined through the brighter galaxies. The reason for that lies in the galaxy mass segregation within voids. In fact, faint galaxies are those living closer to the void centre and, thus, allow to probe the matter distribution even in the innermost part of the voids.

References

Hamaus, N., Sutter, P. M., & Wandelt, B. D., 2014, *PhRvL*, 112, 251302

Lavaux, G. & Wandelt, B. D., 2012, *ApJ*, 754, 109

Quilis, V., 2004, *MNRAS*, 352, 1426

Nadathur, S. & Hotchkiss, S., 2013, arXiv, arXiv:1310.2791

Nadathur, S., Hotchkiss, S., Diego, J. M., Iliev, I. T., Gottlöber, S., Watson, W. A., & Yepes, G., 2014, arXiv, arXiv:1407.1295

Ricciardelli, E., Quilis, V., & Planelles, S., 2013, *MNRAS*, 434, 1192

Ricciardelli, E., Quilis, V., & Varela, J., 2014, *MNRAS*, 440, 601

Sutter, P. M., Lavaux, G., Wandelt, B. D., & Weinberg, D. H., 2012, *ApJ*, 761, 187

Varela, J., Betancort-Rijo, J., Trujillo, I., & Ricciardelli, E., 2012, *ApJ*, 744, 82

The Zeldovich Universe:
Genesis and Growth of the Cosmic Web
Proceedings IAU Symposium No. 308, 2014
R. van de Weygaert, S. Shandarin, E. Saar & J. Einasto, eds.

© International Astronomical Union 2016
doi:10.1017/S1743921316010577

Testing Gravity using Void Profiles

Yan-Chuan Cai[1], Nelson Padilla[2,3] and Baojiu Li[1]

[1]Institute for Computational Cosmology, Department of Physics, University of Durham, South Road, Durham DH1 3LE, UK
email: y.c.cai@durham.ac.uk

[2]Instituto de Astrofísica, Pontificia Universidad Católica, Av. Vicuña Mackenna 4860, Santiago, Chile.
[3]Centro de Astro-Ingeniería, Pontificia Universidad Católica, Av. Vicuña Mackenna 4860, Santiago, Chile.

Abstract. We investigate void properties in $f(R)$ models using N-body simulations, focusing on their differences from General Relativity (GR) and their detectability. In the Hu-Sawicki $f(R)$ modified gravity (MG) models, the halo number density profiles of voids are not distinguishable from GR. In contrast, the same $f(R)$ voids are more empty of dark matter, and their profiles are steeper. This can in principle be observed by weak gravitational lensing of voids, for which the combination of a spectroscopic redshift and a lensing photometric redshift survey over the same sky is required. Neglecting the lensing shape noise, the $f(R)$ model parameter amplitudes $|f_{R0}| = 10^{-5}$ and 10^{-4} may be distinguished from GR using the lensing tangential shear signal around voids by 4 and 8σ for a volume of 1 $(\mathrm{Gpc}/h)^3$. The line-of-sight projection of large-scale structure is the main systematics that limits the significance of this signal for the near future wide angle and deep lensing surveys. For this reason, it is challenging to distinguish $|f_{R0}| = 10^{-6}$ from GR. We expect that this can be overcome with larger volume. The halo void abundance being smaller and the steepening of dark matter void profiles in $f(R)$ models are unique features that can be combined to break the degeneracy between $|f_{R0}|$ and σ_8.

Keywords. gravitational lensing: weak, methods: statistical, gravitation, large-scale structure of Universe

1. Introduction

Scalar-field models of modified gravity can mimic the late-time accelerated expansion of the Universe without invoking a cosmological constant, but usually, extra long-range fifth forces are introduces. These models can still in principle pass local tests of gravity via certain screeening mechanisms. The Vainshtein (Vainshtein 1972) or chameleon (Khoury & Weltman, 2004) mechanisms are two of the typical ones via which the fifth forces are suppressed in high density regions like dark matter haloes and the local Solar system. To distinguish these models from GR, it is therefore important to investigate the under dense regions, where the differences between MG and GR may be larger. A study using spherical evolution model by (Clampitt *et al.* 2013) has shown that in chameleon models, fifth forces in voids are repulsive. Voids are driven by the fifth force to expand faster and grow larger. In this work, we use N-body simulations to investigate observables for this phenomena, using the Hu-Sawicki $f(R)$ models Hu & Sawicki (2007) as an example. We will focus on void profiles in this proceeding. A detailed analysis including void abundances is presented in (Cai *et al.* 2014).

2. Simulations and void finding

Simulations of the Hu-Sawicki $f(R)$ models with model parameter amplitudes of $|f_{R0}| = 10^{-6}$ (F6), 10^{-5}(F5) and 10^{-4}(F4) are employed, where the model F6 has the weakest coupling strength between the scalar field and the density and is the most similar to GR. These simulations are performed using the ECOSMOG code (Li *et al.* 2012) with the same initial conditions and the same background expansion history as that of a GR ΛCDM model, which make it straightforward for comparison. Cosmological parameters of the simulations are: $\Omega_m = 0.24$, $\Omega_\Lambda = 0.76$, $h = 0.73$, $n_s = 0.958$ and $\sigma_8 = 0.80$, where Ω_m is the matter content of the universe, Ω_Λ is the effective dark energy density, h is the dimensionless Hubble constant at present, n_s is the spectral index of the primordial power spectrum and σ_8 is the linear root-mean-squared density fluctuation in spheres of radius 8Mpc/h. The simulations have the boxsize of 1Gpc/h and 1024^3 particles.

Dark matter haloes are found using the spherical overdensity code AHF(Knollmann & Knebe 2009). We use haloes above the minimal mass $M_{\min} \sim 10^{12.8} M_\odot/h$ so that each halo has at least 100 particles. For fair comparison, we adjust this value slightly for different models such that the number density of haloes are the same. This makes sure that the void population found for different models are least affected by possible differences of shot noise. In this sense, any model differences we find for the void properties are perhaps conservative. Voids are defined using an improved version of the spherical overdensity algorithm of Padilla *et al.* (2005).

We define voids in halo fields as they are related to galaxy clusters and groups, which are observational meaningful. We call them halo voids. Using tracers of dark matter to define voids, however, will suffer from the effect of sparse sampling. To overcome this, we introduce the variance of the distances for the nearest four halos from each void center, σ_4 as a free parameter to control the quality of halo voids. Voids with relatively small value (0.2) of σ_4 are chosen such that they are close to spherical in shape. Sub-voids that are 100% overlapped with a main void are excluded by default. Details of void finding can be found in (Padilla *et al.* 2005) and (Cai *et al.* 2014).

3. Void profiles

With void centers found in both $f(R)$ and GR simulations, we measure their halo number density profiles as well as the dark matter density profiles. Results are presented in Fig. 1 and Fig. 2. The left and right panels are void profiles with small radius and large radius, which qualitatively correspond to two different types of voids, void-in-cloud (voids in overdense environment) and void-in-void (voids in underdense environment) (Sheth & van de Weygaert, 2004). Profiles on the left have overdense ridges at 1× void radius $r_{\rm void}$, but not for those on the right. These are qualitatively as expected from the dynamics of void evolution. Small voids are likely to have been through more shell crossings at their walls than large ones. They are also consistent with the expansion velocities of voids. Void-in-cloud has a regime of infall at about $r > 1.5 \times r_{\rm void}$ but the outflow of mass persist at all scales for void-in-void up to 3 times of void radius [see (Cai *et al.* 2014) for more details]. In general, the halo number density profiles are steeper than the dark matter ones at $\sim r_{\rm void}$.

Most interestingly, we find that the halo number density profiles of voids are not distinguishable for different models (top panels of Fig. 1). In contrast, the dark matter density profiles are deeper in $f(R)$ within $r > 1 \times r_{\rm void}$ (Fig. 2). Also, the dark matter overdense ridges at $r \sim r_{\rm void}$, if any, are sharper in $f(R)$ models.

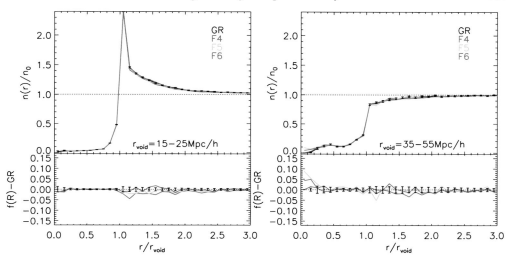

Figure 1. Top panels: the halo number density profiles of voids using haloes above the minimum halo mass of $M_{\min} \sim 10^{12.8} M_\odot/h$ from simulations of different models as labelled in the legend. $M_{\min}$ is slightly different from $10^{12.8} M_\odot/h$ in the $f(R)$ models so that the number of haloes for different models are the same. Error bars shown on the black line (GR) are the scatter about the mean for voids at 15 Mpc/h < $r_{\rm void}$ < 25 Mpc/h (left) and at 35 Mpc/h < $r_{\rm void}$ < 55 Mpc/h (right) found within the $1(\text{Gpc}/h)^3$ volume. There are [6038, 5946, 6096, 6307] (left) and [296, 323, 319, 261] (right) voids in GR, F6, F5 and F4 models passing the selection criteria. Bottom panels: differences of halo number density profiles of voids between $f(R)$ models and GR.

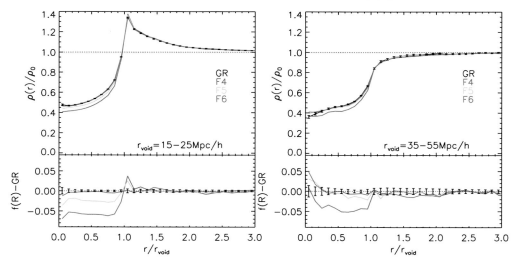

Figure 2. Similar to Fig. 1 but showing void density profiles measured using all dark matter particles from simulations of different models as labelled in the legend. Voids are defined using halo number density fields, which are the same as those being used to make Fig. 1.

It is somewhat counter-intuitive but perhaps not surprising that little model difference is shown in halo number density profiles, because the fifth force is suppressed in overdense regions in $f(R)$. The spatial distribution of massive haloes are perhaps not so different between $f(R)$ and GR. The dark matter void profiles being steeper in $f(R)$ is consistent with the analytical results of (Clampitt *et al.* 2013) that the repulsive fifth force drives voids in $f(R)$ to grow larger and emptier. It is also reasonable that once the growth

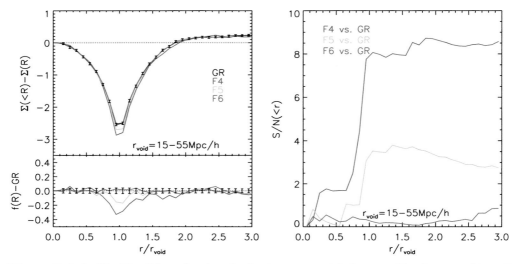

Figure 3. Left: like Fig. 2 but showing the lensing tangential shear profiles from stacking all voids with $15 < r < 55$ Mpc/h. They are projected over two times the void radius along the line of sight. $\Sigma(< R) - \Sigma(R)$ is proportional to the surface mass density within the projected radius of R to which we subtract the surface mass density at R. Right: the corresponding cumulative (from small to large radius) S/N for the differences between GR and $f(R)$ models.

of voids are restricted by their environments, mass start to accumulate at the edges of voids, and the overdense ridges will grow sharper. This effect is stronger in $f(R)$ thanks to the repulsive fifth force in voids.

3.1. *Gravitational lensing of voids*

The differences between $f(R)$ and GR in dark matter void profiles can in principle be observed using gravitational lensing of voids. This requires the overlap of a spectroscopic redshift survey (spec-z) and a (deeper) lensing photometric redshift survey (photo-z) over the same sky. Voids will be identified in the spec-z survey while their gravitational lensing effect on the background galaxy images can be detected in the photo-z survey via stacking of void centres. The lensing tangential shear profiles of the background galaxies are associated with the excess of the projected mass density along the line of sight,

$$\Delta\Sigma(R) = \Sigma(< R) - \Sigma(R), \qquad (3.1)$$

where $\Sigma(R)$ and $\Sigma(< R)$ are the projected surface densities around the centre of a void at the projected distance of R and within R. It can be used to measure the void density profile (Krause *et al.* 2013; Higuchi *et al.* 2013; Clampitt & Jain 2014). This is sensitive to the slope of the mass density. Therefore, the peak of the lensing signal for voids is at $\sim r_{\rm void}$ where the slope of the matter density is the largest, shown in Fig. 3. Neglecting the lensing shape noise, F5 and F4 can be told apart from GR by 4 and 8σ for the $1({\rm Gpc}/h)^3$ volume. F6 is not distinguishable from GR due to the line-of-sight projection of large-scale structure.

We find sub-voids are useful to help increasing the lensing S/N, though the S/N per void is not as great as that of the main voids. With all sub-voids included in our volume, the number of voids increases by approximately 76%. The S/N for F5 and F4 are increased to 7 and 12, but there is no increase of S/N for F6. The S/Ns are degraded if we integrate the projected mass density to larger line-of-sight distances. For example, increasing the line-of-sight projection from 2 to 6 times of void radius decrease the S/Ns by about 30%.

The above forecast may be somewhat optimistic as the lensing shape noise and other systematics are neglected. However, at the relatively large radius ($\sim r_{\rm void}$), which are the most interesting to distinguish $f(R)$ models from GR, the lensing shape noise is expected to be sub-dominant for DEFT Stage IV type of deep imaging survey(Albrecht *et al.* 2006), see (Krause *et al.* 2013; Higuchi *et al.* 2013) for quantitative examples. The above forecast is based on a volume of $1(\mathrm{Gpc}/h)^3$. The current BOSS DR11 CMASS sample has an effective volume of 6.0 $(\mathrm{Gpc}/h)^3$ (Anderson *et al.* 2011; Beutler *et al.* 2013; Sánchez *et al.* 2014). In principle, the significance level should increase by a factor of 2.4 if the BOSS DR11 CMASS sample is used, on condition that deep lensing image data on the same sky is available. The future EUCLID survey (Laureijs *et al.* 2011) is expected to have an effective volume of ~ 20 $(\mathrm{Gpc}/h)^3$, a factor of 4.4 improvement is expected in this case.

4. Conclusions and discussion

Using simulations of $f(R)$ models, we have found that voids are emptier than that in GR in terms of dark matter. However, the void profiles of tracers (haloes) are not necessary distinguishable between $f(R)$ and GR. Moreover, the halo number density profiles of voids are very different from that of dark matter, the former being sharper. This is true even in GR. It rings an alarm that voids found using tracers are not necessary the same as that of dark matter. Note that other authors using different void finding algorithms may conclude differently, see for example (Sutter *et al.* 2014).

We have found that two types of voids, void-in-cloud and void-in-void, are separable using their radii, the latter tend to be larger. Their profiles are different in that the former have developed over dense ridges but not for the latter. From our prospective, it is perhaps unlikely that the void profile takes the same form, and can simply be rescaled only by the void radius for these two different types. For void-in-cloud, an additional parameter is needed to describe the height of the overdense ridge. This seems different from the results of (Nadathur *et al.* 2014), but again, they are using ZOBOV (Neyrinck *et al.* 2005) to find voids. Also, the voids they found are from (mock) LRG galaxies and their void sizes are relatively large compared to ours.

Using halo voids to study their dark matter density profiles has the observational implications in that voids are usually found using tracers. The steepening of the underlying dark matter void profile in $f(R)$ models over that of GR induces stronger lensing tangential shear signals at about $1 \times r_{\rm void}$. Measuring the model differences associated with these voids requires the combination of a spec-z survey and a photo-z survey. This adds value to the idea of combining surveys on top of systematic calibration and canceling of cosmic variance (Zhang *et al.* 2007; McDonald & Seljak, 2009; Bernstein & Cai, 2011; Cai & Bernstein, 2011; Gaztañaga *et al.* 2012; de Putter *et al.* 2013; Kirk *et al.* 2013). Neglecting lensing shape noise, which is expected to be sub-dominant for near future deep imaging surveys, F5 and F4 can be told apart from GR by 4 and 8σ. Line-of-sight projections of large-scale structure set limit on the constraining power. For this reason, it is challenging to distinguish F6 using the $(1\mathrm{Gpc}/h)^3$ volume.

We caution that the steepening of void profiles may also be expected in the same ΛCDM model with a higher σ_8. In this sense, there may be a degeneracy between the $|f_{R0}|$ parameter with σ_8. This may be possible to be broken using the void abundance measurement. In (Cai *et al.* 2014), the halo void abundances are found to be smaller for large voids in $f(R)$ models compared to that in GR. This counters the trend in ΛCDM with a higher σ_8. Therefore, the combination of void abundances and profiles may be a powerful tool for constraining gravity.

References

Albrecht, A., Bernstein, G., Cahn, R., Freedman, W. L., Hewitt, J., Hu, W., Huth, J., Kamionkowski, M., Kolb, E. W., Knox, L., Mather, J. C., Staggs, S., & Suntzeff, N. B., 2006, ArXiv Astrophysics e-prints

Anderson, L., Aubourg, É., Bailey, S., Beutler, F., Bhardwaj, V., Blanton, M., Bolton, A. S., Brinkmann, J., *et al.* 2014, *MNRAS*, 441, 24

Beutler, F., Saito, S., Seo, H.-J., Brinkmann, J., Dawson, K. S., Eisenstein, D. J., Font-Ribera A., Ho, S., *et al.* 2013, ArXiv e-prints

Cai, Y.-C. & Bernstein, G., 2012, *MNRAS*, 422, 1045

Cai, Y.-C., Padilla, N., & Li, B. 2014, arXiv:1410.1510

Bernstein, G. M. & Cai, Y.-C., 2011, *MNRAS*, 416, 3009

Clampitt, J., Cai, Y.-C., & Li, B., 2013, *MNRAS*, 431, 749

Clampitt, J. & Jain, B., 2014, ArXiv e-prints

de Putter R., Doré O., & Takada, M., 2013, ArXiv e-prints

Gaztañaga, E., Eriksen, M., Crocce, M., Castander, F. J., Fosalba, P., Marti, P., Miquel, R., & Cabré A., 2012, *MNRAS*, 422, 2904

Higuchi, Y., Oguri, M., & Hamana, T., 2013, *MNRAS*, 432, 1021

Kirk, D., Lahav, O., Bridle, S., Jouvel, S., Abdalla, F. B., & Frieman, J. A., 2013, ArXiv e-prints

Hu, W. & Sawicki, I., 2007, *Phys. Rev. D*, 76, 064004

Khoury J. & Weltman A., 2004, *Phys. Rev.Lett.*, 93, 171104

Knollmann S. R. & Knebe A., 2009, *ApJ*, 182, 608

Krause, E., Chang, T.-C., Doré O., & Umetsu, K., 2013, *ApJ*, 762, L20

Laureijs, R., Amiaux, J., Arduini, S., Augueres, J.-L., Brinchmann, J., Cole, R., Cropper, M., Dabin, C., *et al.* 2011

Li, B., Zhao, G., Teyssier, R., & Koyama, K., 2012, *J. Cosmo. Astropart. Phys.*, 01, 051

McDonald, P. & Seljak, U., 2009, *JCAP*, 10, 7

Nadathur, S., Hotchkiss, S., Diego, J. M., Iliev, I. T., Gottlöber S., Watson, W. A., & Yepes, G., 2014, ArXiv e-prints

Neyrinck, M. C., Gnedin, N. Y., & Hamilton, A. J. S., 2005, *MNRAS*, 356, 1222

Padilla, N. D., Ceccarelli, L., & Lambas, D., 2005, *MNRAS*, 363, 977

Sánchez A. G., Montesano, F., Kazin, E. A., Aubourg, E., Beutler, F., Brinkmann, J., Brownstein, J. R., Cuesta, A. J., Dawson, K. S., Eisenstein, D. J., Ho, S., *et al.* 2014, *MNRAS*, 440, 2692

Sheth, R. K. & van de Weygaert R., 2004, *MNRAS*, 350, 517

Sutter, P. M., Lavaux, G., Wandelt, B. D., Weinberg, D. H., & Warren, M. S., 2014, *MNRAS*, 438, 3177

Vainshtein, A. I., 1972, *Physics Letters B*, 39, 393

Zhang, P., Liguori, M., Bean, R., & Dodelson, S. 2007, *Physical Review Letters*, 99, 141302

The Zeldovich Universe:
Genesis and Growth of the Cosmic Web
Proceedings IAU Symposium No. 308, 2014
R. van de Weygaert, S. Shandarin, E. Saar & J. Einasto, eds.

© International Astronomical Union 2016
doi:10.1017/S1743921316010589

Void asymmetries in the cosmic web: a mechanism for bulk flows

J. Bland-Hawthorn[1] and S. Sharma[1]

[1]Sydney Institute for Astronomy, University of Sydney,
School of Physics A28, NSW 2006, Australia
email: jbh@physics.usyd.edu.au

Abstract. Bulk flows of galaxies moving with respect to the cosmic microwave background are well established observationally and seen in the most recent ΛCDM simulations. With the aid of an idealised Gadget-2 simulation, we show that void asymmetries in the cosmic web can exacerbate local bulk flows of galaxies. The *Cosmicflows-2* survey, which has mapped in detail the 3D structure of the Local Universe, reveals that the Local Group resides in a "local sheet" of galaxies that borders a "local void" with a diameter of about 40 Mpc. The void is emptying out at a rate of 16 km s^{-1} Mpc^{-1}. In a co-moving frame, the Local Sheet is found to be moving away from the Local Void at ∼ 260 km s^{-1}. Our model shows how asymmetric collapse due to unbalanced voids on either side of a developing sheet or wall can lead to a systematic movement of the sheet. We conjectured that asymmetries could lead to a large-scale separation of dark matter and baryons, thereby driving a dependence of galaxy properties with environment, but we do *not* find any evidence for this effect.

Keywords. Large-scale structure, galaxy surveys, voids, walls, sheets, filaments, bulk flows

1. Introduction

The physics of baryons across the universe is the grandest of all environmental sciences. This drama is played out against a backdrop of evolving dark matter structure from the Big Bang to the present day. Cold dark matter (CDM) simulations without baryons reveal a universe that looks structurally different from the observed universe defined by its baryons. The most recent CDM simulations that include baryons and hydrodynamics emphasise how little we know of baryonic processes throughout cosmic time (Schaye *et al.* 2014; Vogelsberger *et al.* 2014).

This meeting honours the 100th year since the birth of Y.B. Zel'dovich. In 1977, Tallinn, Estonia was the site of the first great conference on large-scale structure. Over the past week, much of the discussion centred on where next for studies of the cosmic web and galaxy redshift surveys. An interesting question is how the ratio of baryons to dark matter by mass ($\langle f_b \rangle$) varies across large-scale structure. Non-standard models do exist which predict baryon to dark matter variations (Malaney & Mathews 1993; Gordon & Lewis 2003). Variations in $\langle f_b \rangle$ of order 10% lead to only few percent variations in the matter power spectrum, but could conceivably lead to observable differences in some local galaxy properties (Nichols & Bland-Hawthorn 2013). In clusters, the baryon fraction approaches the universal average $\langle f_b^o \rangle \approx 15.5\%$ (Planck) with small scatter (Sun *et al.* 2009). For most galaxies in groups, this ratio is more uncertain largely because the warm-hot gas phases are very difficult to detect. In some instances, the majority of the missing baryons may be in a warm circumgalactic medium (Tumlinson *et al.* 2011; Shull *et al.* 2012).

The next generation of large-scale galaxy surveys will seek to associate more of a galaxy's properties with its large-scale environment (Croom *et al.* 2012; Bundy *et al.* 2014; Bland-Hawthorn 2014). While the distinction between clusters and the field is well defined, the dependence of a galaxy's properties on a more graded local density has been hard to establish (e.g. Blanton & Moustakas 2009; Metuki *et al.* 2014). The effects appear to exist only weakly, if at all. These include a weak dependence of the fundamental plane with environment, scatter in the mass-metallicity relation that correlates with environment (Cooper *et al.* 2008), and mean star formation rates showing a trend with environment (Lewis *et al.* 2002; Gomez *et al.* 2003). The weakness of these trends may arise from any of the following: (i) the difficulty of defining environment; (ii) the wrong galaxy parameters are being explored; (iii) a strong local dependence does not exist in nature.

In light of recent simulations where void-void imbalances are observed to push material around (Pichon *et al.* 2011; Codis *et al.* 2012), we look more closely at the prospect of baryon-dark matter variations. While the effects look strong in our 1D toy model, they are essentially non-existent in 3D. But what we do find is a barycentric drift of the collapsed sheet and a possible mechanism for bulk flows in galaxies (e.g. Rubin *et al.* 1976; Burstein et al 1990; Tully *et al.* 2008).

In §2, we introduce our 1D toy model for asymmetric collapse that suggests a strong separation of dark matter and baryons. We investigate this idea further in §3 with a cosmologically motivated 3D model using Gadget-2. In §4, we conclude that there is no general case for baryon-dark matter separation on megaparsec scales, but we establish an interesting mechanism for bulk flows in the presence of void asymmetries in the cosmic web.

2. Toy model

Can baryons and dark matter separate on megaparsec scales? The short answer is yes in specific cases, for example, the Bullet Cluster (e.g. Mastropietro & Burkert 2008) where two massive clusters have passed through each other sweeping out the baryons in both systems. This provides us with our initial motivation for modelling the Local Sheet. To illustrate how the Local Sheet can be kinematically offset from the Local Void, initially, we reduce the the the dynamics of a forming sheet to a one-dimensional problem (cf. Melott 1983). Before exploring a cosmologically motivated N-body simulation in co-moving coordinates, we develop a toy model using the "infinite sheets" approximation (Binney & Tremaine 2008, hereafter BT08). This approach has a long history dating back to theoretical work in plasma physics (Eldridge & Feix 1963) although its relevance to structure formation has been demonstrated in numerous papers (Yamashiro *et al.* 1992). An up-to-date discussion is given by Teles *et al.* (2011) who call for the sheets approximation to be explored in a cosmological context, as we do here.

We simulate the formation and evolution of the Local Sheet using thin sheets of dark matter and a matching set of (initially cospatial) sheets made of gas. The dark matter is treated as collisionless while the gas is assumed to undergo inelastic collisions. In the expanding universe, dark matter and baryons turn around and begin to collapse towards a local density perturbation. At turnaround, when no sheet crossing has occurred, the evolution is described by linear theory. But during the collapse phase, the sheets start to cross each other, and the evolution becomes non-linear. We start the simulation just after turn around. Initially we treat the symmetric case where the sheet separations have a Gaussian distribution in the normal (x-axis) direction. The gas and dark matter sheets are assumed to extend to infinity in the $y - z$ plane such that the force exerted by any sheet is constant at any point

The equation of motion for the sheets along the x axis is given by

$$\ddot{x} = 2\pi G \int_{-\infty}^{\infty} \text{Sgn}(x' - x)\sigma(x')dx' \tag{2.1}$$

or equivalently

$$\ddot{x} = f(x) = 2\pi G(2\Sigma(> x) - \Sigma_{\text{tot}}) \tag{2.2}$$

where $\Sigma(> x)$ is the cumulative surface density and it is assumed that $\Sigma(< x) + \Sigma(> x) = \Sigma_{\text{tot}}$. For a discretized system, one can think of $2N + 1$ sheets distributed in space with i-th sheet having surface density m_i. N must be large enough to render the system "collisionless" as discussed by Yamashiro *et al.* (1992). The mass of the i-th sheet is assumed to be distributed evenly between $(x_{i-1} + x_i)/2$ and $(x_{i+1} + x_i)/2$. Here we assume that all dark matter sheets have the same constant surface density; the gas sheets also have a constant surface density defined to be a factor of 10 lower ($\langle f_b \rangle = 0.1$) than for the dark matter. In our analysis, we set $G = 1$ and choose $\Sigma_{\text{tot}} = \int \sigma(x)dx = 1/(2\pi)$.

Initial conditions. Let $\sigma(\xi)$ $(-1 < \xi < 1)$ be the initial density distribution at shortly after the Big Bang $(t = 0)$. If no shell crossings have happened since the Big Bang, the acceleration is constant and is given by

$$a(\xi) = -2\pi G \int_{-1}^{1} \mathrm{Sgn}(\xi' - \xi)\sigma(\xi')d\xi'. \tag{2.3}$$

Let $v(\xi) = v_{\mathrm{tot}}\xi$ be the initial velocity field. The collapse time is given by

$$t_{\mathrm{collapse}} = -\frac{2v(\xi = 1)}{a(\xi = 1)} = \frac{v_{\mathrm{tot}}}{\pi G\Sigma_{\mathrm{tot}}} \tag{2.4}$$

The position and velocity at a later time τ is given by

$$x(\xi, \tau) = v_{\mathrm{tot}}\xi\tau + \frac{a(\xi)\tau^2}{2} \quad v(\xi, \tau) = v_{\mathrm{tot}}\xi + a(\xi)\tau. \tag{2.5}$$

After BT08, we set $v_{\mathrm{tot}} = 0.75$, which at $\tau = 1$ gives max(v)=max(x)=0.25.

Let the initial sheet distribution be given by a function of form

$$\sigma(\xi) = \frac{k}{1 - a\cos(b\pi\xi)} \quad, -1 < \xi < 1. \tag{2.6}$$

Using $\int_{-1}^{1} \sigma(\xi)d\xi = 1$, the normalization constant is given by

$$k = \frac{\pi b\sqrt{1 - a^2}}{4\tan^{-1}((1 + a)\tan(b\pi/2)/\sqrt{1 - a^2})} \tag{2.7}$$

To sample such a distribution we use the method of inverse transform sampling. Let $F(> \xi)$ be the cumulative distribution, then for u uniformly sampled between 0 and 1, the ξ is given by

$$\xi = F^{-1}(u) \tag{2.8}$$

$$= \frac{2}{b\pi}\tan^{-1}\left[\frac{\sqrt{1 - a^2}}{1 + a}\tan\left(\frac{(b\pi\sqrt{1 - a^2})}{2k}(u - 0.5)\right)\right] \tag{2.9}$$

Here we set $a = 0.3$ and $b = 0.3$ (BT08).

For the asymmetric case, the density is defined as follows

$$\Sigma'(\xi) = \begin{cases} \Sigma(\xi) & \text{if } \xi < 0 \\ 2\Sigma(2\xi) & \text{if } \xi > 0 \end{cases}$$

This is achieved by setting $\xi = \xi/2$ for $\xi > 0$. The position and velocities are calculated as in the symmetric case. Note for the region $\xi > 0$, the collapse time now decreases by a factor of two.

To evolve a system of sheets, we use the kick-drift-kick algorithm. For fixed time step Δt this is given as

$$v_i = v_i + f(x_i)\Delta t \tag{2.10}$$

$$x_i = x_i + v_i\Delta t \tag{2.11}$$

$$v_i = v_i + f(x_i)\Delta t \tag{2.12}$$

A time step of $\Delta t = 10^{-4}\tau$ was employed in all runs. The results were checked for convergence: choosing a lower time step did not yield any difference in results. For collisionless sheets, it is sufficient to just evolve the as shown above, but for gas one has to put in additional physics. We assume the gas sheets undergo fully inelastic collisions, which conserve mass and momentum, but not energy. The prescription to simulate this is as follows. In a given time step, we identify a contiguous set of gas sheets that criss-cross each other as predicted by the equation of motion. This can be easily accomplished by sorting the sheets before and after advancing. If L_1 and L_2 are lists that contain sorted indices, then two gas sheets are set to cross if $L_1 - L_2$ is non-zero. Two (or more) gas sheets that are set to cross in the next step are joined to form a single particle

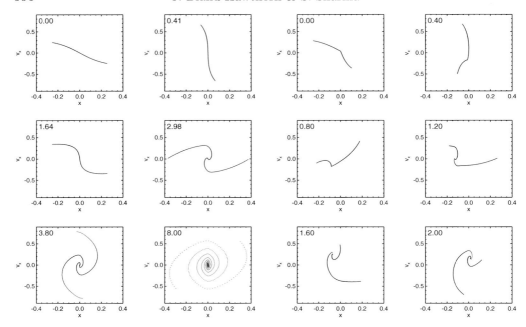

Figure 1. Evolution in phase space of parallel infinite sheets. The time of the Big Bang is $-\tau$ and the units of time are in units of time τ. (Left) Dark matter only, symmetric perturbation. (Right) Dark matter + gas, asymmetric perturbation.

whose position is given by the center of mass of the sheets in the set. The momentum and mass is assumed to be conserved.

In our first test, we run a simulation with dark matter. In Fig. 1 (left), we accurately recover Fig. 9.14 in BT08. The curves are smooth and continuous in phase space, which tells us that the time integration is working correctly. We now study the evolution of sheets when both dark matter and gas are present together. We consider both the symmetric and asymmetric perturbations where the functional form of the perturbation is given by Equation (2.6).

Symmetric perturbation. The first panel in Fig. 2 (Left) shows the variation of the center of mass position and center of mass velocity with time; these are shown separately for gas (x_{gas}, v_{gas}) and dark matter (x_{DM}, v_{DM}). The middle panel shows the dispersion σ_x or spread of the sheets along the x-axis. The dark matter sheets show oscillatory behaviour while the gas sheets stick after the first crossing. The bottom panel shows the variation in the mean kinetic energy. The kinetic energy of the gas sheets is assumed to be lost to internal energy within the sheets through shock heating. We explored different baryon fractions from $f_b = 0.01$ to $f_b = 0.1$. A larger value of f_b increases the time required by dark matter to achieve the second turnaround and collapse. We observe that there is *no* offset between the center of mass of the dark matter and the gas — all of the lines are overlaid (see top left panel).

Asymmetric perturbation. In Fig. 2, we compare our asymmetric collapse model with the symmetric case. Clear differences are evident. The asymmetric distribution of sheets leads to an initial (non-zero) offset in the velocity centroid of the gas and dark matter. As the collapse proceeds, the gas becomes progressively separated from the dark matter. The dark matter starts to exert a force on the gas and, at a later stage, gas tries to move towards the dark matter's centre of mass. A larger value of f_b increases the spatial offset and delays the time required by the baryons to turn around and fall towards the dark matter. The kinematic offset between the gas and dark matter is exaggerated here because of the artificial asymmetry built into our model. Now we turn our attention to sheet collapse in a 3D cosmological context.

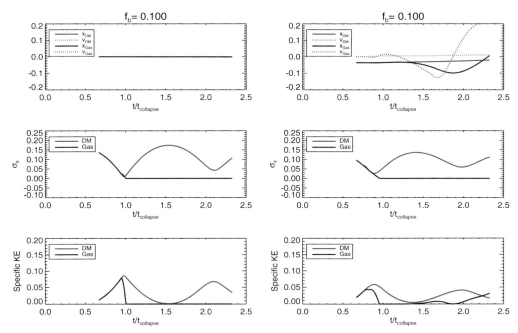

Figure 2. Evolution of position and velocity of the center of mass of dark matter, gas and the whole system. The middle panel shows the dispersion of sheets in space and the bottom panel shows the specific kinetic energy. The simulation explores the evolution of a symmetric perturbation (Left) and an asymmetric perturbation (Right). In the top left figure, all of the curves are overlaid.

3. Simulation with Gadget-2 in a cosmological context

3.1. *Vacuum boundaries*

We explore a collapsing sheet where the perturbation takes the form

$$\rho(z) = A\cos(2\pi z/L). \tag{3.1}$$

Note that the collapse is now along the z axis. The amplitude A was selected so that $\delta\rho/\rho = 1$ at redshift $\mathcal{Z} = \mathcal{Z}_{\text{collapse}} = 2$. This perturbation was evolved using linear theory till $\mathcal{Z} = \mathcal{Z}_{\text{start}} = 6.6$, the point from where the simulation starts. To set up the perturbation, the particles were initially distributed uniformly in a box of size $L = 10$ Mpc h^{-1} and a spherical region was cut from this. We adopt a $\Omega_m = 1$ and $\Omega_b = 0.05$ co-moving cosmology. Here we use $N = 32^3$ particles for the dark matter, and the same for the gas. The displacement field corresponding to the perturbation was calculated, viz.

$$S(z) = \frac{-2\pi z}{LA}\sin(2\pi z/L) \tag{3.2}$$

and the particles were accordingly displaced from the uniform distribution to generate the perturbation. The time integration was done in co-moving coordinates using Gadget-2. The results of the symmetric collapse (Fig. 3) are in good agreement with Dekel (1983).

The asymmetric case was set up by increasing the displacement field by a factor of two for $z > 0$. In Fig. 3, we show the initial $z - v_z$ and density distribution of the particles. At $\mathcal{Z} = 0.9$, there is a bump in the density distribution of dark matter particles at around 400 kpc h^{-1}. This behaviour is not seen in the gas. The gas centre of mass shows a slight displacement with respect to dark matter similar to our idealized 1D simulation, but the displacement is very small and is less than the softening parameter ($\epsilon = 12.5$ kpc h^{-1}). An $N = 64^3$ simulation with

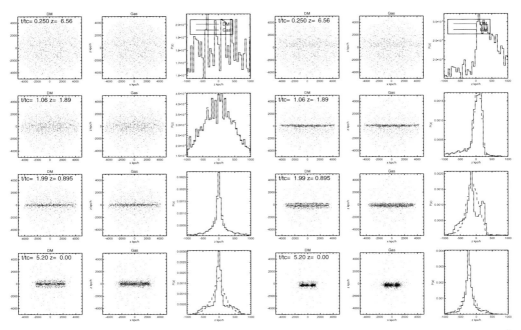

Figure 3. The distribution of particles in $x - z$ space as a function of time for a 10 Mpc h^{-1} simulation with a symmetric cosine perturbation (left) and an asymmetric perturbation (right) along the z-direction. These are evolved with Gadget-2 with no gas cooling. The perturbation is imposed over a uniform sphere.

$\epsilon = 6.25$ kpc h^{-1} also shows similar behaviour. We note that the amount of displacement in the 1D case is much larger than for the 3D set-up in physically motivated co-moving coordinates.

3.2. *Periodic Boundaries*

In earlier simulations, a plane wave perturbation was imposed on a uniform sphere and the asymmetry function had a discontinuity at $z = 0$. We now employ a periodic box and use a better asymmetry function. The symmetric perturbation is a cosine function and we now add a tanh function to make it asymmetric (see §3.3), such that

$$\delta(z) = A \left[\left(1 + \tanh \left(\frac{z2\pi}{\lambda\alpha} \right) \right) \left(1 + \cos \left(\frac{z2\pi}{\lambda} \right) \right) - 1 \right] \qquad (3.3)$$

For $\alpha \gg 1$, this reduces to the plane wave form $\delta(z) = A \cos(2\pi z/\lambda)$. The degree of asymmetry is controlled by the shape parameter α. The smaller the value of α, the higher the asymmetry. The functional form for different α is shown in Fig. 4 (Left). The function and its first derivative are continuous across the boundary of the box; more details are given in § 3.3.

We simulate 7 different types of perturbation: one is a symmetric cosine wave; the others use an asymmetry factor $\alpha = 100, 10, 1, 0.5, 0.25$ and 0.1. These were simulated with and without cooling. A smoothing length of 12.5 kpc h^{-1} was used. $N = 32^3$ particles were used once again for each constituent in the simulation and the cosmology adopted is $\Omega_m = 1$ and $\Omega_b = 0.05$. The amplitude of the perturbation was set so as to make the cosine perturbation collapse at redshift $\mathcal{Z}_{collapse} = 2$. A starting redshift of $\mathcal{Z}_{start} = 6.6$ was used. These last two parameters are the same as in our first Gadget-2 run with vacuum boundary conditions. Here we only show results for the symmetric case and with $\alpha = 1, 0.5$ and 0.25.

In Fig. 5, we show the density distribution of the dark matter and gas along the z-axis (in co-moving coordinates). The type of perturbation and details about cooling are given on top of each figure. To simulate gas cooling, we set an upper bound on the temperature at $1000K$. The results with and without cooling are very similar. A notable feature is the occurrence of

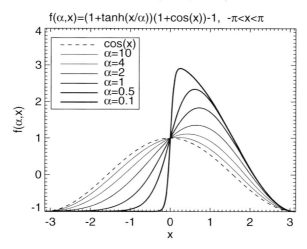

f(α,x)=(1+tanh(x/α))(1+cos(x))-1, -π<x<π

Figure 4. The density profile of the perturbations. A tanh function is used to introduce asymmetry. It has well behaved derivatives across the boundary and is suitable for periodic boundary conditions; see §3.3 for details.

spikes at the edges due to the piling up of particles at turnaround after shell crossing. This was also noticed by Dekel (1983) in his simulations. The gas is not so able to criss-cross and thus forms a central peak. The fact that, even with cooling, the dispersion of the gas does not diminish is interesting. The gas appears to expand as the DM potential becomes shallower which may be why cooling does not seem to have a great effect on the final distribution of gas. For the asymmetric perturbation, the behaviour is similar except for the fact that the spikes are asymmetric.

In the panels of Fig. 6, we plot as a function time the center of mass offset between gas and dark matter– defined as $\Delta_z = \langle z_{DM} \rangle - \langle z_{Gas} \rangle$. The velocity of center of mass of gas and dark matter is also shown on the same panel. All quantities in these set of figures are in physical coordinates, i.e. physical length and physical peculiar velocity (without hubble flow).

For $\alpha < 0.5$, i.e., large asymmetry, one can see that the offset Δ_z reaches to about 40 kpc at redshift zero. The dispersion in z is around 1000 kpc, or an offset that is about 4% of the dispersion. Overall, the offset and velocity of the center of mass are quite small. Nevertheless, it is interesting to explore the cause of the shift. First thing to note is that the offset occurs only after the collapse of the perturbation, i.e., after shell crossing. At this time, the dark matter particles move past each other rapidly and hence the shape of the distribution is also changing rapidly (rapid movement of asymmetric spikes). The gas is less able to criss-cross creates only a central peak and lags behind, thus creating an offset.

In Fig. 5, the vertical lines show the location of the peak in comoving coordinates. If there is no peculiar or bulk velocity associated with peak, then the peak should remain stationary. For the symmetric case this is true. However, for asymmetric case the peak is not stationary. The location of the peak changes rapidly at earlier time, i.e., before the perturbation has collapsed. In Fig. 7, we plot the velocity of the peak, computed as the mean velocity of the particles in and around the peak. For an asymmetric perturbation, a peculiar velocity as large as 260 km s^{-1} can be seen.

3.3. *Asymmetric Perturbation*

The asymmetric perturbation is described by the following functional form.

$$f(\alpha, x) = (1 + \tanh(x/\alpha))\,(1 + \cos x) - 1 \qquad (3.4)$$

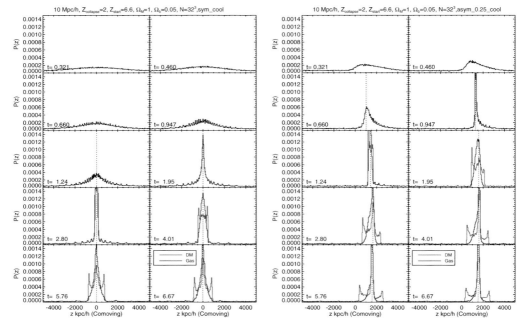

Figure 5. The z distribution of particles as a function of time for a 10 Mpc h^{-1}, $N = 32^3$ simulation. (Left) Symmetric case with cooling; (Right) Asymmetric perturbation $\alpha = 0.25$ with cooling. The results for $\alpha \lesssim 1$ are very similar. The vertical lines show the location of the gas density peak. We do not mark the peak location at very early times, this is because it is difficult to locate the peak when the amplitude of the perturbation is very small.

The properties of this function are very similar to the cos function except that it is asymmetric. Some of the useful properties are as follows. It is defined in range $(-\pi, \pi)$ such that

$$f(\alpha, -\pi) = f(\alpha, \pi) = -1, \tag{3.5}$$

$$\frac{df(\alpha, x)}{dx}\Big|_{x=-\pi} = \frac{df(\alpha, x)}{dx}\Big|_{x=\pi} = 0, \tag{3.6}$$

$$\langle f(\alpha, x)\rangle = \int_{-\pi}^{\pi} f(\alpha, x)dx = 0 \tag{3.7}$$

An asymmetric probability distribution for a periodic box of length l with range $-l/2 < x < l/2$, is given by

$$p(x) = \frac{l}{4\pi^2}\left(1 + Af(\alpha, 2\pi x/l)\right) \tag{3.8}$$

where A is the amplitude of the perturbation. Note $p(x) > 0$ only for $A \leqslant 1$. For $A > 1$, this is the non-linear regime and then $p(x)$ can be negative. The amplitude of a cosine perturbation grows linearly with time. For simplicity, assuming the asymmetric perturbation to also behave in the same way, for our choice of $\mathcal{Z}_c = 2.0$ and $\mathcal{Z}_{\rm start} = 6.36$, the amplitude A from the growth factor is given by 0.3968.

4. Discussion

With apologies to John Donne, no galaxy is an island. Most galaxies are in groups and these accrete from the group environment which in turn accretes from the intergroup medium. This is a more accurate description for most galaxies than a simple statement of that galaxies accrete from the intergalactic medium. It is only in the last few years that modern simulations are able to show how this mechanism operates.

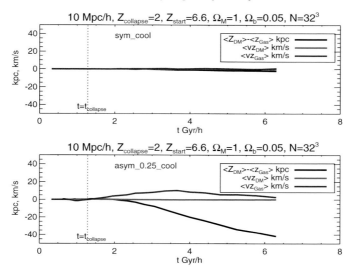

Figure 6. Evolution of the position and velocity (proper) of center of mass of dark mater and gas for a 10 Mpc h^{-1}, $N = 32^3$ simulation with periodic boundaries. (Top) Symmetric case with cooling; (Bottom) Asymmetric perturbation $\alpha = 0.25$ with cooling. Even under large asymmetry, the gas to a good approximation tracks the dark matter.

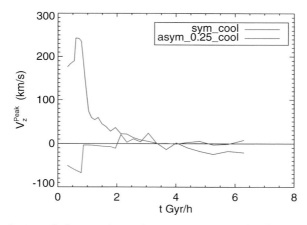

Figure 7. The evolution of the peculiar velocity of the peak (gas) for a 10 Mpc, $N = 32^3$ simulation with periodic boundaries. For the case with the asymmetric perturbation the peak in density distribution is found to move with time. The peculiar velocity of peak is high at early times and then falls off with time.

Our work was motivated by new large, ongoing surveys of galaxies (Bland-Hawthorn 2014; Bundy *et al.* 2014) that seek to understand how the detailed properties of galaxies vary with the local environment. Motivated by the remarkable Bullet Cluster, we conjectured that asymmetries would lead to a large-scale separation of dark matter and baryons, thereby driving a dependence of galaxy properties with environment, but we do not find any evidence for this effect.

We do find, however, a mechanism for generating bulk flows in the galaxy population at a level that could explain the bulk flows of galaxies moving with respect to the cosmic microwave background. With our Gadget-2 simulation, we show that void asymmetries in the cosmic web can exacerbate local bulk flows of galaxies. The *Cosmicflows-2* survey reveals that the Local Group resides in a "local sheet" of galaxies that borders a "local void" with a diameter of about 40 Mpc (Tully *et al.* 2013). The void is found to be emptying out at a rate of 16 km s^{-1} Mpc^{-1}.

In a co-moving frame, the Local Sheet is found to be moving away from the Local Void at ~ 260 km s^{-1}. Our model shows how asymmetric collapse due to unbalanced voids on either side of a developing sheet or wall can lead to a systematic movement of the sheet, and the magnitude of the kinematic offset is (fortuitously) the same, at least in the early stages of the sheet collapse.

Our analysis seeks to honour the memory of Y.B. Zel'dovich whose published work continues to inspire astrophysicists around the world to the present day. We thank the organisers for putting together such an inspiring meeting.

Acknowledgement

JBH acknowledges an ARC Australian Laureate Fellowship and SS is funded by a University of Sydney DVC-R Fellowship. We are grateful to Merton College, Oxford for its hospitality in the final stages of writing up this work. We acknowledge insightful comments from T. Tepper-Garcia.

References

Binney, J. & Tremaine, S. 2008, Galactic Dynamics: Second Edition, ISBN 978-0-691-13026-2, Princeton University Press, Princeton
Bland-Hawthorn, J., in IAU Symp. 309, Galaxy in 3D across the Universe, B. L. Ziegler, F. Combes, H. Dannerbauer, M. Verdugo, Eds. (Cambridge: Cambridge Univ. Press), in press
Blanton, M. R. & Moustakas, J. 2009, *ARAA*, 47, 159
Bundy K. *et al.* 2014, *AJ*, in press
Burstein, D., Faber, S. M., & Dressler, A. 1990, *ApJ*, 354, 18
Codis, S., Pichon, C., Devriendt, J., *et al.* 2012, *MNRAS*, 427, 3320
Cooper, M. C., Tremonti, C. A., Newman, J. A., & Zabludoff, A. I. 2008, *MNRAS*, 390, 245
Croom, S. M, Lawrence, J. S, Bland-Hawthorn, J., Bryant, J. J. *et al.* 2012, *MNRAS*, 421, 872
Dekel, A. 1983, *ApJ*, 264, 373
Eldridge, O. C. & Feix, M. 1963, *Phys. Fluids*, 6, 398
Gómez, P. L., Nichol, R. C., Miller, C. J., *et al.* 2003, *ApJ*, 584, 210
Gordon, C. & Lewis, A. 2003, *Phys. Rev. D*, 67, 123513
Lewis, I., Balogh, M., De Propris, R., *et al.* 2002, *MNRAS*, 334, 673
Malaney, R. A. & Mathews, G. J. 1993, *Phys. Rep.*, 229, 145
Mastropietro, C. & Burkert, A. 2008, *MNRAS*, 389, 967
Melott, A. L. 1983, *MNRAS*, 202, 595
Metuki, O., Libeskind, N. I., Hoffman, Y., Crain, R. A., & Theuns, T. 2014, astro-ph/1405.0281
Nichols, M. & Bland-Hawthorn, J. 2013, *ApJ*, 775, 97
Pichon, C., Pogosyan, D., Kimm, T. *et al.* 2011, *MNRAS*, 418, 2493
Rubin, V. C., Thonnard, N., Ford, W. K., Jr., & Roberts, M. S. 1976, *AJ*, 81, 719
Schaye, J., Crain, R. A., Bower, R. G., *et al.* 2014, arXiv:1407.7040
Shull, J. M., Smith, B. D., & Danforth, C. W. 2012, *ApJ*, 759, 23
Sun, M., Voit, G. M., Donahue, M., *et al.* 2009, *ApJ*, 693, 1142
Teles, T. N., Levin, Y., & Pakter, R. 2011, *MNRAS*, 417, L21
Tully, R. B., Shaya, E. J., Karachentsev, I. D., *et al.* 2008, *ApJ*, 676, 184
Tully, R. B., Courtois, H. M., Dolphin, A. E., *et al.* 2013, *AJ*, 146, 86
Tumlinson, J., Thom, C., Werk, J. K., *et al.* 2011, *Science*, 334, 948
Vogelsberger, M., Genel, S., Springel, V., *et al.* 2014, arXiv:1405.2921
Yamashiro, T., Gouda, N., & Sakagami, M. 1992, *Progress of Theoretical Physics*, 88, 269

The Zeldovich Universe:
Genesis and Growth of the Cosmic Web
Proceedings IAU Symposium No. 308, 2014
R. van de Weygaert, S. Shandarin, E. Saar & J. Einasto, eds.
© International Astronomical Union 2016
doi:10.1017/S1743921316010590

Measuring the growth rate of structure around cosmic voids

A. J. Hawken, D. Michelett, B. Granett, A. Iovino, L. Guzzo and + VIPERS

[1]Osservatorio Astronomico di Brera, via E. Bianchi 46, 23807 Merate, Lc

Abstract. Using an algorithm based on searching for empty spheres we identified 245 voids in the VIMOS Public Extragalactic Redshift Survey (VIPERS). We show how by modelling the anisotropic void-galaxy cross correlation function we can probe the growth rate of structure.

Keywords. cosmology: large scale structure of the universe, observations, cosmological parameters, gravitation

1. Introduction

Different cosmological models, and different theories of gravity, predict that the large scale distribution of matter should be structured in subtly different ways. The light emitted from galaxies can be used as a proxy to trace this weblike structure. The cosmic web can be split into different component structures which show different properties, namely clusters, filaments, walls, and voids. Cosmic voids are the most underdense regions of the universe, and compose most its volume. They are also the most dark energy dominated environments and so are ideal places in which to study the vacuum energy and to search for signatures of modified gravity.

Galaxies used to trace voids are subject to motions apart from the Hubble flow. These motions contribute to the observed redshift of a galaxy and distort its apparent position in space. A galaxy in or close to the edge of a void is likely to be being evacuated away from the void centre, falling onto the surrounding structure (Padilla *et al.* 2005, Dubinski *et al.* 1993). These redshift space distortions (RSD) introduce an anisotropy to the void-galaxy cross correlation function, ξ_{vg}.

Here we give an overview of the search for voids in VIPERS (Guzzo *et al.* 2013) and demonstrate how the anisotropy in the void-galaxy cross correlation function caused by linear redshift space distortions can be observed. Following this we propose that by fitting a model to the observed ξ_{vg}, which includes linear RSD, it is possible to extract a measurement of the growth rate of structure, $f(\Omega)$.

2. The search for voids in VIPERS

VIPERS aims to measure 100,000 redshifts for galaxies out to a redshift of $z \sim 1$ over an area of 24 square degrees. The current public data release contains ~ 57000 spectra. The survey is particularly narrow in declination which makes it difficult to use common void finding techniques such as the watershed algorithm (Platen *et al.* 2007, Neyrinck *et al.* 2008, Sutter *et al.* 2015). We therefore developed an algorithm which searches for voids using empty spheres, which is described in detail in Micheletti *et al.* 2014.

We searched for voids in a volume limited sample of galaxies with a redshift $0.55 < z < 0.9$, from the VIPERS internal data release four. We first thinned the sample by removing galaxies with a third nearest neighbour distance more than 3.5σ from the mean. Removing the most isolated galaxies increases the density contrast between high and low density regions. It also removes galaxies which have possibly been spuriously introduced to voids by non-linear RSD (Hamilton 1992). Furthermore, this provides us with a sample of the most isolated galaxies in VIPERS which can be analysed for galaxy evolution studies.

We then grow empty spheres on a fine regular grid. We limit ourselves to only searching for the most significant empty spheres. This is to avoid selecting spurious under densities generated by masking effects and in the hope that a sample of larger voids will avoid the void-in-cloud problem (Sheth & van de Weygaert 2004). Empty spheres are deemed significant if they have a radius greater than 2.5σ times the mean inter-galaxy separation, in practice this means that the empty spheres we are interested in have a radius $\gtrsim 15 \mathrm{Mpc}h^{-1}$. Our *voids* are then defined as the regions connected by overlapping statistically significant spheres. The subset of statistically significant non-overlapping empty spheres are called *maximal spheres*. We identified 229 maximal spheres in 145 voids in the W1 field of VIPERS, and 159 maximal spheres in 100 voids in W4. The distinction between maximal spheres and voids is illustrated in figure 1.

We measured the cross-correlation function between VIPERS galaxies (including those previously pruned out) and maximal spheres using the Davis and Peebles estimator (Davis & Peebles 1983).

3. A model for the void-galaxy cross correlation function

Following Paz *et al.* 2013, we assume that the line of sight pairwise velocity dispersion approximately follows a Maxwell-Boltzman distribution,

$$1 + \xi_{vg}(\sigma, \pi) = \int \frac{\mathrm{d}w_3}{\sqrt{2\pi}\sigma_v} \exp\left(-\frac{(w_3 - v(r)\frac{r_3}{r})^2}{2\sigma_v}\right)[1 + \xi_{vg}(r)], \qquad (3.1)$$

where $\xi_{vg}(\sigma, \pi)$ is the anisotropic cross correlation function, and w_3 is the line of sight component of the pairwise velocity. Because the densities in the vicinity of voids are small, the linear estimate for the relationship between the density and velocity fields remains valid,

$$v(r) \approx -Hr\Delta(r)\frac{f(\Omega_m)}{3}. \qquad (3.2)$$

The growth factor is commonly parameterised as $f(\Omega_m) = \Omega_m^\gamma$. In standard GR $\gamma \approx 0.55$. Because the densities involved are very low, the gravitational dynamics of galaxies around voids remain in the linear regime. This should be particularly true for our void sample because our voids are relatively large and so are expected to be more linear. A correct description of the velocity field should also consider the impact of galaxy biasing. However, for now, we consider the bias to be constant and equal to unity.

The void galaxy cross-correlation function can be expressed in terms of the void density profile,

$$\xi_{vg}(r) = \frac{1}{3r^2} \frac{\mathrm{d}}{\mathrm{d}r}(r^3\Delta(r)). \qquad (3.3)$$

There are several proposed functional forms for the void density profile in the literature. These can broadly be divided into two categories: phenomenological models which seek to fit the functional form of the void density profile (e.g. Hamaus *et al.* 2014, Paz *et al.* 2013), and theoretically motivated models (e.g. Fimelli *et al.* 2014).

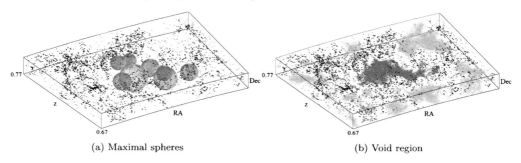

(a) Maximal spheres   (b) Void region

Figure 1: We calculate the cross correlation between significant non overlapping spheres (a) and galaxies. By connecting together spheres which do overlap, our algorithm can be used to find void regions of arbitrary shape (b).

4. Measuring the growth rate

We ran our void finder on 57 mock VIPERS fields generated from the Multidark N-body simulations with a fiducial cosmology of $\Omega_m = 0.27, \Omega_b = 0.0469, \Omega_\Lambda = 0.73, H_0 = 70 \text{km s}^{-1}\,\text{Mpc}^{-1}, n_s = 0.95$, and $\sigma_8 = 0.82$ (Prada *et al.* 2012, de la Torre *et al.* 2013). Using the publicly available Monte-Carlo Markov Chain analysis tool PYMC (Patil *et al.* 2010), we fitted a selection of void profile models to the mean of all the void-galaxy cross correlation functions measured in the mocks, keeping the growth rate fixed to the fiducial cosmology. In an attempt to re-extract the input cosmology we then varied $f(\Omega)$ keeping the other parameters fixed and calculated the most likely value for $f(\Omega)$. The likelihood that a given model is true is related to the χ^2 in the usual way, $\mathcal{L} = \exp(-\chi^2/2)$, where

$$\chi^2 = \sum_{ij} \frac{\mathcal{D}_{ij} - \mathcal{M}_{ij}}{\sigma_{ij}^2} \qquad (4.1)$$

where $\mathcal{D}_{ij}$ is the observed cross correlation function $\xi_{vg}(\sigma, \pi)$ in bin ij, $\mathcal{M}_{ij}$ is the value predicted by the model, and σ_{ij}^2 is the variance of the cross correlation function measured in bin ij. All the models we tested provide a considerably better fit to the mocks when they are normalised so that the central density of the void is fixed to the observed value $\delta_c \sim 0.85$. i.e. $\delta_c > -1$. Failing to account for the fact that the interiors of voids are not completely empty introduces a systematic bias which suppresses the measurement of the growth rate. Keeping the shape parameters fixed to the values measured in the mocks we then fitted the observed $\xi_{vg}(\sigma, \pi)$ as measured in VIPERS, varying $f(\Omega)$. Figure 2 shows the normalised likelihood for different values of $f(\Omega)$ obtained by fitting different models for ξ_{vg}. Also plotted are the conventional measurement of the growth rate from VIPERS, presented in de la Torre *et al.* 2013, and the expected value given a Planck like cosmology Ade *et al.* 2014. There is no significant tension between our most likely value of $f(\Omega)$ and that obtained from a conventional analysis of the VIPERS data.

5. Conclusion

We have demonstrated that it is possible to constrain the growth rate of structure by fitting a model to the anisotropic void-galaxy cross correlation function, as measured in VIPERS. However, the results are highly model dependent and are subject to unknown systematics which may be introducing biases. A paper exploring these issues is currently in preparation.

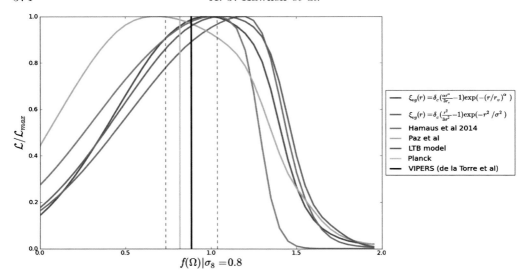

Figure 2: Normalised maximum likelihood for different values of the growth rate $f(\Omega)$, for various model density profiles given the VIPERS data. The solid and dashed black lines represent the measurement of the growth rate, with one-sigma error bars, from VIPERS using standard two-point galaxy clustering statistics. The yellow (pale) vertical line indicates the expected value for $f(\Omega)$ given the best fit Planck cosmological parameters.

References

Ade, P. A. R. & The Planck Collaboration, 2014, *A&A*, 571, A16

Aragon-Calvo, M. A. & Szalay, A. S., 2013, *MNRAS*, 428, 4

Davis, M. & Peebles, P. J. E. 1983, *ApJ*, 267, 465

de la Torre, S., Guzzo, L., Peacock, J. A., *et al.* 2013, *AAP*, 557, A54

Dubinski, J., da Costa, L. N., Goldwirth, D. S., Lecar, M., & Piran, T. 1993, *ApJ*, 410

Finelli, F., Garcia-Bellido, J., Kovacs, A., Paci, F., & Szapudi, I., 2014, ArXiv e-prints:1405.1555

Guzzo, L., The Vipers Team 2013, The Messenger, 151, 41

Hamaus, N., Sutter, P. M., & Wandelt, B. D. 2014, *PRL*, 112

Hamilton, A. J. S. 1992, *APJL*, 385, L5

Kaiser, N. 1987, *MNRAS*, 227

Micheletti, D., Iovino, A., Hawken, A. J., Granett, B., & The VIPERS team, 2014, *A&A* 570, A106

Neyrinck, M. C. 2008, *MNRAS*, 386, 2101

Padilla, N. D., Ceccarelli, L., & Lambas, D. G. 2005, *MNRAS*, 363, 977

Paranjape, A., Lam, T. Y., & Sheth, R. K., 2012, *MNRAS*, 420, 1648

Patil, A., Huard, D., & Fonnesbeck, C. J., 2010, *Journal of Statistical Software*, 35, 4

Paz, D., Lares, M., Ceccarelli, L., Padilla, N., & Lambas, D. G. 2013, *MNRAS*, 436

Platen, E., van de Weygaert, R., & Jones, B. J. T. 2007, *MNRAS*, 380, 551

Prada, F., Klypin, A. A., Cuesta, A. J., Betancort-Rijo, J. E., & Primack, J., 2012 ,*MNRAS*, 423

Regos, E. & Geller, M. J. 1991, *ApJ*, 377, 14

Sheth, R. K. & van de Weygaert, R. 2004, *MNRAS*, 350, 517

Sutter, P. M., Lavaux, G., Hamaus, N., Pisani, A., Wandelt, B. D., Warren, M., Villaescusa-Navarro, F., Zivick, P., Mao, Q., & Thompson, B. B., 2015, *Astronomy and Computing*, 9

The Zeldovich Universe:
Genesis and Growth of the Cosmic Web
Proceedings IAU Symposium No. 308, 2014
R. van de Weygaert, S. Shandarin, E. Saar & J. Einasto, eds.

© International Astronomical Union 2016
doi:10.1017/S1743921316010607

The cosmic web in CosmoGrid void regions

Steven Rieder[1,2], Rien van de Weygaert[1], Marius Cautun[3,1], Burcu Beygu[1] and Simon Portegies Zwart[2]

[1] Kapteyn Instituut, Rijksuniversiteit Groningen, P.O. Box 800, 9700 AV Groningen, The Netherlands, email: `steven@stevenrieder.nl`

[2] Sterrewacht Leiden, Leiden University, P.O. Box 9513, 2300 RA Leiden, The Netherlands,

[3] Department of Physics, Institute for Computational Cosmology, University of Durham, South Road, Durham DH1 3LE, UK

Abstract. We study the formation and evolution of the cosmic web, using the high-resolution CosmoGrid ΛCDM simulation. In particular, we investigate the evolution of the large-scale structure around void halo groups, and compare this to observations of the VGS-31 galaxy group, which consists of three interacting galaxies inside a large void.

The structure around such haloes shows a great deal of tenuous structure, with most of such systems being embedded in intra-void filaments and walls. We use the `Nexus+` algorithm to detect walls and filaments in CosmoGrid, and find them to be present and detectable at every scale. The void regions embed tenuous walls, which in turn embed tenuous filaments. We hypothesize that the void galaxy group of VGS-31 formed in such an environment.

Keywords. large-scale structure of universe, dark matter

1. Introduction

The large-scale structure of the Universe is volume-dominated by voids: enormous regions of space that contain very few galaxies (see Zeldovich *et al.* 1982, and van de Weygaert and Platen 2011 for a review). They are surrounded by walls, filaments and clusters, together forming the *Cosmic Web* (Bond *et al.* 1996). Kreckel *et al.* (2011, 2012) conducted a survey of 60 void galaxies, selected from the SDSS. One of these, VGS-31, was later revealed to be an interacting galaxy system, consisting of three galaxies that appeared to be aligned along an HI filament (Beygu *et al.* 2013). It is a manifestation of void substructure as a result of hierarchical evolution (van de Weygaert and van Kampen 1993; Sheth and van de Weygaert 2004; Aragon-Calvo *et al.* 2010; Aragon-Calvo and Szalay 2013).

2. Simulation and methods

In order to investigate the formation of a system like VGS-31, we study the environment of dark matter haloes in cosmic voids within the high-resolution CosmoGrid ΛCDM suite of simulations (Portegies Zwart *et al.* 2010; Ishiyama *et al.* 2013). CosmoGrid was performed using the GreeM/SUSHI code (Ishiyama *et al.* 2009; Groen *et al.* 2011). The most detailed realisation contains 2048^3 particles in a $(30 \text{ Mpc})^3$ volume, resulting in a mass of 1.28×10^5 M$_\odot$ for each particle. With a softening length ϵ of 175 parsec at $z = 0$ and over 500 snapshots, it has a very high mass-, spatial- and time-resolution.

Due to its high resolution, CosmoGrid is especially suitable for case studies of smaller dark matter haloes (up to Milky Way and Local Group-size). However, its limited volume makes it less suited for studies requiring statistical representation.

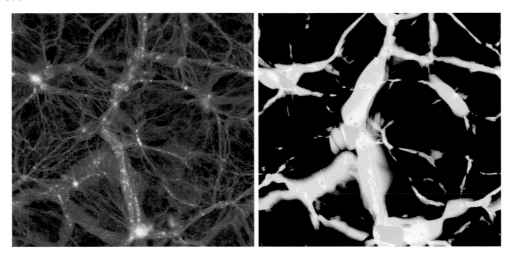

Figure 1. 600 kpc thick slice of the 1024^3-particle CosmoGrid realization at $z = 0$. Left: particles. Right: Nexus+ analysis of the same region. Blue regions represent filaments and clusters, while orange represents the walls.

For the analysis of the CosmoGrid environment, we use the `Nexus+` (Cautun *et al.* 2013, 2014) and `DTFE` (Schaap and van de Weygaert 2000; van de Weygaert and Schaap 2009; Cautun and van de Weygaert 2011) tools and the AMUSE (Portegies Zwart *et al.* 2013; Pelupessy *et al.* 2013) framework. `DTFE` uses the Voronoi and Delaunay tessellations of the particle distribution to construct a volume-weighted linear piecewise density field. The inverse volume of the contiguous Voronoi cells provide a local density estimate, and the Delaunay tessellation are used as an interpolation grid. The resulting field reconstruction retains the anisotropic character and multiscale structure of the particle distribution. We use `DTFE` to determine densities in the volume (smoothed on a scale of 1 h^{-1}Mpc), where we identify regions with a density contrast $\delta < 0$ as void regions.

Additionally, we use `Nexus+` to determine whether a region is part of a wall, a filament or a cluster. `Nexus+` is a technique for the multiscale morphological characterisation of the spatial mass distribution. It is an extension and improvement of the Multiscale Morphology Filter (Aragón-Calvo *et al.* 2007), in which the locally dominant morphological signal is extracted from the 4d scale-space representation of the mass distribution. It dissects the mass distribution into voids, walls, filaments and cluster nodes on the basis of the local signature of the Hessian of the logarithm of the density field. Using this method, we analyse the full volume of CosmoGrid (see figure 1). Due to the multi-scale nature of the cosmic environment, a void region found with a 1 h^{-1}Mpc smoothing scale can still contain walls, filaments and clusters on a smaller scale.

3. VGS-31-like systems in CosmoGrid

3.1. *Candidate systems*

We use CosmoGrid data to investigate possible formation scenarios for the VGS-31 system, and to find out if the filamentary structure seen in the alignment of these galaxies is to be expected or coincidental. To this end, we first search for systems of similar properties to VGS-31 in CosmoGrid. Using the mass, environmental density contrast and size of the system as selection criteria, we find eight candidate systems in CosmoGrid that we name "CosmoGrid Void Systems" (CGV, see table 1 and Rieder *et al.* 2013, for details).

Name	M_{vir}($h^{-1}\mathrm{M}_\odot$)	Last MM (Gyr)	δ	Wall	Filament
CGV-A	3.15×10^{10}	-	-0.68	X	-
CGV-B	3.95×10^{10}	5.24	-0.51	X	X
CGV-C	2.99×10^{10}	1.19	-0.51	X	-
CGV-D	4.60×10^{10}	10.9	-0.63	X	X
CGV-E	1.99×10^{10}	2.44	-0.57	X	X
CGV-F	2.27×10^{10}	-	-0.62	X	X
CGV-G	2.14×10^{10}	5.80	-0.61	X	X
CGV-H	4.63×10^{10}	8.45	-0.50	X	-

Table 1. VGS-31-like systems in CosmoGrid. 1: Virial mass of the main halo, 2: Time of the last major merger, 3: Density contrast, 4: System is embedded in a wall, 5: System is embedded in a filament.

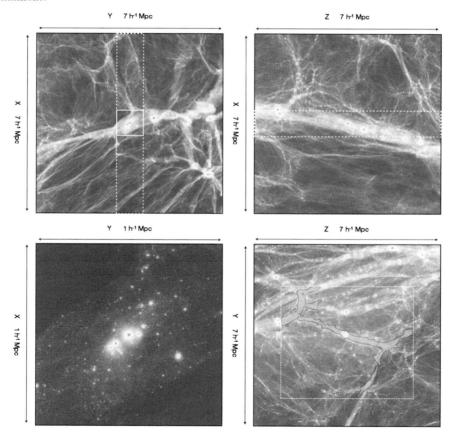

Figure 2. Environment of CGV-G, as seen from different angles, in 1 h^{-1}Mpc thick slices. Lower-right panel includes a **Nexus+** overlay.

While at $z = 0$ these systems appear very similar, their formation histories are quite diverse: some systems continue to experience mergers up to $z = 0$, while others remain virtually unchanged since $z = 6$. Here, we discuss the environment of these haloes.

3.2. *Environment of the systems*

In figure 2, we show the environment of CGV-G as a representative for the void halo systems, at a 7 h^{-1}Mpc scale. While the whole system is in a void, its environment is

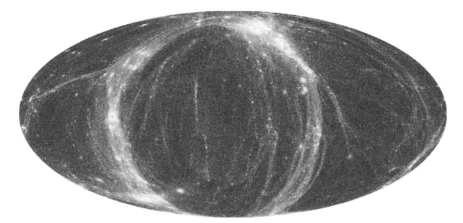

Figure 3. Mollweide projection of the environment around CGV-G at $z = 0$, up to a distance of 2 h^{-1} Mpc from the main halo's centre. The wall manifests itself as a circle, while the filament is visible within the wall as two bright regions on opposite sides (top-right and lower-left).

highly structured. The tenuous wall, which in turn contains a thin filament, is clearly visible in this figure.

While the majority of the CGV systems forms along a tenuous filament, all of the systems are embedded in thin but prominent walls (see table 1). These walls have a typical thickness of around 0.4 h^{-1} Mpc. They show a strong coherence and retain the character of a highly flattened structure out to a distance of at least 3 h^{-1} Mpc from the CGV haloes. The filaments in which five of the systems reside are rather short, not longer than 4 h^{-1} Mpc, with a diameter of around 0.4 h^{-1} Mpc. Compared to the high-density filaments of the larger-scale cosmic web, the environment of void haloes is very feeble. At earlier times, this structure inside the void is much more prominent, becoming more tenuous over time until it can hardly be detected at $z = 0$. However, it remains visible in the alignment of the haloes. The formation histories of these haloes are quite diverse, with the main haloes forming between $z = 6$ and $z = 0$.

In figure 3, we show a Mollweide projection of all the dark matter around CGV-G up to a distance of 2 h^{-1} Mpc. This projection clearly shows the wall in which the system is embedded, while the filament appears as two bright spots on opposite sides. We see very similar structure around the other CGV-systems.

A similar study within a larger volume will provide a more definitive answer on the abundance of VGS-31-like systems. Furthermore, a continuation study with a hydrodynamical simulation will provide insight into the gaseous surroundings of such systems.

Acknowledgements

The authors are grateful to Tomoaki Ishiyama and Bernard Jones, for many insightful discussions. This work was supported by NWO grants IsFast [#643.000.803], VICI [#639.073.803], LGM [#612.071.503] and AMUSE [#614.061.608]), NCF (grants [#SH-095-08] and [#SH-187-10]), NOVA and the LKBF in the Netherlands. SR and RvdW acknowledge support by the John Templeton Foundation, grant [#FP05136-O].

References

M. A. Aragon-Calvo & A. S. Szalay. The hierarchical structure and dynamics of voids. *MNRAS*, 428:3409–3424, February 2013. .

M. A. Aragón-Calvo, B. J. T. Jones, R. van de Weygaert, & J. M. van der Hulst. The multiscale morphology filter: identifying and extracting spatial patterns in the galaxy distribution. *A&A*, 474:315–338, October 2007. .

M. A. Aragon-Calvo, R. van de Weygaert, P. A. Araya-Melo, E. Platen, & A. S. Szalay. Unfolding the hierarchy of voids. *MNRAS*, 404:L89–L93, May 2010. .

B. Beygu, K. Kreckel, R. van de Weygaert, J. M. van der Hulst, & J. H. van Gorkom. An Interacting Galaxy System along a Filament in a Void. *AJ*, 145:120, May 2013. .

J. R. Bond, L. Kofman, & D. Pogosyan. How filaments of galaxies are woven into the cosmic web. *Nature*, 380:603–606, April 1996. .

M. Cautun, R. van de Weygaert, & B. J. T. Jones. NEXUS: tracing the cosmic web connection. *MNRAS*, 429:1286–1308, February 2013. .

M. Cautun, R. van de Weygaert, B. J. T. Jones, & C. S. Frenk. Evolution of the cosmic web. *MNRAS*, 441:2923–2973, July 2014. .

M. C. Cautun & R. van de Weygaert. The DTFE public software - The Delaunay Tessellation Field Estimator code. *arXiv:1105.0370*, May 2011.

D. Groen, S. Portegies Zwart, T. Ishiyama, & J. Makino. High-performance gravitational N-body simulations on a planet-wide-distributed supercomputer. *Computational Science and Discovery*, 4(1):015001–+, January 2011. .

T. Ishiyama, T. Fukushige, & J. Makino. GreeM: Massively Parallel TreePM Code for Large Cosmological N -body Simulations. *PASJ*, 61:1319–, December 2009.

T. Ishiyama, S. Rieder, J. Makino, S. Portegies Zwart, D. Groen, K. Nitadori, C. de Laat, S. McMillan, K. Hiraki, & S. Harfst. The Cosmogrid Simulation: Statistical Properties of Small Dark Matter Halos. *ApJ*, 767:146, April 2013. .

K. Kreckel, E. Platen, M. A. Aragón-Calvo, J. H. van Gorkom, R. van de Weygaert, J. M. van der Hulst, K. Kovač, C.-W. Yip, & P. J. E. Peebles. Only the Lonely: H I Imaging of Void Galaxies. *AJ*, 141:4, January 2011. .

K. Kreckel, E. Platen, M. A. Aragón-Calvo, J. H. van Gorkom, R. van de Weygaert, J. M. van der Hulst, & B. Beygu. The Void Galaxy Survey: Optical Properties and H I Morphology and Kinematics. *AJ*, 144:16, July 2012. .

F. I. Pelupessy, A. van Elteren, N. de Vries, S. L. W. McMillan, N. Drost, & S. F. Portegies Zwart. The Astrophysical Multipurpose Software Environment. *A&A*, 557:A84, September 2013. .

S. Portegies Zwart, T. Ishiyama, D. Groen, K. Nitadori, J. Makino, C. de Laat, S. McMillan, K. Hiraki, S. Harfst, & P. Grosso. Simulating the universe on an intercontinental grid of supercomputers. *IEEE Computer, v.43, No.8, p.63-70*, 43:63–70, October 2010. .

S. Portegies Zwart, S. L. W. McMillan, E. van Elteren, I. Pelupessy, & N. de Vries. Multi-physics simulations using a hierarchical interchangeable software interface. *Computer Physics Communications*, 183:456–468, March 2013. .

S. Rieder, R. van de Weygaert, M. Cautun, B. Beygu, & S. Portegies Zwart. Assembly of filamentary void galaxy configurations. *MNRAS*, 435:222–241, October 2013. .

W. E. Schaap & R. van de Weygaert. Continuous fields and discrete samples: reconstruction through Delaunay tessellations. *A&A*, 363:L29–L32, November 2000.

R. K. Sheth & R. van de Weygaert. A hierarchy of voids: much ado about nothing. *MNRAS*, 350:517–538, May 2004. .

R. van de Weygaert & E. Platen. Cosmic Voids: Structure, Dynamics and Galaxies. *International Journal of Modern Physics Conference Series*, 1:41–66, 2011. .

R. van de Weygaert & W. Schaap. The Cosmic Web: Geometric Analysis. In V. J. Martínez, E. Saar, E. Martínez-González, and M.-J. Pons-Bordería, editors, *Data Analysis in Cosmology*, volume 665 of *Lecture Notes in Physics, Berlin Springer Verlag*, pages 291–413, 2009. .

R. van de Weygaert & E. van Kampen. Voids in Gravitational Instability Scenarios - Part One - Global Density and Velocity Fields in an Einstein - De-Sitter Universe. *MNRAS*, 263:481, July 1993.

I. B. Zeldovich, J. Einasto, & S. F. Shandarin. Giant voids in the universe. *Nature*, 300:407–413, December 1982. .

The Zeldovich Universe:
Genesis and Growth of the Cosmic Web
Proceedings IAU Symposium No. 308, 2014
R. van de Weygaert, S. Shandarin, E. Saar & J. Einasto, eds.
© International Astronomical Union 2016
doi:10.1017/S1743921316010619

The ISW imprints of voids and superclusters on the CMB

S. Hotchkiss[1],† S. Nadathur[2], S. Gottlöber[3], I. T. Iliev[1], A. Knebe[4], W. A. Watson[1] and G. Yepes[5]

[1] Department of Physics and Astronomy, University of Sussex, Falmer, Brighton, BN1 9QH, UK

[2] Department of Physics, University of Helsinki and Helsinki Institute of Physics, P.O. Box 64, FIN-00014, University of Helsinki, Finland

[3] Leibniz-Institute for Astrophysics, An der Sternwarte 16, D-14482 Potsdam, Germany

[4] Departamento de Física Teórica, Modulo C-XI, Facultad de Ciencias, Universidad Autónoma de Madrid, 28049 Cantoblanco, Madrid, Spain

Abstract. We examine the stacked integrated Sachs-Wolfe (ISW) imprints on the CMB along the lines of sight of voids and superclusters in galaxy surveys, using the Jubilee ISW simulation and mock luminous red galaxy (LRG) catalogues. We show that the expected signal in the concordance ΛCDM model is much smaller than the primary anisotropies arising at the last scattering surface and therefore any currently claimed detections of such an imprint cannot be caused by the ISW effect in ΛCDM. We look for the existence of such a signal in the Planck CMB using a catalogue of voids and superclusters from the Sloan Digital Sky Survey (SDSS), but find a result completely consistent with ΛCDM – i.e., a null detection.

Keywords. cosmology: cosmic microwave background, dark energy, large-scale structure of Universe, methods: numerical, methods: data analysis

1. Introduction

The possibility of detecting the ISW imprints on the CMB of voids and superclusters using stacking techniques has received much attention following the work of Granett et al. (2008) (hereafter G08), who claimed a very high-significance ($> 4\sigma$) detection of correlation between superstructure lines-of-sight and corresponding hot and cold spots on the CMB. The magnitude of the claimed signal, $\sim 10 \ \mu$K, has however always been puzzlingly large (Hunt & Sarkar (2010), Granett et al. (2009)). Theoretical estimates (Nadathur et al. (2012)) suggested that the maximum possible stacked ISW signal in a ΛCDM cosmology should be an order of magnitude smaller, a conclusion later supported by estimates based on N-body simulations (Flender et al. (2013), Hernández-Monteagudo & Smith (2013), Cai et al. (2013)).

Various consistency checks of the original detection failed to identify any systematic effects or foreground contamination (Hernández-Monteagudo & Smith (2013)), suggesting a cosmological origin for the effect, which was also found to persist in Planck CMB data (Planck Collaboration (2013a)). The use of an alternative catalogue of voids alone, drawn from independent SDSS data at lower redshift, did not support the original high-significance detection (Ilić et al. (2013), Planck Collaboration (2013a)), but this catalogue was later shown to be flawed (Nadathur & Hotchkiss (2014), hereafter NH14). Another measurement using a better void catalogue (Cai et al. (2013)) also gave inconclusive

† s.a.hotchkiss@sussex.ac.uk

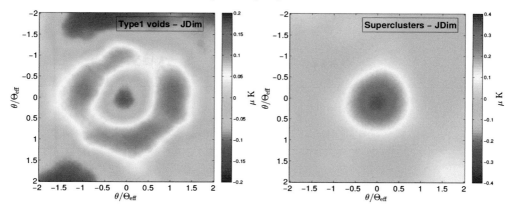

Figure 1. Stacked and rescaled patches from the simulated Jubilee ISW maps along the directions of structures identified in the JDim catalogue. Largest scale modes ($\ell \leqslant 10$) have been removed for clarity. *Left*: Type1 voids; *right*: superclusters.

results, claiming a possible detection of an anomalously large ISW signal from stacked voids, but at low S/N ($\sim 2\sigma$ significance).

We re-examine this issue using the Jubilee ISW simulation as well as the Planck SMICA map (Planck Collaboration (2013b)) and a robust catalogue of structures from the SDSS DR7 LRG samples (NH14). Unlike more recent studies but in the spirit of the original claimed detection, we make use of both voids and superclusters in an attempt to maximize any possible S/N.

2. Method

We make use of the Jubilee ISW simulation, which contains 6000^3 particles in a $(6\ h^{-1}\,\text{Gpc})^3$ box, with a cosmology based on the WMAP5 results (Dunkley *et al.* (2009)): $\Omega_m = 0.27$, $\Omega_\Lambda = 0.73$, $\Omega_b = 0.044$, $h = 0.72$, $\sigma_8 = 0.8$ and $n_s = 0.96$. Initial conditions of the simulation are set at redshift $z = 100$. Full-sky maps of the ISW temperature anisotropy are constructed (see Watson *et al.* (2014) for details) for an observer at the centre of the box by ray-tracing photons through the gravitational potential according to $\Delta T(\hat{\mathbf{n}}) = 2\overline{T}_{CMB} \int \dot{\Phi}(r, \hat{\mathbf{n}}) a \, dr$, where $\dot{\Phi}$ is calculated from the potential Φ using the semi-linear approximation $\dot{\Phi} = -\Phi H(t)\,[1 - \beta(t)]$ (Cai *et al.* (2010)), where $\beta = d\ln D/d\ln a$ and $D(t)$ is the linear growth function. The large size of the Jubilee simulation means that these maps are complete for all structures out to $z \sim 1.4$ without repetition of the box.

Within the Jubilee simulation, halos are resolved on the light cone down to a mass of $\sim 1.5 \times 10^{12} h^{-1} M_\odot$ using a spherical overdensity algorithm. We use the HOD model of Zheng *et al.* (2009) to populate these halos in order to obtain two mock LRG catalogues JDim and JBright between redshifts $z \in (0.16, 0.36)$, $(0.16, 0.44)$, as described by Watson *et al.* (2014), Hotchkiss *et al.* (2014). These are designed to match the SDSS LRG catalogues of Kazin, *et al.* (2010). To identify voids and superclusters within these catalogues we use a modification of the ZOBOV algorithm (Neyrinck (2008)). Our methodology closely follows that outlined by NH14, including the use of buffer particles at the survey boundaries. Selection criteria for the final superstructures are chosen to match those used for Type1 and Type2 voids and superclusters introduced in that paper. Our catalogues of SDSS structures are taken from the public catalogues of NH14†, who found structures in

† Available for download from `www.hip.fi/nadathur/download/dr7catalogue`.

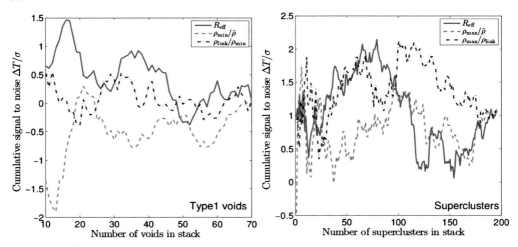

Figure 2. Cumulative S/N values for structures from NH14 and Planck data as the number of structures included in the stack is increased in order of the structure effective radius (blue), maximum density deviation (red dashed) and density contrast (black dot-dashed). *Left*: For Type1 voids; *right*: for superclusters.

the Kazin, *et al.* (2010) samples, labelled *lrgdim* and *lrgbright*. For brevity we discuss only results for Type1 voids and superclusters using JDim (*lrgdim*) here, but results using all the data are described in detail in Hotchkiss *et al.* (2014).

For each superstructure we extract a patch of the simulated ISW map in that direction and filter it using a compensated top-hat filter

$$\Delta T(\theta_R) = \frac{\iint_0^{\theta_R} T(\theta)d\theta d\phi - \iint_{\theta_R}^{\theta_R^*} T(\theta)d\theta d\phi}{\iint_0^{\theta_R} d\theta d\phi}, \qquad (2.1)$$

where θ is the azimuthal angle to the line-of-sight through the centre, θ_R is the filter angle and $\theta_R^* = \arccos(2\cos(\theta_R) - 1)$. As the structures in the catalogues have very different sizes, the filter radius for each is chosen in fixed proportion to the effective angular size it subtends on the sky, $\theta_R = \alpha\Theta_{\rm eff}$. We then determine the average value $\overline{\Delta T}$ for all superstructures. This value is found to be maximized for relative rescaling $\alpha \simeq 0.6$ (Hotchkiss *et al.* (2014)), therefore we fix it to this value henceforth. We experimented with the removal of largest scale modes ($\ell \leqslant 10$) from the map before filtering, and found that removal slightly decreases the total signal amplitude. For superstructures from the SDSS catalogues of NH14 we follow exactly the same procedure except using the Planck SMICA CMB map instead of the simulated ISW-only maps. Errors are estimated by randomizing the lines of sight with the SDSS window (but keeping the same angular size distribution) and calculating the standard deviation over 1000 realizations.

3. Results

Jubilee simulation: Fig. 1 shows stacked and rescaled patches of the ISW maps along directions of Type1 voids and superclusters in the Jubilee JDim sample. Void locations clearly correspond to cold spots and supercluster locations to hot spots, as expected. However, the overall magnitude of the average filtered signal is extremely small, corresponding to $\overline{\Delta T} \simeq 0.15\ \mu K$ for both cases (when $\alpha = 0.6$). This is much smaller than the expected noise in a similar measurement on the CMB due to primary anisotropies at last scattering, indicating that the stacked ISW effect of such structures should not be observable in a ΛCDM cosmology.

At the same time, when the structure catalogues are binned according to physical properties, the simulation results show easily understood trends. In particular $\overline{\Delta T}$ increases with the effective radius R_{eff} of the structure, with decreasing (increasing) minimum (mamimum) density within the void (supercluster), and (more weakly) with increasing *density contrast* of the structure (see Hotchkiss *et al.* (2014) for details). These physical trends should be reproduced for any real ISW-like correlation, even if due to some physics beyond ΛCDM, but will not be seen for spurious detections due to noise.

SDSS data: We repeated the same procedure for the 70 Type1 voids and 196 superclusters from the *lrgdim* catalogues of NH14, obtaining $\overline{\Delta T} = 0.14 \pm 2.8$ μK and $\overline{\Delta T} = 2.05 \pm 1.9$ μK respectively, at the rescaling weight $\alpha = 0.6$. That is, the results are consistent with zero within the measurement error, exactly in accord with the ΛCDM expectation. Fig. 2 shows the effect on the cumulative S/N as the number of structures in each stack is increased from 1 to the final number by adding structures in order of increasing radius, central density deviation and density contrast. In all cases the S/N curves are consistent with random walks due to noise alone, i.e. it is not possible to select any physically motivated subset of the observed structure catalogues which provides a significant S/N. Even allowing the rescaling weight to vary from the optimal value determined from simulation, we do not find an S/N value exceeding 2.5σ at *any* value of α.

4. Conclusions

Our results based on the Jubilee simulation show that the stacked ISW signal of superstructures within ΛCDM is orders of magnitude too small to be observable. This vindicates previous estimates (e.g. Nadathur *et al.* (2012),Flender *et al.* (2013)) which claimed a discrepancy between the ΛCDM expectation and the G08 detection. Any purported detection of the stacked ISW signal of voids and superclusters, if truly cosmological, must therefore be in contradiction with the ΛCDM model.

However, when attempting to reproduce the G08 measurement using an independent catalogue of structures found from SDSS data spanning a lower redshift range than that used by those authors, we fail to detect any significant ISW signal. This is entirely consistent with theoretical expectation. We conclude that whatever the effect seen by G08 was, it does not exist in these independent galaxy samples. Our null result does not definitively exclude the possibility that the original detection was due to some hypothetical new physical effect manifest only at redshifts $\gtrsim 0.5$; however in our opinion it calls into question its likely physical significance.

Finally, although lack of space precludes a full discussion here, in Hotchkiss *et al.* (2014) we have also considered the significance of the tentative detection reported by Cai *et al.* (2013). Given the low significance of the reported detection this is completely consistent with our null result. In addition, we find that it cannot be seen as a confirmation of G08 due to significant differences in methodology. Future observations are required to clear up this issue once and for all, but at present the status of stacked ISW detection claims must remain in doubt.

References

Cai, Y.-C., Cole, S., Jenkins, A., & Frenk, C. S., 2010, *MNRAS*, 407, 201

Cai, Y.-C., Neyrinck, M. C., Szapudi, I., Cole, S., & Frenk, C. S., 2013, *ApJ*, 786, 110

Dunkley, J., *et al.*, 2009, *ApJS*, 180, 306

Flender, S., Hotchkiss, S., & Nadathur, S., 2013, *JCAP*, 1302, 013

Granett, B. R., Neyrinck, M. C., & Szapudi, I., 2008, *ApJ*, 683, L99

Granett, B. R., Neyrinck, M. C., & Szapudi, I., 2009, *ApJ*, 701, 414

Hernández-Monteagudo, C. & Smith, R. E., 2013, *MNRAS*, 435, 1094

Hotchkiss, S. *et al.*, 2014, *MNRAS* in press, arXiv:1405.3552

Hunt, P. & Sarkar, S., 2010, *MNRAS*, 401, 547

Ilić, S., Langer, M., & Douspis, M., 2013, *A&A*, 556, A51

Kazin, E. A. *et al.*, 2010, *ApJ*, 710, 1444

Nadathur, S. & Hotchkiss, S., 2014, *MNRAS*, 440, 1248

Nadathur, S., Hotchkiss, S., & Sarkar, S., 2012, *JCAP*, 1206, 042

Neyrinck, M. C., 2008, *MNRAS*, 386, 2101

Planck Collaboration *et al.*, 2013b, ArXiv e-prints, 1303.5079

Planck Collaboration *et al.*, 2013a, ArXiv e-prints, 1303.5062

Watson, W. A. *et al.*, 2014, *MNRAS*, 438, 412

Zheng, Z., Zehavi, I., Eisenstein, D. J., Weinberg, D. H., & Jing Y., 2009, *ApJ*, 707, 554

The Zeldovich Universe:
Genesis and Growth of the Cosmic Web
Proceedings IAU Symposium No. 308, 2014
R. van de Weygaert, S. Shandarin, E. Saar & J. Einasto, eds.

© International Astronomical Union 2016
doi:10.1017/S1743921316010620

Characterising the local void with the X-ray cluster survey REFLEX II

Chris A. Collins[1], Hans Böhringer[2], Martyn Bristow[1] and Gayoung Chon[2]

[1] Astrophysics Research Institute, Liverpool John Moores University, IC2, Liverpool Science Park, 146 Brownlow Hill, Liverpool L3 5RF, UK
email:c.a.collins@ljmu.ac.uk

[2] Max-Planck-Institut für extraterrestrische Physik, D-85748 Garching, Germany
email: hxb@mpe.mpg.de

Abstract. Claims of a significant underdensity or void in the density distribution on scales out to $\simeq 300$ Mpc have recently been made using samples of galaxies. We present the results of an alternative test of the matter distribution on these scales using clusters of galaxies, which provide an independent and powerful probe of large-scale structure. We study the density distribution of X-ray clusters from the ROSAT-based REFLEX II catalogue, which covers a contiguous area of 4.24 steradians in the southern hempsphere (34% of the entire sky). Using the normalised comoving number density of clusters we find evidence for an underdensity $(30 - 40\%)$, out to $z \sim 0.04$, equivalent to $\simeq 170$ Mpc and with a significance of 3.4σ. On scales between 300 Mpc and 1 Gpc the distribution of REFLEX II clusters is consistent with being uniform. We also confirm recent results that the underdensity has a large contribution from the direction of the South Galactic Cap region, but is not significant in the direction of the Northern Galactic Cap as viewed from the southern sky. Both the limited size of the detected underdensity and its lack of isotropy, argue against the idea that the Type Ia supernovae data can be explained without the need for dark energy.

Keywords. galaxies: clusters, cosmology: large-scale structure of universe, general, X-rays: galaxies: clusters

1. Introduction

There is significant interest in the possibility that we live close to the centre of a large isotropic void that can mimic an accelerating universe, as revelaed by the observations of supernovae (Schmidt *et al.* 1998; Perlmutter *et al.* 1999), without the need for dark energy (e.g., Alexander *et al.* 2009; February *et al.* 2010). In these "minimal-void" scenarios an underdensity of $\sim 40\%$ stretching to ~ 300 Mpc or more can reproduce both the supernovae data and other cosmological constraints without the requirement of an accelerating universe, although how difficult it is for these models not to fail at least one cosmological test remains an open question (e.g., Moss *et al.* 2011). A large void could also explain the tension between measurements of the Hubble constant $(H_0 = 67.3 \pm 1.2\,\mathrm{km\,s^{-1}\,Mpc^{-1}})$ determined from PLANCK measurements of the cosmic microwave background (Planck Collaboration XVI, 2013) and a higher value $(H_0 = 73.8 \pm 2.4\,\mathrm{km\,s^{-1}\,Mpc^{-1}})$ based on galaxies hosting both Type Ia supernovae and Cepheid variable stars (Riess *et al.* 2011). Furthermore, the minimum size of an underdensity large enough to embed the local H_0 measurements and deep enough to induce a 9% difference in H_0 compared to the global mean, a so called "Hubble bubble", is unlikely to occur naturally from density fluctuations in the standard ΛCDM model. If substantiated, a cosmology beyond the standard model may be required (Marra *et al.* 2013).

Two recent observational studies using K-band selected galaxy samples probe the density distribution on scales $300-400\,\mathrm{Mpc}$. Whitbourn & Shanks (2014) use the redshifts of 250,000 galaxies compiled from 6dFGS and SDSS to examine the density distribution in three regions covering a total area of $9,000\,\mathrm{deg}^2$. They find a range of underdensities from $4-40\%$ to a depth of $\sim 200\,\mathrm{Mpc}$, with the most prominent underdensity concentrated on the South Galactic Cap region of the southern sky. Keenan *et al.* (2013) use a sample of 35,000 galaxies with redshifts, covering $600\,\mathrm{deg}^2$, based on the UKIDSS and 2MASS surveys. By fitting the lumonisity function to the galaxy magnitudes the authors provide evidence for a $\simeq 50\%$ reduction in the mass density within a volume of $z=0.07$ or about $300\,\mathrm{Mpc}$. If representative, these results could have significant implications.

2. The Data: REFLEX II

An alternative probe of the underdensity is to use clusters of galaxies. These are well known to be a reliable tracers of the large-scale structure in the universe (e.g., Kaiser 1984) and they easily probe the large volumes necessary to examine the claims of underdensities up to Gpc scales. Furthermore, clusters are biased tracers of the mass distribution compared to galaxies and therefore should amplify any putative density contrast. The study here uses the homogeneous X-ray flux limited cluster survey REFLEX II (Böhringer *et al.* 2013), which covers a contiguous area of $\simeq 4.24$ steradians in the southern sky below a declination of $+2.5^0$ and with a galactic latitide $|b_{II}| \geqslant 20^0$. REFLEX II is based on the RASS X-ray source detections (Trümper 1993; Voges *et al.* 1999). The nominal flux limit of $1.8 \times 10^{-12}\,\mathrm{erg\,cm^{-2}\,s^{-1}}$ in the energy range $0.1-2.4\,\mathrm{keV}$ is reached for 80% of the survey area, with higher flux limits in the remaining 20% which have low exposure or high interstellar absorption. The REFLEX II selection function is fully accounted for in the analysis described below. In total there are 913 clusters and groups in REFLEX II above an X-ray luminosity of $L_X \geqslant 10^{42}\,\mathrm{erg\,s^{-1}}$ $(0.1-2.4\ \mathrm{keV})$; 206 of these lie within $z=0.06$ and 416 out to $z=0.1$. For details of the properties of RE-FLEX II and how it was constructed see Böhringer *et al.* (2013). In this paper we adopt $H_0 = 70\,\mathrm{km\,s^{-1}Mpc^{-1}}$ and $\Omega_m = 0.3$ with a flat ΛCDM cosmology.

3. Results

Here we give an overview of our methodology and results. Full details and discussion can be found in Böhringer *et al.* (2014b). We calculate the relative density distributions for REFLEX II using a best-fit Schechter form for the luminosity function (see Böhringer *et al.* 2014) in order to predict the number of X-ray clusters with redshift and then divide the number of observed clusters in redshift shells by the predicted value. In Figure 1 we show the normalised comoving density of REFLEX II clusters as a function of redshift. There is a clear cluster underdensity of $30-40\%$ in the redshift range below $z=0.04$, equivalent to about $170\,\mathrm{Mpc}$. Table 1 shows the cumulative ratio of the observed to predicted number of clusters, along with the statistical uncertainty using the best-fit Schechter function above the luminosity limit $L_X = 10^{42}\,\mathrm{erg\,s^{-1}}$ $(0.1-2.4\,\mathrm{keV})$. The overall significance of the underdensity averaged over the entire southern sky out to $z \simeq 0.04$ is $3.4\,\sigma$. The void is still present at a significant level if the minimum luminosity limit is increased by a factor 20 to $L_X \geqslant 2 \times 10^{43}\,\mathrm{erg\,s^{-1}}$ $(0.1-2.4\,\mathrm{keV})$ - indicating that both richer clusters and poorer groups follow the same density pattern. More specifically the relative density distributions for REFLEX II volume-limited subsamples over the redshift range $z=0.015$ to $z=0.15$ show remarkably similar density ratios in overlapping redshift shells (see Böhringer *et al.* 2014b). We also note that there

z	0.02	0.03	0.04	0.05	0.06	0.07	0.08
Ratio	0.63 ± 0.11	0.53 ± 0.07	0.61 ± 0.07	0.71 ± 0.06	0.81 ± 0.05	0.88 ± 0.06	0.93 ± 0.05

Table 1. Cumulative ratio of observed to predicted cluster numbers with redshift.

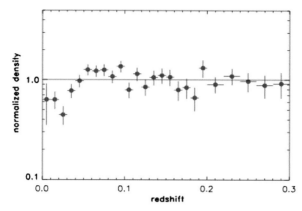

Figure 1. Normalised cluster density distribution from REFLEX II as a function of redshift using a minimum X-ray luminosity of 10^{42} erg s^{-1} $(0.1 - 2.4 \, \text{keV})$. There is a clear signal of an underdensity within a radius of about $z = 0.04$, corresponding to $\simeq 170 \, \text{Mpc}$.

is significant overdensity detected at redshifts between $z = 0.05 - 0.08$. This is due to the dominance of superclusters such as Shapley ($z = 0.046$) and superclusters 42 (part of the Horologium-Reticulum supercluster at $z = 0.065$) and 62 identified in REFLEX II by Chon *et al.* (2013) - see also the contribution by Chon *et al.* in this volume.

Next we restrict our analysis to the identified regions at the South Galactic Cap (SGC) and North Galactic Cap (NGC) where Whitbourn & Shanks (2014) carried out their underdensity analysis using 6dFGS galaxies, as mentioned in Section 1. In Figure 2 we show the normalised comoving density of REFLEX II clusters as a function of redshift for both the SGC and NGC regions, along with the galaxy results from Whitbourn & Shanks (2014). Whereas at the SGC the density falls to 0.45 ± 0.10 at $z = 0.05$ (0.35 ± 0.10 at $z = 0.04$), at the NGC the density is 1.02 ± 0.17 at $z = 0.05$ (0.83 ± 0.18 at $z = 0.04$). This indicates that the detected $30 - 40\%$ underdensity from the full REFLEX II area is not isotropic. In Table 2 we show the comparison of the ratio of observed to predicted numbers of clusters within $z = 0.05$ and $z = 0.1$, providing a direct comparison with Whitbourn & Shanks (2014). Overall there is excellent agreement between the two surveys in both the SGC and NGC regions, even accounting for the relative biasing (see Section 4) and despite the presence of relatively large uncertainties, which reassuringly indicates that both galaxies and clusters are following the same local density variations.

4. Implications

The detection of an overall underdensity of $40 \pm 15\%$ in the cluster distribution implies an underdensity in the matter distribution of $15 \pm 5\%$, based on an average bias of $2.5 - 3.0$, appropriate for REFLEX II clusters at low redshift (Chon *et al.* 2014). In turn this implies a $3 \pm 1\%$ larger Hubble constant locally, based on simple linear theory. This is not sufficient on its own to explain the tension between the local and distant determinations of the Hubble constant which differ by 9% as discussed in Section 1. Furthermore the scale of our detected void falls short of the $\simeq 300 \, \text{Mpc}$ ($z = 0.007$) required to explain supernovae data

		$z < 0.05$		$z < 0.1$	
Region	SGC	NGC	SGC	NGC	
REFLEX II	0.45 ± 0.10	1.02 ± 0.17	0.84 ± 0.09	1.18 ± 0.12	
W&S	0.60 ± 0.05	0.96 ± 0.10	0.75 ± 0.05	0.94 ± 0.07	

Table 2. Relative densities from REFLEX II along with the results of Whitbourn & Shanks 2014. The redshifts $z = 0.05$ and 0.1 are chosen to enable a direct comparison.

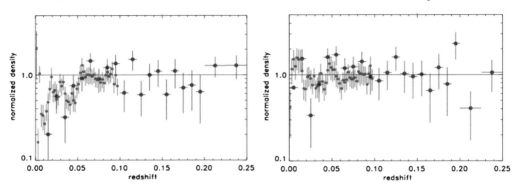

Figure 2. Density distributions for REFLEX II with the galaxy study (smaller points) by Whitbourn & Shanks (2014) for the 6dFGS regions: SGC (left) and NGC (right).

without dark energy induced acceleration. An examination of Figure 1 reveals that from 300 Mpc up to 1 Gpc ($z = 0.25$) the density distribution of REFLEX II is exceptionally uniform. The limited size of the measured underdensity and its lack of isotropy across the southern sky are in contrast to models which place us at the centre of a large void.

References

Alexander, S., Biswas, T.,Notari, A., & Vaid, D. 2009, *JCAP*, 9, 25
Böhringer, H., Chon, G., Collins, C. A., Guzzo, L., *et al.* 2013, *A&A*, 555, A30
Böhringer, H., Chon, G., & Collins. C. A. 2014, *A&A*, 570, A31
Böhringer, H., Chon, G., Bristow, M., & Collins, C. A. 2014b, *A&A*, in press, *arXiv:1410.2172*
Chon, G., Böhringer, H., & Nowak, N. 2013, *MNRAS*, 429, 3272.
Chon, G., Böhringer, H., Collins, C. A., & Krause, M. 2014, *A&A*, 567, A144
February, S., Larena, J., Smith, M., & Clarkson, C. 2010, *MNRAS*, 405, 2231
Kaiser, N. 1984, *ApJ*, 284, L9
Keenan, R. C., Barger, A. J., & Cowie, L. L. 2013, *ApJ*, 775, 62
Marra, L., Amendola, L., Sawicki, I., & Valkenburg, W. 2013, *Phys. Rev. Lett.*, 110, 241305
Moss, A., Zibin, J. P., & Scott, D. 2011, *Phys. Rev. D*, 83, 103515
Perlmutter, S., Aldering, G., Goldhaber, G., Knop, R. A., *et al.* 1999, *ApJ* 517, 565
Planck Collaboration 2013 results XVI 2013, *arXiv 1303.5076v2*
Riess, A. G., Macri, L., Casertano, S.,Lampeitl, H., *et al.* 2011, *ApJ*, 730, 119
Schmidt, B., Suntzeff, N. B., Phillips, M. M., Schommer, R. A., *et al.* 1998, *ApJ*, 507, 46
Trümper, J. 1993, *Science*, 260, 1769
Voges, W., Aschenbach, B., Boller, T., Bräuninger, H., *et al.* 1999, *A&A*, 349, 389
Whitbourn, J. R. & Shanks, T. 2014, *MNRAS*, 437, 2146

The Zeldovich Universe:
Genesis and Growth of the Cosmic Web
Proceedings IAU Symposium No. 308, 2014
R. van de Weygaert, S. Shandarin, E. Saar & J. Einasto, eds.

© International Astronomical Union 2016
doi:10.1017/S1743921316010632

Distinguishing f(R) gravity with cosmic voids

P. Zivick[1]† and P. M. Sutter[1,2,3]

[1]Center for Cosmology and AstroParticle Physics, Ohio State University, Columbus, USA
[2]Sorbonne Universités, UPMC Univ Paris 06, UMR7095, F-75014, Paris, France
[3]CNRS, UMR7095, Institut d'Astrophysique de Paris, F-75014, Paris, France

Abstract. We use properties of void populations identified in N-body simulations to forecast the ability of upcoming galaxy surveys to differentiate models of f(R) gravity from ΛCDM cosmology. We analyze simulations designed to mimic the densities, volumes, and clustering statistics of upcoming surveys, using the public VIDE toolkit. We examine void abundances as a basic probe at redshifts 1.0 and 0.4. We find that stronger f(R) coupling strengths produce voids up to $\sim 20\%$ larger in radius, leading to a significant shift in the void number function. As an initial estimate of the constraining power of voids, we use this change in the number function to forecast a constraint on the coupling strength of $\Delta f_{R0} = 10^{-5}$.

Keywords. cosmology: simulations, cosmology: large-scale structure of universe

1. Introduction

Modifications of gravity provide one way to explain the observed expansion of the universe. One such proposed theory is the *f(R)* class of models, which contain relatively simple modifications to General Relativity (GR). This particular model incorporates the chameleon mechanism (Khoury & Weltman 2004) that screens the fifth force in high density regions while leaving it unscreened in low density regions, strengthening the force of gravity.

Studying these underdense regions, called voids, could provide a way to test *f(R)* gravity. Already voids have been used as a potential diagnostic for examining other models, such as coupled dark energy (Sutter *et al.* 2014). So far, current void-based studies of modified gravity (e.g., Li *et al.* 2012) have only focused on predictions for present-day conditions and ignored realistic survey effects. In this work, we mimic upcoming galaxy redshift surveys such as Euclid (Laureijs *et al.* 2011) and provide an initial estimate of the constraining power of void statistics.

2. Analysis and Results

We analyzed six simulation realizations from Zhao *et al.* (2011). Three models with differing values for structure formation in the universe, expressed by $|f_{R,0}|$ with values 10^{-4} (F4), 10^{-5} (F5), and 10^{-6} (F6), were examined in addition to general relativity (GR). Each simulation box contained 1024^3 dark matter particles and had a cubic volume of 1.5 h^{-1}Gpc per side. For analysis we selected snapshots at redshifts $z = 0.43$ and $z = 1.0$ and subsampled the DM particles to a mean density of $\bar{n} = 4 \times 10^{-3}$ per cubic h^{-1}Mpc. This choice of redshift, density, and volume is designed to represent a typical space-based galaxy survey such as Euclid. Finally, we perturbed particle positions according to their peculiar velocities. We chose to ignore the effects of galaxy bias, as Sutter *et al.* (2013) demonstrated that watershed void properties are relatively insensitive to bias.

Voids were identified using the publicly available Void Identification and Examination (VIDE) toolkit (Sutter *et al.* 2014), which uses a substantially modified version of ZOBOV (Neyrinck

† email: zivick.1@osu.edu

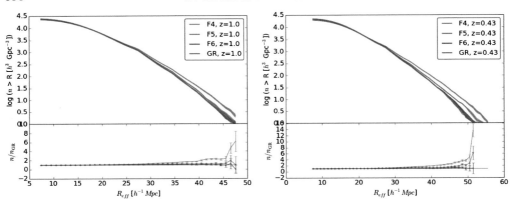

Figure 1. Cumulative void number functions. Shown are the abundances (top) and relative abundances (bottom) for ΛCDM (purple) and modified gravity models F4 (red), F5 (green), and F6 (blue) from realistically subsampled dark matter particle simulations plotted against the effective void radius. The solid lines are the mean number functions of the six realizations, and the shaded regions are the 1σ cosmic variances. Larger values of $|f_{R_0}|$ cause the fifth force to turn on at earlier ages, accelerating the evacuation of matter compared to ΛCDM.

2008). For this work, voids must be larger than the mean particle separation (in our case, $1\ h^{-1}$Mpc) and have central densities higher than 0.2 of the mean particle density $\bar{n}$.

Figure 1 shows the cumulative number function from ΛCDM and $f(R)$ simulations at redshifts $z = 1.0$ and $z = 0.43$. We can see that F4 clearly contains larger voids than in the ΛCDM simulation at both redshifts. With weaker coupling strengths, one will notice that at high redshift, F5 and F6 are not able to separate from GR. At lower redshift, the F5 model becomes distinguishable at roughly the three sigma level from GR at radii as small as $35\ h^{-1}$Mpc. Even the F6 model around $45\ h^{-1}$Mpc separates from GR, albeit by a relatively small amount. The gain in large voids is balanced by a loss of small voids, implying that the fifth force is accelerating the dissipation of interior void walls.

3. Conclusions

These features align with what one would reasonably expect to see from the $f(R)$ models. At higher redshift, the voids have not yet emptied out. Until the local densities pass a low enough threshold, the fifth force will remain screened, making the $f(R)$ models appear identical to GR. Simultaneously, the modified gravity mechanism only affects particle acceleration, and so the differences grow larger with time. Thus the strongest force, F4, produces the greatest number of large voids.

An initial Fisher forecast places the constraint on measuring $|f_{R0}|$ at roughly $\Delta f_{R0} = 10^{-5}$, indicating that for stronger fifth forces, a detection may well be possible, especially at lower redshift where there is more statistical power.

References

Khoury J. & Weltman A., 2004, *Phys. Rev. D*, 69
Laureijs R., Amiaux J., Arduini S., Auguères J. ., Brinchmann J., Cole R., Cropper M., Dabin C., Duvet L., & Ealet A., *et al.*, 2011, ArXiv:1110.3193
Li B., Zhao G.-B., & Koyama K., 2012, *Mon. Not. R. Astron. Soc.*, 421, 3481
Neyrinck M. C., 2008, *Mon. Not. R. Astron. Soc.*, 386, 2101
Sutter P. M., Carlesi E., Wandelt B. D., & Knebe A., 2014, ArXiv e-prints
Sutter P. M., *et al.*, 2014, ArXiv e-prints: 1406.1191
Sutter P. M., Lavaux G., Wandelt B. D. Hamaus N., Weinberg D. H., & Warren M. S., 2013, ArXiv e-prints: 1309.5087
Zhao B., Li B., & Koyama K., 2011, *Phys. Rev. D*, 83

CHAPTER 8B.

Void Galaxies

Yan-Chuan Cai
searching for Chameleons in Voids.

The Zeldovich Universe:
Genesis and Growth of the Cosmic Web
Proceedings IAU Symposium No. 308, 2014
R. van de Weygaert, S. Shandarin, E. Saar & J. Einasto, eds.

© International Astronomical Union 2016
doi:10.1017/S1743921316010644

The Void Galaxy Survey: Galaxy Evolution and Gas Accretion in Voids

Kathryn Kreckel[1], Jacqueline H. van Gorkom[3], Burcu Beygu[2], Rien van de Weygaert[2], J. M. van der Hulst[2], Miguel A. Aragon-Calvo[4] and Reynier F. Peletier[2]

[1]MPIA, Königstuhl 17, 69117 Heidelberg, Germany
email: kreckel@mpia.de

[2]Kapteyn Astronomical Institute, University of Groningen, PO Box 800, 9700 AV Groningen, The Netherlands

[3]Columbia University, MC 5246, 550 W120th St., New York, NY 10027, USA

[4]University of California, Riverside, CA 92521, USA

Abstract. Voids represent a unique environment for the study of galaxy evolution, as the lower density environment is expected to result in shorter merger histories and slower evolution of galaxies. This provides an ideal opportunity to test theories of galaxy formation and evolution. Imaging of the neutral hydrogen, central in both driving and regulating star formation, directly traces the gas reservoir and can reveal interactions and signs of cold gas accretion. For a new Void Galaxy Survey (VGS), we have carefully selected a sample of 59 galaxies that reside in the deepest underdensities of geometrically identified voids within the SDSS at distances of $\sim$100 Mpc, and pursued deep UV, optical, Hα, IR, and HI imaging to study in detail the morphology and kinematics of both the stellar and gaseous components. This sample allows us to not only examine the global statistical properties of void galaxies, but also to explore the details of the dynamical properties. We present an overview of the VGS, and highlight key results on the HI content and individually interesting systems. In general, we find that the void galaxies are gas rich, low luminosity, blue disk galaxies, with optical and HI properties that are not unusual for their luminosity and morphology. We see evidence of both ongoing assembly, through the gas dynamics between interacting systems, and significant gas accretion, seen in extended gas disks and kinematic misalignments. The VGS establishes a local reference sample to be used in future HI surveys (CHILES, DINGO, LADUMA) that will directly observe the HI evolution of void galaxies over cosmic time.

Keywords. large-scale structure of universe, galaxies: evolution, galaxies: ISM

1. Introduction

The large scale clustering of galaxies into structures such as filaments, walls and clusters results in a large volume of the universe with a low density of galaxies. These voids retain imprints from the cosmological forces which shape large scale structure, and can provide unique probes of both gravity and dark energy. While these regions have very low density, they are not empty. The population of void galaxies provides key constraints on our understanding of galaxy formation in a cosmological context, as well as a sample in which to study galaxy evolution in a simplified environment (see e.g. van de Weygaert & Platen 2011 for a review). Simulations only now have reached the level at which they predict not just the dark matter content of halos in a cosmological context, but the evolution of the baryonic components (stars, gas and dust) that are better suited for comparison with observations (Kreckel *et al.* 2011c).

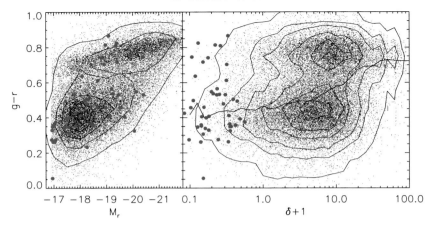

Figure 1. Comparison of global VGS galaxy properties (color, absolute magnitude, environmental density) with a volume limited sample drawn from SDSS. The void galaxies with redder colors are mostly edge-on and gas-rich disks, suggesting that the red colors are due to dust and not indicative of an early type morphology. At fixed luminosity there is no evidence for bluer colors, suggesting that the color-density relation may not extend into the voids.

The galaxies observed to live within voids are distinct from those populating average or overdense environments. There are very few of the most massive galaxies, most are instead lower mass, star-forming, disk galaxies (Grogin & Geller 1999, 2000; Rojas *et al.* 2004, 2005). While they present distinct properties as a population, at fixed stellar mass they are indistinguishable from galaxies living in the 'field' environment. This suggests that in many ways galaxy evolution is largely driven by secular processes that proceed independent of the surrounding large scale environment. This provides important insights into how galaxies evolve, and the role of external processes in driving galaxy evolution.

The lower density environment is expected to result in shorter merger histories and slower evolution of galaxies, allowing a study of how hierarchical merging and gas accretion both contribute to galaxy growth. In particular, cold accretion of gas in the form of filamentary flows is predicted to play an important role for galaxies in low mass halos (Kereš *et al.* 2005). This is expected to be most important for galaxies at high redshift, before the buildup of halos has happened, and in low density void environments at $z = 0$, where this buildup has not yet happened. As a result, void galaxies are the ideal targets for detailed study of the neutral gas in and around galaxies, in order to address general questions of how galaxies get their gas.

Dark matter simulations suggest that the voids should be threaded with low density filamentary substructure. While these may host a substantial number of galaxies, there is some expectation that the galaxies in voids may preferentially lie within these filamentary void substructures (Sheth & van de Weygaert 2004, Aragon-Calvo & Szalay 2013, Rieder *et al.* 2013). Tentative evidence for this has been found in the GAMA redshift survey (Alpaslan *et al.* 2014), which probes to a significantly lower galaxy stellar mass range than existing redshift surveys (e.g. SDSS, 6dF). However, the Local Void, which has been studied in great detail to very low galaxy luminosities, is remarkably empty. Thus, the question remains whether the predicted void substructures can be traced observationally.

We approach the study of void galaxies through the careful selection of 59 void galaxies, which we target with a full multi-wavelength series of observations. Here we present our galaxy sample, and summarize results from our study mapping the neutral gas within and

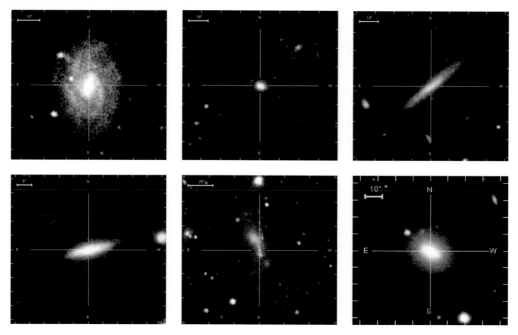

Figure 2. Sample VGS galaxies, all at the same physical scale, display a range of colors and morphologies (Kreckel *et al.* 2012).

around void galaxies. We look for evidence of ongoing gas accretion and void substructure, and find tantalizing clues within our sample.

2. The Void Galaxy Survey

We identified voids within the nearby (z<0.025) universe using topological techniques to identify the surrounding large scale structure (Kreckel *et al.* 2011a). We apply the Delaunay Tessellation Field Estimator (Schaap & van de Weygaert 2000; van de Weygaert & Schaap 2009) to the SDSS galaxy redshift survey to recover the underlying dark matter density field. From this we use the Cosmic Spine formalism (Aragón-Calvo *et al.* 2010) to identify walls and filaments, and the Watershed Voidfinder algorithm (Platen *et al.* 2007) to identify the void boundaries. From these, we selected 59† galaxies which resided in the centers of well defined voids. Comparing the g-r color and $\delta \equiv \rho/\rho_u - 1$ density contrast of our Void Galaxy Survey (VGS) targets to a sample of SDSS galaxies (Figure 1, right), we see the general color-density trend, that red galaxies prefer high density environments while blue galaxies prefer low densities, and find our void galaxies well sample this lowest density population (Kreckel *et al.* 2012).

We have selected our galaxy sample independent of their intrinsic properties (e.g. color, luminosity, morphology), and find that the VGS spans a range of colors and absolute magnitudes (Figure 1, left). The void galaxies with redder colors (g-r > 0.6) are almost all edge-on and gas-rich disks, suggesting that the red colors are due to dust and not indicative of an early type morphology. Upon close examination, only three of the 59 galaxies appear to have a true early type morphology. Figure 2 demonstrates the range

---

† The sample was originally designed to be 60 galaxies, however one was later found to have its redshift mis-identified, and instead falls outside the void.

of colors and morphologies present in the sample. The VGS does not contain any massive galaxies with absolute magnitudes brighter than M_r = -20.4 mag. At fixed luminosity, there is no evidence for bluer colors, suggesting that the color-density relation may not extend into the voids. This sample, however, is too small for a statistical study, and shows the need for careful controls of all factors that correlate with galaxy color.

We have compiled multi-wavelength observations for the full VGS, including GALEX NUV imaging to trace the recent (10-100 Myr) star formation, Hα narrow band imaging to trace the current (<10 Myr) star formation, deep B- and R-band optical imaging to examine the outer stellar disk, Spitzer 3.6 and 4.5 μm imaging of the old stellar component (Beygu(2014)), HI line emission from the extended gas disk and 1.4 GHz radio continuum emission from star formation and AGN. Altogether, these observations will enable us to study in detail the gas and stellar morphology and the star formation history for these galaxies to better understand their evolutionary history. In addition, for select galaxies we have also performed near-IR and WISE 22 μm observations, tracing the old stellar population, and millimeter wavelength CO line observations, tracing the cold molecular gas that is available to form stars.

This sample is ideal for not just understanding the effect of large-scale environment on shaping these galaxies, but also, given their relatively undisturbed evolution, is well suited to study the secular processes that contribute to galaxy evolution.

3. Void Galaxies as Cosmological Probes

We focus in these proceeding on results related to the neutral gas properties in the sample and what they tell us about role of gas accretion on galaxy evolution.

Our full HI survey of the VGS (Kreckel *et al.* 2012) revealed that the galaxies residing in voids typically have extended, gas-rich HI disks. About half exhibit disturbed gas morphologies or kinematics. However, we also find that these HI properties are typical for galaxies at the same stellar mass that are in average density environments. This suggests that while the large scale environment is playing a role in shaping the evolution of galaxies in voids, since they tend to be low mass systems, at fixed stellar mass the galaxy HI properties are largely independent of the environment. However, given the extremely rarified surroundings, we argue that the irregularities in the HI disks show convincing evidence for ongoing gas accretion in these systems that has proven difficult to isolate in general studies of gas in galaxies. In some cases, we see indications of gas accretion directly from out of the void.

3.1. VGS_12 - cold accretion of void gas?

The strongest example we have that presents evidence for accretion out of the void is VGS_12 (Figure 3, Stanonik *et al.* 2009). This small stellar system presents a blue, disk-like morphology that extends to about 5 kpc in diameter. The HI emission also presents a regular disk-like morphology and kinematics, however it is oriented perpendicular to the stellar component, and extends over nearly 30 kpc. This sort of polar disk configuration has been seen before in other galaxies, most notably NGC 4650A. A satellite accretion or interaction is typically understood to be required for the formation of such disjoint kinematic components in one galaxy. Another option is that the material is accreted directly as gas along filaments. In all previously known cases the systems show a polar component that contains stars. However, this is not the case for VGS_12, where we see no evidence for a stellar component in the polar disk through deep R-band imaging. Because of this, the isolated nature of this system, and the apparently undisturbed nature of the central stellar disk, we believe that this void galaxy presents strong evidence for ongoing

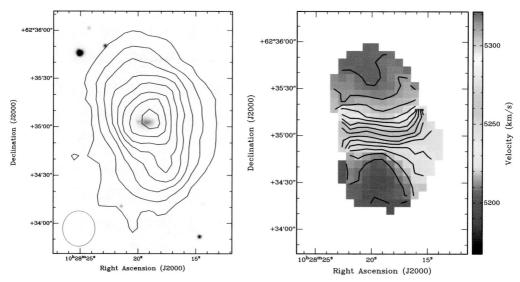

Figure 3. VGS_12, with HI emission shown as contours overlaid on the SDSS g-band image, and HI velocity field (Figure 1 from Stanonik *et al.* 2009, ©AAS. Reproduced with permission). It exhibits an extremely extended polar disk of HI gas that is devoid of stars. Given the pristine gas disk, isolated nature of this system, and apparently undisturbed nature of the central stellar disk, this void galaxy presents strong evidence for ongoing cold accretion of gas onto this system.

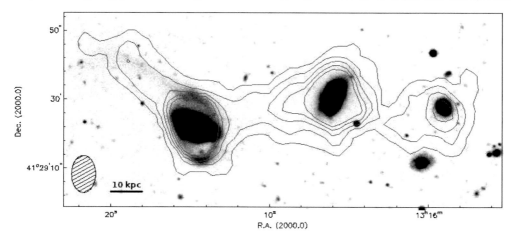

Figure 4. The VGS_31 system, with HI emission shown as contours overlaid on a deep B-band image (Figure 3 from Beygu *et al.* 2013, ©AAS. Reproduced with permission). It shows the possible formation of substructure within a void, as all three systems align linearly and are connected by a tenuous HI bridge.

cold accretion of gas onto this system. This confirms predictions from simulations that such a mode of galaxy growth plays an important role in low mass halos (Kereš *et al.* 2005), but which has proven difficult to isolate in observations.

3.2. *VGS_31 - tracing a void filament?*

Imaging the gas morphology is crucial as it reveals features essential to understanding the formation history of these galaxies. A second example of this is the VGS_31 system, which contains three galaxies that appear to be forming along a filament within the void

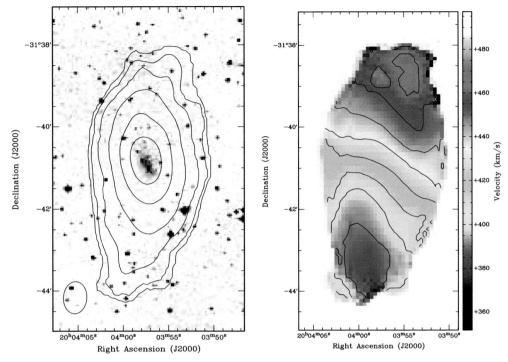

Figure 5. KK246, with HI emission shown as contours overlaid on the B-band Palomar Observatory Sky Atlas image, and HI velocity field (Figure 1 from Kreckel *et al.* 2011b, ©AAS. Reproduced with permission). It is the only galaxy confirmed to reside well within the Local Void, and exhibits an extremely extended and misaligned gas disk. This is suggestive of growth by accretion of gas from out of the void.

(Figure 4, Beygu *et al.* 2013). All three are linearly aligned and at nearly identical velocity, and they are connected by a tenuous HI bridge with relatively smooth gas kinematics connecting all three systems. The entire filamentary configuration extends nearly 100 kpc across the sky. Simulations designed to reproduce such a configuration show that it is likely this system is assembling within a (proto)filament and not just a group of interacting galaxies (Rieder *et al.* 2013). Further observational evidence for this is seen in the low metallicity of an outlying polar disk of material surrounding the eastern-most galaxy, suggesting again that this system presents evidence for cold accretion of gas (Spavone & Iodice 2013). The VGS_31 system presents a fascinating example of the possible formation of substructure within a void.

3.3. *KK246 - a dark void dwarf*

Void substructure is predicted to exist through the filamentary features seen threading the voids in dark matter simulations. However, one clear counter-example is the remarkably empty Local Void (Tully *et al.* 2008). Constraints on this void are particularly strong, since its proximity allows us to identify even very faint ($M_R \sim -10$) galaxies. HI observations of KK246, the only galaxy confirmed to reside well within this Local Void, show it also has an extremely extended and misaligned gas disk (Figure 5, Kreckel *et al.* 2011b). Detailed examination of the kinematics within the gas disk reveal tantalizing evidence for ongoing gas infall onto the system. More sensitive surveys for dwarf galaxies in voids will provide important constraints on the role void substructure plays in the evolution of these systems.

Given the intriguing evidence for ongoing gas accretion in these systems, we plan to use new optical IFU observations for 10 of the more massive VGS galaxies to connect this to possible differences in the growth of the galaxy disks. By examining the radial gradients in metallicity, stellar population and star formation history we will search for further evidence of recent pristine gas accretion. The VGS establishes a local reference sample to be used in current (CHILES, Fernández *et al.* 2013) and future HI surveys (DINGO, LADUMA) that will directly observe the HI evolution of void galaxies over cosmic time.

References

Alpaslan, M., Robotham, A. S. G., Obreschkow, D., *et al.* 2014, *MNRAS*, 440, L106

Aragón-Calvo, M. A., Platen, E., van de Weygaert, R., & Szalay, A. S. 2010, *ApJ*, 723, 364

Aragon-Calvo, M. A. & Szalay, A. S. 2013, *MNRAS*, 428, 3409

Beygu, B., Kreckel, K., van de Weygaert, R., van der Hulst, J. M., a & van Gorkom, J. H. 2013, *AJ*, 145, 120

Beygu, B. 2014, The void galaxy survey : a study of the loneliest galaxies in the universe (Ph.D. Thesis, University of Groningen)

Fernández, X., van Gorkom, J. H., Hess, K. M., *et al.* 2013, *ApJL*, 770, L29

Grogin, N. A. & Geller, M. J. 1999, *AJ*, 118, 2561

—. 2000, *AJ*, 119, 32

Kereš, D., Katz, N., Weinberg, D. H., & Davé, R. 2005, *MNRAS*, 363, 2

Kreckel, K., Platen, E., Aragón-Calvo, M. A., *et al.* 2011a, *AJ*, 141, 4

Kreckel, K., Peebles, P. J. E., van Gorkom, J. H., van de Weygaert, R., & van der Hulst, J. M. 2011b, *AJ*, 141, 204

Kreckel, K., Joung, M. R., & Cen, R. 2011c, *ApJ*, 735, 132

Kreckel, K., Platen, E., Aragón-Calvo, M. A., *et al.* 2012, *AJ*, 144, 16

Platen, E., van de Weygaert, R., & Jones, B. J. T. 2007, *MNRAS*, 380, 551

Rieder, S., van de Weygaert, R., Cautun, M., Beygu, B., & Portegies Zwart, S. 2013, *MNRAS*, 435, 222

Rojas, R. R., Vogeley, M. S., Hoyle, F., & Brinkmann, J. 2004, *ApJ*, 617, 50

—. 2005, *ApJ*, 624, 571

Schaap, W. E. & van de Weygaert, R. 2000, *A&A*, 363, L29

Sheth, R. K. & van de Weygaert, R. 2004, *MNRAS*, 350, 517

Spavone, M. & Iodice, E. 2013, *MNRAS*, 434, 3310

Stanonik, K., Platen, E., Aragón-Calvo, M. A., *et al.* 2009, *ApJL*, 696, L6

Tully, R. B., Shaya, E. J., Karachentsev, I. D., *et al.* 2008, *ApJ*, 676, 184

van de Weygaert, R. & Schaap, W. 2009, *Data Analysis in Cosmology*, 665, 291

van de Weygaert, R. & Platen, E. 2011, *International Journal of Modern Physics Conference Series*, 1, 41

The Zeldovich Universe:
Genesis and Growth of the Cosmic Web
Proceedings IAU Symposium No. 308, 2014
R. van de Weygaert, S. Shandarin, E. Saar & J. Einasto, eds.

© International Astronomical Union 2016
doi:10.1017/S1743921316010656

The Void Galaxy Survey:
Morphology and Star Formation Properties
of Void Galaxies

Burcu Beygu[1], Kathryn Kreckel[2], Thijs van der Hulst[1], Reynier Peletier[1], Tom Jarrett[3], Rien van de Weygaert[1], Jacqueline H. van Gorkom[4] and Miguel Aragón-Calvo[5]

[1] Kapteyn Astron. Inst., Univ. Groningen, PO Box 800, 9700 AV Groningen, the Netherlands
email: `beygu@astro.rug.nl`

[2] MPIA, Königstuhl 17, 69117 Heidelberg, Germany

[3] Dept. Astron., Univ. Cape Town, Private Bag X3, Rondebosch 7701, South Africa

[4] Columbia University, MC 5246, 550 W120th St., New York, NY 10027, USA

[5] University of California, Riverside, CA 92521, USA

Abstract. We present the structural and star formation properties of 59 void galaxies as part of the Void Galaxy Survey (VGS). Our aim is to study in detail the physical properties of these void galaxies and study the effect of the void environment on galaxy properties. We use Spitzer 3.6μm and B-band imaging to study the morphology and color of the VGS galaxies. For their star formation properties, we use Hα and GALEX near-UV imaging. We compare our results to a range of galaxies of different morphologies in higher density environments. We find that the VGS galaxies are in general disk dominated and star forming galaxies. Their star formation rates are, however, often less than 1 M$_\odot$ yr^{-1}. There are two early-type galaxies in our sample as well. In r$_e$ versus M$_B$ parameter space, VGS galaxies occupy the same space as dwarf irregulars and spirals.

1. Voids and void galaxies

Voids are a prominent aspect of the Cosmic Web (see van de Weygaert & Platen (2011) for a review). Surrounded by elongated filaments, sheetlike walls and dense compact clusters, they have formed out of primordial underdensities via an intricate hierarchical process of evolution (Sheth & van de Weygaert (2004), Aragón-Calvo *et al.* (2010), Aragón-Calvo & Szalay (2013), Rieder *et al.* (2013)). Their diluted substructure and population of galaxies remain as fossils of the earlier phases of this void hierarchy.

Within this context, the pristine environment of voids represents an ideal setting for the study of environmental influences on galaxy formation and evolution. Largely unaffected by the complexities and processes that modify galaxies in high-density environments, the characteristics of void galaxies are expected to provide information on the role of environment in galaxy evolution.

A few aspects stand out immediately. The void galaxies are in general gas rich and display a substantial star formation activity. They also have rather low stellar masses, which relates to the finding by theoretical studies that the mass function of galaxies and halos in voids has shifteed considerably towards lower masses (see e.g. Goldberg & Vogeley (2004), Aragón-Calvo (2007), Cautun *et al.* (2014)). What still remains to be understood is the remarkable low abundance of dwarf galaxies in void interiors (Peebles (2001)). Also, we may wonder in how far the gas accretion and outflow history of void

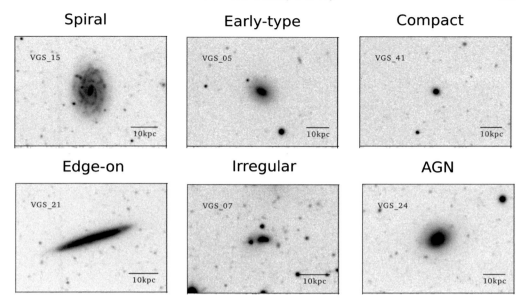

Figure 1. VGS galaxy morphologies. The images are B-band images of 6 VGS galaxies. In each image, the black bar represents a physical scale of 10 kpc.

galaxies has resulted in systematically different objects (e.g. Keres *et al.* (2005), Hoeft & Gottlöber (2010)).

In an attempt to obtain more insight into the properties of void galaxies, in this study we explore the morphology, structural characteristics and star formation properties of void galaxies in the Void Galaxy Survey (VGS).

2. The Void Galaxy Survey

In order to fully study the effect of void environment on galaxy evolution and formation, one needs an unbiased void galaxy sample found in a well defined void environment. The Void Galaxy Survey (VGS) (Kreckel *et al.* (2011), van de Weygaert *et al.* (2011), Kreckel *et al.* (2012), this volume) provides such a sample. This sample has been selected from the Sloan Digital Sky Survey Data Release 7 (SDSS DR7), using geometric and topological techniques for delineating voids in the galaxy distribution and identifying galaxies populating the central interior of these voids (Schaap & van de Weygaert (2000), Platen *et al.* (2007), Aragón-Calvo *et al.* (2010), Kreckel *et al.* (2011)). The typical size of voids in our sample is on the order of 5 to 10 h^{-1} Mpc in radius.

The resulting sample of void galaxies is unbiased and largely independent of intrinsic galaxy properties (except for the the spectroscopic flux limit of 17.7 mag in the r-filter of the SDSS). The VGS galaxies have redshifts in the range $0.02 < z < 0.03$. They have an absolute magnitude in the range of $-20.4 < M_r < -16.1$, colors in between $0.06 < g - r < 0.087$ and a stellar mass $M_* < 3 \times 10^{10} M_\odot$. The Void Galaxy Survey aims to probe the color, morphology, star formation and gas content of the void galaxies. For this we observed 59 VGS galaxies in the 21 cm HI line, in Hα and in the optical B-band. In addition, we acquired GALEX near-UV data, as well as Spitzer 3.6 μm and WISE 22μm imaging. For some VGS galaxies we obtained CO(1-0) observations, which will form the starting point for a study of the relation between their star formation activity and their molecular and atomic gas content.

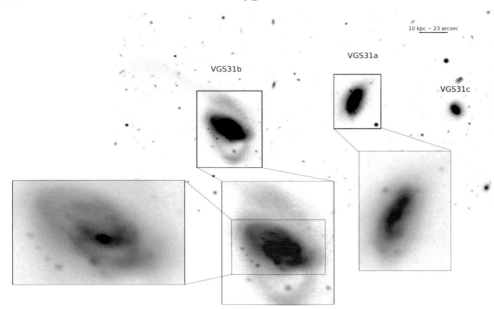

Figure 2. VGS_31. R-band negative image of this configuration of three void galaxies aligned along a tenuous filament inside the void. From left to right: *VGS_31b*: The most remarkable member of the system, a *Markarian* galaxy, has a tail and a ring. Close up images show the inner structures such as the bar. *VGS_31a*: A disk galaxy with a bar structure. *VGS_31c*: Smallest member of the system is optically undisturbed. The black bar on the top-right corner represents 10 kpc (~23″) (Beygu *et al.* (2013)).

So far we have completed the study of the H I properties of 55 VGS galaxies (see Kreckel *et al.* (2012) for details).

3. Morphology & Structural Parameters

The morphological classification of galaxies is rather complex and an accurate classification requires a more elaborate analysis than we are able to provide here. Therefore, instead of carrying out an absolute morphological classification, we seek to classify the morphology of the VGS galaxies in a general way, by eye. We find that the VGS galaxy sample mainly consists of disk galaxies with an occasional bar and spiral structure and sometimes small compact objects, irregulars and two early-types (one is an AGN). In figure 1 we display examples of the different morphological types of galaxies that we find in the VGS sample. Also, we find various peculiar galaxies, such as the dynamically distorted Markarian galaxy VGS_31b in the filamentary VGS31 constellation (figure 2, see Beygu *et al.* (2013)).

3.1. *Structural parameters*

The structural analysis of the VGS galaxies involves the fitting of Sérsic profiles to the light distributions and the determination of the characteristic size (r_e) and surface brightness (μ_e) of the galaxies, along with their total luminosity. The concentration of (stellar) light is quantified by the Sérsic index n. We have compared the determined structural parameter values for the VGS galaxies with the values of the same parameters for a wide range of different galaxies. Amongst these are giant early-type galaxies, dEs and late-type disk galaxies. Figure 3a shows a sketch of where different types of galaxies are

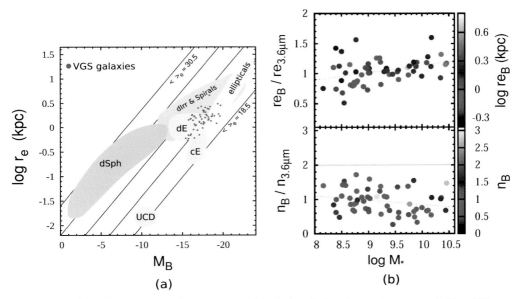

Figure 3. VGS galaxy structural parameters. (a): A sketch showing regions occupied by different types of galaxies and the VGS galaxies in the parameter space of the half-light radii r_e and B-band absolute magnitude M_B adopted from Mo *et al.* (2010). VGS galaxies are shown as red dots. (b): The ratios of the r_e (top) and Sérsic indices (bottom) of the B-band and 3.6μm as a function of stellar mass, $r_{e(B)}$ and Sérsic index n_B.

located in terms of their half-light radii and B-band absolute magnitudes. VGS galaxies (red dots), dIrr and spirals occupy the same space in r_e versus M_B.

Figure 3b shows the $r_{e,B}/r_{e,3.6}$ (top panel) and the $n_B/n_{3.6}$ ratios (lower panel) as a function of stellar mass, $r_{e,B}$ and Sérsic index n_B. The majority of the VGS galaxies have Sérsic indices $n < 2$ in both bands. This confirms that they are disk dominated.

In figure 3b, we see that $r_{e,B}/r_{e,3.6}$ increases as a function of increasing stellar mass and as as well as a function of increasing $r_{e,B}$. In other words, the smaller galaxies have a more concentrated light distribution at 3.6μm than in B. One explanation may be that in smaller galaxies the star formation activity is more concentrated towards the center than in the larger objects. Part of the effect is also caused by extinction. Larger galaxies generally contain more dust in their central regions. Dust affects the light in the B-band more than at 3.6μm. This translates into a larger $(r_e)_B$ than $(r_e)_{3.6}$.

4. Star formation properties

From Hα and near-UV imaging we may conclude that VGS galaxies are galaxies with a substantial star forming activity. Nonetheless, most of them appear to have star formation rates less than 1 M$_\odot$ yr^{-1}. The one exception with a considerably elevated star formation rate is the VGS_31 (Beygu (2014)) system.

We have compared the specific star formation (SFR$_\alpha$/M$_*$) and star formation efficiencies - i.e. SFR$_\alpha$/M$_{HI}$ - of the VGS galaxies to those of galaxies in average density regions (see Beygu (2014) for details). The latter belong to a sample of galaxies that consists of a combination of galaxies defined by three studies (Gavazzi *et al.* (2012), Sánchez-Gallego *et al.* (2012) and Karachentsev *et al.* (2013)). These cover the same stellar mass range, as well as other properties, as the VGS galaxies.

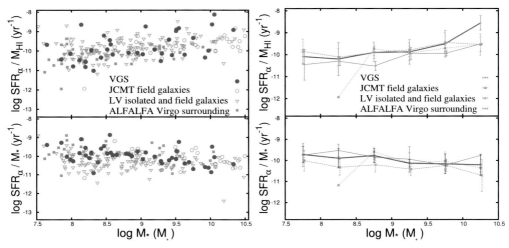

Figure 4. Star formation properties of VGS galaxies. Top left: SFR_α/M_{HI} as a function of the stellar masss M_*. VGS galaxies are indicated by red dots, the comparison sample galaxies are indicated by faint symbols. Bottom left: SFR_α/M_* as a function of stellar mass M_* for the sample sample of galaxies. Right panels: the average of the specific star formation parameters plotted in the corresponding lefthand panels. Note that there does not appear to be a significant difference between the VGS and the comparison sample galaxies.

The left panel of figure 4 reveals a similar dependence of star formation efficiency and specific star formation rate on stellar mass for the VGS galaxies as for the galaxies from the control samples. As a function of stellar mass, the VGS and control sample galaxies show the same weak trends. In other words, we do not seem to detect a significant difference between the VGS galaxies and the control sample galaxies.

5. Conclusion

The voids in our VGS sample do not appear to be populated by a type of galaxy specific for void environments. Voids mainly contain late-type galaxies of different morphologies. Only two VGS galaxies have an early-type morphology. While the void environment expresses itself in the low stellar mass and size of the galaxies (see e.g. Aragón-Calvo (2007), Cautun *et al.* (2014)), there is not evidence for star formation activity that deviates significantly from their peers with a similar mass in the higher denstiy filamentary and cluster-like environments of the cosmic web. It forms an indication for star formation to be a mainly self regulated process, not strongly influenced by the large scale environment. To understand this better, and to answer questions in how far the environment plays a role in initiating star formation, better theoretical understanding of the processes involved will be needed (see e.g. Aragón-Calvo *et al.* (2014).

References

Aragón-Calvo, M. A. 2007 Morphology and Dynamics of the Cosmic Web, PhD thesis, Univ. Groningen

Aragón-Calvo, M. A, van de Weygaert, R., Araya-Melo, P., Platen, E., & Szalay, A. S. 2010 *MNRAS*, Vol. 404, p. 89

&Aragón-Calvo, M. A. and Platen, E., van de Weygaert, R., & Szalay, A. S. 2010, *ApJ*, 723, 364

Aragón-Calvo, M. A &, Szalay, A. S. 2013 *MNRAS*, 428, p. 3409

Aragón-Calvo, M. A., Neyrinck M., & Silk J. 2014, *MNRAS*, subm, arxiv:1412.1119

&Beygu, B. and Kreckel, K. and van de Weygaert, R. van der Hulst, J. M., & van Gorkom, J. H. 2013, *AJ*, 145, 120

Beygu, B. 2014, The void galaxy survey: a study of the loneliest galaxies in the universe (Ph.D. Thesis, University of Groningen)

Cautun, M., van de Weygaert, R. , Jones, B. J.T., & Frenk, C. S. 2014 *MNRAS*, Vol. 441, Issue 4, p. 2923

&Gavazzi, G. and Fumagalli, M. and Galardo, V. and Grossetti, F., Boselli, A., Giovanelli, R., Haynes, M. P., & Fabello, S. 2012, *A&A*, 545, A16

Goldberg, D. M. & Vogeley, M. 2004 *ApJ*, 605, p. 1

Hoeft, M. & Gottlöber, S. *AdvAst*, 2010, 87

Hoyle, F., Rojas, R. R., Vogeley, M. S., & Brinkmann, J. 2005, *ApJ*, 620, 618

Karachentsev, I. D., Makarov, D. I., & Kaisina, E. I. 2013, *AJ*, 145, 101

Keres, D., Katz, N., Weinberg, D. H, & Davé, R. 2005 *MNRAS*, Vol. 363, p. 2

Kreckel, K., Platen, E., Aragón-Calvo, M. A., van Gorkom, J. H., van de Weygaert, R., van der Hulst, J. M., Kovač, K., Yip, C.-W., & Peebles, P. J. E. 2011, *AJ*, 141, 4

Kreckel, K., Platen, E., Aragón-Calvo, M. A., van Gorkom, J. H., van de Weygaert, R., van der Hulst, J. M., & Beygu, B. 2012, *AJ*, 144, 16

Lee, J. C., Gil de Paz, A., Tremonti, C., Kennicutt, Jr., R. C., Salim, S., Bothwell, M., Calzetti, D., Dalcanton, J., Dale, D., Engelbracht, C., Funes, S. J. J. G., Johnson, B., Sakai, S., Skillman, E., van Zee, L., Walter, F., & Weisz, D. 2009, *ApJ*, 706, 599

Mo, H., van den Bosch, F. C., & White, S. 2010, *Galaxy Formation and Evolution*

Peebles, P. J. E. 2001 *ApJ*, 557, 495

Platen, E., van de Weygaert, R., & Jones, B. J. T. 2007, *MNRAS*, 380, 551

Rieder, S., van de Weygaert, R., Cautun, M., Beygu, B., & Portegies Zwart, S. 2013 *MNRAS*, 435, p. 222

Sánchez-Gallego, J. R., Knapen, J. H., Wilson, C. D., Barmby, P., Azimlu, M., & Courteau, S. 2012, *MNRAS*, 422, 3208

Schaap, W. E. & van de Weygaert, R. 2000, *A&A*, 363, L29

Sheth, R. & van de Weygaert, R. 2004 *MNRAS*, 350, p. 517

Stanonik, K., Platen, E., Aragón-Calvo, M. A., van Gorkom, J. H., van de Weygaert, R., van der Hulst, J. M., & Peebles, P. J. E. 2009, *ApJ*, 696, L6

van de Weygaert, R., Platen, E. 2011 *IJMPS*, 1, p. 41

van de Weygaert, R., Kreckel, K., Platen, E., Beygu, B., van Gorkom, J. H, van der Hulst, J. M, Aragón-Calvo, M. A, Peebles, P.J.E., Jarrett, T., Rhee, G., & Kovac, K. 2011 in Environment and the Formation of Galaxies: 30 years later, ASSP, Springer, p. 17

The Zeldovich Universe:
Genesis and Growth of the Cosmic Web
Proceedings IAU Symposium No. 308, 2014
R. van de Weygaert, S. Shandarin, E. Saar & J. Einasto, eds.
ⓒ International Astronomical Union 2016
doi:10.1017/S1743921316010668

Observations of dwarfs in nearby voids: implications for galaxy formation and evolution

Simon A. Pustilnik

Special Astrophysical Observatory of Russian Academy of Sciences,
369167, Nizhnij Arkhyz, Russia
email: sap@sao.ru

Abstract. The intermediate results of the ongoing study of deep samples of $\sim$200 galaxies residing in nearby voids, are presented. Their properties are probed via optical spectroscopy, *ugri* surface photometry, and HI 21-cm line measurements, with emphasis on their evolutionary status. We derive directly the hydrogen mass M(HI), the ratio $M(HI)/L_{\rm B}$ and the evolutionary parameter gas-phase O/H. Their luminosities and integrated colours are used to derive stellar mass M_* and the second evolutionary parameter – gas mass-fraction ($f_{\rm g}$). The colours of the outer parts, typically representative of the galaxy oldest stellar population, are used to estimate the upper limits on time since the beginning of the main SF episode. We compare properties of void galaxies with those of the similar late-type galaxies in denser environments. Most of void galaxies show smaller O/H for their luminosity, in average by $\sim$30%, indicating slower evolution. Besides, the fraction of $\sim$10% of the whole void sample or $\sim$30% of the least luminous void LSB dwarfs show the oxygen deficiency by a factor of 2–5. The majority of this group appear very gas-rich, with $f_{\rm g} \sim$(95–99)%, while their outer parts appear rather blue, indicating the time of onset of the main star-formation episode of less than 1–4 Gyr. Such unevolved LSBD galaxies appear not rare among the smallest void objects, but turned out practically missed to date due to the strong observational selection effects. Our results evidense for unusual evolutionary properties of the sizable fraction of void galaxies, and thus, pose the task of better modelling of dwarf galaxy formation and evolution in voids.

Keywords. galaxies: evolution, galaxies: formation, galaxies: dwarf, galaxies: abundances, galaxies: photometry, galaxies: statistics, large-scale structure of universe, radio lines: galaxies

1. Introduction and Overview

Studies of galaxies in low-density environments were tempting in the hope to probe the basics of galaxy evolution in isolation. The modern concepts suggest however that even the most isolated objects are related to and influenced by the baryon flows of adjacent filaments. Observational studies of galaxy samples in voids, based on large deep surveys (SDSS, 2dRGS), were mostly limited by large distant voids ($D \sim$ 100–200 Mpc). As a consequence, they probed only the upper part of void galaxy luminosity function ($M_{\rm r} < -16$). Only subtle or at most moderate differences with wall galaxies were found on their SFR and colours.

The complementary approach to study tens-hundred less luminous galaxies in nearby voids was suggested in Pustilnik & Tepliakova (2011). The first void intrinisically faint galaxy sample (down to $M_{\rm B} = -11$) was drawn up in the nearby Lynx-Cancer void ($D_{\rm centre} = 18$ Mpc), currently including over 100 objects. Several interesting findings on void galaxy evolution (see below) evidense for importance of this direction and emphasize the need of larger statistics and detailed studies of the least luminous void galaxies.

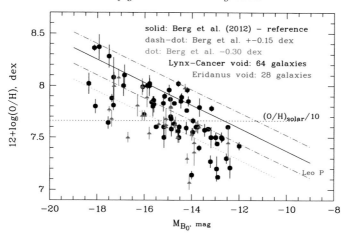

Figure 1. O/H (with error bars) vs M_B relation for 92 galaxies in the Lynx-Cancer (filled octogons) and Eridanus (filled triangles) voids in respect of similar galaxies in denser environment (Berg *et al.* 2012, "reference"). The significant systematic O/H drop in void galaxies is evident as well as the sizable fraction of strong outliers (deficiency of O/H by factor of 2–5) (Pustilnik *et al.* 2011, 2014; Kniazev *et al.* 2014, in prep.)

2. Ongoing project and intermediate results

To significantly encrease the number of faint void galaxies and conduct more reliable statistical study of their properties, we work on the revision of the sample of nearby voids and the sample of galaxies residing in them. In particular, galaxies residing in Monoceros and Cetus voids and in the equatorial part of Eridanus void (Fairall, 1998) are added to the current sample of the Lynx-Cancer void. Some of the important results on the evolutionary status of void galaxies are illustrated below.

In Fig. 1 we summarise determinations of gas-phase O/H in 92 galaxies residing in the Lynx-Cancer (filled octogons) and Eridanus (filled triangles) voids (Pustilnik, Tepliakova & Kniazev, 2011, Pustilnik *et al.* in prep., Kniazev *et al.* in prep.). In the large fraction of studied spectra no [OIII]λ4363 line was detected, and hence semi-empirical method by Izotov & Thuan (2007) was used to determine O/H. These O/H are shown vs absolute blue magnitudes M_B. For comparison we use the sample of similar galaxies from the Local Volume for which the confident O/H, distances and M_B are known (Berg *et al.*, 2012). Their linear regression of 12+log(O/H) and M_B is shown by solid line, while $\pm 1\sigma$ rms scatter in O/H (0.15 dex) are shown by dash-dotted lines. The substantial shift of the whole void O/H data below the 'reference' Berg *et al.* relation is well seen. Moreover, the sizable fraction of void galaxies shows the O/H deficiency of more than by factor of two (up to five).

Another important result comes from mass photometric study of Lynx-Cancer void galaxies based on the SDSS database (Abazajian *et al.* 2009). In particular, we determined *ugri* colours of outer parts for 85 void galaxies and compared them with PEGASE2 (Fioc & Rocca-Volmerange, 1999)) evolutionary tracks for two extreme SF laws: instantaneous and continuous with constant SFR (Fig. 2). While the great majority of void galaxies started their main star formation 7-14 Gyr ago, for about 15% of the sample we have clear indication of the retarded main star formation, commenced 1–5 Gyr ago.

The third direction in study of void galaxies is related to their HI content and structure. Integrated HI data on 96 Lynx-Cancer void galaxies (Pustilnik, Martin, in prep.) indicate that void objects are in average gas-rich, with median $M(\mathrm{HI})/L_B$=1.2, $\sim$40% higher than

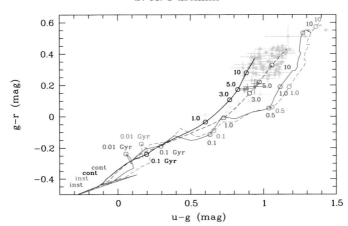

Figure 2. Age indicators: *ugr* colours of outer parts of 85 Lynx-Cancer void galaxies super-imposed on PEGASE2 evolutionary tracks. ~15% show retarded main Star Formation, started only 1–5 Gyr ago (see Perepelitsyna *et al.*, 2014)). Solid lines correspond to Salpeter IMF, while dashed lines - to Kroupa *et al.* IMF.

for the sample of similar galaxies in denser environment. Mapping of their HI with Giant Meterwave Radio Telescope leads to discovery of extremely gas-rich LSB dwarfs with $M(\mathrm{HI})/L_{\mathrm{B}}$=10 and 25, and $M_{\mathrm{gas}}/M_{\mathrm{bary}}$ >0.99 (Chengalur & Pustilnik, 2013; see Fig. 3, left). Such 'unevolved' galaxies are found mostly among the least luminous void galaxies. They can be not rare among void objects with $M_{\mathrm{B}} \gtrsim -11$, but due to severe observational selection effects they escape appearance in common wide-angle spectral surveys. Another interesting result of HI-mapping of three the most isolated void LSBD galaxies shows their disturbed morphology (Fig. 3, right). Probably here we see the most clear cases of cold accretion along void filaments (Chengalur *et al.* in prep.).

3. Implications

Summarising all above and some published findings on properties of galaxies residing in nearby voids, we notice the following implications:

• void galaxies in average show slower chemical evolution, having O/H in average 30%–40% lower than similar galaxies in denser environment; ~10% of void galaxies have O/H lower by 2–5 times, indicating their unusual evolutinory status.

• *ugri* colours of outer parts for ~15% void galaxies indicate the main SF episode started ~1–5 Gyr ago.

• More than a half of void galaxies are gas-rich, with $M(\mathrm{HI})/L_{\mathrm{B}}$ >1. Extremely gas-rich dwarfs, with $M(\mathrm{HI})/L_{\mathrm{B}} = 4 - 25$ already found in voids can be not rare among least luminous dwarfs ($M_{\mathrm{B}} > -11$).

• All together these results imply that evolution of void galaxies in average goes substantially more slowly. In addition, there are indications on that ~10% void galaxies formed with significant delay. This fraction reaches ~30% if we consider the least luminous void LSB dwarfs.

Acknowledgements. SAP is grateful to A. Kniazev, J.-M. Martin, J. Chengalur, A. Tepliakova, Y. Perepelitsyna and E. Safonova for fruitful collaboration and their contribution to the discussed topics. The author acknowledges the partial support of this work through RFBR grant 14-02-00520 and IAU travel grant.

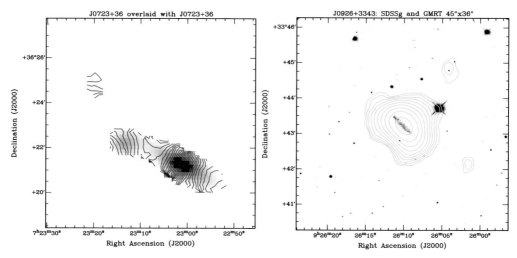

Figure 3. Left panel: Extremely gas-rich dwarf triplet J0723+36 near the centre of Lynx–Cancer void, at $D = 16$ Mpc, with $M(\mathrm{HI})/L_\mathrm{B}$ of $\sim$3, 10 and 25. The two more massive members appear to experience a minor merger, while the least massive and most gas-rich dwarf at the NE is still well separated. HI column density is shown in grey scale, while contures show isovelocity lines with step of 6 km s^{-1}. This finding can be a hint to possible hidden void population of very gas-rich low mass galaxies (Chengalur & Pustilnik, 2013). **Right panel:** Disturbed HI morphology in isolated void LSBD galaxies: the 2-nd most metal-poor LSBD J0926+3343 (Chengalur *et al.*, in prep.). HI column density (in contures) is superimposed on *g*-band SDSS image. Evidence for cold accretion?

References

Abazajian K. N., Adelman-McCarthy J. K., Agüeros M. A. *et al.*, 2009, *ApJS*, 182, 543

Berg D. A., Skillman E. D., Marble A. R., *et al.* 2012, *ApJ*, 754, 98

Chengalur J. N. & Pustilnik S. A. 2013, *MNRAS* 428, 1579

Fairall A., 1998, *Large-Scale Structures in the Universe, Wiley-Praxis, 196 pp.*

Fioc M. & Rocca-Volmerange B. 1999, *arXiv:astro-ph/9912179*

Izotov Y. I. & Thuan T. X. 2007, *ApJ* 665, 1115

Kreckel K., Platen E., Aragon-Calvo M. A., *et al.* 2012, *AJ* 144, 16

Perepelitsyna Y. A., Pustilnik S. A., & Kniazev A. Y. 2014, *Astroph. Bull.* 69, 247 (arXiv:1408.0613)

Pustilnik S. A. & Tepliakova A. L. 2011, *MNRAS* 415, 1188

Pustilnik S. A., Tepliakova A. L., Kniazev A. Y., *et al.* 2010, *MNRAS* 401, 333

Pustilnik S. A., Tepliakova A. L., & Kniazev A. Y. 2011, *Astroph. Bull.* 66, 255 (arXiv:1108.4850)

Pustilnik S. A., Tepliakova A. L., & Kniazev A. Y. 2011, *MNRAS* 417, 1335

Pustilnik S. A., Martin J.-M., Lyamina Y. A., & Kniazev A. Y. 2013, *MNRAS* 428, 1579

The Zeldovich Universe:
Genesis and Growth of the Cosmic Web
Proceedings IAU Symposium No. 308, 2014
R. van de Weygaert, S. Shandarin, E. Saar & J. Einasto, eds.

© International Astronomical Union 2016
doi:10.1017/S174392131601067X

Molecular Gas and Star Formation in Void Galaxies

M. Das[1], T. Saito[2], D. Iono[2], M. Honey[1] and S. Ramya[3]

[1]Indian Institute of Astrophysics, Banaglore, India
email: mousumi@iiap.res.in

[2]Department of Astronomy, The University of Tokyo, Tokyo 113-0033

[3]Shanghai Astronomical Observatory, Shanghai

Abstract. We present the detection of molecular gas using CO(1–0) line emission and followup Hα imaging observations of galaxies located in nearby voids. The CO(1–0) observations were done using the 45m telescope of the Nobeyama Radio Observatory (NRO) and the optical observations were done using the Himalayan Chandra Telescope (HCT). Although void galaxies lie in the most underdense parts of our universe, a significant fraction of them are gas rich, spiral galaxies that show signatures of ongoing star formation. Not much is known about their cold gas content or star formation properties. In this study we searched for molecular gas in five void galaxies using the NRO. The galaxies were selected based on their relatively higher IRAS fluxes or Hα line luminosities. CO(1–0) emission was detected in four galaxies and the derived molecular gas masses lie between $(1-8) \times 10^9$ $M_\odot$. The Hα imaging observations of three galaxies detected in CO emission indicates ongoing star formation and the derived star formation rates vary between from $0.2 - 1.0$ $M_\odot$ yr^{-1}, which is similar to that observed in local galaxies. Our study shows that although void galaxies reside in underdense regions, their disks may contain molecular gas and have star formation rates similar to galaxies in denser environments.

Keywords. ISM: molecules, galaxies: evolution, galaxies: ISM, cosmology: large-scale structure of universe.

1. Introduction

Voids contain a sparse but significant population of galaxies that are usually small, gas rich, late type galaxies (kreckel etal. 2012). The smaller voids are dominated by low surface brightness (LSB) dwarfs and irregular galaxies (karachentsev etal. 1999) but the larger voids also have a population of relatively bright galaxies that are often blue in color. These galaxies have ongoing star formation and are often interacting with companion galaxies in pairs or small groups along filaments in the voids (Beygu *et al.* 2013). Many questions remain regarding star formation in void environments; what is its nature - is it sporadic or continuous, what drives it and how is the star formation related to the location of the galaxies with respect to the filaments, walls and void interiors ?

One of the key elements for supporting star formation in galaxies is the presence of molecular hydrogen (H_2) gas. Although neutral hydrogen has been both detected and mapped in several voids, not much is known about the distribution of H_2 gas in void galaxies. There have been two studies that have detected CO emission and estimated molecular gas masses in a total of five void galaxies (Sage *et al.* 1997; Beygu *et al.* 2013). These results indicate that the H_2 gas masses in void galaxies are comparable to those found in nearby star forming systems. In this study we searched for molecular gas in void galaxies to obtain a larger sample of such H_2 rich galaxies and carried out followup Hα imaging observations of some of the detected galaxies. Our main motivation was to understand how the cold gas masses relate to the star formation properties of these

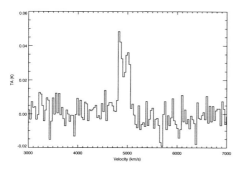 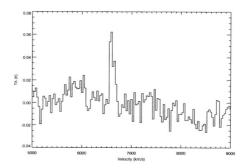

Figure 1. (a) Figure on the left shows the CO(1–0) line emission detected from the galaxy SBS 1325+597 that lies in the void Ursa Minor I. The line has a distinctive double horned profile indicative of a rotating disk and the peak separation is approximately 200 km s^{-1}. (b) Figure on the right shows the CO(1–0) line emission detected from the void galaxy SDSS 153821.22+331105.1. The gas is concentrated in the center of the galaxy and has a line width of ~100 km s^{-1}.

systems. In the following sections we present our observations, results and discuss their implications. For all distances we have used $H_0 = 73\ km\ s^{-1}\ Mpc^{-1}$ and $\Omega = 0.27$.

2. Sample galaxies and observations

Our initial sample comprised of 12 galaxies that were selected based of their relatively high infra-red fluxes or high star formation rates (kreckel et a. 2012; Cruzen *et al.* 2002; Szomoru *et al.* 1996). However due to weather conditions we were able to finally observe only five galaxies from this sample and they are listed in Table 1. Three galaxies have been observed in HI by Kreckel *et al.* (2012) in the Void Galaxy Survey (VGS) (SBS 1325+597, SDSS 143052.33+551440.0, SDSS 153821.22+331105.1) and the remaining two galaxies have been observed in an earlier HI survey of the Bootes void by Szomoru *et al.* (1996). All the galaxies have SDSS data.

The $^{12}CO(J = 1-0)$ emission observations were carried out using the 45 m Nobeyama Radio Telescope during 14 - 25 April, 2013. At the CO rest frequency of 115.271204 GHz, the half-power beam width (HPBW) was 15″ and the main beam efficieny was about 30%. The on source time for the first four galaxies varied between 1 to 1.5 hours; due to poor weather conditions SBS 1428+529 was observed for only 25 minutes. We used the one beam (TZ1), dual polarization, double sideband receiver (TZ) and the digital FX-type spectrometer SAM45, that has a bandwidth of 4 GHz (Nakajima *et al.* 2008). Typical system temperatures were 160 - 260 K. The pointing accuracy was about 2″-4″. Only data with a wind velocity less than 5 km s^{-1} were used and data with winding baselines were flagged. The data was analysed using NRO calibration tool NEWSTAR.

The Hα observations were done using the Himalayan Faint Object Spectrograph Camera (HFOSC) which is mounted on the 2m Himalayan Chandra Telescope (HCT) and were carried out on 2014 April 11 & 25. For SBS 1325+597 the redshift is 0.0165, so the Hα filter (band width ~500 Å) was used to get the Hα line emission. For SDSS 143052.33+551440.0 we used the narrow Hα filter (band width ~100 Å). The galaxy SDSS J153821.22+331105.1 is at a redshift of $z = 0.022$ and the Hα line is shifted to 6714 Å. Hence we used the narrow band [SII] filter (band width ~100 Å and centered around 6724 Å) for this galaxy. To obtain the continuum subtracted Hα images we also obtained broad band images with the R filter centered around the Hα line. The bias frames and twilight flats were used for preprocessing of the images. The data reduction

Table 1. Observed galaxies, molecular gas masses and star formation rates

Galaxy	D_L Mpc	Redshift	CO flux (K km s^{-1})	Molecular Gas Mass(10^9 $M_\odot$)	Hα Flux (10^{-13} ergs s^{-1}cm^{-2})	SFR ($M_\odot yr^{-1}$)
SBS 1325+597	70.4	0.0165	10.7±0.2	1.5±0.03	0.4	0.20
SDSS 143052.33+551440.0	76.6	0.0176	7.0±0.2	1.1±0.03	1.2	0.60
SDSS 153821.22+331105.1	97.6	0.0220	6.4±0.2	1.7±0.05	1.2	1.02
CG 598	248.0	0.0575	5.2±0.1	8.5±0.10	....	.....
SBS 1428+529	191.0	0.0445	< 0.6	< 0.6	....	.....

Notes:
[1] For SBS 1428+529 there was no CO(1–0) detection and no Hα image could be obtained. Upper limits for the molecular gas mass were obtained from the noise which was 0.0024 k and assuming a typical linewidth of 250 km/s.
[2] For CG 598 no Hα image could be obtained.

was done using the standard packages available in IRAF†. The images were corrected for cosmic rays, aligned and corrected for point spread function variations. Flux calibration was done using the spectrophotometric standard star HZ44. The Hα fluxes are listed in Table 1.

3. Results

1. Molecular gas detection : We have detected $^{12}CO(J=1-0)$ emission from four of the five sample galaxies that we observed (Table 1). The non-detection in SBS 1428+529 could partly be due to the short duration of the scan, which was limited by bad weather. Of the four detections, SBS 1325+597 has the most striking line profile; it has a double horned structure indicating a rotating disk of molecular gas (Figure 1a). The velocity separation of the peaks is ~ 200 kms^{-1}; assuming a disk inclination of 59.3° the disk rotation is 116 km s^{-1}. This is similar to the HI rotation speed from Kreckel *et al.* (2012). In SDSS 153821.22+331105.1 the gas is centrally peaked (Figure 1b); probably driven into the center by the bar in the galaxy. In the other two galaxies (SDSS 143052.33+551440.0 and CG 598) the CO line profile is slightly off center from the systemic velocities of the galaxies, which suggests that their gas disks are disturbed, possibly due to interaction with a companion galaxy (Das *et al.* 2014, in preparation).

2. Molecular masses : The CO fluxes in K km s^{-1} were converted to Jy km/s using a conversion factor (Jy/K) of 2.4. The CO line luminosity was determined using the relation $L_{CO} = 3.25 \times 10^7 (S_{CO}\Delta V/Jykms^{-1})(D_L/Mpc)^2(\nu_{res})^{-2}(1+z)^{-1}$ and the molecular gas mass was estimated using the relation $M(H_2) = [4.8 L_{CO}(K kms^{-1})]$ (Solomon & van den Bout 2005). The molecular gas masses lie in the range $(1-8) \times 10^9$ $M_\odot$ which is comparable to that observed from bright galaxies in denser environments.

3. Comparison with previous detections and HI masses : Our molecular gas are similar to that obtained for earlier studies of void galaxies by (Sage *et al.* 1997, Beygu *et al.* 2013) that lie in the range $10^8 - 10^9$ M$_\odot$. Although our study and previous detections indicate relatively large H_2 gas masses and a high detection rate, it must be remembered that the sample was biased towards star forming galaxies and those with high FIR fluxes. The molecular gas masses are comparable to the HI masses of these galaxies (Kreckel *et al.* 2012; Szomoru *et al.* 1996).

4. Hα fluxes and star formation rates (SFR) : The SFRs were calculated from the Hα fluxes using the kennicutt formula SFR = L(Hα)/1.26×10^{41} ergs s^{-1} (Table 1).

† Image Reduction & Analysis Facility Software distributed by National Optical Astronomy Observatories, which are operated by the Association of Universities for Research in Astronomy, Inc., under co-operative agreement with the National Science Foundation.

In SBS 1325+597 the Hα is distributed on either side of the galaxy nucleus, possibly in a ring. The distrbution matches the CO line profile (Figure 1a) which indicates a ring like configuration for the molecular gas. In SDSS 143052.33+551440.0 the Hα emission is concentrated about the nucleus. In SDSS 153821.22+331105.1 it is concentrated along the bar. The SFR is highest for SDSS 153821.22+331105.1; it is probably triggered by gas flowing along the bar in the galaxy although the emission is surprisingly faint in the disk.

5. Are these galaxies interacting? : SBS 1325+597 has a disturbed optical and HI morphology. SDSS 143052.33+551440.0 also has a disturbed optical morphology and its CO profile is not symmetric about its systemic velocity. SDSS 153821.22+331105.1 has a bar that may have been triggered by an interaction. CG 598 appears to be accreting a companion in its SDSS g image and its CO profile is also asymmetric about its systemic velocity. Thus all the detected galaxies in our sample show some signs of interaction.

4. Implications

The main implications of this study is that cold gas and star formation are present in voids, even though the overall environment is underdense. Our sample galaxies also show disturbed morphologies, possibly due to interactions with companion galaxies. Our results can be understood in the hierarchical picture of void evolution, in which voids merge leaving behind a filamentary substructure. Galaxies grow along these filaments and in clusters where filaments intersect (e.g. Sahni *et al.* 1994; Sheth & van de Weygaert 2004; Cautun *et al.* 2014). The presence of both molecular gas and star formation in void galaxies indicates that they are probably evolving within this void substructure. Gas flowing along the filaments can be accreted by these galaxies and will contribute to the accumulation of neutral gas in their disks. High enough gas surface densities will result in the onset of star formation, leading to galaxy evolution within the void environment.

Acknowledgements

This paper was based on observations at the Nobeyama Radio Observatory (NRO) which is a branch of the National Astronomical Observatory of Japan, National Institutes of Natural Sciences. The optical observations were done at the Indian Optical Observatory (IAO) at Hanle. We thank the staff of IAO, Hanle and CREST, Hosakote, that made these obervations possible. This research has made use of the NASA/IPAC Extragalactic Database (NED).

References

Beygu, B., Kreckel, K., van de Weygaert, R., *et al.* 2013, *AJ*, 145, 120
Cautun, M., van de Weygaert, R., Jones, B. J. T. *et al.* 2014, *MNRAS*, 441, 2923
Cruzen, Shawn, Wehr, Tara, Weistrop, Donna *et al.* 2002, *AJ*, 123, 142
Karachentsev, V. E., Karachentsev, I. D., & Richter, G. M. 1999 *A&AS*, 135, 221
Kreckel, K., Platen, E., Aragn-Calvo, M. A., *et al.* 2012, *AJ*, 144, 16
Kreckel, K., Platen, E., Aragn-Calvo, M. A., *et al.* 2011, *AJ*, 141, 4
Nakajima, Taku, Sakai, Takeshi, Asayama, Shin'ichiro, *et al.* 2008, PASJ, 60, 435
Sage, L. J., Weistrop, D., Cruzen, S., & Kompe, C. 1997, *AJ*, 114, 1753
Sahni, Varun, Sathyaprakah, B. S., Shandarin, & Sergei F. 1994, *ApJ*, 431, 20
Sheth, Ravi K. & van de Weygaert, Rien 2004, *MNRAS*, 350, 517
Solomon, P. M. & Vanden Bout, P. A. 2005, *ARA&A*, 43, 677
Szomoru, Arpad, van Gorkom, J. H., Gregg, & Michael D. 1996, *AJ*, 111, 2141

The Zeldovich Universe:
Genesis and Growth of the Cosmic Web
Proceedings IAU Symposium No. 308, 2014
R. van de Weygaert, S. Shandarin, E. Saar & J. Einasto, eds.

© International Astronomical Union 2016
doi:10.1017/S1743921316010681

Surveying for Dwarf Galaxies Within Voids FN2 and FN8

Stephen McNeil[1], Chris Draper[2] and J. Ward Moody[3]

[1] Dept. of Physics, 118 ROM
Brigham Young University Idaho, Rexburg, ID 83460-0520 USA
email: `mcneils@byui.edu`

[2] Dept. of Physics, MS 179
Utah Valley University, Orem, UT 84058 USA
email: `drapechr@uvu.edu`

[3] Dept. of Physics and Astronomy, N283 ESC
Brigham Young University, Provo, UT 86402-4666 USA
email: `jmoody@byu.edu`

Abstract. The presence or absence of dwarf galaxies with $M_{r'} > -14$ in low-density volumes correlates with dark matter halos and how they affect galaxy formation. We are conducting a redshifted Hα imaging survey for dwarf galaxies with $M_{r'} > -13$ in the heart of the well-defined voids FN2 and FN8 using the KPNO 4m Mayall telescope and Mosaic Imager. These data have furnished over 600 strong candidates in a four square degree area. Follow-up spectra finding none of these candidates to be within the void volumes will constrain the dwarf population there to be 2 to 8% of the cosmic mean. Conversely, finding even one Hα dwarf in the void heart will challenge several otherwise successful theories of large-scale structure formation.

Keywords. large-scale structure of universe, galaxies: distances and redshifts, galaxies: dwarf

1. Introduction

Lambda cold dark matter (ΛCDM) models predict the existence of many low-mass dark matter halos in voids (e.g. Dekel & Silk 1986, Peebles 2001, Hoffman, Silk, & Wyse 1992, Tikhanov & Klypin 2009.) If galaxy formation has proceeded in these halos, then voids should have a population of dwarf galaxies in their interior (e.g. Tikhanov & Klypin 2009). But studies like Hoyle *et al.* (2005), do not find them. This problem was termed the "void phenomenon" by Peebles (2001).

To date, void centers have not been systematically searched in large-area optical spectroscopic surveys to levels faint enough to find such dwarfs. For example, the SDSS spectroscopic limit of $r' > 17.77$ is sufficient to locate theoretically interesting dwarf galaxies with $M_{r'} = -14$ only out to a cz of 1700 km/sec. Well defined void centers are not reached until 4,000 to 6,000 km/sec. Therefore the work of ruling out the existence of dwarf galaxies in the centers of voids through such surveys has yet to be accomplished.

2. Technique

In February 2013 we used the KPNO 4m Mayall telescope+Mosaic Imager to examine the heart of voids FN2 (RA = 3h 45m, DEC = +18deg, cz = 4550 km/sec, diam. = 5070 km/sec) and FN8 (RA = 12h 32m, DEC = +71.3deg cz = 4980 km/sec, diam. = 3660 km/sec) from Foster & Nelson (2009). Exposures were through three redshifted Hα filters: H8($\Delta\lambda$ 661.2-669.0 nm), H12($\Delta\lambda$ 665.1-673.0 nm), and H16($\Delta\lambda$ 668.5-676.3 nm).

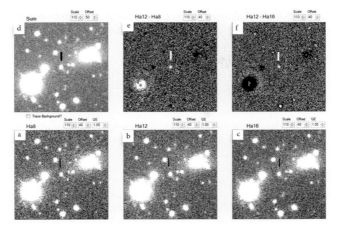

Figure 1. A crowded field showing images through filters a) H8, b) H12, and c) H16. Image d) is the sum of all filters, image e) is H12-H8, and image f) is H12-H16. The marked object stands out in both e) and f) because it has an emission line centered in filter H12. If the emission is Hα, it is a dwarf galaxy in the void center.

Seventeen fields (thirteen in FN8 and four in FN2) covering four square degrees were imaged. Data were summed over all filters for each field and over 400,000 objects were selected from the summed frames using *Source Extractor* (Bertin & Arnouts 1996). Each object was categorized by its value of H12/H8 vs H12/H16. Objects at the distance of the void center having Hα emission greater than 5.0 nm have both H12/H8 and H12/H16 index values greater than 1.1 while objects with no emission have values close to 1.0. Candidates for being inside the voids (over 600) were collected and examined by eye using our own software package *Compare Images* to reject flaws (see Fig. 1).

3. Discussion

The summed signal for an object with $M_{r'} = -13$ is a three to five sigma detection for all fields. At this threshold we are finding many more candidates than expected for the void density and LMF distribution of Hoyle *et al.* (2005). This is in part from detecting objects with strong [OIII]λ5007 in a supercluster at z ∼ 0.3 as judged from maps and object sizes and morphologies. This result is encouraging since it provides a critical measure of the depth of the survey technique and the detection limit of void dwarfs.

The search technique works well and we will be able to find dwarfs at the void center with $M_{r'} > -13$ if they are present. Given the large contamination from the backside, our data are consistent with completely empty voids, but they are also consistent with dwarf galaxies being present. Follow-up spectra alone will determine which candidates are dwarfs and which are not and will resolve the observational side of the void phenomenon.

References

Bertin, E. & Arnouts, S. 1996, *A&AS*, 317, 393
Dekel, A. & Silk, J. 1986, *ApJ*, 303, 39
Foster, C. & Nelson, L. A. 2009, *ApJ*, 699, 1252
Hoffman, Y., Silk, J., & Wyse, R. F. G. 1992, *ApJ*, 388, L13
Hoyle, F., Rojas, R. R., Vogeley, M. S., & Brinkmann, J. 2005, *ApJ*, 620, 618
Peebles, P. J. E. 2001, *ApJ*, 557, 495
Tikhonov, A. V. & Klypin, A. 2009, *MNRAS*, 395, 1915

CHAPTER 9.

Cosmology

Gruber Prize cosmology panel in action:
Moderator Alar Toomre in discussion with
Jaan Einasto, Brent Tully and Rashid Sunyaev

Cosmology in high spirits:
Sergei Shandarin & Jaan Einasto (sitting),
Dick Bond & Rashid Sunyaev (standing)

The Zeldovich Universe:
Genesis and Growth of the Cosmic Web
Proceedings IAU Symposium No. 308, 2014
R. van de Weygaert, S. Shandarin, E. Saar & J. Einasto, eds.

© International Astronomical Union 2016
doi:10.1017/S1743921316010693

Probing the accelerating Universe with redshift-space distortions in VIPERS

Sylvain de la Torre[1] and VIPERS collaboration

[1] Aix Marseille Université, CNRS, LAM (Laboratoire d'Astrophysique de Marseille) UMR 7326, 13388, Marseille, France
email: sylvain.delatorre@lam.fr

Abstract. We present the first measurement of the growth rate of structure at $z = 0.8$. It has been obtained from the redshift-space distortions observed in the galaxy clustering pattern in the VIMOS Public Redshift survey (VIPERS) first data release. VIPERS is a large galaxy redshift survey probing the large-scale structure at $0.5 < z < 1.2$ with an unprecedented accuracy. This measurement represents a new reference in the distant Universe, which has been poorly explored until now. We obtain $f\sigma_8 = 0.47 \pm 0.08$ at $z = 0.8$ that is consistent with the predictions of standard cosmological models based on Einstein gravity. This measurement alone is however not accurate enough to allow the detection of possible deviations from standard gravity.

Keywords. large-scale structure of universe, cosmological parameters, cosmology: observations

1. Introduction

The cosmological model is today well established with regards to our knowledge of the amount of Universe's major constituents (e.g. Planck Collaboration, 2013). This has been achieved thanks to geometrical probes such as cosmic microwave background anisotropies (CMB), baryonic acoustic oscillations in the galaxy clustering (BAO) or type 1a supernovae distance-redshift measurements (SN1a). The later has permitted to evidence that the Universe was experiencing a late-time accelerated expansion due to an unknown dark energy component dominating the total energy density today (Riess *et al.*, 1998; Perlmutter *et al.*, 1999). However, measurements of the expansion history of the Universe from geometrical probes alone, cannot discriminate dark energy from modifications to standard gravity to explain the observed acceleration of expansion. This degeneracy can only be lifted by measuring the growth rate of structure at different epochs $f(z)$ (e.g. Peacock *et al.*, 2006).

Galaxy redshift surveys provide a wealth of cosmological information, mapping the late-time development of the small metric fluctuations that existed at early times, and whose early properties can be viewed in the CMB. The growth of structure during this intervening period is sensitive both to the type and amount of dark matter, and also to the theory of gravity, so there is a strong motivation to make precise measurements of the rate of growth of structure. The main advantage of redshift surveys is that the radial information depends on the expansion history and is corrupted by peculiar velocities. The peculiar velocities induce an anisotropy in the apparent clustering, the *redshift-space distortions*, from which the properties of peculiar velocities can be inferred much more precisely than in any attempt to measure them directly using distance estimators. Because these peculiar velocities on large scales are induced by structure growth, redshift-space distortions is a unique probe of the growth rate of structure.

The possibility of using the redshift-space distortion signature as a probe of the growth rate of structure, together with that of using BAO as a standard ruler to measure the expansion history, is one of the main reasons behind the recent burst of activity in galaxy redshift surveys. The first paper to emphasise this application as a test of gravity theories was the analysis of the VVDS survey by Guzzo *et al.* (2008) and subsequent work using larger surveys has exploited this method to make measurements at $z < 1$ (e.g. Cabré & Gaztañaga, 2009; Beutler *et al.*, 2012). The prime goal of the VIMOS Public Redshift survey (VIPERS) is to provide an accurate measurement of the growth rate of structure at redshift around unity (Guzzo *et al.*, 2014). We report here the initial redshift-space distortions analysis in VIPERS, performed in de la Torre *et al.* (2013).

2. Redshift-space distortions in VIPERS

The VIPERS survey is intended to provide robust and precise measurements of the properties of the galaxy population at an epoch when the Universe was about half its current age. It covers 24 deg^2 on the sky, divided over two areas within the W1 and W4 CFHTLS fields. Galaxies are selected to a limit of $i_{AB} < 22.5$ and by applying a simple and robust gri colour preselection to efficiently remove galaxies at $z < 0.5$. VIPERS spectra has been collected with the VIMOS multi-object spectrograph at moderate resolution (R=210). VIPERS observations are on-going and the dataset used in this analysis corresponds to the VIPERS Public Data Release 1 (PDR-1) catalogue. This catalogue includes 55358 redshifts and represents 64% of the final survey in terms of covered area. The catalogue that we use to perform the redshift-space distortions analysis corresponds to a sub-sample of 45871 galaxies with reliable redshift measurements. A complete description of the survey and the PDR-1 catalogue are given in Guzzo *et al.* (2014) and Garilli *et al.* (2014).

The galaxy clustering in the VIPERS sample is characterized by measuring the two-point statistics of the spatial distribution of galaxies in configuration space, using the two-point correlation function. VIPERS has a complex angular selection function which has to be taken into account carefully when estimating the correlation function. For this, a detailed correction scheme has been developed that is described in de la Torre *et al.* (2013). In order to measure the anisotropy induced by structure growth, we measure the multipole moments of the anisotropic correlation function. We effectively measure the two first non-null moments $\xi_0(s)$ and $\xi_2(s)$, where most of the relevant information is contained.

The measured $\xi_0(s)$ and $\xi_2(s)$ for all galaxies between $0.7 < z < 1.2$ ($z_{med} = 0.8$) in the VIPERS sample are presented in Fig. 1. To derive the growth rate of structure, we fit the Scoccimarro (2004) model to the measurements, defining the galaxy bias parameter times the normalisation of the matter power spectrum $b\sigma_8$, the pairwise velocity dispersion σ_v, and the growth rate of structure times the normalisation of the matter power spectrum $f\sigma_8$, as free parameters. We assume a fixed shape of the mass power spectrum consistent with the cosmological parameters obtained from WMAP9 (Hinshaw *et al.*, 2013) and perform a maximum likelihood analysis. The best-fitting models are shown in Fig. 1 when considering either a Gaussian or a Lorentzian damping function to reproduce the small-scale distortions induced by the Fingers-of-God effect (see de la Torre & Guzzo, 2012). After marginalising over σ_v and $b\sigma_8$ parameters, we obtain a value of $f\sigma_8 = 0.47 \pm 0.08$.

Our measurement of $f\sigma_8$ is shown in Fig. 2 and compared with previous measurements and predictions for various cosmological models. This measurement is consistent with the General Relativity prediction in a flat ΛCDM Universe with cosmological parameters given by WMAP9 and Planck experiments, for which the expected values are $f(0.8)\sigma_8(0.8) = 0.45$ and $f(0.8)\sigma_8(0.8) = 0.46$ respectively. We find that our result is not significantly altered if we adopt a Planck cosmology for the shape of the

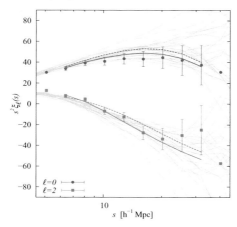

Figure 1. The monopole and quadrupole moments of the redshift-space correlation function, $\xi_0(s)$ and $\xi_2(s)$, as measured in the VIPERS sample. The shallow curves show the results for the 26 mock survey realisations. The points are for the measured VIPERS data at $0.7 < z < 1.2$. The solid and dotted lines correspond to the best-fitting models to the data with different damping functions describing small-scale distortions. *Figure reproduced from de la Torre et al. (2013).*

mass power spectrum. It is also in good agreement with previous measurements at lower redshifts from 2dFGRS (Hawkins *et al.*, 2003), 2SLAQ (Ross *et al.*, 2007), VVDS (Guzzo *et al.*, 2008), SDSS LRG (Cabré & Gaztañaga, 2009; Samushia *et al.*, 2012), WiggleZ (Blake *et al.*, 2012), BOSS (Reid *et al.*, 2012), and 6dFGS (Beutler *et al.*, 2012) surveys. In particular, it is compatible within 1% with the results obtained in the VVDS and WiggleZ surveys at a similar redshift, although WiggleZ measurements tend to suggest lower $f\sigma_8$ values, smaller than expected in standard gravity (but see Contreras *et al.*, 2013). Finally we compare our measurement to the predictions of three of the most plausible modified gravity models, studied in di Porto *et al.* (2012). We consider Dvali-Gabadadze-Porrati (Dvali *et al.*, 2000), $f(R)$, and coupled dark energy models and show their predictions in Fig. 2. We find that our $f\sigma_8$ measurement is currently unable to discriminate between these modified gravity models and standard gravity given the size of the uncertainty, although we expect to improve the constraints with the analysis of the VIPERS final dataset.

3. Conclusions

We have analysed the redshift-space clustering properties of galaxies in the VIPERS survey first data release. The first data release of about 54000 galaxies at $0.5 < z < 1.2$ in the VIPERS survey allows a measurement of the redshift-space clustering of galaxies to an unprecedented accuracy over the redshift range $0.5 < z < 1.2$. The main goal of VIPERS is to provide an accurate measurement of the growth rate of structure through the characterisation of the redshift-space distortions in the galaxy clustering pattern. With the first data release we have been able to provide an initial measurement of $f\sigma_8$ at $z = 0.8$. We find a value of $f\sigma_8 = 0.47 \pm 0.08$ which is in agreement with previous measurements at lower redshifts. This allows us to put a new constraint on gravity at the epoch when the Universe was almost half its present age. Our measurement of $f\sigma_8$ is statistically consistent with a Universe where the gravitational interactions between structures on $10\ h^{-1}$ Mpc scales can be described by Einstein's theory of gravity.

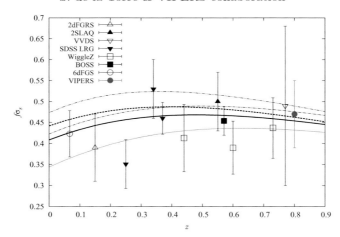

Figure 2. VIPERS $f\sigma_8$ initial measurement compared with a compilation of recent results from various surveys (see inset). The thick solid (dashed) curve corresponds to the prediction for General Relativity in a ΛCDM model with WMAP9 (Planck) parameters, while the dotted, dot-dashed, and dot-dot-dashed curves are respectively Dvali-Gabadaze-Porrati, coupled dark energy, and $f(R)$ model expectations. *Figure reproduced from de la Torre et al. (2013).*

The present dataset represents the half-way stage of the VIPERS project, and the final survey will be large enough to subdivide our measurements and follow the evolution of $f\sigma_8$ out to redshift one. This will allow us to address some issues such as the suggestion from the WiggleZ measurements that $f\sigma_8$ is lower than expected at $z > 0.5$ Our measurement at $z = 0.8$ already argues against such a trend to some extent, but the larger redshift baseline and tighter errors from the final VIPERS dataset can be expected to deliver a definitive verdict on the high-redshift evolution of the strength of gravity.

References

Blake, C., Brough, S., Colless, M., *et al.* 2012, *MNRAS*, 425, 405
Beutler, F., Blake, C., Colless, M., *et al.* 2012, *MNRAS*, 423, 3430
Cabré, A. & Gaztañaga, E. 2009, *MNRAS*, 393, 1183
Contreras, C., Blake, C., Poole, G. B., *et al.* 2013, *MNRAS*, 430, 924
de la Torre, S. & Guzzo, L. 2012, *MNRAS*, 427, 327
de la Torre, S., Guzzo, L., Peacock, J. A., *et al.* 2013, *A&A*, 557, A54
di Porto, C., Amendola, L., & Branchini, E. 2012, *MNRAS*, 419, 985
Dvali, G., Gabadadze, G., & Porrati, M. 2000, *Physics Letters B*, 485, 208
Garilli, B., Guzzo, L., Scodeggio, M., *et al.* 2014, *A&A*, 562, A23
Guzzo, L., Pierleoni, M., Meneux, B., *et al.* 2008, *Nature*, 451, 541
Guzzo, L., Scodeggio, M., Garilli, B., *et al.* 2014, *A&A*, 566, A108
Hawkins, E., Maddox, S., Cole, S., *et al.* 2003, *MNRAS*, 346, 78
Hinshaw, G., Larson, D., Komatsu, E., *et al.* 2013, *ApJS*, 208, 19
Peacock, J. A., Schneider, P., Efstathiou, G., *et al.* 2006, ESA-ESO Working Group on Funda-
    mental Cosmology, Edited by J. A. Peacock *et al.* ESA
Perlmutter, S., Aldering, G., Goldhaber, G., *et al.* 1999, *ApJ*, 517, 565
Planck Collaboration, Ade, P. A. R., Aghanim, N., *et al.* 2013, arXiv:1303.5076
Reid, B. A., Samushia, L., White, M., *et al.* 2012, *MNRAS*, 426, 2719
Riess, A. G., Filippenko, A. V., Challis, P., *et al.* 1998, *AJ*, 116, 1009
Ross, N. P., da Ângela, J., Shanks, T., *et al.* 2007, *MNRAS*, 381, 573
Samushia, L., Percival, W. J., & Raccanelli, A. 2012, *MNRAS*, 420, 2102
Scoccimarro, R. 2004, *PRD*, 70, 083007

The Zeldovich Universe:
Genesis and Growth of the Cosmic Web
Proceedings IAU Symposium No. 308, 2014
R. van de Weygaert, S. Shandarin, E. Saar & J. Einasto, eds.
© International Astronomical Union 2016
doi:10.1017/S174392131601070X

The impact of superstructures in the Cosmic Microwave Background

Stéphane Ilić[1,2], Mathieu Langer[2] and Marian Douspis[2]

[1]Institut de Recherche en Astrophysique et Planétologie,
14 avenue Édouard Belin, 31400 Toulouse, France
email: stephane.ilic@irap.omp.eu

[2]Institut d'Astrophysique Spatiale, Université Paris-Sud , UMR8617, Orsay, F-91405
& CNRS, Orsay, F-91405
emails: mathieu.langer@ias.u-psud.fr, marian.douspis@ias.u-psud.fr

Abstract. In 2008, Granett *et al.* claimed a direct detection of the integrated Sachs-Wolfe (iSW) effect, through the stacking of CMB patches at the positions of identified superstructures. Additionally, the high amplitude of their measured signal was reported to be at odds with predictions from the standard model of cosmology. However, a closer inspection of these results prompts multiple questions, more specifically about the amplitude and significance of the expected signal. We propose here an original theoretical prediction of the iSW effect produced by such superstructures. We use simulations based on GR and the LTB metric to reproduce cosmic structures and predict their exact theoretical iSW effect on the CMB. The amplitudes predicted with this method are consistent with the signal measured when properly accounting the contribution of the non-negligible (and fortuitous) primordial CMB fluctuations to the total signal. It also highlights the tricky nature of stacking measurements and their interpretation.

Keywords. cosmic microwave background, large-scale structure of universe, dark energy

1. Introduction

Dark Energy (DE) is one of the great mysteries of modern cosmology. The integrated Sachs-Wolfe effect (iSW) is an original probe of DE, linked to the large-scale structure of the Universe and the cosmic microwave background (CMB). Indeed, one of the effects of the accelerated expansion is the stretching of gravitational potentials, therefore changing the frequency of the CMB photons that travel through them. This effect is integrated along the whole line of sight of the CMB photons and shifts their temperature by an amount defined by :

$$\delta_T^{\mathrm{iSW}} = 2 \int \dot{\Phi} dt \qquad (\Phi = \mathrm{grav.potential}).$$

Consequently, the iSW effect has a direct but weak impact on the largest scales of the power spectrum of the CMB temperature fluctuations (Kofman & Starobinskii 1985). Since cosmic variance prevents a CMB-only detection of this iSW signal at those low multipoles, the use of external data is therefore required. The conventional approach is to correlate the CMB with a tracer of the matter distribution – usually galaxy surveys. This approach has been attempted numerous times (see Dupé *et al.*2011 for a review), these have yet to give a definitive and unambiguous detection of the iSW effect. This situation is mainly a consequence of the shortcomings of current surveys, not deep enough and/or with too small a sky coverage and therefore not optimised for iSW studies (see Douspis *et al.*2008 for a detailed discussion on the optimisation of such surveys).

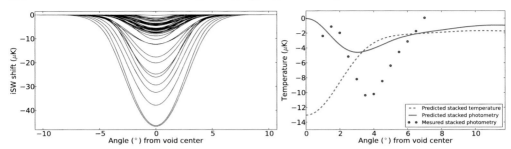

Figure 1. *Left panel:* Predicted temperature profiles of the iSW shift induced by the 50 Granett *et al.* voids. The amplitude of about 10 of these voids clearly stands out from the rest and makes up for most of the predicted signal. *Right panel:* Predicted temperature (dashed red curve) and photometry (solid red) profiles from the stacked iSW signal of the 50 Granett *et al.* voids. The blue points correspond to the real photometry profile measured in the CMB data.

Another approach to the iSW detection is to focus on the individual objects expected to leave the biggest imprint in the CMB, i.e. the largest superstructures in the Universe. While their individual imprint is buried under the primordial CMB fluctuations, we can average patches of the CMB at the locations of many superstructures in order to cancel the random CMB fluctuations while enhancing the iSW signal. In their pioneering work, Granett *et al.* (2008) claimed a $4\,\sigma$ detection of the iSW effect by stacking CMB patches at the positions of 100 superstructures, identified in the DR4 release of the LRGs sample of the SDSS. However, the measured amplitude was reported to be at odds with ΛCDM predictions (e.g. Cai *et al.*2013), while some peculiar features were also noted in the signal (Ilić *et al.*2013) such as the large hot ring around the cold signal from voids.

2. Simulating a superstructure and its impact on the CMB

The use of a stacking approach for the iSW detection raises a number of questions, especially concerning the nature and amplitude of the expected signal. To answer them, we developed an original theoretical prediction of the iSW effect produced by superstructures. The procedure we created is twofold : first, we modelled a structure and its evolution using the Lemaître-Tolman-Bondi (LTB) metric, the most general metric with a spherical symmetry. Using the data from the Granett *et al.* catalogue, we focused on reproducing their 50 voids (size, redshift, etc.) and we derived their evolution history. Secondly, we then computed the exact theoretical iSW effect of these structures on the CMB by solving geodesic equations for photons crossing through the previously computed LTB metric.

As expected, the iSW effect produced by the majority of these voids is indeed small (less than 10μK for the decrement in the temperature of CMB photons) as we can see in the left panel of Fig. 1. However, using my simulations we show here that about one fifth of the Granett *et al.* voids (and, as intuited, the largest ones) create a signal large enough to greatly increase the mean amplitude of the shift. Going further, we were also able to reconstruct the theoretical iSW map associated with these 50 voids, i.e. the temperature shift due to the iSW effect associated to each point in the sky accounting for all the voids present on the corresponding line of sight. Similarly to the analysis of real data, we can perform the stacking procedure on this iSW map, and obtain the expected temperature and photometry signal from these voids (see the right panel of Fig. 1). At a scale of 4 degrees the predicted photometry signal reaches approximatively -4μK, whereas the measured signal peaks at about -10μK at the same scale.

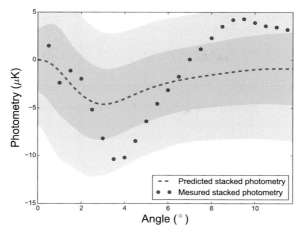

Figure 2. Predicted (dashed red) and measured (blue points) photometry profiles from the stacked iSW signal of the 50 Granett *et al.* voids. From dark to light red, the areas correspond to the $1\,\sigma$ and $2\,\sigma$ levels of the expected photometry profile.

However, this result only shows that the iSW effect of the 50 supervoids cannot solely account for the entirety of the signal measured by stacking. Indeed, it does not invalidate the hypothesis of the presence of an iSW signal in the measurement, that would be mixed with and contaminated by a larger, fortuitous signal coming from the primordial fluctuations of the CMB.

To assess the likelihood of such a scenario and its compatibility with the data, we generated a few thousands Gaussian realisations of the CMB (with the current best-fit cosmological parameters as determined by the *Planck* mission) to which we added our predicted iSW map of the 50 voids. For each of the resulting map, we applied the same stacking procedure as the one used on the real CMB data and obtained the temperature and photometry profiles of the corresponding stacked image.

With the resulting collection of profiles, we estimated that the photometry of the measured stacking signal at 4 degrees stands out at a $1.7\,\sigma$ level compared to its expected value (see Fig. 2). Furthermore, we were also able to compute the reduced χ^2 of the whole photometry profile, which we found to be close to one. Contrary to claims in the literature, it appears here that the signal measured through the stacking procedure is compatible with ΛCDM predictions. The high reported amplitude and the observed peculiarity of the photometry profile seem to be merely due to random (and not particularly rare) fluctuations from the primordial part of the CMB. The measurement from Granett *et al.* (2008) is therefore compatible with the expected iSW from such structures and does not present any significant discrepancy with respect to the ΛCDM paradigm.

References

Cai, Yan-Chuan, Neyrinck, Mark C. & Szapudi, István, Cole, Shaun, Frenk, & Carlos S. 2014, *ApJ*, 786, 110

Douspis, M., Castro, P. G., Caprini, C. & Aghanim, N. 2008, *A&A*, 485, 395-401

Dupé, F.-X., Rassat, A., Starck, J.-L., & Fadili, M. J. 2011, *A&A*, 534, A51

Granett, Benjamin R., Neyrinck, Mark C., & Szapudi, István 2008, *ApJ*, 683, L99-L102

Ilić, Stéphane, Langer, Mathieu, & Douspis, Marian 2013, *A&A*, 556, A51

Kofman, L. A. & Starobinskii, A. A. 1985, *Sv. A. L.*, 11, 271-274

The Zeldovich Universe:
Genesis and Growth of the Cosmic Web
Proceedings IAU Symposium No. 308, 2014
R. van de Weygaert, S. Shandarin, E. Saar & J. Einasto, eds.

© International Astronomical Union 2016
doi:10.1017/S1743921316010711

Faraday rotation in CMB maps

Beatriz Ruiz-Granados[1,2], **Eduardo Battaner**[1,2] **and Estrella Florido**[1,2]

[1]Departamento de Fsica Terica y del Cosmos, Universidad de Granada, Spain

[2]Instituto de Fsica Terica y Computacional Carlos I, Granada, Spain
email: bearg@ugr.es, battaner@ugr.es, estrella@ugr.es

Abstract. WMAP CMB polarization maps have been used to detect a low signal of Faraday Rotation (FR). If this detection is not interpreted as simple noise, it could be produced:
- at the last scattering surface (LSS) (z=1100), being primordial,
- at Reionization (z=10),
- in the Milky Way.

The second interpretation is favoured here. In this case magnetic fields at Reionization with peak values of the order of 10^{-8} G should produce this observational FR.

Keywords. cosmic microwave background, polarization, magnetic fields

1. Overview

A FR signal is detected by using foreground reduced maps of Stokes Q and U parameters provided by WMAP9 (Hinshaw *et al.* (2013)) at frequencies 33, 41 and 61 GHz.

The magnetic field strength at Reionization is obtained taking into account the value of the optical depth at this epoch derived by WMAP9 (i.e. ~ 0.089) (Hinshaw *et al.* (2013)).

Figure 1 shows the observational magnetic field distribution at Reionization from foreground reduced maps at 33, 41 and 61 GHz for a pixel size of 3.6 deg. This plot assumes that FR is produced at this epoch.

Figure 2 shows the angular power spectrum of this observational magnetic field distribution.

In Table 1 it is shown the correlation coefficient of our observational distribution of magnetic field (see Fig.1) and different maps that characterized Recombination and Galactic FR.

The polarized temperatures at Recombination are much lower than the Milky Way intensities but Rotation Measures are higher. This LSS interpretation would require

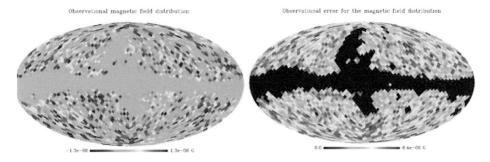

Figure 1. Magnetic field distribution and its corresponding error map.

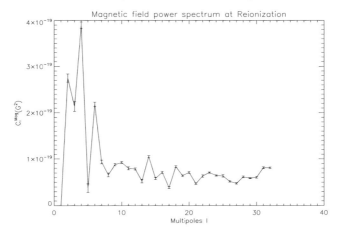

Figure 2. Power spectrum of the magnetic field distribution.

Table 1. Correlation coefficients.

Maps	Correlation coefficient r
CMB temperature	0.054 ± 0.015
Galactic RM (1.4-23 GHz)	-0.014 ± 0.034
Galactic RM (Extragalactic radio sources)	0.068 ± 0.015
Stokes U (23 GHz)	0.048 ± 0.015
Stokes Q (23 GHz)	0.002 ± 0.015
Stokes U (galactic dust)	-0.004 ± 0.015
Stokes Q (galactic dust)	-0.008 ± 0.015
Rotated angle (galactic dust)	-0.018 ± 0.015

magnetic field strengths higher than the limits found by Planck of 3.4×10^{-9} G (Planck Collaboration 2013).

Also, the correlation coefficient with the primordial temperature is very low, as shown in the table.

The correlation coefficients for typical galactic intensities are also very low, which probably indicates that the separation of the galactic component was satisfactorely made.

The multipole range of our spectrum is typical of phenomena taking place at Reionization. Therefore the data are consistent with magnetic fields at Reionization of the order of 1-10 nG.

Acknowledgements

This work was partially supported by projects CSD2010-00064 and AYA2011-24728 of the Spanish Ministry of Economy and Competitiveness (MINECO) and by Junta de Andalucía Grant FQM-108.

References

Hinshaw, G., Larson, D., Komatsu, E., *et al.* 2013, *ApJS*, 208, 19 ; Planck Collaboration, Ade, P. A. R., Aghanim, N., *et al.* 2013, *arXiv:1303.5076*

The Zeldovich Universe:
Genesis and Growth of the Cosmic Web
Proceedings IAU Symposium No. 308, 2014
R. van de Weygaert, S. Shandarin, E. Saar & J. Einasto, eds.

© International Astronomical Union 2016
doi:10.1017/S1743921316010723

Strong limit
on the spatial and temporal variations
of the fine-structure constant

T. D. Le

Department of Physics, Center for Natural Science, Vietnam
email: ldthong@hotmail.com

Abstract. Observed spectra of quasars provide a powerful tool to test the possible spatial and temporal variations of the fine-structure constant $\alpha = e^2/\hbar c$ over the history of the Universe. It is demonstrated that high sensitivity to the variation of α can be obtained from a comparison of the spectra of quasars and laboratories. We reported a new constraint on the variation of the fine-structure constant based on the analysis of the optical spectra of the fine-structure transitions in [NeIII], [NeV], [OIII], [OI] and [SII] multiplets from 14 Seyfert 1.5 galaxies. The weighted mean value of the α-variation derived from our analysis over the redshift range $0.035 < z < 0.281$ $\Delta\alpha/\alpha = (4.50 \pm 5.53) \times 10^{-5}$. This result presents strong limit improvements on the constraint on $\Delta\alpha/\alpha$ compared to the published in the literature

Keywords. galaxies: fundamental parameters (classification, colors, luminosities, masses, radii, etc.); cosmology: observations; methods: data analysis

1. Introduction

The interesting problem of time variation of the fundamental constants such as the finestructure constant,

$$\alpha = \frac{e^2}{4\pi\epsilon_0\hbar c} \tag{1.1}$$

in the expanding Universe was questioned by Dirac and Milne (Dirac 1973; Milne 1937). This subject is current interest because of the new possibilities opened up by the contemporary theories, which lead us to unify the fundamental interactions of nature. Astrophysical estimates of the α is based on the comparison of the line centers in the absorption/emission spectra of astronomical objects and the corresponding laboratory values (Bahcall *et al.* 2004; Dzuba *et al.* 1999; Le 2014). The AD method is based on the fine-structure splitting. The splitting ratios $s(t)$,

$$s(t) = \frac{\lambda_2(t) - \lambda_1(t)}{\lambda_2(0) + \lambda_1(0)} \tag{1.2}$$

at two different epochs, gives the different values in α between them two epochs. It is shown that,

$$\frac{\Delta\alpha}{\alpha}(t) = \frac{1}{2}\left[\frac{(\lambda_2(t) - \lambda_1(t))/(\lambda_2(t) + \lambda_1(t))}{(\lambda_2(0) - \lambda_1(0))/(\lambda_2(0) + \lambda_1(0))} - 1\right], \tag{1.3}$$

where $\lambda_2(t)$ and $\lambda_1(t)$ are the wavelengths obtained from the quasar observations, and $\lambda_2(0)$ and $\lambda_1(0)$ the laboratory wavelengths. The best recent result using AD method applied to quasar absorption lines is $\Delta\alpha/\alpha = (-0.5 \pm 1.3) \times 10^{-5}$ at $0.2 < z < 4.2$ (Bahcall *et al.* 2004).

Another method is the MM method, which was developed by Dzuba *et al.*(1999). This is dependent on the rest wavelengths to the variation of α parameter by using the fitting function,

$$\omega = \omega_0 + q_1 x + q_2 y, \qquad (1.4)$$

where ω_0 and ω are the vacuum wave number measured in the laboratory and in the absorption system at redshift z, and the parameters x and y are defined as

$$x = \left(\frac{\alpha_z}{\alpha_0}\right)^2 - 1$$

$$y = \left(\frac{\alpha_z}{\alpha_0}\right)^4 - 1. \qquad (1.5)$$

The best recent result using MM method applied to KECK/HIRES data were $\Delta\alpha/\alpha = (-0.57 \pm 0.10) \times 10^{-5}$ at $0.2 < z < 3.7$. This method simultaneously analyzes many doublets of many atomic species and due to the large number of linesused, gives an order magnitude improvement in the measurement of $\Delta\alpha/\alpha$ compared with the AD. However, several uncertainties and possible systematics would be some unnoticed systematic effects and might challenge that interpretation and it is very difficult to determine the systematic effects and the true values of $\Delta\alpha/\alpha$.

A new method is needed to improve the disadvantages of these methods to investigate spatial and temporal variation in α using the absorption/ emission lines (Bahcall *et al.* 2004; Dzuba *et al.* 1999).

2. Astrophysical data tests of $\Delta\alpha/\alpha$

In spectra of QSOs systems, they include almost resonance lines of the ions, and others corresponding to the transitions ground state. The separation between energy levels caused by fine- splitting is proportional to α^4 with leading to term of energy level being proportional to α^2. The spectroscopic observations of gas clouds present ideal laboratories in which to search for any spatial and temporal of α, which is based on the fact that the energy of each line transition depends individually on a change in α.

Therefore, we applied the new method that was developed by Le (2014) to improve the disadvantages of the AD and MM methods to search for spatial and temporal variation of the fine structure constant. This method allows us to estimate α values at early stages of the Universe evolution by using the fine-splitting doublets and multiplets seen in quasar spectra and has the advantage that it is more transparent and less subject to systematics is approximately written as

$$\frac{\Delta\alpha}{\alpha} = \frac{1}{2}\left(\frac{\frac{1}{2}\left(\frac{\lambda_2(t)}{\lambda_1(t)}\right) - 1}{\frac{1}{2}\left(\frac{\lambda_2(0)}{\lambda_1(0)}\right) - 1} - 1\right) \qquad (2.1)$$

provided that $\Delta\alpha/\alpha$ is very small.

Thus, comparing the relative of measured wavelengths $\lambda_1(t)$ and $\lambda_2(t)$ in an absorption (or emission) system at the redshift z with the corresponding laboratory values, one can directly infer the possible variation of α at different epochs, space-time points, regions of the Universe and the present value.The observed wavelengths from quasars will be determined through the cosmological redshift parameteras $1 + z = \lambda_{observ}/\lambda_{emiss}$, where λ_{emiss} is the wavelength at the moment of emission (absorption) by a distant source in the past and λ_{observ} is the observed value. Thus, this method can be used to study not only possible effect of the time variability of the fine-structure constant at present but also at earlier epochs of the Universe, when they were causally disconnecte.

In this study, we have applied a new method to analysis the optical spectra of the fine-structure transitions in [NeIII], [NeV], [OIII], [OI] and [SII] multiplets from 14 Seyfert 1.5 galaxies to constrain past variations in α (Grupe *et al.* 2005). This approach eliminates the largest systematic errors present in other determinations of α and provides estimates of the remaining statistical and systematic errors. This method includes α-independent line ratios which can identify the true size of statistical and systematic errors.

3. Conclusion

Our result is $\Delta\alpha/\alpha = (4.50 \pm 5.53) \times 10^{-5}$ over the redshift range $0.035 < z < 0.281$. Our result corresponds to two orders of magnitude more sensitive than that using the same data derived by Grupe *et al.* (2005). This result is nearly two order magnitude improvements on the constraint on / compared with the AD method Bahcall *et al.* (2004) and an order compared with MM method (Dzuba *et al.* 1999). Our results consist with the results coming from Oklo natural nuclear reactor (Shylakher 1976). The spatial gradient in values of α are in progress.

References

Bahcall J. N. *et al.* 2004, *ApJ* 600, 520
Dirac A. M. 1973, *Nature* 139, 323
Dzuba V. A. *et al.* 1999, *PRL* 82, 888
Milne E. A. 1937, *Proc. R. Soc. Ser. A.* 158, 324
Le T. D. 2014, *AIP Conf. Proc.* 1594, 23
Grupe D. *et al.*, 2005, *AJ* 130, 255
Shylakher A. I., 1976, *Nature* 264, 340

CHAPTER 10.

Miscellaneous

Celebrating the Symposium's Success:

under the direction of Enn Saar
Rien van de Weygaert and Sergei Shandarin
follow a traditional Estonian star dance.

The Zeldovich Universe:
Genesis and Growth of the Cosmic Web
Proceedings IAU Symposium No. 308, 2014
R. van de Weygaert, S. Shandarin, E. Saar & J. Einasto, eds.
© International Astronomical Union 2016
doi:10.1017/S1743921316010735

Evolution of low-frequency contribution in emission of steep-spectrum radio sources

Alla P. Miroshnichenko

Institute of Radio Astronomy of NAS of Ukraine,
UA-61002, Kharkov, Ukraine
email: `mir@rian.kharkov.ua`

Abstract. We consider evolution properties of galaxies and quasars with steep radio spectrum at the decametre band from the UTR-2 catalogue. The ratios of source's monochromatic luminosities at the decametre and high-frequency bands display the dependence on the redshift, linear size, characteristic age of examined objects. At that, the mean values of corresponding ratios for considered galaxies and quasars have enough close quantities,testifying on the unified model of sources. We analyse obtained relations for two types of steep-spectrum sources (with linear steep spectrum (S) and low-frequency steepness after a break (C+)) from the UTR-2 catalogue.

Keywords. Steep radio spectrum, decametre emission, galaxy, quasar

1. Introduction

We continue to study the properties of the steep-spectrum sources from the Grakovo decametre survey (UTR-2 catalogue) within the frequency range 10 to 25 MHz (Braude *et al.* 1978, Braude *et al.* 1979, Braude *et al.* 1981a, Braude *et al.* 1981b, Braude *et al.* 2003). This peculiar class of radio sources (the value of low-frequency spectral index exceeds 1) corresponds to conception of the long evolution, when the critical frequency of the synchrotron emission can displace to values less than 10 MHz. Befor (Miroshnichenko 2012a, Miroshnichenko 2012b, Miroshnichenko 2013) we received estimates of the main physical parameters of quasars and galaxies with steep radio spectrum over the sample of objects at the decameter band (at the frame of the Lambda-CDM model of the Universe). The sample of objects with linear steep spectrum (type S) includes 78 galaxies and 55 quasars with flux density more than 10 Jy at the frequency 25 MHz. The sample of objects with break steep spectrum (type C+) contains 52 galaxies and 36 quasars with flux density more than 10 Jy at the frequency 25 MHz. The optical and high-frequency data for examined sources have been got from the NED database. The redshift range of objects forms 0.017-3.570. Note, our calculations show that galaxies and quasars with steep low-frequency spectra have the great luminosity (by order of 10^{28} W/(Hz ster) at the frequency 25 MHz) and very extended radio structure with linear size by order of 1 Mpc, and characteristic age by order of 100 million years (Miroshnichenko 2012a, Miroshnichenko 2012b, Miroshnichenko 2013).

2. Contribution of decametre emission in sources with S-type and C+ - type of steep spectra

It is known that the decametre emission of galaxies and quasars corresponds to emission of their extended regions, outlying from source's core. At the same time, the source's emission at the higher frequencies, mainly, is connected with the emission from the central

Table 1. Mean values of the ratios of monochromatic luminosities at the different bands for quasars and galaxies with steep spectrum.

Mean value of ratio	Quasars	Galaxies
$< lg(S_25/S_5000) >$	1.69 ± 0.08	1.74 ± 0.05
$< lg(S_25/S_I R) >$	4.30 ± 0.11	3.67 ± 0.19
$< lg(S_25/S_opt) >$	5.00 ± 0.10	5.15 ± 0.12
$< lg(S_25/S_X) >$	7.78 ± 0.17	7.89 ± 0.33
$< lg(S_I R/S_X) >$	3.54 ± 0.20	4.68 ± 0.47

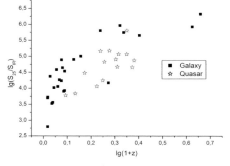

Figure 1. The ratio of monochromatic luminosities of examined sources at the decametre and infrared bands versus the redshift (for type S).

Figure 2. The ratio of monochromatic luminosities of examined sources at the infrared and X-ray bands versus the linear size (for type S).

region of radio source. The ratio of low-frequency and high-frequency luminosities may be as characteristic of different substructures of objects indicating some features of their evolution. Moreover, one is not influenced by the Universe model. So, we determine the ratios of the flux densities of emission in the different bands: decametre (25 MHz), centimetre (5000 MHz), infrared (IR, K-band), optical (opt, V),X-ray (1 keV) band for quasars and galaxies from the steep-spectrum sample. These are identical to the ratios of the corresponding monochromatic luminosities (Table 1). It is important that the mean values of the corresponding luminosity's ratios for considered quasars and galaxies in Table 1 have enough close quantities. Thus, the obtained characteristics of sources with steep radio spectrum are in concordance with the unified model of sources.

We have received the relations for derived luminosity ratios of quasars and galaxies with spectrum S and C+ versus the redshift, linear size, characteristic age (see Fig.1-4). These relations evidence for the essential cosmological evolution of luminosities of steep-spectrum sources. The interesting picture is displayed in the relation of infrared and X-ray luminosities versus the linear size of sample objects (Fig. 2). The founded two branches in this relation may testify on the recurrence of the nucleus activity in galaxies and quasars with steep radio spectrum.

3. Conclusions

Galaxies and quasars with steep radio spectrum display the essential cosmological evolution. The relative contribution of the decametre emission in steep-spectrum sources increases for more extended objects. The revealed two branches in relation of the ratio

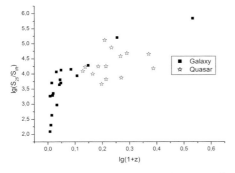

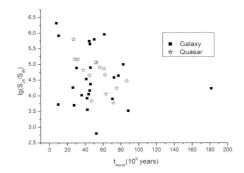

Figure 3. The ratio of monochromatic luminosities of examined sources at the decametre and infrared bands versus the redshift (for type C+).

Figure 4. The ratio of monochromatic luminosities of examined sources at the decametre and infrared bands versus the characteristic age (for type S).

of infrared and X-ray luminosity versus the linear size of steep-spectrum galaxies and quasars may indicate on the activity recurrence of sources. Mutual similarity of the structure and the physical parameters of steep-spectrum galaxies and quasars corresponds to the unified model of sources.

References

Braude, S., Megn, A., Rashkovski, S., Ryabov, B. *et al.* 1978, *Ap & SS*, 54, 37

Braude, S., Megn, A., Sokolov, K., Tkachenko, A., & Sharykin, N. 1979, *Ap & SS*, 64, 73

Braude, S., Miroshnichenko, A., Sokolov, K., & Sharykin, N. 1981a, *Ap & SS*, 74, 409

Braude, S., Miroshnichenko, A., Sokolov, K., & Sharykin, N. 1981b, *Ap & SS*, 76, 279

Braude, S., Miroshnichenko, A., Rashkovski, S., & Sidorchuk, K. *et al.* 2003, *Kinematics & Physics of Celestial Bodies*, 19, 291

Miroshnichenko, A. 2012a, *Radio Physics & Radio Astronomy*, 3, 215

Miroshnichenko, A. 2012b, *Odessa Astronomical Publications*, 25, 197

Miroshnichenko, A. 2013, *Odessa Astronomical Publications*, 26, 248

The Zeldovich Universe:
Genesis and Growth of the Cosmic Web
Proceedings IAU Symposium No. 308, 2014
R. van de Weygaert, S. Shandarin, E. Saar & J. Einasto, eds.

© International Astronomical Union 2016
doi:10.1017/S1743921316010747

The cosmic web and microwave background fossilize the first turbulent combustion

Carl H. Gibson[1] and R. Norris Keeler[2]

[1]MAE and SIO, Univ. of Cal. San Diego,
La Jolla, CA, 92093-0411, USA
email: cgibson@ucsd.edu

[2]6652 Hampton Park Court,
McLean VA 22101, USA
email: rnkeeler@verizon.net

Abstract. Collisional fluid mechanics theory predicts a turbulent hot big bang at Planck conditions from large, negative, turbulence stresses below the Fortov-Kerr limit ($< -10^{113} Pa$). Big bang turbulence fossilized when quarks formed, extracting the mass energy of the universe by extreme negative viscous stresses of inflation, expanding to length scales larger than the horizon scale ct. Viscous-gravitational structure formation by fragmentation was triggered at big bang fossil vorticity turbulence vortex lines during the plasma epoch, as observed by the Planck space telescope. A cosmic web of protogalaxies, protogalaxyclusters, and protogalaxysuperclusters that formed in turbulent boundary layers of the spinning voids are hereby identified as expanding turbulence fossils that falsify $\Lambda CDMHC$ cosmology.

Keywords. turbulence, cosmology, dark matter, dark energy.

1. Introduction

Yakov Borisovich Zeldovich pioneered the difficult field of turbulent combustion, and would have appreciated the complexities that arise when collisional fluid mechanics, general relativity, self-gravitational-stratification, fossil turbulence, fossil turbulence waves, and beamed zombie turbulence maser action mixing chimneys (journalofcosmology.com volume 21) are combined in the first turbulent combustion of the big bang (JoC volumes 15-24), Gibson (2004), Gibson (2005), Keeler and Gibson (2012).

Space telescopes show distinctive fossil turbulence patterns in the cosmic web and in the cosmic microwave background that confirm a big bang turbulent combustion mechanism, where turbulence is defined by inertial vortex forces and negative Fortov-Kerr pressures [Fortov (2012)] extract mass-energy from turbulence needed to trigger inflation by gluon viscous stresses of the strong force freeze-out. Such turbulence always cascades from small scales to large (see journalofcosmology.com volumes 15-23) and leaves patterns termed fossil turbulence in a variety of hydrophysical fields that preserve information about the previous turbulence. For astrophysical sample calculations and physical constants, see Gibson (2000). Photon viscosity during the plasma epoch explains the fragmentation of protogalaxies, and the observations showing all galaxies have the same mass, Tyson, J. A. & P. Fischer (1995).

The cosmic microwave background spectrum reveals fossil turbulence patterns at large wavelengths (now $> 10^{25}$ m) fossilizing big bang turbulent combustion, and smaller wavelengths fossilizing viscous gravitational fragmentation of the plasma epoch at 10^{12} seconds to produce $\approx 10^{24}$ m Superclusters and Superclustervoids in the turbulent boundary layers of spinning SuperSuperclustervoids, now $\approx 2 \times 10^{25}$ m in diameter. See JoC vol. 24

for details. Protogalaxies formed at 10^{12} s serve as fluid particles for the weak boundary layer turbulence.

The CMB spectral turbulence pattern is a single peak reflecting a highly concentrated vortex, and two secondary peaks reflecting transverse secondary vortices at right angles that stretch the primary vortex by inertial vortex forces into a tubular shape about one part per million of the Kolmogorov space time at the Planck conditions of the big bang. Fortov-Kerr negative stresses less than $-10^{113}Pa$ extract mass-energy from the vacuum.

The plasma turbulence peak and its two harmonics has been misinterpreted as sonic oscillations of trapped baryons in cold dark matter potential wells. $\Lambda CDMHC$ cosmology is generally falsified by the fluid mechanically based cosmology and observations presented, where cold dark matter, dark energy and hierarchical clustering of CDM clumps is questioned.

Schild (1996) proved the dark matter of galaxies is Earth-mass planets of hydrogen–helium, confirming fluid mechanical predictions of Gibson (1996) that plasma super-clusters and clusters of protogalaxies fragmented into Jeans mass clumps of primordial planets at the plasma to gas transition at time 10^{13} seconds. Stars and larger planets are formed in these proto-globular-star-cluster clumps by binary mergers. Since $\approx 10^{80}$ dark matter planets are produced by the big bang, numerous habitats for the formation of life are provided by HGD cosmology, Gibson, Schild, & Wickramasinghe (2011).

2. Discussion

Turbulence and gravitational structure formation are absolutely unstable random processes. Several critical length scales (Schwarz scales) distinguish different structure formation regimes and hydrogravitational phase diagrams, Gibson (1996). The most important length scale for cosmological structure formation is the viscous–gravitational Schwarz scale $L_{SV} \approx [\gamma\nu/\rho G]^{1/2}$. The rate-of-strain γ is $\approx 1/t$. The kinematic viscosity ν during the plasma epoch is huge (the photon viscosity), so the fragmention mass (10^{44} kg) is that of a galaxy, Gibson (2000). After transition to gas at 10^{13} s, ν decreases by a factor of 10^{13}, giving a planetary (10^{24} kg) fragmentation mass. This matches that observed by quasar microlensing of a foreground galaxy, Schild (1996).

3. Conclusions

Evidence of empty SuperSuperVoids supports the hypothesis of big bang turbulence. The cosmic web is identified as linear structures characteristic of fossil turbulent vortex lines in the turbulent boundary layers of these spinning big bang fossil turbulence vortex lines that have triggered structure formation during the plasma epoch.

References

Fortov, V. E. 2012, *J. of Cosmology*, 18, 14, 8105–8138.
Gibson, C. H. 1996, *Appl. Mech. Rev.*, 49, 299
Gibson, C. H. 2000, *J. Fluids Eng.*, 122, 830–835. JoC, 23, 11, 11206–11212.
&Gibson, C. H., R. E. Schild, & N. C. Wickramasinghe 2011, *Int. J. of Astrobiology*, 10 (2): 83–98. The origin of life from primordial planets.
Gibson, C. H. 2004, *Flow Turbul. Combust.*, 72, 161-179.
Gibson, C. H. 2005, *Combust. Sci. and Tech.*, 177, 1049-1071. arXiv:astro-ph/0501416
Keeler, R. N. & C. H. Gibson 2012, *J. of Cosmology*, 18, 13, 8095–8104.
Schild, R. E. 1996, *ApJ*, 464, 125.
Tyson, J. A. & P. Fischer 1995, *ApJ*, 446, L55.

Author index

Printed in the United States
by Baker & Taylor Publisher Services